SAP PRESS e-books

Print or e-book, Kindle or iPad, workplace or airplane: Choose where and how to read your SAP PRESS books! You can now get all our titles as e-books, too:

- By download and online access
- For all popular devices
- And, of course, DRM-free

Convinced? Then go to www.sap-press.com and get your e-book today.

Treasury and Risk Management with SAP S/4HANA®

SAP PRESS is a joint initiative of SAP and Rheinwerk Publishing. The know-how offered by SAP specialists combined with the expertise of Rheinwerk Publishing offers the reader expert books in the field. SAP PRESS features first-hand information and expert advice, and provides useful skills for professional decision-making.

SAP PRESS offers a variety of books on technical and business-related topics for the SAP user. For further information, please visit our website: *www.sap-press.com*.

Stoil Jotev
Configuring SAP S/4HANA Finance (3rd Edition)
2025, 744 pages, hardcover and e-book
www.sap-press.com/5920

Aylin Korkmaz
Financial Reporting with SAP S/4HANA
2022, 707 pages, hardcover and e-book
www.sap-press.com/5416

Dirk Neumann, Lawrence Liang
Cash Management with SAP S/4HANA (2nd Edition)
2021, 561 pages, 2nd edition hardcover and e-book
www.sap-press.com/5169

Ryan, Bala, Raghav, Mohammed, Kukreja
Group Reporting with SAP S/4HANA:
The Financial Consolidation Guide (2nd Edition)
2024, 484 pages, hardcover and e-book
www.sap-press.com/5872

Anand Seetharaju, Mayank Sharma
General Ledger Accounting with SAP S/4HANA
2023, 886 pages, hardcover and e-book
www.sap-press.com/5630

Luke Carlson, Andrew Carlson, Jeffrey Lasecki

Treasury and Risk Management with SAP S/4HANA®

The Comprehensive Guide

Editor Meagan White
Acquisitions Editor Emily Nicholls
Copyeditor Doug McNair
Cover Design Graham Geary
Photo Credit AI-generated with midjourney.com
Layout Design Vera Brauner
Production Hannah Lane
Typesetting III-satz, Germany
Printed and bound in the United States of America, on paper from sustainable sources

ISBN 978-1-4932-2610-8
1st edition 2025

© 2025 by:
Rheinwerk Publishing, Inc.
2 Heritage Drive, Suite 305
Quincy, MA 02171
USA
info@rheinwerk-publishing.com

Represented in the E.U. by:
Rheinwerk Verlag GmbH
Rheinwerkallee 4
53227 Bonn
Germany
service@rheinwerk-verlag.de

Library of Congress Cataloging-in-Publication Control Number: 2024046344

All rights reserved. Neither this publication nor any part of it may be copied or reproduced in any form or by any means or translated into another language, without the prior consent of Rheinwerk Publishing.

Rheinwerk Publishing makes no warranties or representations with respect to the content hereof and specifically disclaims any implied warranties of merchantability or fitness for any particular purpose. Rheinwerk Publishing assumes no responsibility for any errors that may appear in this publication.

"Rheinwerk Publishing" and the Rheinwerk Publishing logo are registered trademarks of Rheinwerk Verlag GmbH, Bonn, Germany. SAP PRESS is an imprint of Rheinwerk Verlag GmbH and Rheinwerk Publishing, Inc.

All screenshots and graphics reproduced in this book are subject to copyright © SAP SE, Dietmar-Hopp-Allee 16, 69190 Walldorf, Germany.

SAP, ABAP, ASAP, Concur Hipmunk, Duet, Duet Enterprise, ExpenseIt, SAP ActiveAttention, SAP Adaptive Server Enterprise, SAP Advantage Database Server, SAP ArchiveLink, SAP Ariba, SAP Business ByDesign, SAP Business Explorer (SAP BEx), SAP BusinessObjects, SAP BusinessObjects Explorer, SAP BusinessObjects Web Intelligence, SAP Business One, SAP Business Workflow, SAP BW/4HANA, SAP C/4HANA, SAP Concur, SAP Crystal Reports, SAP EarlyWatch, SAP Fieldglass, SAP Fiori, SAP Global Trade Services (SAP GTS), SAP GoingLive, SAP HANA, SAP Jam, SAP Leonardo, SAP Lumira, SAP MaxDB, SAP NetWeaver, SAP PartnerEdge, SAPPHIRE NOW, SAP PowerBuilder, SAP PowerDesigner, SAP R/2, SAP R/3, SAP Replication Server, SAP Roambi, SAP S/4HANA, SAP S/4HANA Cloud, SAP SQL Anywhere, SAP Strategic Enterprise Management (SAP SEM), SAP SuccessFactors, SAP Vora, TripIt, and Qualtrics are registered or unregistered trademarks of SAP SE, Walldorf, Germany.

All other products mentioned in this book are registered or unregistered trademarks of their respective companies.

No part of this book may be used or reproduced in any manner for the purpose of training artificial intelligence technologies or systems. In accordance with Article 4(3) of the Digital Single Market Directive 2019/790, Rheinwerk Publishing, Inc. expressly reserves this work from text and data mining.

Contents at a Glance

1	Introduction to Treasury and Risk Management with SAP	25
2	Master Data	53
3	Core Configuration Elements	103
4	Contract-Specific Configuration	201
5	General Contract Processes	321
6	Treasury Management–Specific Processes	369
7	Exposure Management	485
8	Cash Flow Hedging	515
9	Balance Sheet Hedging	599
10	Correspondence	647
11	SAP Treasury Analyzers	705
12	Integration with Other Areas	743
13	Reports, Key Performance Indicators, and Alerts	763

Contents

Preface 19

1   Introduction to Treasury and Risk Management with SAP   25

1.1   Changes in Technology and the Evolving Corporate Treasury Function 26
1.2   Treasury and Risk Management with SAP 27
     1.2.1   Current Scope and Use 28
     1.2.2   Core Treasury Components 29
1.3   SAP Fiori User Experience 32
     1.3.1   Maintaining a User Profile 33
     1.3.2   Customizing the User Screen 37
     1.3.3   Creating Custom Views 41
     1.3.4   Creating Custom SAP Fiori Tiles 43
     1.3.5   Sharing Custom SAP Fiori App Views 45
1.4   Supported Treasury Contracts 46
     1.4.1   Money Market Debt and Investments 46
     1.4.2   Foreign Exchange 47
     1.4.3   Securities 48
     1.4.4   Derivatives 49
     1.4.5   Trade Finance Instruments 50
1.5   Summary 51

2   Master Data   53

2.1   Company Code 54
2.2   Bank Master 55
2.3   House Bank 57
2.4   House Bank Account/Account ID 59
2.5   Business Partners 60
     2.5.1   Business Partner Maintenance 62
     2.5.2   Assigning Business Partner Payment Details 67
     2.5.3   Assigning Business Partner Authorizations 71

2.6		Derivative Contract Specification	74
2.7		Calendars	84
2.8		Traders	86
	2.8.1	Defining User Data	86
	2.8.2	Assigning Traders	88
2.9		Market Data	89
	2.9.1	Foreign Exchange Rates	90
	2.9.2	Swap Rates	92
	2.9.3	Security Prices	93
	2.9.4	Reference Interest Rates	94
	2.9.5	Yield Curves	96
	2.9.6	Commodity Prices	98
2.10		Market Data Interfaces	99
2.11		Summary	102

3   Core Configuration Elements 103

3.1		Company Code Additional Data	104
3.2		Accounting Codes	106
3.3		Valuation Areas	107
3.4		Valuation Class	108
3.5		General Valuation Class	109
	3.5.1	Define General Valuation Class	109
	3.5.2	Assign General Valuation Class	111
3.6		Position Management Procedure	113
	3.6.1	Define and Assign Position Management Procedure	113
	3.6.2	Customizing Valuations	118
3.7		User Data	131
3.8		Product Categories	134
	3.8.1	Money Market	135
	3.8.2	Foreign Exchange	136
	3.8.3	Derivatives	137
	3.8.4	Securities	139
	3.8.5	Trade Finance	141
3.9		Product Types	141
3.10		Number Ranges	145

3.11	Transaction Types		146
3.12	Field Selection		148
3.13	Flow Types		151
	3.13.1	Defining Flow Types	152
	3.13.2	Assigning Flow Types to Transaction Types	154
3.14	Update Types		156
	3.14.1	Defining Update Types and Assigning Usages	156
	3.14.2	Assigning Flow Types to Update Types	159
	3.14.3	Indicate Update Types Relevant to Posting	160
3.15	Condition Types		161
	3.15.1	Defining Condition Types	162
	3.15.2	Assigning Condition Types to Transaction Type	164
3.16	Accruals and Deferrals		165
3.17	Portfolios		170
3.18	Administration Fields		171
3.19	Workflow		172
	3.19.1	Defining the Release Procedure	173
	3.19.2	Adjusting the Workflow Template	174
3.20	Account Symbols		176
3.21	Posting Specifications		178
	3.21.1	Defining Posting Specifications	178
	3.21.2	Assigning Update Types to Posting Specifications	183
3.22	Account Assignment Reference		184
	3.22.1	Defining Account Assignment References	185
	3.22.2	Determining Account Assignment References	186
	3.22.3	Assigning General Ledger Accounts to Posting Keys	192
	3.22.4	Allocating Additional Account Assignments to Account Assignment References	193
3.23	Position Transfers		194
3.24	Summary		199

4 Contract-Specific Configuration 201

4.1	Debt and Investments		201
	4.1.1	Money Market Contracts	203
	4.1.2	Intercompany Loans	213

Contents

	4.1.3	Credit Facilities	226
	4.1.4	Letters of Credit	239
4.2		**Foreign Exchange**	256
	4.2.1	Product Types	257
	4.2.2	Transaction Types	258
	4.2.3	Assigning Foreign Exchange Attributes	259
	4.2.4	Assigning Fixing Spreads	261
	4.2.5	Defining Flow Types	262
	4.2.6	Assigning Flow Types to Update Types	264
	4.2.7	Position Management Procedures	264
	4.2.8	Set Effects of Update Types on Position Components	265
	4.2.9	Assigning Update Types for Valuation	265
	4.2.10	Assigning Update Types for Derived Business Transactions	266
	4.2.11	Alternative Update Types for Position Outflows	268
	4.2.12	Defining Nondeliverable Currencies	269
	4.2.13	Foreign Exchange Mirroring	270
4.3		**Securities**	275
	4.3.1	General Configuration	276
	4.3.2	Transaction Management	283
	4.3.3	Position Management	286
4.4		**Derivatives**	293
	4.4.1	Product Types	293
	4.4.2	Transaction Types	295
	4.4.3	Defining Flow Types	296
	4.4.4	Assigning Flow Types to Update Types	299
	4.4.5	Defining Condition Types	300
	4.4.6	Assigning Condition Types to Transaction Types	300
	4.4.7	Position Management Procedure	301
	4.4.8	Assigning Update Types for Valuation	302
	4.4.9	Assigning Update Types for Derived Business Transactions	303
4.5		**Commodity Derivatives**	305
	4.5.1	Commodity Types	305
	4.5.2	Specifying Commodities	306
	4.5.3	Specifying Market Identifier Codes	307
	4.5.4	Defining the Exchange	308
	4.5.5	Assigning Exchanges to Market Identifier Codes	309
	4.5.6	Defining Product Types	310
	4.5.7	Defining Transaction Types	311
	4.5.8	Defining and Assigning Flow Types	312
	4.5.9	Defining Update Types and Assigning Usages	315

		4.5.10	Updating Types for Position Updates	317
		4.5.11	Indicating that Update Types Are Relevant to Posting	318
4.6	Summary			319

5 General Contract Processes — 321

5.1	Create Financial Transaction App			321
	5.1.1	High-Level Process Flow		322
	5.1.2	Initial Screen		323
	5.1.3	Header Information		325
	5.1.4	Structure Tab		326
	5.1.5	Additional Tabs		332
5.2	Contract Settlement			342
5.3	Transaction Posting			344
	5.3.1	Account Assignment Reference		345
	5.3.2	Post Flows		346
	5.3.3	Run Accrual/Deferral		350
	5.3.4	Run a Key Date Valuation		354
	5.3.5	Transfer from Long Term to Short Term		358
5.4	Payment Processing			364
5.5	Summary			368

6 Treasury Management–Specific Processes — 369

6.1	Debt and Investments			369
	6.1.1	Process Flow		370
	6.1.2	Borrowing or Invest		371
	6.1.3	Risk-Free Interest Rates		379
	6.1.4	Interest Rate Adjustments		383
	6.1.5	Planned Records		387
6.2	Facilities			390
	6.2.1	Process Flow		390
	6.2.2	Facility Entry Screen		391
	6.2.3	Bilateral Facility Entry		397
	6.2.4	Syndicated Facility Entry		398
	6.2.5	Facility Drawings		399

Contents

6.3	**Intercompany Loans**		402
	6.3.1	Process Flow	402
	6.3.2	Contract Creation	404
	6.3.3	Mirroring	406
	6.3.4	Settlement	411
	6.3.5	Changes	413
6.4	**Securities**		415
	6.4.1	Process Flow	416
	6.4.2	Accounts	417
	6.4.3	Classes	418
	6.4.4	Contracts	422
	6.4.5	Accounting Postings	428
	6.4.6	One-Off Postings	434
6.5	**Foreign Exchange**		438
	6.5.1	Process Flow	439
	6.5.2	Contract Creation	440
	6.5.3	Contract Posting	441
	6.5.4	Monthly Processing	442
	6.5.5	Transaction Settlement	450
	6.5.6	Other Foreign Exchange Processes	452
6.6	**Interest Rate Swaps**		459
	6.6.1	Process Flow	460
	6.6.2	Entry Screen	461
	6.6.3	Interest Rate Adjustments	464
	6.6.4	Cash Flows	465
	6.6.5	Mark-to-Market Valuation	465
6.7	**Trade Finance**		468
	6.7.1	Process Flow	468
	6.7.2	Letter of Credit Contract Creation	468
6.8	**Commodity Contracts**		476
	6.8.1	Process Flow	476
	6.8.2	Contract Creation	477
	6.8.3	Cash Settlement	480
	6.8.4	Contract Settlement	483
6.9	**Summary**		484

7 Exposure Management — 485

7.1	**Raw Exposures**	486
7.2	**Raw Exposures Apps**	491
	7.2.1 Process Raw Exposures – Collective Processing App	491
	7.2.2 Import Raw Exposures – Spreadsheet App	494
7.3	**Releasing Raw Exposures**	495
7.4	**Exposure Configuration**	497
	7.4.1 Defining Global Settings	498
	7.4.2 Defining Periods	498
	7.4.3 Defining Exposure Types	499
	7.4.4 Settings for Free Attributes	502
	7.4.5 Maintaining Release Procedures	503
	7.4.6 Assigning Users and Roles to Release Steps	504
	7.4.7 Defining Product Types for Exposures	509
	7.4.8 Defining Exposure Position Types	510
	7.4.9 Defining the Derivation Strategy for Exposure Fields	510
	7.4.10 Defining Exposure Origins	513
7.5	**Summary**	514

8 Cash Flow Hedging — 515

8.1	**Hedge Management Configuration**	517
	8.1.1 Defining Hedging Classifications	518
	8.1.2 Defining Target Quota Types	519
	8.1.3 Defining Authorization Groups for Hedging Areas	522
	8.1.4 Defining Hedge Request Reasons	522
8.2	**Hedge Accounting for Positions Configuration**	523
	8.2.1 Defining Designation Types	523
	8.2.2 Defining Product Types for Exposure Subitems	524
	8.2.3 Assigning Update Types to Product Types for Exposure Subitems	525
	8.2.4 Assigning General Valuation Classes to Product Types	526
	8.2.5 Defining Hedge Accounting Calculation Types	526
	8.2.6 Effectiveness Test	531
	8.2.7 Defining Hedging Profiles	534
	8.2.8 Defining Update Types and Assigning Usages	538
	8.2.9 Assigning Update Types to Hedging Business Transactions per Product Type	539
	8.2.10 Defining and Activating Groups	540

Contents

8.3	**Hedging Area**		**541**
	8.3.1	Main Data	542
	8.3.2	General Settings	545
	8.3.3	Currencies	547
	8.3.4	Filters for Exposures	548
	8.3.5	Filters for Hedges	550
	8.3.6	Target Quotas	551
	8.3.7	FX Hedge Request	552
	8.3.8	Hedge Accounting I	554
	8.3.9	Hedge Accounting II	557
	8.3.10	Administration	558
8.4	**Cash Flow Hedging Process**		**558**
	8.4.1	Taking a Snapshot	559
	8.4.2	Hedge Management Cockpit	561
	8.4.3	Processing Hedge Requests	578
	8.4.4	Creating a Financial Transaction	580
	8.4.5	Managing Hedging Relationships	582
8.5	**Period End Closing**		**591**
	8.5.1	Calculating Net Present Values	591
	8.5.2	Running the Valuation	592
	8.5.3	Executing Classification	592
8.6	**Contract Close**		**594**
	8.6.1	Dedesignating Hedging Relationships	595
	8.6.2	Posting Cash Flows	595
	8.6.3	Posting Derived Business Transactions	596
	8.6.4	Executing Reclassification	597
8.7	**Summary**		**597**

9   Balance Sheet Hedging   599

9.1	**Reviewing Balance Sheet Risk**		**600**
9.2	**Taking a Foreign Exchange Snapshot for Balance Sheet Risk**		**602**
	9.2.1	Taking a Snapshot for Balance Sheet Foreign Exchange Risk	602
	9.2.2	Scheduling Treasury Middle Office Jobs	605
9.3	**Balance Sheet Risk Overview Reporting**		**608**
9.4	**Processing Hedge Requests**		**609**
	9.4.1	Process Hedge Requests – Balance Sheet FX Risk App	609
	9.4.2	Schedule Treasury Middle Office Jobs App	610

9.5	SAP Trading Platform Integration		613
	9.5.1	Manage Trade Requests App	615
	9.5.2	Manage Trades App	624
	9.5.3	Manage Block of Trade Requests App	627
	9.5.4	Counterparty Limit Utilization	628
	9.5.5	Configuration	629
9.6	Master Data Setup		636
9.7	Summary		645

10 Correspondence 647

10.1	Configuration		648
	10.1.1	Defining Communication Channels	649
	10.1.2	Defining Format Metatypes	650
	10.1.3	Defining Formats	651
	10.1.4	Defining Correspondence Recipient Types	652
	10.1.5	Defining the Correspondence Class	652
	10.1.6	Defining the Correspondence Class for Inbound Process	653
	10.1.7	Defining Communication Profiles	654
	10.1.8	Assigning Formats	658
	10.1.9	Defining the Correspondence Partner	659
	10.1.10	Assigning Format Mapping for Outbound and Inbound Process	663
10.2	Inbound and Outbound Process Settings		663
	10.2.1	Inbound Process	664
	10.2.2	Outbound Process	665
10.3	Additional Correspondence Settings		667
	10.3.1	Defining Correspondence Activity	667
	10.3.2	Defining Start and End Fields for Sequences in SWIFT Messages	669
	10.3.3	Dynamic Table Assignment for Configuration	669
	10.3.4	Defining BIC Codes and Accounts for Business Partners	670
	10.3.5	Setting Up Number Ranges	671
10.4	Correspondence Mapping Rules		671
	10.4.1	SAP GUI	671
	10.4.2	SAP Fiori	672
10.5	Outbound Messaging		683
	10.5.1	Correspondence from a Financial Transaction	683
	10.5.2	Manual Correspondence	684

10.6 Inbound Messaging ... 687
- 10.6.1 Automatic Inbound Processing ... 687
- 10.6.2 Manual Inbound Processing ... 688
- 10.6.3 Types of Incoming Messages ... 689

10.7 Correspondence Matching ... 690
- 10.7.1 Configuration ... 690
- 10.7.2 Automatic Matching ... 691
- 10.7.3 Manual Matching ... 692

10.8 Display Correspondence ... 693
- 10.8.1 Displaying Correspondence in the Financial Transaction ... 694
- 10.8.2 Correspondence Monitor ... 694

10.9 Alerts ... 699
- 10.9.1 Configuration ... 699
- 10.9.2 Monitoring Alerts ... 702

10.10 Summary ... 703

11 SAP Treasury Analyzers   705

11.1 Credit Risk Analyzer ... 705
- 11.1.1 Defining Limits ... 706
- 11.1.2 Single Transaction Checks ... 707
- 11.1.3 Interim Limits ... 708
- 11.1.4 End-of-Day Processing ... 709
- 11.1.5 Review Limit Utilizations App ... 710
- 11.1.6 Reviewing Bank Risk ... 713
- 11.1.7 Reviewing the Deal Default Risk Limit ... 714
- 11.1.8 Configuration ... 715

11.2 Market Risk Analyzer ... 724
- 11.2.1 Market Data Collection ... 725
- 11.2.2 Mark-to-Market Valuation ... 727
- 11.2.3 Sensitivity Analysis and Simulations ... 728
- 11.2.4 Value at Risk ... 733
- 11.2.5 Configuration ... 734

11.3 Summary ... 741

12 Integration with Other Areas 743

12.1 Cash Management 744
12.1.1 Assigning Planning Levels 746
12.1.2 Specifying Update Types for Cash Management 747
12.1.3 Basic Settings for Cash Management Integration 747
12.1.4 Managing Substitution and Validation Rules: Treasury Flows 748
12.1.5 Cash Flow Analyzer 751
12.2 Accounting 752
12.3 Payments 753
12.3.1 Flow Types 754
12.3.2 Posting Specifications 754
12.3.3 Assigning General Ledger Accounts 755
12.3.4 Contract: Payment Details 756
12.3.5 Payment Requests and Payment Execution 758
12.4 In-House Cash and In-House Banking 759
12.5 Summary 761

13 Reports, Key Performance Indicators, and Alerts 763

13.1 Standard SAP Fiori Reports 763
13.1.1 Treasury Position History App 764
13.1.2 Treasury Position Values App 765
13.1.3 Display Treasury Position Flows App 766
13.1.4 Display Treasury Posting Journal App 767
13.1.5 Interest Rate Overview App 769
13.1.6 Debt and Investment Maturity Profile App 772
13.1.7 Debt and Investment Analysis App 773
13.1.8 Foreign Exchange Overview App 774
13.1.9 Market Data Overview App 775
13.1.10 Credit Line Analysis App 776
13.1.11 Display Treasury Alerts App 777
13.1.12 Cash Flow Analyzer App 780
13.2 SAP Fiori Analysis Reports 782
13.2.1 Filtering and Designing Analysis Reports 782
13.2.2 Treasury Position Analysis Apps 791

Contents

13.3 **SAP GUI Reports** 794
    13.3.1 Treasury: Journal of Financial Transactions Report 794
    13.3.2 Transaction Release: Work Item Overview and Status of all Transactions Report 794
    13.3.3 Treasury: Change Documents for Transaction Report 795
    13.3.4 Journal: Transactions with Cash Flows Report 797
    13.3.5 Rate/Price Adjustment Schedule Report 798
    13.3.6 Facilities: Credit Lines and Utilization Report 798
    13.3.7 Facilities: Lines of Credit, Drawing, and Fees Report 799
    13.3.8 Treasury Position Flows Report (Classic View) 800
    13.3.9 Treasury Posting Journal Report (Classic View) 800
    13.3.10 Money Market: Collective Processing Report 801
    13.3.11 FX: Collective Processing Report 802

13.4 **Summary** 803

Conclusion 805

The Authors 809

Index 811

Preface

For the past eighteen years, we've consulted in the treasury and risk management space and witnessed firsthand the evolution of SAP's powerful solution. We've seen treasury and risk management grow from the very early versions of SAP ERP into a comprehensive, sophisticated functionality within SAP S/4HANA that's capable of addressing the unique needs and requirements of some of the most complex treasury organizations. Our journey of implementing the treasury and risk management solution for a vast array of Fortune 500 clients has given us a unique perspective on the advantages and limitations of the solution. During our time as consultants, we've celebrated the solution's advancements and innovations but at the same time become well aware of the challenges and complexities involved in successfully implementing it. Our experience has also allowed us to recognize the areas where treasury and risk management excels, such as its robust treasury capabilities, and where it falls short, such as the steep learning curve that can overwhelm new users.

Objective of the Book

Our treasury and risk management consulting career began at a time when SAP was not the household name it is today and high-quality resources were hard to come by. In the early days, finding reliable information on treasury and risk management was very difficult. There were no comprehensive guides, online forums, or hands-on classes. Unlike today, when you can hardly attend a sporting event without seeing the SAP logo, back then, SAP's visibility and resources were minimal, especially in the area of treasury. Despite the challenges of learning a new discipline, we persevered by learning through hands-on experience, trial and error, and a relentless pursuit of knowledge. Today, while information on SAP is more accessible, we still see significant gaps in comprehensive and practical treasury and risk management resources.

These gaps underscore the need for a single source of truth that not only explains the system's full capabilities but also offers actionable insights that users can put into practice. All too often, we receive calls and emails from companies that struggle to navigate the complexities of the system, leading to inefficient use and underutilization of its capabilities. Also, many existing resources also gloss over the key details or are too technical for practical day-to-day use—and that sparked our idea to write a comprehensive guide that would not only explain the functionalities of treasury and risk management but would also provide practical insights and strategies for implementation and hands-on application. Our goal was to write a book that would serve as a comprehensive manual for consulting professionals and end users, guiding them through the complex world of treasury and risk management with clarity and confidence.

Learning treasury and risk management for the first time felt like building a house with an incomplete blueprint. Each piece of information was like a separate piece of the house—important on its own but not sufficient without the context of the plan for the entire house. When we're focusing solely on one small area, such as the framing or the electrical wiring, it's hard to step back and see that that we're actually building an entire house. Our goal in this book is to provide the full picture. We aim to discuss in detail the various areas of treasury and risk management and also provide the context of how it all fits together into one unified solution. Our objective with this approach is to make sure that users understand the individual components in detail—but it's also extremely important that they understand how all the components fit together and integrate to form a comprehensive treasury management system.

We still vividly remember the countless hours we spent trying to understand technical treasury and risk management terms and the trial and error involved in fumbling through configuration to create new instruments. This book is a product of our journey, and it's written from the perspective of novices who've become experts. We've distilled our experiences and lessons learned into a format that we believe will be most helpful to our readers. By sharing our own consulting experience and approach, we aim to make the learning curve less steep and the implementation process more manageable for anyone embarking on the treasury and risk management journey. Writing this book from the perspective of once struggling consultants ensures that it's not only informative but also relatable and applicable to real-world scenarios that you'll encounter in every project.

Our long-term goal is that this book will empower more companies to harness the full potential of treasury and risk management and increase awareness of the powerful capabilities the treasury and risk management solution has to offer. We've structured the book to address real-world scenarios, providing step-by-step instructions, tips, and examples that reflect the common challenges we've encountered. Whether you're a seasoned consultant looking to deepen your understanding or a new consultant just starting your journey, this book is designed to be your go-to resource for mastering treasury and risk management. By bridging the gap between theory and real-world treasury scenarios, we aim to provide you with the insights and confidence you need to effectively navigate the complex world of treasury and risk management.

Content and Structure of the Book

The content this book covers is designed to focus on the most common scenarios and financial instruments used in treasury management today. Rather than covering every possible instrument available for configuration in the system, this book intends to build a baseline understanding by focusing on the most commonly used instruments that we see in treasury departments. In other words, we concentrate on addressing the 80% of scenarios that are most frequently encountered, rather than spending time on

Content and Structure of the Book

less common edge cases. By focusing on these frequently used elements, we aim to provide practical and relevant insights that align with the day-to-day operations of treasury departments. Our approach is grounded in real-world scenarios, reflecting the challenges and situations that most treasury professionals face.

In **Chapter 2** and **Chapter 3**, we explain the configuration and setup required for all financial instruments within treasury and risk management. You'll find that many of the treasury and risk management configuration steps are foundational and are required whether you're setting up a simple commercial paper contract or a more complex derivative instrument. Understanding the foundational knowledge is important as it establishes the baseline configuration necessary for all instruments, regardless of type. By understanding these core setup requirements, you can gain a comprehensive view of the system's flexibility and insight into how to tailor the base configuration settings based on unique requirements. These chapters serve as the groundwork for more advanced configurations and ensures that users have a solid starting point.

In **Chapter 4**, we dive into the configuration settings that are common to all instruments, and in **Chapter 5**, we move on to the configuration settings that are instrument specific. Each instrument type has variability within the configuration that is built on top of the baseline configuration. We provide step-by-step instructions, starting from the initial instrument creation to the last step, in which we define the accounting postings for each instrument. This includes detailed screenshots of the configuration screens to guide you through the entire process. Each section begins with an overview of the financial instrument and its application in real-world treasury operations. We then move into the specifics of the initial setup, highlighting the key nodes and settings that need to be configured. Alongside the technical steps, we incorporate practical insights and tips drawn from our consulting experience, helping readers understand not just the "how" but also the "why" behind each configuration step. To enhance the instrument-specific configuration examples, we include a variety of practical examples that mirror the scenarios treasury departments encounter in their daily operations. By focusing on these commonly used instruments, we ensure that readers can configure and implement them effectively before moving on to more complex derivations of the instruments.

In **Chapter 6**, we turn our attention to the user side of the process. We walk through the steps of creating and processing contracts within treasury and risk management, providing an in-depth look at the daily operational tasks that users will encounter in their job while navigating the system. This chapter offers detailed instructions on how to execute the day-to-day tasks, such as entering contract details, managing approvals, processing associated journal entries, and performing month-end activities. Providing practical examples and detailed screenshots, we ensure that users can follow along with ease and gain a thorough understanding of how to manage contracts from inception to completion. Heavily focusing on treasury users' day to day activities in SAP S/4HANA helps us bridge the gap between system configuration and practical

21

applications, ensuring that users can efficiently navigate the solution and make the most of treasury and risk management's capabilities.

Furthermore, **Chapter 6** covers common issues and troubleshooting tips related to contract management, offering solutions to potential problems that users may face. We discuss best practices for entering and maintaining contracts, monitoring contract performance, and ensuring compliance with regulatory requirements. By spending time explaining the user side of treasury and risk management, we aim to equip users with the knowledge and skills they need to confidently handle complex contract management scenarios.

In **Chapter 7**, **Chapter 8**, and **Chapter 9**, we turn our focus to the analysis and risk management capabilities within the treasury management process. Starting with hedge and exposure management, we explore how treasury and risk management helps organizations identify, measure, and manage their exposure to various financial risks. We cover the steps involved in setting up and using hedge management tools within treasury and risk management, ensuring that users can implement effective hedging strategies to protect against market volatility. These chapters include practical examples of hedge accounting and risk mitigation techniques, highlighting the system's capabilities in managing complex financial instruments.

Chapter 10 covers the functionality and processes involved in managing correspondence within the system. It covers how SAP S/4HANA enables companies to automate and track communication with counterparties, trading platforms, and other stakeholders, using templates, output types, and triggers to generate and send documents such as trade confirmations and notifications. In **Chapter 11**, we'll dive into the Credit Risk Analyzer and Market Risk Analyzer, which are essential tools for assessing and managing the financial risks associated with credit exposures and market fluctuations. Our coverage includes step-by-step guides on configuring these analyzers, interpreting their outputs, and integrating their insights into broader risk management strategies.

In **Chapter 12**, we discuss how treasury and risk management fits into the overall SAP architecture and interacts natively with other areas, such as SAP S/4HANA Finance, the general ledger, and payment processing. Treasury and risk management's integration with other areas of SAP S/4HANA is extremely important and ensures seamless data flow and cohesive functionality across the enterprise. We illustrate how treasury and risk management's native integration with other areas enables seamless integration without the need for a custom interface.

Finally, in **Chapter 13**, we explore the robust reporting capabilities of treasury and risk management. Effective reporting is crucial for providing transparency, supporting compliance, and facilitating strategic decision-making, and we cover the various reporting tools available within treasury and risk management, including standard reports and customizable options. This chapter also provides guidance on generating and interpreting reports that offer insights into net positions, risk exposures, and financial performance. By understanding the reporting tools available, users can ensure that they can meet the necessary reporting requirements using standard treasury and risk management reporting.

Acknowledgements

We would like to extend our gratitude to several individuals and teams who have made significant contributions to the creation of this book.

First and foremost, we would like to thank Meagan White and Emily Nicholls of SAP PRESS for their support and guidance throughout the publishing process. Your expertise and dedication have been invaluable in bringing this project to fruition.

We would also like to recognize Christian Mnich, Haresh Chhaya, Arif Esa, and Michael Lucente from SAP for their steadfast dedication to the long-term vision of SAP S/4HANA Finance for treasury and risk management. Their commitment to driving new innovations and shaping the future of treasury solutions ensures that SAP will continue to meet the evolving needs of corporations. Their work has been pivotal in advancing the capabilities of treasury and risk management, ensuring its sustained success in a dynamic financial landscape.

Thank you all for your contributions to and belief in this project. We hope that the knowledge and insights shared within these pages will empower readers to navigate the complexities of treasury and risk management with confidence.

Chapter 1
Introduction to Treasury and Risk Management with SAP

In the fast-paced and ever-evolving world of corporate finance, efficient treasury and risk management has become key to sustaining business growth and financial stability. This is a playbook for professionals navigating the intricacies of this complex functionality, and it will provide you with detailed insights, practical applications, and strategic approaches. Whether you are new to SAP treasury, a seasoned consultant, or an end user, this book will equip you with the knowledge and tools you need. Join us on this journey to mastering one of the most sophisticated treasury management software packages in the industry.

In recent years, the world has undergone profound changes due to the COVID-19 pandemic, geopolitical tensions, and rising interest rates, all of which have accelerated the digital transformation of many organizations. These changes have reshaped financial priorities and highlighted the need for reliable treasury and risk management solutions. Companies now face increased complexity in managing liquidity, optimizing debt and investments, and mitigating financial risks. As markets become more volatile and financial instruments more diverse, the need for a sophisticated treasury management system becomes greater than ever. This is where *SAP S/4HANA Finance for treasury and risk management* (which we will call *treasury and risk management* for short hereafter) comes into play. It provides organizations with a comprehensive solution for how to manage their treasury operations effectively.

In this chapter, Section 1.1 examines how advances in technology are transforming treasury management from a primarily operational focus to a strategic role within organizations. It highlights the impact of automation, real-time analytics, and integrated systems on decision-making processes. Section 1.2 provides a comprehensive overview of treasury and risk management, outlining its core functionalities, including hedge and exposure management, the market and credit risk analyzer, and transaction execution. Section 1.3 explores the SAP Fiori user experience, which enhances the usability of treasury and risk management. Lastly, Section 1.4 covers supported treasury contracts, detailing the various financial instruments and contracts that treasury and risk management accommodates, such as foreign exchange contracts, money market instruments, bonds, and derivatives.

1 Introduction to Treasury and Risk Management with SAP

1.1 Changes in Technology and the Evolving Corporate Treasury Function

The role of corporate treasury within the organization has experienced significant changes and transformations in recent years. Advances in technology have primarily driven these changes, raising treasury professionals' expectations for software and technology. Fifteen or twenty years ago, most treasury departments didn't use specialized treasury software to manage their operations—the running joke in treasury consulting was that Microsoft Excel was the most widely used treasury management software in the world. Every client we interacted with seemed to be managing their entire treasury department using spreadsheets, and it was very common for the entire treasury department to rely on a slew of spreadsheets to manage its day-to-day operations. The small percentage of treasury departments that used treasury management software had very low expectations for the solution. This is in stark contrast to today's treasury departments, which see sophisticated treasury management systems as necessary for conducting daily treasury operations.

From our perspective, a modern treasury management system must prioritize three main things:

- **Real-time data and analytics**
  Efficiently managing contracts in a system is great, but how can the data be reported on and used to make quick decisions?
- **Native integration with third-party treasury providers**
  Most treasury departments conduct trading activity by using a trading portal that must integrate with their treasury software.
- **Standardized processes and controls**
  Treasury software should create efficiency across the organization, but more importantly, it should instill standard processes and controls.

One of the most dramatic changes in the industry has been the shift toward leveraging reliable data for decision-making. In the past, treasury departments often relied on fragmented data sources, leading to inconsistencies and inefficiencies in financial reporting and analysis. Today, integrated treasury management systems provide a single source of truth, enabling treasurers to access accurate and consistent data across all financial operations. This shift hasn't only improved the accuracy of financial information, but it has also enhanced the ability to make informed decisions quickly and confidently.

With advances in technology in recent years, treasury departments now expect to have real-time, accurate information at their fingertips. Enhanced reporting and analytical capabilities are now must-haves for the modern treasury organization. Modern treasury management systems offer advanced reporting tools that provide detailed insights into cash flow, liquidity, and risk exposures, and these tools allow treasurers to generate comprehensive reports that meet the needs of internal stakeholders and comply with regulatory requirements. The ability to customize reports and dashboards has empowered treasury departments to present data in a more meaningful way, facilitating better communication with senior management and other key stakeholders. This has led to more strategic discussions about financial health and risk management, ultimately contributing to more effective treasury operations.

The evolution of corporate treasury over the last decade has been marked by a move toward data-driven decision-making, enhanced reporting, and real-time information. These changes haven't only improved the operational efficiency of treasury departments but have also elevated the departments' strategic importance within organizations. By leveraging reliable data, advanced reporting tools, and real-time insights, corporate treasurers are now better equipped to manage financial risks, optimize liquidity, and contribute to the overall financial stability and growth of their organizations.

In this rapidly changing environment, treasury and risk management stands out as a natural fit for organizations for several reasons. One of the primary advantages of treasury and risk management is its ability to provide real-time data from all the other areas of SAP S/4HANA. Unlike traditional systems that rely on manual interfaces and disparate data sources, treasury and risk management integrates seamlessly with the broader SAP ecosystem. This means that all the data needed for informed decision-making is housed in one centralized location within SAP S/4HANA. This integration eliminates the inefficiencies associated with manual data handling, ensuring that treasurers have access to the most up-to-date and reliable information at all times. This real-time data availability is crucial for managing liquidity, monitoring cash flows, and assessing financial risks promptly and accurately.

1.2 Treasury and Risk Management with SAP

Before we jump into the system setup and configuration, we need to make clear what treasury and risk management entails. Treasury and risk management is designed to help organizations efficiently manage their treasury transaction operations. It provides comprehensive tools for various treasury functions, including financial contract management, hedge and exposure management, and risk mitigation. By leveraging treasury and risk management, treasury departments can gain better control over their financial assets and liabilities, streamline processes, and ensure compliance with regulatory reporting requirements. Treasury and risk management is crucial for organizations aiming to optimize their treasury operations and mitigate financial risks, and the core components of treasury and risk management are designed to address the diverse needs of the modern treasury department. We further define the core components in this section.

By utilizing treasury and risk management, treasury departments can achieve enhanced control over their financial assets and liabilities, streamline complex processes, and ensure adherence to regulatory reporting requirements. This capability is particularly vital for organizations that aim to optimize their treasury operations while minimizing financial risks in an increasingly dynamic and complex financial environment.

In this section, we'll start by walking through the current scope and use of treasury and risk management. Then, we'll explore its core components in detail, laying the foundation for understanding how they contribute to the overall effectiveness of the treasury function. This foundational knowledge will be essential as we proceed to the more technical aspects of system setup and configuration, ensuring that you're well-equipped to leverage treasury and risk management to its full potential.

1.2.1 Current Scope and Use

Treasury and risk management is incredibly powerful and has undergone significant enhancements in usability, particularly with the rollout of SAP S/4HANA and the introduction of the new user interface of SAP Fiori. SAP S/4HANA's in-memory computing capabilities allow for faster data processing and real-time analytics, enabling treasurers to make quick and informed decisions. The SAP Fiori interface has revolutionized the overall SAP user experience by providing a more intuitive, user-friendly solution. With SAP Fiori, users can access tailored dashboards, personalized reports, and real-time insights, making the management of treasury operations more efficient and effective. This combination of powerful functionality and improved usability ensures that organizations can leverage treasury and risk management to its full potential, optimizing their treasury operations in a user-centric environment.

From its inception, treasury and risk management has been designed to meet the evolving needs of corporate treasury departments. The first iterations of the solution laid the groundwork by providing essential tools for managing financial transactions, liquidity, and risk. As the demands of corporate treasury have grown more complex, SAP has continuously updated and expanded its functionalities to keep pace. This ongoing evolution has ensured that treasury and risk management remains relevant and capable of addressing contemporary challenges faced by treasury professionals.

One of the key strengths of treasury and risk management is its ability to continually evolve to keep up with the continually changing treasury environment and industry best practices. With the introduction of SAP S/4HANA, treasury and risk management leveraged in-memory computing to offer real-time data processing and analytics, drastically enhancing the speed and accuracy of financial reporting and risk assessment. The adoption of the SAP Fiori user interface further improved usability, making the system more intuitive and accessible for users. These enhancements demonstrate SAP's commitment to keeping its treasury solution at the forefront of innovation.

A major advantage of treasury and risk management over other treasury management systems is its longstanding presence and stability in the treasury solutions market. First introduced in the early 1990's, treasury and risk management has been a cornerstone of the corporate treasury landscape for decades. Over the years, the treasury management systems landscape has seen significant disruptions, with some companies being acquired or undergoing major transformations. Despite these changes, treasury and risk management has remained a constant and reliable solution for organizations worldwide.

Treasury and risk management is a natural fit for organizations in today's dynamic environment, due to its real-time data availability, powerful functionality enhanced by SAP S/4HANA and SAP Fiori, and its long-standing presence in the market. These attributes make treasury and risk management a comprehensive, reliable, and future-proof solution for corporate treasury departments.

1.2.2 Core Treasury Components

Let's now dive into the core components of treasury and risk management. We'll encounter each of the following components again and again throughout the book:

- **Transaction manager**

  We use the *transaction manager* to create and manage treasury contracts in SAP S/4HANA. The solution encompasses a wide range of instruments used by most treasury departments, such as debt and investments, foreign exchange contracts, trade finance, and derivatives. It ensures that every transaction that is entered into the system is meticulously recorded, monitored, and posted to the general ledger.

  The process starts with the creation of financial contracts. We enter the details of each trade manually or by using an interface with a trading platform. This includes everything from entering the initial contract details, confirming the counterparty payment details, and verifying the associated cash flows. Once we have entered the contracts into the system, we move on to the downstream activities of processing the contract to maturity. These downstream activities include general ledger postings, in which we ensure that all financial movements are correctly posted to the general ledger.

  In this book, we also cover month-end activities, which are essential for closing financial periods accurately. This involves reconciling accounts, validating the accuracy of recorded transactions, and preparing financial statements. These tasks are critical for providing a true and fair view of the organization's financial position at the end of each period.

  Moreover, we dive into the continuous monitoring and reporting processes that follow initial contract creation. This includes tracking the performance of investments, managing debt obligations, and monitoring foreign exchange (FX) positions to mitigate risks and optimize financial outcomes. We highlight the reporting tools within treasury and risk management, showcasing how we can utilize them to generate insightful reports that aid in decision-making and strategic planning.

- **Market Risk Analyzer**

  The *Market Risk Analyzer* is a vital component of treasury and risk management, offering a suite of tools designed to assess and manage various market risks effectively. It enables organizations to thoroughly analyze market data, accurately value financial instruments, and continuously monitor risk exposures.

  By utilizing the Market Risk Analyzer, businesses can gain a deeper understanding of their exposure to market variables such as interest rate fluctuations, currency exchange rate changes, commodity price shifts, and more. This comprehensive analysis allows companies to identify potential risks early and develop strategies to mitigate them, thereby safeguarding their financial stability.

  It provides advanced valuation methods and analytical techniques, which are crucial for calculating the market value of financial instruments. These methods ensure that organizations can assess their positions accurately, reflecting the true market conditions. This is particularly important for derivatives and other complex financial instruments, where precise valuation is essential for effective risk management. In addition to valuation, the Market Risk Analyzer offers robust tools for monitoring risk exposures. Organizations can set up alerts and thresholds to detect when risk levels exceed acceptable limits, enabling proactive risk management.

  The Market Risk Analyzer also supports scenario analysis and stress testing, allowing organizations to simulate various market conditions and evaluate their potential impacts on the portfolio. This helps treasury professionals understand the resilience of financial strategies under different market scenarios and in preparing contingency plans.

- **Credit Risk Analyzer**

  The *Credit Risk Analyzer* is an important component of treasury and risk management, offering comprehensive tools for evaluating and managing credit risks. It provides organizations with the ability to assess creditworthiness, monitor credit exposures, and implement effective risk mitigation strategies.

  It allows treasury organizations to analyze credit data from various sources, enabling a thorough evaluation of counterparties and financial instruments. By leveraging advanced analytical techniques, the Credit Risk Analyzer helps treasury professionals identify potential credit risks early, ensuring that organizations can take proactive measures to mitigate them.

  One of the key features of the Credit Risk Analyzer is its ability to monitor credit exposures in real time. Organizations can set limits and thresholds for credit exposures, and the system will automatically alert users when these limits are approached or breached. This continuous monitoring is essential for maintaining control of credit risks and avoiding unexpected financial losses. The Credit Risk Analyzer also supports reporting capabilities, allowing organizations to generate detailed reports on credit exposures, counterparty risk assessments, and overall credit risk profiles. These reports provide valuable insights that can inform strategic decision-making and enhance the organization's risk management framework.

Additionally, the credit risk analyzer facilitates stress testing and scenario analysis, enabling businesses to evaluate the potential impact of adverse credit events on their financial position. This helps them understand the resilience of credit risk strategies and prepare for potential challenges. By integrating with other areas of SAP S/4HANA, the Credit Risk Analyzer ensures seamless data flow and cohesive risk management across the organization. This integration allows for a holistic view of credit risks, incorporating data from various financial processes and ensuring that risk management strategies are aligned with overall business objectives.

- **Exposure management**
  Exposure management is another important component within treasury and risk management. It's designed to help organizations identify, measure, and manage their exposure to various financial risks. These risks encompass a broad range of exposures, including market risk, interest rate risk, and FX risk. The primary objective of exposure management is to offer a comprehensive view of the financial exposures a company faces and to facilitate the development of effective strategies to mitigate these risks.

  Exposure management provides tools to help us accurately capture and assess exposure data from different financial transactions and operations within the organization. By consolidating this data, exposure management helps treasury professionals gain a holistic understanding of the company's risk profile. It helps them see how various risk factors interact and impact the overall financial health of the organization.

  One of the key functionalities of exposure management is its ability to measure and quantify the extent of financial risks. Through advanced analytical models and risk assessment techniques, exposure management helps treasury professionals calculate potential losses under different scenarios. This quantitative analysis is essential for informed decision-making and developing strategies that can protect the organization from adverse financial outcomes. Exposure management also supports the implementation of risk mitigation strategies. By providing a clear picture of where the risks lie, it allows organizations to take proactive measures such as hedging, diversifying investments, or adjusting their financial strategies to minimize risk exposure. Exposure management can simulate various risk scenarios and their potential impacts, helping businesses prepare for different market conditions and economic environments.

  Exposure management seamlessly integrates with other treasury and risk management components, ensuring that risk data is consistent and comprehensive across the organization. This integration supports a unified approach to risk management, enhancing the organization's ability to coordinate risk mitigation efforts across different financial functions and departments.

- **Hedge management**
  Hedge management in treasury and risk management is a comprehensive process for managing and mitigating financial risks, including currency, interest rate, and commodity price fluctuations. It begins with identifying risk exposures, allowing companies to analyze and categorize potential financial impacts. Based on this analysis, organizations design and implement appropriate hedging strategies by using financial instruments such as forwards, options, and swaps to offset adverse market movements.

  Treasury and risk management facilitates the execution of these hedging transactions, ensuring they are accurately recorded and compliant with regulatory requirements. The system's real-time monitoring and reporting capabilities enable organizations to continuously assess the effectiveness of their hedges, making necessary adjustments to maintain optimal risk management.

Treasury and risk management's integration with other areas of SAP S/4HANA, such as financial accounting and controlling, ensures a unified view of financial data, enhancing decision-making and operational efficiency. Additionally, treasury and risk management supports detailed documentation and hedge accounting, aligning hedging activities with international accounting standards like the International Financial Reporting Standards (IFRS) and Generally Accepted Accounting Principles (GAAP), thus ensuring transparency and consistency in financial reporting.

By integrating these core components, treasury and risk management provides a holistic solution for managing treasury operations and financial risks. It enables organizations to achieve greater financial transparency, improve decision-making, and enhance their ability to respond to market changes and financial uncertainties.

1.3 SAP Fiori User Experience

In 2013, SAP introduced SAP Fiori, a completely redesigned interface that changed how users interact with SAP. Moving from the traditional transaction codes that had been a staple of SAP since its inception, SAP Fiori was designed with a focus on simplicity and adaptability. It shifted the power from IT departments to end users, enabling them to customize reports and modify tiles directly within the interface. SAP Fiori introduced a new level of customization and usability, setting a benchmark for enterprise software interfaces. This approach not only streamlined everyday tasks within SAP but also empowered users by giving them more control over how they interact with SAP.

SAP Fiori is a big shift away from the original SAP GUI, so we want to cover the basics of how to navigate SAP Fiori. In the following sections, we'll explore the core functionalities of SAP Fiori, providing a clear understanding of how it enhances the user experience and the ways users can tailor it to meet specific business needs. Specifically, we'll walk through maintaining the user profile, customizing the SAP Fiori launchpad, creating custom views, creating custom tiles, and sharing custom views.

1.3.1 Maintaining a User Profile

Each SAP user maintains a unique user profile within the system, allowing them to personalize settings that cater to their specific needs and preferences. The user profiles ensure that customized settings, such as display preferences, shortcuts, and default parameters, are applied each time a user logs in to SAP S/4HANA. SAP Fiori offers a range of user profile settings that users can customize to further streamline navigation and improve productivity. These settings allow users to configure their user screens according to their roles and tasks, enabling them to access the most relevant information and tools quickly. Figure 1.1 shows a standard SAP Fiori home screen. Note the user profile icon in the upper right-hand corner of the screen.

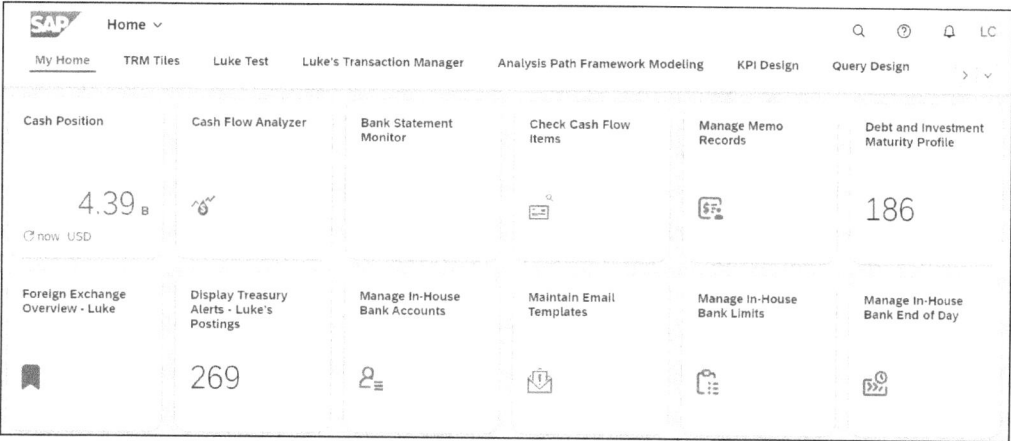

Figure 1.1 Standard SAP Fiori Home Screen

On each SAP Fiori home screen, a small icon at the top right of the screen contains the user profile menu. This icon commonly contains the initials of the logged-in user but may have a different presentation depending upon individual system settings and the theme of the SAP Fiori screen. Regardless, this menu contains the options that a user can modify per their preferences. Clicking this icon will reveal the menu displayed in Figure 1.2.

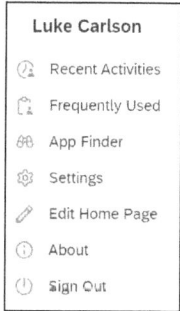

Figure 1.2 User Profile Menu Presented after Clicking User Profile Icon

The user profile menu shows many different options that the user can modify to suit their needs. Each item on this list has a different function, as follows:

- **Recent Activities**
  This item presents a list of past user actions across the SAP Fiori landscape. This function can be particularly useful for quickly recalling a previously executed action or re-executing a previous search for a particular phrase or application.

- **Frequently Used**
  This quick-action item displays frequently used activities. Once again, this can be useful if a user is completing a related task and is looking for a quick way to repeat navigation within SAP Fiori.

- **App Finder**
  A user can employ this function to search for a particular SAP Fiori app within the user's universe of available apps. It's important to note that each app's availability is controlled by the security roles and profiles that are assigned to each user. If an app is not present in **App Finder**, it doesn't mean that the application doesn't exist; it may simply be unavailable to a particular user. SAP maintains a list of all available SAP Fiori apps in the SAP Fiori apps reference library. Users can access this library at *https://fioriappslibrary.hana.ondemand.com/*.

- **Settings**
  This option allows the user to customize settings specific to the user's SAP Fiori experience. The settings menu contains many options, several of which are as follows:

  - **Appearance**
    This setting allows the user to customize the visual appearance of their SAP Fiori screen. It's important to note that these settings, like all SAP Fiori settings, are user specific and only persist for each individual user. Changing this setting won't modify the appearance of user screens across the enterprise.

  - **Home Page**
    This setting provides the user with options for how to view their initial SAP Fiori screen, otherwise known as the homepage. The user has two choices here:

    - **Show all content**
      All available and assigned SAP Fiori apps and SAP Fiori app groups (including custom groups) are presented on the homepage. This allows a user to scroll through all available apps and groups. Figure 1.3 shows the SAP Fiori homepage with the **Show all** content option selected.

    - **Show one group at a time**
      The homepage will only display one app group at a time (including custom groups). Users need to select additional app groups from the app group tabs to display the relevant apps. Figure 1.4 shows the SAP Fiori homepage with the **Show one group at a time** option selected. Note how the individual app group is selected from the group tab bar.

1.3 SAP Fiori User Experience

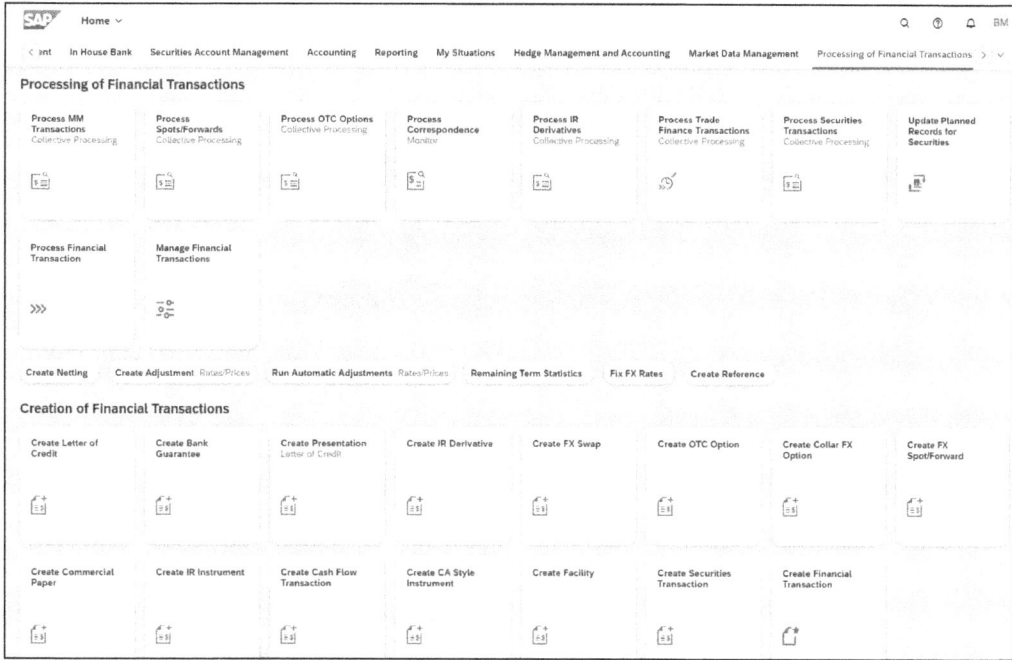

Figure 1.3 SAP Fiori Homepage with Show All Content Option Selected

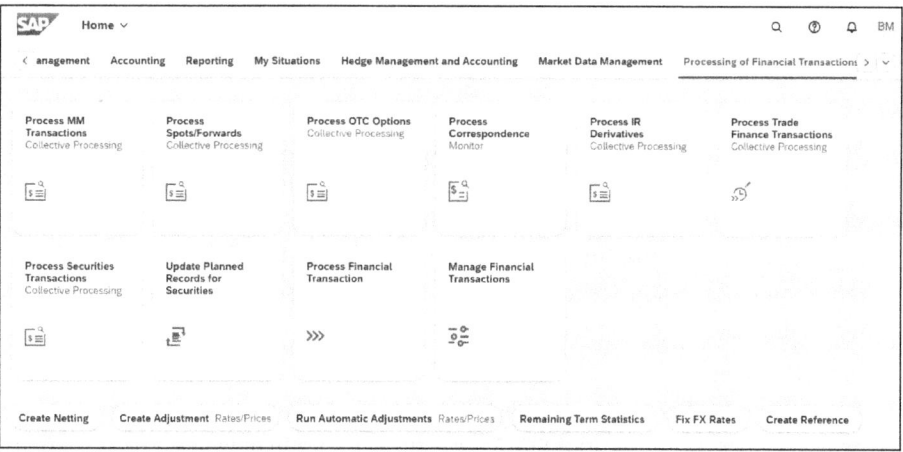

Figure 1.4 SAP Fiori Homepage with Show One Group at a Time Option Selected

- **Spaces and Pages**

    This setting changes the way the apps are presented on the homepage. Along with SAP Fiori, SAP has introduced the concept of *spaces*, which are customized groupings of SAP Fiori apps grouped by business user function. In theory, the apps that are most relevant to a specific business function are grouped together in a space. Selecting this option will group apps in this manner on the user's SAP Fiori homepage, as shown in Figure 1.5.

1 Introduction to Treasury and Risk Management with SAP

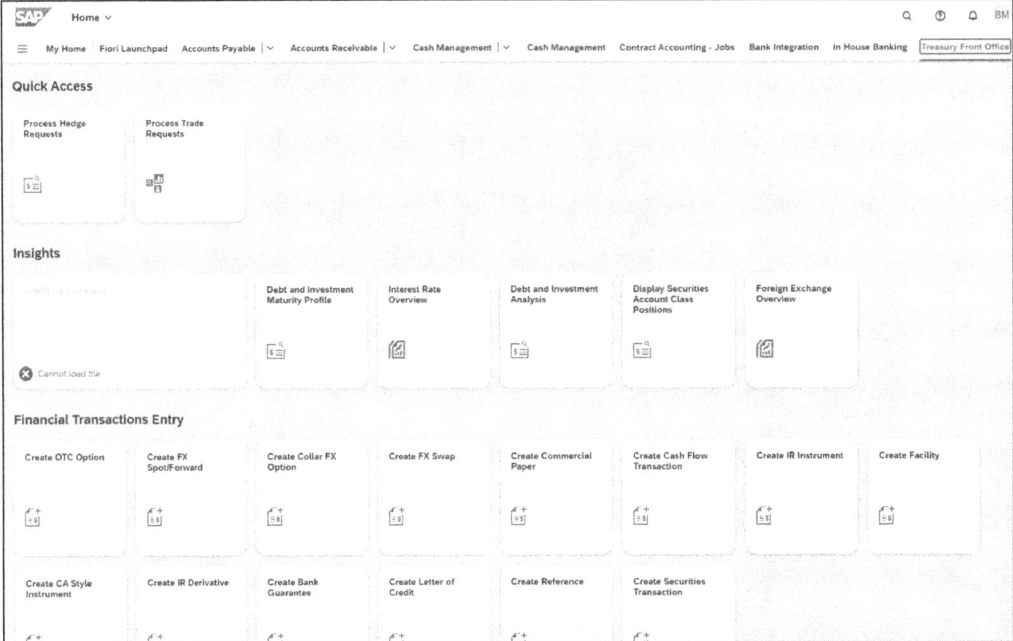

Figure 1.5 SAP Fiori Homepage with Use Spaces Option Enabled

- **User Activities**
  This is a simple toggle to enable or disable the tracking of recent activity and frequently used apps.

- **Language and Region**
  These settings customize the SAP Fiori user experience to match regional parallels. Here, the user can select the language in which the SAP Fiori menus are presented, and they can also modify their preferred systemwide date and time formats, the time zone, the week numbering, and the decimal format that the system should use. Again, it's important to note that these settings are unique to the individual user.

- **Default Values**
  The settings in this section (see Figure 1.6) allow the user to establish customized default values for the various SAP Fiori screens they will navigate. There are many fields available in this section to which to assign default values. This setting can be especially useful if a user is focused on only one particular area of a business or if the scope of business operations is relatively simple. For example, perhaps a user's role is to work exclusively with one specific company code. Using these settings, the user could assign a default value to the **Company Code** field, and that value would populate in each SAP Fiori screen that requires the input of a company code. This can save an active user significant time they would otherwise spend frequently reentering the same data.

1.3 SAP Fiori User Experience

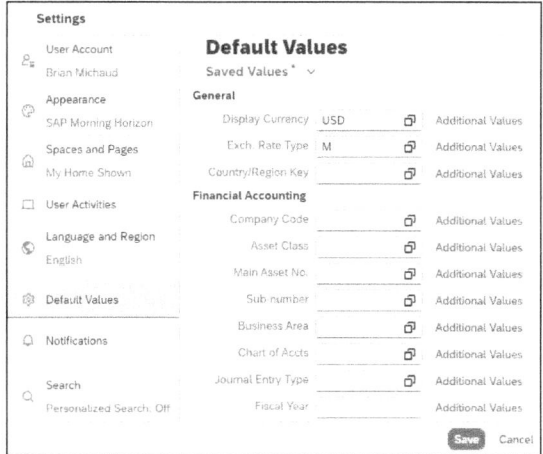

Figure 1.6 Default Values Assignment Screen in User Profile Settings Area

- **Notifications**
  This user setting allows the user to toggle on or off notifications for high-priority alerts.
- **Search**
  This setting allows the user to control personalized search functions, delete previously recorded search data, and enable or disable the personalized search scope.

- **Edit Home Page**
  Users can select this function to modify the contents of their SAP Fiori homepage. We detail the major features of this functionality in the next several sections.
- **About**
  This function presents the user with relevant information about the SAP system they are currently navigating.
- **Sign Out**
  This function allows a user to sign out of SAP Fiori.

In the next several sections, we discuss additional functions of SAP Fiori that are designed to enhance the user experience.

1.3.2 Customizing the User Screen

Whereas user settings customize the systemwide settings for each individual user, there's an additional functionality within SAP Fiori that allows users to customize the way the homepage is structured. This customization provides several benefits by allowing each user to organize their homepage to best suit their workflow. In the following sections, we'll specifically walk through adding and removing apps, creating custom user groups, and rearranging apps.

1 Introduction to Treasury and Risk Management with SAP

Adding and Removing Applications

The SAP Fiori homepage is delivered with sets of curated apps that are displayed depending upon each user's security access to the SAP system. While these apps are intended to be organized into commonly used groupings, the requirements of each user may differ from the predelivered structure. Luckily, users can modify the apps on their homepage to match their requirements.

To do this, a user selects the **Edit Home Page** function within the SAP Fiori user profile menu. Once they select this, the homepage displays an editing screen that allows the user to modify the homepage's appearance, as shown in Figure 1.7.

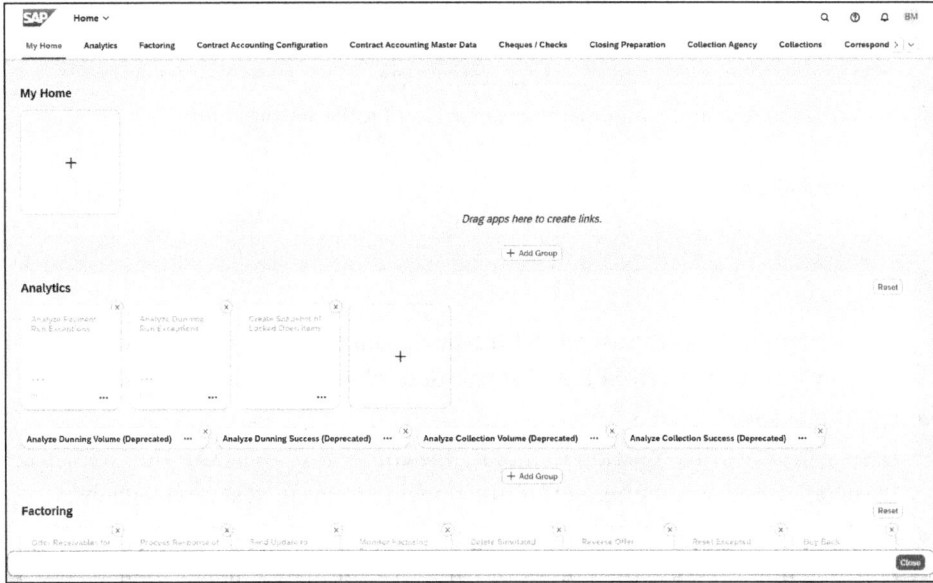

Figure 1.7 Homepage Editing Screen That Is Displayed after User Selects Edit Homepage Function

After entering the editing screen, the user can remove applications from the defined app groups by simply clicking the X in the upper right-hand corner of the tile of each app they want to remove. To add an app to an existing app group, the user selects the empty tile containing the **+** sign. This brings the user to the **App Finder** screen, as shown in Figure 1.8. Using this screen, the user can navigate to the app(s) they desire to add to the selected app group. After the user locates the app(s), they can add the app(s) to the selected app group by simply clicking the pushpin icon in the bottom right corner of each tile.

After the user clicks the pushpin icon, a small message will notify the user that the selected app has been added to the desired app group. The pushpin icon will also change color, indicating that the app has already been added to the intended app group.

1.3 SAP Fiori User Experience

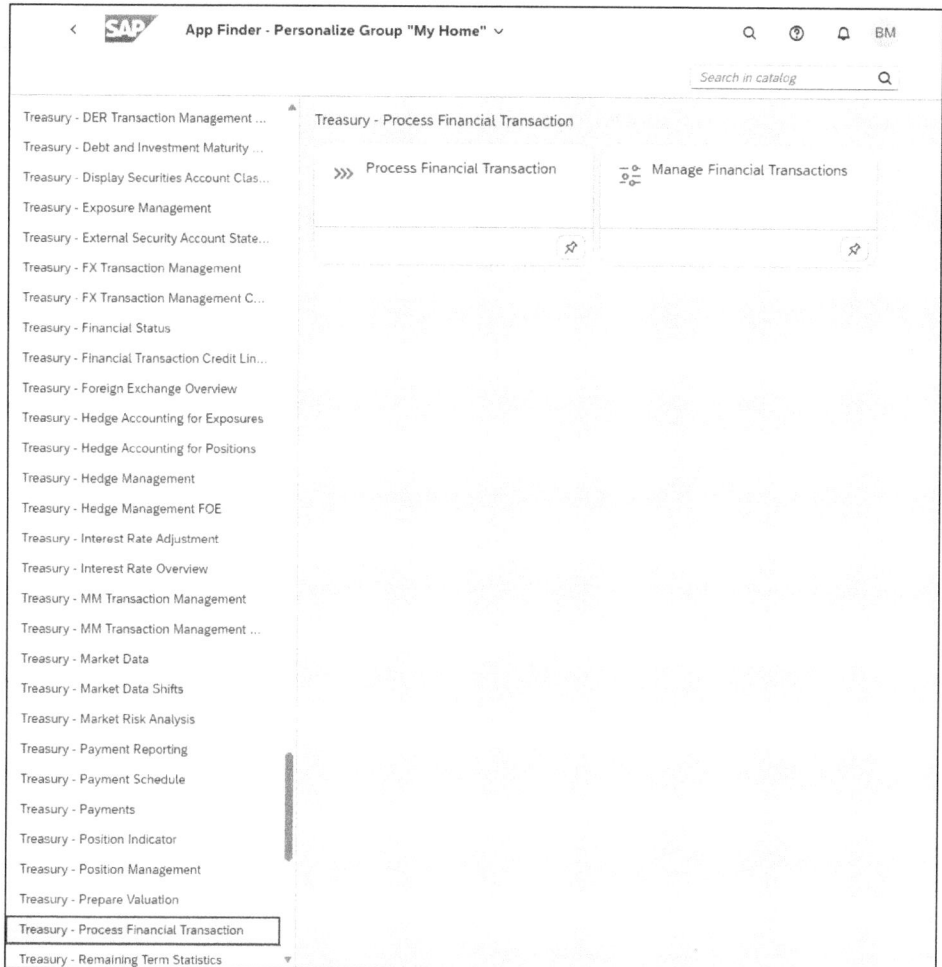

Figure 1.8 App Finder Screen Showing Pushpin Icon That Is Used to Add SAP Fiori Apps to App Groups

Creating Custom User Groups

Each SAP Fiori homepage is delivered with an additional standard group titled **My Home**. The intent of this group is to house and organize the user's most popular apps. It provides quick access to the most essential and most frequently used applications. This group is originally delivered empty, but users can easily populate it by using the steps outlined in this section.

In addition to the standard **My Home** group, an SAP Fiori user can create custom app groupings and assign them relevant titles. Again, any custom app grouping that an individual user creates is unique to that user. Using the details in this section, a user can essentially fully customize the look and feel of their homepage to optimize SAP Fiori to suit their individual needs.

1   Introduction to Treasury and Risk Management with SAP

On the homepage editing screen, there's a button titled **Add Group**. After the user clicks this button, a new group row will appear between the two groups where the **Add Group** button originally appeared. The user can then rename the newly created group with the title they need. Figure 1.9 shows a created custom group on the homepage editing screen.

Figure 1.9 Homepage Editing Screen with New Custom Group Created

Additionally, on the homepage editing screen, a user can modify groupings in several other ways:

- The user can arrange groups to suit their preferences by dragging the title of a group into the space they desire. It's important to note that the **My Home** group will always remain at the top of the screen and can't be rearranged.
- The user can rename a group by selecting the title of the group and changing the text of the title. Again, the **My Home** group can't be renamed.
- The user can reset a predelivered group to its original settings by clicking the **Reset** button in the chosen app group's row.

1.3 SAP Fiori User Experience

- The user can delete custom app groups by clicking the **Delete** button within the custom app group's row.

Any changes made to the homepage are saved in real-time. Once the user has made any changes, they will need to click the **Close** button to exit the homepage editing screen.

Rearranging SAP Fiori Applications

Users can add and remove SAP Fiori applications to and from various app groups by using the steps described above, and users can also rearrange these applications. For example, if a user wants to create a logical flow of apps that they utilize every morning when they arrive for work, they could theoretically arrange the necessary apps from left to right, one after the other, to create a simple workflow that's easy to follow. That's one simple use case for this type of functionality. Users can also organize the SAP Fiori screen further by simply rearranging the order of the apps on the screen. They can accomplish this by simply dragging and dropping apps into the preferred order on the homepage editing screen.

1.3.3 Creating Custom Views

To further enhance usability within SAP Fiori, the user has the option to create and save custom views and filters for later recall when running a report or executing a transaction. Akin to a transaction variant, the custom view functionality allows the user to manage preferred filters and customized views independently. This could prove especially useful to a user who consistently recalls the same data from a report or wants to routinely see specific information that's not present in the standard report views.

To utilize this functionality, the user will first populate the information that corresponds to the view in the filter area of a transaction (see Figure 1.10). Note the asterisk next to **Standard**, indicating that the currently displayed view has been modified. After the user appropriately populates the data, they can click on the **Select View** dropdown list to manage the view that they've just constructed.

Figure 1.10 Filter Area of SAP Fiori Application

The user has two options after electing to manage the current view:

- **Save As**
  This option allows the user to save the current view. Each view with a transaction

must be given a unique name, and if the user has modified a view that has already been created and named, the current settings will be overwritten (as with the **Save** function in Microsoft Word). The **Save As** menu also provides several other options that can be beneficial to the user:

- **Set As Default**
  This option sets the current view as the default view for the user.
- **Public**
  This option marks the view as public, meaning other users can incorporate it into their own view selections. This is a useful way to share views that may be applicable to several other users without each user needing to create their own version of the same view.
- **Apply Automatically**
  This option automatically applies the settings of the view the next time the user opens the transaction.

Figure 1.11 shows the **Save As** options for an SAP Fiori view.

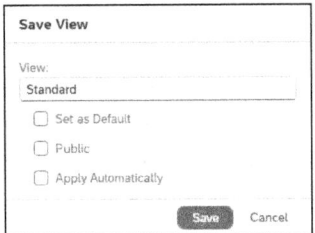

Figure 1.11 Save As Menu Options

- **Manage**
  This option allows the user to manage previously saved views, and views that are marked as public will also appear when the user selects this option. From this menu, the user can set a view as the default view, apply a view automatically, delete a view, and favorite a view. Views created by the user are automatically favorited, and the user can favorite public views created by other users on this screen and incorporate them into the user's view dropdown list for future use. Figure 1.12 shows the **Manage Views** screen and related functionalities.

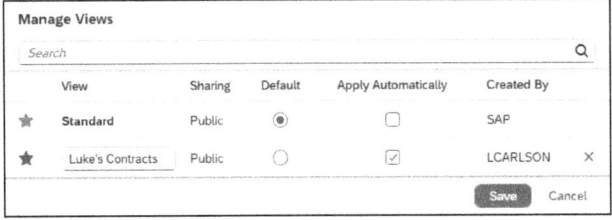

Figure 1.12 The Manage Views Menu Options

1.3.4 Creating Custom SAP Fiori Tiles

Now that we've outlined the concept of creating a custom view, we can explore how users can use those custom views to create customized SAP Fiori tiles that can reside directly on the home screen. This eliminates the need to navigate into a transaction and apply a custom view. Instead, users can select the custom tile for a transaction directly from the SAP Fiori homepage, giving them direct access to their most frequently used view and transaction combinations. This functionality can be particularly useful for high-volume users. To save a view as a custom tile, users select the **Share** menu at the top right of the screen, directly beneath the user profile icon. Figure 1.13 shows the expanded share menu.

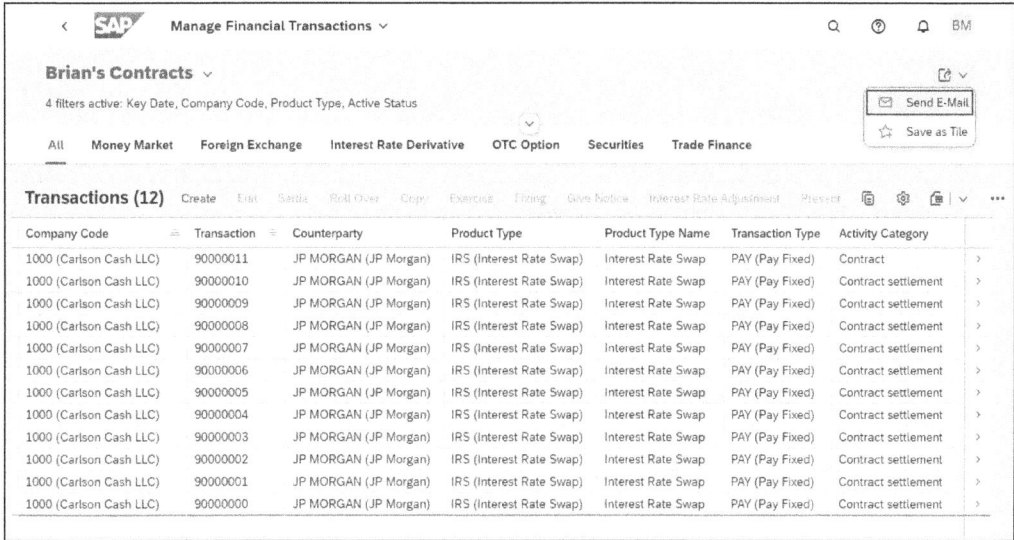

Figure 1.13 SAP Fiori Transaction Screen with Share Menu Expanded

The **Share** menu has two options for the user to select from, as follows:

- **Send E-Mail**
  This option allows the user to send the current view as an email to a separate party. We discuss this functionality in the next section.

- **Save as Tile**
  This option saves this view as a custom tile on the user's SAP Fiori homepage. A custom tile allows for quick recall of both the transaction and the saved view in a single click. When saving a view as a tile, the user can customize the look and feel of the tile. The user can assign a custom name to the tile, add a subtitle and a description, and even select the group that the tile will reside in. The user can also manually rearrange the custom tile by using other standard SAP Fiori tile rearrangement methods. Figure 1.14 shows the custom tile on the SAP Fiori homepage.

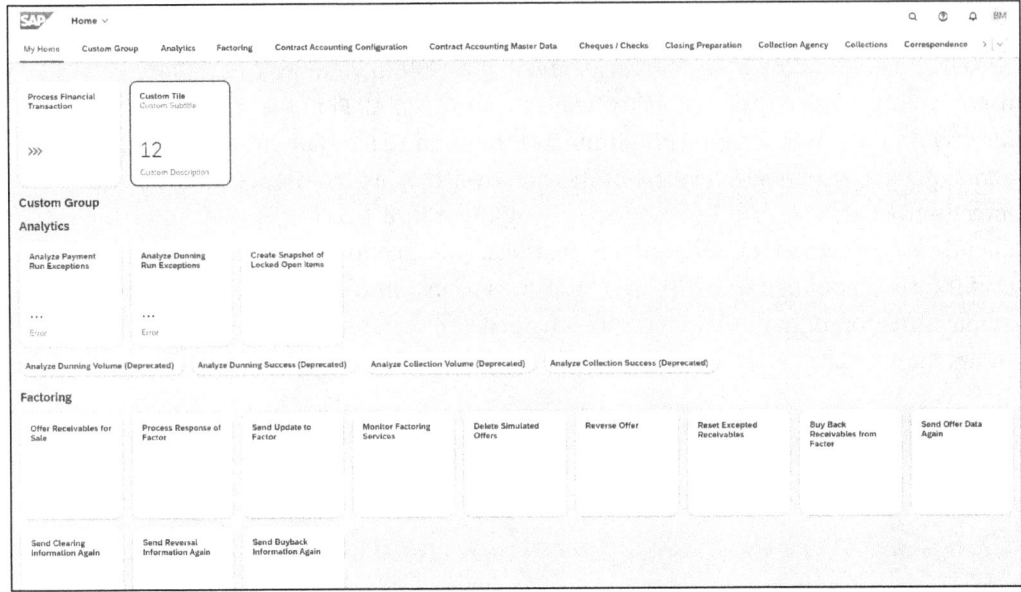

Figure 1.14 SAP Fiori Homepage Showing Custom-Created Tile

In addition to the static tiles created from a transaction's custom view, SAP Fiori allows the user to create live updating tiles that can display small amounts of real-time data. This functionality is commonly used in the cash management area when a cash manager or treasury professional may want to see a cash balance in a particular currency or for a particular company code or region.

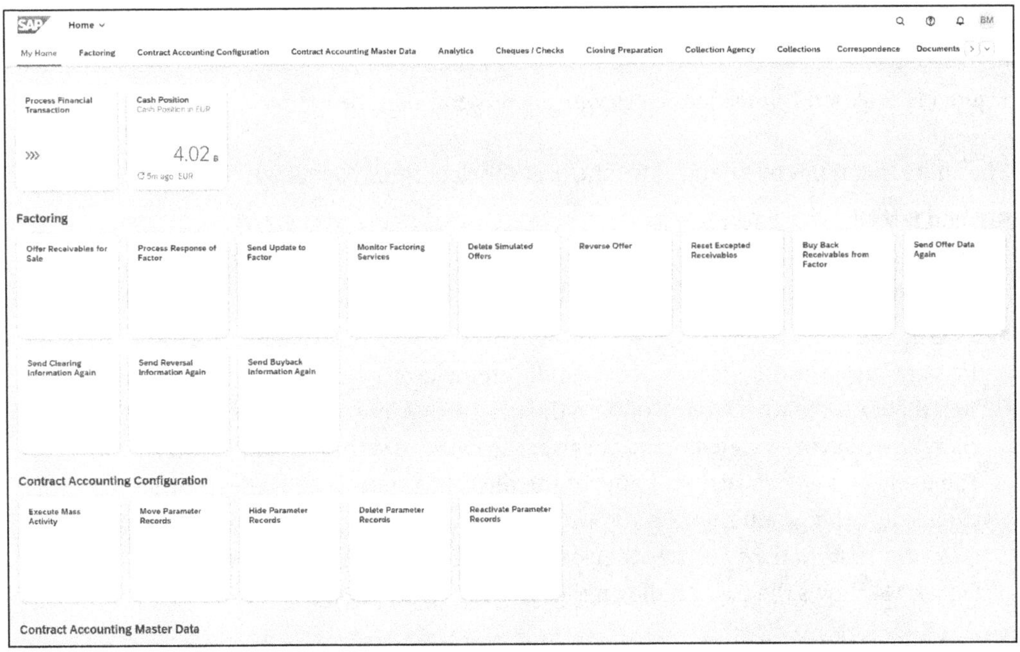

Figure 1.15 SAP Fiori Homepage Showing Custom-Created Live-Updating Tile

The user can save a tile from the app, and that tile will be a live-updating tile reflecting the desired data in real time. There's no need to reexecute the application. The data on the live tile will consistently refresh, as with a widget in many of today's mobile phone operating systems. Live tiles are created in the same way as a static tile, but only a subset of applications can generate a live tile. Figure 1.15 shows a customized live tile displaying today's cash balance in euros instead of the standard US dollars.

Custom-created tiles will automatically refresh and utilize the saved view parameters as well as available system data to provide small pieces of information in real time directly on the user's SAP Fiori homepage.

1.3.5 Sharing Custom SAP Fiori App Views

In addition to customizable views, custom static and live-updating tiles, and the ability to create custom groups and rearrange tiles as the user sees fit, SAP Fiori has an easy way to share custom views with additional users. Creating a public view is one way to share information with an additional party, and SAP Fiori also allows users to send emails with direct links to the content on the screen. This greatly improves the ability to share a replica of the screen the user is viewing with someone else.

After generating the view and staging the data the way the user intends, the user clicks the **Share** button and chooses the **Send E-Mail** option from the menu. This automatically engages the user's preferred email client and generates an email with a direct link to the transaction and view that the user shared. The best part of this functionality is that once the recipient receives the message, they can click the shared link and be taken directly to the same transaction the original user shared with the same view parameters engaged. This functionality is especially useful when two colleagues may need to review the same contract. Rather than attempting to match the view parameters, one colleague can simply share a link to the contract with the other. After one click, they can be looking at the same data. Figure 1.16 shows an example of an email generated from the **Send E-Mail** share option.

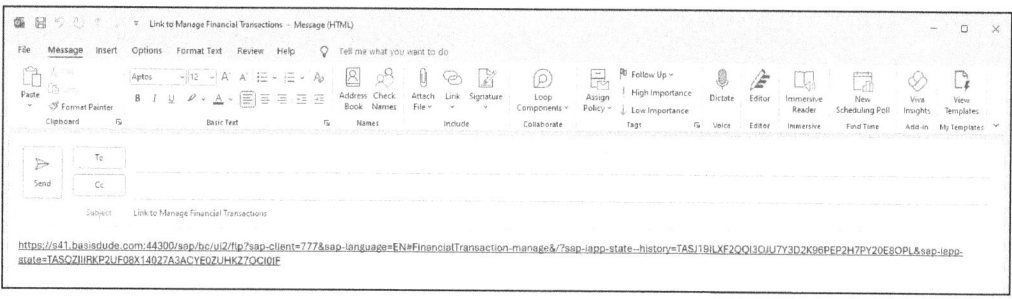

Figure 1.16 Example of Email Generated from Send E-Mail Share Option

1.4 Supported Treasury Contracts

One of the key strengths of treasury and risk management is its ability to manage a broad spectrum of financial instruments. When selecting a treasury software solution, it's important to consider how well it supports the various instruments used by the treasury department. In this section, we cover in detail the financial instruments that treasury and risk management can accommodate. By gaining an understanding of the instruments that treasury and risk management can manage, you can better evaluate how well it aligns with your organization's specific needs and unique business requirements, ensuring that the solution effectively supports your treasury management objectives.

The solution supports a range of financial instruments, which are organized into five primary categories: money market, FX, derivatives, securities, and trade finance. Each category encompasses a variety of specific instruments tailored to meet diverse treasury management needs. In the following sections, we provide a description of each of these areas and explore the types of instruments it includes.

1.4.1 Money Market Debt and Investments

Money market debt and investment contracts are among the most widely used types of instruments in corporate treasury departments. They are typically low risk and highly liquid, and they enable organizations to effectively manage short-term cash needs and park surplus liquidity in interest-earning investments. Here are some of the money market instruments most commonly used by treasury departments:

- **Interest rate instruments**
  Interest rate instruments include instruments with a fixed term that have specific repayment methods, including final, installment, or annuity repayments. Supporting both fixed and variable interest rates, the system supports the full lifecycle of the contract from creation to contract maturity.

- **Intercompany loans**
  Intercompany loans are commonly used with corporate treasury, and they function very much like debt or investment contracts. The key difference is that instead of transacting with an external counterparty, the loan is set up between two internal entities and is used to manage the borrowings that take place internally.

- **Time deposits**
  Time deposits are fixed-term deposits offered by financial institutions that pay interest over a specified period. Treasury and risk management allows users to manage time deposits efficiently, including tracking their maturity dates, interest payments, and renewal options. Debt and investment contracts support the full lifecycle of time deposits, from initiation through to maturity.

- Commercial paper
  Commercial paper is a short-term, unsecured debt instrument issued by corporations to meet immediate funding needs. Treasury and risk management facilitates the management of commercial paper, including issuance, tracking, and settlement.
- Deposits at notice
  Deposits at notice are savings accounts that require prior notice before withdrawal. Treasury and risk management supports the management of these deposits, including tracking notice periods, interest accruals, and withdrawal requests.
- Credit facilities
  Credit facilities are agreements between a borrower and a lender that allow the borrower to draw funds up to a specified limit. Treasury and risk management manages credit facilities by tracking utilization, interest payments, and covenant compliance. Treasury and risk management provides comprehensive tools for monitoring and managing credit lines and the associated fees.

1.4.2 Foreign Exchange

FX instruments are used by treasury departments to manage currency risk, optimize foreign currency cash flows, and facilitate international transactions. Global corporations often operate in multiple currencies, which exposes them to the risk of fluctuations in exchange rates. FX instruments are used to hedge this risk, maintain predictable cash flows, and ensure that exchange rate volatility doesn't significantly impact profitability. The most common FX instruments are as follows:

- Spot contracts
  Spot contracts are agreements to exchange one currency for another at the current market rate, with settlement typically occurring within two business days. Treasury and risk management provides robust capabilities for managing FX spot contracts, including tracking transaction details, managing settlements, and ensuring accurate accounting. FX allows users to enter and monitor spot contracts with real-time exchange rates, providing visibility into current and historical transactions.
- Forward contracts
  Forward contracts are agreements to exchange currencies at a predetermined rate on a future date. These contracts are essential for hedging against currency fluctuations and managing future cash flows. Treasury and risk management supports the full lifecycle of FX forward contracts, from contract initiation through to settlement. FX facilitates the configuration of forward contracts, including setting forward rates, tracking contract performance, and managing associated risks.
- Non-deliverable forward contracts
  Non-deliverable forward (NDF) *contracts* are a type of forward contract used in the FX market. They are used for specific currencies that are not freely traded due to regulatory restrictions. Unlike standard forward contracts, NDFs don't have physical cash

flow of the foreign currency. Instead, there's a cash settlement at the end of each contract, and the cash settlement is the difference between the agreed-upon exchange rate and the spot rate at the time of contract maturity.

- **Intercompany foreign exchange contracts**
  Intercompany FX contracts are commonly used in corporate treasury to trade FX among internal entities. Many times, internal FX contracts are used after an aggregate external FX contract is placed from a central trading entity. Internal FX contracts are then placed to distribute the foreign currency from the trading entity to all other entities within the organization.

1.4.3 Securities

Securities play a key role in corporate treasury operations, providing a way to manage long-term liquidity, optimize investment returns, and secure financing. Treasury departments often deal with various types of securities, including bonds, stocks, and government debt, to achieve financial goals such as preserving capital, generating income, and funding corporate activities. The type of securities used varies with the company's risk tolerance, cash flow requirements, and investment horizon. In the following list, we detail the most common securities used by treasury departments:

- **Equities**
  Equities represent ownership shares in a corporation and are fundamental components of investment portfolios. Treasury and risk management provides comprehensive tools for managing equity investments, including tracking purchase and sale transactions, managing dividends, and monitoring stock performance. Securities enable users to record and manage equity holdings, generate reports on portfolio performance, and ensure compliance with accounting standards.

- **Bonds**
  Bonds are debt securities issued by corporations or governments to raise capital, offering fixed interest payments over a specified period. Treasury and risk management supports the management of both issued and purchased bonds by tracking their issuance, interest payments, maturity dates, and trading activities. Treasury and risk management provides features for managing bond portfolios, including calculating accrued interest, tracking bond ratings, and assessing credit risk.

- **Warrants**
  Warrants are financial instruments that give holders the right, but not the obligation, to buy or sell an underlying asset at a predetermined price by a specific date. Treasury and risk management manages warrants by facilitating their issuance, tracking their performance, and handling the associated accounting. Treasury and risk management supports the configuration of warrant terms, including strike prices and expiration dates, and it provides tools for monitoring their value and impact on financial positions.

- **Investment funds**
  Investment funds (or *mutual funds*) are financial vehicles that pool money from multiple investors to invest in a diversified portfolio of assets like stocks, bonds, and other securities. Within treasury, the most common investment funds used are money market funds, due to their liquid nature and relatively low risk.

1.4.4 Derivatives

Treasury and risk management provides extensive capabilities for managing various derivative instruments, which are essential for hedging, speculation, and managing financial risks. This section covers key derivatives, as follows:

- **Caps and floors**
  Caps and floors are financial derivatives used to limit the range of interest rate fluctuations. Caps provide protection against rising interest rates by setting a maximum rate, while floors offer protection against falling rates by establishing a minimum rate. Treasury and risk management supports the management of both caps and floors by facilitating their setup, tracking their performance, and calculating their impact on financial positions.

- **Swaps**
  Swaps are contracts in which two parties agree to exchange cash flows based on different financial instruments or rates. Treasury and risk management manages various types of swaps, including interest rate swaps and currency swaps.

- **Forward rate agreements**
  Forward rate agreements (FRAs) are contracts that fix the interest rate for a future period. Treasury and risk management supports the management of FRAs by tracking their terms, calculating the difference between the agreed-upon rate and the prevailing market rate, and managing settlements.

- **Currency barrier options**
  Currency barrier options are exotic options that either become active or expire, depending on whether the underlying currency hits a specific barrier level. Treasury and risk management manages these options by facilitating their configuration, tracking their performance, and analyzing their impact on currency risk.

- **FX options**
  FX options provide the right, but not the obligation, to buy or sell a currency pair at a predetermined rate. Treasury and risk management supports the management of FX options by handling their setup, tracking their performance, and calculating their impact on FX exposure.

- **Repurchase agreements**
  Repurchase agreements (repos) are short-term borrowing arrangements in which securities are sold with an agreement to repurchase them at a later date. Treasury

and risk management manages repos by tracking their terms, calculating interest payments, and ensuring accurate settlement.

- **Futures**
  Futures are standardized contracts to buy or sell an asset at a future date for a predetermined price. Treasury and risk management supports the management of futures contracts, including by tracking contract details, managing margin requirements, and monitoring performance.

- **Forward contracts**
  Forward contracts are agreements to buy or sell an asset at a future date for a price agreed upon today. Treasury and risk management manages forward contracts by tracking their terms, monitoring performance, and ensuring accurate settlements.

- **Interest rate swaps**
  Interest rate swaps involve exchanging fixed interest rate payments for floating-rate payments or vice versa. Treasury and risk management supports the management of interest rate swaps by facilitating their configuration, tracking cash flows, and calculating their impact on interest rate exposure.

1.4.5 Trade Finance Instruments

Treasury departments use *trade finance instruments* to bridge the gap between exporters and importers, ensuring that both parties can manage the financial aspects of their trade deals efficiently. Trade finance instruments are commonly used within treasury operations to support and facilitate trade by providing solutions to mitigate risk and enable smooth cross-border transactions. The most common trade finance contracts are as follows:

- **Letters of credit**
  Letters of credit are financial instruments issued by a bank guaranteeing payment to a seller on behalf of a buyer, provided that the seller meets specific terms and conditions. Treasury and risk management supports the management of letters of credit by facilitating their issuance, tracking compliance with terms, and managing payments.

- **Bank guarantees**
  Bank guarantees are commitments from a bank to cover financial obligations if parties fail to fulfill their contractual duties. Treasury and risk management manages bank guarantees by supporting their issuance, tracking their terms and conditions, and monitoring their execution.

1.5 Summary

Now that you have a clear understanding of the purpose and objectives of this book, we'll shift our focus to master data, which is an essential component of treasury and risk management. *Master data* serves as the critical foundation for effective treasury management in SAP S/4HANA, providing the necessary reference information for the accurate processing and reporting of financial instruments. In the upcoming chapter, we'll explore the various aspects of treasury and risk management master data, including its structure, significance, and configuration. This detailed walkthrough will help us set up and manage key master data elements, ensuring that the system is set up with the necessary master data before we dive into the configuration.

Chapter 2
Master Data

Master data in treasury and risk management represents the core of the solution, the consistent information that underpins various treasury processes across the system. It serves as the foundation for accurate and efficient transactions, providing a single source of truth for all treasury activities within SAP S/4HANA.

Master data in SAP S/4HANA is crucial as it forms the foundation upon which the rest of the system relies to process transactions effectively. In this section, we'll cover the various types of master data that are required for the various processes within SAP S/4HANA Finance for treasury and risk management. This chapter will cover master data creation from end to end and will highlight which processes use master data. The master data covered in this chapter includes the following:

- Company code
- Bank master
- House bank
- House bank account/account ID
- Business partners
- Derivative contract specifications
- Calendar
- Traders
- Market data
- Market data interfaces

In summary, there are many types of master data that we must create and maintain as a prerequisite for contract creation in treasury and risk management. Due to the amount of master data needed for treasury and risk management, we must establish a solid process for master data creation and maintenance so that we can have clear and accurate data. In the following sections, we'll dive deeper into master data, and we'll cover how to enter this data into SAP S/4HANA and any of the design considerations we should think through when creating master data.

2 Master Data

2.1 Company Code

Before we can create any financial transaction, we must assign it to a company code. Each corporate entity has a different company code, and depending on the location of the entity, there could be a different set of accounting rules that is assigned to the transaction. Once a transaction has been assigned to a company code, it's locked in and will stay in place for the life of the transaction.

The company code in SAP represents a financial entity within SAP and includes the specific settings for each entity. These include the name, legal name, country, and currency of the entity, along with additional settings specific to the company code. We configure company codes in the **Financial Accounting • Financial Accounting Global Settings • Global Parameters for Company Code • Enter Global Parameters** menu path. As shown in Figure 2.1, we've defined our main company code as 1000 and assigned it to a country, currency, and chart of accounts.

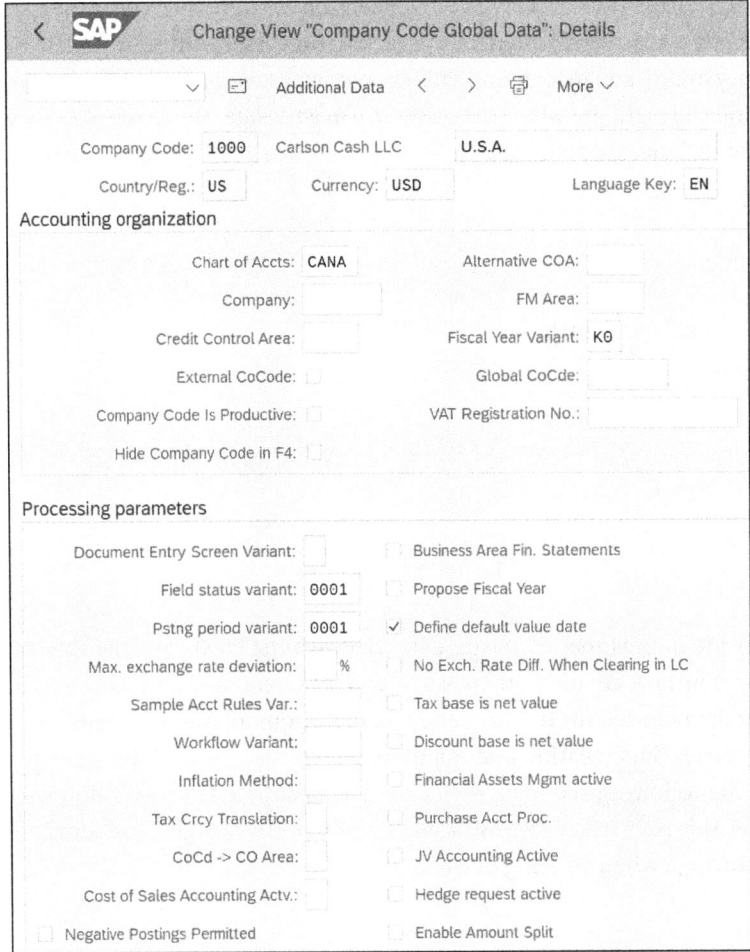

Figure 2.1 Defining Company Code Global Data

2.2 Bank Master

All data related to a specific bank needs to be maintained in the *bank master*, which is a series of data elements that comprise information the SAP system needs to process various financial operations. The bank master maintains all essential information about banks that the SAP S/4HANA system will conduct business with. Several of the key data elements maintained in the bank master include the bank's address, SWIFT code/Bank Identifier Code (BIC), and routing number/American Bankers Association (ABA) number. Each of these data elements is critical to correctly processing treasury payments out of the SAP system, as well as processing information from electronic bank statements. Any bank that the SAP system will encounter will need to be maintained in this bank master, which is primarily stored in the BNKA table, as shown in Figure 2.2.

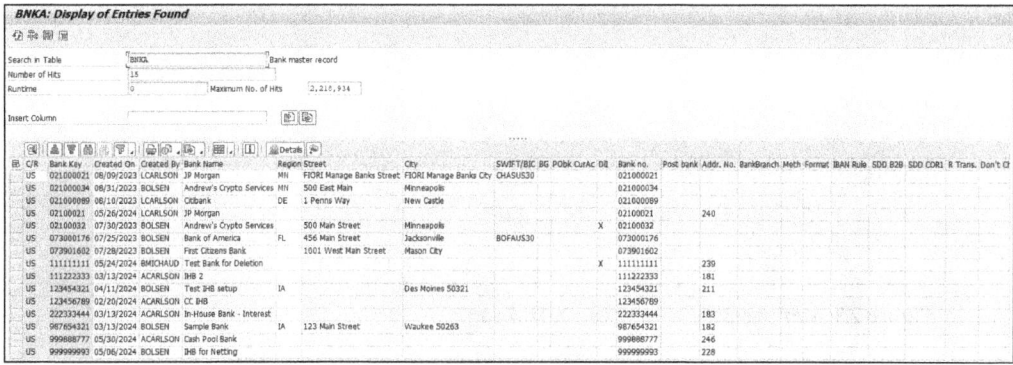

Figure 2.2 Excerpt from BNKA Bank Master Table

There are several SAP transactions that we can use to maintain bank master data. First, we can create bank master data by utilizing the Manage Banks – Master Data app (shown in in Figure 2.3) or Transaction FI01.

To create a new bank entry, we click the **Create** button shown in Figure 2.3, and selecting an individual record on this screen allows us to manage that selected entry. When creating a new bank entry, we will be required to input both the country and the region of the new bank entry, as well as an SAP identifier called **Bank Key**. The bank key is a unique identifier that the SAP S/4HANA system uses to identify a specific bank record within the bank master. Many times, the bank key is composed of corresponding information that already uniquely identifies a bank—such as routing number or bank number. Ultimately, the combination of region and bank key needs to be unique for each bank entry in the bank master.

To maintain an existing bank key record, we can utilize the same Manage Banks – Master Data app or Transaction FI02, the latter of which will allow us to input a specific bank key record for recall and management. We can modify bank address information and SWIFT/BIC data, but we can't change the bank key. In Figure 2.4, the bank master record has been created, and all of the fields are available for us to make changes, with the exceptions of the bank key and bank number fields.

2   Master Data

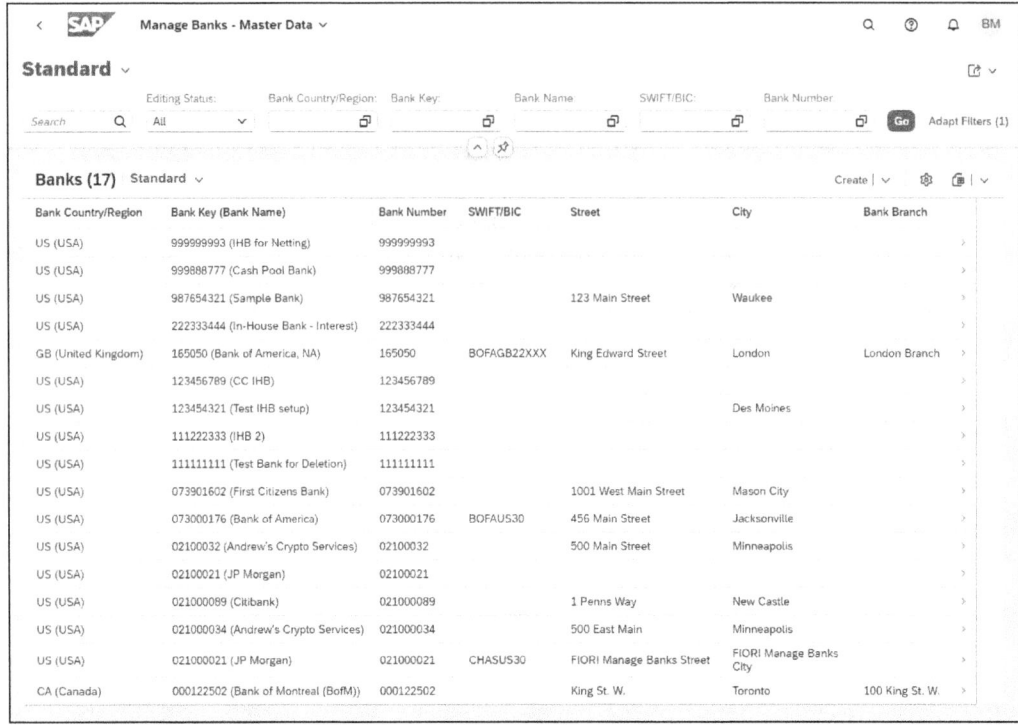

Figure 2.3 Manage Banks – Master Data App

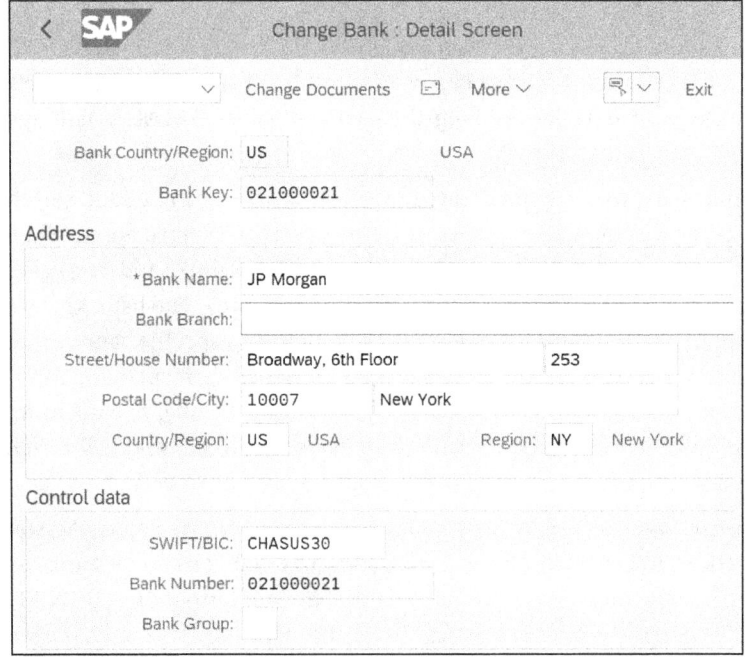

Figure 2.4 Bank Master Data Using Transaction FI02 to Change Bank

Bank master data is difficult to delete, and rarely is this type of data completely removed from SAP databases. The information is retained to maintain past records and to maintain the integrity of past financial data within the SAP S/4HANA system. Instead of allowing deletion, SAP will allow us to mark or flag a particular bank master record as unnecessary, thus removing it from use going forward. We do this by using Transaction FI06. We'll check the **Deletion Indicator** box as shown in Figure 2.5, and then, we'll be able to delete the bank master record by archiving or reorganizing records.

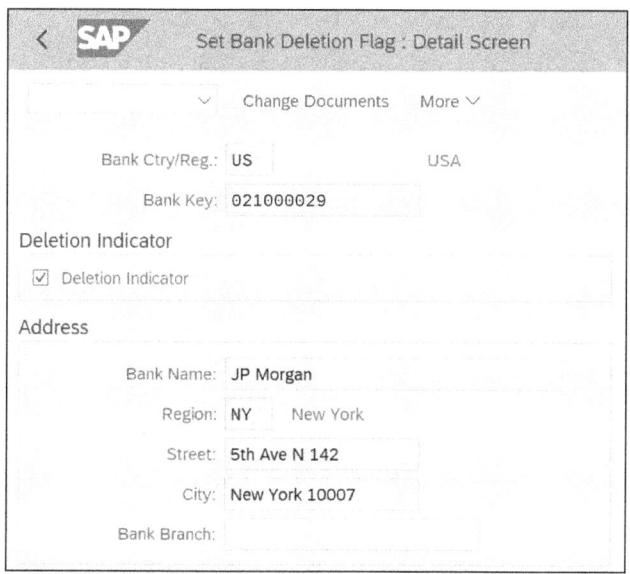

Figure 2.5 Set Bank Deletion Flag Screen in Transaction FI06

2.3 House Bank

The house bank must be unique in each company code, and it's equivalent to a bank branch within SAP S/4HANA. This is the first piece of information you'll assign when determining an account for the incoming and outgoing payments within treasury and risk management.

When we create a corporation's accounts in SAP S/4HANA for various purposes, we need to create them with the correct details. We can use these accounts in SAP S/4HANA to determine an account for accounts payable payments, treasury payments, or any type of bank statement processing. Prior to creating the account, we need to create a house bank. The house bank holds all information relevant to the bank branch where the account is held, and this includes the bank country, bank key, SWIFT code, and address information for the bank branch. In an SAP ERP system, we create the house banks in Transaction FI12, but in an SAP S/4HANA system, we can create them in a couple of places. The new transaction code we use to create house banks in SAP S/4HANA is Transaction FI12_HBANK.

2 Master Data

Alternatively, we can use the Manage Banks or Manage Banks – Cash Management app to create the house bank. Note that in SAP S/4HANA 2022 and earlier, the app used for creating house banks is Manage Banks. In SAP S/4HANA 2023 and later, we must first create the bank in the Manage Banks – Master Data app, and then, we can create the house bank in the Manage Banks – Cash Management app.

In the Manage Banks – Cash Management app, we'll see a list of all banks that have been created in the SAP S/4HANA system. We can click into any of those banks to view additional details, and Figure 2.6 shows the initial screen with the list of available banks.

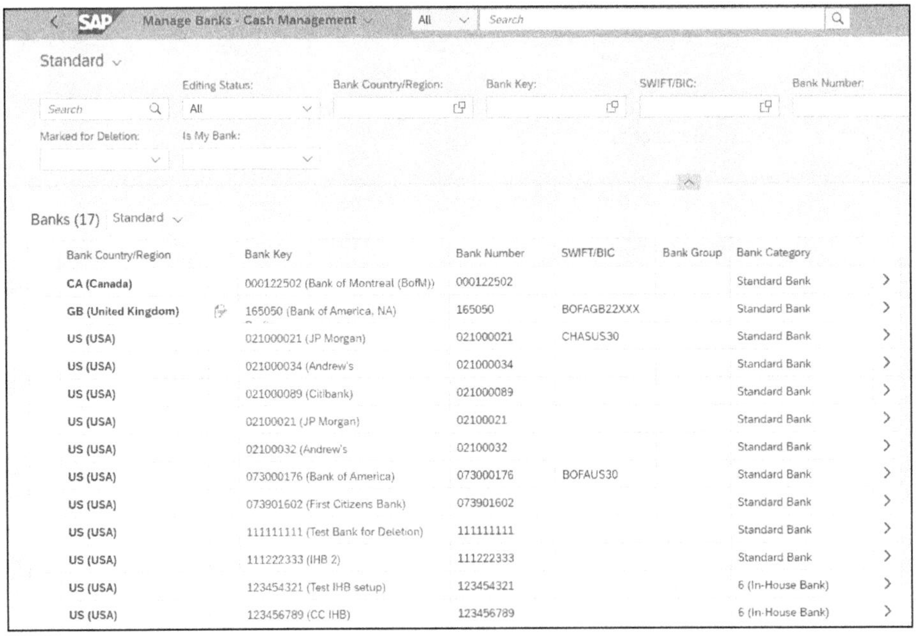

Figure 2.6 Manage Banks – Cash Management App

After drilling down into a bank, we can scroll down to the **House Banks** section (shown in Figure 2.7) and view the house banks that have been created in this area. To add any new house banks, we click the **Edit** button and add assignments of house banks by clicking the **Create** button.

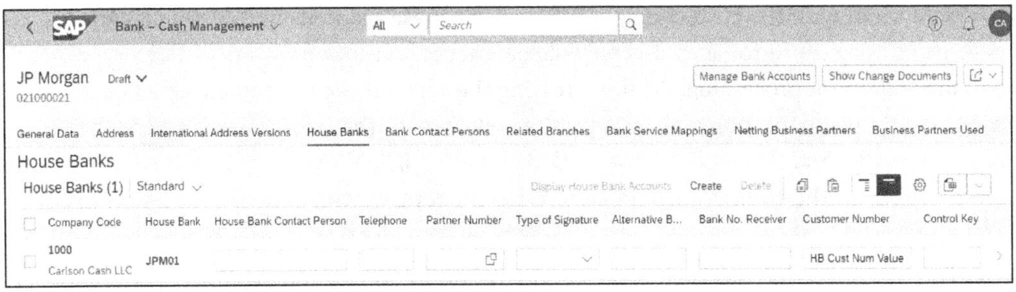

Figure 2.7 Creating Bank Account Master Record

2.4 House Bank Account/Account ID

On the screen that appears, the key information for the house bank is the company code, house bank, bank country, and bank key, but as seen in Figure 2.8, we can add more information.

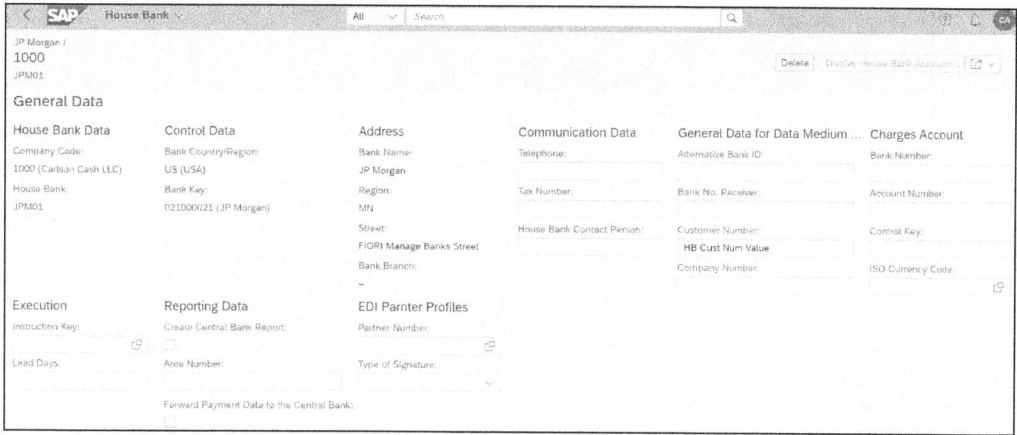

Figure 2.8 House Bank Creation Screen for House Bank JPM01

Now that we've created the house bank, we can look into the setup for the account ID.

2.4 House Bank Account/Account ID

The next master data we'll look at is the house bank account or account ID. Depending on the area and version of SAP S/4HANA you are in, the name is labeled differently. The account ID is unique in each company code, and we should treat it as equivalent to a bank account number. When we select any payment or account details to assign an account to the payments function, we'll assign the account ID.

After we create the house bank, we can assign accounts to it. The account ID is SAP S/4HANA's representation of an individual account at a bank, and each account that we hold at a bank will have an account ID. The key details held at the account ID are a description of the account and an assignment of a general ledger account for the bank account. We assign the account ID in the Manage Bank Accounts app. We must first create an account, and then we can assign the house bank and account ID. We can assign these details in the House Bank Account Connectivity app. In the example in Figure 2.9, the house bank, account ID, and general ledger account have all been assigned to the account.

2 Master Data

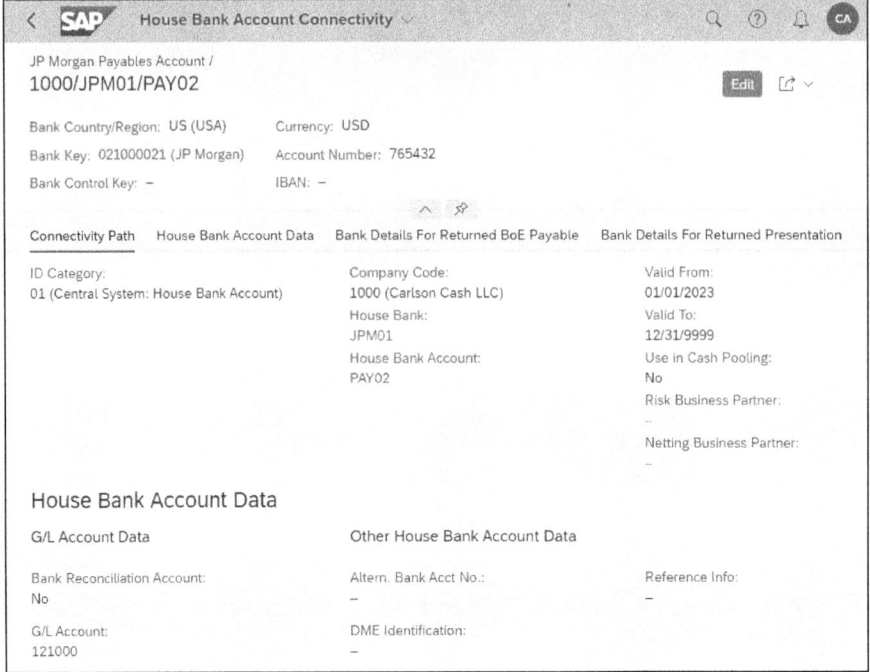

Figure 2.9 House Bank Account Connectivity Linkage

Now that we've assigned the house bank and account ID, we can show how the accounts are assigned in the business partner standing instructions.

2.5 Business Partners

When we create a financial transaction, we need to define which counterparty the transaction is held with. This helps drive reporting in this module, and functionally, standing payment instructions are assigned to the business partner to set up the default payment instructions. Defining this data correctly helps us avoid redundant work and having to assign payment details in each transaction. In the following sections, we'll cover the various business partner roles within treasury and risk management and when to use each of them.

In SAP S/4HANA, a business partner is a foundational piece of master data that represents an individual or organization with whom a business engages in various transactions. It serves as a centralized master data object that consolidates information about customers, vendors, and other parties involved in business processes. Unlike traditional master records, which are maintained separately, the business partner concept in SAP S/4HANA integrates various roles into a single master record. General data is maintained for the record, and additional roles are assigned to and maintained for the business partner.

In treasury and risk management, the primary purpose of business partners is for use in financial contracts. There are two primary business partner types we use:

- **External business partners**
  These include external entities with which the organization engages in various transactions, including debt and investments, derivatives, and FX contracts.

- **Internal business partners**
  These constitute internal entities linked to specific company codes within the organization. These counterparts play a pivotal role, primarily in facilitating intercompany loan transactions, wherein funds are transferred or borrowed between different divisions, subsidiaries, or affiliated entities within the same organization.

In treasury and risk management, business partners serve multiple roles, and we can utilize them for different purposes. A single business partner can fulfill a wide variety of functions within the organization, and we can integrate it into various financial contracts based on the assigned role. Upon the establishment of the general data of the business partner, we can assign a variety of roles to the general business partner based on the specific requirements and functionalities of the treasury contracts. The key roles relevant to treasury and risk management include the following:

- **Counterparty**
  The counterparty role is the cornerstone of treasury management, serving as the primary role of all business partners utilized in creating contracts within transaction manager. We must maintain the counterparty role in a business partner to create a treasury contract using that business partner.

- **Issuer**
  The role of the issuer is primarily associated with issuing securities contracts in transaction manager. The assignment of the issuer role to the respective business partner takes place before the issuance of a bond involving that business partner.

- **Payer**
  The payer business partner role is utilized by an entity responsible for making payments in financial transactions. We assign this role to a business partner who is obligated to settle financial obligations.

- **Depository bank**
  We use the depository bank business partner role in scenarios where a bank acts as a custodian or holder of funds or assets on behalf of another entity. This role is typically utilized in financial contracts involving a deposit of funds, for example, in the case of a securities contract.

We'll now shift our focus to the creation of a treasury business partner, ensuring the proper assignment of the counterparty role. We'll follow this by updating the relevant standing instructions and transaction authorizations, all of which are prerequisites for facilitating transaction processes in the treasury system.

2 Master Data

2.5.1 Business Partner Maintenance

Now that we've discussed the purpose of business partners in treasury and risk management, we can move on to the creation and maintenance of business partners within SAP S/4HANA. To create a business partner, we'll use the Maintain Business Partner app, as shown in Figure 2.10.

Figure 2.10 Maintain Business Partner App

Once we're in the app, there are a number of options at the top of the screen. To create a new business partner, we click **Organization**. This is the initial step when creating a business partner that we'll use in transaction manager. Clicking this button will kick off the creation process for the business partner, as shown in Figure 2.11.

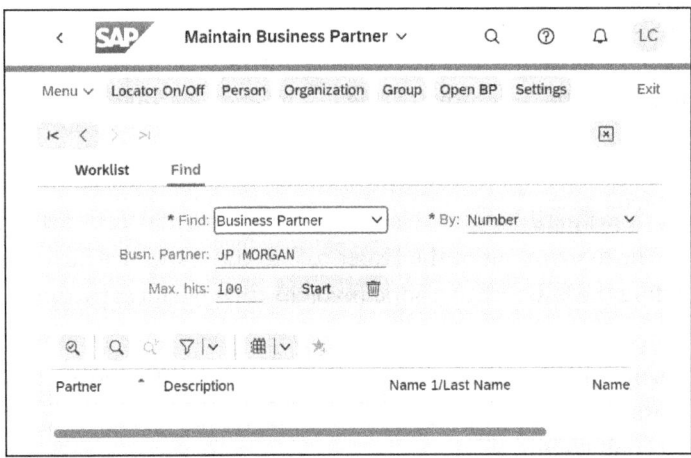

Figure 2.11 Maintain Business Partner App Initial Screen

We start by entering the name of the new business partner, and we then choose a grouping that is relevant to the business partner. There are two primary groupings that we can use. Depending on the requirements of the project, we can choose to create the business partner using either an internal or an external number range, as follows:

- **External grouping**
  Similar to other external number ranges in SAP S/4HANA, the system doesn't generate an identification number for the business partner internally. The name of the business partner is assigned externally by the user creating the business partner.

Externally assigned number ranges are commonly used for business partners as a way to easily distinguish between one another.

- **Internal grouping**
  In this grouping, the system generates identification numbers internally, based on predefined number ranges configured within the SAP S/4HANA system. These numbers are assigned automatically by the system and are typically sequential or based on a predefined pattern.

The next step is to click on the **Create Organization** button at the top of the screen. After that, the business partner name in the **Business Partner** field will carry over to the next screen. We can then chose our external number assignment grouping as shown in Figure 2.12.

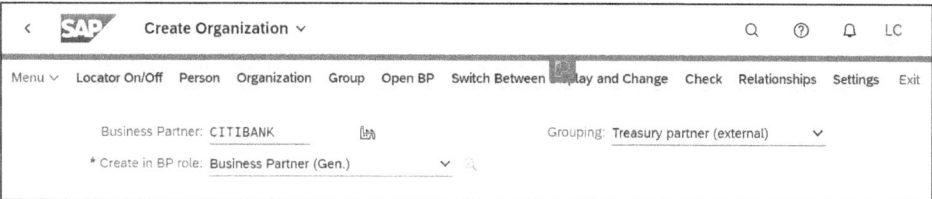

Figure 2.12 Creating Treasury Business Partner for Citibank

We can now add the applicable information on the **Address** tab of the business partner. The address information contains the details of the address of the counterparty or bank. In the address tab, we fill out all the required fields, including the following (see Figure 2.13), plus a complete address:

- Name
- Search Term 1/2

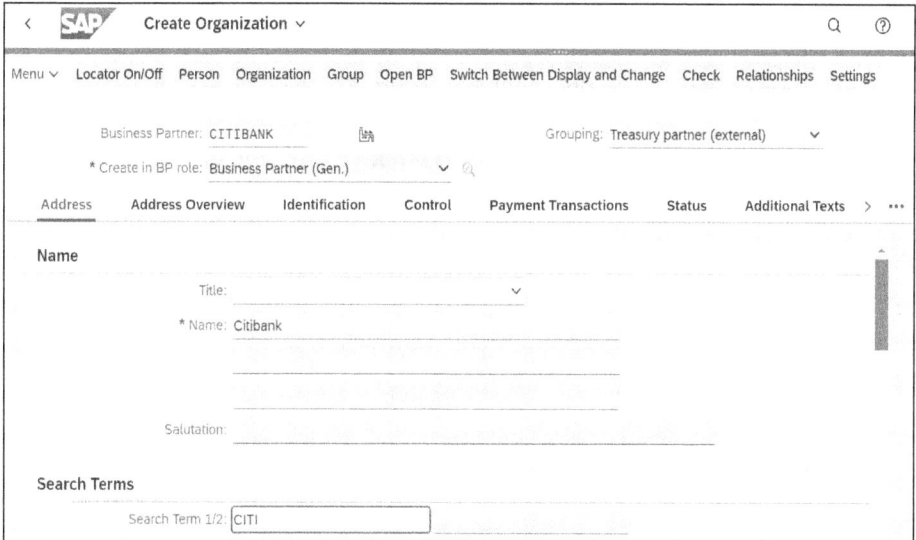

Figure 2.13 Business Partner Creation: General Address Tab

2 Master Data

If we scroll down in the **Address** tab, we can see the additional address fields, as shown in Figure 2.14. We can then populate the following required fields related to the address of the counterparty:

- Street/House Number
- Postal Code/City
- Country/Reg.
- Region
- Language

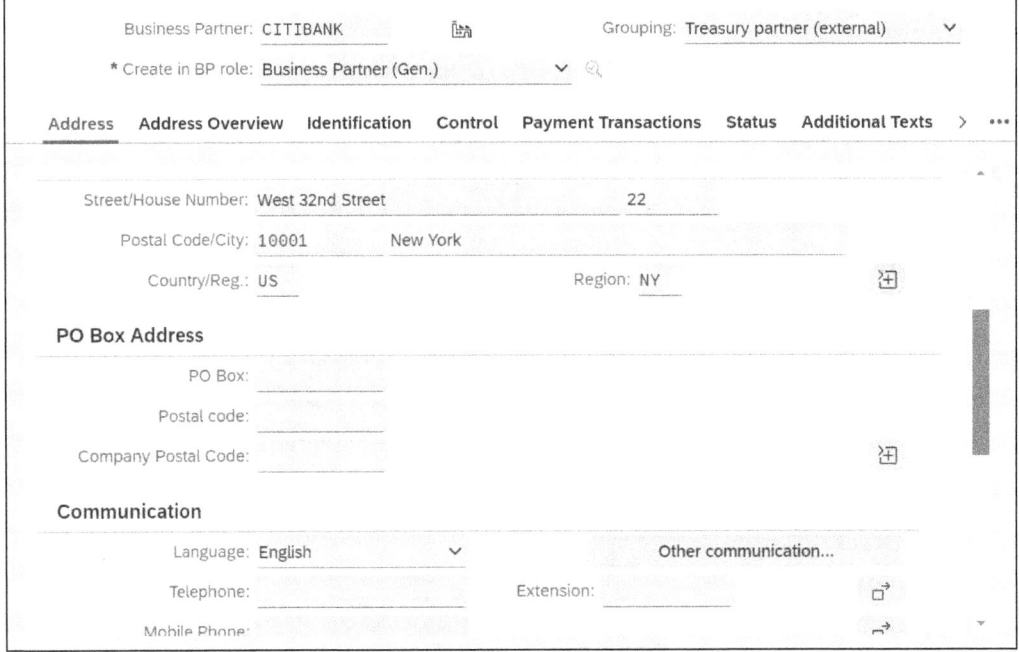

Figure 2.14 Business Partner Creation: Additional Address Details

At this point, we've entered everything that is required on the **Address** tab of the business partner. However, there are a number of additional fields in the business partner record that can store more information related to the address of the counterparty. An important set of fields to keep in mind is the address validity as shown in Figure 2.15. We can assign an address valid from and to date, and this is important to keep in mind as this date must fall within the dates on the contracts created within the transaction manager.

Now that we've entered the address information for the business partner, we can view the address information overview in the next tab: **Address Overview**. We can see our business partner's address information and the validity dates of the address; this is an important element because a business partner may have various addresses that have changed over time. In Figure 2.16, we can see the address information we entered into the **Address** tab of the business partner.

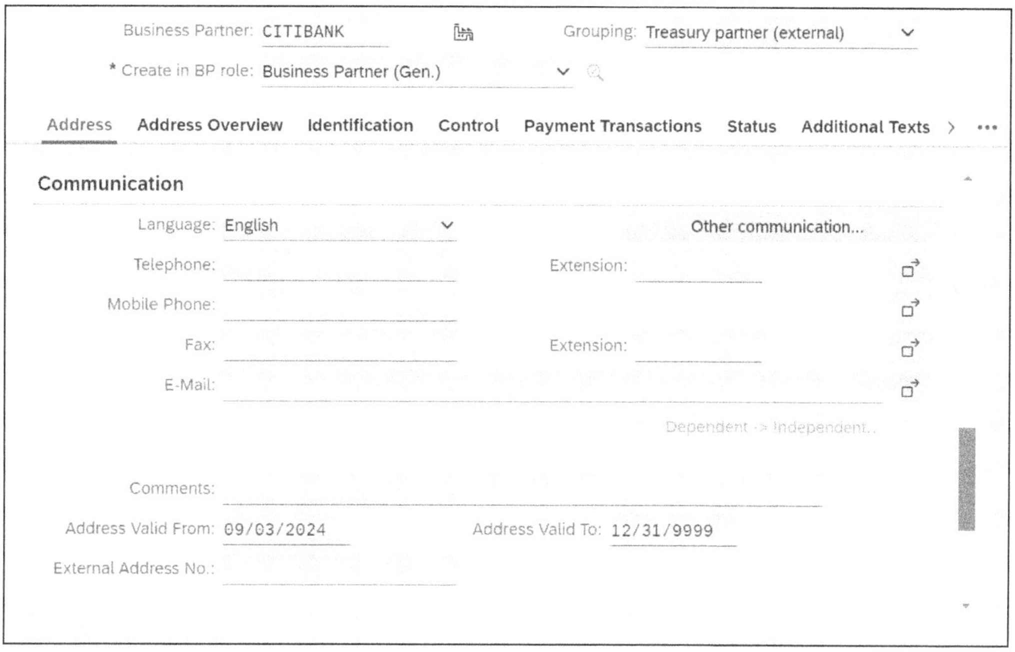

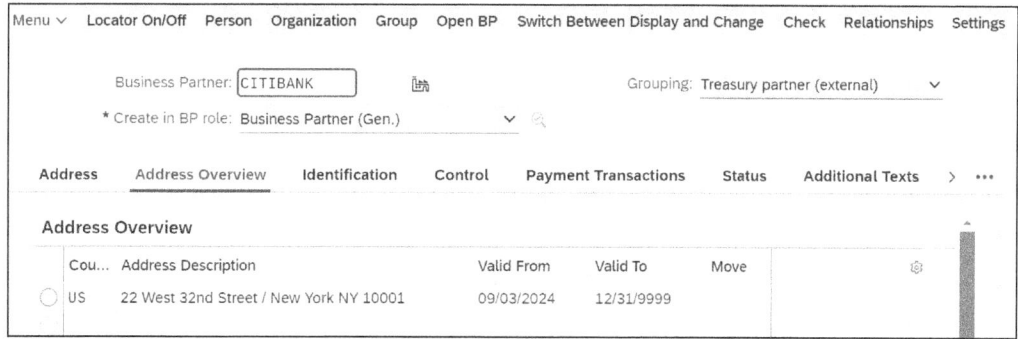

Figure 2.15 Business Partner Creation: Maintaining Address Validity Period

Figure 2.16 Address Overview Maintenance

Now that we've entered all the address information, the next tab that is applicable to creating a counterparty is the **Payment Transactions** tab (see Figure 2.17), which is where we'll store the bank details related to the counterparty's bank account. In other words, when we pay this business partner, we must determine the bank account to

2 Master Data

which we'll make the payment. This is where that information on the business partner is stored. In this tab, we'll add the following bank information:

- **ID** (used to define the set of payment instructions for the business partner)
- Country code (**C/R**)
- **Bank Key**
- **Bank acct**
- **IBAN**

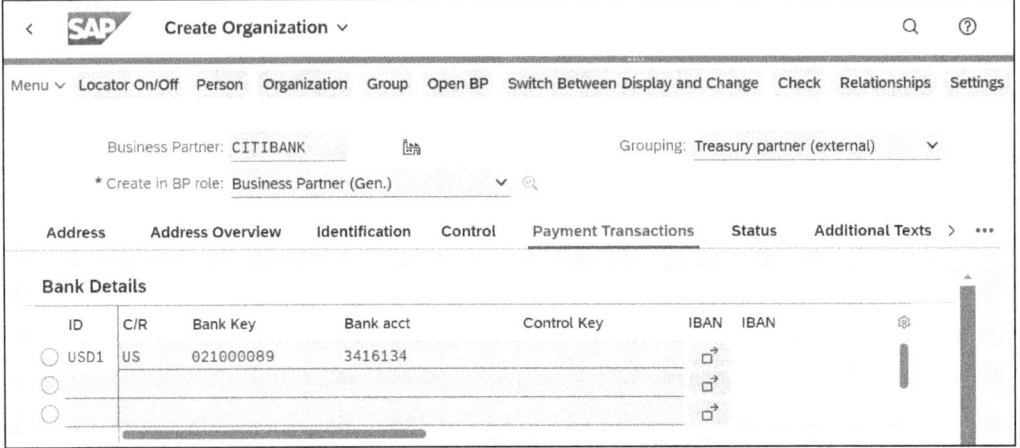

Figure 2.17 .17: Maintaining Business Partner Bank Details

As with the address validity date, the bank accounts associated with a business partner can change over time. Therefore, there's a way to track the validity of the bank account details, as shown in Figure 2.18. To change the validity of bank details, we click the radio button to the left of the bank details and then click the **Validity** button. This is important to keep in mind as the bank details must be valid and align with the dates when we're creating a transaction manager contract.

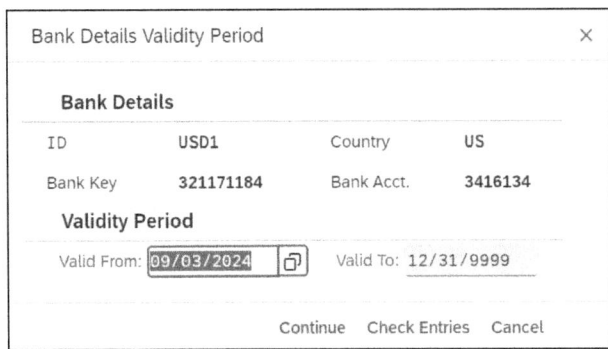

Figure 2.18 Business Partner Bank Details Validity Period

At this point in the process, we've created all of the information related to the general business partner (see Figure 2.19). Before moving on to the creation and maintenance of the counterparty role, we should save the information contained within the business partner by clicking the **Save** button in the lower right corner of the screen. At this point, we've created and saved the general business partner, and we can move on to creating the counterparty role.

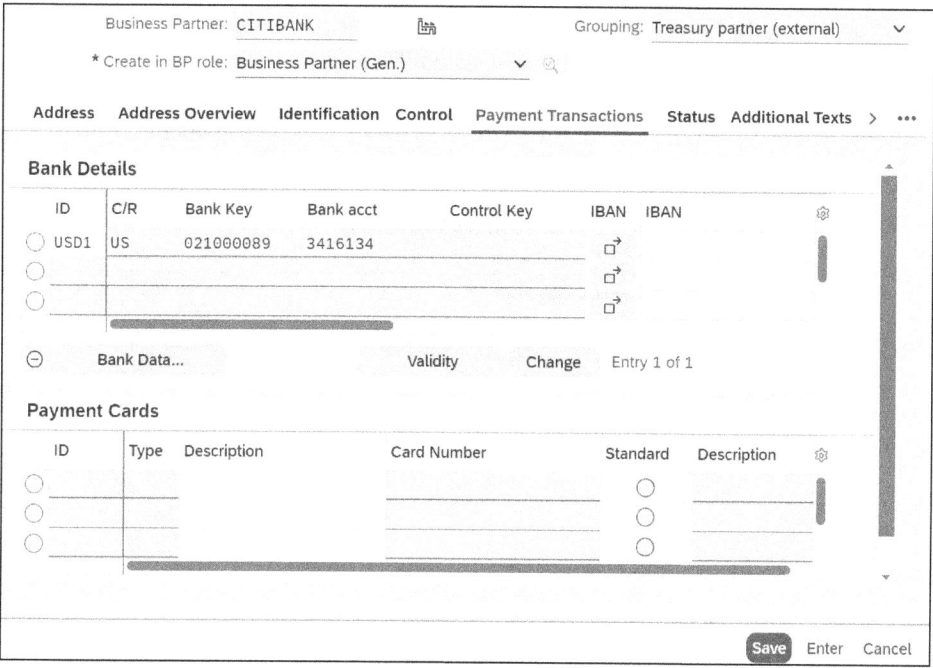

Figure 2.19 Maintaining Business Partner Bank Details

2.5.2 Assigning Business Partner Payment Details

After we initially create the business partner's general information, we can assign the various business partner roles applicable to the business partner. As we mentioned earlier in this chapter, the counterparty role is the foundational business partner role that virtually every business partner will use. The counterparty role contains the linkage and relationship between the business partner and the transaction manager contracts, and it's where we define the relationship between the business partner and the types of treasury and risk management contracts that we can create using the business partner.

The first step to establish a new role of a business partner is to go to the **Change in BP role** field and switch it from **Business Partner (Gen)** to the role that we'd like to add to the business partner, which in this case is the **Counterparty (New)** role. In the example in Figure 2.20, we've added the counterparty role to the business partner. The next step is to click the **Company Code** button located at the top of the screen.

2 Master Data

Figure 2.20 Creating Business Partner Counterparty Role

As shown in Figure 2.21, now that we're in the counterparty role, the screen has changed to display the counterparty role and the associated fields. In this section, we'll focus on the business partner tabs within the counterparty role that we primarily use and that are required before we can create a contract with treasury and risk management, as follows:

- **Payment Details**
- **Authorizations** (which we will cover in Section 2.5.3)

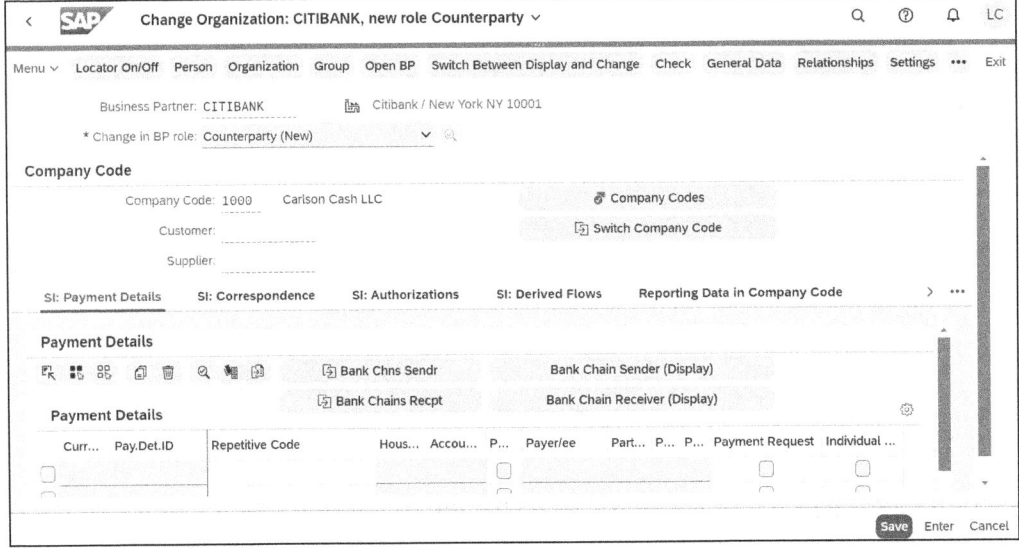

Figure 2.21 Counterparty Role Standing Instructions Payment Details

When we're managing business partners within the counterparty role, our first step is to define default payment details that will flow into the treasury and risk management contract upon its creation. We do this by using the **Payment Details** tab, and during this process, we define default payment details for each of the contracts that we've configured for transaction manager. We can always change payment details on an individual basis within the contract, but the purpose of this step is to define the default payment details that make contract entry much quicker. Within this tab, we'll define the following fields:

- **Currency**

  This represents the currency of the contract, and it should align with the currency in which the contract is created. If the contract we create doesn't have an entry in this table for the currency in which the contract is created, the payment details won't flow into the contract correctly. In this example, we'll use US dollars.

- **Payment Details ID**

  We'll use this field to name the payment details to differentiate them from other payment details assigned to other types of contracts. An example of a naming convention commonly used for payment detail IDs is as follows:
  - The first two or three characters represent the product type, and they're followed by a dash or underscore.
  - The following three characters represent the currency of the contract.

  Examples are this are as follows:
  - IRI_USD
  - FX_EUR
  - FX_GBP
  - ICL_USD
  - ICL_SEK

- **House Bank**

  This field represents the originating house bank or payment house bank.

- **Account ID**

  This field represents the originating account ID or payment account ID—in other words, what bank account we're sending this payment from.

- **Payer/Payee**

  This field represents the business partner we're paying. In most cases, this should be the same business partner.

- **Partner Bank**

  We use this field to select the bank details we've previously created on the bank details screen of the general business partner. The **Partner Bank** represents the bank account details that should be used when making payments to a specific counterparty. It stores the necessary information, such as the bank account number and bank details, ensuring that payments are directed to the correct account.

- **Payment Request**

  We check this box if we're planning to send a payment and create a payment request for this contract.

- **Payment Methods**

  We use this field to map in the default payment methods for the contract. In most cases, we would map in both an incoming and an outgoing payment method.

2 Master Data

As shown in Figure 2.22, we've populated the key fields that are required, and these values will populate into our contract upon its creation.

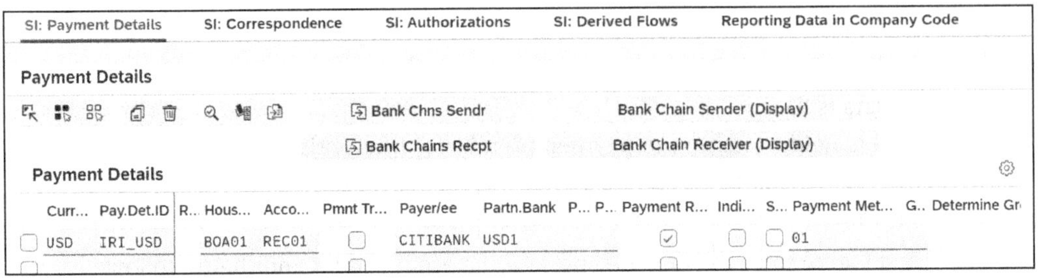

![Figure 2.22]

Figure 2.22 Maintaining Standing Instructions Payment Details

Once the payment details have been fully populated, the next step is to assign the payment details to particular contract types. Without this step, the payment details contained within this setup have no association with a specific contract type. It's in this step that we'll assign the payment details to specific product and transaction types. To start this process, we check the box next to the **Payment Details** line and click the **Assign** button as shown in Figure 2.23.

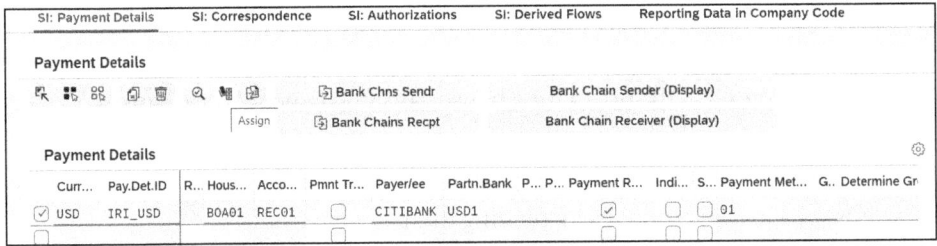

Figure 2.23 Assigning Business Partner Payment Instructions

Once we're in the assignment screen, we can select the appropriate product types and whether these payment details pertain to inbound payments, outbound payments, or both. We do this by toggling the **Select/Deselect** buttons. We then click on the type of transaction we want to assign the business partner payment details to and then click on the transaction type. Next, we assign the transaction to the incoming and the outgoing sides by clicking the **Select** buttons as shown in Figure 2.24.

2.5 Business Partners

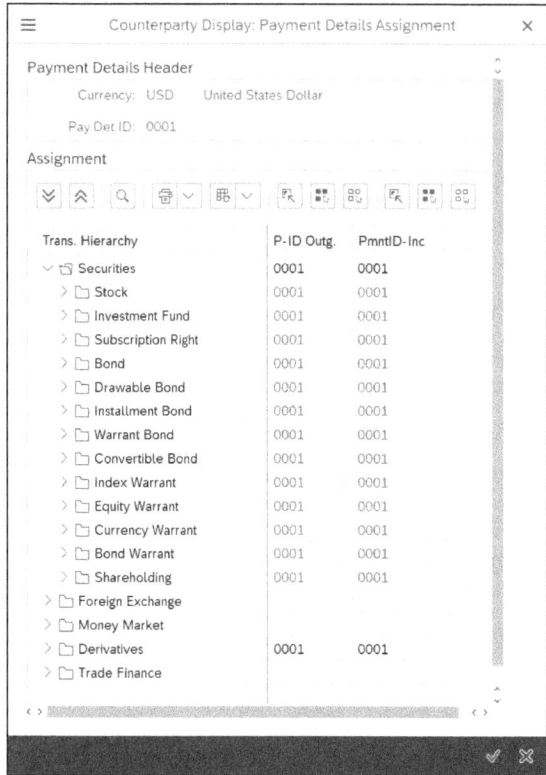

Figure 2.24 Assigning Counterparty Payment Details

Once we've assigned the correct types of contracts, we must make sure to click **Continue** to save the entries. We must complete this process for all entries on the **Payment Details** tab.

2.5.3 Assigning Business Partner Authorizations

We can also authorize business partners within the counterparty role to only transact in specific product types. This helps us isolate which business partners can maintain certain types of contracts. To authorize a business partner for a product type or a series of product types, we first need to authorize that business partner via the **Authorizations** tab in the counterparty business partner company code settings.

We'll then select the **Authorizations** tab and authorize it for the specific contracts that we'll use this business partner for. In this step, we can be as granular as desired, or we can authorize transactions at a higher level. The goal of the authorizations is to only authorize business partners for the types of transactions they'll engage in. As shown in Figure 2.25, we've authorized the business partner for all transaction types by checking the boxes in the **Auth.** column.

71

2 Master Data

Figure 2.25 Authorizing Business Partner Standing Instructions

As mentioned, we can also authorize transactions at a much more granular level. As shown in Figure 2.26, we can expand each of the categories and authorize individual product and transaction types.

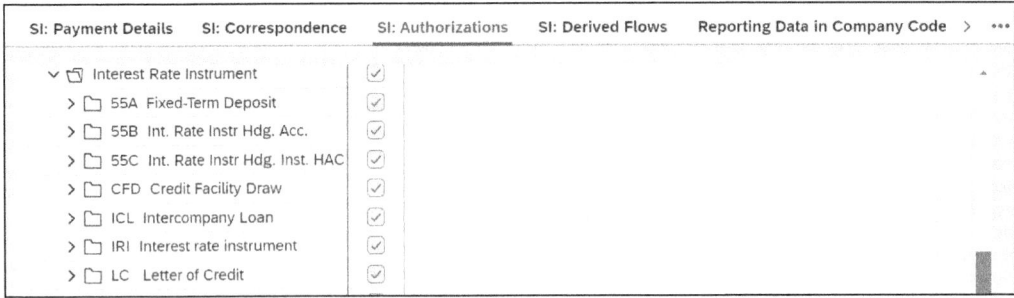

Figure 2.26 Assigning Business Partner Standing Instructions at Product Type Level

> **Tip**
> Each business partner has three dates that we must change to align with the dates of our treasury contracts. Having these dates out of synch can cause issues when we're creating contracts if the contract dates lie outside of the business partner dates. We've outlined the date fields in this section.

We then click the button to the right of the business partner role, as shown in Figure 2.27.

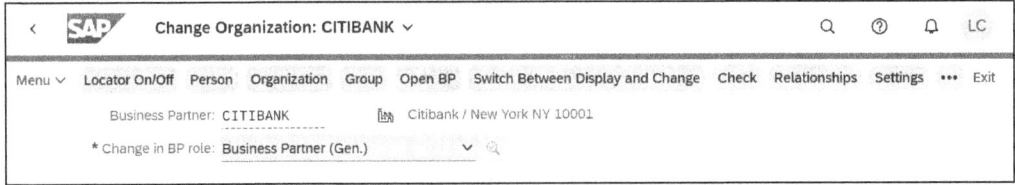

Figure 2.27 Validity Dates for Business Partner Counterparty Role

We'll then see the effective date of the counterparty role, and we'll change the **Vld From** and **Valid To** dates as shown in Figure 2.28. Then, we click **Enter**.

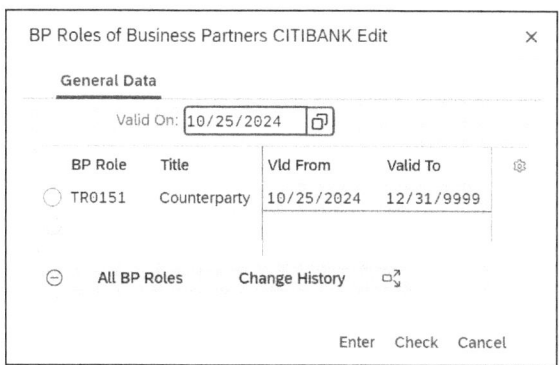

Figure 2.28 Maintaining Counterparty Role Validity Dates

On the **Address** tab, we enter the **Address Valid From** and **Address Valid To** dates, as shown in Figure 2.29.

Figure 2.29 Counterparty Address Validity Dates

On the **Payment Transactions** tab, we select the bank details and click the **Validity** button, as shown in Figure 2.30.

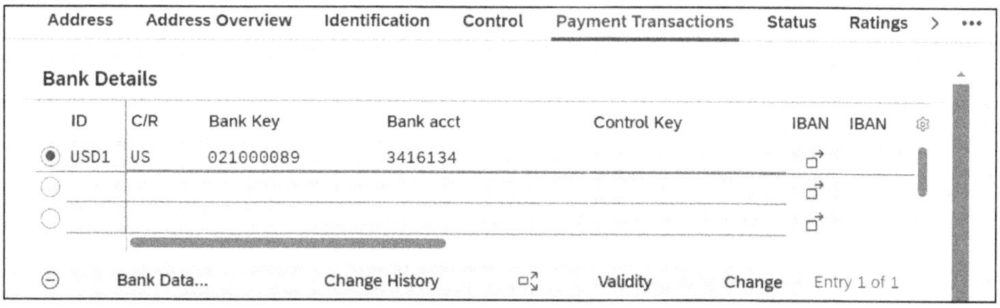

Figure 2.30 Validity of Payment Transaction Bank Details

We then ensure that the validity date of the payment details in the **Valid From** field is changed, as shown in Figure 2.31.

2 Master Data

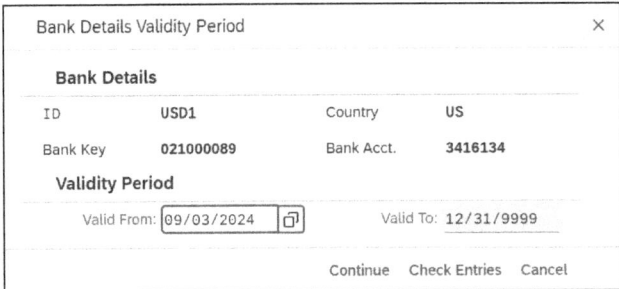

Figure 2.31 Updating Bank Details Validity Period

2.6 Derivative Contract Specification

This data represents the type of contract at a trading exchange, and it is required for commodity trades and creates the overall structure of the commodity. This data also includes any of the contract conditions and quotations for the commodity.

A unique master data element specific to commodity derivatives is the derivative contract specification, which is an important piece of master data that we must have before we can create any commodity derivative. It defines how a commodity is traded on a particular exchange, it's linked to a specific commodity type and market identifier code (MIC), and it encompasses significant information related to the traded commodity.

The derivative contract specification includes trading-specific details such as the unit of measure, quotation currency, lot size, minimum tick value, and tick size. These details are crucial for defining the terms and conditions of a commodity derivative contract. Because these elements are captured in the derivative contract specification as a master data element, there's no need to repeatedly enter this information for each new contract. Instead, we reference the derivative contract specification during the contract creation process, automatically populating the contract with the necessary trading details.

This approach provides consistency and accuracy among all commodity derivative contracts because all of the relevant data is predefined in the derivative contract specification. It also streamlines the contract creation process, reducing the risk of errors as contracts are manually entered into SAP S/4HANA. The comprehensive nature of the derivative contract specification makes it an important component in managing commodity derivatives within treasury and risk management.

We'll now walk through the setup required when creating a derivative contract specification in SAP S/4HANA. We use Transaction FDCSO1 to create, view, or change a derivative contract specification. The first step is to define a *derivative contract specification ID*, which we will use along with the description to easily identify the derivative contract specification. We then must define a derivative category for the derivative contract specification; the following options are available:

- Commodity Futures
- Listed Options
- Commodity Forward Index

The final step on the main creation screen is to define the **Valid From** date, which specifies the starting date from which the derivative contract specification is valid and can be utilized within a contract.

Setting the **Valid From** date is important because doing so makes sure that the derivative contract specification is only applied to contracts created on or after this specified date, thereby maintaining the integrity of the master data. This step provides a clear temporal boundary for the applicability of the derivative contract specification, preventing its use in contracts that predate its validity. Additionally, defining this date helps us manage changes over time. If we update the trading parameters for a commodity, we can create a new derivative contract specification with a different **Valid From** date, allowing the system to distinguish among contracts created under different trading conditions.

As shown in the example in Figure 2.32, we've created a derivative contract specification for aluminum using the **Commodity Futures** derivative category.

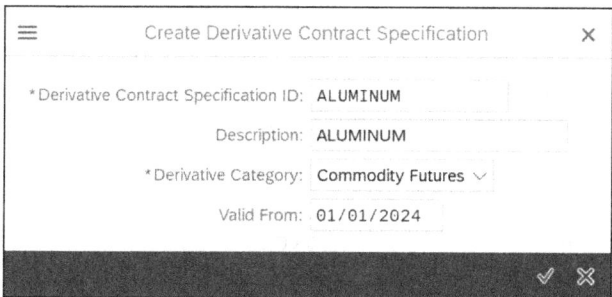

Figure 2.32 Derivative Contract Specification Creation Screen

The **Basic Data** tab of the derivative contract specification screen captures most of the key details of the commodity, including the following:

- Product Symbol
  This is an optional field, and we use it to store the product symbol that the exchange uses to identify that particular commodity.
- Contract Specification URL
  This is also an optional field, and we can use it to store the URL of the commodity specifications for that commodity.
- Commodity
  This is a required field, and we should populate it with the commodity that the derivative contract specification represents.

- **Unit of Measure**
  This is a required field on the derivative contract specification screen, and it represents the unit of measure that the commodity uses to trade at the exchange. It will automatically populate within the derivative contract specification, using the unit of measure defined when creating the commodity.

- **Period Determination**
  This field is optional, and it's required if we're planning to populate the various periods related to the commodity using system logic. If we don't populate this field, we'll have to manually maintain the periods within the derivative contract specification manually.

- **Maturity Code Determination**
  We can manually maintain the maturity codes of the commodities, or the system can automatically populate them in the derivative contract specification using logic based on the security ID and a combination of other system data:

  - **MAN: Manual Input**
    We should use this option if we'll be maintaining the maturity manually.

  - **SISO: Security ID with pattern <ps>YYYYMMDD**
    This option will populate the maturity codes using a combination of the product symbol, year, month, and day.

  - **SMIC: Security ID with pattern <ps><mic>MMYY**
    This option will populate the maturity codes using a combination of the product symbol, MIC, month, and year.

  - **SMY: Security ID with pattern <ps>MY**
    This option will populate the maturity codes using a combination of the product symbol, month, and last digit of the year.

  - **SMYY: Security ID with pattern <ps>MYY**
    This option will populate the maturity codes using a combination of the product symbol, month, and last two digits of the year. We'll see this in action later in Figure 2.38.

- **Expiration Date Logic**
  The expiration date logic is a required field, and we use it to help determine the grid points of the commodity curve. We use the last date of the defined period to identify the period for that derivative. The available values within the dropdown are as follows:

  - Last Contract Date
  - Last Trading Date
  - Last Quotation Date

2.6 Derivative Contract Specification

- **Reporting Date Logic**
  After we've created an exposure with floating prices in a derivative contract specification, we use its due date to automatically identify the corresponding future contract based on the expiration date logic. This logic ensures that the identified future contract is assigned to the correct reporting period. Most commonly, this period is defined by the last day of the contract period of the identified future. The available values in the dropdown list are as follows (see Figure 2.33):
  - Last Contract Date
  - Last Contract Settlement Date
  - Last Physical Settlement Date
  - Last Trading Date
  - Last Quotation Date
  - Shifted by Time to Maturity

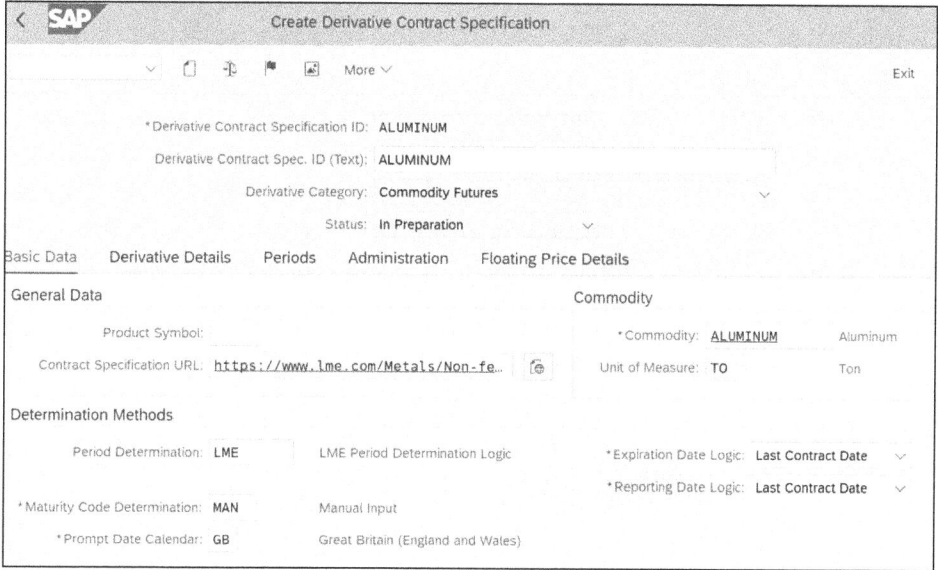

Figure 2.33 Basic Data of Derivative Contract Specification

- **Market Identifier Codes**
  This is a required field, and every derivative contract specification must contain at least one MIC. After we've created the MIC, we assign it to the derivative contract specification here.

- **Archiving Market Data – Retention Period**
  This field (see Figure 2.34) represents the retention period during which the market data related to the derivative contract specification is available in the system before it's archived.

2 Master Data

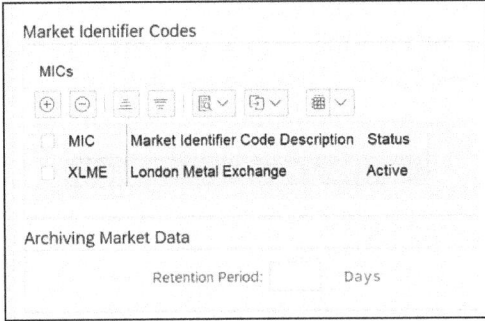

Figure 2.34 Derivative Contract Specification Market Identifier Codes

Now that we've walked through the information in the **Basic Data** tab, we can move to the **Derivative Details** tab of the derivative contract specification. This tab contains the contract-specific data related to the commodity. We'll cover each of the fields in detail in the following list:

- **Valid From**
  This is a required field, and we should be populate it with the date from which the derivative contract specification should be valid (see Figure 2.35). On this screen, it's very easy to create a new version of the derivative contract specification with a new validity date. When we click the **Create New Version** button, we're prompted to enter a new validity date for the new derivative contract specification version.

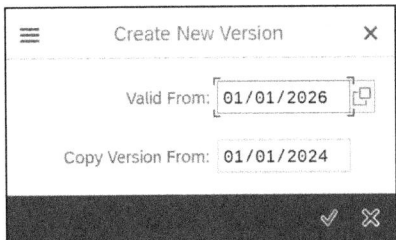

Figure 2.35 Derivative Contract Specification: Create New Version

When we click the green checkmark, it creates a new version of the derivative contract specification, as shown in Figure 2.36. We then have a new version of the derivative contract specification with a validity date that we specified when we created the new version.

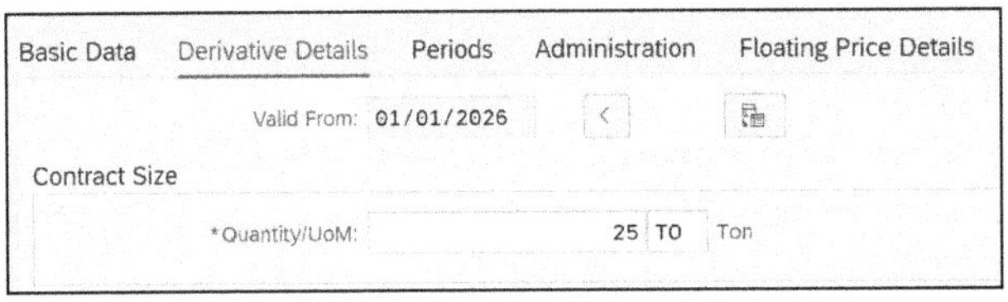

Basic Data	Derivative Details	Periods	Administration	Floating Price Details
	Valid From: 01/01/2026	<		
Contract Size				
	*Quantity/UoM:		25 TO Ton	

Figure 2.36 Derivative Contract Specification: Derivative Details Tab

- **Quantity/Unit of Measure**
  This is a required field and one of the most important elements when creating the derivative contract specification. In this field, we specify both the quantity and the unit of measure for a contract relating to a specific commodity. Commodities are traded on exchanges in standard lot sizes, which can vary greatly from one commodity to another. The purpose of this derivative contract specification section is to define the standard contract or lot size for each traded commodity, making sure that all contracts created in SAP S/4HANA adhere to these predefined standards.

  Accurately specifying the quantity and unit of measure is very important as it impacts how the commodity is traded, reported, priced, and settled. For example, commodities such as block cheese are traded in lots of 20,000 pounds, while metals such as aluminum are traded in lots of 25 metric tons. Each commodity's standard lot size is tailored to market conventions and trading practices of that exchange, and specifying the lot size within the derivative contract specification enables smooth and consistent transaction processing when we're tracking them in SAP S/4HANA.

  In Chapter 6, we'll see how this lot size specification plays an important role in the creation of commodity derivative contracts. It ensures that all derivative contracts are aligned with market standards, providing accurate pricing and reporting. The lot size influences additional aspects of the commodity contract, including margin requirements, settlement processes, and exposure calculations.

- **Currency**
  This is a required field, and it represents the quotation currency of the currency that we use when quoting the commodity price.

- **Currency Unit**
  This is a required field, and it represents the currency unit we use to define the price based on a currency using a defined ratio. The majority of the time, the currency for this field is the same as the quotation currency.

- **Decimal Places**
  This is a required field, and it represents the number of decimal places that are used when the commodity prices are quoted by the exchange.

- **Quotation Unit of Measure**
  This is a required field, and it represents the unit of measure that is used when the price is quoted.

- **CPE Unit of Measure**
  This field represents the retention period during which the market data related to the derivative contract specification is available in the system before it's archived.

- **Tick Size**
  This is a required field that we populate in the derivative contract specification, and it represents the minimum tick size of the commodity (see Figure 2.37).

2   Master Data

- **Tick Value**
  Based on the tick size and the lot size, the minimum tick value is calculated and automatically populated within the derivative contract specification.

Figure 2.37 Derivative Contract Specification: Derivative Details Tab

The last piece of the derivative contract specification that we'll cover includes the various periods related to the traded commodity. The periods are very important as they relate to the various dates for each commodity. The periods that we can define in the derivative contract specification are as follows:

- **Contract Period**
  We use this to define the term of the commodity contract.

- **Trading Period**
  We use this to define the period when the derivative can be traded.

- **Quotation Period**
  We use this to define the period when quotations are available for this derivative.

- **Physical Settlement Period**
  We use this to define the period when the contract can be physically settled or a party can take delivery of it.

- **Cash Settlement Period**
  We use this to define the period when the cash settlement of the derivative can take place.

- **Expiration Period**
  We use this to define the period when the option expires if it hasn't been exercised.

- **Quotation Period Forward**
  We use this to define the period when the forward prices are provided by our market data provider.

- **Quotation Period Settlement**
  We use this to define the period when the settlement prices are provided by the market data provider.
- **Averaging Period**
  We use this to define the period when the exchange prices of the underlying derivative are read and used to calculate the settlement price.

The type of contract that is created from the derivative contract specification drives the specific periods that are available and also drives what period types are mandatory. Table 2.1 clarifies the period types that are available and which ones are mandatory for each contract type. Those marked with asterisks (**) have a mandatory quotation period forward or quotation period settlement.

Period Type	Commodity Future and Forward	Listed Options	Commodity Forward Index
Contract Period	Required	Required	Required
Trading Period	Required	Required	
Quotation Period	Required	Required	
Cash Settlement Period	Available	Required	
Physical Settlement Period	Available	Required	
Expiration Period		Required	
Averaging Period		Required	
Quotation Period Forward			Required**
Quotation Period Settlement			Required**

Table 2.1 Derivative Contract Specification Period Options

In Figure 2.38, we can see that the periods are automatically populated in this tab based on the period determination method that we specified in the **Basic Data** tab of the derivative contract specification. We can change the start date in this tab, and the periods will automatically populate from the new start date.

At this stage, we've provided all the necessary information for the derivative contract specification. This includes the commodity type, market identifier code, unit of measure, quotation currency, lot size, minimum tick value, tick size, and populated periods. With all these details in place, the derivative contract specification is complete, with all of the required information (see Figure 2.39). Now, we can proceed to save the derivative contract specification.

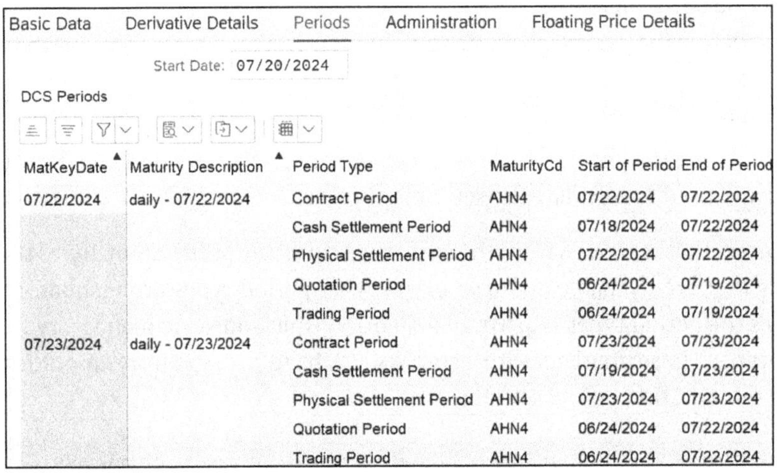

MatKeyDate	Maturity Description	Period Type	MaturityCd	Start of Period	End of Period
07/22/2024	daily - 07/22/2024	Contract Period	AHN4	07/22/2024	07/22/2024
		Cash Settlement Period	AHN4	07/18/2024	07/22/2024
		Physical Settlement Period	AHN4	07/22/2024	07/22/2024
		Quotation Period	AHN4	06/24/2024	07/19/2024
		Trading Period	AHN4	06/24/2024	07/19/2024
07/23/2024	daily - 07/23/2024	Contract Period	AHN4	07/23/2024	07/23/2024
		Cash Settlement Period	AHN4	07/19/2024	07/23/2024
		Physical Settlement Period	AHN4	07/23/2024	07/23/2024
		Quotation Period	AHN4	06/24/2024	07/22/2024
		Trading Period	AHN4	06/24/2024	07/22/2024

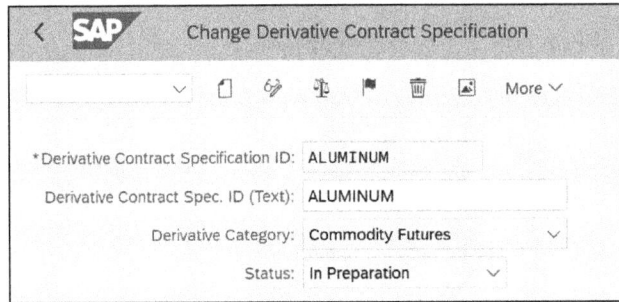

Figure 2.38 Derivative Contract Specifications

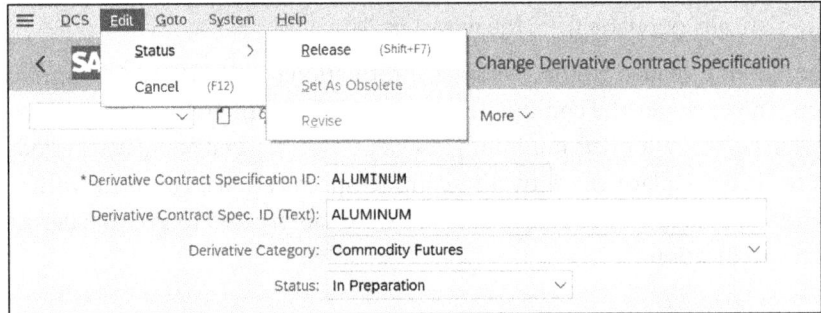

Figure 2.39 Derivative Contract Specification in Preparation Status

Saving the derivative contract specification finalizes the setup, but it's not yet available for use in creating and managing commodity derivative contracts. As shown in Figure 2.39, the derivative contract specification is still in the status of **In Preparation**, so we must release it before it's considered final and ready to use when creating a commodity derivative. To release the derivative contract specification, we access the menu ribbon on the top of the screen and navigate to **Edit • Status • Release**, as shown in Figure 2.40.

Figure 2.40 Release Procedure for Derivative Contract Specification

2.6 Derivative Contract Specification

Once we've fully released the derivative contract specification, it will appear with a status of **Released**, as depicted in Figure 2.41. This indicates that the derivative contract specification is now finalized and ready for use in creating a commodity derivative contract. At this point, all of the important elements of the commodity are contained within the derivative contract specification and locked in, and we can leverage them when creating a contract.

It's important to note that if we want to make any changes to the derivative contract specification, we must change the derivative contract specification status to **In Revision** before we can make those changes. This ensures that any adjustments are tracked and managed systematically within SAP S/4HANA. Once we've made the changes, we must release the derivative contract specification again so that we can use it in a contract. This process of revising and rereleasing the derivative contract specification ensures that all changes are properly vetted and approved before they impact trading operations.

Additionally, although a derivative contract specification is a form of master data, we can't create it directly in each SAP S/4HANA environment like other master data elements. The best practice is to initially create the derivative contract specification in a lower, nonproduction environment. This approach allows for thorough testing and validation of the derivative contract specification before it's moved through the various environments, ultimately reaching the production environment through a transport. The inability to change the derivative contract specification in production helps prevent errors and ensures that the derivative contract specification functions correctly in the live trading environment, maintaining the integrity and reliability of the master data in all SAP S/4HANA environments.

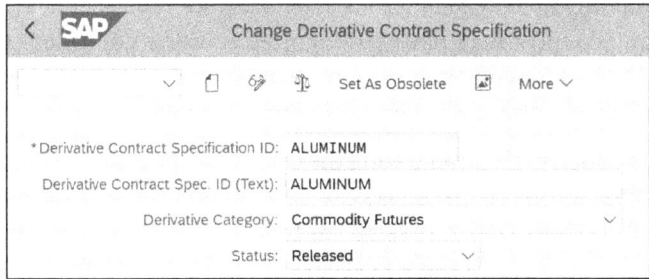

Figure 2.41 Derivative Contract Specification Release

Commodity Contract Month Codes

It's important to understand how to quote commodities based on their maturity month, instead of representing each month with a number or typing out the full month maturity codes. The commodity month maturity codes are standardized codes used across all exchanges to represent the specific delivery months of commodity futures contracts. The codes are very important in the trading and settlement of commodity futures because they indicate the month in which a contract is due for delivery.

Each commodity month code consists of a letter that represents a specific month, which makes it easy to reference and communicate the commodity contract details. The standard commodity month maturity codes are as follows:

- F: January
- G: February
- H: March
- J: April
- K: May
- M: June
- N: July
- Q: August
- U: September
- V: October
- X: November
- Z: December

2.7 Calendars

In numerous interactions with treasury and risk management, users will frequently encounter the concepts of dates, working days, and holidays. Properly managing financial calendars is imperative to ensuring that contracts are being valued appropriately and that any payments and related accounting entries are being made on allowable days. In the SAP S/4HANA system, we can create and maintain calendars to ensure that SAP is aligned with the calendar of financial systems around the world.

There are two primary types of calendars that are maintained in the SAP S/4HANA system: holiday calendars and factory calendars. A *holiday calendar* maintains a listing, by country, of all the holidays that are recognized in that region. This can include both religious and banking holidays, and SAP S/4HANA will treat each day listed in this calendar as a nonworking day. Maintaining holiday calendars is essential since certain holidays and banking holidays in one country may not be recognized in other countries or regions of the world. SAP S/4HANA is agnostic to this information, and thus, we'll need to instruct it on which days to recognize and which ones to treat normally.

We use a *factory calendar* to establish the days during the week that should be treated as working days. For example, in the United States, except for recognized holidays, Monday, Tuesday, Wednesday, Thursday, and Friday are treated as working days. In Qatar, working days during the week are Monday, Tuesday, Wednesday, Thursday, and Sunday. These settings can vary between countries and regions, and we must therefore maintain them independently. It's important to note that before we can assign or attribute any holiday to a particular holiday calendar, we must first create it as a public holiday. We maintain all calendar information in Transaction SCAL, and in the example in Figure 2.42, we can see the calendars that have been defined for the United States.

2.7 Calendars

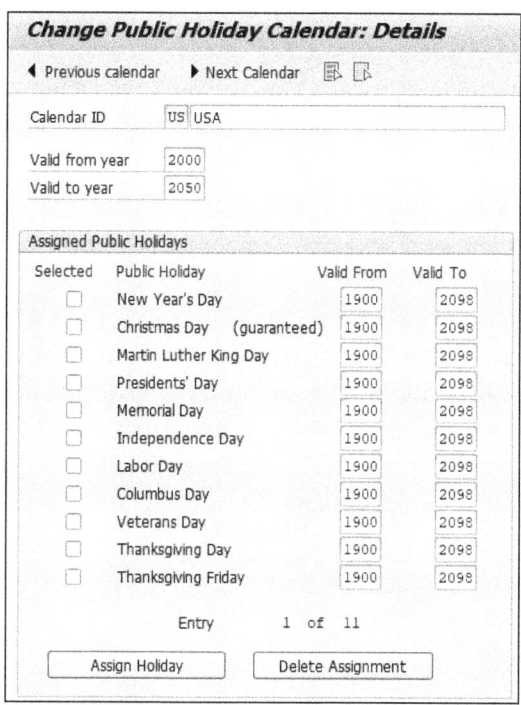

Figure 2.42 US Holiday Calendar with Applicable Holidays

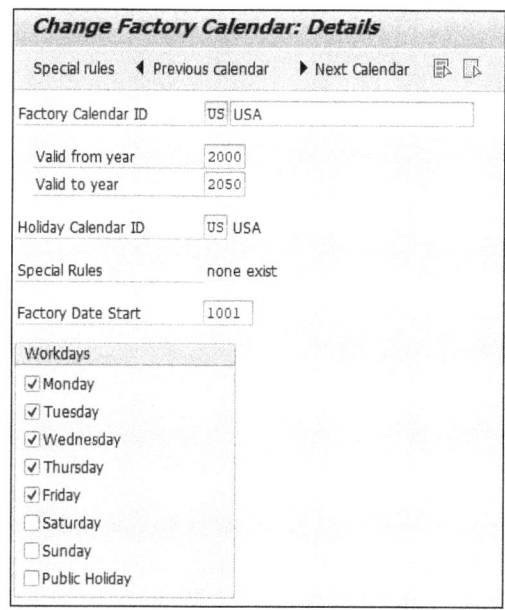

Figure 2.43 Standard-Delivered US Factory Calendar

2.8 Traders

Assigning traders is an optional step within treasury and risk management. Different traders within a company can have different responsibilities, and if we want to allow only certain treasury and risk management users to create certain types of transactions, then we can assign different traders to their allowed product types and transaction types. Assigning traders in SAP S/4HANA also allows us to further track who is executing the trades in treasury and risk management.

We can create traders via the **Financial Supply Chain Management • Treasury and Risk Management • Transaction Manager • General Settings • Organization • Define Traders** menu path. In this configuration, we create the traders by company code. We can assign the traders in this area, but keep in mind that the field is limited to twelve characters. To create a new trade, we click the **New Entries** button and then define the various traders required. As shown in Figure 2.44, we've defined the necessary traders that will be creating treasury contracts.

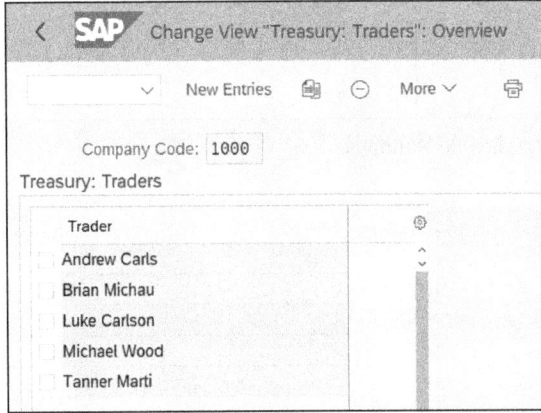

Figure 2.44 Creating Treasury Traders

Now that we've defined the traders that we will use in our treasury contracts, the next step is to further define traders and associated user data as well as authorize the traders for specific treasury contract types.

2.8.1 Defining User Data

After we've created the traders, we can assign them to a user ID in treasury and risk management. We assign them to a user ID by following the menu path **Financial Supply Chain Management • Treasury and Risk Management • Transaction Manager • General Settings • Organization • Define User Data** menu path. In Figure 2.45, we can see that the SAP user ID is now mapped to the corresponding trader.

2.8 Traders

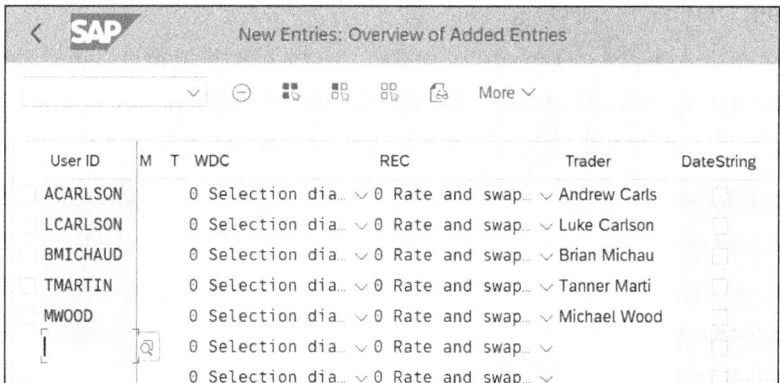

Figure 2.45 Defining User Data and Assigning SAP User IDs to Traders

There are a few other details in this configuration that we should highlight:

- **M**
  This is the one-character abbreviation for "one million." In this column, we can assign this abbreviation, which we do when creating a contract using the Create Financial Transaction app. For example, if we define *M* as our abbreviation, then we can type "1M" in the Create Financial Transaction app instead of typing out "1,000,000."

- **T**
  This is the one-character abbreviation for "one thousand," and it functions the same as *M* except that it's for one thousand instead of one million.

- **WDC**
  This is the working day check indicator. When we have a date assigned in treasury and risk management and it falls on a nonworking day, we can define the system behavior here. The options for this are in the dropdown list. We can also choose to create an error message that stops us from assigning dates on the weekend, or we can bypass the working day check altogether.

- **REC**
  This is the rate entry check indicator, and it works like the WDC except that it's for FX rates and swap rates. If you enter a rate that differs from the current rate in SAP S/4HANA by a certain amount, the system can give you a warning. You can define this to check both FX rates and swap rates, only check FX rates, or only check swap, rates or you can designate that you don't want to run a check at all.

- **DateString**
  If we check this box, then SAP S/4HANA calculates and interprets the dates entered in the value date fields based on the spot value date (rather than the contract conclusion date) when we're entering FX or currency options.

2.8.2 Assigning Traders

After we've created the traders, we can authorize them for each of the types of contracts via Transaction TBT1. We can define this by **Contract Type**, **Product Category**, **Product Type,** and **Transaction Type**. Once we select one of the options, we can click the **Enter** button to go to the next page. In Figure 2.46, we're filtering by **Contract Type** "4," which is for FX transactions. The following contract types are available in the **Contract Type** dropdown list:

- 2: Securities
- 4: Foreign Exchange
- 5: Money Market
- 6: Derivatives
- T: Trade Finance

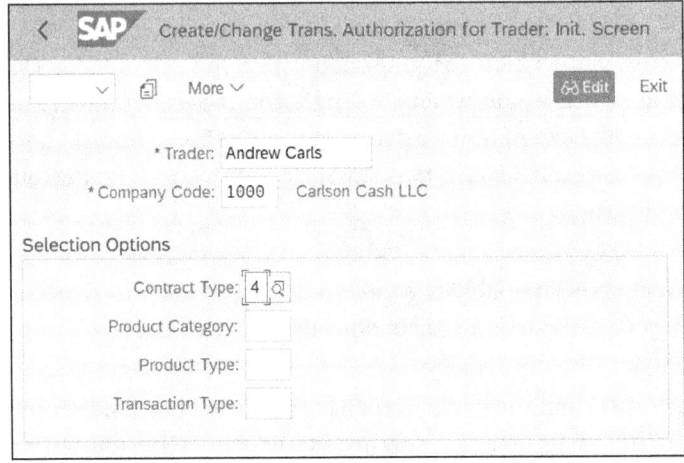

Figure 2.46 Authorizing Traders Using Transaction TBT1

On the next screen, we can check and uncheck the boxes in the **Auth.** column to select which transactions the user is and is not authorized to create. In Figure 2.47, we can see that this user is authorized to enter all types of transactions with the FX product type but is not allowed to enter transactions with any other FX product type.

As a result, if a user goes to the Create Financial Transaction app and tries to enter one of the transactions they are not authorized to enter, they'll receive an error message. The transaction will allow the user to go to the entry screen as shown in Figure 2.48, but as soon as they start entering data, an error will appear.

We can execute this activity for all types of treasury and risk management contract types, including securities, FX, money markets, derivatives, and trade finance.

2.9 Market Data

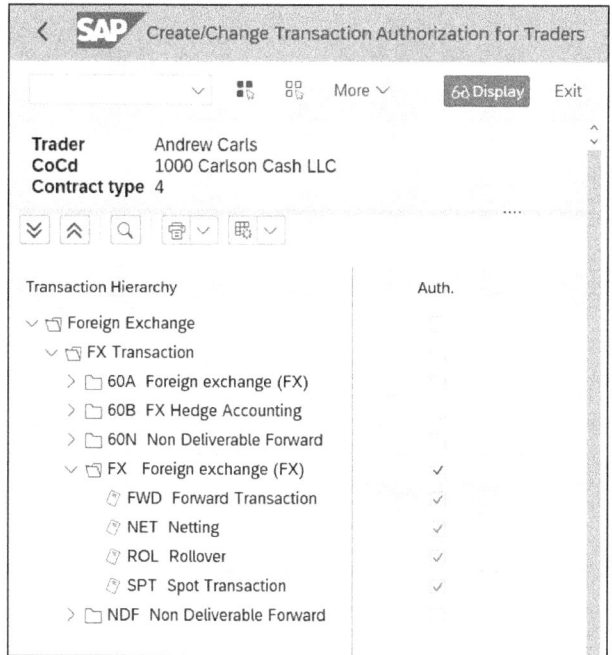

Figure 2.47 Authorizing Specific Product and Transaction Types Using Transaction TBT1

Figure 2.48 Attempting to Create Treasury Contract without Transaction TBT1 Authorization

2.9 Market Data

Many financial transactions that are entered in treasury and risk management are dependent on market data. For example, if a debt contract is tied to a variable interest rate, the variable interest rate needs to be available in SAP for the debt contract to calculate the interest amounts correctly. Also, if a transaction requires a month end mark-to-market, then yield curves would be required to calculate this valuation. Since many of the financial transactions are dependent on this data, we'll be covering the entry and usage of the market data. Examples of the market data we'll cover include FX rates, swap rates, security prices, reference interest rates, yield curves, commodity prices, and indexes.

Most financial transactions in treasury and risk management also have cash flows or calculations that are dependent on some type of market data. For example, FX rates are required when you have contracts in a foreign currency or in FX transactions to correctly calculate the values in the company code's local currency. Additionally, many transactions require a valuation at the end of the month, for the valuation to calculate correctly, a yield curve and interest rates must be in SAP S/4HANA. This section will cover the various types of market data that we can bring into the treasury and risk management module, and it will also highlight why the data is required.

2.9.1 Foreign Exchange Rates

We use FX rates in more areas of SAP S/4HANA than just treasury transactions. FX rates are relevant for any accounting entry that has more than one currency represented and are also relevant for any currency conversions. Before covering how to enter FX rates, we'll cover some configuration that drives how the currencies operate in SAP S/4HANA. Since we use many configuration tables to set up the currencies and they are standard delivered tables, we'll focus on the tables that impact the FX rate functionality in treasury and risk management.

Defining the Leading and Following Currencies

We need to enter the leading and following currencies for treasury and risk management to read the exchange rates correctly. We can find this configuration in the **Financial Supply Chain Management • Treasury and Risk Management • Transaction Manager • General Settings • Transaction Management • Currencies • Define Leading Currency** menu path. Figure 2.49 shows many of the most common FX currency pairs.

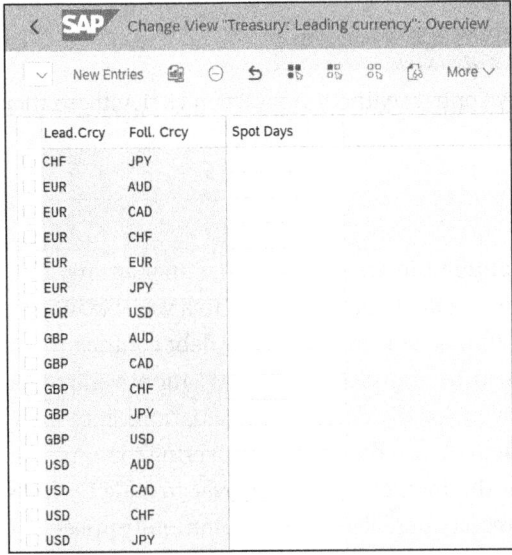

Figure 2.49 Defining Leading and Following Currency Pairs for Foreign Currency Pairs

2.9 Market Data

The first step when defining a currency pair is to define the leading currency in the **Lead.Crcy** column, and we then define the following currency by populating the **Foll. Crcy** column.

Rate Types

Another group of settings that is useful for currency exchange rates is the rate type configuration. Depending on the activities within treasury and risk management, we may create a new exchange rate type. Frequently, a company has a rate type for monthly FX rates and a different exchange rate type for daily FX rates, and we can create these rate types in the configuration path of **Financial Supply Chain Management • Treasury and Risk Management • Basic Functions • Market Data Management • Master Data • Currencies • Check Rate Types** or via Transaction OB07.

Once in this configuration, we can see the settings that we can assign to the rate type. One of the most useful settings in the rate types that we frequently need in implementations is the inverted exchange rate (**Inv**). We check this box (see Figure 2.50) when we might not have a complete set of FX rates in the system. If we have an exchange rate from US dollars to British pounds in the rate tables but the system is trying to figure out the translation from pounds to dollars, it won't be able to correctly translate the amount for that entry. If we check this box, SAP S/4HANA can invert the exchange rate to correctly figure out the rate for pounds to dollars.

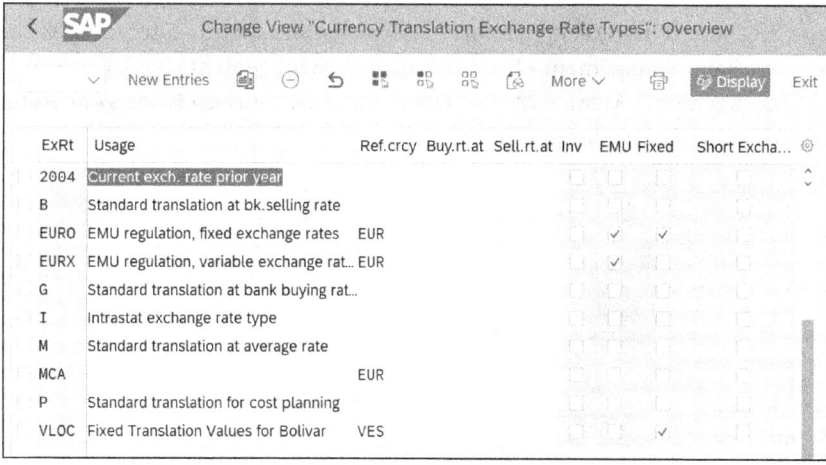

Figure 2.50 Creating Exchange Rate Types

Entering Foreign Exchange Rates

We can maintain FX by using Transaction OB08 or by following the user menu path of **Accounting • Financial Supply Chain Management • Treasury and Risk Management • Basic Functions • Market Data Management • Manual Market Data Entry • Currency • Enter Exchange Rates**. This table can also be populated by using a market data interface. The entries in this table show how we can assign rates to different exchange rate types, and we can enter rates using either a direct quotation or an indirect quotation. If we enter a rate using the opposite quotation from that which is configured, this table will still allow the entry, but the entry will be highlighted. Figure 2.51 shows an example of this table with populated values using both direct and indirect quotations for FX rates. One thing to note is that SAP S/4HANA will use the most recent FX rate, and if we have a transaction on a certain date and an FX rate is not available for that date, then it will look to the past for the most recent FX rate.

ExRt	ValidFrom	Indir.quot	X	Ratio(from)	From	=	Dir.quot.	X	Ratio (to)	To
M	01/01/2001		X	1	CAD	=	0.66800	X	1	ARS
M	01/01/2001		X	1	CAD	=	1.30000	X	1	BRL
M	01/01/2001		X	1	CAD	=	385.00000	X	1	CLP
M	01/01/2001		X	1	CAD	=	1.48000	X	1,000	COP
M	01/01/2001		X	1	CAD	=	76.70000	X	1	JPY
M	01/01/2001		X	1	CAD	=	6.50000	X	1	MXN
M	01/01/2001		X	1	CAD	=	2.35000	X	1	PEN
M	01/01/2023	0.75580	X	1	CAD	=		X	1	USD
M	01/01/2001		X	1	CAD	=	468.00000	X	1	VEB

Figure 2.51 Currency Exchange Rate Maintenance Table

2.9.2 Swap Rates

FX swap rates are located in the same area as the FX rates, and we can enter them in the **Financial Supply Chain Management • Treasury and Risk Management • Basic Functions • Market Data Management • Manual Market Data Entry • Currencies • Enter Swap Rates** menu path or by using Transaction TMDFXFP.

Effective From	Exchange Rate Type	From	To	Term(Days)	Swap Rate
05/01/2024	M	AUD	CAD	7	0.20400
05/01/2024	M	EUR	HUF	7	27.27000
05/01/2024	M	EUR	SGD	30	3.28000-
05/01/2024	M	EUR	USD	1	0.60000
05/01/2024	M	EUR	USD	7	1.71000
05/01/2024	M	EUR	USD	30	7.67000
05/01/2024	M	EUR	USD	90	23.23000
05/01/2024	M	EUR	USD	180	51.09000
05/15/2024	M	AUD	HKD	1	0.00007-
05/15/2024	M	AUD	HKD	7	0.00050-
05/15/2024	M	AUD	HKD	30	0.00117-
05/15/2024	M	AUD	HKD	90	0.00051-
05/15/2024	M	AUD	HKD	180	0.00004
05/15/2024	M	AUD	HKD	270	0.00224
05/15/2024	M	AUD	HKD	360	0.00105
05/15/2024	M	AUD	HKD	720	0.01955-
05/15/2024	M	AUD	JPY	1	0.01250-
05/15/2024	M	AUD	JPY	7	0.08670-
05/15/2024	M	AUD	JPY	30	0.39110-
05/15/2024	M	AUD	JPY	90	1.19070-
05/15/2024	M	AUD	JPY	180	2.31800-
05/15/2024	M	AUD	JPY	270	3.47270-

Figure 2.52 Swap Rate Maintenance

Depending on the settings in the configuration, we'll need the FX swap rates for the calculation of the NPV and for the key date valuation of FX transactions. As shown in Figure 2.52, the swap rates are maintained similar to FX rates and are stored in the table by currency pair and the term represented in days.

2.9.3 Security Prices

Treasury and risk management can incorporate real market data into the system for multiple purposes, and one primary reason for integrating this type of market data into the system is to generate a value for certain contracts. A bond contract, for example, could draw on real market data that has been integrated into SAP S/4HANA to calculate a realistic, market-driven value for that specific contract.

The first step in security price market data integration is the creation and/or maintenance of a price type, and we manage this configuration by using the following menu path: **Financial Supply Chain Management • Treasury and Risk Management • Basic Functions • Market Data Management • Master Data • Securities • Define Security Price Types**. As shown in Figure 2.53, we first define a price type in the **Price Type** column, and we'll use this as the descriptor to represent the specific price type. Then we can define the **Long name** and **Short Name** to further describe the price type.

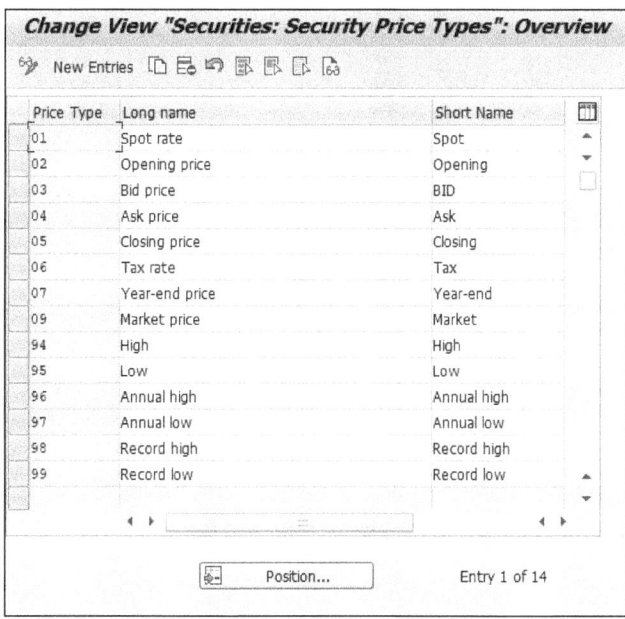

Figure 2.53 Configuration Screen for Price type

The system can then utilize each configured price type in other areas of configuration as well as when uploading market data security prices into SAP S/4HANA. After we've configured the respective price types, market data can be integrated into the system.

2 Master Data

There are two primary methods for integrating security price data into the system. First, we can import the security price data with Transaction TBD4/TBDM via a file interface, and second, we can maintain the security price data manually by using Transaction FW18 or by following the **Financial Supply Chain Management • Treasury and Risk Management • Basic Functions • Market Data Management • Manual Market Data Entry • Securities and Indexes** menu path. The user will need to input the security class and the corresponding exchange that the market data pertains to by populating the **Exchange** field. Once the user has entered this data, they will be able to adjust the security price data as necessary. Figure 2.54 shows the manual market data entry screen for a selected security class.

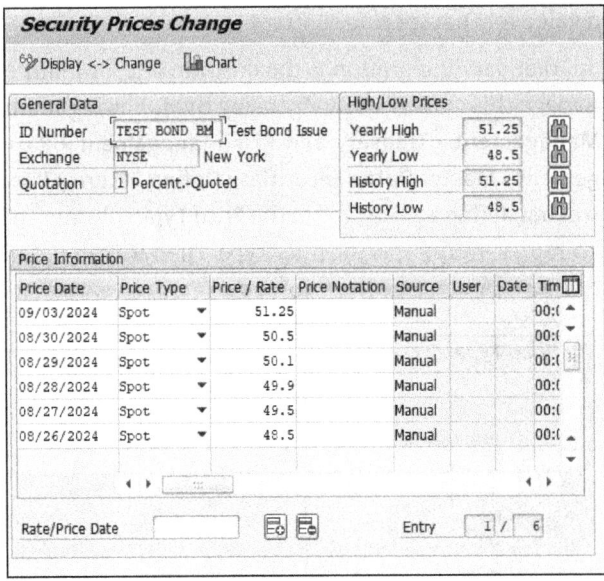

Figure 2.54 Manual Market Data Entry Screen for Selected Security Class

2.9.4 Reference Interest Rates

Reference interest rates are key pieces of market data that are generally required in any treasury and risk management implementation. We use them to save the interest rate data points for each individual day, and once we've done that, the financial transactions can use these rates for the variable interest calculations and for month-end valuations. In this section, we'll cover how to assign reference interest rates to yield curves—but first, we'll cover some of the relevant configuration steps required for the rates.

Configuration

Before entering reference interest rates, we need to create the rate type in the configuration. We can find this in the **Financial Supply Chain Management • Treasury and Risk**

Management • Basic Functions • Market Data Management • Master Data • Settings for Ref Interest Rates and Yield Curves • Define Reference Interest Rates menu path. We assign the settings for each of the required reference interest rates as shown in Figure 2.55. Some of these fields are only informational, so we'll only cover the more functional properties in this configuration:

- **Currency**
  This field ties the reference interest rate to a currency, and we use it to assign rates to a yield curve. The yield curve is dependent on a currency, and we can't assign interest rates that are assigned to a different currency than the yield curve.

- **Interest Calculation Method**
  In this field, we assign a default interest calculation method. When creating a financial transaction, we can assign a different interest calculation method, but SAP will generate a warning letting you know that the transaction's interest calculation method differs from the default calculation method for the reference interest rate.

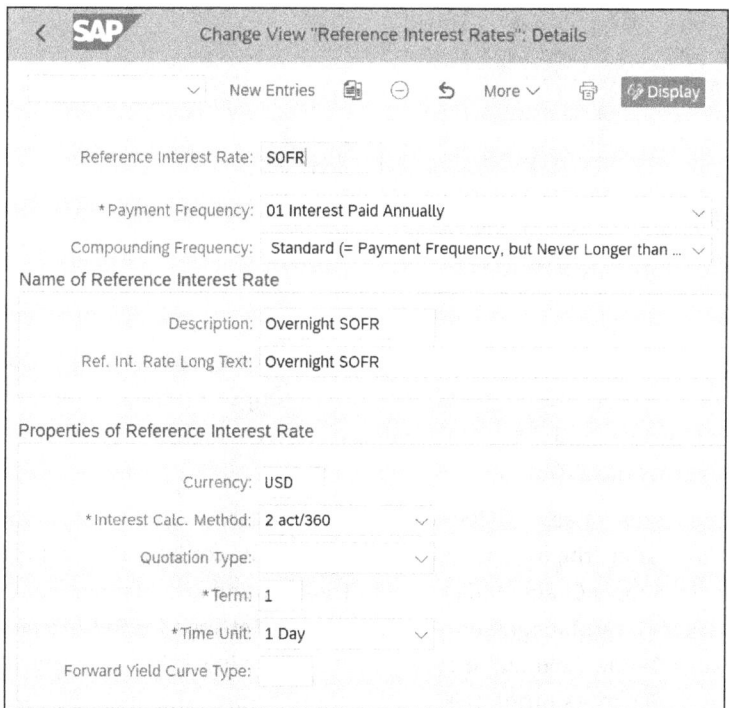

Figure 2.55 Creating Reference Interest Rate

Entry

We can enter reference interest rates in SAP S/4HANA by using Transaction JBIRM or by following the user menu path **Financial Supply Chain Management • Treasury and Risk Management • Basic Functions • Market Data Management • Manual Data Entry • Interest • Enter Reference Interest Rates**.

The interest rates have a few fields where we need to enter data for each rate: the **Reference** interest rate, the **Valid From date,** and the actual interest rate (**Int. Rate**). One thing to note is that this functionality works differently from how SAP S/4HANA looks up the most recent FX rate. When we're using these rates to calculate the interest rate amounts, a rate needs to be available on the fixing date because SAP S/4HANA won't look for the most recent rate when calculating the interest. We can see some examples of saved interest rates in Figure 2.56.

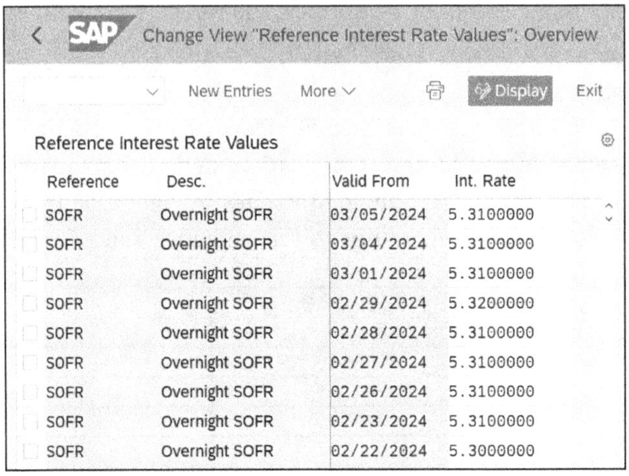
Change View "Reference Interest Rate Values": Overview
New Entries
More
Display
Exit
Reference Interest Rate Values
Reference
Desc.
Valid From
Int. Rate
SOFR Overnight SOFR 03/05/2024 5.3100000
SOFR Overnight SOFR 03/04/2024 5.3100000
SOFR Overnight SOFR 03/01/2024 5.3100000
SOFR Overnight SOFR 02/29/2024 5.3200000
SOFR Overnight SOFR 02/28/2024 5.3100000
SOFR Overnight SOFR 02/27/2024 5.3100000
SOFR Overnight SOFR 02/26/2024 5.3100000
SOFR Overnight SOFR 02/23/2024 5.3100000
SOFR Overnight SOFR 02/22/2024 5.3000000

Figure 2.56 Maintaining Reference Interest Rates

2.9.5 Yield Curves

We use yield curves in SAP S/4HANA to determine the mark-to-market or NPVs for financial transactions. The yield curves are created from groups of reference interest rates with varying date values, and these rates are used to create a yield curve that we can leverage when creating the NPVs.

As with the reference interest rates, we'll look at the configuration for the yield curves to show how the settings affect the functionality. We can find the configuration for the yield curves in **Financial Supply Chain Management • Treasury and Risk Management • Basic Functions • Market Data Management • Master Data • Settings for Ref Interest Rates and Yield Curves • Define Yield Curve Types.** The first step in setting up the yield curve is to define the yield curve's properties, and Figure 2.57 shows the fields we need to fill out: **Payment Frequency**, **Compounding Frequency**, **Extrapolation Method**, **Quotation Type**, and **Maximum Age**.

For each yield curve type, we can create different yield curve settings for different currencies. To do this, we click the **New Currency** button while in edit mode, which will bring us to the screen shown in Figure 2.58. To define details for each currency, we double-click into the desired line, and we can then update the currency, calendar, interest calculation method, and reference interest rates applicable to that specific currency.

2.9 Market Data

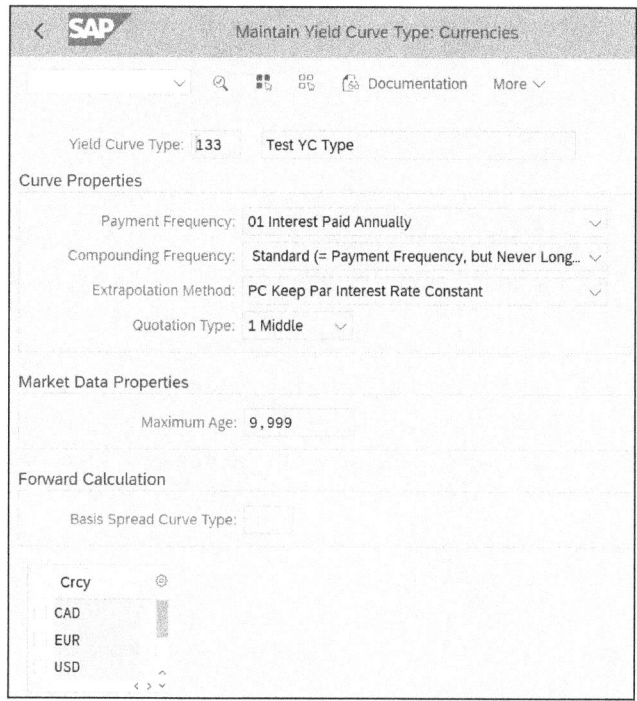

Figure 2.57 Yield Curve Creation

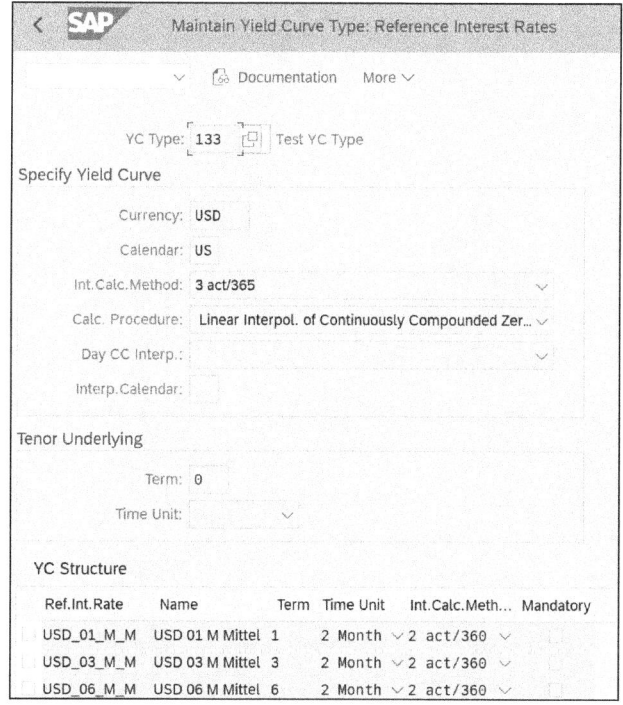

Figure 2.58 Yield Curve Structure Maintenance for US Dollar Currency

Once in the currency, we can see the yield curve structure at the bottom of the screen. This is where we assign different reference interest rates and assign the corresponding time structure, all of which serve as different data points for the yield curve.

2.9.6 Commodity Prices

Commodity price data in SAP S/4HANA is a crucial master data element for companies involved in trading, manufacturing, and procurement of raw materials. The commodity price data represents the foundational information that drives accurate contract pricing, valuation, and risk management for commodities within the system. By setting up reliable commodity price master data, companies can streamline the tracking of market prices, manage fluctuations, and integrate pricing seamlessly across the management of commodity contracts. We'll discuss how to maintain and track commodity prices in SAP S/4HANA.

We maintain commodity prices by using Transaction FDCS17, with a combination of the derivative contract specification and the market identifier code. As shown in Figure 2.59, we're maintaining commodity prices for aluminum traded on the London Metals Exchange (XLME). After we enter the **DCS ID** as well as the **Market Identifier Code**, we can maintain the pricing data by clicking **Execute**.

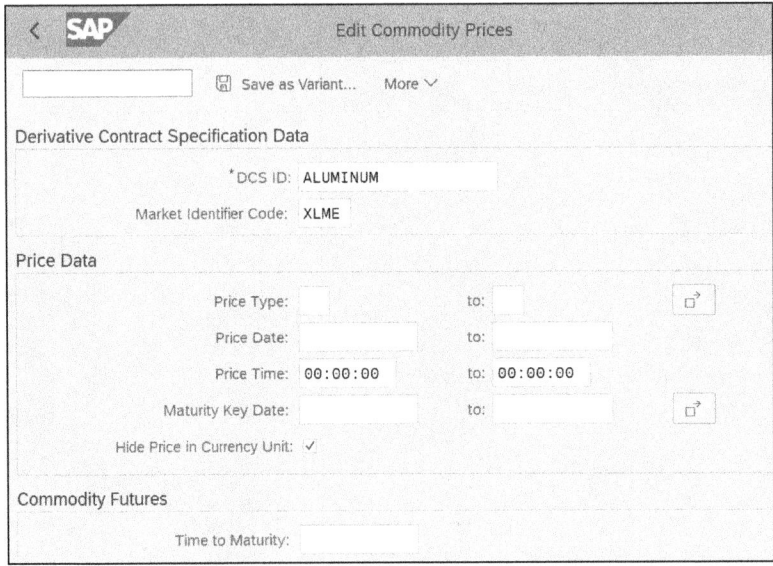

Figure 2.59 Editing Commodity Prices Using Transaction FDCS17

After we press **Execute,** we're directed to the screen shown in Figure 2.60, where we can maintain the commodity prices for the specific commodity. The first step is to define the **Price Date** with the date when a commodity was quoted at a particular price. We then maintain the **Price Type** along with the **MatKeyDate**. The **Quotation Price** is then populated with the commodity price. Values such as currency and unit of measure are automatically populated from the **Derivative Contract Specification.**

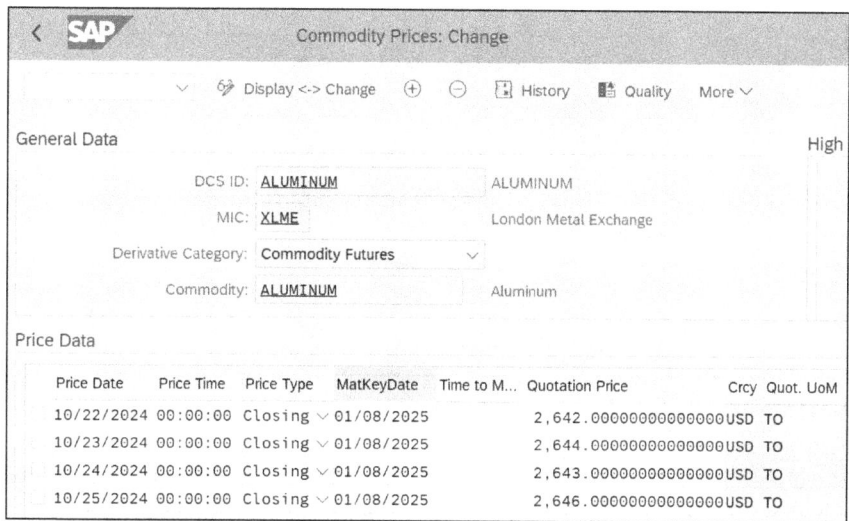

Figure 2.60 Commodity Price Maintenance for Aluminum Futures

2.10 Market Data Interfaces

Now that we've covered the various types of market data that SAP S/4HANA can consume, it's time to discuss how to import this data into the system. We can integrate market data through market rate management or via external data sources. Common providers for this external market data include Bloomberg and Refinitiv, though SAP S/4HANA also supports data feeds from a wide range of other providers. Whether you're utilizing market rate management or external feeds, having reliable access to market data is essential for accurate financial and risk management processes in SAP S/4HANA.

Although SAP S/4HANA has a number of standard programs that can import market data, customized importation of market data into SAP S/4HANA is very common in many projects. This is due to the fact that many market data providers deliver market data in a format that SAP S/4HANA can't natively consume, and customization is often required to massage or change the market data file into a format that SAP S/4HANA can consume. After the market data file is changed into the correct format, we can upload it using the standard Transaction TBD4, but before that, we must perform a series of configuration steps to enable the program to run correctly.

The first step to enable Transaction TBD4 is to define the data feed name by following the menu path **Financial Supply Chain Management • Cash and Liquidity Management • General Settings • Market Data • Datafeed • Technical Settings • Define Datafeed Name**. Within this configuration node, we establish the data provider that will be providing the market date. We first enter the name of the data provider in the **Data Provider** column,

2 Master Data

as shown in Figure 2.61, and we can then add a description of the data provider and the type of data that will be provided. We should also check the **Usage Log** and **FeedActive?** boxes and provide a usage log when the transaction is executed.

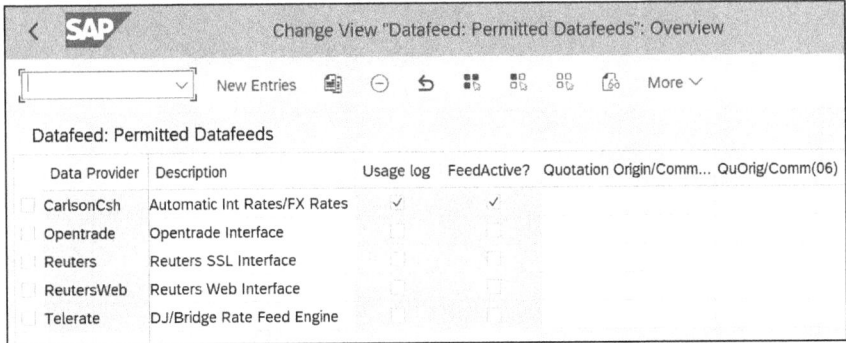

Figure 2.61 Defining Data Feed Name Configuration

Now that we've defined the data feeds, we can assign data feed remote function call destinations for the data provider. To start the configuration, we navigate to **Financial Supply Chain Management • Cash and Liquidity Management • General Settings • Market Data • Datafeed • RFC Settings for External Partner Program • Assign Datafeed RFC Destination**. In this step, we map the data provider (which we set up in the prior step) to the program associated with that data provider. We also establish the operating mode for the data, which we can define as synchronous or asynchronous:

- **Synchronous**
  In this operating mode, SAP S/4HANA waits until the external partner program has delivered all the data.

- **Asynchronous**
  In this operating mode, SAP S/4HANA doesn't wait until the external partner program has delivered all the data and switches control back to the calling program.

As shown in Figure 2.62, we've mapped our data provider (**CarlsonCSh**) using a **Synchronous** operating mode, and we've then mapped it to the applicable program.

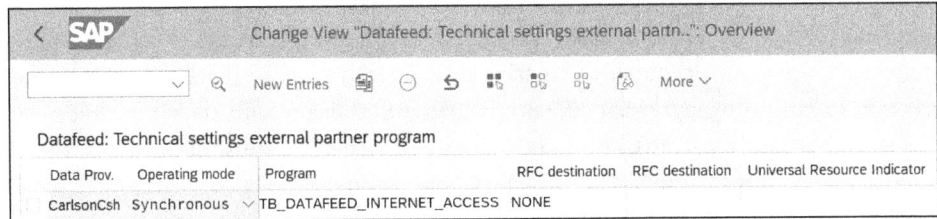

Figure 2.62 Configuration to Assign Data Feed Remote Function Call Destination

The next step is to define internet settings for web server access, which we can do in the **Financial Supply Chain Management • Cash and Liquidity Management • General**

Settings • Market Data • Datafeed • Technical Settings • Internet Settings for External Partner Program • Define Internet Settings for Web Server Access menu path. In this step, we define the settings for access to the data partner's web server, which will enable real-time data feed connectivity between SAP S/4HANA and the data provider. The first step is to define the data provider in the work area, as shown in Figure 2.63.

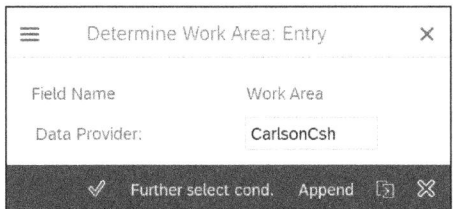

Figure 2.63 Defining Internet Settings for Web Server Access

After entering the data provider, we'll be navigated to the screen shown in Figure 2.64. In this configuration node, we map the URL of the market data service provider by populating the **URL** field, and we then enter the **User name** and **Password** provided by the market data service provider.

Figure 2.64 Internet Settings for Web Server Access URL, User Name, and Password Configuration

The final step is to define the types of data that will be imported using the market data provider. In this example, we'll walk through the setup for FX rates, but the process is similar for other types of market data. To define the currency pairs, we navigate to **Financial Supply Chain Management • Cash and Liquidity Management • General Settings • Market Data • Datafeed • Translation Table • Define Currencies**. In this step, we map the various exchange rate currency pairs to the external instrument by rate type, as defined by the market data provider. The first step is to define the **From** and **To** currencies along with the **RType,** which will then be mapped to the **Instrument** for that specific rate that should be brought into SAP S/4HANA. The **Instrument** is the internal descriptor used by the market data provider to distinguish that currency pair, as shown in Figure 2.65.

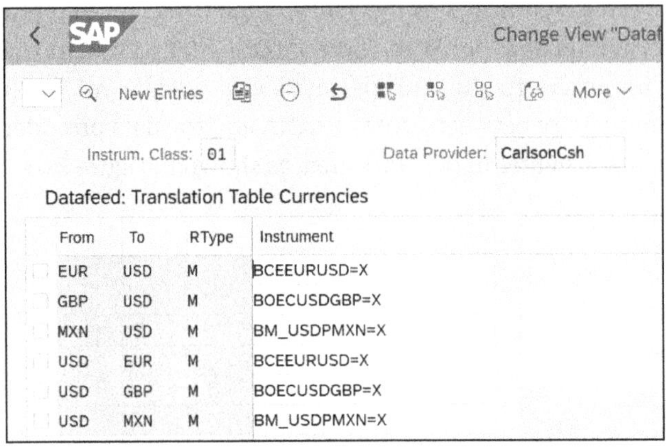

2   Master Data

From	To	RType	Instrument
EUR	USD	M	BCEEURUSD=X
GBP	USD	M	BOECUSDGBP=X
MXN	USD	M	BM_USDPMXN=X
USD	EUR	M	BCEEURUSD=X
USD	GBP	M	BOECUSDGBP=X
USD	MXN	M	BM_USDPMXN=X

Instrum. Class: 01   Data Provider: CarlsonCsh
Datafeed: Translation Table Currencies

Figure 2.65 Mapping Currency Pairs to Market Data Provider Instrument Codes

2.11 Summary

Now that we've walked through the various master data components, we can appreciate how master data in SAP S/4HANA serves as the foundational element enabling efficient transaction processing across the system. This chapter provides a comprehensive overview of the essential master data types required for various treasury and risk management processes within SAP S/4HANA Finance for treasury and risk management. Covering the complete end-to-end process of master data creation, it emphasizes how each process depends on specific master data. The chapter includes detailed insights into key master data categories relevant to treasury operations, establishing a clear understanding of their roles and applications in supporting end-to-end financial processes.

Chapter 3
Core Configuration Elements

We invite you on a journey through the intricate landscape of SAP S/4HANA configuration, where clarity and precision give you the knowledge and skills vital for seamless implementation and optimization. Whether you are a novice navigating the complexities of SAP S/4HANA or a seasoned professional seeking to refine your expertise, this chapter offers a comprehensive roadmap to success in instrument configuration within the treasury landscape.

Now that we've explored the key master data elements of treasury and risk management, we'll transition our focus to the core configuration that serves as the backbone of SAP S/4HANA Finance for treasury and risk management. Recognizing the differences in the functionalities of instruments, it becomes evident that configuration requirements can differ greatly from one another. Thus, our primary goal in this chapter is to discuss in detail the fundamental configuration essentials applicable to all instrument types within treasury and risk management.

Irrespective of the contract type, it's important to address and understand the foundational elements of configuration. The journey of creating a new instrument in treasury and risk management unfolds through a series of logical steps. Regardless of the instrument type that we're creating, there are core configuration steps that apply to all instruments within the setup of treasury and risk management. These steps commence with the establishment of baseline settings and encompass crucial aspects such as accounting codes and valuation configuration, which lay the groundwork for subsequent configurations. This process finishes with us configuring the integration with the general ledger, ensuring a seamless alignment between treasury and risk management and the broader SAP S/4HANA landscape.

In this chapter, we dive into the foundational aspects necessary for configuring any instrument type within treasury and risk management. By starting with the fundamental configuration elements, we can establish a baseline before moving on to additional instrument-specific configuration. It's important to understand the hows and whys behind the primary configuration nodes, as we will use many of the same nodes again and again, regardless of the instrument type we're creating.

3   Core Configuration Elements

3.1   Company Code Additional Data

When configuring any instrument within treasury and risk management, we must start by performing a series of steps to set the stage before we create the individual instrument types. The first step we'll cover is the creation of additional company code data. This step is important because it's the first step required for all company codes in scope for treasury and risk management. We may have a variety of company codes configured throughout our organization, but this node is required to activate the company codes that we plan to build out for treasury instrument creation.

To start the configuration of this node, we navigate to the following menu path: **Financial Supply Chain Management • Treasury and Risk Management • Transaction Manager • General Settings • Organization • Define Company Code Additional Data**. Upon entering the node we'll notice that many of the configuration fields are grayed out and we can't enter data there. The majority of the settings within this node are automatically populated by the underlying company code settings for each of the company codes, and we can start the configuration by entering a company code at the top part of the screen. As we can see in Figure 3.1, as soon as the **Company Code** field is populated, the **Currency**, **FY Variant**, **Chart of Accts** and **Fiscal Year End** date are automatically populated based on the company code settings already established.

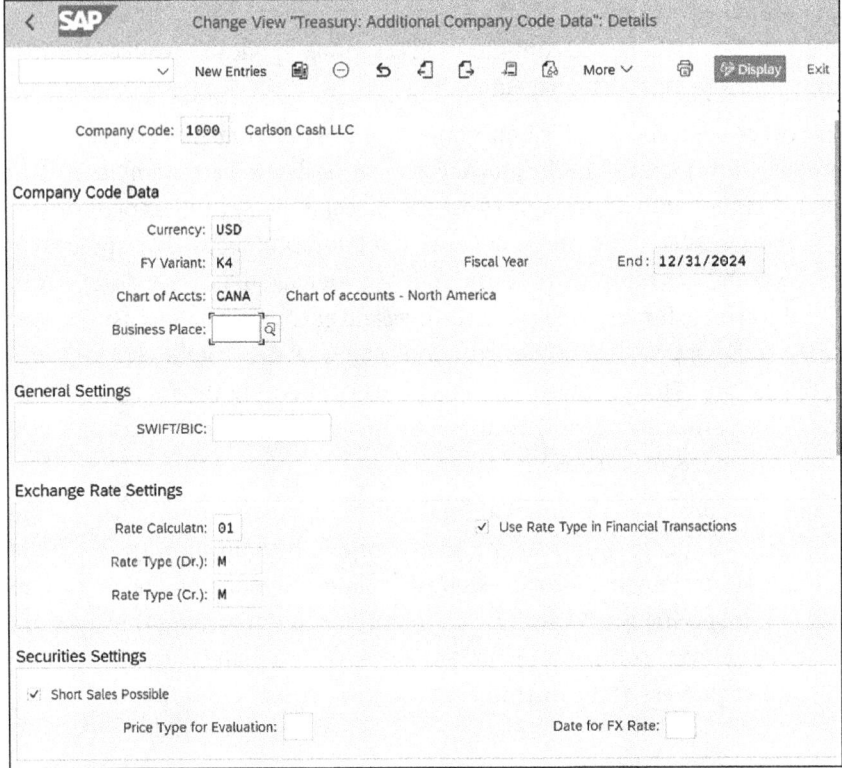

Figure 3.1 Company Code Additional Data Configuration Screen

3.1 Company Code Additional Data

Once we've configured the company code data, the next piece of configuration within this node that's worth discussing is the **Exchange Rate Settings**, with which we define the exchange rate type that we use for both the debit and credit entries when we're posting treasury contracts. We define the rate calculation type by navigating to the following menu path: **Financial Supply Chain Management • Treasury and Risk Management • Transaction Manager • General Settings • Organization • Define Calculation Indicator**. It's very common to use the M rate for both the debit and the credit, but in theory, we could use any rate type. The first step is to create a two-character **RateCalc.ID** and then assign a **RateType(D)** for the debit rate type and a **RateType(C)** for the credit rate type. In the example in Figure 3.2, we've configured rate calculation ID 01 to use the M rate for both the debit and the credit.

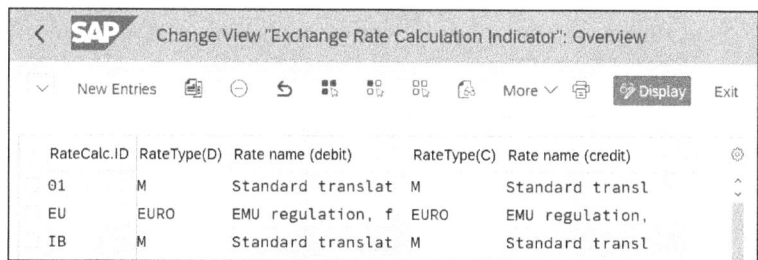

Figure 3.2 Exchange Rate Calculation Indicator

It's important to note that this configuration is only applicable to securities transactions; for it to apply to all other financial transactions, we must check the **Use Rate Type in Financial Transactions** box.

The other important element of this configuration node is determining how variable interest rates, commodity prices, and security prices appear in the cash position. As we can see in Figure 3.3, we can treat the settings for interest rates, commodity prices, and security prices independently, and they can appear differently in the cash position. The options for these three sections are as follows:

- Settings for Variable Interest Rates
  - Zero update
  - Update with interest rates maintained automatically
  - Update types with interest rates maintained manually
  - Update with current interest rates
  - Update with interest rates maintained automatically/manually
- Settings for Variable Commodity Prices
  - Zero update
  - Update with commodity prices maintained automatically
  - Update with commodity prices maintained manually
  - Update with current commodity prices

3 Core Configuration Elements

- Update with commodity prices maintained automatically/manually
- Update with commodity prices according to commodity curve
- Settings for Variable Security Prices
  - Zero update
  - Update with security prices maintained automatically
  - Update with security prices maintained manually
  - Update with current security prices
  - Update with security price maintained automatically/manually

Figure 3.3 Settings for Variable Prices

3.2 Accounting Codes

The next key step that relates to the company code is the creation of the accounting codes. Establishing the accounting code is a crucial step that is necessary to link the company codes to position management within treasury and risk management. The accounting codes should align closely with the company codes defined within SAP and should ideally maintain a one-to-one relationship with them. Furthermore, it's important that the naming conventions of the accounting codes mirror what was used during the initial setup of company codes.

Each company code included within the scope of treasury and risk management requires the completion of this configuration to establish a link between the internal accounting code defined within treasury and the already established company code in SAP S/4HANA. Despite its seeming like a repetitive step, configuring accounting codes is a required element in the setup process, ensuring consistency and alignment of the company code with the accounting code within treasury and risk management.

To create the account codes, we follow the menu path **Financial Supply Chain Management • Treasury and Risk Management • Transaction Manager • General Settings • Accounting • Organization • Define Accounting Codes**. The first step is to define the accounting code, which will mirror the company code in the **AC** field. Next, we enter the name of the accounting code or company code in the **Accounting code** field. The

106

accounting code name should be the same as the name of the company code and should maintain a one-to-one relationship with the names of the company codes. The last step is to define the company code that the accounting code is mapped to in the **CoCode** field. As we can see in Figure 3.4, we've configured the accounting code to equal the company code.

AC	Accounting code	Co...
0001	Head Office	0001
1000	Carlson Cash LLC	1000
2000	Carlson Cash LLC	2000
3000	Carlson Cash GB LLC	3000
4000	Carlson Cash CA LLC	4000

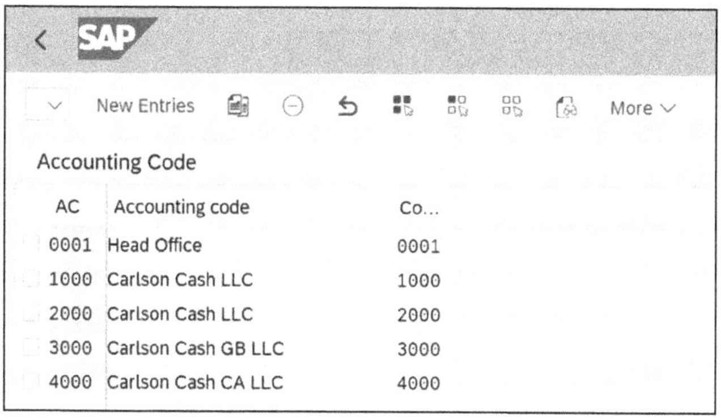

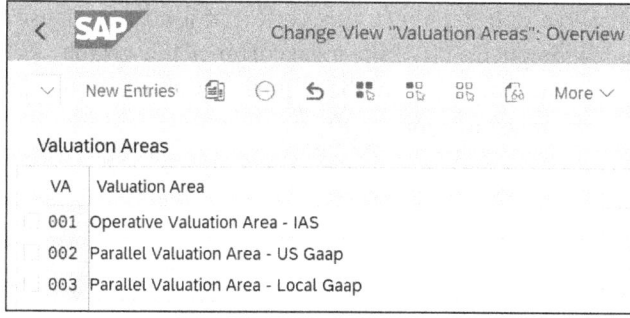

Figure 3.4 Accounting Code Configuration

3.3 Valuation Areas

A *valuation area* in treasury and risk management represents a specific accounting principle. Each accounting principle in the scope of a valuation area must be maintained within a configuration, and common accounting principles such as the International Financial Reporting Standards and the US generally accepted accounting principles and are supported. To set up the valuation area, we navigate to the following menu path: **Financial Supply Chain Management • Treasury and Risk Management • Transaction Manager • General Settings • Accounting • Organization • Define Valuation Areas**. In the first step, we define the valuation area with a name and a unique set of characters. We will assign the functions of the valuation area in subsequent customizing steps. As shown in Figure 3.5, we map the valuation area (**VA**), and then we define the **Valuation Area** name.

VA	Valuation Area
001	Operative Valuation Area - IAS
002	Parallel Valuation Area - US Gaap
003	Parallel Valuation Area - Local Gaap

Figure 3.5 Defining Valuation Areas

3   Core Configuration Elements

Then, we must create the accounting code, assign it to a valuation area, and link the valuation area with the accounting code and company code. As illustrated in Figure 3.6, we connect the accounting code to the previously established valuation area in the **Assign Accounting Codes and Valuation Areas** by following this menu path: **Financial Supply Chain Management • Treasury and Risk Management • Transaction Manager • General Settings • Accounting • Organization • Assign Accounting Codes and Valuation Areas**.

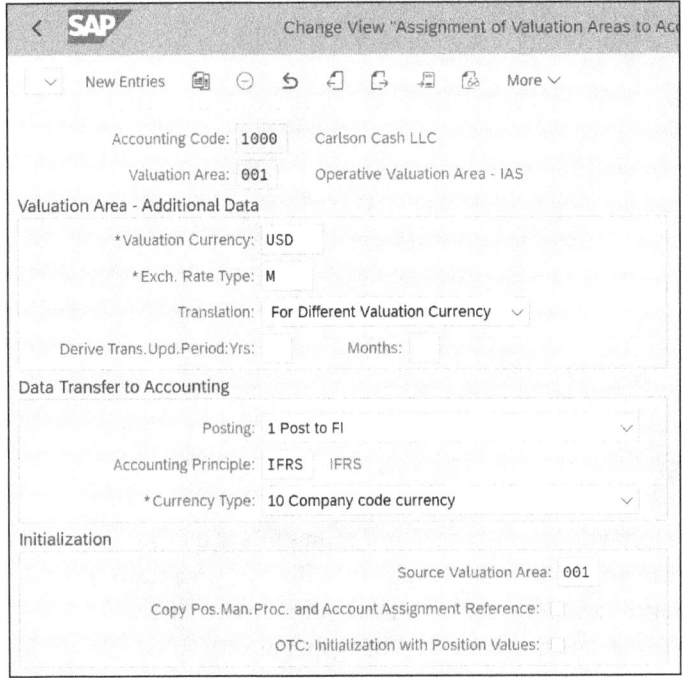

Figure 3.6 Assigning Valuation Area to Accounting Code

The key attributes we configure to complete this assignment are as follows:

- The **Valuation Currency** field should reflect the currency in which the financial statements are reported in this valuation area. This is generally the local currency of the company.
- **Exch. Rate Type** is the default rate that translates postings in the position currency into the valuation currency. We only reference this setting if a posting doesn't already have a local currency assigned to it. If the local currency is available, SAP S/4HANA will simply use that amount and won't be required to translate the currency based on the defined exchange rate type.

3.4   Valuation Class

We'll now move on to the configuration of the *valuation class*, which is a required configuration activity that we use to map the valuation area we created in the previous section. The valuation class setup categorizes the various valuation areas by function and maps them to a respective valuation class. For example, it may make sense to break out the valuation classes based on the type of instrument because the valuation method will be different based on the type of transaction, such as FX, debt/investment, or derivatives.

To configure this node, follow this menu path: **Financial Supply Chain Management • Treasury and Risk Management • Transaction Manager • General Settings • Accounting • Settings for Position Management • Define and Assign Valuation Classes**. The first step in creating a valuation class is to specify the valuation area (in the **VA** column). Then, we map the valuation area to a valuation class, as shown in Figure 3.7. We specify the valuation class with a numeric value in the **ValCl** column and then define a valuation class descriptor in the **Name of Valuation Class** field.

Valuation Class			
VA	Name of Valuation Area	ValCl	Name of Valuation Class
001	Operative Valuation Area - IAS	1	At Fair Value Through Profit or Loss
001	Operative Valuation Area - IAS	2	Available for Sale
001	Operative Valuation Area - IAS	3	Held to Maturity / Loans and Receivables
001	Operative Valuation Area - IAS	4	Liabilities
001	Operative Valuation Area - IAS	5	Foreign Exchange
001	Operative Valuation Area - IAS	11	Debt - Short Term
001	Operative Valuation Area - IAS	12	Debt - Medium Term
001	Operative Valuation Area - IAS	13	Debt - Long Term
002	Parallel Valuation Area - US Gaap	1	Trading
002	Parallel Valuation Area - US Gaap	2	Available for Sale
002	Parallel Valuation Area - US Gaap	3	Held to Maturity
002	Parallel Valuation Area - US Gaap	4	Liabilities
003	Parallel Valuation Area - Local Gaap	1	Current Assets

Figure 3.7 Defining Valuation Class

3.5 General Valuation Class

The *general valuation class* is a required configuration component that's independent of the valuation class, and it's a reference to a specific valuation class within each valuation area. In the following sections, we'll walk through how to both define and assign a general valuation class.

3.5.1 Define General Valuation Class

To define a general valuation class, we navigate to the following menu path: **Financial Supply Chain Management • Treasury and Risk Management • Transaction Manager • General Settings • Accounting • Settings for Position Management • Define and Assign Valuation Classes**. The first step is to define the descriptor that we will use to identify the general valuation class in the **GVC** column. We can then define the name of the general valuation class in the **Name of gen valn cl.** column; the name should add further detail on the type of contract that will roll up into the general valuation class. The last step is to assign the general valuation class to the valuation class that we defined in Section 3.4, and we do this this by populating the valuation class in the **ValCl** column as shown in Figure 3.8.

VA	Name of Valuation Area	GVC	Name of gen valn cl.	ValCl	Name of Valuation Class
001	Operative Valuation Area - IAS	1	Short-term investments	1	At Fair Value Through Profit or Loss
002	Parallel Valuation Area - US Gaap	1	Short-term investments	1	Trading
003	Parallel Valuation Area - Local Gaap	1	Short-term investments	1	Current Assets
001	Operative Valuation Area - IAS	2	Mid-term investments	2	Available for Sale
002	Parallel Valuation Area - US Gaap	2	Mid-term investments	2	Available for Sale
003	Parallel Valuation Area - Local Gaap	2	Mid-term investments	1	Current Assets
001	Operative Valuation Area - IAS	3	Long-term investments	3	Held to Maturity / Loans and Receivables
002	Parallel Valuation Area - US Gaap	3	Long-term investments	3	Held to Maturity
003	Parallel Valuation Area - Local Gaap	3	Long-term investments	2	Fixed Assets
001	Operative Valuation Area - IAS	4	Liabilities	4	Liabilities
002	Parallel Valuation Area - US Gaap	4	Liabilities	4	Liabilities
003	Parallel Valuation Area - Local Gaap	4	Liabilities	4	Liabilities
001	Operative Valuation Area - IAS	5	Foreign Exchange	5	Foreign Exchange

Figure 3.8 Assigning General Valuation Class to Valuation Class

The general valuation class also defines whether a transaction can be designated as a current portion of long-term debt. The settings for this are shown in Figure 3.9 and are referenced in the following list. There must be at least two general valuation classes for the transfer. The following settings are required for us to run a valuation class transfer for the current portion of long-term debt:

❶ In the figure, general valuation class 13 is set for **Long-term debt**. When we create any debt contracts, they are assigned this general valuation class by default.

❷ General valuation class 11 has been defined as **Short-term debt**. When we execute the valuation class transfer, this will be the destination general valuation class.

❸ In the **Current Portion Handling** column, we need to select one of the **Current Portion** options for the **Short-term debt** general valuation class. We have two options, as follows:

- Current Portion: Nominal-Based Interest Assignment
  We determine the nominal amount of all current and long-term positions, and the interest payments that are transferred are distributed based on the assignments of the nominal amounts. In summary, the interest generated based on the current portion is transferred to the *short-term* general valuation class, and the interest generated based on the long-term position stays in the source general valuation class.

– Current Portion: Date-Based Interest Assignment
  The valuation class transfer looks at all interest payments between the key date and the horizon date and transfers them all to the short-term general valuation class.

Functionally, this will change the entry screen in Transaction TPM15M (Valuation Class Transfer). When we select source general valuation class 13 to transfer to destination general valuation class 11, an additional horizon date field will appear where we can determine which transactions to change to short-term accounts.

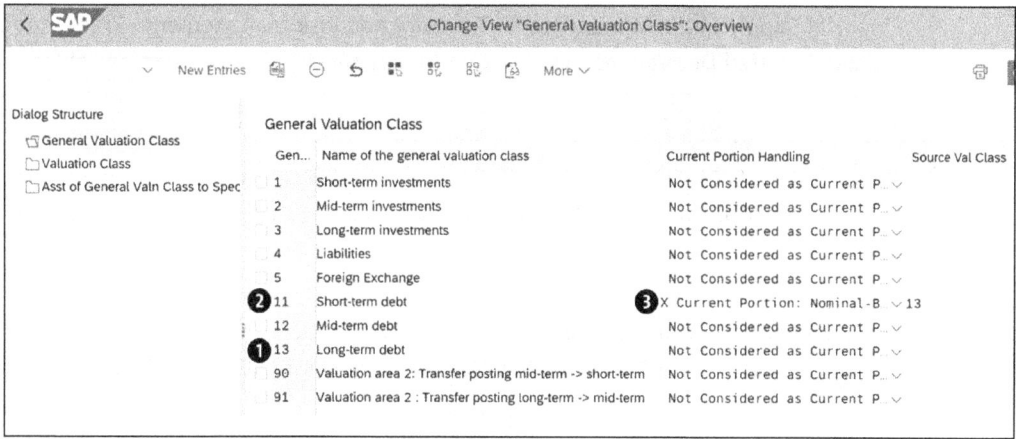

Figure 3.9 General Valuation Class Configuration

3.5.2 Assign General Valuation Class

For every company code, product type, and transaction type combination, we want to designate a universal valuation class to serve as the default value when we're initiating a financial transaction. This general valuation class serves as the foundation, allowing for the derivation of specialized valuation classes across various valuation areas. This is not a required configuration element, as in each contract, we have the ability to manually choose the general valuation class. The default value within the contract will default into the contract during creation, and we can adjust it within the contract if required during contract creation. By defining a general valuation class tailored to the specific blend of company code, product type, and transaction type, we ensure that there will be an efficient default value when we're inputting a transaction type. This approach facilitates streamlined financial transaction processing and ensures consistency in valuation across all of the product and transaction types.

We can find the configuration to assign a default general valuation class in the customizing, and the menu path to it varies based on the type of contract we are defining, as follows:

- **Money market**
  Financial Supply Chain Management • Treasury and Risk Management • Transaction Manager • Money Market • Transaction Management • Assign General Valuation Class

- **FX**
  Financial Supply Chain Management • Treasury and Risk Management • Transaction Manager • Foreign Exchange • Transaction Management • Assign General Valuation Class

- **Securities**
  Financial Supply Chain Management • Treasury and Risk Management • Transaction Manager • Securities • Transaction Management • Assign General Valuation Class

- **Listed derivatives**
  Financial Supply Chain Management • Treasury and Risk Management • Transaction Manager • Listed Derivatives • Transaction Management • Assign General Valuation Class

- **OTC derivatives**
  Financial Supply Chain Management • Treasury and Risk Management • Transaction Manager • OTC Derivatives • Transaction Management • Assign General Valuation Class

- **Trade finance**
  Financial Supply Chain Management • Treasury and Risk Management • Transaction Manager • Trade Finance • Transaction Management • Assign General Valuation Class

Once we follow any of these menu paths, we can click on **New Entries,** enter the **Company Code**, product type (**PTyp**), and transaction type (**TTyp**) combination, and assign it to a default **General Valuation Class**. We must maintain every combination of company code, product type, and transaction type in this table because all of the fields are required and don't allow for masking. In Figure 3.10, we can see the various entries that have been mapped within the configuration.

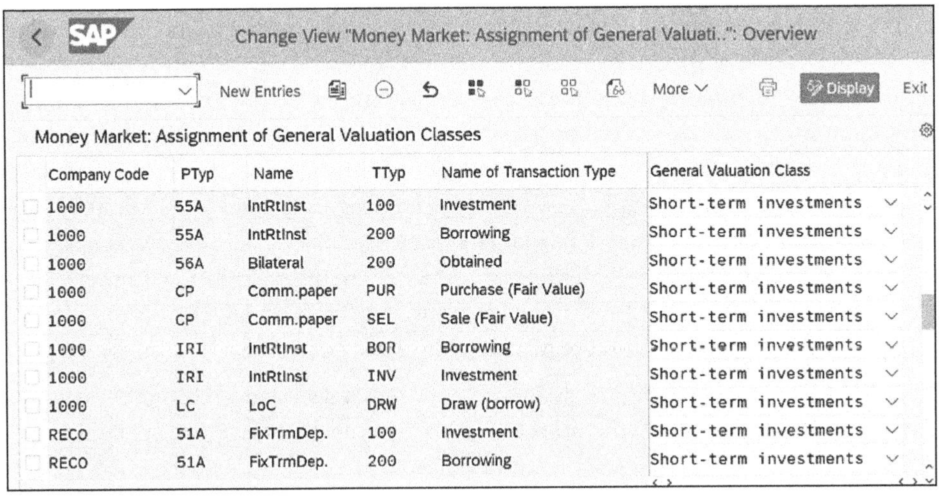

Company Code	PTyp	Name	TTyp	Name of Transaction Type	General Valuation Class
1000	55A	IntRtInst	100	Investment	Short-term investments
1000	55A	IntRtInst	200	Borrowing	Short-term investments
1000	56A	Bilateral	200	Obtained	Short-term investments
1000	CP	Comm.paper	PUR	Purchase (Fair Value)	Short-term investments
1000	CP	Comm.paper	SEL	Sale (Fair Value)	Short-term investments
1000	IRI	IntRtInst	BOR	Borrowing	Short-term investments
1000	IRI	IntRtInst	INV	Investment	Short-term investments
1000	LC	LoC	DRW	Draw (borrow)	Short-term investments
RECO	51A	FixTrmDep.	100	Investment	Short-term investments
RECO	51A	FixTrmDep.	200	Borrowing	Short-term investments

Figure 3.10 Assignment of General Valuation Class

3.6 Position Management Procedure

Next, we move on to creating the position management procedure, which serves as a pivotal attribute. It governs the intricacies of internal position management, dictating the methods used for position valuation. Additionally, this attribute plays an important role in computing price fluctuations and delineating gains or losses incurred during outflows. Moreover, it outlines the specific components managed by the system for each position, ensuring comprehensive oversight and management of position-related activities.

There are two main areas we'll focus on for the position management procedure, and these configuration activities work together to manage positions. In Section 3.6.1, we'll cover the creation of the position management procedure, which determines how we treat positions for all financial transactions managed in treasury and risk management. In Section 3.6.2, we'll cover the different types of valuations. Since the management of the position and any valuations are closely related, we assign the different types of valuations to the position management procedure. In this section, we'll describe the reasons for different types of valuations and how to assign them.

3.6.1 Define and Assign Position Management Procedure

To create a position management procedure, we navigate to the following menu path: **Financial Supply Chain Management • Treasury and Risk Management • Transaction Manager • General Settings • Accounting • Settings for Position Management • Define Position Management Procedure**.

We're free to customize the position management procedure according to our preferences. Along with providing a description, it's common to assign such values to a category—known as **the Position Management Category** field—that governs crucial properties. This category predominantly influences the key figures accessible for position inquiries. Since many components listed in this section are relevant only to specific financial instruments, the position management categories are structured based on these instruments. These categories drive many aspects of the position management of a transaction, but most importantly, they drive the types of flows that we can calculate and the types of key date valuations that we can perform. The available categories include the following:

- Securities/Loans/Money Market (without index linked bonds)
  This position management category is appropriate for the financial instruments we delineated earlier, as well as for routinely managed listed options. We advise refraining from utilizing it in cases where the financial instrument incorporates installment repayments.

- Securities/Loans with Installment Payments
  This position management category primarily mirrors its predecessor, offering the repayments component tailored for financial instruments featuring installment

repayments. It adequately fulfills the criteria for amortization as per both US generally accepted accounting principles and International Financial Reporting Standards, exclusively through this position management category. Additionally, this category is applicable to financial instruments with a single final repayment, and we should use it for bonds falling under the installment with repayment product category.

- **Index Linked Bonds**
  This position management category is specifically designed for index-linked bonds. As a result, it includes the index valuation key figure in addition to the key figures associated with the primary position management category. Additionally, users have the option to select whether they prefer to use the book value or nominal values as the basis for valuation.

- **Futures**
  This position management category is appropriate for futures and listed options handled similarly to futures.

- **Foreign Exchange Transactions**
  This position management category is only appropriate for FX transactions, and it includes the key date valuations to appropriately value the FX between the traded currencies.

- **Forwards/Repos**
  This position management category is appropriate for use with forwards and repo contracts.

- **OTC Derivatives (profit/loss posting)**
  This position management category is applicable to all OTC derivatives. The inclusion of profit and loss posting signifies that the book value of the option is recorded in the profit and loss statement upon physical exercise of the option.

- **OTC Options (transfer posting to underlying)**
  This position management category aligns with the previous one. Upon physical exercise of an OTC option, the book value is transferred to the underlying asset and doesn't get cleared into the profit and loss statement.

Within the position management procedure, we also establish the method by which an assigned ledger position undergoes valuation. This valuation process involves up to five distinct valuation steps. Once we've designated the **Transfer Category** of the step (as depicted in Figure 3.11), based on the position management procedure we've selected, the screen will exclusively offer the valuation steps that are applicable, and the relevant valuation procedures that are available will align with the valuation step that we select. It's important to note that these steps not only impact the valuation process but also contribute to determining the price gains and losses incurred during outflows.

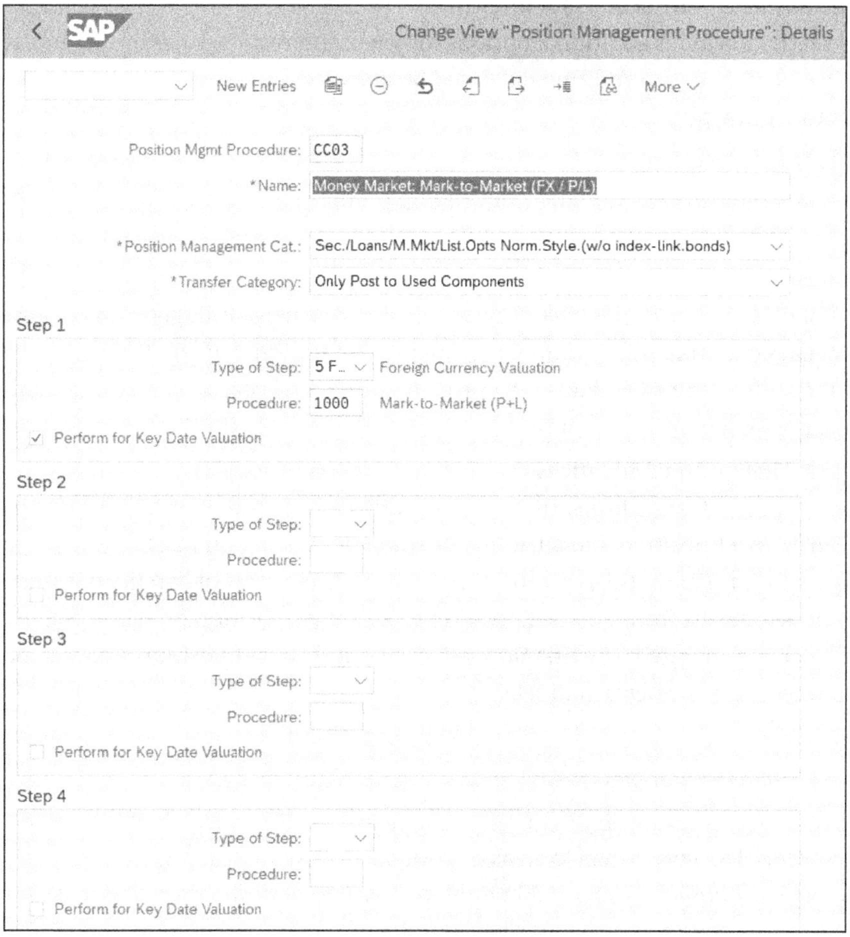

3.6 Position Management Procedure

Figure 3.11 Position Management Configuration

Additional settings are available in the position management procedure in the bottom section that follows the valuation steps (see Figure 3.12). These settings are as follows:

- **Valuation for Position Outflow**
  We only use this setting for FX transactions. By default, we don't execute a valuation at the settlement of the transaction, but selecting this option allows for the valuation to be run. This valuation would follow the settings that we have previously set up in the valuation steps of the position management procedure. If an alternative rate type is required for the valuation, we can determine it in this section.

- **Hedge Accounting for Positions**
  The settings in this area for **Evaluation Type** and **CVA/DVA Type** are defaults that we can set in the position management procedure. If we do not define these settings, we'll need to select them when running the NPV calculation with Transaction TPM60CVA.

- **Derived Business Transactions for Interest**
  This selection includes additional options for when interest postings are generated and if we need to create additional derived business transactions to match specific accounting principles.

- **Liability/Asset Position**
  Since this option works differently depending on the type of financial instrument, we need to separate the definition into two sections.

- **Securities/Loans/Money Markets**
  If the financial transaction is a liability, we must define it accordingly with the **Position Managed as Liability** option. We can leave the indicator blank for asset positions. If we don't define this indicator correctly, an error will occur when we generate the first position flow for a contract.

- **Foreign Exchange and Derivatives**
  In these types of transactions, there are scenarios where the book value can vary between an asset position and a liability position depending on market conditions for the valuation. Due to this, we can't simply define these asset or liability positions. When there's a shift from an asset position to a liability position or vice versa, two flows must be generated. The first flow will clear out the asset position, and the second flow will post the new liability book value.

Foreign Exchange Example

Let's go through an example of what FX valuations look like when the book value changes from a liability position to an asset position. In this example, let's say that the book value is determined to be –500 on the first valuation of the transaction. We'll write down the book value with the following entries:

- DR: Unrealized Loss
- CR: Position Liability

Now, let's say that in the second month, the book value is 200. This would require a writeup of 700. Since these balances might need to go to separate general ledger accounts, this change in book value would require two flows. The first would create a transaction to clear the unrealized loss:

- DR: Position Liability
- CR: Unrealized Loss

Then, we would generate a second flow to post the gain. An example of this posting would be as follows:

- DR: Position Asset
- CR: Unrealized Gain

- **Transfer to Underlying (OTC Derivatives)**
  When we physically execute an OTC derivative, there are a few options for the value that is transferred to the underlying. There are options for when the entire book value is transferred or only the premium is transferred. Additionally, if we need to transfer the premium and valuations but they need to post to separate accounts, then we should select the **Transfer Premium and Valuations on Separate Components** option.

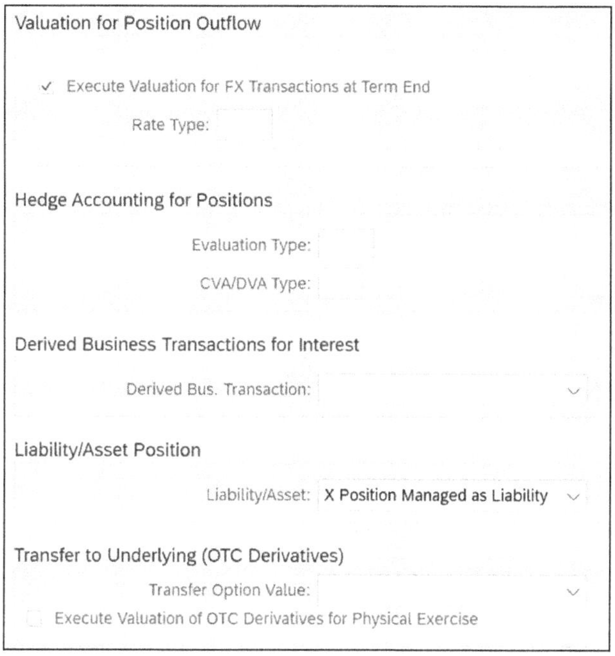

```
Valuation for Position Outflow

    ✓  Execute Valuation for FX Transactions at Term End
              Rate Type:


Hedge Accounting for Positions

              Evaluation Type:
              CVA/DVA Type:


Derived Business Transactions for Interest

              Derived Bus. Transaction:                          ∨


Liability/Asset Position

              Liability/Asset:   X Position Managed as Liability  ∨


Transfer to Underlying (OTC Derivatives)
              Transfer Option Value:                             ∨
    ☐  Execute Valuation of OTC Derivatives for Physical Exercise
```

Figure 3.12 Valuation Procedure Configuration

Once we've fully configured the position management procedure, we must assign it to the relevant transactions. We do this in the customizing in the **Financial Supply Chain Management** • **Treasury and Risk Management** • **Transaction Manager** • **General Settings** • **Accounting** • **Settings for Position Management** • **Assign Position Management Procedure** menu path.

In this customizing, we can select the position management procedure by **Accounting Code, Valuation Area, Valuation Class, Product Category Product Type, Transaction Type, Portfolio,** and securities account group (**Sec. Acct Grp**). In the example in Figure 3.13, we are assigning the **Position Management Procedure** directly to each **Product Type**. We should use the additional fields in this configuration if more specific assignments are required for the position management procedures.

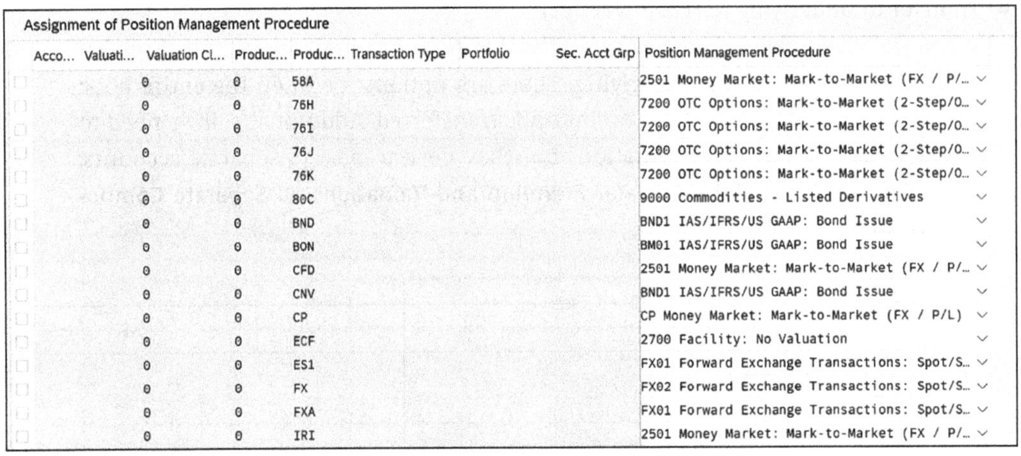

Assignment of Position Management Procedure								
Acco...	Valuati...	Valuation Cl...	Produc...	Produc...	Transaction Type	Portfolio	Sec. Acct Grp	Position Management Procedure
☐		0	0	58A				2501 Money Market: Mark-to-Market (FX / P/... ⌄
☐		0	0	76H				7200 OTC Options: Mark-to-Market (2-Step/O... ⌄
☐		0	0	76I				7200 OTC Options: Mark-to-Market (2-Step/O... ⌄
☐		0	0	76J				7200 OTC Options: Mark-to-Market (2-Step/O... ⌄
☐		0	0	76K				7200 OTC Options: Mark-to-Market (2-Step/O... ⌄
☐		0	0	80C				9000 Commodities - Listed Derivatives ⌄
☐		0	0	BND				BND1 IAS/IFRS/US GAAP: Bond Issue ⌄
☐		0	0	BON				BM01 IAS/IFRS/US GAAP: Bond Issue ⌄
☐		0	0	CFD				2501 Money Market: Mark-to-Market (FX / P/... ⌄
☐		0	0	CNV				BND1 IAS/IFRS/US GAAP: Bond Issue ⌄
☐		0	0	CP				CP Money Market: Mark-to-Market (FX / P/L) ⌄
☐		0	0	ECF				2700 Facility: No Valuation ⌄
☐		0	0	ES1				FX01 Forward Exchange Transactions: Spot/S... ⌄
☐		0	0	FX				FX02 Forward Exchange Transactions: Spot/S... ⌄
☐		0	0	FXA				FX01 Forward Exchange Transactions: Spot/S... ⌄
☐		0	0	IRI				2501 Money Market: Mark-to-Market (FX / P/... ⌄

Figure 3.13 Assignment of Position Management Procedure

3.6.2 Customizing Valuations

To further detail the settings for the position management procedure, we must go back to the settings in the valuation steps. We assign the valuation customizing to each of the position management procedures. The definition of the valuation steps is a key configuration for the position management procedure, and if we need to execute any amortizations, valuations, or impairments, then we need to define each transaction with relevant valuations to be executed. One key detail to note is that the position management category will drive the types of valuation that are available within a position management procedure. Now that we know this, the full list of available valuation steps for transactions are as follows:

- Foreign currency valuation
- Security valuation
- One-step price valuation
- Index valuation
- Rate valuation for forward exchange transactions
- Amortization
- Impairment procedure
- Current portion transfer procedure

To assign the valuation steps to the position management procedure, we must first define them. We can find the series of customizing nodes for these valuation procedures in the **Financial Supply Chain Management • Treasury and Risk Management • Transaction Manager • General Settings • Accounting • Settings for Position Management • Key Date Valuation** menu path. One thing to keep in mind is that there are additional settings behind these valuations in SAP S/4HANA. Once we run the valuations with the applicable transaction, additional flows are generated, and we must assign the associated update types and accounting. We can see in Figure 3.14 that there's a specific tab available for each of the types of valuation.

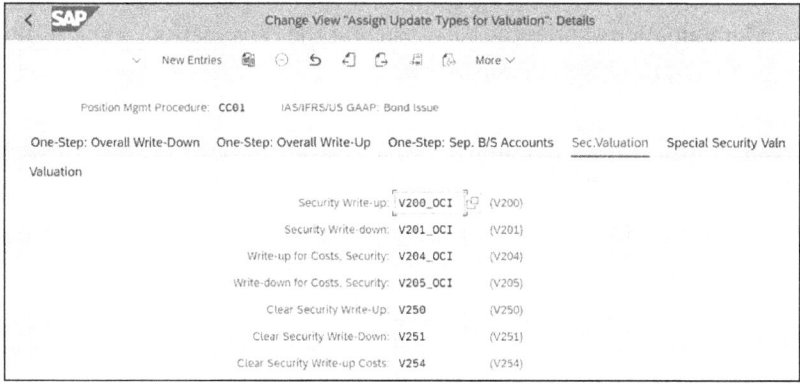

Figure 3.14 Assigning Update Types for Valuation

Foreign Currency Valuation

The foreign currency valuation is relevant when the position currency of a transaction is different from the valuation currency. In this scenario, we run a valuation to reflect the difference between these two currencies. We'll look at how the position value changes according to the valuation currency and the defined exchange rate. The settings for the foreign currency valuation can be seen in Figure 3.15.

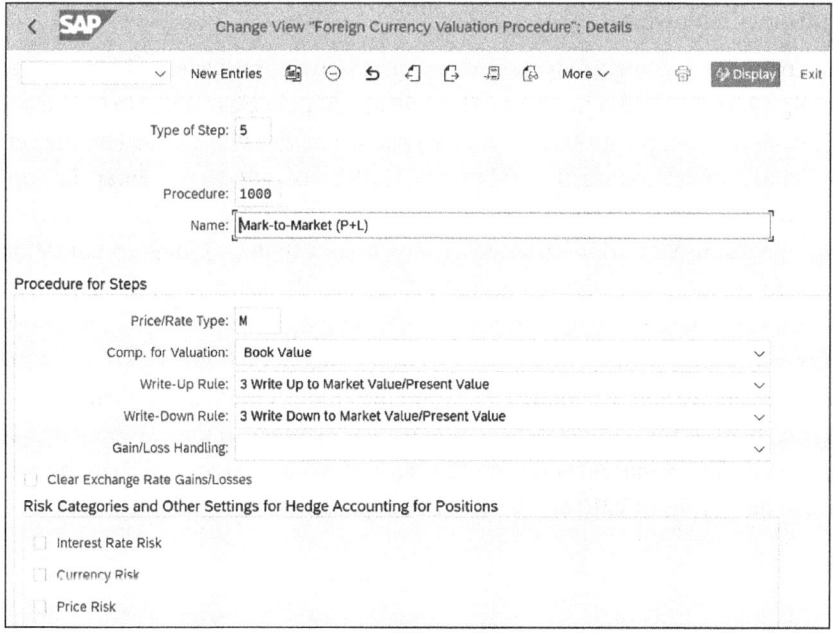

Figure 3.15 Foreign Currency Valuation Procedure

In the **Price/Rate Type** field, we can define the exchange rate type that we'll use for the valuation. We can define different rate types with a daily or monthly rate, and we can also have bid, ask, middle, or other types of rates. Business requirements will define which rate type we should use in this setting.

The **Component for Valuation** defines the base value for the FX valuation; we generally use **Book Value**. Other options include the following:

- **Variation Margin**
  We only use this setting for futures and listed options.
- **Amortized Acquisition Value**
  If we define an amortization for the position, the base value will change throughout the transaction, so we need to factor the amortized value into the calculation.
- **Amortized Acquisition Value, only at Valuation**
  We can use this amortized value for the calculation of the derived business transaction or only for a key date valuation. If the derived business transactions should look at the book value and not at the amortized value, we should select this option.
- **Hedge Adjustment**
  We use this component when the transaction is related to hedge management.

We set up write-ups and write-downs individually, and we also determine the amount used for the related flows. The write-up and write-down rules are generally set up to look at the valuations and will use the market value/present value. This will compare the position currency to the valuation currency at the designated key date and then determine the present value. Based on this calculation, a flow will be generated to post the write-up or write-down.

If there's a chance that the gain or loss could switch positions, we should select the **Clear Exchange Rate Gain/Loss** indicator. This will adequately clear the gain or loss for the change from an asset position to a liability position or vice versa. Based on the settings in this valuation procedure, the update types will be assigned in the **Financial Supply Chain Management • Treasury and Risk Management • Transaction Manager • Accounting • Key Date Valuation • Update Types • Assign Update Types for Valuation** menu path. The update types that are called depend on the settings in the valuation procedure. For a standard write-up or write-down in a foreign currency valuation, we'll assign **Forex Write-Up** "V202" to the write-up and **Forex Write-Down** "V203" to the write-down as in Figure 3.16. The update types we use in this configuration will change depending on whether the component for valuation is changed. If the component was assigned to the **Variation Margin Forex Write-Up** (or **Write-Down**) option, then the corresponding update types of V210 and V211 will be applicable.

3.6 Position Management Procedure

Figure 3.16 Assigning Update Types for Valuation

Security Valuation

The security valuation takes into account the fluctuation of market prices. Using this information, treasury and risk management can determine the fair value of a contract, and subsequently, we must record the change in fair value. Many financial instruments need to be valued with the security valuation. Figure 3.17 shows the configuration screen where we set up the security valuation.

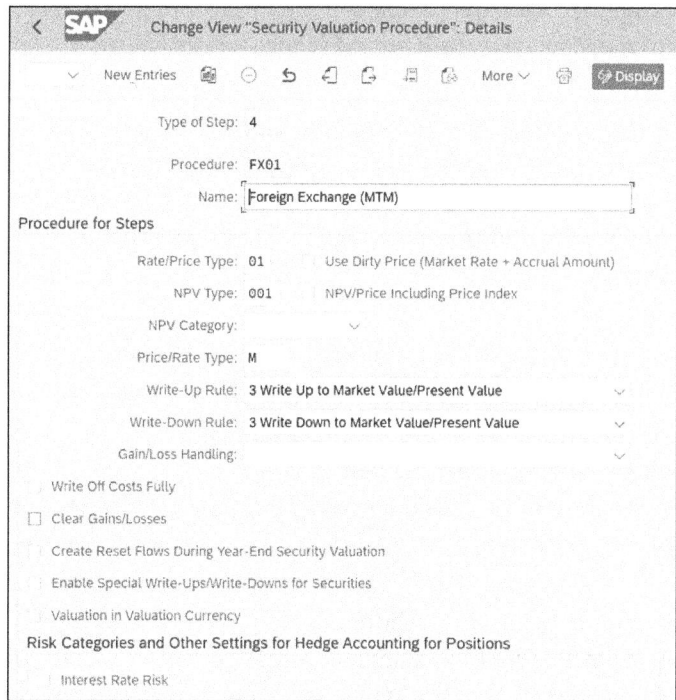

Figure 3.17 Security Valuation Procedure

3 Core Configuration Elements

The following are the options in the security valuation:

- **Rate/Price Type**
  As with the exchange rate type for foreign currency valuations, we determine the type of market price to use for the security valuation. Different market price types can be the opening price, the closing price, and other variations on the market price.

- **Use Dirty Price**
  If we check this box, the accrual amount is added to the calculation of the market rate to determine the *dirty price* as the basis of the valuation.

- **NPV type**
  We use this to categorize the price type of the valuation.

- **NPV Category**
  We can determine whether to use the dirty or clean price for the net present value in OTC transactions. The price/rate type in this section is for an FX rate type. Ideally, the NPV determined for a contract is in the position currency and we can post it in that currency. If the contract is in a different currency, then the amount is translated into the position currency based on this exchange rate type. If this rate is not populated, then the exchange rate determined in the accounting code is used to determine the position currency amount.

- **Write-Up Rule/Write-Down Rule**
  The write-up and write-down rules work like the FX valuations that we covered in the previous section, except that a different section of the configuration assigns the update types and different update types are assigned to the security valuation.

- **Write Off Costs Fully**
  This indicator is relevant to capitalized costs on a contract. If the valuation results in a write-down, checking this box will ensure that the costs are fully written off. If we leave this box unchecked, then the write-down for the costs will only partially apply to the cost amount. The distribution of the cost write-off will be proportionally applied to the cost and the book rate.

- **Clear Gains/Losses**
  This indicator works the same as the FX valuation setting with the same name. We should check this box to properly reset the gains or losses if the valuation could change from positive to negative and vice versa.

- **Create Reset Flows During Year-End Security Valuation**
  Some accounting principles require us to generate reset flows yearly for the security valuation. Checking this indicator allows the yearly valuation to generate the reset flow.

- **Enable Special Write Ups/Write Downs for Securities**
  If we check this box, we can manually enter and post valuations outside of SAP. We enter manual valuations in the Enter Book Values for Manual Valuation app or Transaction TPM74.

3.6 Position Management Procedure

- **Valuation in Valuation Currency**
  If we need to run the valuation in a different currency than the position currency, then we can check this box to make the alternate currency valuation occur.

From a posting perspective, the security valuation uses different update types to facilitate the correct postings. Generally, the security write-up and write-down will use update types V200 and V201. We can assign additional update types depending on the type of security valuation and whether a reset is required for the valuation. The available settings for the update type assignment for security valuations can be seen in Figure 3.18. For more information about this topic, see Chapter 4, Section 4.2.9.

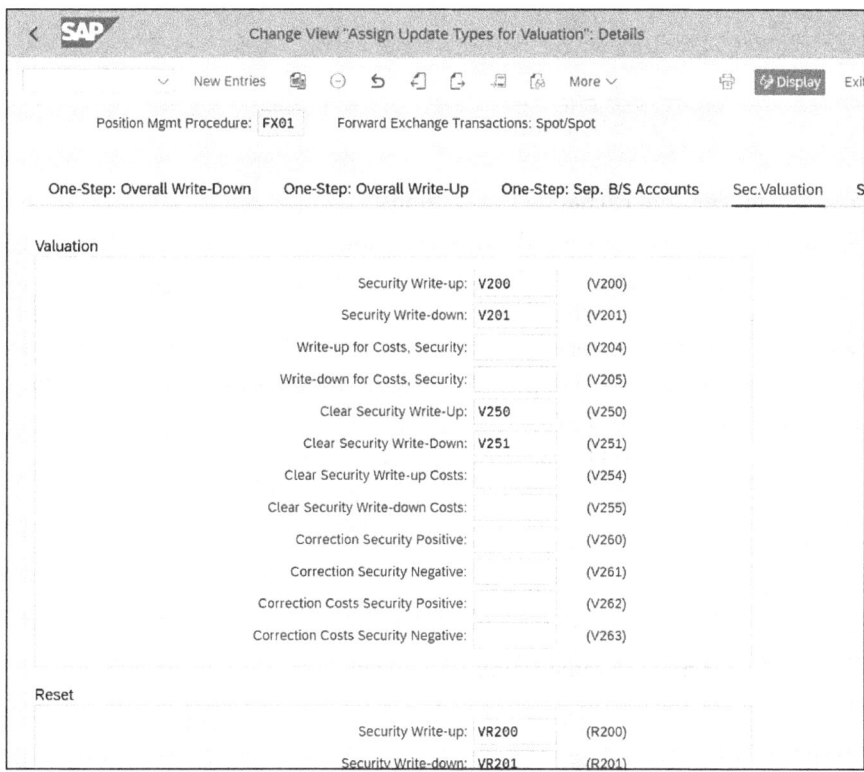

Figure 3.18 Assigning Update Types for Valuation

One-Step Price Valuation

The *one-step price valuation* is a single valuation that we assign to the position management procedure, but it is actually a combination of the security valuation and the FX valuation. The valuation includes two steps to determine the full write-up or write-down amount. First, we complete the valuation in the position currency and translate it into the valuation currency using the defined exchange rate. Then, the valuation compares that value to the new book value in the valuation currency to determine the FX value. Each of the relevant fields shown in Figure 3.19 was covered in the previous two sections.

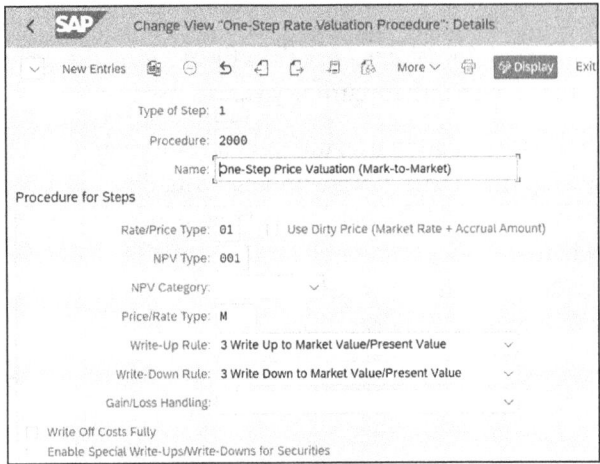

Figure 3.19 One-Step Rate Valuation Procedure

Rate Valuation for Forward Exchange Transactions

The next valuation that we'll cover is the rate valuation for FX transactions. This valuation is specific to FX forwards, and we calculate it differently than the FX valuation. We use this valuation procedure to create a valuation by comparing the FX market rates to the FX rates in the financial transaction. The market rates we use are the current spot rates and the swap rates/forward points. This valuation procedure will only use those data points and doesn't factor in the interest rate yield curves. Let's now look at the unique fields for this valuation procedure, which can be seen in Figure 3.20.

The **Forward Category** has four options that we can select, as follows:

- Market spot rate for forward rate in transaction
- Market forward rate for forward rate in transaction
- Market spot rate for spot rate in transaction
- Valuation with net present value from risk module

When we run a valuation, the system displays the amount in the valuation currency. If we check the **Cross-Valuation** box, each side of the valuation will be shown separately for the valuation result. The separate flows will also post to two different update types, so the accounting will post separately for each leg of the valuation result.

The **Netting** indicator is relevant for rollovers and premature settlements when we create a second transaction to net the flows of the original FX forwards. The netting transaction in this scenario creates an offsetting financial transaction with the same due date as the original trade. The point of this transaction is to clear the original payment flows, and the system also creates a new financial transaction with the updated due date. From a valuation perspective, SAP S/4HANA will look at these three trades as completely separate transactions that need a valuation. By default, the valuation uses the spot and forward rates on the day the transaction concludes, and the valuations from the original transaction and the netting transaction won't offset each other. If we check the **Netting** box, the system applies the same rates from the original transaction to the netting transaction.

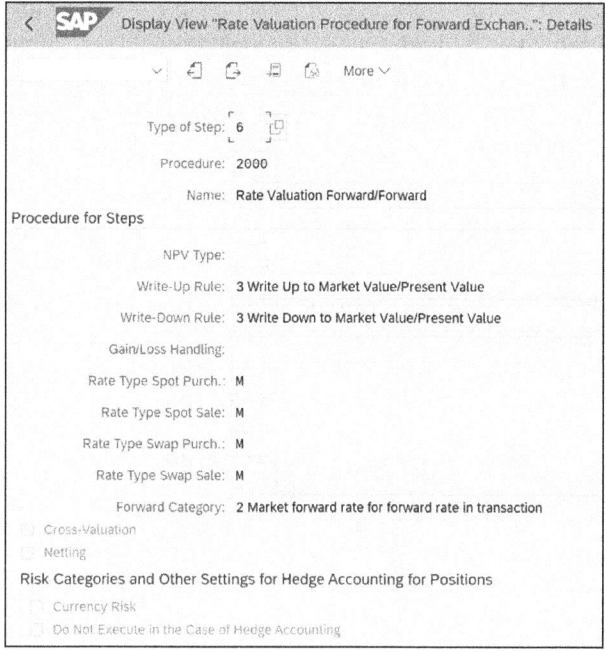

Figure 3.20 Rate Valuation Procedure for FX

From an accounting perspective, we can determine accounting entries for both the purchase side and the sell side of the FX forward. As we can see in Figure 3.21, there are separate update types for the buy and sell sides of the transaction to facilitate this separation when we check the **Cross-Valuation** box. We'll cover this topic in more detail in Chapter 4, Section 4.2.9.

Valuation		
	Write-up purch. currency:	(V500)
	W-down purch. currency:	(V501)
	Clear write-up purchase currency:	(V502)
	Clear write-down purchase currency:	(V503)
	Write-up sales currency:	(V504)
	Write-down sales currency:	(V505)
	Clear write-up sales currency:	(V506)
	Clear write-down sales currency:	(V507)
Reset		
	Write-up purch. currency:	(R500)
	W-down purch. currency:	(R501)
	Clear write-up purchase currency:	(R502)
	Clear write-down purchase currency:	(R503)
	Write-up sales currency:	(R504)

Figure 3.21 Rate Valuation for Forward Exchange Transactions

Amortization

The amortization step differs from the valuation steps since instead of making a valuation, we're distributing the premium or discount of a financial transaction. The premium or discount on a transaction is the difference between the nominal value of the contract and the amount actually paid. These differences are due to market conditions that can make the transaction more or less valuable than the nominal value. The amortization of this premium or discount is designed to allocate an amount of this value to the income statement. Since the premium and discount are included in the cost of the financial instrument, the costs are generally amortized throughout the life of the contract. There are many different ways to calculate and recognize the amortization, so we'll cover the customizing of these settings in detail to highlight the key setting that we can select to impact both the calculation and the accounting entries for the amortization key date valuation. In Figure 3.22, we can see the customizing of the amortization procedure.

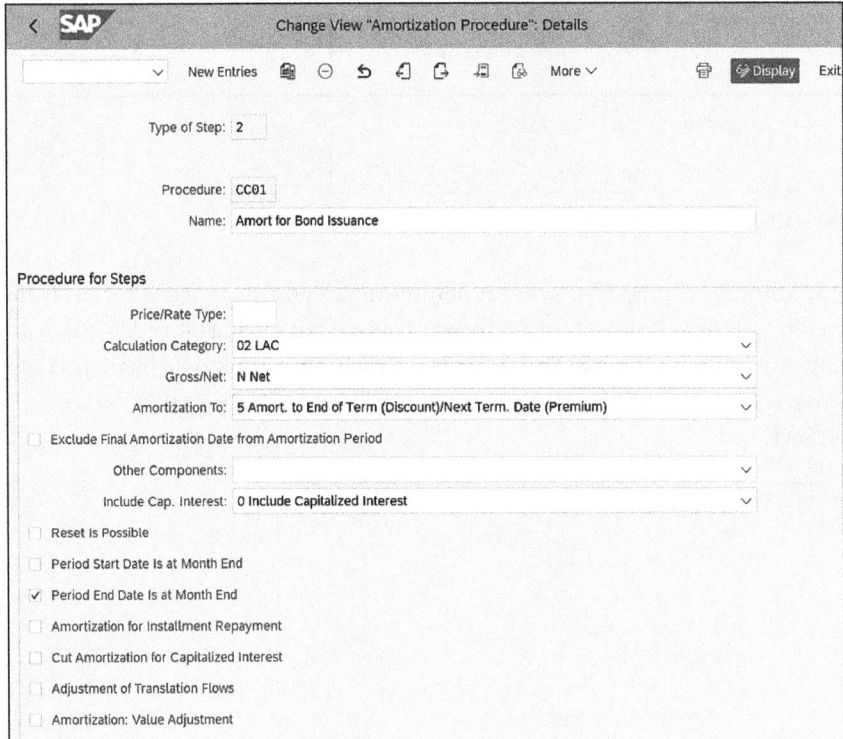

Figure 3.22 Amortization Procedure Configuration

Let's now discuss the main settings in detail:

- **Price/Rate Type**
  We can assign an exchange rate type to determine the local currency amount of the amortization if it's in a foreign currency.

- **Calculation Category**
  The calculation category drives how the amortization is calculated. The following options are available:
  - **Linear amortized cost**
    We use this to evenly distribute the amortization over the periods of the transaction.
  - **Scientific amortized cost**
    This is a calculation procedure that looks at the effective interest rate of the transaction and then discounts the future cash flows to determine the amortization amount. The amount varies in each calculation since the amortized acquisition value is the basis of each period's calculation.
  - **Effective interest method**
    In the United States, this allows for the amortization to use the effective interest method. Additional details for the calculation are defined in the **Effective Interest Rate Treatment** field.
- **Gross/Net**
  There are a few options in this area for posting the transactions either gross or net. The purpose of this setting is to drive the accounting entries for both the creation of the transaction and the subsequent amortizations. In the net procedure, the acquisition value of the transaction is posted. As the premium or discount is amortized, the value reflected in accounting will reflect the updated amortized book value.

  Using the gross procedure, the nominal value and the premium or discount are posted separately. This is useful if we want to post the nominal value and the premium or discount to different general ledger accounts. Then the amortizations will post accordingly in their respective posting periods.
- **Amortization To**
  This option drives how the final amortization date is determined. There are various options in this area for ensuring that the correct calculation occurs for the final amortization.
- **Effective Interest Rate Treatment**
  This selection appears if we set the **Calculation Category** to **Effective Interest Method (U.S.)**. This drives how the effective interest rate is applied. The standard calculation is to adjust the effective interest rate for each period, but alternative selections are available. For example, we can determine a constant effective interest rate for the amortization instead of applying a new effective interest rate each period.
- **Exclude Final Amortization Date from Amortization Period**
  If there should be no amortization on the final period of the amortization, we should check this box.
- **Other Components**
  This selection determines any additional flows or calculations that need to be considered for the amortization.

- **Deferral Item, Purchase Value**
  This option creates an additional position component for the deferral of the purchase value. The *position component* is the difference between the purchase value and the intrinsic value of the bond. The intrinsic value reflects the issuance yield curve on the purchase date of the bond, and this flow is also deferred on a linear basis for amortization.

- **Negotiation Spread Linear**
  During the amortization, we split the amount into two amounts: the issue spread and the negotiation spread. Before going into each of these amounts, we need to cover the issue amount for the bond or loan. The *issue amount* is the nominal amount of the transaction multiplied by the issue rate/price. The *issue spread* is the difference between the repayment amount and the compounded issue amount on the purchase date. The *negotiation spread* is the difference between the compounded issue amount on the purchase date and the purchase amount of the bond or loan. These amounts are amortized in different ways: the issue spread amortizes based on the settings in the amortization procedure, and the negotiation spread follows a linear amortization.

- **Negotiation Spread Exponential**
  We use the same definition as detailed in the previous bullet point for the issue spread and the negotiation spread. In this method, the issue spread follows the amortization procedure settings, and the negotiation spread is amortized on an exponential basis.

- **Negotiation Spread without Amortization**
  In this method, the issue spread is amortized based on the amortization procedure, and the negotiation spread clears as rate/price gains.

- **Include Interest**
  This selection appears if we set the calculation category to **Scientific Amortized Cost**. This selection determines how we treat interest for the Scientific Amortized Cost interest calculation category. We can ignore the interest component, include the interest with adjustment for what has been accrued, or use the International Accounting Standard 39 cash flow method to include the interest in the calculation without the interest rate adjustment.

- **Include Cap Interest**
  This indicator defines whether capitalized interest should be considered in the amortization process.

- **Reset is Possible**
  We select this option if it's possible that the amortization will use a book and reset method when creating the accounting entries.

- **Period Start Date is at Month End**
  Given how the calendar days work with months, checking the month-end indicator box can ensure that the correct dates are selected for calculation if the date is a

month end. This is especially an issue when the month end is in February. If the start date of the period falls on a month end day, we check this box.

- **Period End Date is at Month End**
  As with **Period Start Date is at Month End**, we check this box if the end date of the period falls on a month end day.

- **Amortization for Installment Repayment**
  If we check this box, a derived business transaction is generated for each installment repayment. This could result in the amortization taking a very long time to run. If we do not check this box, the derived business transactions won't be generated.

- **Cut Amortization for Capitalized Interest**
  We should check this box if the transaction will have capitalized interest. If we don't check it, a derived business transaction can be generated for every capitalized interest flow, so the amortization calculation can take a very long time. Checking this box allows the system to only create the amortization result when running the key date valuation or for position outflows and inflows.

Impairment Procedure

The impairment procedure allows for the write-off of the value of a contract if the value has decreased to a point where there's no reasonable expectation that we'll recover the asset. In this case, we must write off a value to an expense account. When this occurs, we use the impairment procedure in Figure 3.23. The relevant settings are as follows:

- **Reset Valuation**
  There are four options for the reset valuation. If we need to reset the other key date valuations at the time we're impairing an asset, then we should set the applicable indicators.
  - Reset Unrealized Foreign Currency Valuation
  - Reset Realized Security Valuation
  - Reset Realized Foreign Currency Valuation
  - Reset Index Valuation

- **Amortization After Impairment**
  We check this box if we still want to create amortization flows after the impairment is recorded.

- **Stop Index Valuation After Impairment**
  If we no longer want to run an index valuation after the impairment, we should check this box.

- **Record FX Impairment**
  If the transaction is FX relevant, the system can create an FX impairment. Checking this box will allow us to select the **One-Step Impairment** option. This type of impairment is similar to the one-step valuation that combines a foreign currency and security valuations into one step. Just as with the one-step valuation, there's a calculation

that looks at the fluctuations of price and exchange rate that could offset each other. This could result in either a positive or a negative impairment for securities using this method.

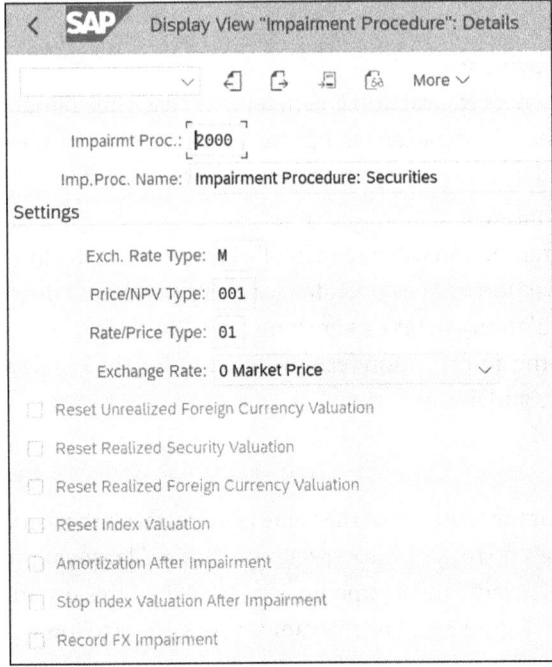

Figure 3.23 Impairment Procedure Configuration

Current Portion Transfer Procedure

The current portion transfer procedure is relevant when there are payments within a short-term period that need to be repaid on long-term assets or liabilities. When this scenario occurs, a current portion step of the valuation can occur to separate the long-term balance from the short-term balance. The settings required on this configuration are shown in Figure 3.24:

- **Price/Rate Type**
  This field is where we determine the rate type used for currency.

- **Period (Months)**
  This field is where we determine how many months should be considered as the current portion.

- **Base Amount**
  Here, we determine whether to transfer the nominal amount or the book value.

- **Conversion Category**
  If we are executing a currency conversion, we use this dropdown list to determine whether it will use the **Book Rate** or the **Market Rate** in the market data tables.

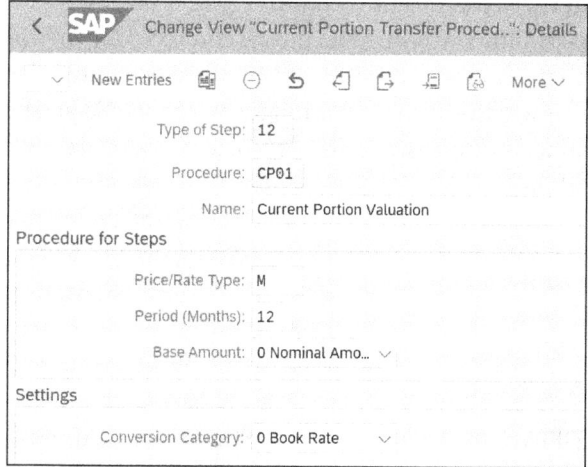

Figure 3.24 Current Portion Transfer Procedure Settings

3.7 User Data

Most of the configuration nodes in this chapter are centered on essential configuration that is required for all product types. However, this piece of the configuration is an exception to that and is an optional step we use to make the user experience more streamlined. When creating contracts within treasury and risk management, we may find that it's easier to use various shortcuts as the contracts are created to speed up the process. We can use this node to drive various shortcuts during the contract entry phase.

To define user data, we navigate the following menu path: **Financial Supply Chain Management • Treasury and Risk Management • Transaction Manager • General Settings • Organization • Define User Data.** This node is driven by our SAP user ID, and we can make changes that are specific to a user ID; in other words, this can be configured differently for each user. Let's look at our options, which are shown in Figure 3.25:

- **Abbreviations for thousand and million (in the M and T columns)**
  When creating contracts in treasury and risk management, we can define abbreviations when entering amounts into the contract. Traditionally, we use *T* is for thousand and *M* for million, but within this configuration, we can define any characters to use for thousand and million.

- **Workday check (in the WDC column)**
  During the contract creation process, we can define how the workdays are checked when entering specific dates into the contract. Users can choose and define the following options:
  - Warning
  - Error Message
  - No Check

3   Core Configuration Elements

- **Exchange rate and SWAP input check (in the REC column)**
  When we're entering FX contracts into SAP S/4HANA, the system can compare the rate and swap rates entered into the contract with the current market data stored in the rate tables. This setting drives whether or not a warning message should appear if the data entered into the contract is inconsistent with the current market data. We can configure the following options:
  - Rate and swap entry check
  - Only rate entry check
  - Only swap entry check
  - No check

- **Default value for Trader**
  Once we create a contract, we'll have to manually populate the trader field in the contract with the trader. Leveraging this configuration, we can link the user's ID to their SAP trader. Then, every time that user ID creates a contract within treasury and risk management, the trader field will automatically be populated with the correct trader. As we can see in Figure 3.25, we can map our user ID to the trader that we have created.

- **Date string from spot value date (with the DateString checkbox)**
  When we are creating FX or currency option contracts, this checkbox comes into play. After we check this box, SAP S/4HANA will calculate the dates used in the value date fields from the spot value date and not from the contract conclusion date.

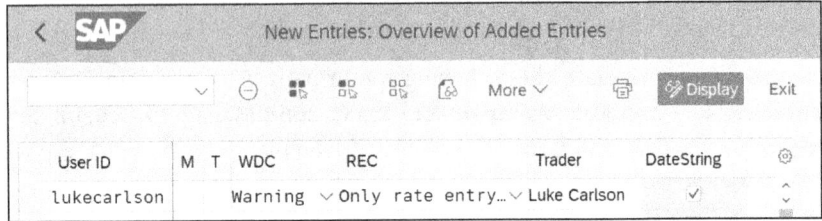

Figure 3.25 Defining User Data Configuration

In addition to these options, SAP S/4HANA allows us to apply default values to contract entry screens for more efficient processing when creating contracts. A common concern of treasury users is the number of button clicks or inputs they must perform when creating treasury contracts. Within the user data, we can apply specific default parameters to the contract entry screens and reduce the number of inputs users must make.

To define specific default parameters per user, we navigate to the SAP menu and go to **System • User Profile • User Data**, as shown in Figure 3.26.

3.7 User Data

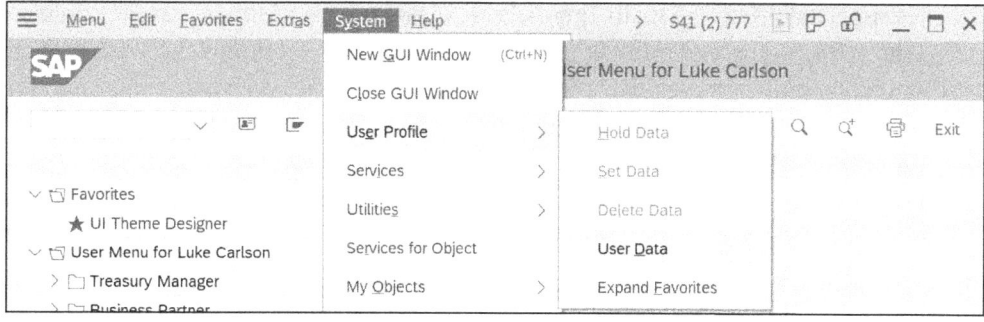

Figure 3.26 Navigation to Change User Data

On the **Maintain User Profile** page, we can modify the default parameters that the system utilizes. The process begins with us identifying the specific **Parameter ID** for which we want to set a default value. Once we've selected the **Parameter ID**, we can establish the desired default value that will automatically populate within the relevant transaction. We do this by entering the desired value into the **Parameter value** field, as illustrated in Figure 3.27.

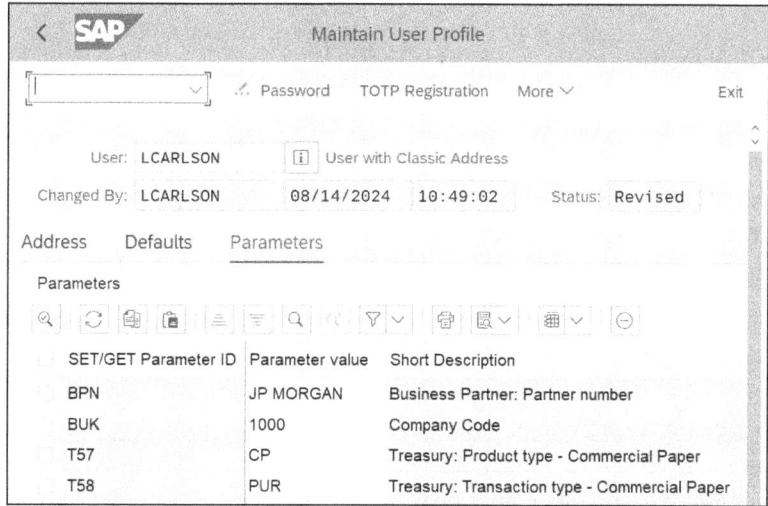

Figure 3.27 Maintaining User Profile for Treasury Contract Parameters

---

TIP

To determine the parameter ID of any field, we navigate to the user screen and press F1 to view the technical information. In the technical information, the **Parameter ID** for each field will be available in the **Field Data** section.

Having established the default parameters for creating a commercial paper contract, we can now streamline the data entry process significantly. We've specified all the necessary values on the main entry screen, including the business partner, company code, product type, and transaction type. These predefined settings ensure that whenever we initiate the creation of a commercial paper contract, the system will automatically load these default parameters into the contract, reducing manual inputs and minimizing the potential for errors. This automated process is visually demonstrated in Figure 3.28 where we can see how the default values populate the contract fields automatically.

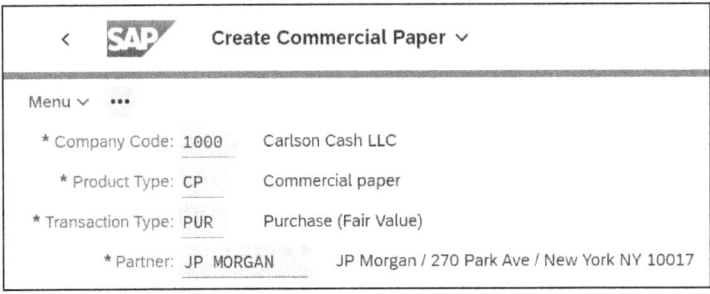

Figure 3.28 Creating Commercial Paper Default Values

3.8 Product Categories

Treasury and risk management was designed to meet the diverse needs of all companies, regardless of industry, by accommodating a wide range of financial instruments. SAP provides a robust and adaptable framework for setting up and configuring a variety of financial instrument types. Its flexibility is driven by the concept of product categories, which are the key to SAP's adaptability in managing different types of financial instruments.

Product categories are important as they reflect the fundamental differences in processing various financial contract types. By organizing financial instruments into distinct product categories based on their underlying structures, SAP ensures that each type of instrument is accurately represented and managed within the system. These predefined product categories form the essential building blocks with which we can construct any required treasury contract and configure it to meet specific business needs.

The underlying product category not only defines the structure of the instrument but also provides the elements necessary for its creation, operation, and ongoing management within the SAP S/4HANA environment.

These product categories break down into five major financial instruments, each of which is based on the specific function it serves. This allows users to navigate the configuration process by instrument category and configure all instruments within a category all in one place within the configuration tree. The standard delivered product categories are systematically divided into these five financial instruments. This provides an organized approach to configuration by placing each category in a separate configuration area within the menu path. The five main financial instruments are broken out within the configuration tree as follows:

- Money market
- Foreign exchange
- Derivatives
- Securities
- Trade finance

3.8.1 Money Market

We use the money market product categories when configuring the short- to mid-term money market–type transactions that most treasury departments use on a daily basis. The money market product categories are arguably the most commonly used product categories, due to the fact that most treasury organizations use money market–type instruments to invest and borrow on a short-term basis. It is important to note that we use these product categories for money market transactions and not money market funds. Let's now look at each of the product categories for money market instruments:

- **Fixed-term deposit: Product category 510**
  A *fixed-term deposit* is one of the most basic product categories available when we're configuring a money market transaction. It's characterized by a fixed term with agreed-upon contract start and end dates that can't be changed once the contract is created. The interest structure is fixed as well and doesn't support the ability to use a variable or reference interest rate. This product category can also be rolled over with a new end date and a change in principal.

- **Deposits at notice: Product category 520**
  The deposit-at-notice product category is very similar to the fixed term-deposit in that it uses a start date and a fixed interest rate. The biggest difference is that a deposit at notice does not have a fixed term. Instead, we define a notice period along with the interest payment date or the frequency of the interest payments.

- **Commercial paper: Product category 530**
  The commercial paper product category operates differently from fixed-term deposits and deposits at notice. Unlike the latter two, for which we establish a start date, an end date, and an interest rate upon contract initiation, commercial paper entails a unique structure tailored to facilitating specific transactions. These transactions involve issuance at face value and subsequent trading at a discount to face value. The purchase price of the commercial paper contract is the nominal value of the contract minus the interest, and there are no interest payments made throughout the life of the contract. Instead, the interest is added to the face value of the contract and is repaid in a lump sum at the end of the contract.

- **Cash flow transaction: Product category 540**
  The cash flow transaction category is the most flexible product category available. It's generally used for transactions that follow an atypical structure and when the requirements of the instrument can't be met by any of the other money market product categories. Using this product category, we have the flexibility to freely add incoming and outgoing flows to a contract to build the structure of the contract and achieve the type of contract we are trying to create.

- **Interest rate instrument: Product category 550**
  The interest rate instrument product category is the most frequently used money market product category. All of the structural requirement needs for product categories 510 and 520 are contained in this product category as well. Additionally, this category allows for both fixed- and variable-interest contracts. We can also create additional interest conditions to change the interest rate throughout the lifecycle of a contract. Capitalization of interest is also possible in this product category. Repayment of principal and interest can take place at contract maturity or at intervals we define when creating the contract. The maturity date of the contract can also be rolled over or extended as the contract changes.

- **Facility: Product category 560**
  The facility product category is structured differently from the other money market product categories. We use it when configuring a credit line or facility that has been prenegotiated with a counterparty. The structural components of the facility, such as start date, maturity, and any associated fees, are predefined within the product category.

  The facilities can be categorized into bilateral and syndicated facilities. *Bilateral facilities* involve a borrower and a single lender, whereas *syndicated facilities* involve a borrower and multiple lenders that have their own credit lines. Within syndicated facilities, there are primary lines of credit that operate independently alongside secondary sublines of credit. We can track and report on the primary line as well as the secondary sublines independently.

  It's important to note that we use this product category when defining the structure of the facility, the associated fees, and the total credit line. We would configure another product category, such as 550, to draw or borrow on the facility.

3.8.2 Foreign Exchange

We use the FX product categories when configuring contracts related to FX management. We use these categories to handle transactions such as spot trades, forward contracts, and NDFs. FX product categories are frequently employed in treasury and risk management to manage FX positions and hedge currency risk.

There's really only one important product category—FX Transaction: Product category 600. This product category is flexible enough that we can use it for FX spots, FX forwards, and NDFs. Within the configuration using the FX product category, we define whether the FX contract is settled with cash, is physically settled, or functions as a non-deliverable currency. When setting up an NDF, we determine the FX gain or loss at the fixing date of the contract. Upon conclusion of the contract, instead of physically settling the foreign currency, we settle only the gain or loss that has been recognized at that time.

3.8.3 Derivatives

We use the derivatives product categories when configuring all types of derivatives contracts, which treasury departments widely use to manage financial risk. These product categories are particularly important for companies looking to hedge against fluctuations in interest rates, commodity prices, or currency exchange rates. Derivatives such as swaps, options, and futures are crucial tools in a treasury department's risk management strategy and are supported by the following product categories:

- **Cap and floor: Product category 610**
  We use the cap and floor product category to configure a series of interest rate options that are exercised when the option falls outside of the predetermined cap or floor specified within the contract.

- **Interest rate swap: Product category 620**
  We can use the interest rate swap product category for all interest rate swap transactions, and typically, we use it to hedge long-term debt contracts. We can also configure the interest rate swap product type as a traditional interest rate swap or a cross-currency interest rate swap. The floating leg of the interest rate swap uses a reference interest rate maintained within the master data that then flows automatically into the contract. In addition to cross-currency interest rate swaps, both discount swaps and compound swaps are supported in this product category.

- **Forward rate agreement: Product category 630**
  The forward rate agreement product category supports a financial derivative contract between two parties that allows them to lock in an interest rate for a future period of time. One party agrees to pay a fixed interest rate, while the other party agrees to pay a floating interest rate based on a specified reference rate.

- **Total return swap: Product category 640**
  The total return swap product type is structured as a contract between two parties where one party agrees to pay the total return of a specific asset to the other party in exchange for regular cash flows based on a floating interest rate.

- **Future: Product category 700**
  This product category is designed to track a *futures contract*, which is an agreement to buy or sell assets—such as commodities, currencies, or financial instruments—at a predetermined future price on a specified future date. The contracts are standardized in terms of the quantity and delivery date, and they are traded on an exchange and most of the time are traded on margin.

- **External underlying: Product category 710**

  An external underlying options contract is an options contract in which the underlying asset is an external financial instrument or commodity, rather than an asset directly traded on the exchange where the options contract is listed.

- **Security as underlying: Product category 712**

  A security as an underlying option is an options contract in which the underlying asset is a specific security, such as a stock or a bond. In options trading, the underlying asset is the financial instrument on which the value of the option is based, and it determines the rights and obligations of the option contract.

- **Repo: Product category 730**

  We use the repo product category when setting up repurchase agreements. A *repurchase agreement*, also referred to as a *repo*, is usually a short-term transaction in which a party sells securities to another party with an agreement to repurchase them at a later date, usually at a marginally higher price.

- **Forward securities transaction: Product category 740**

  A forward securities transaction functions very much like an FX forward, the difference being that instead of purchasing a currency at an agreed-upon date in the future, one of the two parties buys or sells a specified quantity of securities at a price agreed upon up front.

- **Listed option: Product category 750**

  We use this product category for any listed option instruments that are traded on an exchange. Commonly referred to as an *exchange-traded option*, a listed option gives the holder the right but not the obligation to buy or sell a specific asset at a predetermined strike price. The contracts are standardized because they are traded on an exchange.

- **Over-the-counter option: Product category 760**

  This is very similar to product category 750; the biggest difference is that we use it when setting up over-the-counter (OTC) options or any options that are not traded on an exchange. Due to the customized and nonstandardized nature of these contracts, they are unique, and we should set them up using this product category.

- **Securities lending: Product category 770**

  This product category supports financial agreements that allow investors to lend their securities to other market participants in exchange for a fee. It involves the temporary transfer of securities from the lender that owns the securities to the borrower with an understanding that the borrower will return the securities at a later date along with a specified fee.

- **Forward: Product category 780**

  Similar to a futures contract, a *forward* is a financial agreement between two parties to buy or sell an asset at a predetermined price in the future on a future expiration date. Unlike standardized futures contracts traded on organized exchanges, forward contracts are agreed upon between two parties, are typically tailored to meet the specific needs of the parties involved, and are traded directly between the two parties OTC.

- **Forward loan: Product category 790**
  A *forward loan*, also known as a *forward-start loan* or *forward-commitment loan*, is a type of loan agreement in which the disbursement of funds and the commencement of interest accrual are scheduled to occur at a future start date, known as the *forward-start date*. Unlike traditional loans, in which funds are disbursed and interest begins accruing immediately upon agreement, in a forward loan, the disbursement and interest accrual are delayed until a predetermined future date.

- **Commodity forward: Product category 800**
  When setting up a commodity forward, we must use a specific product category outside of the traditional forwards product category. The commodity forward contract is agreed upon by two parties, and the terms of the agreement are defined based on the requirements of both parties. Unlike with commodity futures contracts, the terms of the agreement are not standardized, and thus, commodity forwards are not traded on an exchange.

- **Commodity swap: Product category 810**
  A *commodity swap* is a financial derivative contract in which two parties agree to exchange cash flows based on the price movements of underlying commodities or commodity indexes. Commodity swaps allow the investor to manage their exposure to commodity price risk, speculate on commodity price movements, and gain exposure to commodity markets without owning physical commodities.

3.8.4 Securities

The securities transaction categories are fundamental for configuring contracts related to the purchase and sale of financial securities, such as bonds, stocks, and fixed-income instruments. These categories are particularly important for treasury departments managing an organization's investment portfolio, ensuring that both short-term and long-term investments are effectively handled within treasury and risk management. The following security-related product types are supported:

- **Stock: Product category 010**
  We should use the stock product type when setting up the purchase and sale of equity contracts by an organization. We should only use this product category when purchasing and selling shares because it doesn't support the issuance of a company's own stock.

- **Investment fund: Product category 020**
  We use the investment fund product type within securities to track the purchase and sale of an investment fund such as a money market fund. Purchases and redemptions are supported, as well as investments or dividends to purchase additional shares of the fund.

- **Subscription rights: Product category 030**
  The subscription rights product category gives the stockholder the option to purchase shares of an offering at a discounted price.

- **Bond: Product category 040**
  We use the bond product category when issuing or investing in bonds with either a fixed or a variable interest rate. Zero-coupon bonds are also supported within this product category. The bond is issued and regular coupon payments are made over the life of the contract, with the final repayment taking place at the end of the contract.

- **Installment bond: Product category 042**
  We use this product category when setting up an *installment bond*, sometimes referred to as a *serial bond*. The biggest difference between product categories 042 and 040 is that an installment bond requires periodic payments of both principal and interest over the life of the bond, rather than a single lump-sum payment at maturity. The installments are made at regular intervals, such as monthly or annually, until the bond reaches its maturity date

- **Warrant bond: Product category 060**
  A *warrant bond* is a type of bond that comes with detachable warrants. These *warrants* give the bondholder the right to purchase additional shares of the issuer's stock at a predetermined strike price.

- **Convertible bond: Product category 070**
  The convertible bond category functions very similar to product category 040, except that the owner can convert the entire bond or part of the bond to equity based on the convertible terms specified within the bond. SAP S/4HANA only supports the purchase of convertible bonds; the issuance of convertible bonds is not available.

- **Index warrant: Product category 111**
  An *index warrant* is a financial derivative that gives the holder the right but not the obligation to buy or sell a specific stock market index financial instrument at a predetermined strike price within a specified period of time. An index warrant functions like a stock option, but instead of an individual stock, it's based on an index.

- **Equity warrant: Product category 112**
  An equity warrant functions very much like an index warrant. It gives the holder the right but not the obligation to buy or sell a specific stock at a predetermined strike price over a specific period of time.

- **Currency warrant: Product category 113**
  A *currency warrant* is a derivative that gives the holder the right but not the obligation to buy or sell a specific currency at a predetermined exchange rate within a specified period of time.

- **Bond warrant: Product category 113**
  A *bond warrant* is a financial instrument that gives the holder the right but not the obligation to purchase a specific bond at a predetermined price within a specific period of time.

3.8.5 Trade Finance

We use the trade finance transaction categories to configure contracts, such as letters of credit and bank guarantees, that are related to the financing of trade activities. These categories are particularly important for treasury departments managing an organization's trade-related financial obligations, and they ensure that both import and export transactions are efficiently handled within treasury and risk management. The following trade finance–related product types are supported:

- **Letter of credit: Product category 850**
  The letter of credit product category allows for the ability to track letters of credit issued by a bank or financial institution. Documentary, standby, revocable, and irrevocable letters of credit are all supported by this product category.

- **Bank guarantee: Product category 860**
  The bank guarantee product category allows for the setup of a bank guarantee issued by a bank or financial institution, which serves as a promise from the bank to the beneficiary that the holder will fulfill their contractual obligations.

3.9 Product Types

Now that we've explored the nuances of setting up the base configuration of treasury and risk management, it's time to dive into the essential core configuration required for creating new contract types from scratch. At the forefront of this process lies the creation of product types. This initial setup is extremely important because it lays the groundwork for how contracts will operate within the system. Creating product types involves not only defining the characteristics and functionalities of the contracts but also assigning them to appropriate product categories. This step is very important as it directly influences the behavior and functionality of the contracts. The accurate determination of the product category ensures that the contract aligns with the desired design and functionality, while inaccuracies in this setup phase will impact subsequent processing of the contract and how the contract entry screens are populated. Thus, we must pay careful attention to detail during product type creation and product category assignment to ensure the intended contract functionality.

Product types serve as the fundamental building blocks of treasury contracts. Creating them is the first step when creating a new contract type, and it plays a crucial role in how we classify the type of financial instrument we're configuring. We use the product type to represent the type of instrument we're configuring. For example, commercial paper, FX, and interest rate instruments are examples of product types that we can configure during this step. Treasury contracts exhibit differences in their overall purpose and core functionalities, prompting SAP to categorize them based on their general functions. For example, a debt or investment contract operates differently from an FX contract. To accommodate the differences between the various instruments, SAP has structured the configuration process into six primary financial instrument types. By delineating these, SAP ensures that users can efficiently navigate the configuration process by breaking up the configuration activities of like instruments into specific areas. This approach not only simplifies the setup process but also facilitates alignment between the contract type and its intended function. The six financial instrument types are broken out and the configuration for each one resides in a separate menu path, as follows:

- **Money market**
  Financial Supply Chain Management • Treasury and Risk Management • Transaction Manager • Money Market

- **Foreign exchange**
  Financial Supply Chain Management • Treasury and Risk Management • Transaction Manager • Foreign Exchange

- **Securities**
  Financial Supply Chain Management • Treasury and Risk Management • Transaction Manager • Securities

- **Listed derivatives**
  Financial Supply Chain Management • Treasury and Risk Management • Transaction Manager • Listed Derivatives

- **OTC derivatives**
  Financial Supply Chain Management • Treasury and Risk Management • Transaction Manager • OTC Derivatives

- **Trade finance**
  Financial Supply Chain Management • Treasury and Risk Management • Transaction Manager • Trade Finance

> **Note**
> SAP offers a set of predefined configurations for product types as a standard feature. While these configurations are available for use, it's common for us to create new product types during projects to better align them with the specific business needs of the client. By creating new product types beyond the standard offerings, we gain the flexibility to define the naming conventions according to our preferences. This allows for the use of more descriptive text, making it easier for users to identify and distinguish product types when creating contracts or generating reports.

As we start the configuration for a product type within SAP S/4HANA, we'll quickly notice a common thread running through the structure of each financial instrument type's configuration nodes. Despite their similarities, it's important to note that the specific configuration nodes we navigate through will differ based on the type of contract we are in the process of configuring. They may seem very similar, but the nodes within the financial instrument types vary based on the unique requirements of each instrument. This underscores the flexibility of SAP S/4HANA's configuration framework, allowing for nuanced adjustments tailored to the unique requirements of each contract type. By navigating through the different nodes, users can fine-tune settings and parameters to align with the characteristics and functionalities of the contract at hand.

To start the configuration of a money market product type, we follow the **Financial Supply Chain Management • Treasury and Risk Management • Transaction Manager • Money Market • Transaction Management • Product Types • Define Product Type** menu path. We can then add and configure a new product type based on the unique requirements of our organization.

When we create a new product type, the screen shown in Figure 3.29 will appear. The main configuration elements are the following:

- **Product Type**
  We should define the product type with any three characters that will easily identify the product. This is important because this will be the value that users see as they are creating the contract.

- **Text**
  This is the text used to identify this product type, and it will appear when we are creating the contract as well as in any reporting in the future.

- **Prod. Category**
  This is the most important piece of the configuration within the product type creation process. We populate this field with one of the product categories we covered in Section 3.8, and it will drive the screen layout and instrument functionality in the subsequent configuration settings.

- **Int. Calc. Method**
  This serves as the default interest calculation method integrated into the contract upon creation. Whenever a contract requires an interest calculation method, we have the opportunity to predetermine its default value by populating this specific field with the appropriate parameter. Although we can modify the value during contract creation, establishing a default interest calculation method streamlines the entry process. This practice ensures that the most frequently utilized interest calculation method is automatically applied when we generate contracts of similar types. The interest calculation methods that are supported are the following:
  - 360E/360
  - 360E/365

3 Core Configuration Elements

- 360E/actY
- 360/360: International Swaps and Derivatives Association (ISDA)
- 360/365: ISDA
- 360/360: Equal to 30/360
- 365/360
- 365/365
- act/360
- act/364
- act/365
- act/365.25
- act/366
- act/365P
- act/actP: International Capital Markets Association (ICMA)
- act/actY: ISDA
- act/actE: Association Francaise de Banques (AFB)
- act/actEP: AFB
- act/365L
- act/365Y
- actW/252 (W = working days)
- act/act[M] (M = monthly)

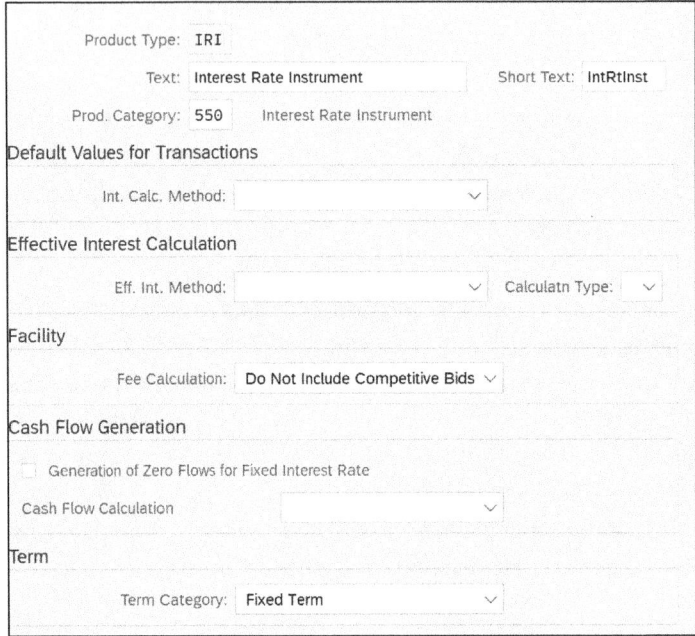

Figure 3.29 Creating Product Types

3.10 Number Ranges

Now that we've defined the product type, we'll focus on how we want to view the specific transactions once the contracts are created. Every contract that we create within treasury and risk management is assigned a number that is unique within the company code. The number assigned to the contract is very important as it's the number that will forever be tied to the contract, it's also the primary way we'll identify the contract in the future. In Section 3.11, we'll assign a number and a number range to each contract at the transaction type level. The contract numbers can be up to thirteen digits in length and can be either a combination of numeric and alpha characters or sequential numbers.

We can find the number range definition using Transaction TAC1 or the following menu path: **Financial Supply Chain Management • Treasury and Risk Management • Transaction Manager • Money Market • Transaction Management • Transaction Types • Define Number Ranges for Financial Transactions**. Our first step when defining the number range is to distinguish which **Company Code** we'll be creating the number ranges for (see Figure 3.30), since number ranges are defined at the company code level. We'll then click on **Create Intervals**.

Figure 3.30 Initial Screen for Creating Financial Transaction Number Range

Before we define a number range within the company code, it's important for us to determine whether it will be an internally or externally assigned number range. The difference between an internal and an external number range is as follows:

- **Internal number range**
  This is a numerical sequence allocated automatically upon contract creation. The number is assigned to the contract as part of the contract creation in the background and is only visible once the contract has been saved. The number is internally assigned to the contract, and the system uses the next sequential number available within the range. We most commonly use this method to define number ranges for high-volume treasury contracts.

- **External number range**
  An external number range is a numerical sequence designated by the user during contract creation. When entering the contract details, users have the flexibility to

specify the alphanumeric identifier that distinguishes the contract. We most commonly use an externally assigned number range for contracts with low transactional frequency, where an alphanumeric identifier proves helpful for identification such as in the case of facilities or credit lines. Additionally, using an external number range can be useful for aligning with a counterparty's contract number or another value provided by the counterparty. We can use this approach as a way to link the external contract with the contract stored within treasury and risk management, and it also eliminates the need to store the value within another field in the contract.

When defining the number range, we start with a two-digit character that will identify that number range, followed by the range of numbers we would like to use for that specific number range (see Figure 3.31). The number range number corresponds with the first digit of the number range; this is only for ease of use and is not a requirement. If we want to use external number ranges, we check the **External** box, which allows for alphanumeric values within the number range.

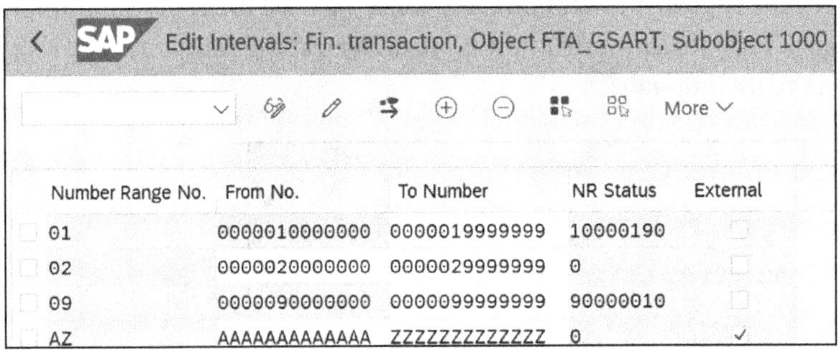

Number Range No.	From No.	To Number	NR Status	External
01	0000010000000	0000019999999	10000190	
02	0000020000000	0000029999999	0	
09	0000090000000	0000099999999	90000010	
AZ	AAAAAAAAAAAA	ZZZZZZZZZZZZ	0	✓

Figure 3.31 Number Range Creation

3.11 Transaction Types

The next step within the configuration process of setting up a treasury and risk management contract is the setup of the transaction type. Within the configuration, we can find it in the **Financial Supply Chain Management** • **Treasury and Risk Management** • **Transaction Manager** • **Money Market** • **Transaction Management** • **Transaction Types** • **Define Transaction Types** menu path. A transaction type further differentiates the characteristics of a product type by specifying the types of transactions available when we're entering a contract using that product type.

For instance, consider an interest rate instrument product type, which can be subdivided into distinct transaction types representing the debt and investment aspects of the contract. The relationship between the product type and its associated transaction type forms a unique identifier that shapes the functionality of the contract. The distinction between transaction types becomes apparent when considering factors such as cash flow direction. Configuring the transaction type entails delineating the nuances between different variations of each product type within scope and configuring corresponding transaction types to accurately represent these transactions. In Figure 3.32, we can see a list of the most common money market transaction types configured within treasury and risk management.

Money Market: Transaction Types			
PTyp	Prod.Type Desc.	TTyp	Name of Transactio...
CFD	Credit Facility Draw	BOR	Borrowing
CP	Commercial paper	PUR	Purchase (Fair Value)
CP	Commercial paper	SEL	Sale (Fair Value)
ECF	External Credit Facility	BOR	Borrowing
IRI	Interest rate instrument	BOR	Borrowing
IRI	Interest rate instrument	INV	Investment
LC	Letter of Credit	DRW	Draw (borrow)
SCF	Syndicated Facility	BOR	Borrowing

Figure 3.32 Money Market Transaction Type Configuration

The following configuration elements are important to understand when creating a transaction type, as shown in Figure 3.33:

- **Transaction Type**
  Once we've selected the appropriate product type that we configured during product type setup, we must specify a three-character transaction type that serves to further distinguish the instrument. When naming the transaction type, we must ensure that the values we choose are easily discernible, since both the product type and transaction type will be utilized together whenever contracts are generated. This ensures clarity and precision in the contract creation processes, allowing for efficient differentiation and identification of various instrument types.

- **Name of transaction type**
  This text further defines the transaction type and will populate in the contract when we're creating and reporting on contracts.

- **Transaction Cat**
  The transaction category selection is the most important element in the transaction type setup screen. When populating this field, we have predefined transaction categories that are tied to the product category that we used during product type configuration. They drive the direction of the cash flows within the contract, so using the correct transaction category is crucial within the configuration. The transaction categories should also correspond to the type of transaction we're creating. For example, product category 550 – Interest Rate Instrument has two transaction categories that we can use: 100 – Investment or 200 – Borrowing.

- **Number ranges - Transactions**
  We populate the number range that we should use when creating a contract.

3   Core Configuration Elements

- **Processing Cat.**
  The processing category is extremely important since it defines how users process and create contracts. When a user creates a contract, a second user can perform an additional settlement step. Whether or not a contract requires an additional settlement step is defined by the processing category. The settlement step is a way to enforce a dual-control principle when creating contracts in SAP S/4HANA. If the settlement step is in place, one user creates the contract, but that contract will be blocked for additional processing until the contract has been settled.

- **Automatic posting release**
  Following the creation of a contract and its subsequent settlement (if applicable), there's an option to control the release of the accounting postings. Enabling this feature imposes an additional step that involves releasing accounting postings before they are eligible for posting. We commonly employ this practice to differentiate the responsibilities between accounting and treasury functions, facilitating a two-step posting process. Accounting departments may also opt to release postings prior to their integration into the general ledger, allowing for enhanced oversight and alignment with internal control processes.

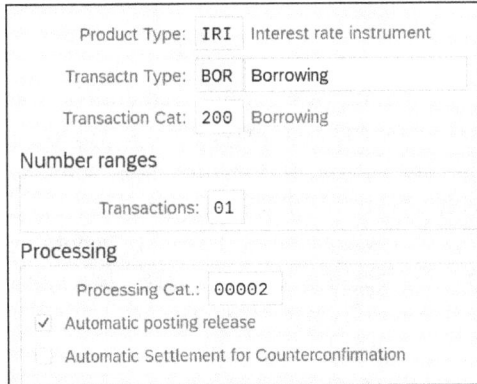

Figure 3.33 Define Transaction Types Configuration Screen

3.12   Field Selection

Now that we've defined the product and transaction types, we'll turn our attention to the usability of the selection screen when we're entering contracts into SAP S/4HANA. When users are creating a contract in SAP S/4HANA, the fields that are visible to them on the entry screen are predefined to include all of the possible fields that are available for that specific product category. It's not a required step within the configuration, but we can modify the specific fields that are available to users during the creation, settlement, and editing of a contract. This is a convenient way to streamline the contract entry process and eliminate the fields on the contract entry screen that only pertain to that specific instrument. The configuration can be found in the **Financial Supply Chain Management • Treasury and Risk Management • Transaction Manager • General Settings • Transaction Management • Define Field Selection** menu path.

Our first step when setting up the field selections is to define what fields we would like users to see when they are creating, editing, or settling a contract. By clicking on the **Field selection** button shown in Figure 3.34, we can determine which fields and tabs will be available to users.

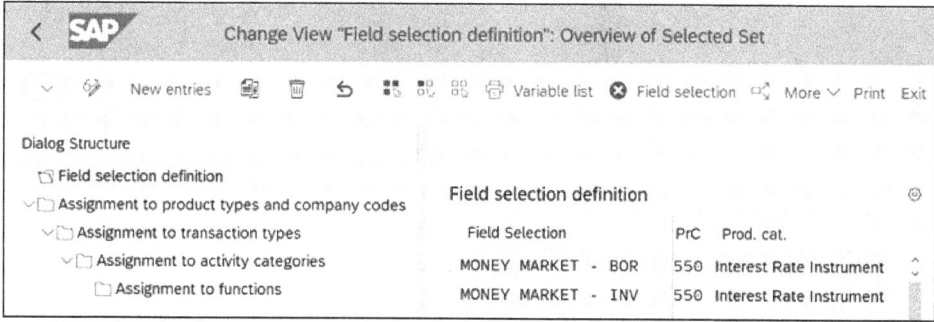

Figure 3.34 Field Selection Definition

When determining what fields and tabs are available, we have the following options, as shown in Figure 3.35:

- **Hide**
  Clicking this radio button will hide the field or tab so that users can't see it. This can be very helpful when we want to eliminate a field or tab that users would never populate as they enter a contract.

- **Req. entry**
  Clicking this radio button makes this field visible onscreen and also signifies that populating this field is required, meaning users can't save a contract without first making an entry in this field.

- **Opt. entry**
  Clicking this radio button lets users see the field within the contract entry screen without making it a requirement to populate this field to save the contract.

- **Display**
  Clicking this radio button makes the field visible to users, but they won't be able to enter a value in the field.

- **Not spec.**
  When we click **this** radio button, the settings that are defined out of the box will remain unchanged for each field or tab. We click this radio button when we want to leave the standard delivered SAP S/4HANA settings in place.

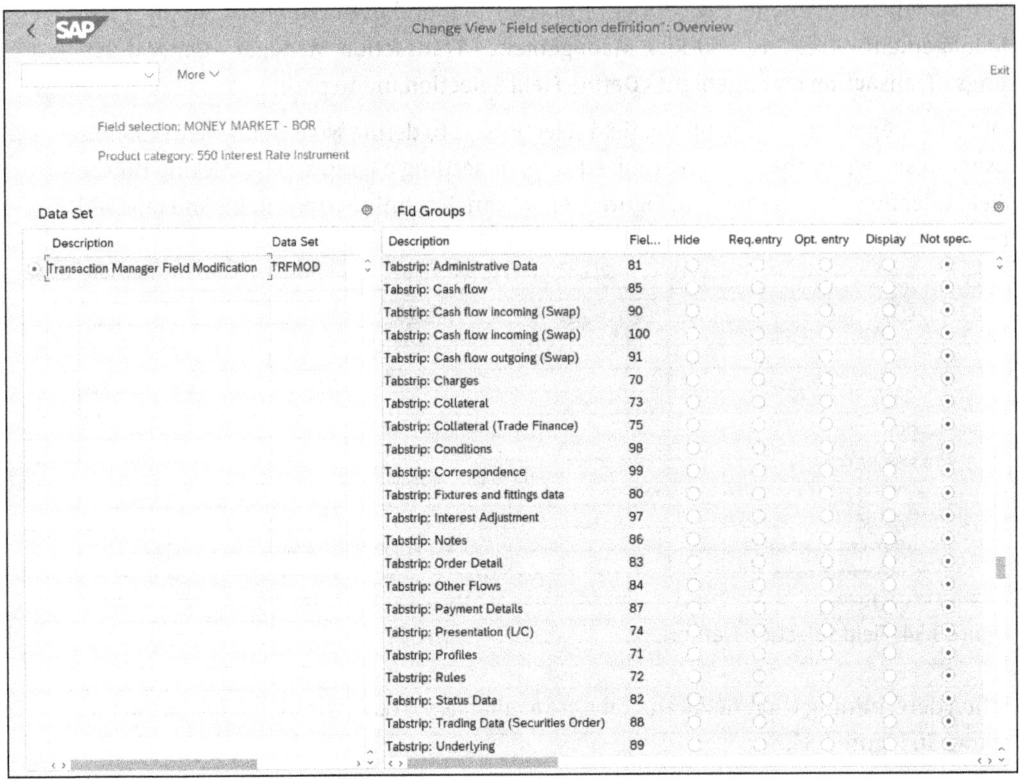

3 Core Configuration Elements

Figure 3.35 Field Selection Options

We can assign the field selection name along with the underlying settings at various levels within the contract, based on how granular we want the field selections to be. We can assign the field selection at the product type and company code levels, as shown in Figure 3.36. To do so, we click on the **Assignment to product types and company codes** folder and then map our product type (**PTyp**) and company code (**CoCode**) to the **Field Selection**.

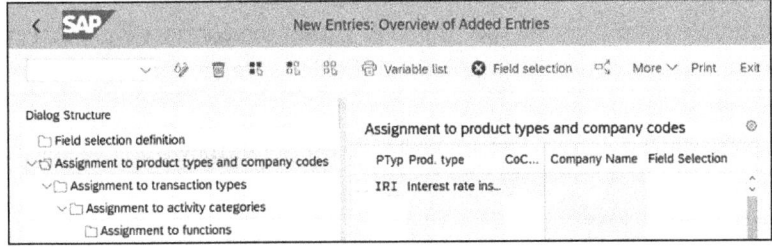

Figure 3.36 Assignment of Field Selection to Product Types and Company Codes

We can also assign the field selections at a more granular level and at the transaction type level as shown in Figure 3.37. To do so, we click on the **Assignment to transaction types** folder and then map our transaction type (**TTyp**) to the **Field Selection**.

3.13 Flow Types

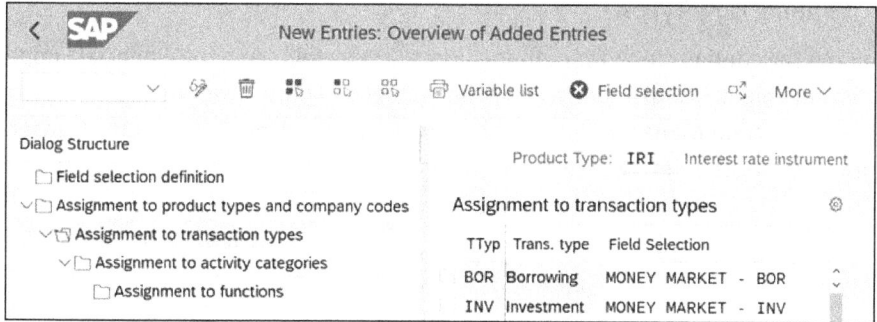

Figure 3.37 Assignment of Field Selection to Transaction Types

Depending on the unique requirements of the project, we can also be even more specific and assign the field selection at the contract activity level as shown in Figure 3.38. To do so, we click on the **Assignment to activity categories** folder and then map our activity (**ACat**) to the **Field Selection**.

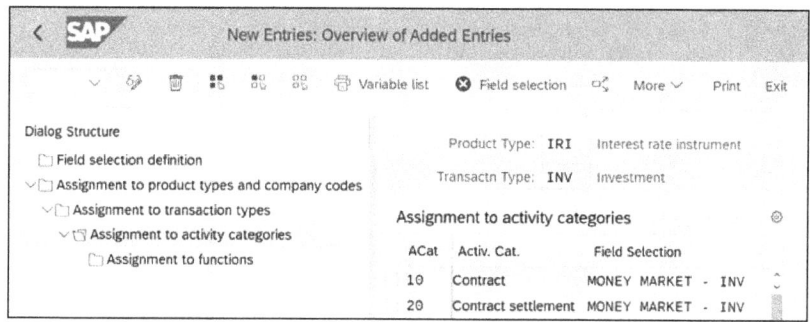

Figure 3.38 Assignment of Field Selection by Activity Category

3.13 Flow Types

Once we've established the product type and corresponding transaction types, the next step involves defining the cash flows associated with each product and transaction type. These cash flows are configured within the system as flow types, and each flow type serves to clarify the purpose of cash movements within the contract. While many flow types are linked to cash flows, it's important to note that they can also encompass noncash movements.

As the contract progresses, these flow types eventually transition into accounting flows that generate accounting documents posted to the general ledger. Therefore, it's important to configure flow types for every flow that involves accounting throughout the duration of the contract's lifecycle. This step ensures comprehensive accounting coverage and accurate financial reporting across all stages of the contract.

In the following sections, we'll walk through how to both define and assign flow types.

3.13.1 Defining Flow Types

We can find the configuration for defining flow types by following the **Financial Supply Chain Management • Treasury and Risk Management • Transaction Manager • Money Market • Transaction Management • Product Types • Define Flow Types – MM Transactions** menu path. The key configuration components of the flow type are shown in Figure 3.39.

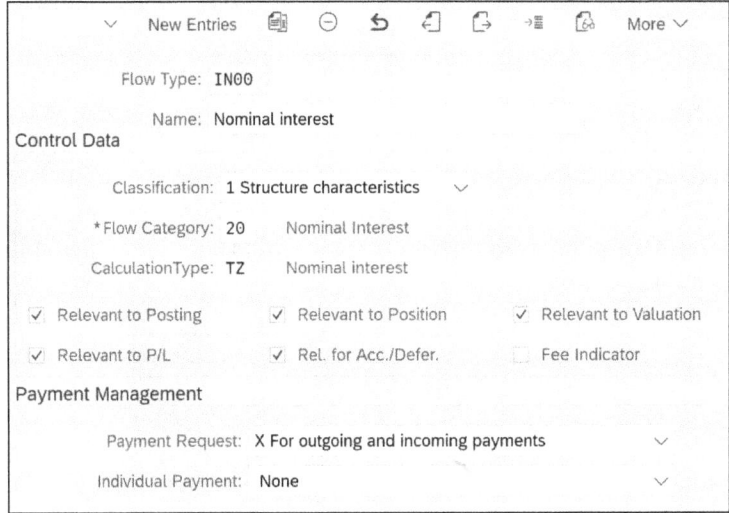

Figure 3.39 Defining Flow Types

The key fields are as follows:

- **Flow Type**
  A *flow type* is characterized by a blend of any four alphanumeric characters. SAP S/4HANA's standard flow types are exclusively defined with numeric characters. The flow type is accessible to users during contract entry and, in certain instances, manually specified within the contract. Therefore, establishing a naming convention that aligns the flow type with the cash flow can add to the usability of the solution. For instance, utilizing "IN00 – Nominal Interest" to denote an interest cash flow within the contract provides clarity regarding the type of flow being represented.

- **Name**
  The name further defines the type of flow that has been configured. We should use a name that is easily recognizable as this name will appear whenever the flow is used within a contract.

- **Classification**
  The classification of the flow is the most important aspect of the flow type creation. It defines how the flow should be used downstream within the contract. If we don't define this with the correct classification, the flow type won't appear correctly in the contract creation process. The options that we can choose are predefined as follows:

- Structure characteristics
- Accrual/deferral
- Valuation
- Transfer

- **Flow Category**

  The flow category is the second most important configuration element of a flow type. The flow category defines at what point in the contract process a flow should be utilized. In this example, IN00 – Nominal Interest is tied to flow category 20: Nominal Interest. The following flow categories are predefined for use within a money market transaction, but they will vary among the various product categories.

  - 10: Principal Increase
  - 11: Principal Decrease
  - 12: Final Repayment
  - 13: Installment Repayment
  - 14: Annuity Repayment
  - 15: Interest Capitalization
  - 16: Capitalized Interest Payment
  - 20: Nominal Interest
  - 21: Interest Rate Adjustment
  - 24: Amounts Equivalent to Interest
  - 25: Payment Rate
  - 26: Price Index Adjustment
  - 27: Accrued Interest Rate
  - 29: Split Event
  - 60: Accumulating
  - 70: Inflow (generic, Fiduciary)
  - 71: Outflow (generic, Fiduciary)
  - 72: Withdrawal
  - 73: Extension
  - 90: Other Flow/Condition

- **Control data checkboxes**

  We use the various control data checkboxes to drive the behavior of the flow type within the contract. Based on the purpose of the flow type, we make different selections for each of the flows:

  - Relevant to Posting

    We should check this box for any flow type that is relevant to post to the general ledger.

- **Relevant to Position**
  We should check this box when the flow is relevant to a change in the position.
- **Relevant to Valuation**
  We should check this box when the flow type is relevant to a valuation and should therefore be used during the valuation process.
- **Relevant to P/L**
  We should check this box for any expenses or revenue-related cash flows.
- **Relevant for Acc./Defer.**
  We should check this box if the flow should be accrued or deferred.
- **Fee Indicator**
  We should check this box for any flow that is a fee.

■ **Payment Management**
The payment management section of the flow type configuration relates to whether or not the various flows create payment requests and how the payment requests are handled within the treasury payment program.

- **Payment Request**
  Within the flow, we can determine whether or not the flow will generate a payment request for outgoing payments, incoming payments, or both incoming and outgoing payments.
- **Individual Payment**
  If a payment request is created as a result of the flow type, we can determine how the payment request is paid when the payment is made to the same counterparty payment instructions. By default, all payments within a payment run will net together, but we can drive the behavior within the update type. We can consolidate the payment requests by using the following logic:
  - **For outgoing payments**
    All outgoing payments won't net and will create individual payments during the payment run.
  - **For incoming payments**
    All incoming payments won't net and will create individual payments during the payment run.
  - **For outgoing and incoming payments**
    Both incoming and outgoing payments won't net and will create individual payments during the payment run.

3.13.2 Assigning Flow Types to Transaction Types

After we establish the various flow types, the subsequent stage involves associating each flow type to the relevant product and transaction type combination. We must map each product and transaction type combination to the relevant flow types that should be utilized though the life of the contract. This step is essential as it ensures that all relevant flow types are properly configured for each of the product and transaction type configured. Not assigning the correct flow type in this node will lead to its omission during the contract creation and subsequent posting process. Therefore, we must focus our attention on this configuration to ensure accurate representation of all cash flows that are associated with a contract type.

We can find the configuration to assign the flow types to the various transaction types within the following menu path: **Financial Supply Chain Management • Treasury and Risk Management • Transaction Manager • Money Market • Transaction Management • Product Types • Assign Flow Types to Transaction Type – MM Transactions**. It's important to note that this configuration activity spans all of the various financial instruments within treasury and risk management and must be set up within the correct financial instrument. Start by clicking on **New Entries**, we can enter the **Product Type** and **Transactn Type** and then assign the corresponding **Flow Type** applicable to this product and transaction type combination, as shown in Figure 3.40.

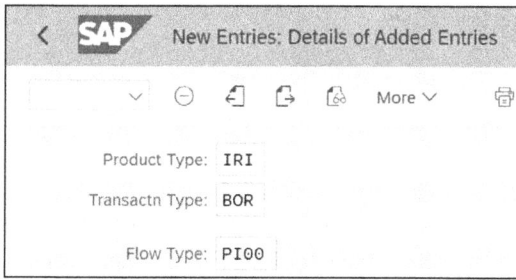

Figure 3.40 Assigning Flow Type to Transaction Type

After saving the entry, we can see all of the entries that reside in this table (as shown in Figure 3.41). An important detail that is often overlooked is that only flow types that are not associated with a condition type must be maintained in this table.

> **Tip**
>
> Any flow type that is used as a result of a condition type is already assigned to the product and transaction type combination via the **Assignment of Condition Types to Product Types** node covered in in Section 3.15.2. Therefore, we don't have to dual-maintain that flow type within this configuration. Adding such flow types to this configuration table won't negatively impact the processing of the contract, and we may decide to maintain all flow types within this configuration for posterity, but it's not required for the contract to function correctly.

PTyp	Prod.type desc.	TTyp	Name of Trans...	FTyp
IRI	Interest rate instrume.	BOR	Borrowing	CH00
IRI	Interest rate instrume.	BOR	Borrowing	IN00
IRI	Interest rate instrume.	BOR	Borrowing	PD00
IRI	Interest rate instrume.	BOR	Borrowing	PF00
IRI	Interest rate instrume.	BOR	Borrowing	PI00
IRI	Interest rate instrume.	INV	Investment	IN00
IRI	Interest rate instrume.	INV	Investment	PD00
IRI	Interest rate instrume.	INV	Investment	PF00
IRI	Interest rate instrume.	INV	Investment	PI00
LC	Letter of Credit	DRW	Draw (borrow)	IN00
LC	Letter of Credit	DRW	Draw (borrow)	PFN0
LC	Letter of Credit	DRW	Draw (borrow)	PIN0
SCF	Syndicated Facility	BOR	Borrowing	CH00
SCF	Syndicated Facility	BOR	Borrowing	FC00
SCF	Syndicated Facility	BOR	Borrowing	FC01
SCF	Syndicated Facility	BOR	Borrowing	FC02
SCF	Syndicated Facility	BOR	Borrowing	FC03

Figure 3.41 Assigning Flow Types to Transaction Types

3.14 Update Types

We now will transition to the maintenance of update types. Update types are where we start to see the accounting take shape within treasury and risk management. We create them to define the cash movements and valuation impacts of the financial instrument, outlining how each transaction will impact the position. Once we define these update types, they are assigned to the previously configured flow types, which represent the different cash flows—such as interest, repayments, or fees—associated with the instrument. By linking flow types with update types, the system ensures that every movement of cash or change in value is accurately captured on the general ledger.

3.14.1 Defining Update Types and Assigning Usages

The next step when configuring a contract is to configure the update types by following the **Financial Supply Chain Management • Treasury and Risk Management • Transaction Manager • Money Market • Transaction Management • Update Types • Define Update Types and Assign Usages** menu path.

The update types serve as an extension of the flow types within the system, offering a deeper insight into the trajectory of accounting entries as they transition from contracts to financial postings in financial accounting. These update types build upon the foundation laid by flow types, providing a directional aspect to cash flows. In most

3.14 Update Types

cases, cash flows, as represented by flow types, represent both incoming and outgoing cash movements.

For instance, an interest payment can denote either incoming or outgoing cash, depending upon whether the contract pertains to a debt or investment contract. To capture this dynamic, we establish update types to specify the directionality of each cash flow. While many flow types may entail two update types, reflecting both incoming and outgoing flows, this isn't mandatory. Update types are primarily designed to reflect the orientation of cash flows, and depending on the instrument's scope, certain flow types may require only a single corresponding update type instead of a pair. Importantly, not all update types are directly linked to flow types; specifically, those associated with valuations operate independently. Valuation transactions mandate the creation of an update type, although not directly tied to a flow type within the configuration framework. As shown in Figure 3.42, we've configured our update types required for a debt or investment contract. We define the **UpdateType** followed by the text of the **UpdateType**, and the text should describe the type of flow associated with the update type. The update type and update type text will appear in any journal entries that are posted using the update type, so it's important to use descriptive text that characterizes the update type accurately.

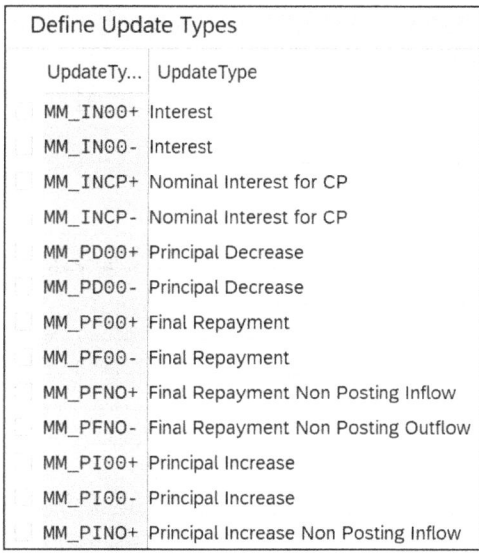

Figure 3.42 Defining Update Types

Considering the wide range of update types managed within configuration, SAP S/4HANA offers a standard naming convention for their creation. The initial two characters serve to further specify the update type, as follows:

- MM: Money market area
- FX: Foreign exchange trading area

157

3   Core Configuration Elements

- DE: OTC derivatives, futures, and listed options area
- SE: Securities transactions area
- CML: Loans area
- SAM: Securities account management
- SAT: Securities account transfer
- RHT: Security right
- CA: Corporate action
- V: Valuation
- VR: Reset valuation
- AD: Accrual/deferral
- DBT: Flow of a derived business transaction
- AAR: Account assignment reference transfer
- VT: Valuation class transfer

When we're creating update types directly associated with a flow type, incorporating the four-character flow type name into the naming convention of the update type can be immensely beneficial. We can follow this by specifying the direction of the flow. As demonstrated in Table 3.1, implementing a naming convention helps users distinguish between update types, and it facilitates a clear understanding of the update types' function and direction. It also enables users to identify the corresponding flow type. This approach streamlines the process of managing the configuration, helps users understand how the various flow types correlate with their respective update types, and makes for easy maintenance of the configuration in the future.

Area	Flow Type	Direction	Update Type
MM: Money Market	Nominal Interest – IN00	Inflow	MM_IN00+
MM: Money Market	Nominal Interest – IN00	Outflow	MM_IN00-

Table 3.1 Update Type Naming Convention Relationships

After we create the update types, all of them are assigned to specific usages. This categorization is based on the intended purposes of the update types and indicates when they will be utilized in further configuration settings. We should select the update type usage (**Upd. Ty. Usage**) of **Transaction Management** for all structural update types within a contract. To configure this node, we first choose the appropriate update type usage from the dropdown menu. We can then populate the update type in the **Update-Type** column, as shown in Figure 3.43. Once we've populated the update type, the text from the update type will automatically populate in the **Update Type Description**. As shown below, we've created all the required update types and assigned them to the **Transaction Management** update type usage.

Assign Update Type to Usages		
Upd. Ty. Usage	UpdateType	Update Type Description
1 Transaction Management	MM_IN00+	Interest
1 Transaction Management	MM_IN00-	Interest
1 Transaction Management	MM_INCP+	Nominal Interest for CP
1 Transaction Management	MM_INCP-	Nominal Interest for CP
1 Transaction Management	MM_PD00+	Principal Decrease
1 Transaction Management	MM_PD00-	Principal Decrease
1 Transaction Management	MM_PF00+	Final Repayment
1 Transaction Management	MM_PF00-	Final Repayment
1 Transaction Management	MM_PFN0+	Final Repayment Non Posting Inflow
1 Transaction Management	MM_PFN0-	Final Repayment Non Posting Outflow
1 Transaction Management	MM_PI00+	Principal Increase
1 Transaction Management	MM_PI00-	Principal Increase
1 Transaction Management	MM_PIN0+	Principal Increase Non Posting Inflow

Figure 3.43 Assigning Update Types to Usages

3.14.2 Assigning Flow Types to Update Types

The next step after creating the update types is assigning additional attributes to each of the update types. We can find the configuration for this step in the **Financial Supply Chain Management • Treasury and Risk Management • Transaction Manager • Money Market • Transaction Management • Update Types • Assign Flow Types to Update Types** menu path. In this configuration step, we add three distinct attributes to the various update types.

First, we assign the correct contact type to the update type based on how the update type will be used in the contract type (**Cont.Type**) field (see Figure 3.44). The contract types align with the five main contract types that are created within treasury and risk management. The contract types available in the dropdown list are as follows:

- Money Market
- Securities
- Foreign Exchange
- Derivatives
- Trade Finance

The next attribute that we configure within this node is the assignment of the update type to its corresponding flow type. Within the configuration node, we first define the flow type (**FTyp**) that we are mapping, and we then map the flow type to the specific **UpdateType** that it's associated with. This step establishes the link between the flow type and the update type that should be used for that particular cash flow. As we can see in Figure 3.44, each of the update types is mapped to its corresponding flow type.

The last attribute we configure during this step is the assignment of a cash flow **Direction** (either **+ Inflow** or **– Outflow**) to each update type. Up until this point, we haven't defined the direction of the cash movement within the update type. The orientation in the name of the update type is nothing more than a naming convention, and we have to further define the direction of the cash flow for each update type by assigning a direction to each update type. As we can see in Figure 3.44, we've assigned a direction to each update type that corresponds to the direction used in our update type naming convention.

Assignment of Business Flow Type to Update Type					
Cont.Type	FTyp	Name	Direction	UpdateType	Update Type Description
5 Money Market	IN00	Nominal interest	+ Inflow	MM_IN00+	Interest
5 Money Market	IN00	Nominal interest	- Outflow	MM_IN00-	Interest
5 Money Market	INCP	Nominal interest for CP	+ Inflow	MM_INCP+	Nominal Interest for CP
5 Money Market	INCP	Nominal interest for CP	- Outflow	MM_INCP-	Nominal Interest for CP
5 Money Market	PD00	Principal Decrease	+ Inflow	MM_PD00+	Principal Decrease
5 Money Market	PD00	Principal Decrease	- Outflow	MM_PD00-	Principal Decrease
5 Money Market	PF00	Final repayment	+ Inflow	MM_PF00+	Final Repayment
5 Money Market	PF00	Final repayment	- Outflow	MM_PF00-	Final Repayment
5 Money Market	PFNO	Final Repayment - No Post	+ Inflow	MM_PFNO+	Final Repayment Non Pos
5 Money Market	PFNO	Final Repayment - No Post	- Outflow	MM_PFNO-	Final Repayment Non Pos
5 Money Market	PI00	Principal Increase	+ Inflow	MM_PI00+	Principal Increase
5 Money Market	PI00	Principal Increase	- Outflow	MM_PI00-	Principal Increase

Figure 3.44 Assignment of Flow Types to Update Types

3.14.3 Indicate Update Types Relevant to Posting

Prior to configuring the accounting settings for specific update types, it's important for us to identify the update types applicable for posting. Not all of the update types we configure will constitute postings to the general ledger, so we must define the specific update types within each contract that should generate an accounting posting. We can find the configuration node for this step in the **Financial Supply Chain Management • Treasury and Risk Management • Transaction Manager • General Settings • Accounting • Link to Other Accounting Components • Indicate Update Types Relevant to Posting** menu path.

To ensure that every update type we expect to trigger a posting within SAP S/4HANA Finance does so, we must configure each relevant update type within this configuration table. Within this setup, we select the update types alongside the corresponding **Valuation Areas**, and we flag them as relevant for posting by checking the **Rel.** boxes as shown in Figure 3.45. There's an option to leave the **Valuation Area** field blank, indicating that the update type is applicable across all valuation areas. This configuration step ensures that the specific update types that should be posted to the general ledger are in fact specified to post to the general ledger.

Posting Relevant Update Types				
UpdateTy...	Update Type Description	Valuation Area	Name of Valuation Area	Rel.
MM_AE00+	Accrued Expense			✓
MM_AE00-	Accrued Expense			✓
MM_AI00+	Accrued Interest			✓
MM_AI00-	Accrued Interest			✓
MM_CH00+	Charges			✓
MM_CH00-	Charges			✓
MM_CM00+	Commission			✓
MM_CM00-	Commission			✓
MM_FC00+	Facility charges: Not utilized			✓
MM_FC00-	Facility charges: Not utilized			✓
MM_FC01+	Facility charges: Utilized			✓
MM_FC01-	Facility charges: Utilized			✓
MM_FC02+	Facility charges: Overdrawn			✓

Figure 3.45 Indicating Update Types Relevant to Posting

Tip

This configuration table is often overlooked, and many times, that's the reason why a contract does not post as expected. It's very important to configure this table with all update types that we expect to trigger an accounting posting because only the update types configured in this table will trigger a posting to the general ledger.

3.15 Condition Types

Now that we've determined which update types are relevant for posting, we can turn our attention to the configuration of condition types. In treasury and risk management, a condition type is required for any flow types that rely on a calculation to derive the amount of the cash flow. Any time a calculation occurs within a contract, a condition type must be created and assigned to the corresponding flow type. While some flows in contracts directly define the value or amount within the contract, several others require calculations before their cash flow values can be determined and posted to financial accounting. For instance, in a debt or investment contract, the user defines the initial principal is defined in the contract, and the contract doesn't require a calculation. In contrast, values such as interest and the final repayment of principal are not readily available in the contract, and the user must calculate them based on the contract's underlying values. These calculated values require the setup and configuration of a condition type.

In the following sections, we'll walk through defining and assigning condition types.

3.15.1 Defining Condition Types

We can find the configuration node for defining condition types by using the following menu path: **Financial Supply Chain Management • Treasury and Risk Management • Transaction Manager • Money Market • Transaction Management • Condition Types • Define Condition Types - MM Transactions**. Each condition type within treasury and risk management is identified by a unique four-character numeric code, complemented by a title describing the condition type. These attributes serve as the key elements for defining the characteristics of each condition type. As we can see in Figure 3.46, we've configured many of the most common condition types used during the setup of a money market contract.

CTyp	Name	Condition category
1010	Final repayment	Final Repayment
1120	Final repayment	Final Repayment
1130	Instalment repayment	Installment Repayment
1140	Annuity repayment	Annuity Repayment
1150	Interest capitalization	Interest Capitalization
1152	Interest capitalization (580)	Interest Capitalization
1200	Nominal interest	Nominal Interest
1203	Nominal interest for CP	Nominal Interest
1204	Facility charges: Not utilized	Amounts Equivalent to Interest
1205	Facility charges: Utilized	Amounts Equivalent to Interest
1206	Facility charges: Overdrawn	Amounts Equivalent to Interest
1207	Facility charges: Credit line	Amounts Equivalent to Interest
1210	Interest rate adjustment	Interest Rate Adjustment
1270	Accrued Interest Rate	Accrued Interest Rate
1901	Charges	Other Flow/Condition

Figure 3.46 Defining Money Market Condition Types

Within the configuration of the condition types, there are two important elements to pay attention to. Both the classification and the condition category act as pivotal drivers of how the condition type will behave.

When it comes to classification, there are four distinct categories available. These classifications serve as the guiding principles for organizing and categorizing condition types based on their intended purposes within the contract:

- **Structure Characteristics**
  We use these for conditions representing structural flows within the contract.
- **Accrual/Deferral**
  We use this for conditions related to accrual and deferral flows.
- **Valuation**
  We use this for conditions related to valuation cash flows.
- **Transfer**
  We use this for conditions related to the transfer of positions.

The **Condition category** is a subset of the classification, and it further defines the type of condition. Based on the classification that we choose within the condition type, the available options within the **Condition category** dynamically change to represent the condition categories that correspond to the classification. In the following, we've outlined the condition categories available for each classification:

- Structure Characteristics
    - 10: Principal Increase
    - 11: Principal Decrease
    - 12: Final Repayment
    - 13: Installment Repayment
    - 14: Annuity Repayment
    - 15: Interest Capitalization
    - 16: Capitalized Interest Payment
    - 20: Nominal Interest
    - 21: Interest Rate Adjustment
    - 24: Amounts Equivalent to Interest
    - 25: Payment Rate
    - 26: Price Index Adjustment
    - 27: Accrued Interest Rate
    - 29: Split Event
    - 60: Accumulating
    - 70: Inflow (generic, Fiduciary)
    - 71: Outflow (generic, Fiduciary)
    - 72: Withdrawal
    - 73: Extension
    - 90: Other Flow/Condition
- Accrual/Deferral
  We use these categories for conditions related to accrual and deferral flows:
    - 10: Accrual
    - 15: Reset of Accrual
    - 20: Deferral
    - 25: Reset of Deferral
    - 30: Accrual/Deferral for Diff.Process
- Valuation
  We use these categories for conditions related to valuation cash flows:
    - 10: Realized Forex Loss
    - 11: Realized Forex Gain

3 Core Configuration Elements

- 16: Realized Loss: Security
- 17: Realized Gain: Security
- 40: Forex Write-Up
- 41: Forex Write-Down
- 42: Security Write-Up
- 43: Security Write-Down
- 90: Loss in Transaction Currency
- 91: Gain in Transaction Currency

- **Transfer**
  We use this category for conditions related to the transfer of positions:
  - 5: Transfer in Transact. Currency

We can now set up an example condition type to use in our interest rate instrument. As we can see in Figure 3.47, when setting up a condition type for a final repayment flow, we choose **Structure characteristics** as the **Classification** and "12" (**Final Repayment**) as the **Cond.category**. Setting up this condition type facilitates the calculation and determines the value that should be derived when we post the final repayment flow type. Within the condition type, we then define the flow type that should be used as a result of this condition type. The **Generated FlTyp** field within the condition is populated with the flow type that should be used when this condition is used within a contract (in this example, **PF00 – Final Repayment**).

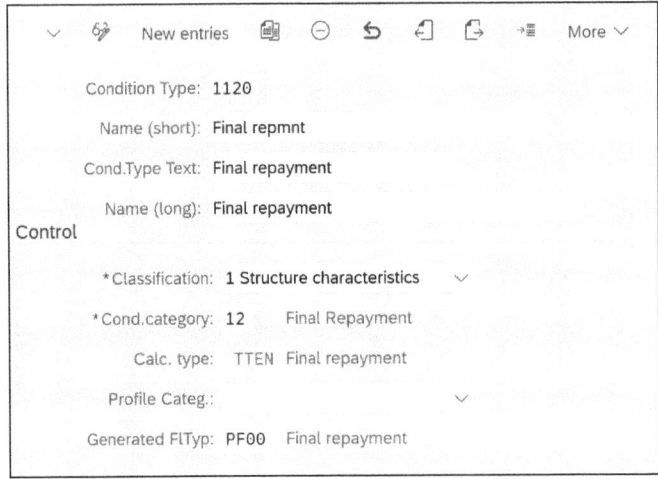

Figure 3.47 Defining Condition Types: Final Repayment

3.15.2 Assigning Condition Types to Transaction Type

Now that we've defined our various condition types, we have to assign them to the applicable product and transaction types. We have to determine which types of contracts the condition types should trigger because without this step, the condition types won't be used within the instruments. To start the configuration, we navigate to the **Financial Supply Chain Management • Treasury and Risk Management • Transaction Manager • Money Market • Transaction Management • Condition Types • Assign Condition Types to Transaction Types - MM Transactions** menu path.

Within this node, we'll map the applicable product and transaction types to the correct condition types. We first map our product type (**PTyp**) and transaction type (**TTyp**) combinations, and then, we assign them to the applicable condition types (**CTyp**) in the far-right column. As we can see in Figure 3.48, our product types have been mapped to the correct condition types.

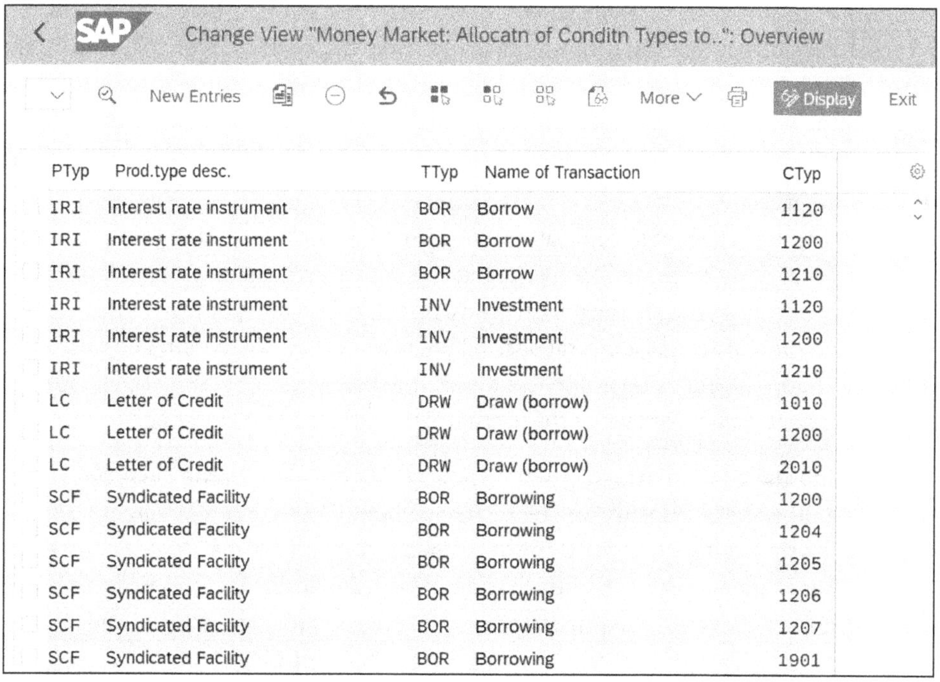

PTyp	Prod.type desc.	TTyp	Name of Transaction	CTyp
IRI	Interest rate instrument	BOR	Borrow	1120
IRI	Interest rate instrument	BOR	Borrow	1200
IRI	Interest rate instrument	BOR	Borrow	1210
IRI	Interest rate instrument	INV	Investment	1120
IRI	Interest rate instrument	INV	Investment	1200
IRI	Interest rate instrument	INV	Investment	1210
LC	Letter of Credit	DRW	Draw (borrow)	1010
LC	Letter of Credit	DRW	Draw (borrow)	1200
LC	Letter of Credit	DRW	Draw (borrow)	2010
SCF	Syndicated Facility	BOR	Borrowing	1200
SCF	Syndicated Facility	BOR	Borrowing	1204
SCF	Syndicated Facility	BOR	Borrowing	1205
SCF	Syndicated Facility	BOR	Borrowing	1206
SCF	Syndicated Facility	BOR	Borrowing	1207
SCF	Syndicated Facility	BOR	Borrowing	1901

Figure 3.48 Assigning Condition Types to Transaction Types

3.16 Accruals and Deferrals

In treasury and risk management, accruals and deferrals play a crucial role in ensuring accurate financial reporting and adherence to accounting standards. SAP S/4HANA provides functionalities to manage these processes efficiently, enabling organizations to maintain proper accounting practices and ensure proper recognition of the profit and loss impact of interest or fees.

Accruals and deferrals within accounting ensure that revenues and expenses are recognized in the appropriate accounting periods, matching them with the related revenues or expenses. *Accruals* are revenues or expenses that are recognized before the related cash transaction occurs, while *deferrals* are revenues or expenses that are recognized after the related cash transaction has occurred. These are very important concepts to understand as they are very common accounting practices for treasury departments when booking interest and fees for treasury contracts.

Treasury and risk management uses two primary methods of booking and calculating accruals and deferrals. The two options standardly available within SAP S/4HANA are as follows:

- **Book and reset**
  Book and reset involves the calculation and posting of the accrual or deferral at month end, followed by a subsequent posting the first day of the following month to reset the accrual posting that was made at month end. The following month, the accrual/deferral is posted again, and the entry then contains the accrual/deferral from the previous month as well as the current month. The entry is then reset or reversed the first day of the following month.

- **Incremental**
  The incremental accrual/deferral method involves the calculation and posting of the accrual or deferral at month end. Instead of resetting the accrual/deferral posting the following month, the posting stays on the books, and at the next month end, the incremental accrual/deferral posting takes place. The calculation and posting of the accrual/deferral value is done on a monthly basis and represents the incremental change in value from when the accrual/deferral was last completed.

We can find the configuration activities for accruals/deferrals in the **Financial Supply Chain Management • Treasury and Risk Management • Transaction Manager • General Settings • Accounting • Accrual/Deferral • Update Types • Define Update Types and Assign Usages** menu path. We must first define the applicable update types. In Figure 3.49 we've defined the four required update types needed for most accrual and deferral scenarios, using the same procedure we covered in Section 3.14.1:

- Accrued Expense MM_AE00+
- Accrued Expense MM_AE00-
- Accrued Interest MM_AI00+
- Accrued Interest MM_AI00-

There's nothing different about the update types for accrual and deferral when compared to other update types, except that the update types for accrual and deferral are noncash movement update types and thus don't require the assignment of a corresponding flow type.

3.16 Accruals and Deferrals

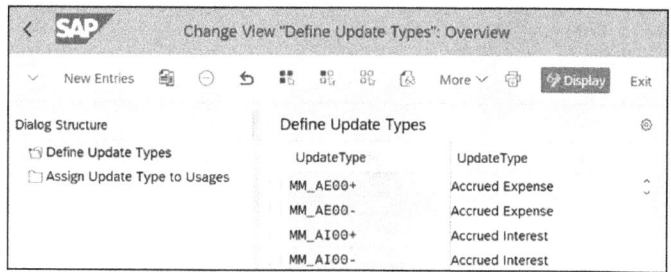

Figure 3.49 Defining Update Types

As with all update types, once we have created them, we must assign them to the correct usage. We must assign any accrual or deferral update types to the **Upd. Ty Usage** of **Accrual/Deferral**, as shown in Figure 3.50. This is an important step, as we must assign the accrual and deferral update types to the correct usage so we can further configure them for accrual and deferrals in the next step.

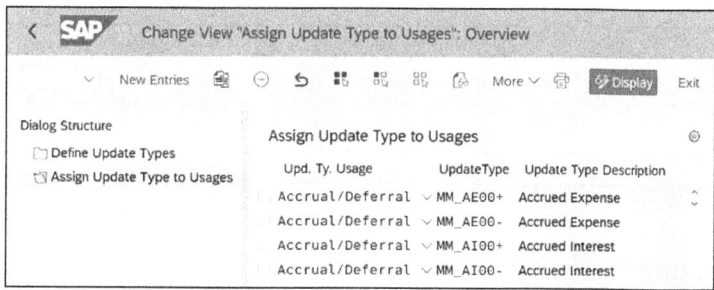

Figure 3.50 Assigning Update Types to Usages

Now that we've configured the necessary update types, we can move on to the next step, which involves utilizing the update types and assigning them to the applicable contracts. We can find the configuration activities for accruals/deferrals in the **Financial Supply Chain Management • Treasury and Risk Management • Transaction Manager • General Settings • Accounting • Accrual/Deferral • Update Types • Assign Update Type for Accrual/Deferral** menu path.

The first step is to establish a structure area and determine how granular we want to be with our accrual definition. At the most generic level, we can set up this step at the product group level, without a company code and product type. Alternatively,, we can set it up at a very granular level by accounting code and product type, which is what we've set up in our example in Figure 3.51. Once we've established the structure, we can further define the activities within the structure by checking the box next to the entry and clicking into the **Update Types** folder.

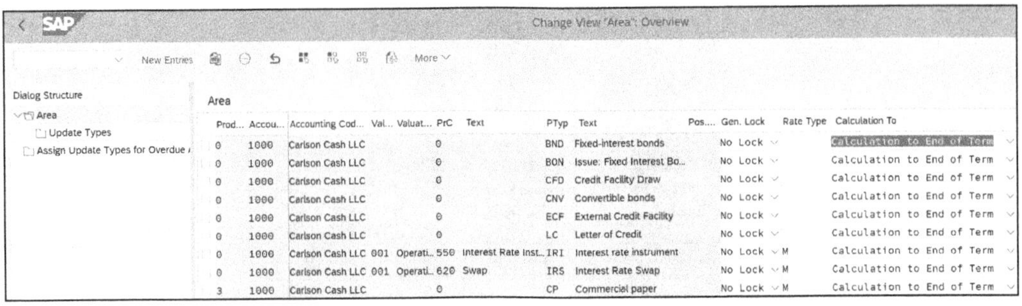

3 Core Configuration Elements

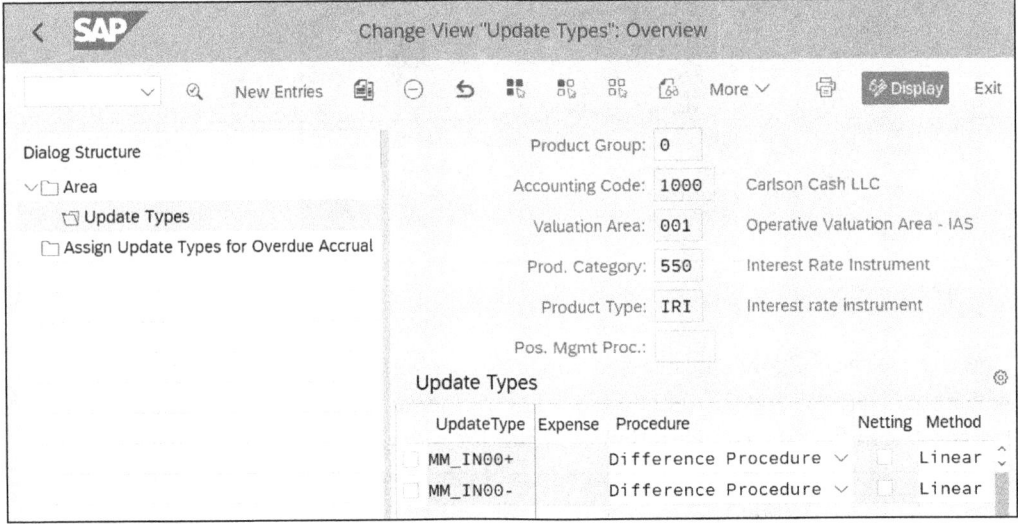

Figure 3.51 Configuring Accrual and Deferral

Now, we can further define the applicable update types for accrual and deferral. Within this step, we determine what update types are applicable to accrual and deferral, and we define the applicable update types that should trigger an accrual and deferral. As we can see in our example (see Figure 3.52), we've defined update types MM_IN00+ and MM_IN00– as our applicable update types.

Figure 3.52 Defining Interest Rate Instrument Accrual and Deferral

The most important element of this configuration node is determining the correct **Procedure**. The options within this node are the **Difference Procedure** and the **Reset Procedure**. Also, the accrual/deferral **Method** is important to consider as it will impact the amounts of the accruals and deferrals. The options available are as follows:

- **Linear**
  Most companies use the linear method most often. It simply takes the total amount divided by the number of days that the accrual is being run for.
- **Pro Rata Temporis**
  In this method, we determine the accrual or deferral amount by using a formula that

168

calculates the interest flow to be accrued or deferred. We can base this calculation on either a linear or an exponential approach.

- **Complete**
  In this method, the entire income amount is accrued or deferred, irrespective of the key date. This approach is necessary, for instance, to accrue or defer dividends in compliance with the US generally accepted accounting principles.

- **Pro Rata with Linear Discounting**
  This method is only used for commercial paper. The total amount is first accrued/deferred over the calculation period and then discounted on a linear basis.

The final step of fully setting up the accruals and deferrals is to assign the update types we should use for each of the update types that are being accrued or deferred. By clicking into each update type entry, we navigate to a new configuration screen where we can further define how the update type is accrued or deferred, as shown in Figure 3.53.

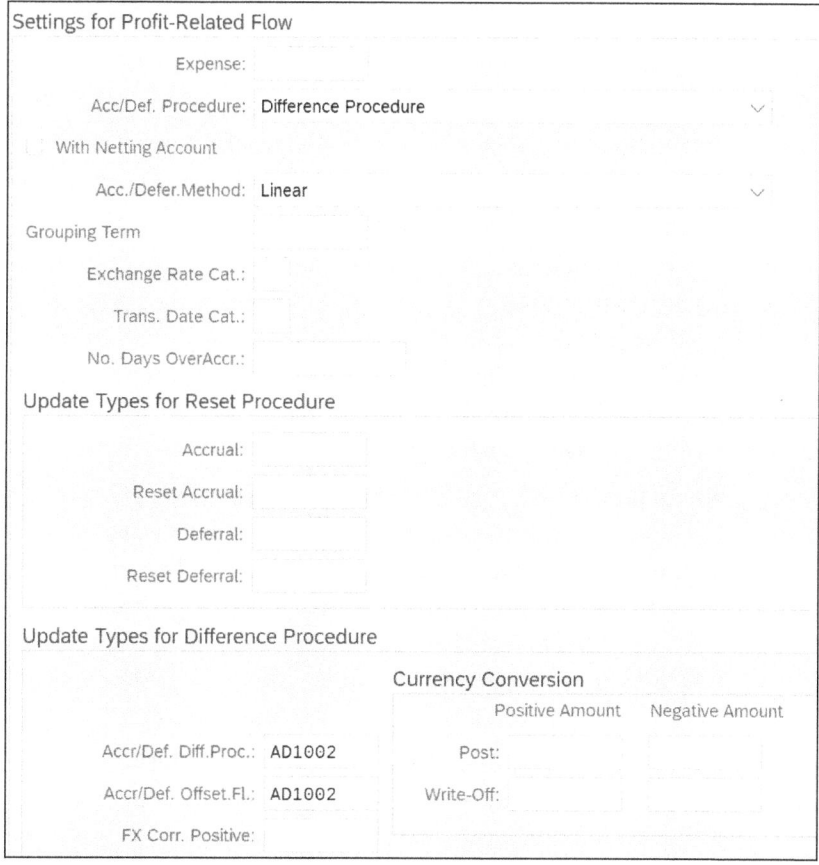

Figure 3.53 Accrual and Deferral Difference Procedure

Let's look at two sections in more detail:

- **Update Types for Reset Procedure**
  We use this section when the accrual or deferrals are using a book-and-reset method. We use the **Accrual** configuration field to define the update type we want to use during the accrual, while the **Reset Accrual** field is where we define the update type we want to use to reset the previously posted accrual. Conversely, the **Deferral** field is where we define the update type we want to use for a deferral posting, and the **Reset Deferral** field is where we define the update type we want to use as the reset of the deferral.

- **Update Types for Difference Procedure**
  When configuring this configuration node for a difference procedure accrual and deferral, we should define both the **Accr/Def. Diff. Proc** and **Accr/Def. Offset. FI** with the update type that we use to post the accrual or deferral. An offset update type is not required because the posting is not reset and an incremental posting will be made each month using the same update.

3.17 Portfolios

A *portfolio* in treasury and risk management is a configured tag that we can assign to a contract or a group of contracts. The portfolio field is not a required configuration element, but we configure and assign it to contracts for a few key reasons.

The most common use of a portfolio is to segregate or bucket treasury and risk management contracts for reporting purposes. We can also use it within a contract to further differentiate specific transactions, and the portfolio values will then be populated within the contract and any reports that are generated. We can also use it to drive various reporting functions. Once we've entered the portfolio field in a transaction, we can view all transactions within the same portfolio in all of the available standard reports.

One commonly used application of the portfolio feature is to categorize contracts based on geographical regions. Rather than reporting on a company code or a cluster of company codes, organizations create and allocate portfolios to all contracts within particular business units or regions. This approach allows users to generate reports encompassing all contracts, with the added benefit of segmenting or subtotaling them by region for enhanced visibility and analysis.

Another primary feature of portfolios is their capacity to facilitate alternative accounting treatments for specific contracts. The **Portfolio** field acts as a customizable parameter that users can leverage during contract creation, offering a dynamic means of differentiation. This functionality proves particularly useful in scenarios where distinct product types necessitate varying accounting procedures, thereby eliminating the need for additional product types. For example, consider a situation where a product and transaction type exhibit similar functionalities but differing accounting requirements. In these cases, we can establish multiple portfolios, each representing a unique accounting treatment option. These portfolios provide users with the flexibility to select the most suitable accounting approach when creating contracts.

The portfolio's use extends beyond categorization; it can serve as a determinant for the account assignment reference in each contract. By aligning the account assignment reference with the chosen portfolio, we can effectively direct the application of accounting treatments tailored to the selected portfolio during contract creation. We discuss this further when we cover the account assignment reference configuration in Section 3.22.2.

We can create a portfolio using the following menu path: **Financial Supply Chain Management • Treasury and Risk Management • Transaction Manager • Organization • Define Portfolio.** The initial step in portfolio creation involves selecting the company code for which the portfolio will be established. We establish portfolios at the company code level and implement them in every company code within the organization. It's also important to understand that portfolios span all product and transaction types within treasury and risk management. This means that any portfolio we create can be applied to any contract type, and there are no restrictions on its usage based on specific product and transaction type combinations. It's also important to highlight that we must define the portfolio within configuration before we can use it within a contract; it's not a freeform field like many of the other administration fields that we will cover in Section 3.18.

In the example in Figure 3.54, four distinct portfolios have been established based on regional criteria. The first step in creating a portfolio is to define the actual **Portfolio**, which is the value that will appear in the dropdown list when creating a contract. The next step is to define the **Portfolio Name**, which should be a descriptor that further defines the portfolio. Once configured, these portfolios become available options for selection and utilization during the contract creation phase.

Company Code:	1000
Treasury: Portfolio	
Portfolio	Portfolio Name
APAC	Asia Pacific
EU	Europe
LATAM	Latin America
NA	North America

Figure 3.54 Defining Portfolio

3.18 Administration Fields

A common requirement for many organizations implementing treasury and risk management is the ability to add data to contracts for identification and reporting purposes. In addition to the portfolio field detailed in the previous section, SAP S/4HANA provides three fields that we can used to store information when creating treasury contracts. These values are populated and stored on the contracts and can then be used for reporting purposes in all standard reports. The following fields are available for customization:

- Assignment
- Characteristics
- Internal Reference

The biggest difference between the additional fields and the **Portfolio** field is that similar to the portfolio field, we can configure the additional fields as values in the customizing, and they are available as dropdown list selections when we're creating contracts. In addition, we can use the fields as freeform fields, and we can populate them with values outside of the configured values. This means that besides selecting from preconfigured values, users have the freedom to input values other than the predefined options.

We can find the configuration for these fields in the **Financial Supply Chain Management • Treasury and Risk Management • Transaction Manager • Values for Fields on Administration Tab** menu path. To create a new value, our first step is to define the name that will appear in the dropdown list when we're creating a contract in the **Internal Reference** field. Our next step is to add a **Value Description** that further describes the value (see Figure 3.55).

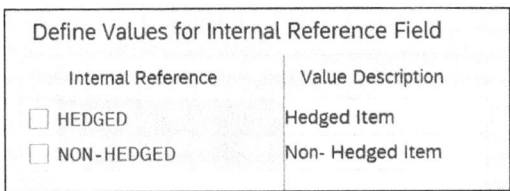

Figure 3.55 Internal Reference Field Configuration

3.19 Workflow

In addition to the standard process of limiting approvers by security, we can use a workflow to define the release procedure for financial transactions. We can set up the workflow to have as many as three release steps, and we can set it up for the specific situations that we want to trigger the workflow. For example, we could set it up so that the workflow can only be triggered during the creation of the contract, during changes to contracts, or during any create/change event for the contract. To facilitate this process, we use business object BUS2042 to trigger the workflow. There are a few customization nodes that we need to configure for this process to work, and we detail them in the following sections.

3.19.1 Defining the Release Procedure

The first step in setting up the workflow is located in the **Financial Supply Chain Management • Treasury and Risk Management • Transaction Manager • General Settings • Transaction Management • Release • Define Release Procedure** menu path.

In the **Release procedure** folder, we can identify which product types and transaction types initiate a workflow. This configuration is set up by company code, so when going into this configuration, we first identify which **Company Code** we are setting up for the workflow. In Figure 3.56, we determine which product types and transaction types are relevant to requiring a workflow release. In this example, product type IRI and transaction type BOR are set up to use the release procedure since the **Rel. Proc.** box is checked in the **Release procedure** section.

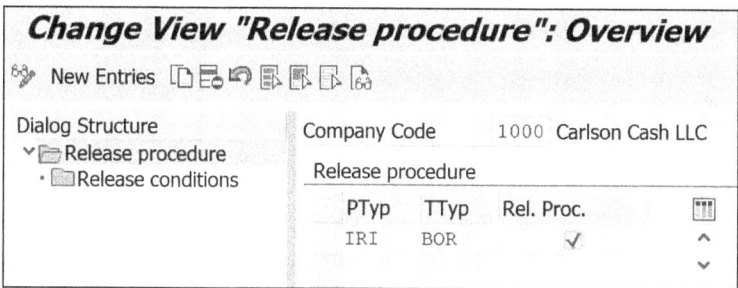

Figure 3.56 Workflow Release Procedure Configuration

For each combination of product type and transaction type, we highlight that row and double-click the **Release Conditions** folder to drill into the release conditions. In this area, there are attributes that we can use to drive the scenarios that kick off a workflow. The options in the release conditions are as follows:

- **Activity Category**
  The activity category describes the status of the contract, and we can set approvals depending on each activity category (e.g., **10: Contract**, **20: Contract Settlement**).

- **Activities**
  We can set the approval by activity type, depending on whether the contract is being created or changed. Options for this field are as follows:
  - 01: Add or Generate
  - 02: Change
  - 85: Reverse

- **Release Require**
  We check this box to require a workflow release.

- **Release Steps**
  In this field, we determine the number of release steps required. The number in this column determines the number of releases. For example, if we require one release

step, then one user will create or change the contract and a different user will need to execute the release.

- **Transaction Release**
  We use this field to determine the change protocol for transactions that are in the workflow and not fully approved. The options for this are as follows:

  – 1

  This option restricts any changes to a contract when it's in the workflow process.

  – 2

  With this option, changes are permitted when the transaction is in workflow, but the workflow status will be reset.

  – 3

  With this option, additional changes are possible when the transaction is in workflow, but the change will trigger an additional workflow to approve the new change.

  – 4

  With this option, ongoing transactions are processed and a change to the contract will result in the workflow resetting.

We add a line in this area to assign all the activities needed to initiate the workflow. The create and change activities were set for our IRI and BOR product type and transaction type to require a release for all create and change activities if the contract is in a contract or settlement status. This is shown in Figure 3.57.

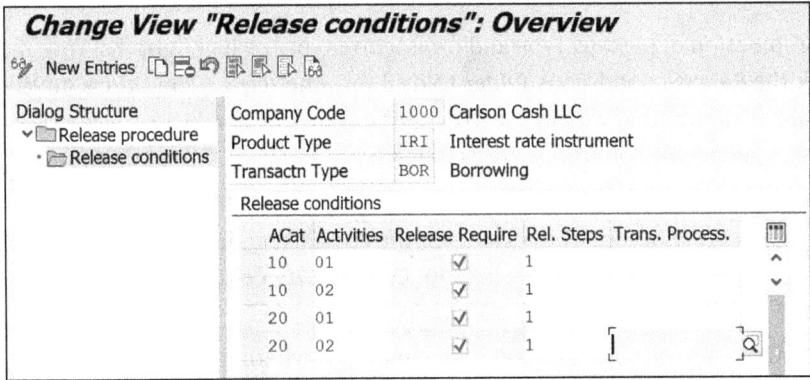

Figure 3.57 Workflow Release Conditions Configuration

3.19.2 Adjusting the Workflow Template

The second step for the workflow is in the **Financial Supply Chain Management • Treasury and Risk Management • Transaction Manager • General Settings • Transaction Management • Release • Adjust/Copy Workflow Template** menu path.

First, a popup will display when we go into this transaction. The first option is to first go to the **Define Standard Rule** activity. This setting will drive the assignment of the approvers to the workflow. The standard responsibility available is 20000034, but we can copy and adjust it to meet our requirements. Once in this area, we can click the create button to create a responsibility. An example of how to set up the responsibility for our IRI-BOR transactions is in Figure 3.58. We can define each attribute we want to use as a filter in the **Responsibility Specs**. In the figure, we determine this rule for **Company code 1000**, **Product type IRI**, and **Transaction type BOR**. In the **Release appr. levels** field, we determine how many approvals are needed for the workflow, and in the **Current appr. level** field, we determine which step we are defining for the approval. This allows us to create multiple responsibilities in this area and determine different approvers for each level of the approval.

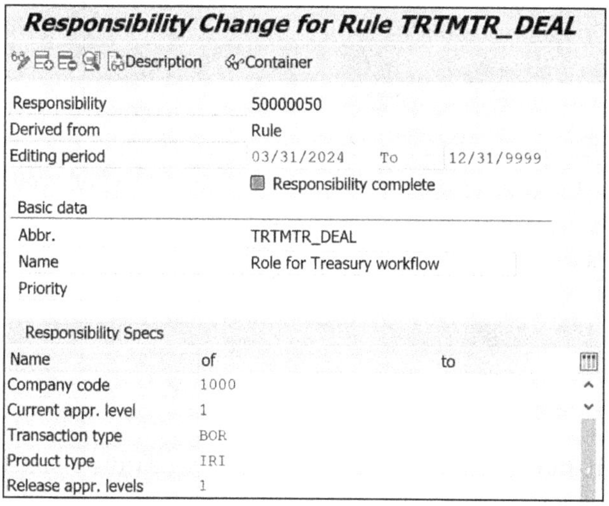

Figure 3.58 Assignment of Release Procedure

After we save this responsibility, we can assign users to the treasury workflow. To do this, we highlight the **Role for Treasury Workflow** line for the responsibility and click the **Insert Agent Assignment** button. This lets us select who we want to assign to the responsibility. After we complete the assignments, the screen will display the assigned users as in Figure 3.59.

Figure 3.59 Assigning Users to Treasury Workflow

The second step in this configuration is in **Transaction release: Adjust Workflow**. We navigate to this by going into the same configuration node and selecting the second option. We can use standard workflow templates 20000138 and 20000139, or we can edit them to meet our specific requirements. For the standard workflow templates and any copied templates, we need to make the event linkage active. We can do this in this transaction or in Transaction SWETYPV (Maintain Event Type Linkages). To activate it, we find the applicable line and check the **Linkage Activated** box. The active object for the two standard workflows is displayed Figure 3.60.

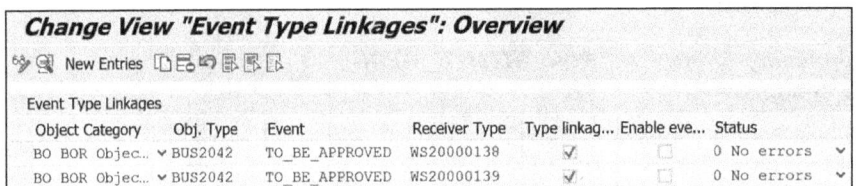

Figure 3.60 Reviewing Event Type Linkages for Treasury Workflow in Transaction SWETYPV

Once these steps are complete, a workflow item will be created for the contract creation or changes as defined in the release procedure configuration.

3.20 Account Symbols

Now that we've walked through the creation steps to configure a new treasury and risk management contract, we can move on to the accounting configuration in which we tie the contract to the debit and credit entries that will flow through to financial accounting. To begin the accounting setup, we navigate to the **Financial Supply Chain Management • Treasury and Risk Management • Transaction Manager • General Settings • Accounting • Link to Other Accounting Components • Define Account Determination for Treasury and Risk Management** menu path. Within this configuration node is a series of substeps for building out the accounting for the treasury contracts. We complete the substeps in order as follows:

1. **Definition of Account Symbols**
2. **Definition of Posting Specifications** (Section 3.21.1)
3. **Assignment of Update Types to Posting Specifications** (Section 3.21.2)
4. **Assignment of G/L Accounts to Account Symbols** (Section 3.22.3)

The first step in crafting the accounting framework for treasury and risk management involves defining account symbols, which we will eventually use to build the debit and credit within the posting of the document. However, in this initial step, we are creating account symbols that represent a posting to a general ledger account, and we must align these account symbols with the specific types of general ledger accounts that are intended for each transaction. It's important to recognize that account symbols and general ledger account mappings don't operate as a simplistic one-to-one mapping. Consequently, when creating account symbols, we should adopt a generic approach that reflects the type of posting rather than one that's narrowly tailored to individual general ledger accounts.

Consider a scenario where we have different interest expense accounts based on the type of contract. Instead of creating numerous interest expense account symbols, we can use a more streamlined approach that involves the creation of a single interest expense account symbol. We can then flexibly map this symbol to diverse general ledger accounts as dictated by the specific requirements of each contract type.

In our experience, it has proven beneficial to establish posting keys according to the nature of general ledger activities. As depicted in Figure 3.61, we've devised account symbols and corresponding text that reflect the intended general ledger account type for posting purposes. Employing a concise two-character naming convention for these posting symbols can significantly streamline the configuration process and enhance the clarity of the postings in the upcoming steps. The first step is to define an account symbol that easily distinguishes the type of activity, and we then further elaborate on the account symbol by assigning **Account Symbol Text** to describe the account symbol.

Definition of Account Symbols		
Acc...	Account Symbol Text	Posting Category
(AE)	Accrued Int Exp/ Deferred Inc	6 Other G/L Posting in Position Currency
(AI)	Accrued Int Income/Deferred Exp	6 Other G/L Posting in Position Currency
(BF)	Bank Fees / Misc Third Party Expenses	6 Other G/L Posting in Position Currency
(BK)	Bank	3 Bank Posting in Payment Currency
(DE)	Deferred Expense	6 Other G/L Posting in Position Currency
(DI)	Deferred Income	6 Other G/L Posting in Position Currency
(EN)	Euro Netting	6 Other G/L Posting in Position Currency
(GL)	Realized Gain / Loss	5 Profit-Related Posting in Payment Curren
(GR)	Gain - Receivable	6 Other G/L Posting in Position Currency
(IE)	Interest Expense	6 Other G/L Posting in Position Currency
(II)	Interest Income	6 Other G/L Posting in Position Currency
(IR)	Interest Receivable	6 Other G/L Posting in Position Currency
(LP)	Loss - Payable	6 Other G/L Posting in Position Currency
(MC)	Commission/charges	6 Other G/L Posting in Position Currency
(OE)	Other expense	6 Other G/L Posting in Position Currency
(PA)	Position Asset	6 Other G/L Posting in Position Currency

Figure 3.61 Definition of Account Symbols

Once we have defined the account symbol and account symbol text, we can map each account symbol to a posting category by choosing a posting category from the dropdown list. The posting category we use depends on the function of the account symbol and the type of document that will be posted as a result of the posting. The following posting categories are available options when assigning the account symbols:

3 Core Configuration Elements

- Position Posting (Book Value) in Position Currency
- Subledger Posting in Payment Currency
- Bank Posting in Payment Currency
- Profit-Related Posting in Position Currency
- Profit-Related Posting in Payment Currency
- Other G/L Posting in Position Currency
- Other G/L Posting in Payment Currency
- Currency Swap Between Position Currency/Payment Currency
- Tax Posting in Payment Currency
- Central Clearing Posting in Payment Currency

3.21 Posting Specifications

Now that we've defined the posting symbols, the next key step is to combine these symbols to build a posting specification. In the following sections, we'll first define posting specifications and then assign update types to them.

3.21.1 Defining Posting Specifications

A *posting specification* dictates both the debit and the credit entries that constitute the journal entry posting. Posting specifications are identified by up to five characters, and their details are further described in the specification text. We've found that a straightforward method of creating posting specifications involves concatenating the debit and credit account symbols to establish the posting specification name. Building posting specifications in this way is an easy way to quickly identify the debit and credit sides of the entry.

In Figure 3.62, the posting specifications have been created by a concatenation of accounting symbols, resulting in the creation of a four-character posting specification. Within this specification, the first two characters denote the debit side of the entry, while the last two characters signify the credit side of the entry. It's important to note that while this naming convention is commonly employed, it's not a mandatory requirement. Posting specifications can adopt any naming convention, but by sticking to a structured naming convention as delineated, we can effortlessly discern the intended postings associated with each specification. This naming convention approach enhances clarity and efficient organization of posting specifications.

3.21 Posting Specifications

Definition of Posting Specifications	
Post.s...	Specs text
AEBK	DR ACCRUE EXP (AE) / CR BANK (BK)
AEII	DR ACCRUE EXP (AE) / CR INT INC (II)
AIBK	DR ACCRUE INT (AI) / CR BANK (BK)
AIEN	CR ACCRUE INC (AI) / DR Euro Netting (EN)
AIII	DR ACCRUE INC (AI) / CR INT INC (II)
AIIR	DR ACCRUE INC (AI) / CR INT REC (IR)
AIPO	DR ACCRUE INC (AI) / CR POSITION (PO)
BFBK	DR BANK FEES (BF) / CR BANK (BK)
BKAI	DR BANK (BK) / CR ACCRUE INC (AI)
BKBF	DR BANK (BK) / CR BANK FEES (BF)
BKII	DR BANK (BK) / CR INTEREST INCOME (II)
BKIR	DR BANK (BK) / CR INT REC (IR)
BKPO	DR BANK (BK) / CR POSITION (PO)
ENAI	DR Euro Netting (EN) / CR ACCRUE INC (AI)
IEAE	DR INT EXP (IE) / CR ACCRUE EXP (AE)
IEAI	DR INT EXP (IE) / CR ACCRUE INC (AI)

Figure 3.62 Definition of Posting Specifications

Within the configuration, crafting posting specifications involves a two-step process. Initially, after assigning a name and description to the posting specification, we proceed to define it in detail to specify the account symbols required for both debit and credit postings. By delineating the document type to be utilized during document creation, we establish the foundation for how the general ledger posting will be constructed. Once we have populated the posting specification and the posting specification text and assigned it to the document type, we can begin to build out the specifics of the posting specification. The debit side and the credit side of the entries are configured independently, but the debit and credit configurations involve the same configuration elements. The configuration elements we use to define the posting specification are as follows:

- **Posting key**
  We use the posting key to describe the type of transaction that is being posted and whether or not it's a debit or credit posting. The most common posting key we use for a treasury debit posting is **40 – Debit Entry**, which we used when posting a debit to a general ledger account. When posting a credit entry, the most common posting key we use is **50 – Credit Entry**, which we use when posting a credit to a general ledger account. Although posting keys 40 and 50 are most commonly used, we can use any of the posting keys listed in Table 3.2 and Table 3.3.

- **Account Symbol**
  The account symbol field is populated with the account symbols that we created in the previous step.

- **Account Symbol Text**
  The account symbol text will automatically populate with the account symbol text that was assigned to the account symbol.

- **Posting Category**
  The posting category will also automatically populate with the posting category we defined when creating the account symbol.

- **Special G/L Indicator**
  If we're posting with the intention of also posting to a special general ledger indicator, we can define it here. This is an optional field, and we should only use it when posting to a special general ledger indicator.

Posting Key	Posting Key Name	Account Type
01	Invoice	D: Customer
02	Reverse credit memo	D: Customer
03	Expenses	D: Customer
04	Other receivables	D: Customer
05	Outgoing payment	D: Customer
06	Payment difference	D: Customer
07	Other clearing	D: Customer
08	Payment clearing	D: Customer
09	Interest Receivable	D: Customer
0A	CH Bill.doc. Deb	D: Customer
0B	CH Cancel.Cred.memoD	D: Customer
0C	CH Clearing Deb	D: Customer
0X	CH Clearing Cred	D: Customer
0Y	CH Credit memo Cred	D: Customer
0Z	CH Cancel.BillDocDeb	D: Customer
21	Credit memo	K: Vendor
22	Reverse invoice	K: Vendor
24	Other receivables	K: Vendor
25	Outgoing payment	K: Vendor

Table 3.2 Debit Posting Keys

Posting Key	Posting Key Name	Account Type
26	Payment difference	K: Vendor
27	Clearing	K: Vendor
28	Payment clearing	K: Vendor
29	Reverse b/e payable	K: Vendor
40	Debit entry	S: General Ledger
70	Debit asset	A: Asset
80	Stock initial entry	S: General Ledger
81	Costs	S: General Ledger
83	Price difference	S: General Ledger
84	Consumption	S: General Ledger
85	Change in stock	S: General Ledger
86	GR/IR debit	S: General Ledger
89	Stock inwrd movement	M: Materials

Table 3.2 Debit Posting Keys (Cont.)

Posting Key	Posting Key Name	Account Type
11	Credit memo	D: Customer
12	Reverse invoice	D: Customer
13	Reverse charges	D: Customer
14	Other payables	D: Customer
15	Incoming payment	D: Customer
16	Payment difference	D: Customer
17	Other clearing	D: Customer
18	Payment clearing	D: Customer
19	Reverse interest due	D: Customer
1A	CH Cancel.Bill.docDe	D: Customer
1B	CH Credit memo Deb	D: Customer

Table 3.3 Credit Posting Keys

3 Core Configuration Elements

Posting Key	Posting Key Name	Account Type
1C	CH Credit memo Deb	D: Customer
1X	CH Clearing Cred	D: Customer
1Y	CH Cancel.Cr.memo C	D: Customer
1Z	CH Bill.doc. Cred	D: Customer
31	Invoice	K: Vendor
32	Reverse credit memo	K: Vendor
34	Other payables	K: Vendor
35	Incoming payment	K: Vendor
36	Payment difference	K: Vendor
37	Other clearing	K: Vendor
38	Payment clearing	K: Vendor
39	Bill of exch.payable	K: Vendor
50	Credit entry	S: General Ledger
75	Credit asset	A: Asset
90	Stock initial entry	S: General Ledger
91	Costs	S: General Ledger
93	Price difference	S: General Ledger
94	Consumption	S: General Ledger
95	Change in stock	S: General Ledger
96	GR/IR credit	S: General Ledger
99	Stock outwd movement	M: Materials

Table 3.3 Credit Posting Keys (Cont.)

In Figure 3.63, we can see that the debit entry corresponds to the first two characters of the posting specification naming convention and that the last two characters correspond to the credit side of the entry. In this example, when we post the journal entry, a debit will post using **(IE) Interest Expense** and a credit will post using **(AE) Accrued Int. Exp/ Deferred Inc.**

3.21   Posting Specifications

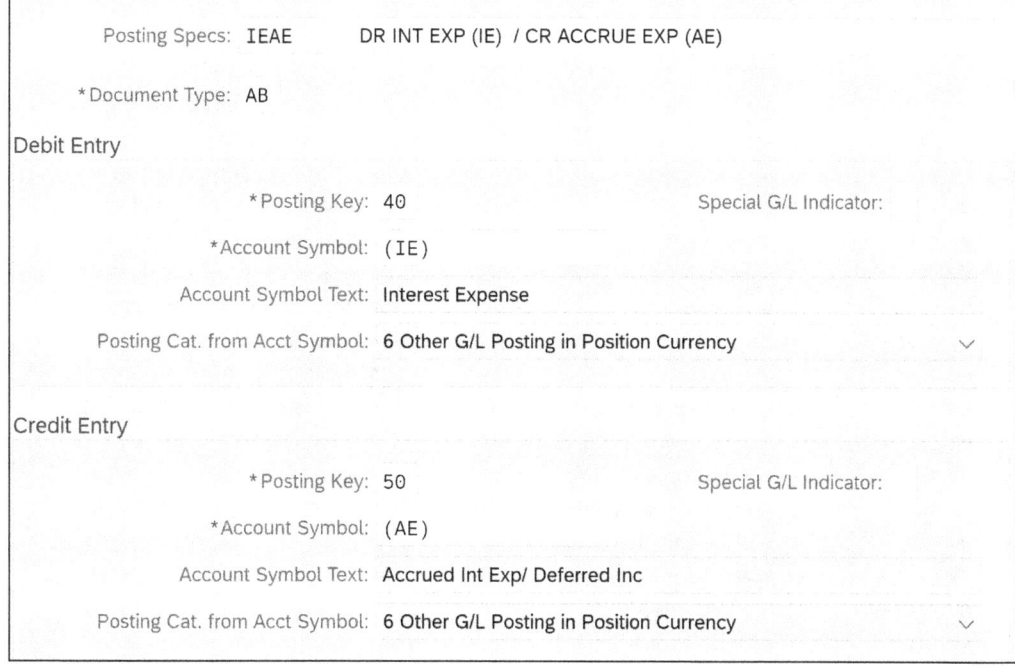

Figure 3.63 Defining Posting Specifications

3.21.2   Assigning Update Types to Posting Specifications

After we establish the posting specifications, our next step is to align the posting specifications with their respective update types in the following menu path: **Financial Supply Chain Management • Treasury and Risk Management • Transaction Manager • General Settings • Accounting • Link to Other Accounting Components • Define Account Determination for Treasury and Risk Management • Assignment of Update Types to Posting Specifications**.

The purpose of this configuration step is to assign each update type that will result in a financial posting to a set of posting specifications that will drive the accounting posting.

First, we define the update type that will be posted, and then, we assign the respective posting specification to that update by populating the four-character posting specification in the **Posting Specs** column. Figure 3.64 shows the mapping of **UpdateType** to its associated posting specification. For instance, update type **MM_AE00-** is linked to **IEAE**. Consequently, when this update type is executed, a debit entry will be posted to **Interest Expense** and a credit entry will be posted to **Accrued Expense**.

3 Core Configuration Elements

Assignment of Update Types to Posting Specs				
UpdateType	P	Update Type Description	Posting Specs	Posting Specifications Text
MM_AE00-		Accrued Expense	IEAE	DR INT EXP (IE) / CR ACCRUE EXP (AE)
MM_AI00+		Accrued Interest	AIII	DR ACCRUE INC (AI) / CR INT INC (II)
MM_AI00-		Accrued Interest	IEAI	DR INT EXP (IE) / CR ACCRUE INC (AI)
MM_CH00+		Charges	BFBK	DR BANK FEES (BF) / CR BANK (BK)
MM_CH00-		Charges	BKBF	DR BANK (BK) / CR BANK FEES (BF)
MM_CM00+		Commission	BFBK	DR BANK FEES (BF) / CR BANK (BK)
MM_CM00-		Commission	BKBF	DR BANK (BK) / CR BANK FEES (BF)
MM_IN00+		Interest	BKAI	DR BANK (BK) / CR ACCRUE INC (AI)
MM_IN00-		Interest	AEBK	DR ACCRUE EXP (AE) / CR BANK (BK)
MM_INCP+		Nominal Interest for CP	BKAI	DR BANK (BK) / CR ACCRUE INC (AI)
MM_INCP-		Nominal Interest for CP	AEBK	DR ACCRUE EXP (AE) / CR BANK (BK)
MM_PD00+		Principal Decrease	BKPO	DR BANK (BK) / CR POSITION (PO)
MM_PD00-		Principal Decrease	POBK	DR POSITION (PO) / CR BANK (BK)
MM_PF00+		Final Repayment	BKPO	DR BANK (BK) / CR POSITION (PO)
MM_PF00-		Final Repayment	POBK	DR POSITION (PO) / CR BANK (BK)
MM_PI00+		Principal Increase	BKPO	DR BANK (BK) / CR POSITION (PO)
MM_PI00-		Principal Increase	POBK	DR POSITION (PO) / CR BANK (BK)

Figure 3.64 Assignment of Update Types to Posting Specs Configuration Screen

3.22 Account Assignment Reference

As you're reading through the accounting section, you might be wondering how the various posting specs tie to a specific general ledger account. This is where the account assignment reference comes into play. Each contract we create within treasury and risk management is assigned an account assignment reference, which serves as the cornerstone of the contract's accounting postings and plays a pivotal role in determining the accounting treatment specific to that contract. The ability to handle diverse posting requirements across contract types, entities, and regions is all made possible through the creation of distinct account assignment references.

In this section, we'll detail the steps required to create unique account assignment references for each of our contract types. We'll then address how we can create rules to derive the account assignment references from our different contract types based on key elements within the contract, using derivation rules. We'll then cover how to assign the account assignment references to the derivation rules, and we'll cover the checks that we can configure to ensure that the correct account assignment reference is assigned to each contract.

3.22.1 Defining Account Assignment References

We can craft the various account assignment references in a number of ways, based on the unique requirements of the project. Sometimes, we'll find that it's sufficient to create an account assignment reference per product type, while other times, to achieve the correct accounting treatment, we must create the account assignment reference at a much more granular level. How granular the account assignment reference must depend on the complexity of the accounting posting requirements. For instance, if all contracts of a specific type post the exact same way regardless of the company code or region, there's no need to break out the account assignment references at a more granular level, and creating an account assignment reference specific to each product type is sufficient. At the opposite end of the spectrum, if the accounting postings differ based on the company code, the region, or a specific element within the contract, we'll have to create unique account assignment references to drive the specific postings for each of the different posting scenarios.

When defining the account assignment reference, we're only defining the **Acct Assignm** and the **Name of Account Assignment Reference**. The value we use for the **Acct Assignm** reference is the value we'll see populated on the instrument. We can find the configuration for the account assignment reference in the following menu path: **Financial Supply Chain Management • Treasury and Risk Management • Transaction Manager • General Settings • Accounting • Link to Other Accounting Components • Define Account Assignment References**. As shown in Figure 3.65, we've created account assignment references that link one-to-one with our various product types.

Account Assignment Reference	
Acct Assignm	Name of Account Assignment Reference
BOND ISSUE	Bond Issue
BONDS	Bonds
CFD	Credit Facility Draw
COMMODITIES	Commodities
CP_PUR	Commercial Paper Purchase
CP_SEL	Commercial Paper Sell
ECF	External Credit Facility
FX	Foriegn Exchange
FX_CFH	FX - Cash Flow Hedging
IRI	Interest Rate Instrument
IRI_ST	Interest Rate Instrument
IRS	Interest Rate Swaps
LOC	Letter of Credit

Figure 3.65 Creating Account Assignment References

3.22.2 Determining Account Assignment References

Now that we've created a variety of account assignment reference values, our next task is to decide how to apply each of these references to specific contracts. This involves deriving the intended account assignment reference based on the contract elements and placing the account assignment reference within the contract. This determination process consists of two steps. First, we establish an account assignment reference derivation rule, and then, we utilize this rule, assigning the appropriate criteria within it to ensure the correct assignment of the account assignment reference for each contract. Using this approach, we ensure accurate alignment between contract elements and their corresponding account assignment references.

We can find the configuration for the account assignment reference determination in the following menu path: **Financial Supply Chain Management • Treasury and Risk Management • Transaction Manager • General Settings • Accounting • Link to Other Accounting Components • Define Account Assignment Reference Determination**. The account assignment reference configuration is a two-part process. First, we create the various derivation procedures used to determine the correct account assignment reference, and then, we use the derivation procedure and populate the account assignment reference values that should be used during the derivation procedure.

Creating the Account Assignment Reference Derivation Rule

The process of assigning the correct account assignment reference entails two primary steps within the configuration: we establish the derivation rule and then utilize it. This derivation rule serves as a set of guidelines and criteria configured to determine a specific account assignment reference based on the stored values within the treasury contract. The primary objective of this configuration step is to ensure the accurate assignment of the account assignment reference to the transaction upon its creation, subsequently dictating the appropriate accounting postings.

To achieve this, we must understand the variances in accounting postings among different instruments. Often, the accounting requirements remain largely consistent among product and transaction types within the same company code. In such instances, the product type and transaction type serve as the primary parameters driving the account assignment reference determination. However, in more intricate scenarios where accounting requirements vary based on factors like company code, country, region, or business partner, we need to use a derivation rule to accommodate these diverse needs.

First, we click on the **Create** button to create a new account assignment reference derivation rule, and then, we can choose to determine what type of step we're planning to create. The most common practice is to use a **Derivation rule** (as shown in Figure 3.66), which is a set of defined parameters that we can use to populate the intended account assignment reference. It's also the most dynamic way to assign the account assignment reference.

3.22 Account Assignment Reference

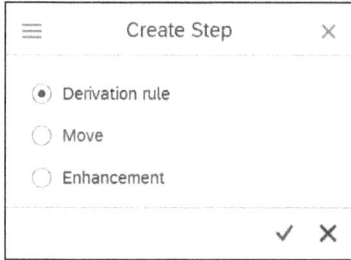

Figure 3.66 Create Step for Account Assignment Reference Determination

To start the process, we outline the different source fields we intend to use within the derivation rule. These source fields encompass the set of fields to be utilized within the rule, and they serve as the key fields we query when executing the derivation function. The first step is to define the source fields that we'll use to derive the account assignment reference; in this example, we're deriving the account assignment reference by leveraging the combination of product type and transaction type (see Figure 3.67). We then determine what value we want to map by using our defined source fields. The value that is being mapped is the target field, and the target field in this example is the account assignment reference field. In the **Target Fields** section, we define the **Origin** as "COMMON DATA," and we define the target field **Name** as "AA_REF," which is the account assignment reference field. When we're using a derivation rule to populate an account assignment reference, this value will always be "AA_REF."

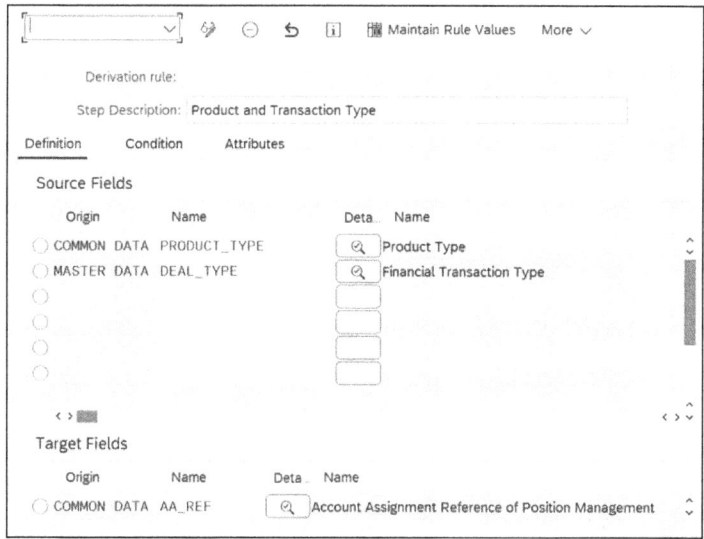

Figure 3.67 Creating Derivative Rule

Conditions serve as additional tools for refining the derivation rule. Although they are not mandatory in configuration, conditions can enhance the specificity of a derivation rule. We establish conditions to provide further differentiation within the rule, and it's important to note that we must meet all conditions listed within the condition set or the derivation rule will not be applied. To create conditions, we click the **Condition** tab (shown in Figure 3.68) and then define **Origin** and **Name** of the field that we want to use as the condition. The next step is to define the field **Value** that should apply to the condition. The derivation rule will only apply when the condition is met; in this example, the derivation rule will only apply to contracts using business partner JP Morgan.

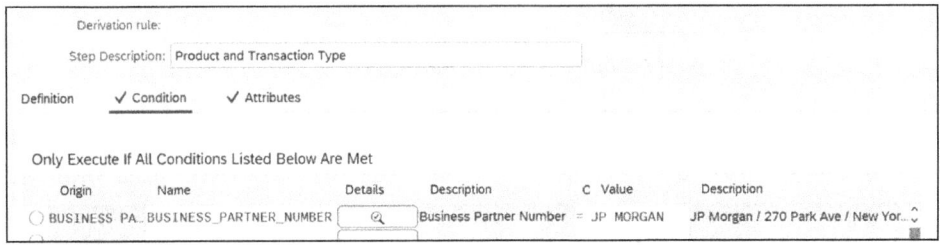

Figure 3.68 Creating Conditions

We can also add attributes to the derivation rule on the **Attributes** tab, as shown in Figure 3.69. These are not required configuration elements, but we can use them to add controls around the derivation process. The two main checkboxes we can use to do this are as follows:

- **Issue error message if no value found**
  If we check this box, an error message will appear if an account assignment reference was not derived upon creation of the contract.

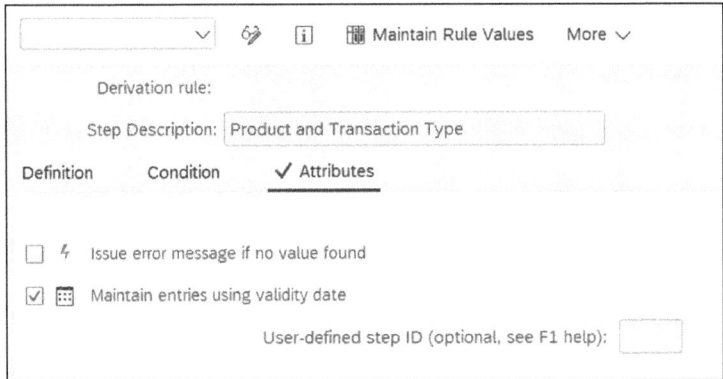

Figure 3.69 Derivation Rule Attributes

- **Maintain entries using validity date**
  If we check this box, an additional validity date will appear when we're maintaining the rule values. This can be useful when we're maintaining a value that's only applicable for a specific time period, as shown in in Figure 3.70.

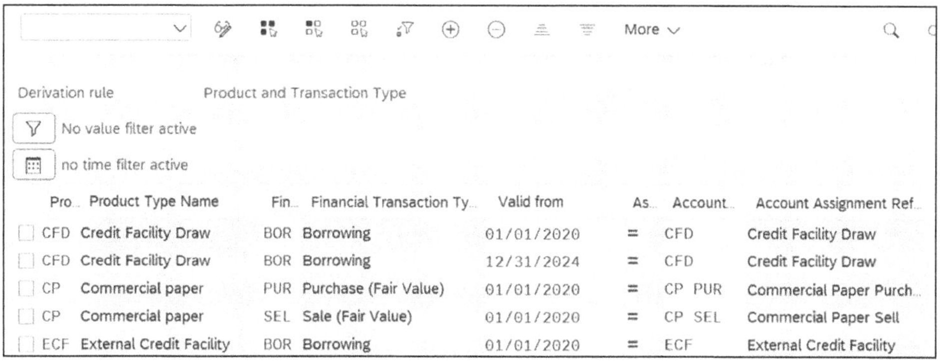

3.22 Account Assignment Reference

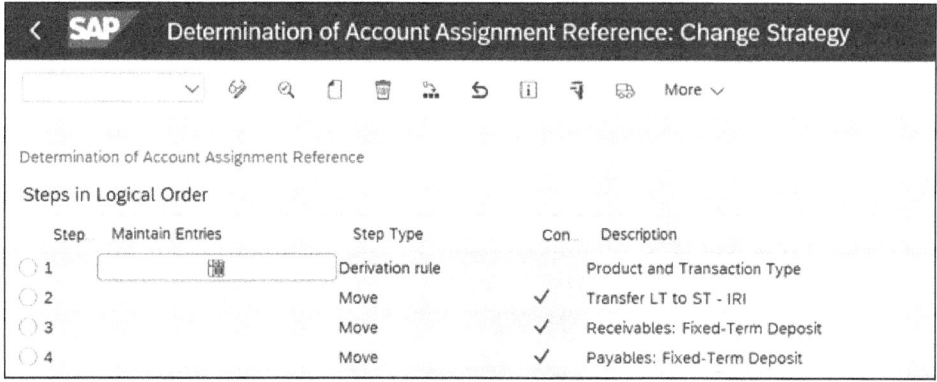

Figure 3.70 Maintaining Entries Using Validity Dates

Once we've established the derivation rule, it will appear as shown in Figure 3.71. These derivation rules are sequentially processed as transactions are generated, so we must consider the granularity of the rules we create. When a transaction is initiated within treasury and risk management, it undergoes processing through the derivation rule designated for the initial step. If an account assignment reference is not determined at this stage, then the transaction proceeds to the subsequent step, and so forth. If an account assignment reference is assigned by a derivation rule, then the processing stops, preventing further progression through the remaining derivation rules. The rationale behind this sequence is that it ensures that more detailed derivation rules are prioritized over less detailed ones, as we can see in Figure 3.71.

Figure 3.71 Determination of Account Assignment Reference

Account Assignment Reference Derivation Rule: Maintain Values

Now that we've defined the account assignment reference derivation rules, the next step in the process is to maintain the values within the derivation rule. First, we must specify the product type and transaction types that should use this derivation rule, and then, we can assign the account assignment reference to the product and transaction type combination by populating the account assignment reference in the **Account**

Assignment Reference column. As shown in Figure 3.72, **Product Type** and **Financial Transaction Type** are used to derive the account assignment reference for each product and transaction type combination.

> **Note**
>
> We must either manually maintain the account assignment reference table in each SAP S/4HANA environment or save the entries to a workbench transport and move it through the production path.

Product Type	Product Type Name	Financial Tran...	Financial Transaction Ty...	Assigned	Account Assignm...	Account Assignment Reference of ...
CFD	Credit Facility Draw	BOR	Borrowing	=	CFD	Credit Facility Draw
CP	Commercial paper	PUR	Purchase (Fair Value)	=	CP_PUR	Commercial Paper Purchase
CP	Commercial paper	SEL	Sale (Fair Value)	=	CP_SEL	Commercial Paper Sell
ECF	External Credit Facility	BOR	Borrowing	=	ECF	External Credit Facility
FX	Foreign exchange (FX)	FWD	Forward Transaction	=	FX_CFH	FX - Cash Flow Hedging
FX	Foreign exchange (FX)	SPT	Spot Transaction	=	FX	Foriegn Exchange
IRI	Interest rate instrument	BOR	Borrowing	=	IRI	Interest Rate Instrument
IRI	Interest rate instrument	INV	Investment	=	IRI	Interest Rate Instrument
IRS	Interest Rate Swap	PAY	Pay Fixed	=	IRS	Interest Rate Swaps
IRS	Interest Rate Swap	REC	Recieve Fixed	=	IRS	Interest Rate Swaps
LC	Letter of Credit	DRW	Draw (borrow)	=	LOC	Letter of Credit

Figure 3.72 Defining Account Assignment Reference Determination

As highlighted in Section 3.17, an alternative approach to deriving the account assignment reference involves leveraging the combination of the product type, the transaction type, and the portfolio field. Given that a user can manually define a portfolio field in the contract by in real time, this approach provides a convenient way to establish varied accounting treatments based on user-defined input, eliminating the need for the user to create additional product and transaction types. In Figure 3.73, a similar derivation rule has been formulated; it incorporates the portfolio field as a source value. The derivation values are now maintained through a combination of the product type, transaction type, and portfolio field, and the advantage of this approach is that we can drive the accounting postings dynamically by a value in the contract that is driven by the users as opposed to in the background.

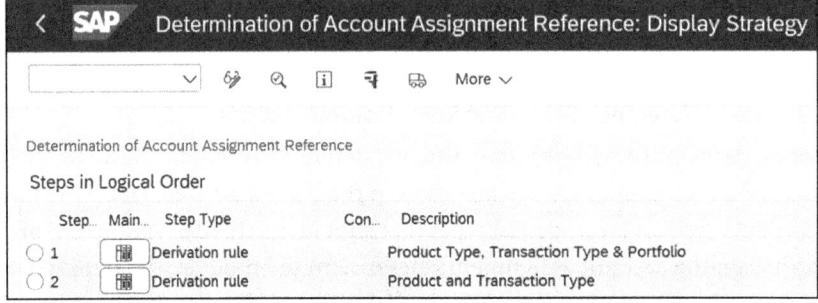

Figure 3.73 Derivation Rule: Product Type, Transaction Type, and Portfolio

Figure 3.74 demonstrates how to set up the derivation rule using the portfolio as a key value in the determination. We can do this by selecting the **PRODUCT_TYPE** and **DEAL_TYPE** origin fields and then, in the third line of the derivation rule definition, pulling **PORTFOLIO** into the source fields we use for the derivation. The system will then take into account the product type, transaction type, and portfolio when deriving the account assignment reference.

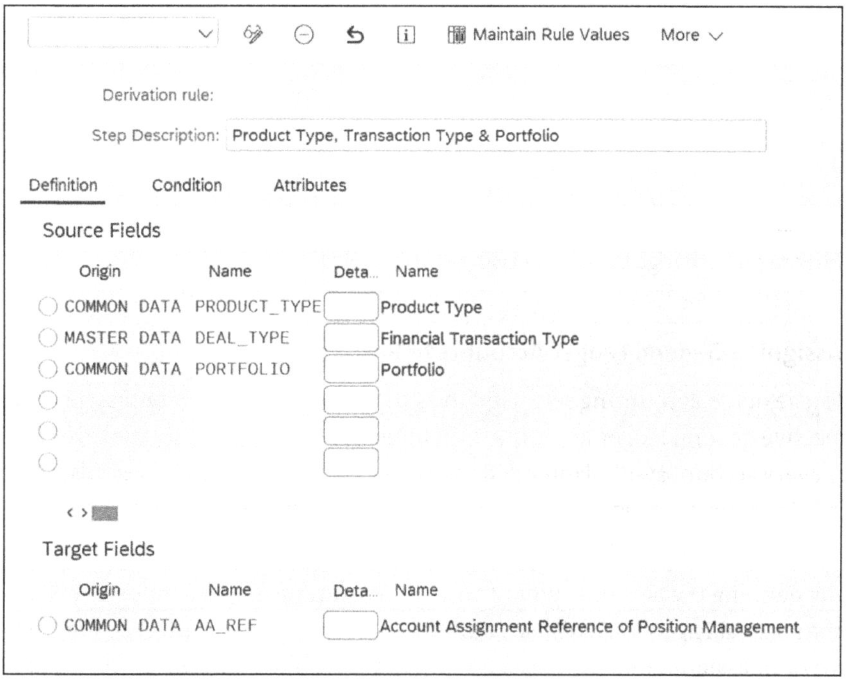

Figure 3.74 Product Type, Transaction Type, and Portfolio Derivation Rule

Figure 3.75 shows the fully configured table, which has been set up using a combination of product type, transaction type, and portfolio. This configuration is a critical step in ensuring that the system can accurately derive the appropriate account assignment reference based on the values the user populates in the portfolio. By leveraging this configuration, the system can dynamically determine the account assignment reference based on user input, which is a very common requirement for most organizations.

Furthermore, the configuration allows us to implement region-specific accounting treatments. The portfolio's regional definition plays a role in this process, allowing the system to apply different accounting treatments to FX contracts based on the region associated with the portfolio. This capability is particularly valuable in a global business environment, where accounting standards and practices may vary among different regions. By configuring the system in this manner, we ensure that the FX contracts are processed in accordance with applicable regional accounting requirements. This is just one example of the flexibility of the account assignment reference determination, and it underscores our tailored approach when managing complex accounting requirements.

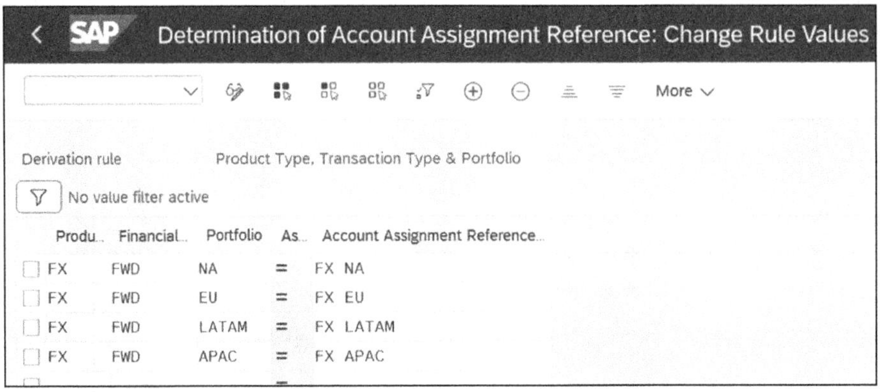

Figure 3.75 Foreign Exchange Derivation Using Portfolio Field

3.22.3 Assigning General Ledger Accounts to Posting Keys

The final stage of the accounting setup involves the assignment of account symbols to their respective general ledger accounts designated for posting. In this step, we must configure every account symbol intended for use during the posting process. It's essential to establish a direct association between each account symbol and its corresponding account assignment reference with the targeted general ledger account.

We start by defining the account symbol and the account assignment reference. Then, we map the applicable general ledger account that we want to use for that account symbol and account assignment reference combination. As shown in Figure 3.76, **Account Symbol (AE) Accrued Int Exp/ Deferred Inc** is an example of such a linkage, and it extends to three distinct general ledger accounts based on varying account assignment references. It's important to note that the configuration within this node can be influenced by specific valuation areas or currency parameters; this adds further depth and customization to the accounting setup process.

To streamline the entries within this configuration table, we are not required to create an entry for every account assignment reference when a general ledger account is consistently linked to an account symbol. Instead, we can directly map account symbols to a consistent general ledger account, independent of the account assignment reference. To do this, we define the account symbol but leave the **AcAsRef** field blank, and then, we map a general ledger account into the **G/L Acct** column. In other words, the configuration entry will include only the account symbol and its corresponding general ledger account. This ensures that all postings using that account symbol will apply the same general ledger account across the board, regardless of the specific account assignment reference.

Assignment of G/L Accounts to Account symbols							
Account Symbol	Account Symbol Text	VA	Valuation Area	AcAsRef	Acct Assignmnt Ref.	Currncy	G/L Acct	Short Text
(AE)	Accrued Int Exp/ Deferred Inc			BOND ISSUE	Bond Issue		221010	ACCRUED INT EXP
(AE)	Accrued Int Exp/ Deferred Inc			CFD	Credit Facility Draw		221010	ACCRUED INT EXP
(AE)	Accrued Int Exp/ Deferred Inc			CP SEL	Commercial Paper Sell		221010	ACCRUED INT EXP
(AE)	Accrued Int Exp/ Deferred Inc			IRS	Interest Rate Swaps		221010	ACCRUED INT EXP
(AE)	Accrued Int Exp/ Deferred Inc			LOC	Letter of Credit		221010	ACCRUED INT EXP
(AI)	Accrued Int Income/Deferred Exp			50500	Issues		113110	Bank1 (forgn crcy A)
(AI)	Accrued Int Income/Deferred Exp			BOND ISSUE	Bond Issue		221005	ACCRUED INT INC

Figure 3.76 Assignment of General Ledger Accounts to Account Symbols

> **Note**
> There are many schools of thought on how to define the account assignment reference structure and assign the appropriate general ledger accounts. The approach we use in this table, where each product type has its own account assignment reference, makes for very easy maintenance if a general ledger account changes at some point in the future. We can also swap out the general ledger account, knowing that doing so will only impact the specific product type tied to that account assignment reference. The drawback of this approach is that this table can become very long, with redundant entries that would not be required if we were using a broader account assignment reference approach.

3.22.4 Allocating Additional Account Assignments to Account Assignment References

Now that we've explored the steps required to post a treasury contract to the general ledger, let's dive into the option of incorporating supplementary account assignments into treasury postings. When posting a contract, we might need to allocate it to additional accounts without altering the primary financial settings within SAP S/4HANA. In treasury and risk management, we can modify these account assignments within the treasury accounting configuration. We can adjust parameters such as **Business Area**, **Cost Center**, and **Functional Area** according to the assigned account assignment reference for each specific contract. Irrespective of the treasury function or the program executing the posting within treasury and risk management, we will utilize and apply this configuration during the posting process.

To configure this functionality, we navigate to the **Financial Supply Chain Management • Treasury and Risk Management • Transaction Manager • General Settings • Accounting • Link to Other Accounting Components • Allocate Additional Account Assignments to Account Assignment References** menu path. We then click **New Entries** and enter the **Company Code** and **Account Assignment Ref.** that the additional account assignments pertain to. We can then maintain the corresponding **Business Area**, **Cost Center**, and **Functional Area** that we want to use for the posting. As with the previous accounting setup steps, the account assignment reference assigned to the contract is the key determining

3 Core Configuration Elements

factor we use when configuring this configuration node. As shown in Figure 3.77, we use a **Company Code** and **Account Assignment Ref.** combination to define the additional account assignments that we will use when we carry out the posting.

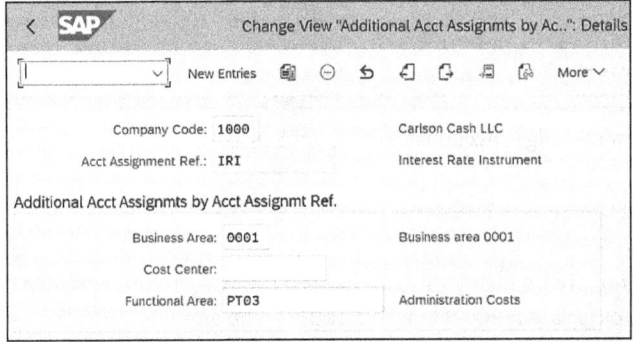

Figure 3.77 Allocating Additional Account Assignments by Account

It's worth mentioning that not every entry requires values in the **Business Area**, **Cost Center**, and **Functional Area** fields; the configuration is adaptable and mandates populating only the necessary values for executing the posting. As depicted in Figure 3.78, the **Account Assignment Reference** field is optional, which offers flexibility in setup. This configuration proves useful when we must post all account assignment references for a specific company similarly; it lets us make fewer redundant entries into the table by maintaining a single entry.

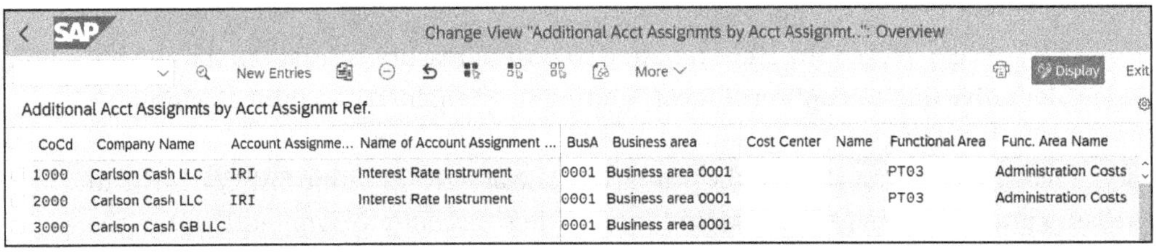

Figure 3.78 Additional Account Assignments for Interest Rate Instruments

3.23 Position Transfers

During the lifecycle of a treasury contract, there might arise a need to initiate transfers or adjustments to accounting positions midway through the contract's duration. Such instances often require the relocation of positions from one general ledger account to another, particularly for contracts that have already undergone posting. Alternatively, there might be scenarios where there's a requirement to designate a new general ledger account for every subsequent posting that occurs in the future. Many organizations commonly encounter a business requirement to transition accounting positions from a long-term to a short-term general ledger account when a contract approaches maturity within twelve months, signifying its reclassification as a short-term debt or investment. The transfer categories are as follows:

- **Post to Same Components**
  In this transfer category, the elements of the source position are directly moved to the target position, regardless of whether the target position typically utilizes these elements. This behavior remains consistent even in cases where a transfer category hasn't been specified within the system.

- **Only Post to Used Components**
  In this transfer category, when the source position contains values for components not handled by the target position, those specific components from the source position are transferred to the corresponding existing components of the target position.

- **Reverse Security and Foreign Exchange Valuations**
  In this transfer category, the valuations related to securities and foreign exchange aren't directly moved to the target position; instead, they are reset.

- **Reverse Security Valuation, Transfer Foreign Exchange Valuations**
  In this transfer category, the foreign currency valuation is shifted to the target position, while the security valuation undergoes a reset.

- **Transfer Security Valuation, Reverse Foreign Exchange Valuation**
  In this transfer category, the valuation of securities is moved, while the valuation of foreign currency is reset.

- **Transfer Unrealized Valuations**
  This transfer category aligns with the **Only Post to Used Components** category, with one exception: the valuations of securities and FX, which don't impact the profit and loss, are consistently transferred and cleared throughout the remaining term.

- **Post Book Value to Purchase Value Component**
  In this transfer category, the individual components remain untouched, and instead, the entire book value of the source position is recorded in the target position as the purchase value.

The ultimate goal of a position transfer within treasury and risk management is to change the defining element driving the account position of the contract. By changing this, we can drive the future general ledger position and facilitate the desired future accounting schematic. SAP offers a few ways to fulfill this accounting requirement, through the modification of three different standard position attributes, including valuation class, portfolio, and account assignment reference.

We'll start by taking a look at the account assignment reference transfer. This functionality is one of the most commonly used methods of transferring an accounting position with treasury and risk management. As we discussed in previous sections, we assign the account assignment reference to every contract, and it serves as the backbone for how the accounting is driven. The account assignment reference transfer function allows us to change the account assignment reference originally assigned to a contract and move it to a new account assignment reference midway through the life of the contract. By changing the account assignment reference, we can transfer positions with their book value from one general ledger account to a new general ledger account.

To facilitate the account assignment reference transfer process, we must perform some additional setup within configuration. When the account assignment reference transfer is carried out, a two-step procedure takes place. In the first step, the system clears the initial posting in the general ledger account and moves the opposite side to a clearing account. In the second step, the system clears the clearing account that the balance resided in and shifts the other side of the posting to a new general ledger account specified by the new account assignment reference. This effectively moves the position from one general ledger account to another. It's worth noting that this feature doesn't undo previous postings; instead, it transfers the initial postings to different general ledger account specified by the new account assignment reference.

The benefit of an account assignment reference transfer is that we can assign different accounting treatments to a contract midway through the life of the contract. This gives us enhanced visibility by allowing us to track changes to accounting within the actual trade, as opposed to making manual journal entries.

We'll now detail the steps required to set up an account assignment reference transfer. To start the configuration process, we go to the **Financial Supply Chain Management • Treasury and Risk Management • Transaction Manager • General Settings • Accounting • Transfer Account Assignment Reference** menu path and then proceed through the following steps in order:

1. We define an update type that's required to transfer the position out of the original position. We create and use these update types for the sole purpose of clearing out the old position within the contract, as with any other update type that we've previously created. We can see the new update type that clears out the old position in Figure 3.79. In our example, we've created two update types: one for the debt contracts and the other for investments.

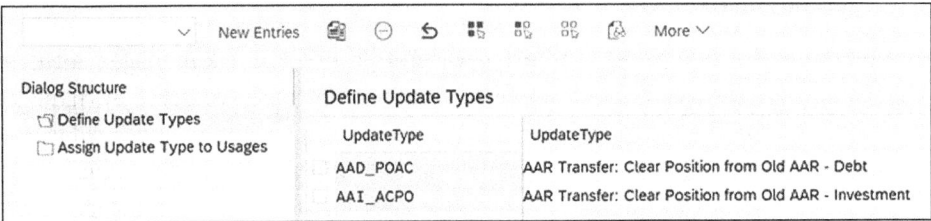

Figure 3.79 Clearing Position from Old Account Assignment Reference

2. We define the update types required to post the new posting. As we can see in Figure 3.80, we've defined two update types: one for debt contracts and another for investment contracts. At this point we've defined all of the update types required to successfully make a transfer.

3. We define the new update types required to post the new position balance.

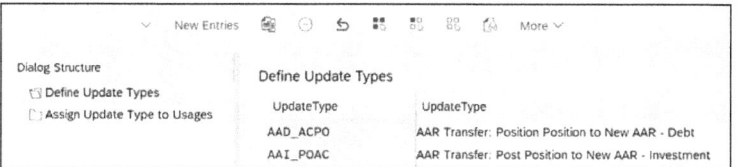

Figure 3.80 Posting Position of New Account Assignment Reference

4. As with all update types, we now assign the update types to an update type usage. The update type usage that we'll always use for this function is the **Account Assignment Transfer**, as shown in in Figure 3.81. This is extremely important because without the correct update type usage, the update type won't be triggered during the account assignment reference transfer process.

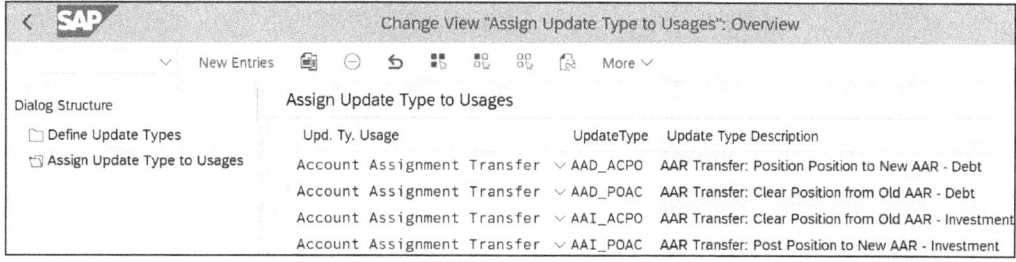

Figure 3.81 Assign Update Types to Usages

5. We utilize the update types that were created in the previous step and assign them to the account symbol that we intend to change during the account assignment reference transfer process. We should assign the update types to the fields in Figure 3.82 in the following way:
   – **Clear (D) and Post (D)**
   These two fields refer to the clearing of both the incoming and the outgoing debit position. In these two fields, we assign the two update types we use for the clearing and posting of the debit position.
   – **Clear (C) and Post (C)**
   These two fields refer to the clearing of both the incoming and the outgoing credit position. In these two fields, we assign the two update types we use for the clearing and posting of the credit position.

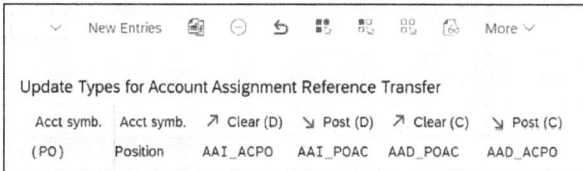

Figure 3.82 Assigning Update Types for Account Assignment Reference Transfer

6. Once we've confirmed the account assignment reference transfer, we can move on to the accounting to facilitate the move from long term to short term. As the position moves from a long-term to a short-term general ledger account, a clearing account is required to act as a technical clearing account and a pass-through account that is posted to during the transfer process. In our example, we refer to this as the "AAR Clearing Account," and we've defined the account symbol in Figure 3.83.

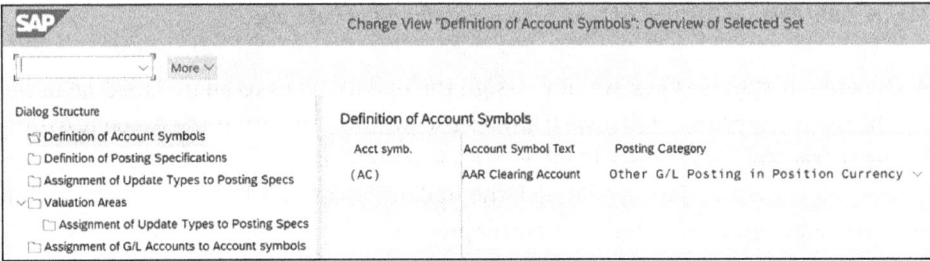

Figure 3.83 Defining Account Symbol for Account Assignment Reference Clearing Account

7. Now that we've assigned the account symbol, we can define the posting specifications used for both the clearing of the old position and the posting of the new position. In our example in Figure 3.84, we've defined two sets of posting specs: one for the debt scenario and another for the investment scenario.

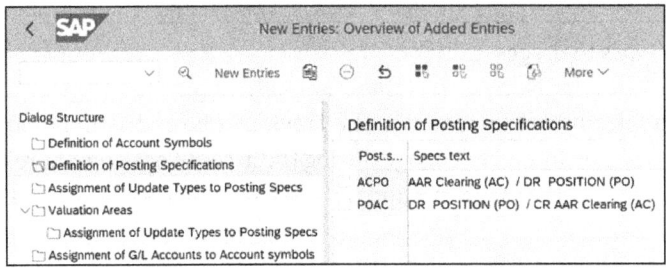

Figure 3.84 Definition of Posting Specifications

8. Now that we've completed the posting specifications, we can assign them to the update types we defined previously in this section. In both examples in Figure 3.85, the posting specifications are designed to clear out the old position, post to the AAR clearing account, and then post the new position to the next general ledger account.

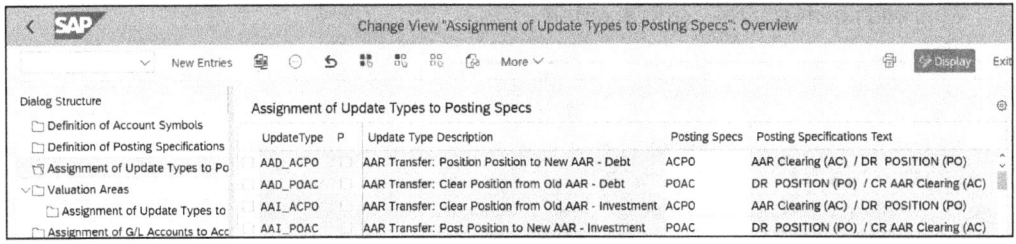

Figure 3.85 Assignment of Update Types to Posting Specifications

9. The key element of the account assignment reference transfer process it the transferring of the position via the account assignment reference transfer. To facilitate that process, we need both a long-term and a short-term account assignment reference (see Figure 3.86). When we create the contract, we assign the "LT Debt" account assignment reference to the contract, and during the account assignment reference transfer process, we can assign it to a new general ledger account.

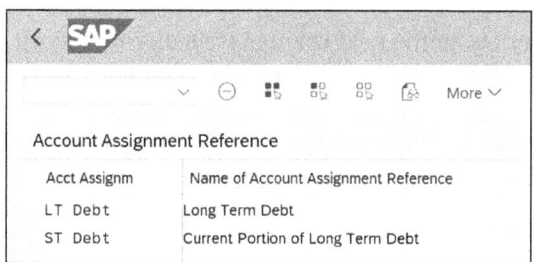

Figure 3.86 Account Assignment Reference for Short-Term and Long-Term Debt

10. Finally, we assign the account symbols to their respective account assignment references and assign the corresponding general ledger accounts. In Figure 3.87, we can see that account symbol (PO) has been assigned to both the long-term and the short-term account assignment references. Also, the long-term account assignment reference is assigned to long-term general ledger account 191009, while the short-term account assignment reference is assigned to account 191008, which is the short-term general ledger account that will be posted.

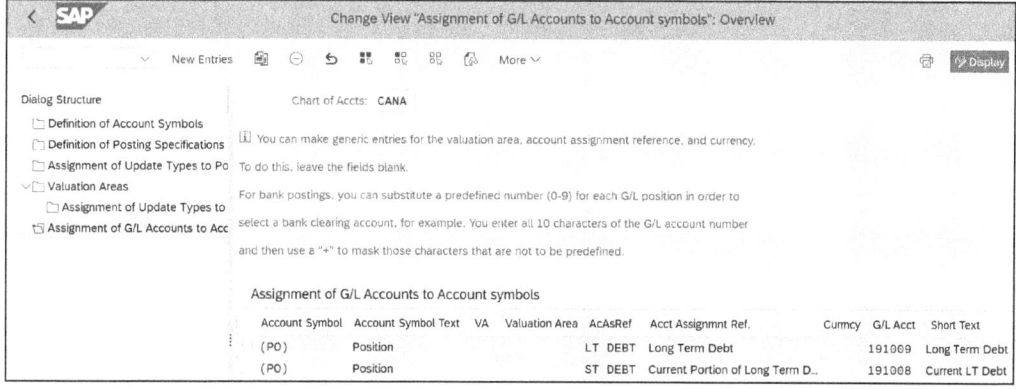

Figure 3.87 Assignment of General Ledger Accounts to Account Symbols

3.24 Summary

This chapter discussed the foundational configuration required for all financial instruments within treasury and risk management. It covered the fundamental setup steps that apply universally, regardless of the specific contract type, including product type, transaction type, flow type, update types, and setting up the accounting integration. By focusing on these core configuration elements, users can build a strong configuration framework, which is necessary before they can tailor specific instruments with unique business requirements.

Hopefully, by now, you feel that you're equipped with a thorough understanding of the core configuration steps that serve as the backbone of all treasury instruments. With this knowledge, you are well prepared to move forward confidently in configuring more complex, instrument-specific setups. In the next chapter, we'll dive deeper into contract-specific configuration, exploring how to customize treasury and risk management for different financial instruments based on their unique requirements.

Chapter 4
Contract-Specific Configuration

Contracts deployed by treasury organizations can vary greatly from one to another, and it's important to understand the contract-specific configuration within treasury and risk management. This chapter provides readers with the knowledge needed to tailor their treasury and risk management contracts to meet the unique requirements of the most common instruments used in corporate treasury departments.

Now that we've covered the core configuration elements of SAP S/4HANA Finance for treasury and risk management, we'll shift our focus to the contract-specific configuration that underpins the system's ability to handle a variety of financial instruments. This chapter provides a deep dive into the setup required for managing specific contracts related to debt, investments, intercompany loans, FX, and derivatives. Each of these financial instruments comes with its own set of nuances as it relates to the setup of the transaction type, flow types, and update types. Understanding these contract-specific elements is important to treasury professionals' ability to ensure that each instrument is aligned with the unique requirements of the contract constructed in SAP S/4HANA. By mastering these configuration steps, treasury professionals can optimize their use of treasury and risk management and understand the unique contractual elements of each instrument type. This chapter will guide you through how to set up each instrument, providing a detailed step-by-step approach to customizing treasury and risk management to meet the requirements of your company's treasury operations. The instruments that we'll cover are as follows:

- Debt and investments
- Foreign exchange
- Securities
- Derivatives
- Commodity derivatives

4.1 Debt and Investments

We'll start by turning our attention to the set up and configuration of a money market debt and investment contract. This is a logical starting point for demonstrating the

4 Contract-Specific Configuration

contract creation process from start to finish. A money market contract is the most straightforward and basic contract within treasury and risk management, making it an ideal foundation to understand before setting up more complex contract types. By understanding how to configure a money market debt and investment contract, we'll gain insights into the structure and elements required for more complex contracts we discuss in later chapters.

A money market contract refers to money market instruments, such as commercial paper, time deposits, and credit facilities. These instruments are widely used by corporate treasury departments as short-term financing and investment tools. By starting with money market debt and investment contracts, we can build a solid base understanding of the contract configuration before moving on to more complex treasury contracts as we progress through the book.

Before we jump into the money market configuration, it's important to understand the configuration required to set up the contract from start to finish and to understand the interconnectivity among the product type and transaction types and how the individual cash flows manifest themselves in eventual postings to the general ledger. In Figure 4.1, we can see this relationship portrayed in how we set up an interest rate instrument for both a debt contract and an investment contract. We start with the product and transaction types configuration, and then we create the flow types and associated update types, and eventually, we assign the update types to posting rules. We then use the posting rules when posting to the general ledger and provide the debit and credit entries of the journal entry that is posted.

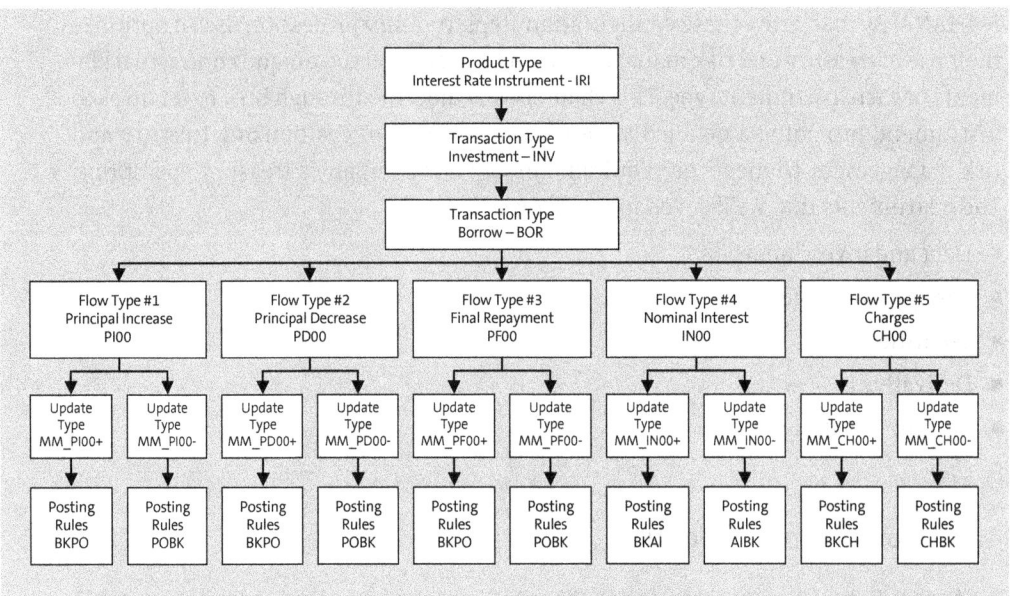

Figure 4.1 Relationship Between Product and Transaction Types and Posting Rules

In the following sections, we'll walk through four types of debt/investment instruments: money market contracts, intercompany loan configurations, credit facilities, and letters of credit.

4.1.1 Money Market Contracts

Creating a money market contract in treasury and risk management begins with the creation of the product type, which involves selecting the appropriate product category, which defines the general characteristics of the financial instrument. Next, we create the transaction type to specify the nature of the deal, such as whether it's a debt or an investment. Then, we configure flow types to outline the cash flows associated with the contract, like interest payments or repayments of principal. Finally, we create update types to ensure that these flows are correctly posted to the relevant accounts in financial accounting. All these components are interconnected, so we'll now dive into the first step, which is the creation of the product type.

Product Types

We assign the product type to an associated product category. The product type provides the overall structure of the contract, and the supported product categories that we can configure within the money market contract are as follows:

- 510: Fixed Term Deposit
- 520: Deposit at Notice
- 530: Commercial Paper
- 540: Cash Flow Transaction
- 550: Interest Rate Instrument
- 560: Facility
- 570: Fiduciary Deposit
- 580: Current Account Style Instrument

We can find the configuration node to configure the product types for money market contracts in the **Financial Supply Chain Management • Treasury and Risk Management • Transaction Manager • Money Market • Transaction Management • Product Types** menu path. In this example, we'll use **Product Category** "550," **Interest Rate Instrument**, and then provide a **Text** and a **Short Text** to describe the product type, as shown in Figure 4.2. At this point, we've completed the essential fields within the product type.

4 Contract-Specific Configuration

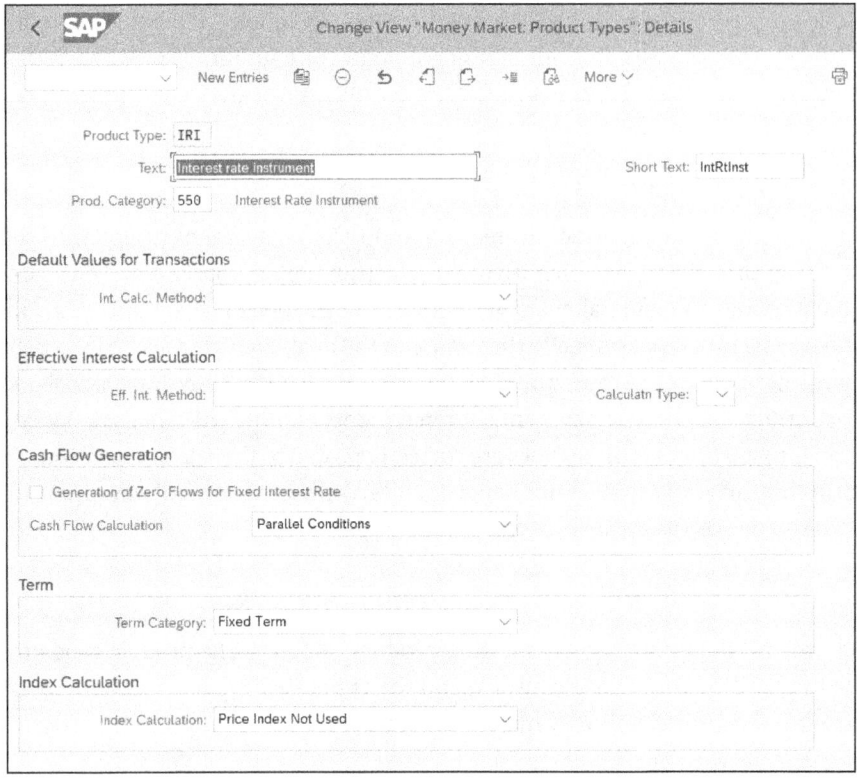

Figure 4.2 Creating Interest Rate Instrument Product Types

Transaction Types

With the product type now fully configured for our interest rate instrument, we can proceed to the next step: configuring the transaction type. This configuration is essential for defining how different transactions will be processed within the system. To begin this process, we'll need to navigate to **Financial Supply Chain Management • Treasury and Risk Management • Transaction Manager • Money Market • Transaction Management • Transaction Types**. This is where we'll configure the necessary transaction types associated with an interest rate instrument.

The two options available when using product category 550 for the **Transaction Cat** field are as follows:

- 100: Investment
- 200: Borrowing

When naming product types, it's important to choose labels that are intuitive and easily recognizable by users. Names like "Debt and Investment" or "Invest and Borrow" are commonly used to clearly represent the debt and investment aspects of a transaction type. For the purposes of this example, we'll use "Borrow" (see Figure 4.3) and "Investment" (see Figure 4.4) to differentiate the two distinct product types. These names should reflect the nature of the transactions they represent and be easily understood by those who will be working with them.

Beyond adhering to naming conventions, it's equally important to select the appropriate transaction category (**Transaction Cat**). The category we choose must align with the transaction type's naming, ensuring consistency within the setup between the name of the transaction type and the transaction category. This alignment is important for maintaining coherence in how transactions are categorized and ultimately posted to the general ledger.

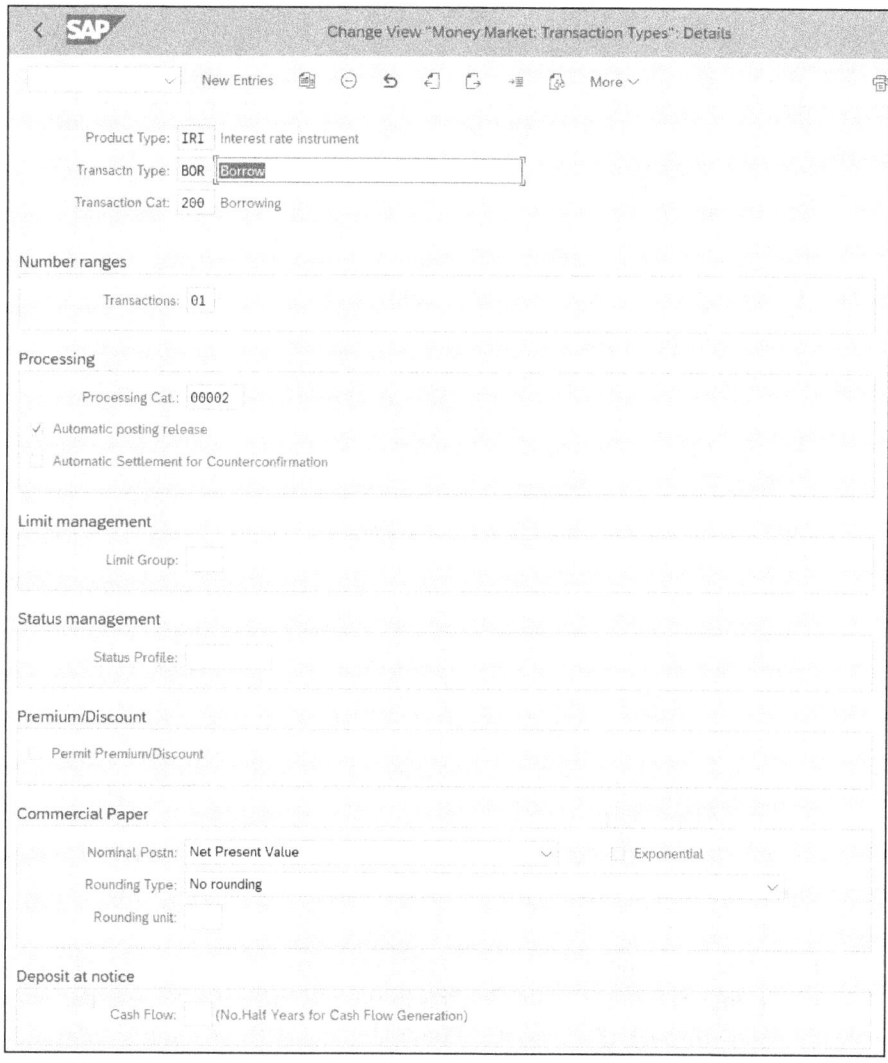

Figure 4.3 Creating Interest Rate Instrument Transaction Type for Borrowing Contract

4   Contract-Specific Configuration

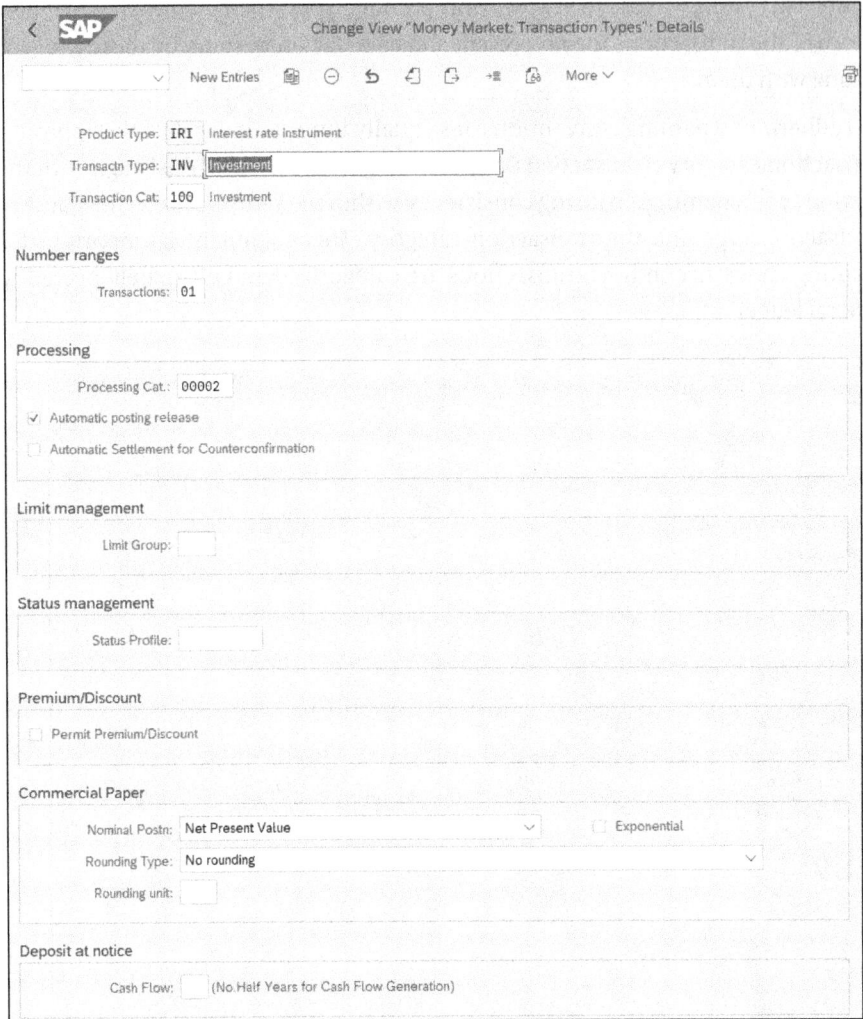

Figure 4.4 Creating Interest Rate Instrument Transaction Type for Investment Contract

Defining Flow Types and Assigning Them to Transaction Types

As mentioned at the beginning of this chapter, a money market debt and investment contract is a great place to start to understand how treasury and risk management contracts are configured during each step of the process. The cash flows associated with a simple debt and investment contract are straightforward, and for the most part, they all follow the same structure. The process typically begins with the initial transfer of principal, known as a principal increase. Following this initial exchange, the loan is repaid over a designated period through the accrual of nominal interest. Throughout the duration of the loan, the balance may be subject to changes, with potential increases or decreases in the principal amount. The loan then concludes with a final repayment at a predetermined time or with a series of payments throughout the life of the contract.

4.1 Debt and Investments

It's at this stage of the configuration that we'll define the various cash flows that take place throughout the life of the contract and assign the cash flows to the product and transaction types. We can find the configuration in the **Financial Supply Chain Management • Treasury and Risk Management • Transaction Manager • Money Market • Transaction Management • Flow Types • Define and Assign Flow Types** menu path.

To account for the cash flows within a money market debt and investment contract, we must configure several flow types. For step-by-step instructions for configuring flow types, refer to Chapter 3, Section 3.13. The key flow types required during this step are as follows:

- **Principal Increase (PI00)**
  This flow type captures the initial disbursement of funds from the lending entity to the borrowing entity, as shown in Figure 4.5. It records the increase in the principal amount of the loan.

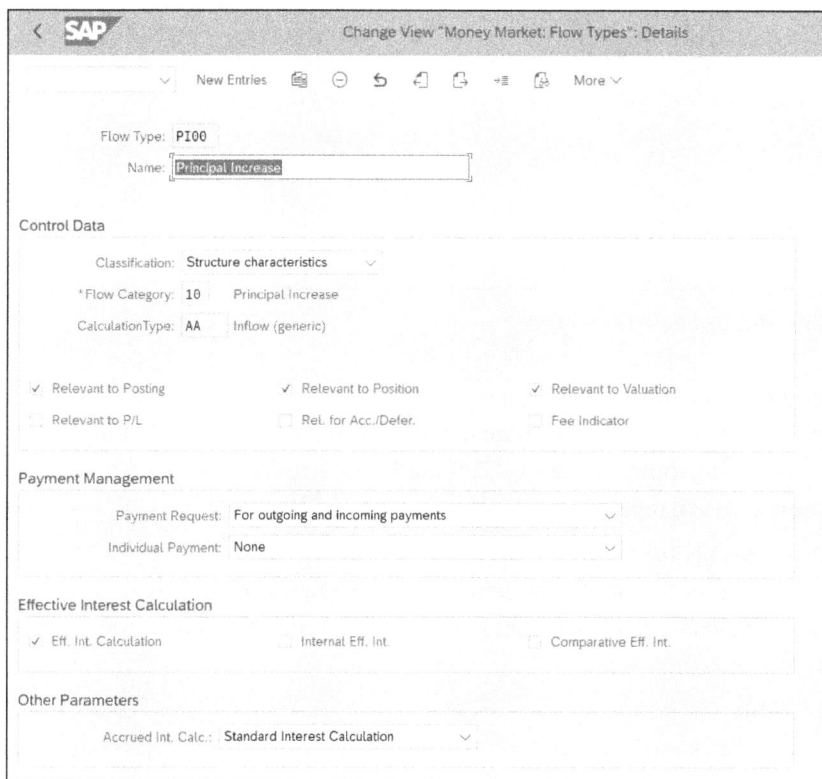

Figure 4.5 Principal Increase Flow Type

- **Principal Decrease (PD00)**
  This flow type accounts for any reductions in the principal amount over the course of the loan, reflecting repayments or adjustments made during the loan's lifecycle, as shown in Figure 4.6.

207

4   Contract-Specific Configuration

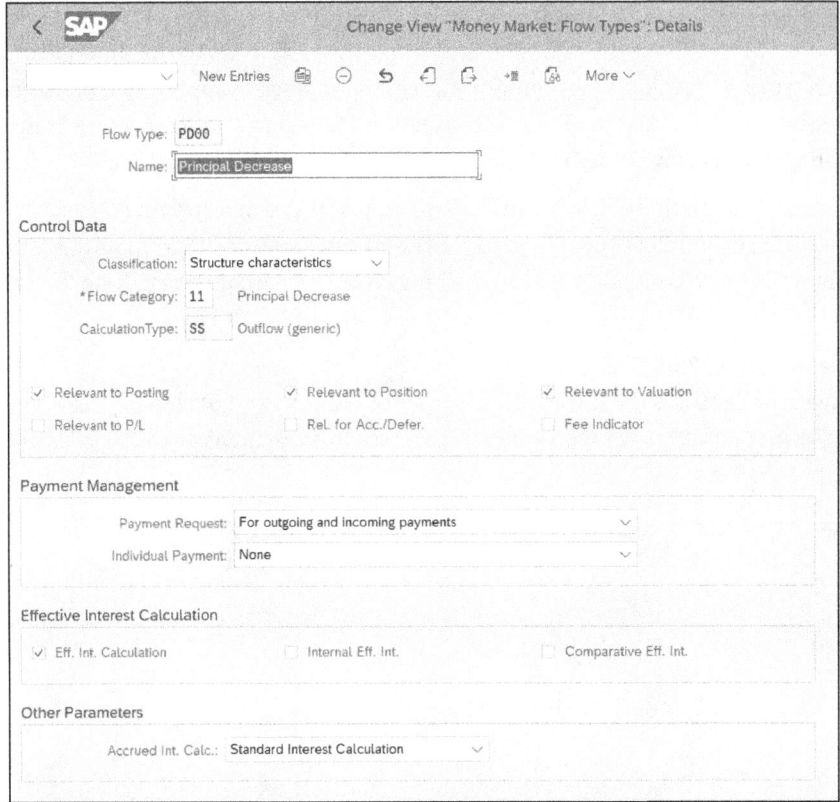

Figure 4.6 Principal Decrease Flow Type

- **Final Repayment (PF00)**
  This flow type documents the ultimate repayment of the loan's principal at the end of the loan term, as shown in Figure 4.7. It marks the closure of the loan agreement by recording the final repayment amount.

- **Nominal Interest (IN00)**
  We use this flow type to track the periodic interest payments made by the borrowing entity, as shown in Figure 4.8. It records the interest accrued based on the loan agreement's terms and conditions.

- **Charges (CH00)**
  We use this to track and account for any charges that will be tied to this particular contract, as shown in Figure 4.9.

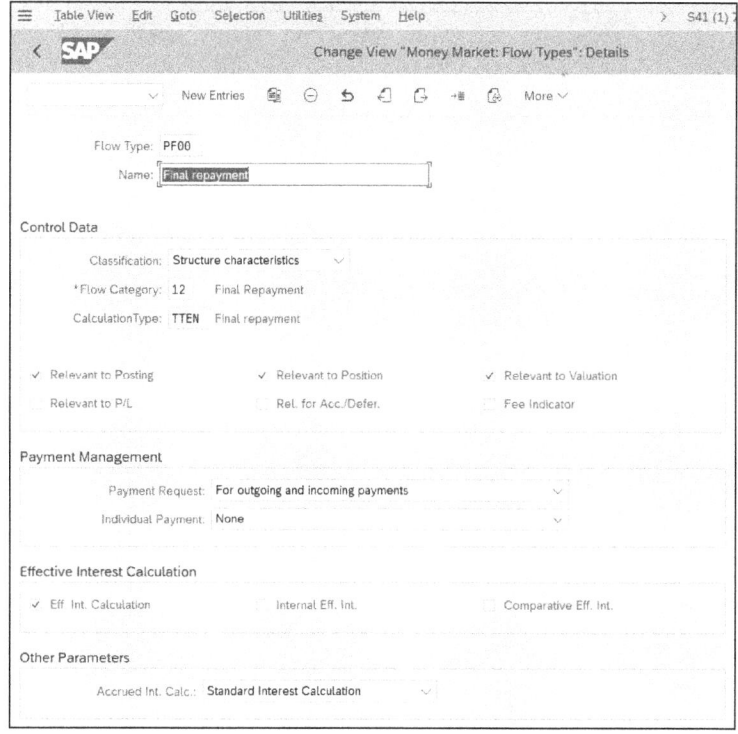

Figure 4.7 Final Repayment Flow Type

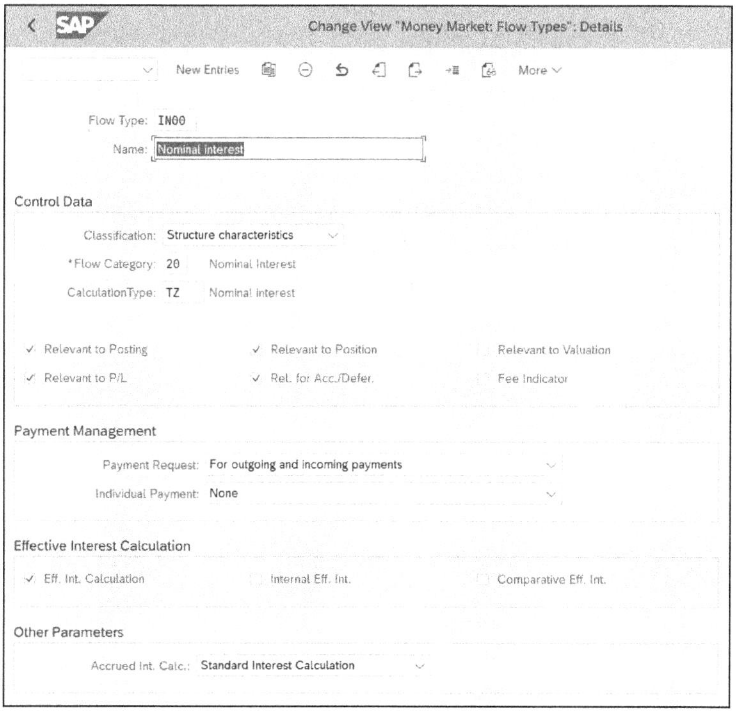

Figure 4.8 Nominal Interest Flow Type

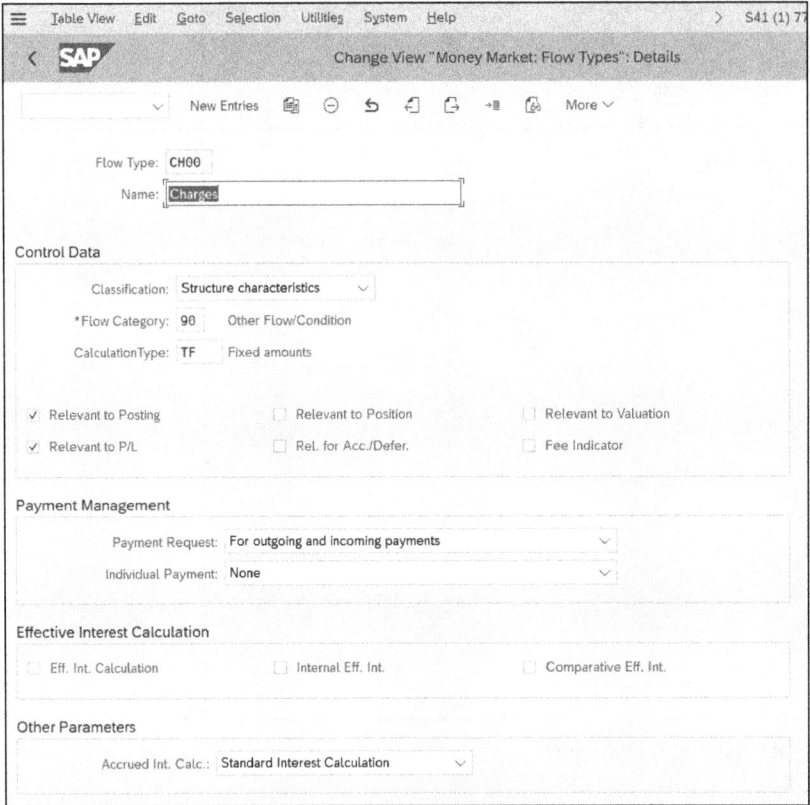

Figure 4.9 Charges Flow Type

For interest rate instruments, once we've defined these flow types, it's important to assign them to their product and transaction type combinations (as discussed in Chapter 3, Section 3.13.2), as shown in Figure 4.10. The assignment ensures that each flow type is linked to the correct interest rate instrument product and transaction type, facilitating the correct selection of flow types in the background as transactions are created in SAP S/4HANA. Proper configuration of these flow types and their assignment to product and transaction types enables the system to handle various debt and investment scenarios accurately through the lifecycle of the contract. Every flow type that is applicable to the product and transaction type must be specified in this table so that we can use the flow types when creating the contract.

> **Note**
>
> We don't have to maintain the flow types that are associated with a condition type in this configuration table as long as we maintain the condition type and assign it to the product and transaction type combination. Although it's not required, we can dually maintain and assign them to the product type and transaction type combination within this table, and it will not impact the functionality.

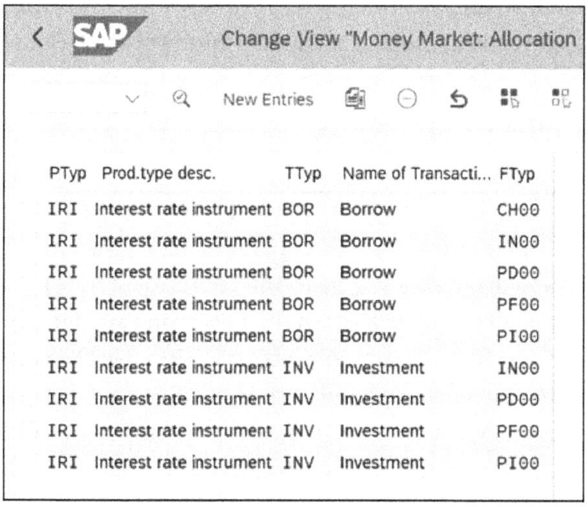

4.1 Debt and Investments

Figure 4.10 Assigning Product and Transaction Types to Flow Types

Condition Types and Condition Type Assignment

The final step in the base configuration of an interest rate instrument involves creating and assigning the necessary condition types that apply to the financial calculations associated with the instrument. Condition types are required for any flow type that involves a calculation to derive the flow amount. For interest rate instruments, these calculations typically pertain to the accrual of interest and the execution of the final settlement. If the interest rate instrument features a variable interest rate, an additional condition type is required to handle the calculation of interest rate adjustments.

To start the configuration, we navigate to **Financial Supply Chain Management • Treasury and Risk Management • Transaction Manager • Money Market • Transaction Management • Condition Types • Define Condition Types**. Specifically, we should establish the following condition types and align them with their respective flow types to accurately reflect the cash movements of an interest rate instrument:

- **Final Repayment (1120)**
  This condition type is critical for the calculation of the final repayment amount, which occurs at the end of the loan's term. It ensures that the full principal amount or any remaining balance is accurately computed and accounted for during the final repayment process.

- **Nominal Interest (1200)**
  We use this condition type to calculate the periodic interest payments made by the borrowing entity. It's essential for accurately determining the interest charges based on the terms of the loan agreement and ensuring these amounts are properly recorded and reported.

- **Interest Rate Adjustment (1210)**
  This condition type becomes necessary if the intercompany loan has a variable

211

interest rate. It handles the calculation of adjustments to the interest rate, ensuring that fluctuations in the rate are correctly applied to the loan's interest calculations.

Creating these condition types involves defining the specific parameters and calculation rules that apply to each type. This ensures that the system can automatically calculate and apply the correct amounts for interest accrual, interest rate adjustments, and final repayment based on the established loan terms. Once we've created these condition types, it's important to assign them to the corresponding flow types within the SAP S/4HANA system. This links each condition type to the appropriate flow type, ensuring that the calculations are applied correctly to both the borrowing and lending sides of the contract. Proper assignment allows for the correct calculations to take place for the condition types, corresponding cash flows related to the final repayment, interest, and interest rate adjustments. As shown in Figure 4.11, the interest rate instrument (IRI) product type has been mapped to the appropriate condition types.

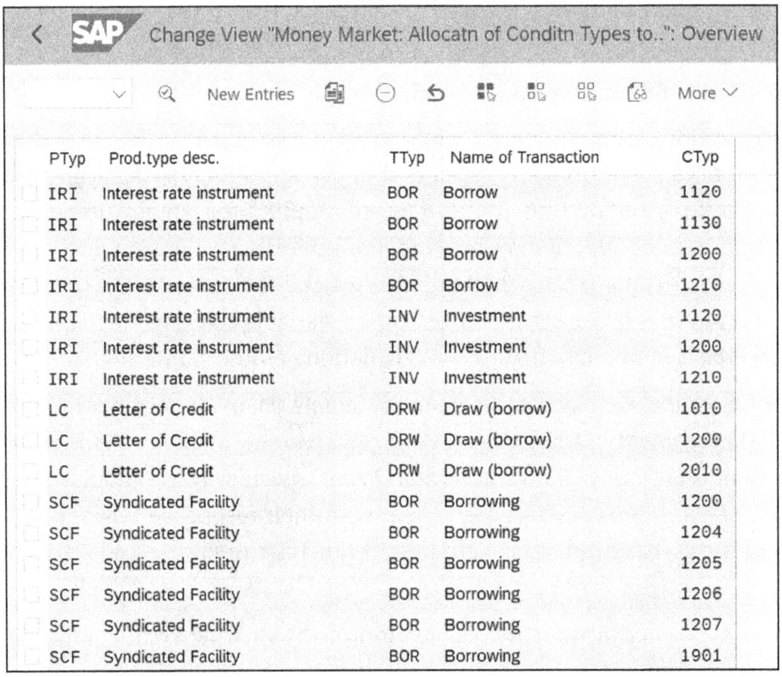

PTyp	Prod.type desc.	TTyp	Name of Transaction	CTyp
IRI	Interest rate instrument	BOR	Borrow	1120
IRI	Interest rate instrument	BOR	Borrow	1130
IRI	Interest rate instrument	BOR	Borrow	1200
IRI	Interest rate instrument	BOR	Borrow	1210
IRI	Interest rate instrument	INV	Investment	1120
IRI	Interest rate instrument	INV	Investment	1200
IRI	Interest rate instrument	INV	Investment	1210
LC	Letter of Credit	DRW	Draw (borrow)	1010
LC	Letter of Credit	DRW	Draw (borrow)	1200
LC	Letter of Credit	DRW	Draw (borrow)	2010
SCF	Syndicated Facility	BOR	Borrowing	1200
SCF	Syndicated Facility	BOR	Borrowing	1204
SCF	Syndicated Facility	BOR	Borrowing	1205
SCF	Syndicated Facility	BOR	Borrowing	1206
SCF	Syndicated Facility	BOR	Borrowing	1207
SCF	Syndicated Facility	BOR	Borrowing	1901

Figure 4.11 Assigning Product and Transaction Types to Condition Types

Assigning Flow Types to Update Types

After we create the update types for all the corresponding flow types, as with all product types, the next step is to assign the flow types to their corresponding update types. We can find the configuration at **Financial Supply Chain Management • Treasury and Risk Management • Transaction Manager • Money Market • Transaction Management • Update Types • Assign Flow Types to Update Types**.

As we can see in Figure 4.12, it's important to assign the **Money Market** contract type (**Cont.Type**) as this is a money market transaction. If we assign the incorrect contract type, the update type will not be called when we process the transaction. Also notice that in the example, the orientation of the update type corresponds with the value in the direction column.

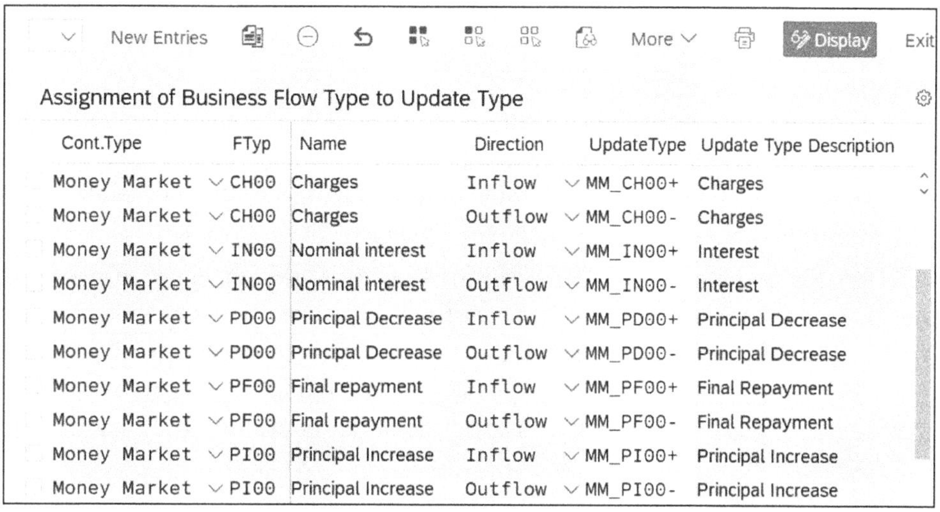

Cont.Type	FTyp	Name	Direction	UpdateType	Update Type Description
Money Market	CH00	Charges	Inflow	MM_CH00+	Charges
Money Market	CH00	Charges	Outflow	MM_CH00-	Charges
Money Market	IN00	Nominal interest	Inflow	MM_IN00+	Interest
Money Market	IN00	Nominal interest	Outflow	MM_IN00-	Interest
Money Market	PD00	Principal Decrease	Inflow	MM_PD00+	Principal Decrease
Money Market	PD00	Principal Decrease	Outflow	MM_PD00-	Principal Decrease
Money Market	PF00	Final repayment	Inflow	MM_PF00+	Final Repayment
Money Market	PF00	Final repayment	Outflow	MM_PF00-	Final Repayment
Money Market	PI00	Principal Increase	Inflow	MM_PI00+	Principal Increase
Money Market	PI00	Principal Increase	Outflow	MM_PI00-	Principal Increase

Figure 4.12 Assigning Flow Types to Update Types

4.1.2 Intercompany Loans

Having successfully configured the product types and transaction types for interest rate instruments, we can now shift our focus to the next key instruments used by most corporations: intercompany loans. This section involves configuring the specific parameters and settings required to manage loans among different entities within a corporation. Let's dive into the process of configuring these loan arrangements, focusing on the differences in the setup of an intercompany loan compared to a traditional debt or investment contract.

Product type configuration of an intercompany loan product type within treasury and risk management in SAP S/4HANA follows a process that is very similar to the setup of other debt or investment product types. The configuration relies on the same principles for managing traditional debt and investment contracts, but there are distinct differences during the actual creation of the intercompany loan contract.

While setting up a conventional debt or investment contracts involves an external business partner such as a financial institution or another external counterparty, the creation of an intercompany loan is very different. Instead of connecting with an outside entity, the intercompany loan contract utilizes an internal business partner that represents the corresponding internal company code or entity within the same organization.

In other words, when setting up an intercompany loan, the internal business partner serves as the business partner for what would ordinarily be an external counterparty. This internal partner represents the company code within the organization, and the creation process involves selecting or defining this internal entity as part of the intercompany loan agreement, ensuring that the loan reflects the internal financial transactions accurately.

Configuration of the internal business partner in SAP S/4HANA involves additional steps to ensure it is properly linked to the correct company codes and reflects the internal business partner relationship within the organization. Understanding and implementing this internal business partner mechanism is important for the accurate representation and management of intercompany loans, allowing for seamless integration within the broader treasury and risk management framework.

We start the configuration of intercompany loans by using the same configuration nodes as we used with other debt and investment contracts. We can find the nodes at **Financial Supply Chain Management • Treasury and Risk Management • Transaction Manager • Money Market.** We'll now look at how to set up product types, transaction types, flow types, condition types, and how to assign them. We'll also walk through intercompany loan mirroring in detail.

Product Type

As with all contract types within treasury and risk management, the first step in setting up an intercompany loan involves creating the relevant product type. This foundational step is crucial for categorizing and managing the loan throughout its lifecycle. In our example, we'll designate the **Product Type** as "ICL" to signify the intercompany loan. We can find the configuration node to set up the product types at **Financial Supply Chain Management • Treasury and Risk Management • Transaction Manager • Money Market • Transaction Management • Product Types.**

To achieve the highest level of flexibility and functionality when creating and managing intercompany loan contracts, we recommend utilizing product category (**Prod. Category**) "550," which is classified as an **Interest Rate Instrument**, as shown in Figure 4.13. This product category is specifically designed to handle various aspects of interest rate management, and it offers extensive flexibility in terms of configuring and executing the loan contracts. Selecting product category 550 ensures that the system can accommodate a wide range of scenarios and variations in intercompany loan agreements, providing the necessary options to customize the contracts according to specific business requirements such as variable interest rates, the ability to change the end date of a contract, etc.

While it's possible to select other product categories for intercompany loans, doing so might limit our flexibility in managing an intercompany loan's requirements. Product category 550 is a good fit and useful for an intercompany loan because it integrates well with the complex requirements required for most intercompany loan agreements, such as variable interest rate structures, different repayment schedules, and intricate financial terms. Using this category also facilitates a flexible approach to interest rate management within the contracts, allowing us to use either a fixed or a variable interest rate, which is often a required component of many intercompany loan arrangements.

By starting with the appropriate product type and category, such as "ICL" under product category 550, companies can ensure that their intercompany loan contracts are set up in a manner that maximizes the usability of the solution. This foundational setup paves the way for accurate and effective management of intercompany loans when borrowing and lending money within the organization.

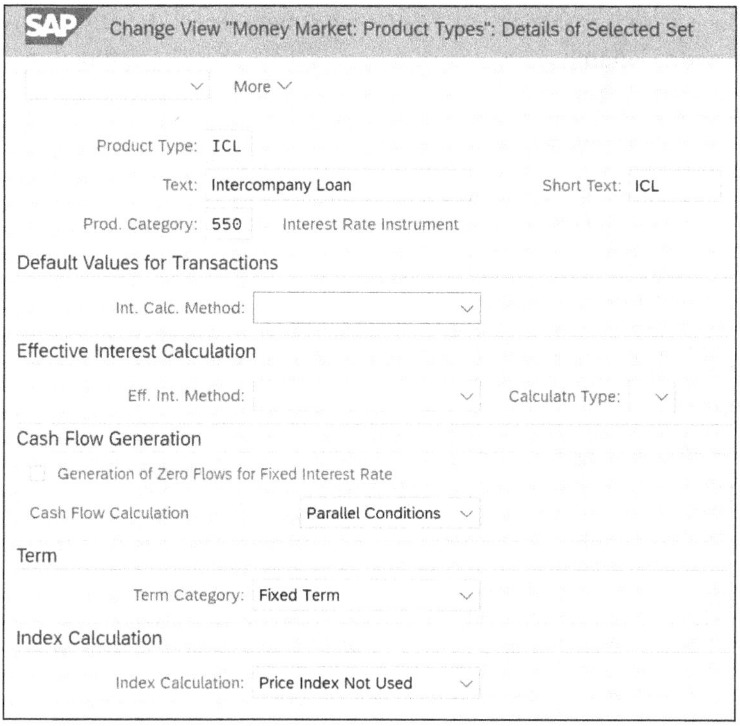

![Configuration of Product Types for Intercompany Loans]

Figure 4.13 Configuration of Product Types for Intercompany Loans

Transaction Types

After we've successfully created the product types, the next step in the process is to set up the transaction types specific to intercompany loans. This involves creating distinct transaction types that will represent both the borrowing and the lending aspects of the intercompany loan. For a comprehensive view of the full intercompany loan, we'll need to create two separate transaction types: one for the lending side and another for the borrowing side of the transaction. To start the configuration for the transaction type, we navigate to the **Financial Supply Chain Management • Treasury and Risk Management •**

Transaction Manager • **Money Market** • **Transaction Management** • **Transaction Types** menu path.

It's common to use terms such as *lend and borrow* and *invest and borrow* when defining these transaction types. For the purpose of this example, we'll use "BORROW" and "Investment" to clearly distinguish between the two sides of the intercompany loan. The "BORROW" transaction type will represent the entity receiving the loan, while the "Investment" transaction type will represent the entity providing the loan.

To illustrate this, consider Figure 4.14 and Figure 4.15, which show the creation of these two transaction types. Each transaction type is configured to accurately reflect its role in the loan agreement, and a crucial aspect of this configuration is setting the transaction category, which ensures that each transaction type is appropriately mapped to its respective category within the system. For the lending or investing side, we should map the transaction type to **Transaction Cat.** "100": **Investment**. This category encompasses activities related to lending or investing funds, and it aligns with the functionality needed for intercompany loans.

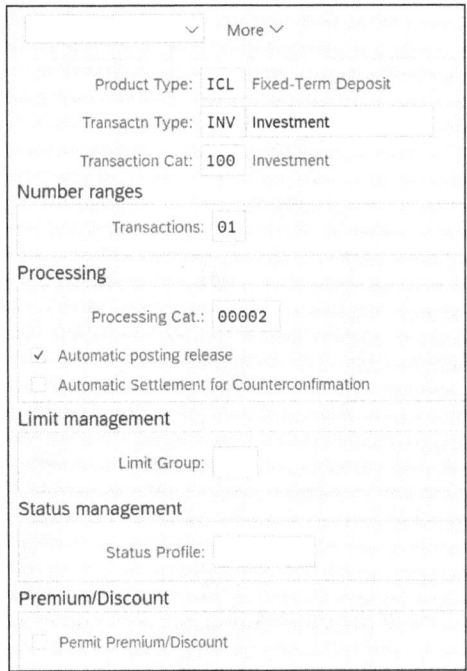

Figure 4.14 Creating Intercompany Loan Transaction Type for Investments

Conversely, we should map the borrowing side of the transaction to **Transaction Cat.** "200": **Borrowing**. This category is designed to handle activities associated with borrowing funds, and it accurately reflects the obligations and financial arrangements of the borrowing entity. The correct mapping of these transaction types is essential for maintaining the integrity and accuracy of the intercompany loan within the SAP S/4HANA system, as it dictates how transactions are recorded, processed, and reported.

By defining the "BORROW" and "Investment" transaction types and mapping them to their respective categories, we can ensure that the intercompany loan transactions are properly classified and managed. This setup not only aids in the accurate tracking of each side of the loan, but it also supports the subsequent steps in the configuration process, in which we link the "BORROW" and "Investment" transaction types to a single unified intercompany loan.

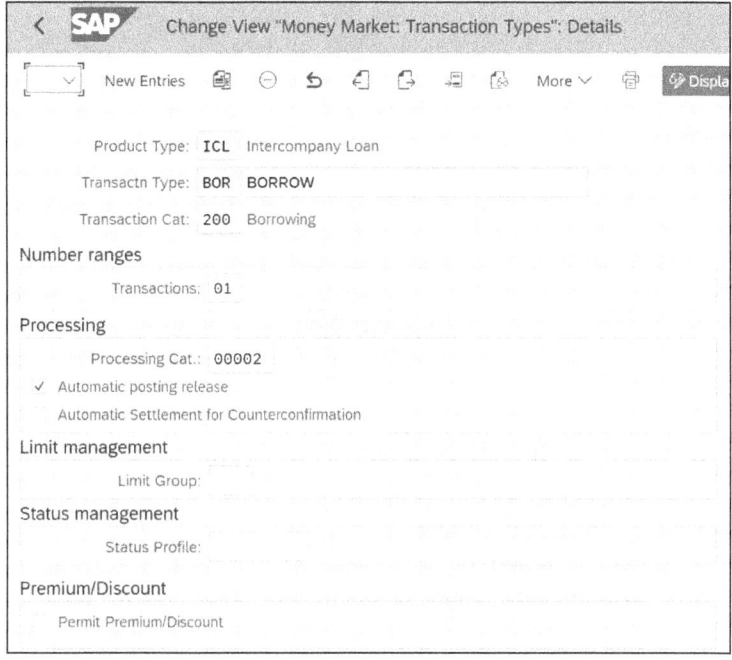

Figure 4.15 Creating Intercompany Loan Transaction Type for Borrowing

Defining Flow Types and Assigning Them to Transaction Types

Structurally, an intercompany loan is very similar to an external debt or investment contract. Therefore, the flow types necessary for managing an intercompany loan closely resemble those used in debt and investment contracts. These flow types are integral to capturing the various stages and activities throughout the lifecycle of the loan. When considering the lifecycle of an intercompany loan, the process typically begins with the initial transfer of principal, which is known as a *principal increase*. Following this initial exchange, the loan is repaid over a designated period through the accrual of nominal interest. Throughout the duration of the loan, the balance often will show increases or decreases in the principal amount as borrowing needs change. Finally, the loan concludes with a final repayment at a predetermined time. We can find the configuration at **Financial Supply Chain Management • Treasury and Risk Management • Transaction Manager • Money Market • Transaction Management • Flow Types • Define and Assign Flow Types**.

To effectively manage these activities, we should configure several key flow types for intercompany loans, mirroring the standard flows seen in debt and investment contracts. These flow types include the following:

- **Principal Increase (PI00)**
  This flow type captures the initial disbursement of funds from the lending entity to the borrowing entity. It records the increase in the principal amount of the loan.
- **Principal Decrease (PD00)**
  This flow type accounts for any reductions in the principal amount over the course of the loan, reflecting repayments or adjustments made during the loan's lifecycle.
- **Final Repayment (PF00)**
  This flow type documents the ultimate repayment of the loan's principal at the end of the loan term. It marks the closure of the loan agreement by recording the final repayment amount.
- **Nominal Interest (IN00)**
  This flow type is used to track the periodic interest payments made by the borrowing entity. It records the interest accrued based on the loan agreement's terms and conditions.

These flow types are essential to accurately tracking and managing the financial transactions associated with intercompany loans. Once we've defined these flow types in the **FTyp** field, as shown in Figure 4.16, it's important to assign them to the appropriate product and transaction type combinations. This ensures that each flow type is correctly linked to the respective intercompany loan product and transaction type, thus facilitating the correct selection of flow types in the background as intercompany loans are created in SAP S/4HANA.

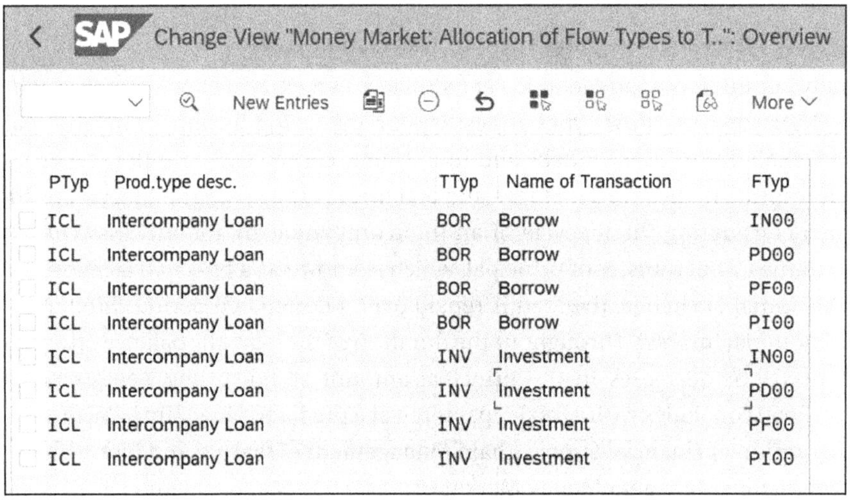

PTyp	Prod.type desc.	TTyp	Name of Transaction	FTyp
ICL	Intercompany Loan	BOR	Borrow	IN00
ICL	Intercompany Loan	BOR	Borrow	PD00
ICL	Intercompany Loan	BOR	Borrow	PF00
ICL	Intercompany Loan	BOR	Borrow	PI00
ICL	Intercompany Loan	INV	Investment	IN00
ICL	Intercompany Loan	INV	Investment	PD00
ICL	Intercompany Loan	INV	Investment	PF00
ICL	Intercompany Loan	INV	Investment	PI00

Figure 4.16 Assigning Intercompany Loan Contracts to Their Flow Types

Proper configuration of these flow types and their assignment to product and transaction types enables the system to handle various loan scenarios accurately, and this includes tracking the initial disbursement of funds, recording any changes in the loan balance, calculating interest, and documenting the final repayment.

Condition Types and Condition Type Assignment

The final step in the base configuration of an intercompany loan involves creating and assigning the necessary condition types that govern the financial calculations associated with the loan. Condition types are required for any flow type that involves a calculation to derive the flow amount. For intercompany loans, these calculations typically pertain to the accrual of interest and the execution of the final repayment. If the intercompany loan features a variable interest rate, an additional condition type is required to handle the calculation of interest rate adjustments. To start the configuration, we navigate to **Financial Supply Chain Management** • **Treasury and Risk Management** • **Transaction Manager** • **Money Market** • **Transaction Management** • **Condition Types** • **Define Condition Types**.

Specifically, we should establish the following condition types and align them with their respective flow types to accurately reflect the financial dynamics of the intercompany loan:

- **Final Repayment (1120)**
  This condition type is critical for the calculation of the final repayment amount, which occurs at the end of the loan's term. It ensures that the full principal amount or any remaining balance is accurately computed and accounted for during the final repayment process.

- **Nominal Interest (1200)**
  This condition type is used to calculate the periodic interest payments made by the borrowing entity. It's essential for accurately determining the interest charges based on the terms of the loan agreement and ensuring these amounts are properly recorded and reported.

- **Interest Rate Adjustment (1210)**
  This condition type becomes necessary if the intercompany loan has a variable interest rate. It handles the calculation of adjustments to the interest rate, ensuring that fluctuations in the rate are correctly applied to the loan's interest calculations.

Creating these condition types involves defining the specific parameters and calculation rules that apply to each type. This ensures that the system can automatically calculate and apply the correct amounts for interest accrual, interest rate adjustments, and final repayment based on the established loan terms. Once we've created these condition types, it's important to assign them to the corresponding flow types within the SAP S/4HANA system, as shown in Figure 4.17. This links each condition type to the appropriate flow type, ensuring that the calculations are applied correctly to both the borrowing and lending sides of the contract.

It's important to ensure that these condition types are consistently assigned to both the borrowing and investment aspects of the intercompany loan. This dual assignment guarantees that the calculations for interest, interest rate adjustments, and final repayment are accurately reflected on both sides of the transaction, providing a comprehensive and balanced view of the intercompany loan activity during all stages of the loan management process in SAP S/4HANA.

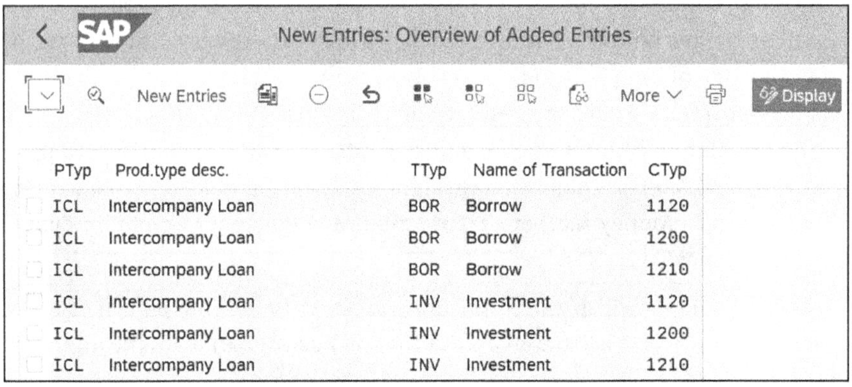

PTyp	Prod.type desc.	TTyp	Name of Transaction	CTyp
ICL	Intercompany Loan	BOR	Borrow	1120
ICL	Intercompany Loan	BOR	Borrow	1200
ICL	Intercompany Loan	BOR	Borrow	1210
ICL	Intercompany Loan	INV	Investment	1120
ICL	Intercompany Loan	INV	Investment	1200
ICL	Intercompany Loan	INV	Investment	1210

Figure 4.17 Assigning ICL Product Types to Necessary Condition Types

Intercompany Loan Mirroring

Intercompany loans in treasury and risk management operate in a manner very similar to other debt and investment contract types, but with a distinct feature: they represent a financial arrangement between two entities within the same organization. What sets intercompany loans apart is that each loan is essentially one unified agreement, yet it's represented by two separate contracts within the SAP S/4HANA system.

This dual-contract approach is unique because it allows each side of the loan—the borrowing side and the lending side—to be managed and viewed independently. On the one hand, we create a contract to reflect the borrowing side of the loan, detailing the terms and conditions associated with the debt. On the other hand, we establish a separate contract to represent the lending side, outlining the investment aspects of the arrangement.

Take for example a scenario in which company code 1000 is lending to company code 2000. Two individual trades are required to represent each side of the transaction in the respective company codes: (1) a contract in company code 1000 lending to or investing with company code 2000 and (2) another contract in company code 2000 borrowing from company code 1000. These two contracts together represent the full intercompany loan, each contract has a unique contract number, and there are separate contracts in SAP S/4HANA that are linked together.

For the intercompany loans to create both sides of the loan in two company codes, we must configure *mirroring*, which is a function within treasury and risk management where upon creation of the contract, both sides of the contract are created simultaneously. Two unique contract numbers are created when we save the contract in SAP S/4HANA, but they are tied together by a unique identification number. Going forward, we can change the intercompany loan on either the borrowing side or the lending side of the transaction, and the changes or adjustments will take place in both contracts without the need to maintain both sides independently. If we do not enable the mirroring functionality, the user will be required to create two contracts for each intercompany loan. The user will have to create both the lending and the borrowing side of the loan manually, and any changes that take place in the loan will then be dually maintained.

The configuration required to set up mirroring is a five-step process. We start it by activating the treasury and risk management mirroring functionality, and we then define the types of contracts and relationships that should be mirrored when creating a contract. In the final steps, we define the internal company code and assign it to its corresponding business partner number. We can find the mirroring configuration by using the following menu path: **Financial Supply Chain Management • Treasury and Risk Management • Transaction Manager • General Settings • Organization • Transaction Management • Intercompany Trading: Distribution of Mirror Transactions**.

Specifying the Mirroring Mode for the Processing of Financial Transactions
The first step in setting up mirroring within treasury and risk management is to activate the mirroring feature. This configuration is essential because it's a prerequisite for enabling mirroring within treasury and risk management. It's crucial to complete this step before proceeding with any subsequent configuration tasks, and activating mirroring ensures that the system can correctly replicate transaction details in all related contracts. We can find the mirroring configuration in the following menu path: **Financial Supply Chain Management • Treasury and Risk Management • Transaction Manager • General Settings • Organization • Transaction Management • Intercompany Trading: Distribution of Mirror Transactions • ICT - Specify Mirroring Mode for Processing of Financial Transactions**.

Within the configuration node, we'll notice that the following three distinct options available (see Figure 4.18). Each of them controls how contracts are mirrored:

- **Mirror Processing of ICo Transactions incl. Settlement**
  This option enables mirroring of intercompany transactions during both the creation and the settlement of the contract. Upon contract creation, two contracts are created representing both sides of the contract, and upon contract settlement, one side of the contract is settled and the other side is settled at the same time. Any changes to the contract will also be reflected automatically on both sides of the contract.

- **Mirror Processing of Intercompany Transactions**
  This option is very similar to the previous option, and it enables mirroring of intercompany transactions during the creation of the contract. The biggest difference is

that the settlement step is not mirrored to the other side of the contract, so after contract creation, we need to settle both sides of the contract independently.

- **Do Not Mirror Processing of Intercompany Transactions**
  The final option available is to not mirror intercompany transactions, and we should use this setting if we don't want to mirror intercompany transactions. With this option, we create, settle, and maintain both sides of the loan independently.

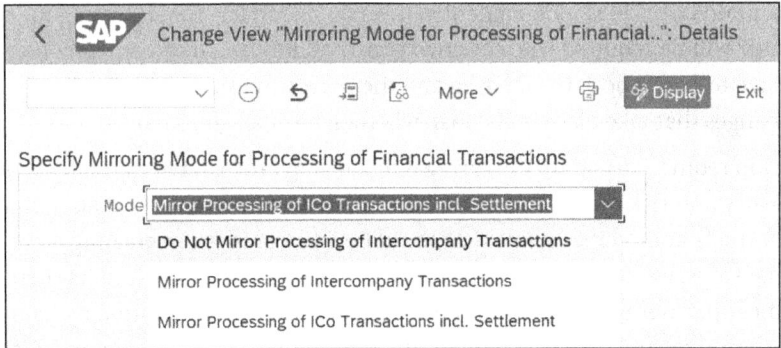

Figure 4.18 Specifying Mirroring Mode for Processing of Financial Transactions

Maintaining Relevant Product Types and Transaction Types

Next, we must define the types of contracts that should be mirrored upon creation and what activities should trigger mirroring. We can find the configuration at **Financial Supply Chain Management** • **Treasury and Risk Management** • **Transaction Manager** • **General Settings** • **Transaction Management** • **Intercompany Trading: Distribution of Mirror Transactions** • **ICT - Maintain Relevant Product Types and Transaction Types**.

The first step is to specify the company code (**CoCode**), product type (**Prod. Type**), and transaction type (**Trans.Type**) that should be mirrored. We then specify the type of activities (**Activity**) and the activity category (**Activ.Cat.**). The activity type and category drive when a mirrored transaction should take place, and the activity is the function that is performed in the contract. The activities that are applicable to intercompany transactions are the following:

- **01: Add or Create**
  This activity is the creation of contracts, and we should always use it to enable the mirroring of intercompany transactions.

- **02: Change**
  This activity is a change in contracts, and we should use it to enable mirroring related to changes in a contract.

We specify the activity category to indicate during what process the intercompany activity should be mirrored. If we want intercompany activities to be mirrored during both the contract creation process and the settlement process, we'll ensure that both the **Activ.Cat** "10" (contract) and "20" (contract settlement) are configured in this table.

In the last column within this configuration node, we can indicate the internal business **Partner** that we will use with this combination of company code, product type, transaction type, activity, and activity category. It's important to map in this table every loan pair that we intend to be mirrored. As we can see in Figure 4.19, we've mapped company code (**CoCode**) 1000 to business partner (**Partner**) ICL-2000 and company code 2000 to ICL-1000. We must maintain both sides for the transaction to mirror correctly.

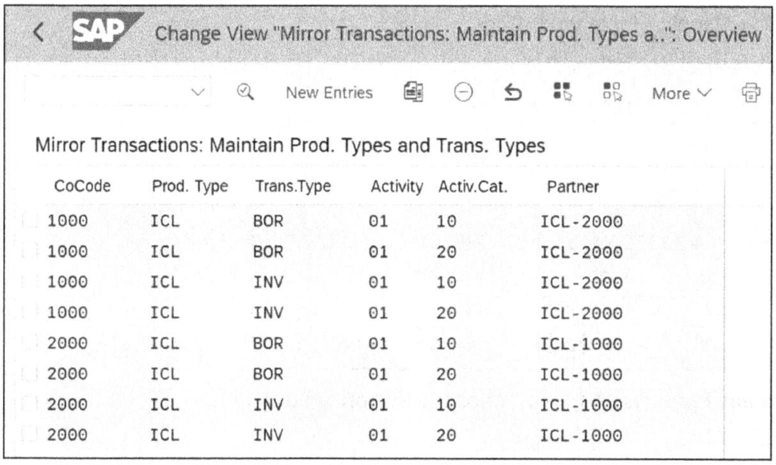

CoCode	Prod. Type	Trans.Type	Activity	Activ.Cat.	Partner
1000	ICL	BOR	01	10	ICL-2000
1000	ICL	BOR	01	20	ICL-2000
1000	ICL	INV	01	10	ICL-2000
1000	ICL	INV	01	20	ICL-2000
2000	ICL	BOR	01	10	ICL-1000
2000	ICL	BOR	01	20	ICL-1000
2000	ICL	INV	01	10	ICL-1000
2000	ICL	INV	01	20	ICL-1000

Figure 4.19 Mirroring Transactions: Maintaining Product Types and Transaction Types

Mapping Product Types and Transaction Types

The next step is to specify how we're mapping the original contract that's created with the mirrored contract that's created automatically in the background. We can find the configuration at **Financial Supply Chain Management • Treasury and Risk Management • Transaction Manager • General Settings • Transaction Management • Intercompany Trading: Distribution of Mirror Transactions • ICT - Map Product Types and Transaction Types**.

A contract that a user creates is called the *outgoing* side of the intercompany relationship, while the mirrored side of the intercompany relationship that is created automatically is called the *incoming* side of the transaction. We use this configuration table to map the outgoing transaction to a corresponding incoming transaction by using a unique **MetaText**, which we use as an identifier to link the outgoing side of the contract to the incoming side of the contract. SAP S/4HANA then knows what contract it should create in the background as a result of the manual contract that we input.

In Figure 4.20, we can see that we have a number of entries in the table between **CoCode** "1000" and **CoCode** "2000." Each outgoing entry has a corresponding incoming entry, and both the outgoing and incoming entries are tied together using the **MetaText**. For example, when we create a contract with the details 1000-ICL-BOR, a corresponding contract with the details 2000-ICL-INV is created using **MetaText** 1000/2000 – 1000 BOR. Additionally, when we create contract 2000-ICL-INV, a corresponding contract 1000-ICL-BOR is created using **MetaText** 1000/2000 – 2000 INV. Although the contracts are

4 Contract-Specific Configuration

exactly the same (**CoCode** 2000 is lending money to **CoCode** 1000), the difference comes down to what company code the contract is manually created on. If we'd like the contracts to mirror regardless of what side of the contract is created or changed, we must maintain this table from the creation perspective on both sides of the contract. In other words, for each company code pair, we must make a total of eight entries to configure this table to completion.

CFM: Maint. View Mapping Table Prod./Trans.Type Mirr.Trans.					
CoCode	Prod. type	Trans.ty...	MetaText	Direction	Function
1000	ICL	BOR	1000/2000 - 1000 BOR	Outgoing	Mirror Image
1000	ICL	BOR	1000/2000 - 2000 INV	Incoming	Mirror Image
1000	ICL	INV	2000/1000 - 1000 INV	Outgoing	Mirror Image
1000	ICL	INV	2000/1000 - 2000 BOR	Incoming	Mirror Image
2000	ICL	BOR	2000/1000 - 1000 INV	Incoming	Mirror Image
2000	ICL	BOR	2000/1000 - 2000 BOR	Outgoing	Mirror Image
2000	ICL	INV	1000/2000 - 1000 BOR	Incoming	Mirror Image
2000	ICL	INV	1000/2000 - 2000 INV	Outgoing	Mirror Image

Figure 4.20 Mapping Products and Transactions to Mirroring Functions

The following fields are required within the table:

- CoCode
- Prod. Type
- Trans.Type (transaction type)
- MetaText
- Direction
- Function

Processing Incoming Data

After we've mapped the product and transaction types, the next step is to process incoming data. We can find the configuration table at **Financial Supply Chain Management • Treasury and Risk Management • Transaction Manager • General Settings • Transaction Management • Intercompany Trading: Distribution of Mirror Transactions • ICT – Process Incoming Data**.

In this configuration step, we maintain the incoming intercompany loan pairs from the perspective of the receiving company code. As with the previous configuration node, if our intent is to have the transactions mirror regardless of whether the contact is created or changed on the borrowing or investment side of the trade, we'll have to maintain each company code combination within this table. The first step is to maintain the **CoCode**, **Prod. Type** and **Trans.Type** combination along with the **Activity** and **Activ.Cat.**, as shown in Figure 4.21. We then map this to the corresponding business partner/**Counterparty** that is associated with that company code. The **Function** column value will always be set to **Mirror Image** since that's the only function supported in this table for intercompany loans.

As with the previous configuration node, for each loan pair in scope, we must make a total of eight entries in this table for every company code pair.

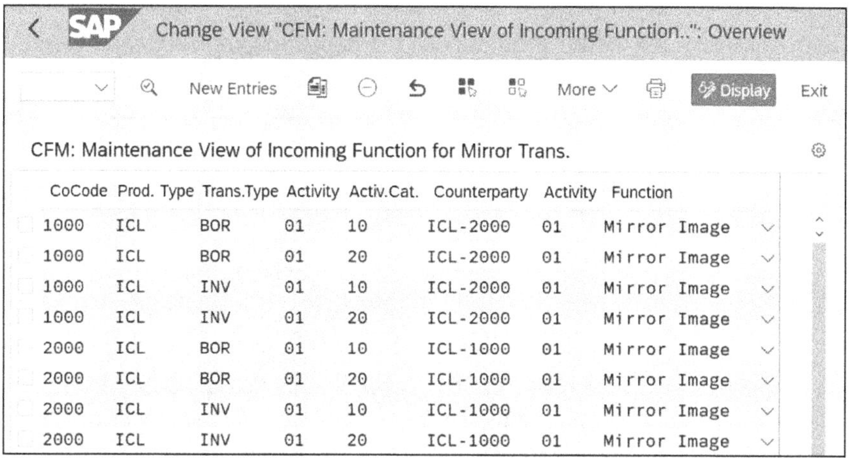

CoCode	Prod. Type	Trans.Type	Activity	Activ.Cat.	Counterparty	Activity	Function
1000	ICL	BOR	01	10	ICL-2000	01	Mirror Image
1000	ICL	BOR	01	20	ICL-2000	01	Mirror Image
1000	ICL	INV	01	10	ICL-2000	01	Mirror Image
1000	ICL	INV	01	20	ICL-2000	01	Mirror Image
2000	ICL	BOR	01	10	ICL-1000	01	Mirror Image
2000	ICL	BOR	01	20	ICL-1000	01	Mirror Image
2000	ICL	INV	01	10	ICL-1000	01	Mirror Image
2000	ICL	INV	01	20	ICL-1000	01	Mirror Image

Figure 4.21 Maintaining Incoming Functions for Mirror Transactions

Assigning Company Codes to Partners

The final configuration step required for mirroring of intercompany loans is to assign the company code to a partner. We can find this by using the following menu path: **Financial Supply Chain Management • Treasury and Risk Management • Transaction Manager • General Settings • Transaction Management • Intercompany Trading: Distribution of Mirror Transactions • ICT – Assign Company Code to Partner**. This configuration node maps the SAP S/4HANA **CoCode** to the business **Partner** that the **CoCode** corresponds to within the master data setup. In this example, **CoCode** 1000 is mapped to ICL-1000 and **CoCode** 2000 is mapped to business partner ICL-2000 (see Figure 4.22). This is a critical setup within the mirroring functionality, and without this configuration, SAP S/4HANA would have no idea what company codes correspond with which business partners. Every company code in scope for intercompany loans requires an entry in this table.

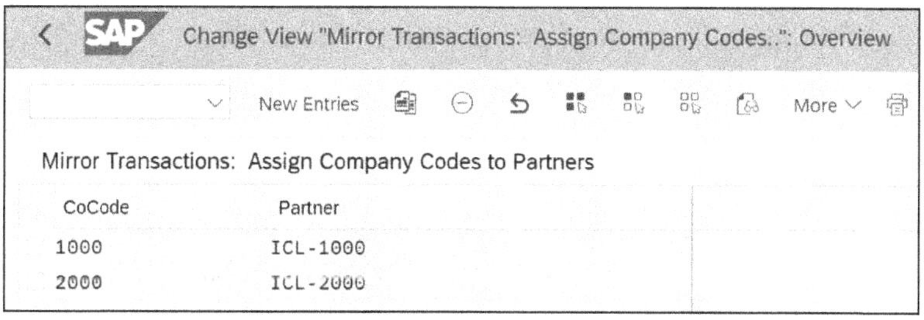

CoCode	Partner
1000	ICL-1000
2000	ICL-2000

Figure 4.22 Mirror Transactions: Assigning Company Codes to Partners

4.1.3 Credit Facilities

The majority of money market contracts are set up similarly to one another, with the exception of credit facilities, which are distinct from traditional debt or investment instruments. A credit facility is a loan arrangement between a counterparty and a borrower that provides access to a predetermined amount of funds for various operational purposes, such as obtaining working capital, making capital expenditures, and meeting other financial needs. This arrangement allows the borrower to draw funds up to a specified limit, repay them, and draw again as needed during the term of the facility. Credit facilities are common tools within treasury departments, and they ensure the necessary liquidity to manage cash flow, finance growth, and handle unexpected cash outlays.

Unlike traditional debt or investment contracts, credit facilities function as lines of credit for businesses, and they feature unique structural characteristics and cash flows. In SAP S/4HANA, the process begins with creating the credit facility, which defines the structure of the contract, the overall credit line, credit facility syndicates, and any associated fee structures. Once we've established the facility, we manage any drawings on the facility by creating a new contract that specifies the term, drawing amount, and interest rate. We then attach this newly created contract to the facility.

We maintain both the facility and the drawing objects as separate contracts, but we link them together when creating the drawing object. Understanding this concept is important before starting configuration, and it will provide clarity as we build out the configuration for a credit facility. We'll now jump into the configuration of a credit facility, starting with product type configuration.

Product Type

As with all money market instruments within treasury and risk management, the very first step of building out any contract is to start with the product type. We navigate to the product type configuration via the **Financial Supply Chain Management • Treasury and Risk Management • Transaction Manager • Money Market • Transaction Management • Product Types** menu path. For the most part, the configuration is straightforward and similar to that of other debt and investment product types. The product category we use for all facilities is **Prod. Category** "560" (**Facility**).

There are two types of facilities that we can set up under product category 560: a bilateral or a syndicated facility. A bilateral facility is originated by a single counterparty or lender, and it represents a straightforward relationship between the borrower and the lender. This type of facility is typically easier to manage due to the involvement of only two parties. In contrast, a syndicated facility involves multiple lenders or counterparties. These lenders pool their funds to provide a larger credit facility to the borrower, spreading the credit risk among all of the counterparties involved. A syndicated facility is usually organized by one or more lead arrangers, who handle the administrative tasks and coordinate the group of lenders. This type of facility is more complex, as all of the counterparties involved in the credit facility must be maintained while a bilaterial facility involves a single counterparty.

When creating the product type, we determine whether to set it up as a syndicated or a bilateral facility. In Figure 4.23, we can see that this facility is being set up as a **Syndicated Facility** in the **Facility Category**. The **Fee Calculation** dropdown list is applicable only to syndicated facilities and doesn't apply to bilateral facilities. The two options available for this field are as follows:

- **Do Not Include Drawing Objects Assigned to Partners**
  In this option, the fee calculation is applied exclusively to drawing objects associated with credit lines that don't have specific partners assigned. When we create a drawing object within a credit facility, the system calculates fees based on whether the drawing object is linked to a designated credit line. If the drawing object is not tied to a specific partner within the credit line, the system generates a fee according to predefined criteria.

- **Include Drawing Objects Assigned to Partners**
  In this option, the fee calculation encompasses all drawing objects assigned to credit lines, regardless of whether specific partners have been chosen or not.

The fee process is designed to be comprehensive and ensure that all drawing objects linked to credit lines are subject to the relevant fee structures. This includes drawing objects that have designated partners as well as those that don't.

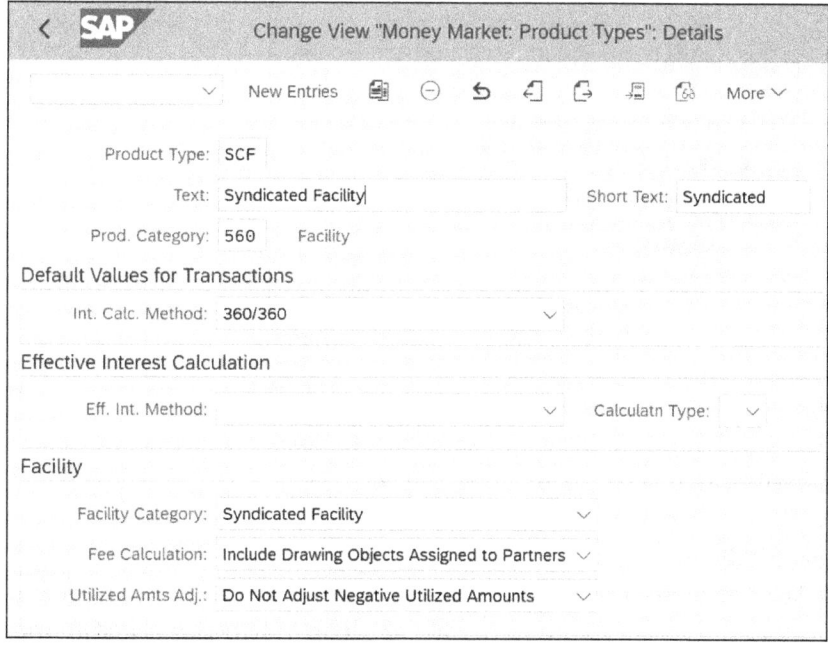

Figure 4.23 Creating Product Type for Syndicated Facility

4   Contract-Specific Configuration

Transaction Types

Moving on to transaction type configuration for both a facility and a syndicated facility, we navigate to **Financial Supply Chain Management • Treasury and Risk Management • Transaction Manager • Money Market • Transaction Management • Transaction Types**. The two transactions categories available are as follows:

- 100: Assigned
- 200: Obtained

Transaction category 200 is the most frequently used category because it represents a facility that has been obtained by or extended to a borrower. This category is the most common to most transactions within treasury operations, and it reflects standard lending or borrowing practices. On the other hand, transaction category 100, labeled as **Assigned**, pertains to the issuance of a facility by an organization. This scenario is very uncommon and typically doesn't occur within most treasury organizations, so transaction category 100 is not frequently utilized in standard treasury processes. As shown in Figure 4.24, we're using **Transaction Cat** "200" (**Obtained**).

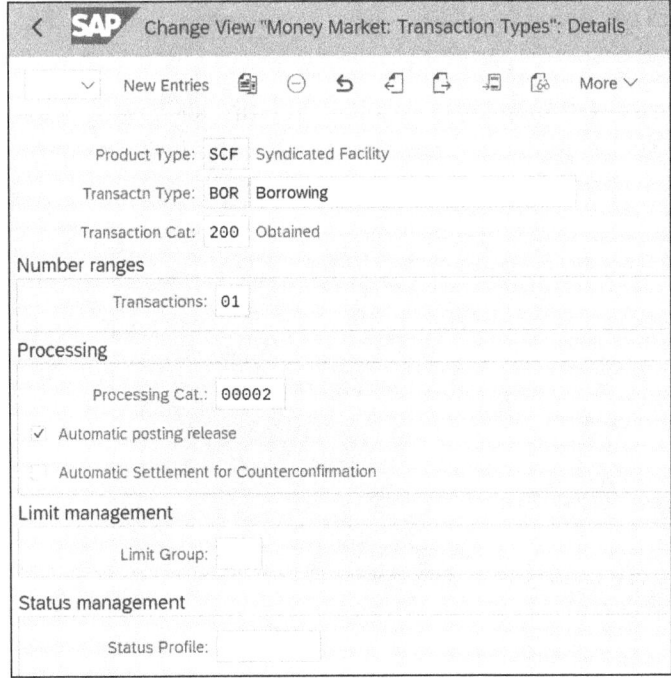

Figure 4.24   Creating Syndicated Facility Transaction Type

Defining and Assigning Flow Types

One of the key functions of the credit facility in SAP S/4HANA is providing precise tracking and management of the overall credit line, along with the associated drawing objects. This involves a monitoring system that keeps a close eye on the available credit, the portions that have been utilized, and the specific transactions linked to these drawdowns. This is how SAP ensures that borrowers maintain a clear and up-to-date understanding of their credit line's status at all times throughout the life of the facility.

Additionally, the SAP S/4HANA credit facility plays an important role in managing and tracking the various fees associated with the credit agreement. These fees can be complex, often involving calculations based on both the total credit line and the amount drawn from it. The system's ability to accurately handle these complex fee structures is essential for ensuring transparency and compliance with the terms of the credit agreement. By integrating these functions, SAP S/4HANA's credit facility provides a robust framework for managing credit lines, offering real time-insights and control concerning the health of the facility arrangement.

There are four main fees commonly associated with most credit facilities:

- **Unutilized fee**
  An unutilized fee, also known as a commitment fee, is charged on the unused portion of the credit facility. This fee is typically a fixed percentage rate applied to the undrawn amount of the credit facility

- **Utilized fee**
  A utilized fee is charged on the portion of the credit facility that the borrower actually draws. It's typically calculated as a fixed percentage and is charged periodically, either monthly or quarterly.

- **Overdrawn fee**
  An overdrawn fee is assessed when the borrower exceeds the agreed-upon credit limit of the facility. This fee is usually higher than other fees as it is not commonly assessed and is meant to penalize borrowers from exceeding the credit line.

- **Total credit line fee**
  The total credit line fee is assessed based on the total amount of the credit line, regardless of whether drawings on the facility take place. The fee is assessed as a fixed or floating rate and applied to the entire credit line.

We can now create flow types associated with the credit facility fees by navigating to **Financial Supply Chain Management • Treasury and Risk Management • Transaction Manager • Money Market • Transaction Management • Flow Types • Define and Assign Flow Types**. To account for the cash flows associated with the credit facility fees, we'll create a variety of flow types. For more information about creating flow types, see Chapter 3, Section 3.13. The key **Flow Type**(s) required during this step are as follows:

- **Facility Charges: Not Utilized (FC00)**
  This flow type captures fees associated with the unutilized portion of the credit facility, as shown in Figure 4.25.

- **Facility Charges: Utilized (FC01)**
  This flow type captures fees associated with the utilized or drawn portion of the credit facility, as shown in Figure 4.26.

4   Contract-Specific Configuration

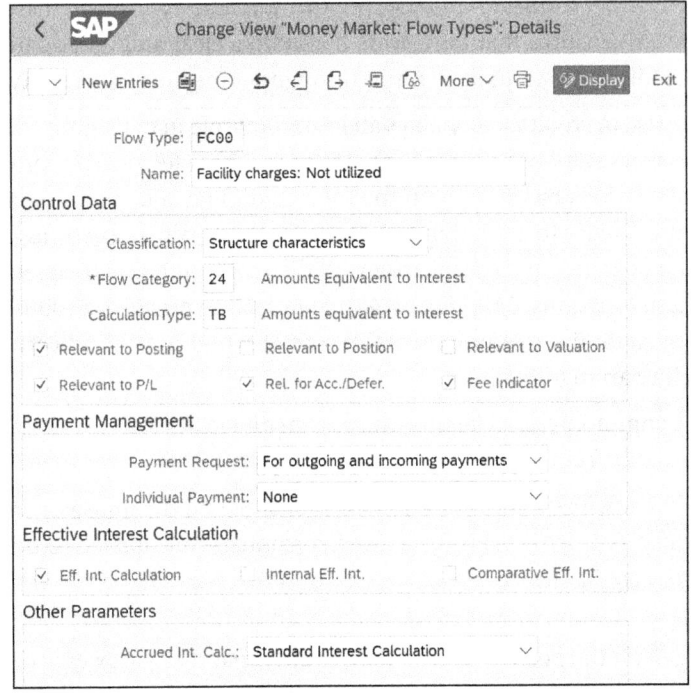

Figure 4.25 Facility Charge FC00

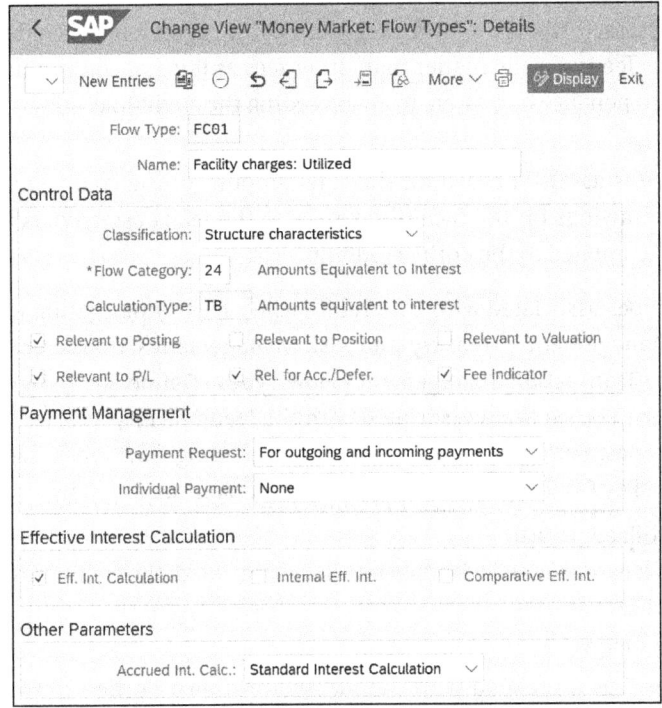

Figure 4.26 Facility Charge FC01

- **Facility Charges: Overdrawn (FC02)**

    This flow type captures fees associated with overdrafts on the facility, as shown in Figure 4.27.

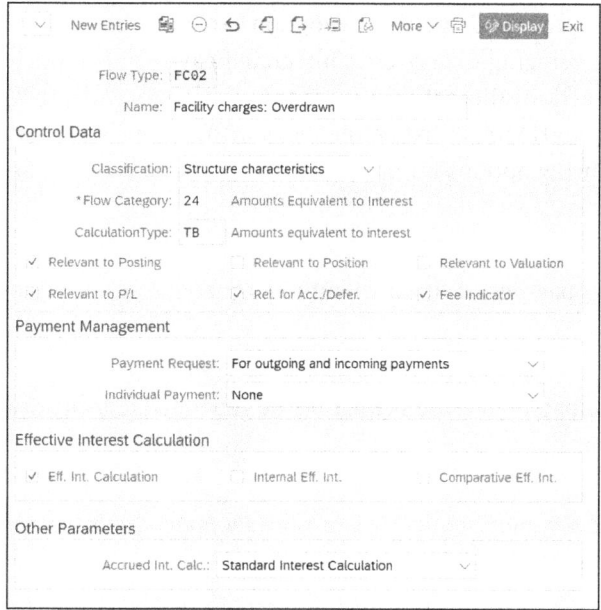

Figure 4.27 Facility Charge FC02

- **Facility Charges: Credit Line (FC03)**

    This flow type captures the fees associated with the overall credit line assessed by the counterparty, as shown in Figure 4.28.

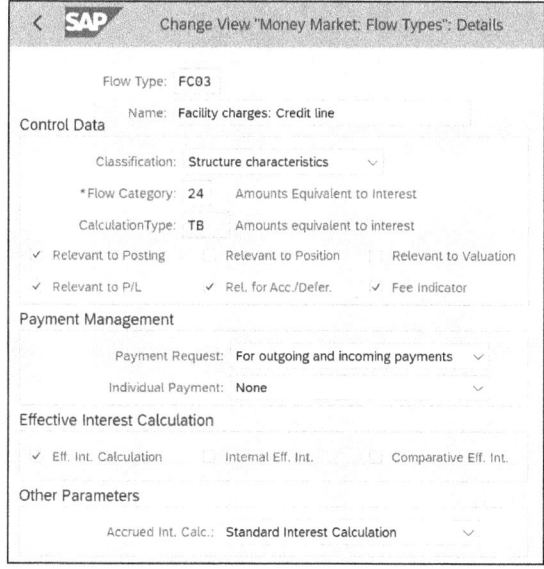

Figure 4.28 Facility Charge FC03

4   Contract-Specific Configuration

As shown in Figure 4.29, setting up the fee related to the facility follows a process similar to configuring any other fee within treasury and risk management. An important consideration is whether the fee will trigger a **Payment Request**. Fees associated with credit facilities can be structured in multiple ways, providing us with the flexibility to determine how payments will be processed. We have the option to handle the payment directly within SAP S/4HANA, leveraging the system's integrated payment process to automate and manage the transaction. Alternatively, we can choose to process the payment externally, outside of SAP S/4HANA, if that better aligns with our organization's specific requirements. The following options for how to handle payment requests are available in a dropdown list:

- **None**
  Selecting this option means a payment request will not be generated as a result of this flow.

- **For outgoing payments**
  Selecting this option means a payment request will only be generated for outgoing payments using this flow type.

- **For incoming payments**
  Selecting this option means a payment request will only be generated for incoming payments using this flow type.

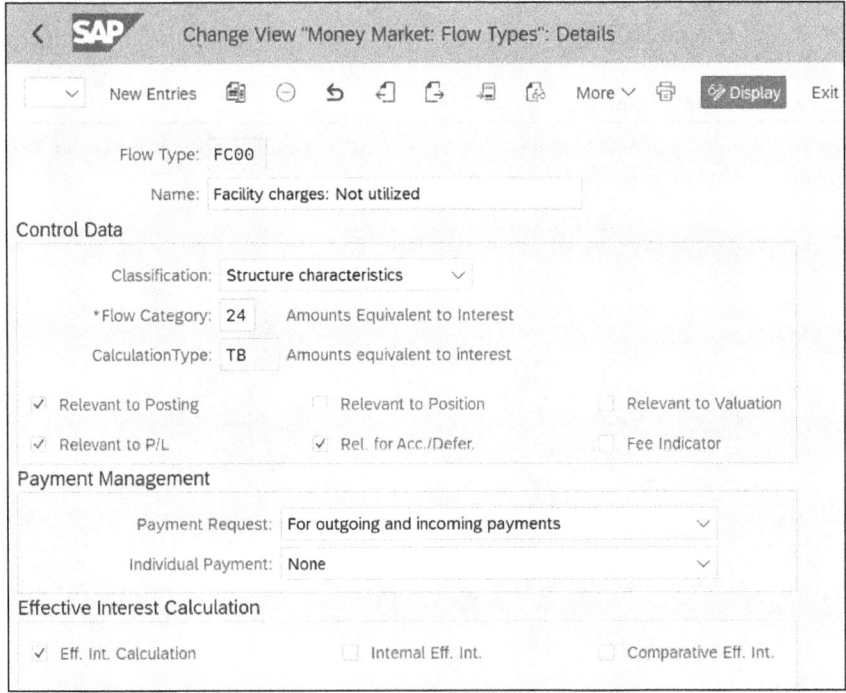

Figure 4.29 Creating Facility Charges Flow Types

- **For outgoing and incoming payments**
  Selecting this option means a payment request will be generated for both incoming and outgoing payments using this flow type.

This flexibility is valuable in scenarios where different fee structures or payment schedules are in place, and it allows us to tailor the process to meet the unique requirements of the credit facility. Whether we opt for internal or external handling of the fees, SAP S/4HANA provides the necessary tools to ensure that payments are accurately tracked, recorded, and reported.

With all the fee-related flow types now established, we can proceed to the next step of assigning them to the appropriate product and transaction type combinations. This step is crucial as it ensures that the fee structures we've defined are correctly linked to the specific financial products and transactions within SAP S/4HANA as the fees occur within the facility. By performing this step, we create a seamless integration that allows for accurate tracking, management, and reporting of fees across the facilty as the balance changes within the facility. The assignment of the flow types is shown in Figure 4.30 (again, refer to Chapter 3, Section 3.13, for step-by-step instructions).

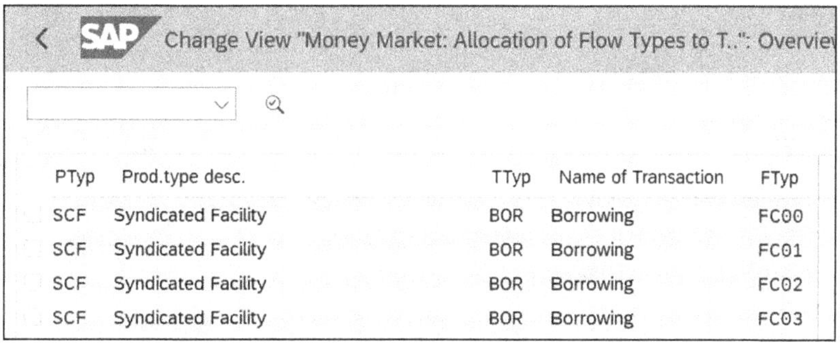

PTyp	Prod.type desc.	TTyp	Name of Transaction	FTyp
SCF	Syndicated Facility	BOR	Borrowing	FC00
SCF	Syndicated Facility	BOR	Borrowing	FC01
SCF	Syndicated Facility	BOR	Borrowing	FC02
SCF	Syndicated Facility	BOR	Borrowing	FC03

Figure 4.30 Assigning Product Type SCF to Applicable Flow Types

Defining Update Types and Assigning Usages

Now that we've created all the flow types for the various fees associated with a facility, we can create specific update types for each flow type. In this example, we're setting up a common scenario where the business is establishing a borrowing facility issued by a counterparty. This means that all the fees associated with the facility are assessed by the issuer, resulting in outgoing amounts. Thus, we'll configure our fees to reflect this by creating outbound cash flow update types for each flow type created in the previous step.

Typically, we would set up two update types for each flow type, which are applicable if we're creating two transaction types that require both inbound and outbound flows. However, in this example, since the facility is issued by a counterparty, the fees will always be outbound cash flows. We can find this configuration at **Financial Supply Chain Management • Treasury and Risk Management • Transaction Manager • Money Market • Transaction Management • Flow Types • Define Update Types and Assign Usages**. We previously defined update types in Chapter 3, Section 3.14.

In Figure 4.31, we've created the four outbound update types that correspond to the four flow types that we've created. In Figure 4.32, we then assign the four update types to the correct usage (**Upd. Ty. Usage**), which is **Transactions Management**.

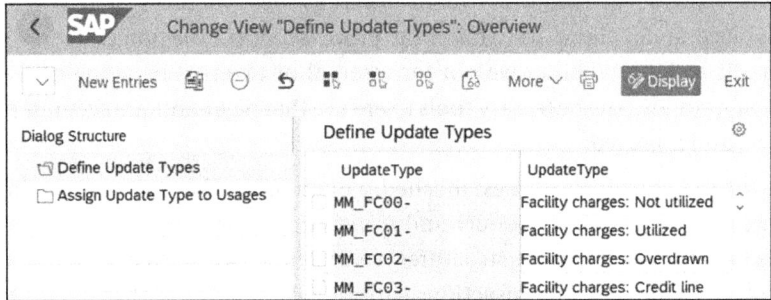

Figure 4.31 Defining Update Types for Facility Charges

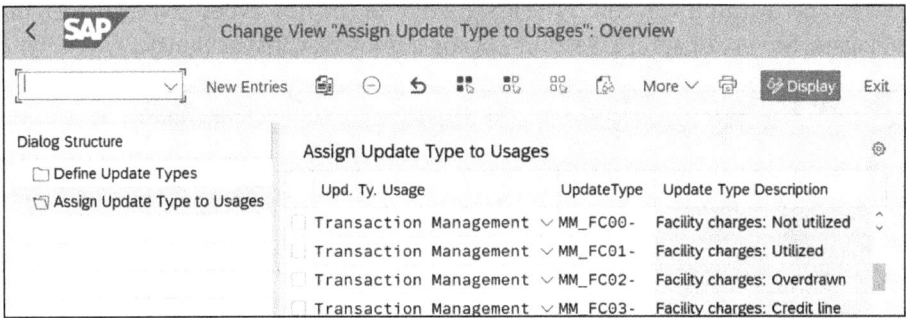

Figure 4.32 Assigning Update Types to Usages

After creating and assigning the usages for the four update types, the final key step is linking these update types to their corresponding flow types, as shown in Figure 4.33. This linkage is important because it ensures that the financial transactions within the SAP S/4HANA system are accurately recorded and processed according to the specific update types we've defined during the fee calculation and posting process.

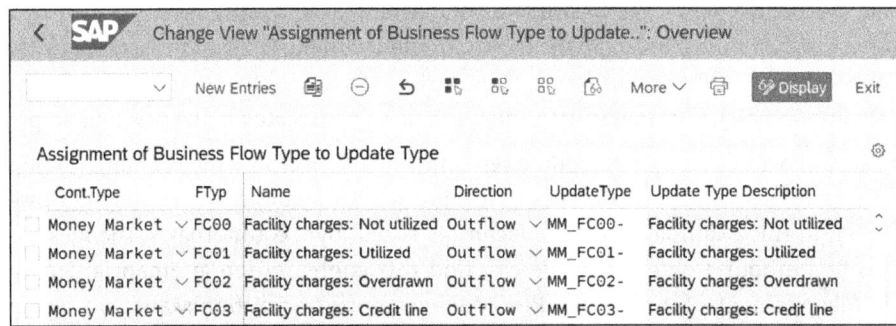

Figure 4.33 Assigning Credit Facility Flow Types to Update Types

Condition Types and Condition Type Assignment

As discussed earlier in this chapter, condition types are essential for any flow type that requires a calculation to determine the flow amount. This is the case for all the fees associated with a credit facility, as these fees are derived and calculated based on the total credit line and the amount of drawings linked to the facility. Because of this, each credit facility fee must have a condition type assigned to the relevant flow type to ensure accurate fee calculation.

Given the complexity and importance of accurately calculating these facility fees, it's crucial to create specific condition types and align them with their respective flow types. Using these condition types is necessary to effectively track cash payments related to facility fees. The configuration of condition types is especially important because it involves selecting the appropriate profile category. The profile category is a key element that determines how the fee is calculated, ensuring that the system correctly reflects the specific fee that is being calculated.

To properly manage and track these fees, we should establish four distinct condition types and link them to their corresponding flow types. This setup will enable the proper calculation and reporting of the fees tied to the credit facility, allowing visibility into the facility fees and ensuring compliance with the terms of the credit agreement. The profile categories we choose for each condition type will drive the calculation method, ensuring that each fee is accurately computed according to the specific financial parameters of the credit facility. We should use the following profile categories to align with the corresponding flow types one to one to avoid any confusion in the configuration and setup. The four main condition types used within a facility are as follows:

- Utilized
- Not Utilized
- Overdrawn
- Total Credit Line

To start the configuration process, we navigate to **Financial Supply Chain Management • Treasury and Risk Management • Transaction Manager • Money Market • Transaction Management • Condition Types • Define Condition Types**. Then, we need to set up the following condition types:

- **Facility charges: Not utilized (1204)**
  We use this condition to calculate the unutilized fee on the facility, and it's mapped to the unutilized fee flow type (**Generate FlTyp**) FC00. The **Profile Categ.** we use to calculate the unutilized fee is **Not Utilized**, as shown in Figure 4.34.
- **Facility charges: Utilized (1205)**
  We use this condition to calculate the utilized fee on the facility, and it's mapped to the utilized fee flow type (**Generate FlTyp**) FC01. The **Profile Categ.** we use to calculate the unutilized fee is **Utilized**, as shown in Figure 4.35.

4   Contract-Specific Configuration

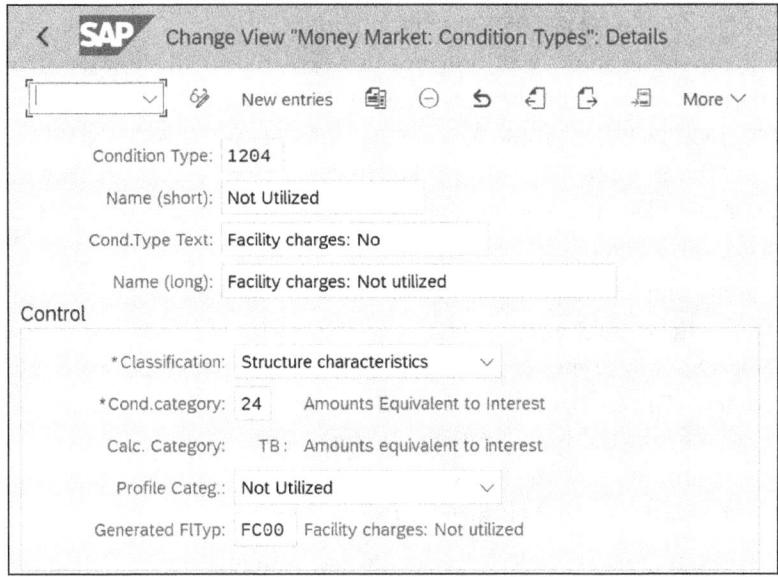

Figure 4.34 Creating Condition Type 1204

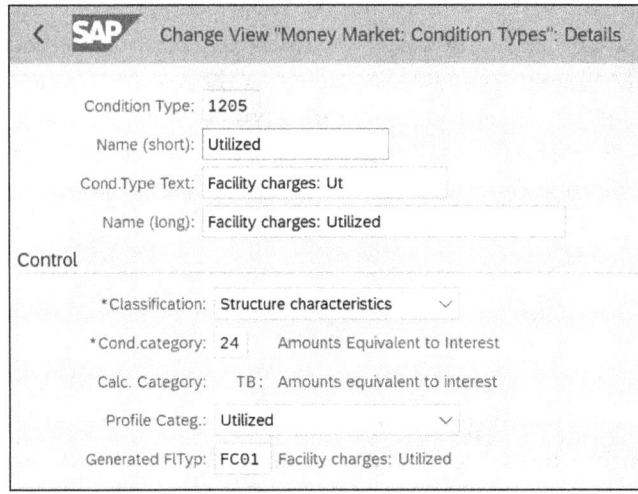

Figure 4.35 Creating Condition Type 1205

- **Facility charges: Overdrawn (1206)**
  We use this condition to calculate the overdrawn fee on the facility and it's mapped to the overdrawn fee flow type (**Generate FlTyp**) FC02. The **Profile Categ.** we use to calculate the overdrawn fee is **Overdrawn**, as shown in Figure 4.36.

4.1 Debt and Investments

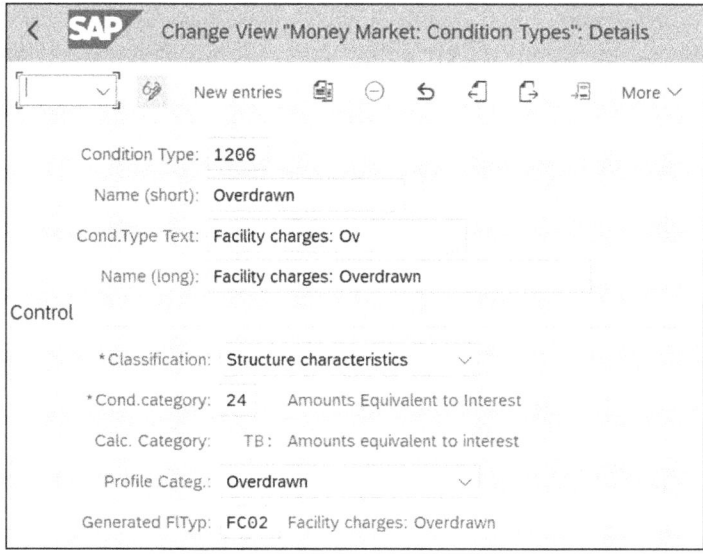

Figure 4.36 Creating Condition Type 1206

- **Facility charges: Credit Line (1207)**
  We use this condition to calculate the total credit line on the facility, and it's mapped to the credit line flow type (**Generate FlTyp**) FC03. The **Profile Categ.** we use to calculate the unutilized fee is **Total Credit Line**, as shown in Figure 4.37.

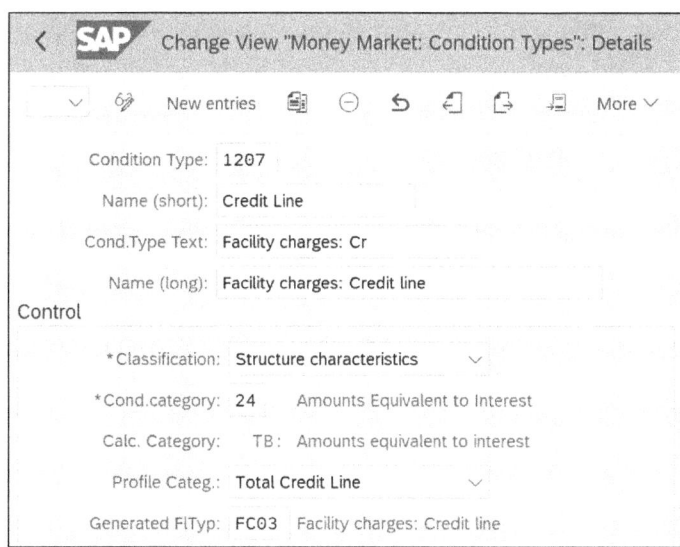

Figure 4.37 Creating Condition Type 1207

Now that we have successfully created the condition types, we can move on to assigning the condition types to the product and transaction types for our syndicated facility (as we detailed in Chapter 3, Section 3.15.2). As we can see in Figure 4.38, we've assigned all of the previously created condition types to our product type SCF with transaction type (**TTyp**) "BOR."

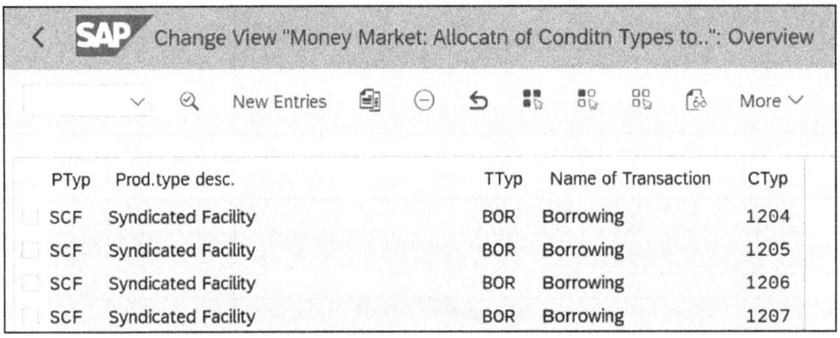

PTyp	Prod.type desc.	TTyp	Name of Transaction	CTyp
SCF	Syndicated Facility	BOR	Borrowing	1204
SCF	Syndicated Facility	BOR	Borrowing	1205
SCF	Syndicated Facility	BOR	Borrowing	1206
SCF	Syndicated Facility	BOR	Borrowing	1207

Figure 4.38 Assigning Product Types to Condition Types for Syndicated Facility

Defining Partner Rank: Syndicated Facilities

Next, we can define ranks for partners. These ranks include specific information pertaining to the specific tasks and responsibilities that each partner plays in relation to the facility. By assigning ranks, we can differentiate the roles of various partners involved in the issuance of the facility. This can become useful when there are a number of syndicates within a facility, and it gives the users the ability to differentiate various facility syndicates. This configuration is important because it clearly defines each partner's duties as explicitly outlined and understood within the context of the facility contract.

To start the configuration, we navigate to **Financial Supply Chain Management • Treasury and Risk Management • Transaction Manager • Money Market • Transaction Management • Syndicated Facilities • Define Partner Rank – Syndicated Facilities.**

As shown in Figure 4.39, we've established two rankings to represent the different roles within the facility. **Ranking** 0001, which we've designated as the "Syndicate Manager," will be assigned to the primary relationship bank that organizes the facility and maintains the main relationship. Additionally, we've defined **Ranking** 0002 to represent the syndicate banks participating in the facility. This ranking will be assigned to all other banks involved in the syndicated facility.

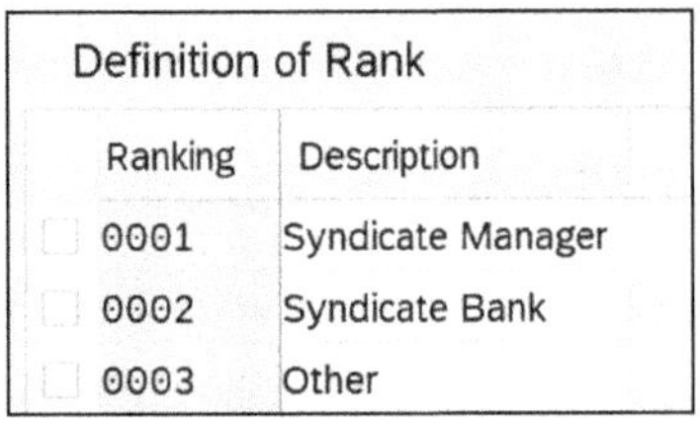

Definition of Rank	
Ranking	Description
0001	Syndicate Manager
0002	Syndicate Bank
0003	Other

Figure 4.39 Defining Ranks for Syndicated Facilities

4.1.4 Letters of Credit

Configuring a letter of credit product type within treasury and risk management closely follows how we configure a debt and investment product type. The setup process relies on principles similar to those we employ for traditional financial products. However, there's a key difference between the structure of a letter of credit and that of other financial products. Unlike standard debt instruments, letters of credit involve complex relationships and compliance requirements. These requirements pertain not only to the beneficiary and the issuing bank but also to the terms and conditions outlined within the letter of credit itself.

Setting up a letter of credit involves defining these terms and conditions in SAP S/4HANA to ensure they accurately reflect the contractual obligations among the buyer, the seller, and the financial institution. The process includes establishing the payment terms, shipping details, and documentation requirements, all of which must align with the commercial contract governing the transaction. This setup is important for facilitating seamless execution and monitoring of the letter of credit, ensuring that all parties adhere to the agreed terms and that the transaction complies with international trade standards.

Letters of credit within treasury and risk management are handled within the trade finance area of the configuration. We can find the configuration to set up a contract at **Financial Supply Chain Management • Treasury and Risk Management • Transaction Manager • Trade Finance**. The configuration of a letter of credit contract includes additional steps not found in other product types within treasury and risk management, and we'll detail those steps in this section.

Product Type

Similar to other contracts within treasury and risk management, the initial step in setting up a letter of credit involves creating the relevant product type. In our example, we'll designate the **Product Type** as "LCI" to signify the letter of credit. We can find the configuration node to set up the product types at **Financial Supply Chain Management • Treasury and Risk Management • Transaction Manager • Trade Finance • Transaction Management • Product Types • Define Product Types – Trade Finance**.

A letter of credit contract functions very differently from a debt or investment contract, so we recommend using **Prod. Category** "850," which is specifically classified as a letter of credit. Prior to the addition of the trade finance functionality in treasury and risk management, it was common to create a letter of credit using an interest rate instrument. This is still possible, but we'll find that using **Prod. Category** "850" gives us additional functionality unique to letters of credit. This product category is tailored to handling the complex requirements of letter of credit transactions, and it offers extensive configurability for executing these contracts. Selecting **Prod. Category** "850" ensures that the system can accommodate various scenarios and variations in letter of credit agreements, providing the necessary options to customize the contracts according to specific business requirements, such as different types of letters of credit, varied documentation requirements, and unique payment conditions.

While it's possible to select other product categories for letters of credit, doing so might limit our flexibility in managing the unique aspects of these financial instruments. **Prod. Category** 850 integrates well with the complex needs of letter of credit agreements, such as managing multiple stages of document verification, handling various types of letters of credit, and supporting intricate financial terms and conditions. In Figure 4.40, we're creating a product type "LCS" to represent letters of credit.

It's important to keep in mind what type of letter of credit we're creating; the system supports two trade finance categories (**TF Cat.**), and we must define the category when we create the product type, as follows:

- Letter of Credit
- Standby Letter of Credit

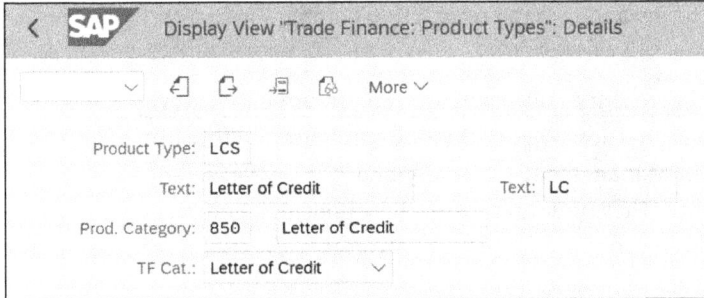

Figure 4.40 Creating Product Type for Letter of Credit Contract

Transaction Type

After we create the product type for letters of credit, the next step involves setting up the transaction types specific to the letter of credit. Letters of credit can be set up and issued by a corporation, or they can be received. Depending on the business requirements and whether they are issued or received, we may need to create two distinct transaction types: one for the issuance and another for the receipt of the letter of credit. To begin configuring these transaction types, we navigate to the **Financial Supply Chain Management • Treasury and Risk Management • Transaction Manager • Trade Finance • Transaction Management • Transaction Types • Define Transaction Types – Trade Finance** menu path.

Common terminology used when defining these transaction types includes *issue* and *receive*. For this example, we'll use those terms to clearly differentiate the transaction types. The "Issue" transaction type will represent the entity issuing the letter of credit, while the "Receive" transaction type will represent the entity receiving or benefiting from it. We'll map the "Issue" transaction to **Transaction Cat** "100" and the "Receive" transaction to **Transaction Cat** "200," as shown in Figure 4.41 and Figure 4.42. By defining the "ISS" and "REC" transaction types and mapping them to their respective product categories, we ensure that the letter of credit transactions are properly classified and managed.

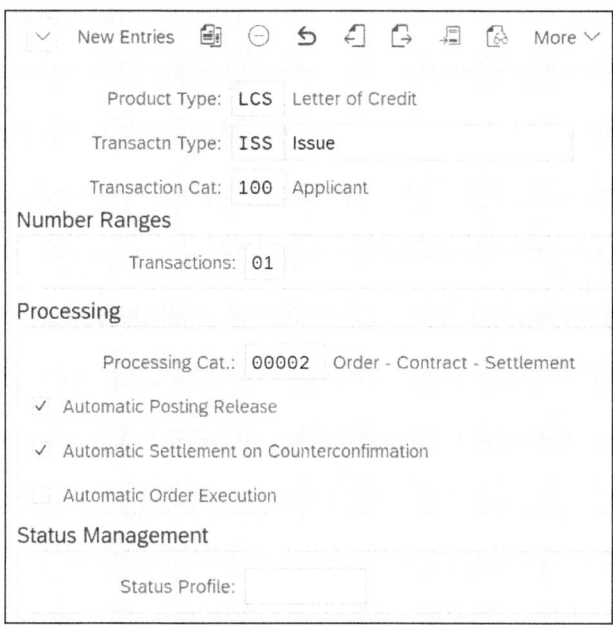

Figure 4.41 Issue Transaction Type for Letter of Credit

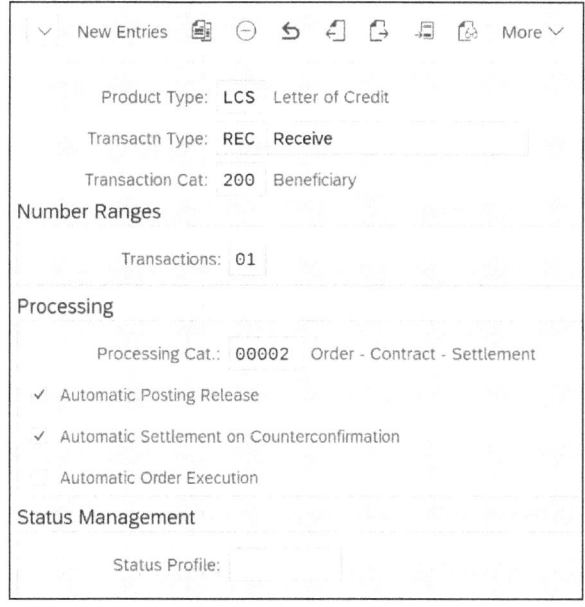

Figure 4.42 Receive Transaction Type for Letter of Credit

4   Contract-Specific Configuration

Flow Type

Much like the configuration process for other contract types within treasury and risk management, the setup of flow types follows a structured approach. To begin configuring flow types for trade finance transactions, we navigate to this menu path: **Financial Supply Chain Management • Treasury and Risk Management • Transaction Manager • Trade Finance • Transaction Management • Flow Types • Define Flow Types – Trade Finance Transactions**. This pathway provides access to the necessary configuration settings where we can define and customize flow types according to the specific requirements of trade finance transactions.

Within the configuration node for flow types, users can define various flow types that correspond to different aspects of trade finance transactions. These flow types encompass the financial movements and events associated with trade finance activities, including the disbursement of funds, the receipt of payments, and the handling of fees and charges. Commonly used **Flow Type**(s) required for letters of credit include the following:

- **Nominal Decrease (PD00)**
  This flow type represents the decrease of the letter of credit balance, as shown in Figure 4.43. With a letter of credit, there's no physical exchange of cash related to the principal amount, so a payment requestion for this flow type is not relevant.

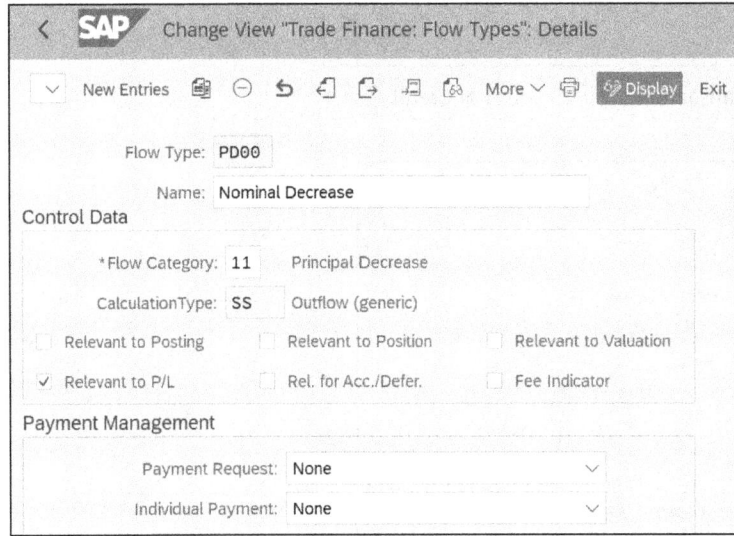

Figure 4.43 Letter of Credit Flow Type PD00

- **Nominal Increase (PI00)**
  This flow type represents the increase of the letter of credit balance, as shown in Figure 4.44. As with the principal decrease flow type, there's no physical exchange of cash related to the principal amount, so a payment requestion for this flow type is not relevant.

4.1 Debt and Investments

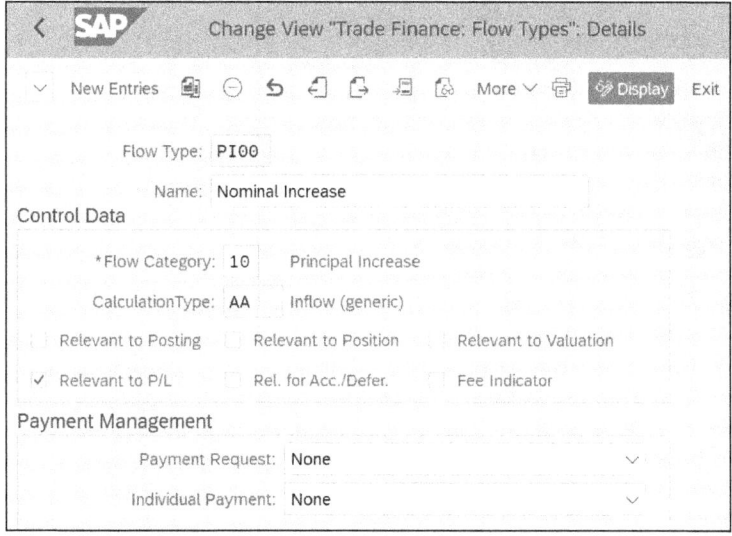

Figure 4.44 Letter of Credit Flow Type PI00

- **Acceptance Payment (AP00)**
  This flow type represents the acceptance payment of the letter of credit, as shown in Figure 4.45.

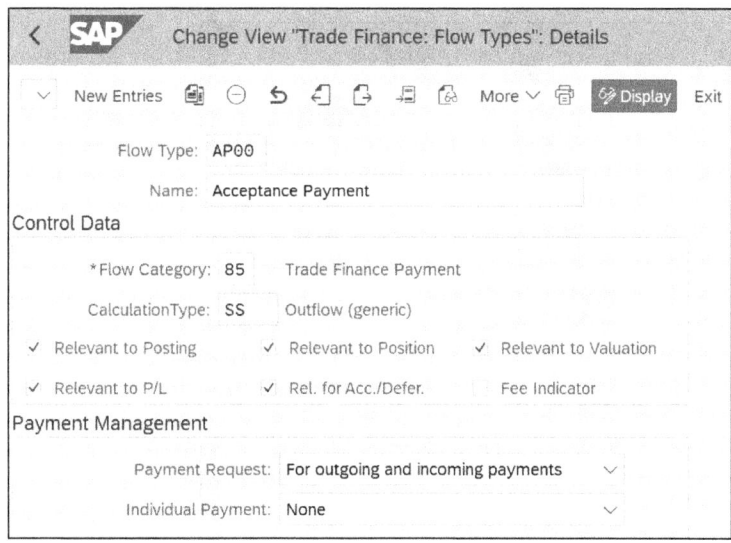

Figure 4.45 Letter of Credit Flow Type AP00

- **Charges (CH00)**
  We use this flow type to represent any agreed-upon charges associated with the letter of credit, as shown in Figure 4.46.

4   Contract-Specific Configuration

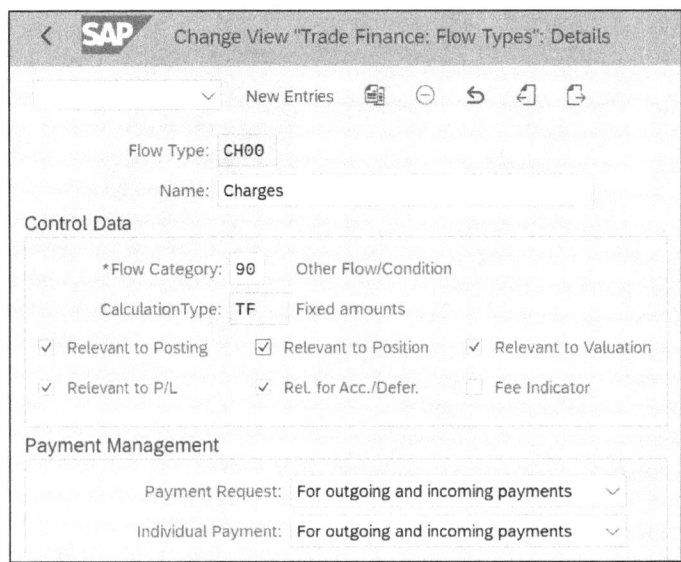

Figure 4.46 Letter of Credit Flow Type CH00

- **Fee on Total Amount (FE01)**
  This is the fee applied to the total amount of the letter of credit, as shown in Figure 4.47.

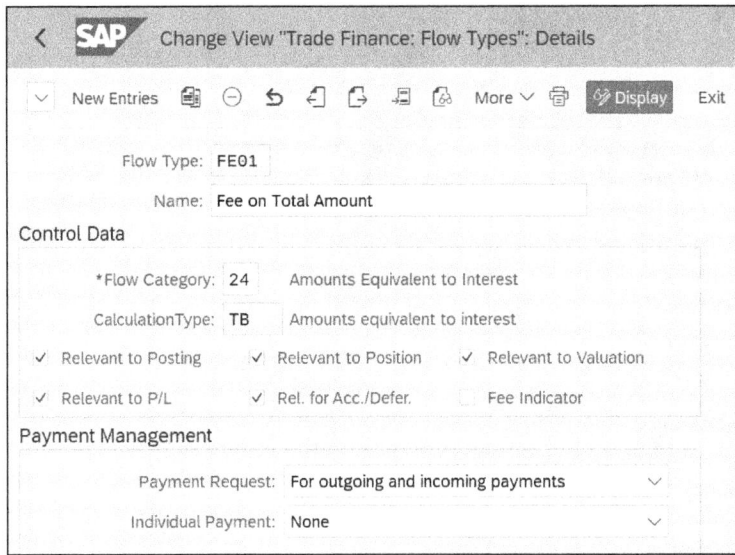

Figure 4.47 Letter of Credit Flow Type FE01

- **Fee on Presented Amount (FE02)**
  This is the fee on the presented or utilized amount of the letter of credit, as shown in Figure 4.48.

4.1 Debt and Investments

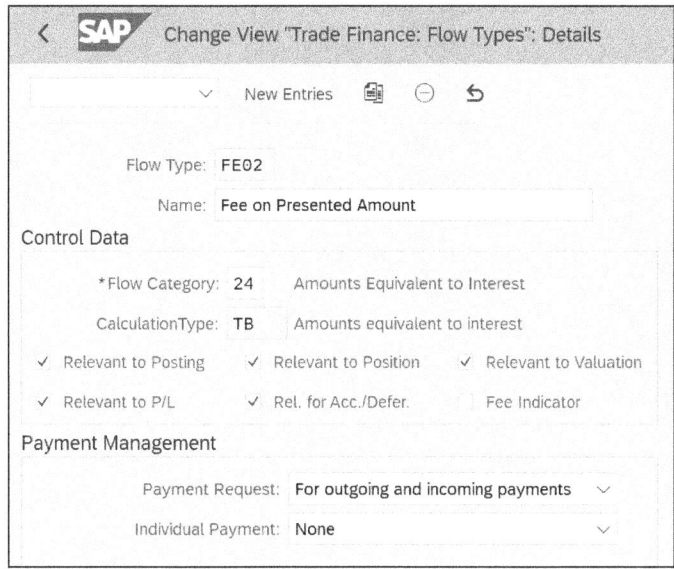

Figure 4.48 Letter of Credit Flow Type FE02

- **Fee on Available Amount (FE03)**
  This is the fee on the available amount of the letter of credit, as shown in Figure 4.49.

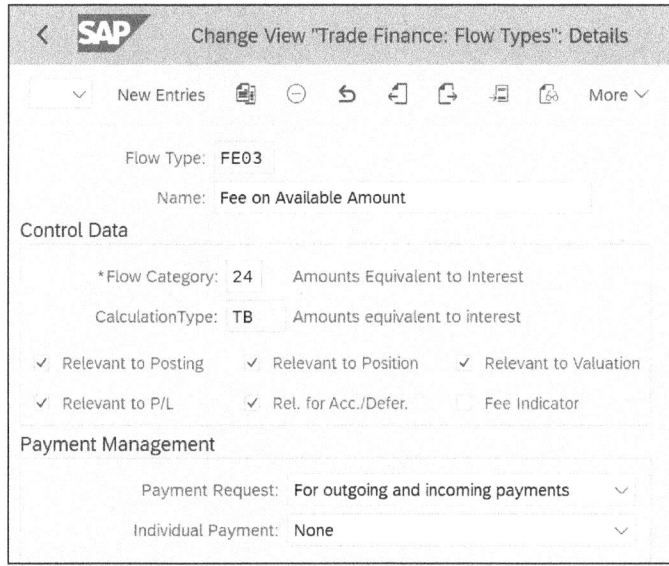

Figure 4.49 Letter of Credit Flow Type FE03

- **Fee on Overdraft Amount (FE04)**
  This is the fee on the portion of the credit facility that is overdrawn, as shown in Figure 4.50.

245

4   Contract-Specific Configuration

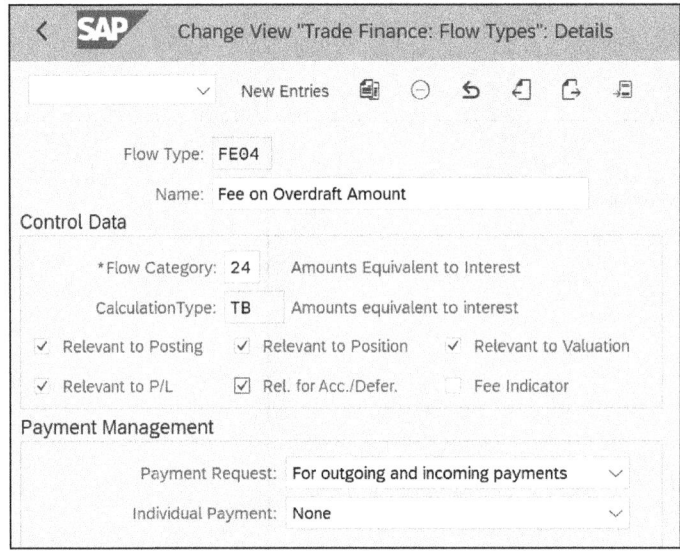

Figure 4.50 Letter of Credit Flow Type FE04

- **Payment Obligation (PO00)**
  This is the obligation the issuing bank has to pay the beneficiary, as shown in Figure 4.51.

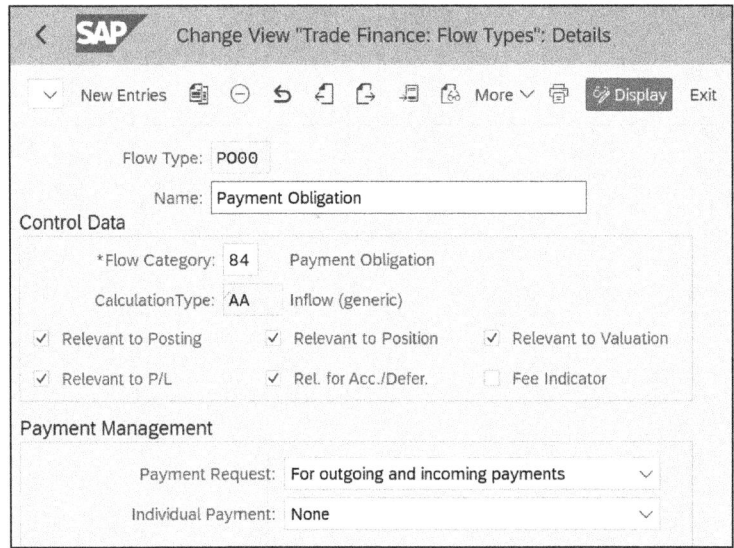

Figure 4.51 Letter of Credit Flow Type PO00

- **Remaining Credit Amount (RC00)**
  This flow type represents the remaining credit left on the letter of credit, as shown in Figure 4.52.

4.1 Debt and Investments

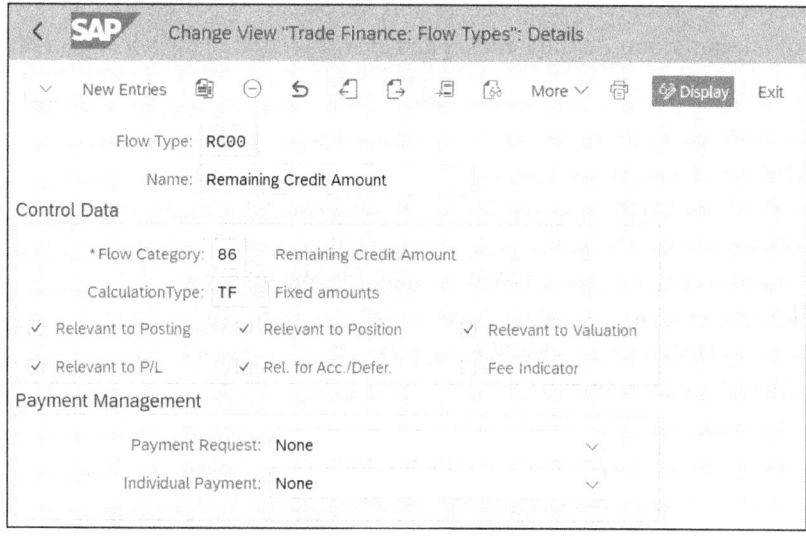

Figure 4.52 Letter of Credit Flow Type RC00

After we create the flow types, the subsequent step involves assigning them to the relevant components of the letter of credit product and transaction types. To execute this task, we navigate to the designated menu path: **Financial Supply Chain Management • Treasury and Risk Management • Transaction Manager • Trade Finance • Transaction Management • Flow Types • Assign Flow Types to Transaction Types – Trade Finance Transactions**. Within this configuration node, we assign the created flow types to specific product types and transaction types associated with the letter of credit, as shown in Figure 4.53.

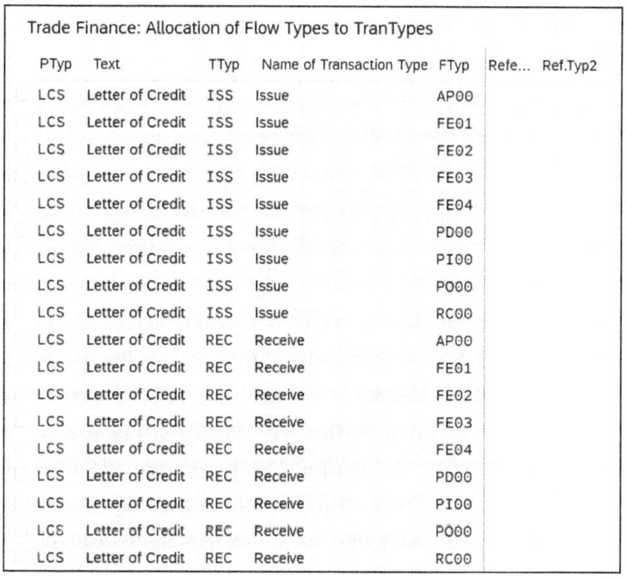

PTyp	Text	TTyp	Name of Transaction Type	FTyp	Refe...	Ref.Typ2
LCS	Letter of Credit	ISS	Issue	AP00		
LCS	Letter of Credit	ISS	Issue	FE01		
LCS	Letter of Credit	ISS	Issue	FE02		
LCS	Letter of Credit	ISS	Issue	FE03		
LCS	Letter of Credit	ISS	Issue	FE04		
LCS	Letter of Credit	ISS	Issue	PD00		
LCS	Letter of Credit	ISS	Issue	PI00		
LCS	Letter of Credit	ISS	Issue	PO00		
LCS	Letter of Credit	ISS	Issue	RC00		
LCS	Letter of Credit	REC	Receive	AP00		
LCS	Letter of Credit	REC	Receive	FE01		
LCS	Letter of Credit	REC	Receive	FE02		
LCS	Letter of Credit	REC	Receive	FE03		
LCS	Letter of Credit	REC	Receive	FE04		
LCS	Letter of Credit	REC	Receive	PD00		
LCS	Letter of Credit	REC	Receive	PI00		
LCS	Letter of Credit	REC	Receive	PO00		
LCS	Letter of Credit	REC	Receive	RC00		

Trade Finance: Allocation of Flow Types to TranTypes

Figure 4.53 Assigning Letter of Credit Transaction Types to Applicable Flow Types

4   Contract-Specific Configuration

By aligning flow types with product type LCS and transaction types ISS and REC, we ensure that all relevant financial movements and events related to the letter of credit are accurately captured and recorded when we create the contracts.

Update Types

As with all other product types in treasury and risk management, we now have to create update types that correspond to each of the previously created flow types. To start this process, we navigate to **Financial Supply Chain Management • Treasury and Risk Management • Transaction Manager • Trade Finance • Transaction Management • Update Types • Define Update Types and Assign Usages**. For each flow type, we create two update types that correspond to the cash flow directions, as shown in Figure 4.54.

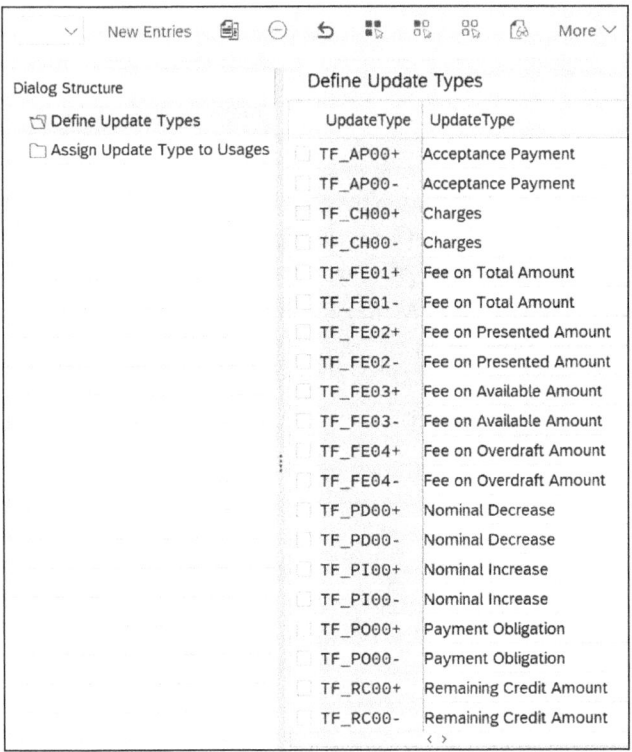

Figure 4.54 Creating Trade Finance Update Types

Then, we allocate the update types to the designated update type usage (**Upd. Ty. Usage**), namely **Transaction Management**, as shown in Figure 4.55. This phase of the process is crucial because it's important to ensure that we appropriately assign all transaction types within this node. By associating the update types (**UpdateType**) with the **Transaction Management** usage, organizations establish a direct linkage between the defined update types and the specific transactional activities managed within this context.

4.1 Debt and Investments

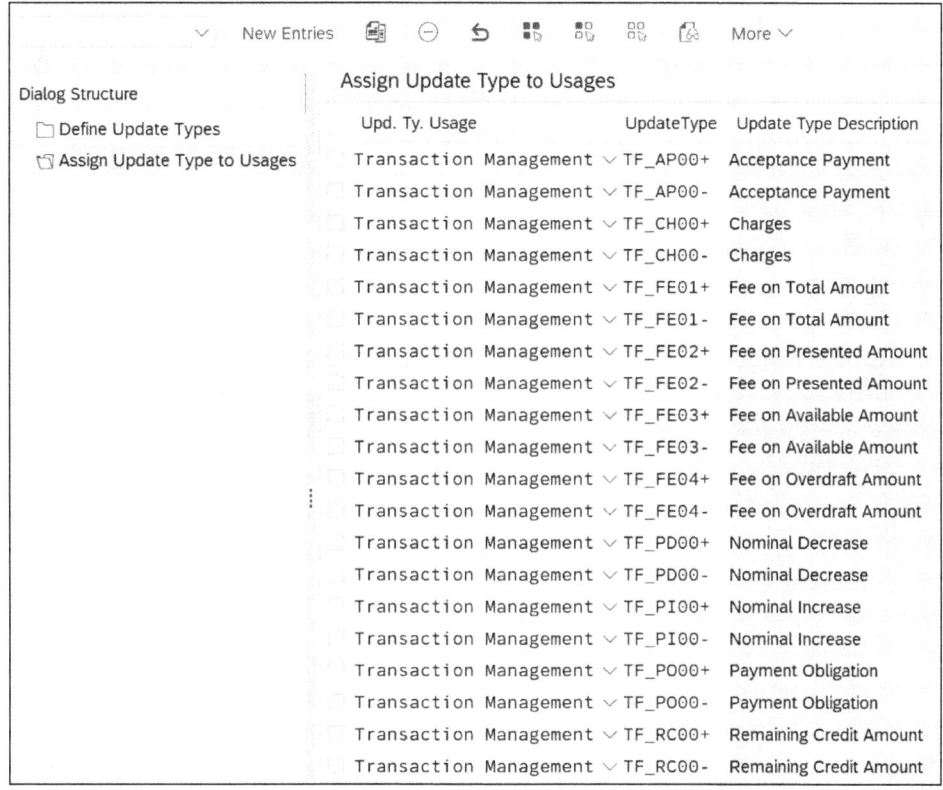

Figure 4.55 Assigning Trade Finance Update Types to Update Type Usage

Cont.Type	FTyp	Name	Direction	UpdateType	Update Type Description
Trade Finance	AP00	Acceptance Payment	Inflow	TF_AP00+	Acceptance Payment
Trade Finance	AP00	Acceptance Payment	Outflow	TF_AP00-	Acceptance Payment
Trade Finance	CH00	Charges	Inflow	TF_CH00+	Charges
Trade Finance	CH00	Charges	Outflow	TF_CH00-	Charges
Trade Finance	FE01	Fee on Total Amount	Inflow	TF_FE01+	Fee on Total Amount
Trade Finance	FE01	Fee on Total Amount	Outflow	TF_FE01-	Fee on Total Amount
Trade Finance	FE02	Fee on Presented Amount	Inflow	TF_FE02+	Fee on Presented Amount
Trade Finance	FE02	Fee on Presented Amount	Outflow	TF_FE02-	Fee on Presented Amount
Trade Finance	FE03	Fee on Available Amount	Inflow	TF_FE03+	Fee on Available Amount
Trade Finance	FE03	Fee on Available Amount	Outflow	TF_FE03-	Fee on Available Amount
Trade Finance	FE04	Fee on Overdraft Amount	Inflow	TF_FE04+	Fee on Overdraft Amount
Trade Finance	FE04	Fee on Overdraft Amount	Outflow	TF_FE04-	Fee on Overdraft Amount
Trade Finance	PD00	Nominal Decrease	Inflow	TF_PD00+	Nominal Decrease
Trade Finance	PD00	Nominal Decrease	Outflow	TF_PD00-	Nominal Decrease
Trade Finance	PI00	Nominal Increase	Inflow	TF_PI00+	Nominal Increase
Trade Finance	PI00	Nominal Increase	Outflow	TF_PI00-	Nominal Increase
Trade Finance	PO00	Payment Obligation	Inflow	TF_PO00+	Payment Obligation
Trade Finance	PO00	Payment Obligation	Outflow	TF_PO00-	Payment Obligation
Trade Finance	RC00	Remaining Credit Amount	Inflow	TF_RC00+	Remaining Credit Amount
Trade Finance	RC00	Remaining Credit Amount	Outflow	TF_RC00-	Remaining Credit Amount

Figure 4.56 Assigning Trade Finance Flow Types to Corresponding Update Types

4 Contract-Specific Configuration

The final step in configuring update types is assigning the flow types we created to their corresponding update types. To do this, we navigate to **Financial Supply Chain Management • Treasury and Risk Management • Transaction Manager • Trade Finance • Transaction Management • Update Types • Assign Flow Types to Update Types**. While this is similar to the configuration for other contract types within treasury and risk management, the distinction lies in using the **Trade Finance** contract type (**Cont.Type**) for letters of credit, as shown in Figure 4.56.

Condition Types

Now that we've fully set up the update types for the letter of credit, we can move on to configuring the condition types. To do this, we navigate to **Financial Supply Chain Management • Treasury and Risk Management • Transaction Manager • Trade Finance • Transaction Management • Condition Types • Define Condition Types – Trade Finance Transactions**. Condition types are important, and there are four main condition types that we must create for a letter of credit: **Fee on Total Amount, Fee on Presented Amount, Fee on Available Amount**, and **Fee on Overdraft Amount**. These condition types enable the system to calculate fees based on different criteria, such as the total amount involved in the transaction, the amount presented, the available funds, and any overdraft amounts. By configuring these condition types, organizations can customize the calculation and application of fees within their letter of credit contracts an ensure that the associated fees are calculated correctly when posting the flows associated with the fees. In the following, we've outlined the condition type to flow type mapping for each condition type:

- **Fee on Total Amount (1204)**
  This condition type is critical for the calculation of the fee on the total amount of the letter of credit.
  - Flow Type: FE01
  - Profile Category: Total Credit Line
- **Fee on Presented Amount (1205)**
  This condition type is used to calculate the fee on the presented amount of the letter of credit.
  - Flow Type: FE02
  - Profile Category: Utilized
- **Fee on Available Amount (1206)**
  This condition type is used to calculate the fee on the available amount left on of the letter of credit, and we should map it to flow type FE03.
  - Flow Type: FE03
  - Profile Category: Not Utilized
- **Fee on Overdraft Amount (1207)**
  This condition type is used to calculate the fee on the overdraft amount of the letter of credit, and we should map it to flow type FE04.
  - Flow Type: FE04
  - Profile Category: Overdrawn

4.1 Debt and Investments

Assigning Condition Types

To assign product type LCS and transaction types ISS and REC to the condition types within the letter of credit framework, we follow this menu path: **Financial Supply Chain Management • Treasury and Risk Management • Transaction Manager • Trade Finance • Transaction Management • Condition Types • Assign Condition Types to Transaction Types – Trade Finance Transactions**. Within this configuration node, we map our associated **Condition Type**(s)—such as fee on total amount, fee on presented amount, fee on available amount, and fee on overdraft amount—to the designated **Product Type** and **Transaction Type**, as shown in Figure 4.57. Assigning these condition types to **Product Type** "LCS" and **Transaction Types** "ISS" and "REC" ensures that the system accurately calculates and applies fees based on the specified criteria for letter of credit transactions.

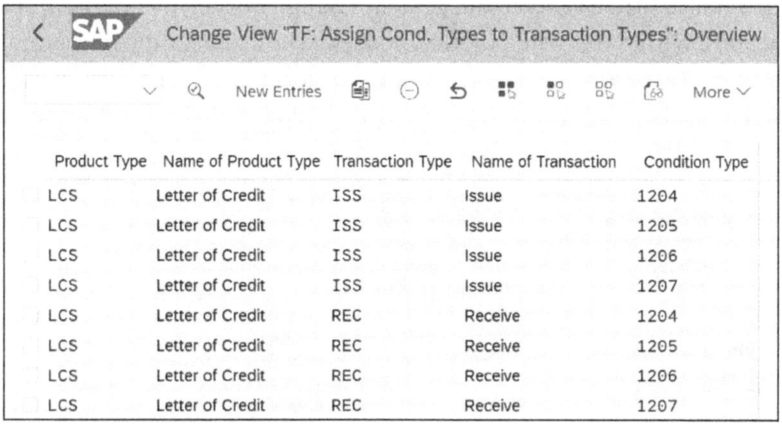

Product Type	Name of Product Type	Transaction Type	Name of Transaction	Condition Type
LCS	Letter of Credit	ISS	Issue	1204
LCS	Letter of Credit	ISS	Issue	1205
LCS	Letter of Credit	ISS	Issue	1206
LCS	Letter of Credit	ISS	Issue	1207
LCS	Letter of Credit	REC	Receive	1204
LCS	Letter of Credit	REC	Receive	1205
LCS	Letter of Credit	REC	Receive	1206
LCS	Letter of Credit	REC	Receive	1207

Figure 4.57 Assigning Letter of Credit Condition Types to Transaction and Product Types

Defining Document Types

In the realm of letters of credit and trade finance contracts within SAP S/4HANA, a unique configuration element that stands out is the use of document types. These document types are uniquely tailored to the specific requirements of trade finance instruments and their role in ensuring compliance and the smooth execution of international trade transactions.

Document types in this context are designed and structured to meet the various stipulations set forth by letters of credit or bank guarantees. Assigning their configuration to trade finance contracts within SAP S/4HANA is essential for aligning with the financial and operational requirements of international commerce.

Each document type serves as a basis for verifying the fulfillment of conditions specified by the letter of credit or bank guarantee, facilitating the approval process and payment execution. Document types represent different facets of the transaction, providing evidence that each step aligns with the agreed-upon terms between the involved parties. By ensuring accuracy and completeness, these documents safeguard the integrity of the transaction and support compliance with international trade standards. Common examples of documents used within trade finance include the following:

- Bill of Lading
- Packing List
- Inspection Certificate
- Transport Documents
- Certificate of Origin
- Insurance Certificate

We can configure these and other document types by navigating to the following path: **Financial Supply Chain Management • Treasury and Risk Management • Transaction Manager • Trade Finance • Transaction Management • Define Document Types – Trade Finance Transactions**. Within this configuration node, we start by defining the various document types (**Doc. Type**). We then further detail each document type with a specific **Document Description** that outlines its characteristics and intended use within trade finance transactions, as shown in Figure 4.58.

Document Type Definition	
Doc. Type	Document Description
PL	Packing List
IS	Inspection Certificate
TD	Transport Documents
IC	Insurance Certificate

Figure 4.58 Trade Finance Document Type Definitions

Defining Document Templates

After we create the document types, the next step is to define the document templates by navigating to this node: **Financial Supply Chain Management • Treasury and Risk Management • Transaction Manager • Trade Finance • Transaction Management • Define Document Templates – Trade Finance Transactions**. The document template serves as the format for how the various documents are handled within the contract. When assigning document types to a document template, we use the appropriate checkboxes to choose to use them for the creation of a letter of credit or for the presentation. Additionally, we can specify the number of documents required by specifying the number of original documents and copies of the document. When we generate a letter of credit in SAP S/4HANA or perform a presentation for a letter of credit, we can select a document template to automatically display all of the associated document types. This methodology eliminates the need to individually enter each related document type and streamline the process.

The first step is to define the various document type templates by choosing the **Template Names** folder, and we can then specify the **Doc. Template Name**. In this example, we're creating a single document type template called "Letter of Credit Documents" that we can use for all document types related to the letter of credit, as shown in Figure 4.59. We can create this at a more granular level, or we can do it at a higher level like we're doing here and assign all document types to a single template.

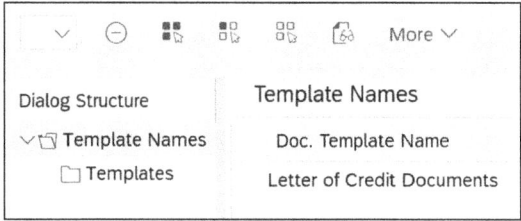

Figure 4.59 Trade Finance Document Templates

Once we've established the document template name, the next step involves configuration of the document types associated with that specific template. This process includes not only adding the relevant document types but also determining their role and timing within the lifecycle of the letter of credit or trade finance transaction. First, we choose the template name within the dialog structure by highlighting the template name, and then we click on the **Templates** folder. We can then begin to add the specific document types that are applicable to this document type template.

One aspect of this configuration is specifying the exact point in the transaction process when each document should be presented, as shown in Figure 4.60. This involves defining whether the documents are needed in the letter of credit creation stage or in the presentation stage, when documents are submitted for review and approval. By determining this timing, we align the documentation process with the procedural requirements of the letter of credit, ensuring that the right documents are available at the right time in the process. We can then define the quantity requirements of the documents, detailing how many originals and copies are needed to satisfy the terms of the letter of credit and meet the demands of all involved parties.

Dialog Structure	Template Name:	Letter of Credit Documents				
∨ Template Names	Templates					
Templates	Doc. Type	Document Description	Crtn-Rel.	Pres.-Rel.	Originals	Copies
	IC	Insurance Certificate	✓	✓	1	2
	IS	Inspection Certificate		✓	1	3
	PL	Packing List		✓	1	2
	TD	Transport Documents		✓	1	1

Figure 4.60 Creating Letter of Credit Document Template

4 Contract-Specific Configuration

Defining Conditions for Payment and the Presentation Period

The system comes preconfigured with two default period conditions: **Shipment Date** for the presentation period and **Presentation Date** for the payment period, as shown in Figure 4.61. These defaults serve as markers and guidelines for when to present documents and when payments are due. To cater to the unique needs of our business processes or specific transaction requirements, we have the flexibility to define additional presentation and payment period conditions within the following node: **Financial Supply Chain Management • Treasury and Risk Management • Transaction Manager • Trade Finance • Transaction Management • Define Conditions for Payment and Presentation Period**.

Period for Payment and Presentation			
Cond.	PerCondTy.	Field name	Period Condition Description
01	Presentation Period Condition	VTBFHAPO-SHIPMENT_DATE	Shipment Date
02	Payment Period Condition	VTBFHAPO-PRESENT_DATE	Presentation Date

Figure 4.61 Defining Conditions for Payment and Presentation Terms

The conditions are configured based on the details within the contract and the standard fields within the contract. The standard fields and logic that we can use are as follows:

- **Term From**
  This condition specifies the start date for the terms of the trade finance agreement or letter of credit. It marks the beginning of the contractual period during which the agreed-upon terms and conditions are in effect. It also establishes the timeline for the execution of the contract, ensuring that all parties adhere to the stipulated start date for activities related to the trade finance transaction.

- **Payment or Delivery Date**
  This condition specifies the deadline by which payment must be made or goods must be delivered, and it also sets a date for fulfilling the financial or delivery obligations outlined in the trade finance agreement.

- **Presentation Date**
  This condition establishes the timeframe within which payment must be made, which is typically linked to the date when documents are presented for payment. It provides a definitive timeline for the settlement of financial obligations, ensuring that payment schedules are clearly aligned with the document submission process.

- **Acceptance Date**
  This condition specifies the date by which documents must be accepted or approved, marking the point at which the terms of the trade finance transaction or letter of credit are formally acknowledged. It establishes the timeline for the review and acceptance of documents, ensuring that the approval process is completed within the specified period.

- **Shipment Arrival Date**
  This condition is a customizable period that sets the timeline for the presentation of documents based on the arrival of goods at their destination. It provides flexibility in aligning document submission with the receipt of goods, ensuring that the documentation reflects the actual delivery status.

Defining Reasons for Rejection of Letter of Credit Presentation

The successful operation of letters of credit hinges on the strict compliance of the presented documents with the stipulated terms and conditions. Deviations from these requirements often result in rejections, which can impede the transaction process, cause delays in payment, and lead to disputes between the parties involved. Common reasons for rejection of a letter of credit include the following:

- Quantity Mismatch
- Description Inconsistency
- Value Discrepancy
- Incorrect Dates

To establish reasons for rejection of letter of credit presentations, we navigate this menu path: **Financial Supply Chain Management • Treasury and Risk Management • Transaction Manager • Trade Finance • Transaction Management • Define Rejection Reasons for Letter of Credit Presentation**. This process entails the creation and definition of reasons for rejection to ensure clarity when identifying why a letter of credit presentation was rejected.

The initial step involves generating a two-character code to represent the reason for rejection in the **RRe** field. This code serves as an identifier for the cause of rejection. We can define a more detailed and descriptive reason for rejection in the **Long name** field, as shown in Figure 4.62. By defining the reasons for rejection through this menu path, organizations can effectively enhance their ability to pinpoint and address issues related to letters of credit presentations. The detailed categorization of reasons for rejection enables corporations to track trends, identify recurring issues, and implement measures to mitigate future rejections.

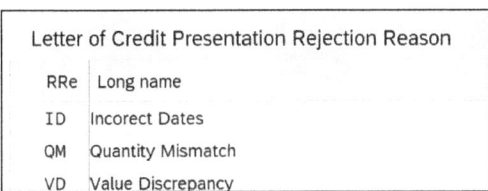

Figure 4.62 Configuring Reasons for Rejection

Defining Bank Guarantee Types

With regard to bank guarantees, it may make sense to establish various bank guarantee types utilized in trade finance. This is an entirely optional step, but we can use it to further differentiate bank guarantees. To configure these distinct guarantee categories, we can navigate this menu path: **Financial Supply Chain Management • Treasury and Risk Management • Transaction Manager • Trade Finance • Transaction Management • Define Bank Guarantee Types**. In this setup, the initial step involves defining a four-character bank guarantee type in the **BG Type** field, followed by providing a detailed description of the characteristics and purposes of the bank guarantee type in the **BG Type Desc.** field, as shown in Figure 4.63.

By defining these bank guarantee types, we can add further granularity within a bank guarantee without creating additional product or transaction types to highlight the different types of guarantees used by an organization. These categorized guarantee types serve as standardized templates, facilitating efficient creation and management of bank guarantees within treasury and risk management.

Bank Guarantee: Bank Guarantee Type	
BG Type	BG Type Desc.
ADVP	Advance Payment Guarantee
BILL	Bill of Lading Guarantee
CRED	Credit Facilities Guarantee
CUST	Customs Guarantee
LEAS	Lease Guarantee
PAYM	Payment Guarantee
PGCO	Performance Guarantee (Warranty Obligation)
PGDO	Performance Guarantee (Delivery Obligation)
PGWO	Performance Guarantee (Contractual Obligation)
TEND	Tender Guarantee

Figure 4.63 Configuring Bank Guarantee Types

4.2 Foreign Exchange

We now shift our focus to the process of configuring FX contracts within treasury and risk management. FX contracts, particularly spot and forward contracts, are essential instruments used by corporate treasury departments to manage foreign currency risk by locking in exchange rates for future transactions. These contracts are fundamental tools for hedging currency exposure, ensuring that fluctuations in exchange rates don't adversely affect an organizations financial position.

This section will guide you through the setup and configuration of both FX spot and forward contracts, covering the key elements required to effectively manage these transactions within the treasury and risk management framework. Unlike configuring money market instruments, which is relatively straightforward, configuring FX contracts involves additional layers of complexity. This is due to the need to account for

4.2 Foreign Exchange

various factors, such as the two sides of the trade in different currencies, the calculation of forward rates, and the management of the settlement process that takes place at a future date.

By mastering the setup of FX contracts, you'll gain a deeper understanding of the intricacies involved in managing currency risk and be better equipped to handle more advanced financial instruments as we progress through the subsequent chapters. Despite the differences related to FX contracts, we'll use the same general principles and that we applied when configuring other instruments. First, we'll jump into the creation of the FX product types, and then, we'll move onto the creation of the transaction types, flow types, and update types. Finally, we'll move on to the unique configuration that is required for an FX contract.

4.2.1 Product Types

We can find the core configuration for FX contracts in the **Financial Supply Chain Management • Treasury and Risk Management • Transaction Manager • Foreign Exchange** menu path. This section covers FX-specific configuration, so we won't cover every configuration field in detail, and we'll only cover the ones related to FX. The first configuration to complete in this area is to define the FX product type, and we find it in the configuration path at **Transaction Management • Product Types • Define Product Type – Foreign Exchange**.

The purpose of the product type is to differentiate among different types of financial products and, in this case, different FX transactions. Figure 4.64 shows the product type creation screen. We can assign a text, a short text, and a product category to the product type. The product category that we must use for all FX transactions is "600."

Figure 4.64 Creating FX Product Type

Once we've defined the **Text**, **ShortText**, and **Prod. Category**, we can look at the various settlement options available for FX contracts (see Figure 4.65). This is an additional setting that we use to determine the settlement type used for the FX transaction. The settlement types are as follows:

4   Contract-Specific Configuration

- **Physical Exercise**
  This settlement type makes sure that both sides of the FX transaction are exchanged.
- **Cash Settlement**
  This settlement type allows the transaction to have a cash settlement option that settles only the differential amount on the settlement date of the transaction.
- **Non-Deliverable Forward**
  This settlement type also only settles the difference between the values of the two currencies at the exercise date of the transaction, but we should only use it when configuring an NDF currency.

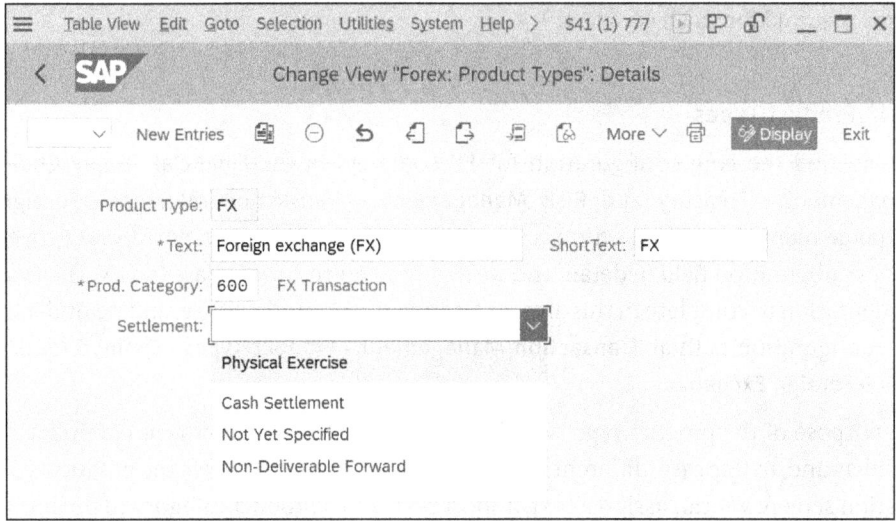

Figure 4.65 FX Settlement Options

4.2.2 Transaction Types

As we've previously discussed in this book, the transaction type plays a crucial role in further defining and categorizing transactions within treasury and risk management. When configuring both forward and spot FX transactions, we start by creating a product type specifically for FX. Within this product type, we then set up distinct transaction types: one for forwards and another for spots.

For all FX transaction types, including both forwards and spots, we'll assign the transaction category as "100." This assignment is exactly the same across both types of transactions, reflecting the fact that a spot and a forward are very similar contracts in how they function. However, there's one key distinction in the configuration: for spot transactions, we need to check the **Spot Transaction** box (checkbox shown but not checked in Figure 4.66). Checking this box indicates that the transaction is a spot trade, which typically settles within two business days, as opposed to a forward trade that settles on a future date.

The reason for setting up separate transaction types for spot and forward trades is primarily to enhance visibility and reporting capabilities. By distinguishing between these two types of FX transactions, we can more effectively manage and analyze our trading activities in all transaction types. This differentiation may allow for clearer reporting and better tracking of our financial positions, but if such distinctions are not required, we have the option to create a single transaction type that encompasses both spot and forward trades. This approach simplifies the setup process but may limit the granularity of our transaction tracking and reporting.

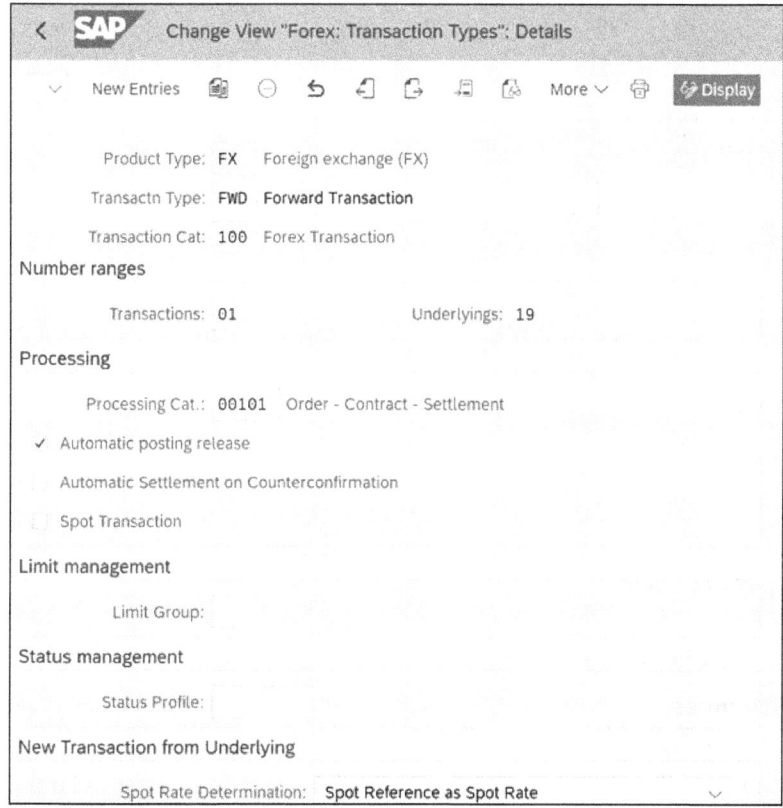

Figure 4.66 Setting Up FX Transaction Types

4.2.3 Assigning Foreign Exchange Attributes

We can add attributes to each of the FX product types and transaction types. This configuration is located in **Financial Supply Chain Management • Treasury and Risk Management • Transaction Manager • Foreign Exchange • Transaction Management • Transaction Types • Assign Forex Attributes**. We use this customizing predominantly for assigning attributes to FX rollovers and premature settlements. The entry screen is shown in Figure 4.67.

4 Contract-Specific Configuration

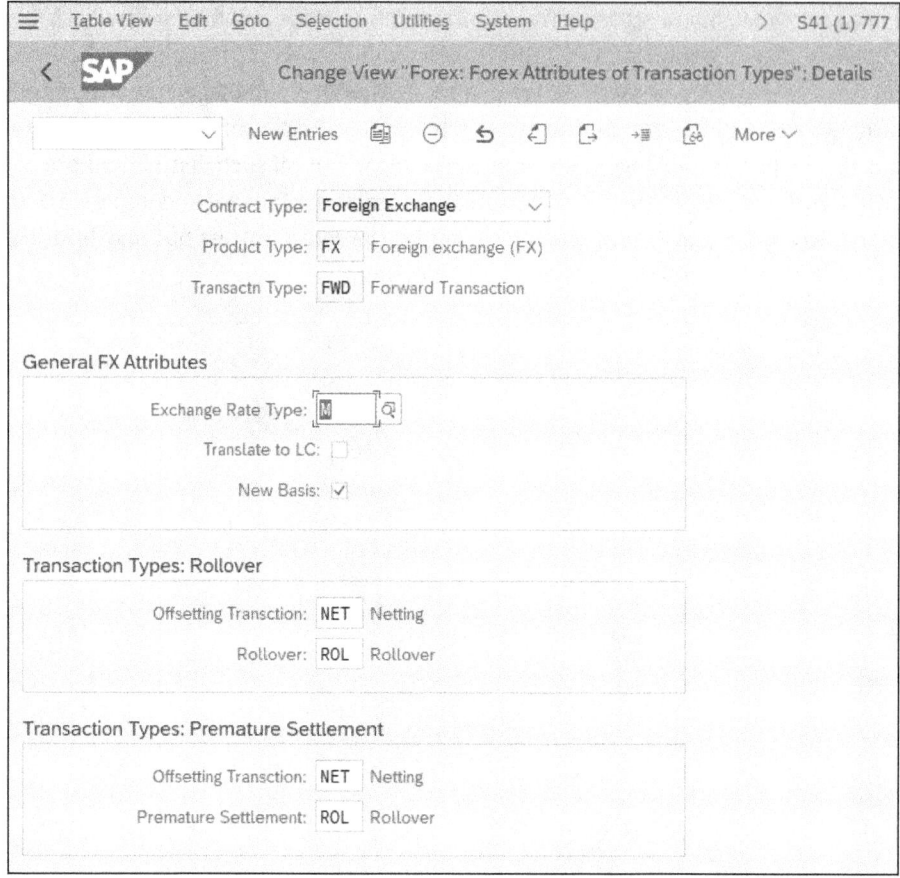

Figure 4.67 Assign FX Attributes

The key fields to enter in this customizing are as follows:

- **General FX Attributes**
  - **Exchange Rate Type**
    This is the rate for the translation of FX amounts for this product type and transaction type.
  - **Translate to LC**
    Checking this box lets us save a local currency amount and rate in the foreign currency flows. We can use this to avoid recognizing profits and losses for a transaction when running the realized gain and loss.
  - **New Basis**
    If we don't check this box and we then execute a rollover or premature settlement, then no gain or loss will be recognized, and the offsetting spot transaction will use the exact same FX rate as the original forward. If we do check this box, we can enter the current FX rate.

- **Transaction Types: Rollover**

  Two transaction types are created as a result of a rollover. One transaction is created to offset the maturity of the FX forward: an offsetting spot is created, and that nets out against the original forward. This can either net out exactly, or most likely, there will be a difference in one of the currencies if the FX rates vary.

  – **Offsetting Transaction**

  This transaction type is used to assign to the new spot transaction that offsets the original FX forward.

  – **Rollover**

  This transaction type creates a new FX forward to roll over this transaction to a new maturity date.

- **Transaction Types: Premature Settlement**

  – **Offsetting Transaction**

  This transaction type is used to offset the original FX forward with a spot.

  – **Premature Settlement**

  A new transaction type is determined to create a new FX forward with the premature settlement date. This is how we settle a contract at a different date than the original maturity date.

4.2.4 Assigning Fixing Spreads

In certain transactions, it may be necessary to add a *fixing spread*. This typically arises when there are discrepancies among the bid, ask, and middle rates. Such variations can impact the accuracy of financial transactions, so it's crucial to account for these differences by applying a fixing spread.

The customization options available within the system allow us to set a default fixing spread based on several key factors, including product type (**PTyp**), transaction type (**TTyp**), business **Partner**, and currency pair for the lead and following currencies (**Lead.Crcy** and **Foll.Crcy**). By configuring these settings, we ensure that the appropriate fixing spread is automatically applied to relevant transactions, thereby enhancing the precision and consistency of our financial operations.

These default settings are particularly useful in scenarios where market fluctuations or specific transaction conditions require a specific approach to rate management. By predefining fixing spreads, we can streamline the transaction process, reduce the likelihood of manual errors, and maintain alignment with our organization's financial policies.

In Figure 4.68, we provide detailed examples of sample settings for fixing spreads, illustrating how we can effectively implement them to address the challenges associated with bid, ask, and middle rate discrepancies. These examples will guide us in customizing our system to handle such scenarios efficiently.

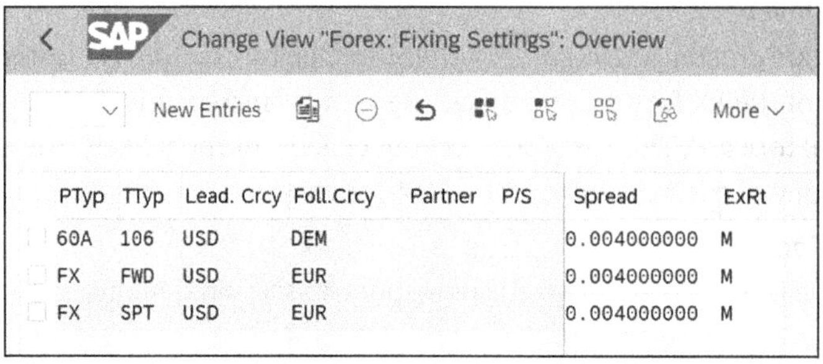

4 Contract-Specific Configuration

PTyp	TTyp	Lead. Crcy	Foll.Crcy	Partner	P/S	Spread	ExRt
60A	106	USD	DEM			0.004000000	M
FX	FWD	USD	EUR			0.004000000	M
FX	SPT	USD	EUR			0.004000000	M

Figure 4.68 Assigning Fixing Spreads

4.2.5 Defining Flow Types

Turning our attention to the flow types associated with an FX transaction, it's important to understand that much like other financial instruments, flow types are responsible for driving the contract cash flows within FX contracts. However, a key difference sets FX contracts apart from the other contracts we've discussed so far: the presence of two primary cash flows. These two flows correspond to the buy and sell currencies involved in the FX contract.

In an FX contract, each of these currencies has its own set of cash flows, which we need to accurately track and manage to ensure that each leg of the contract is tracked accurately. The dual cash flow structure adds a layer of complexity to the configuration process, as we must carefully align in the system both sides of the transaction representing the exchange of one currency for another. We can find the configuration at **Financial Supply Chain Management • Treasury and Risk Management • Transaction Manager • Money Market • Transaction Management • Flow Types • Define and Assign Flow Types**.

To account for the cash flows within an FX contract, we must configure several flow types. The key flow types required during this step are as follows:

- **Buy foreign exchange (PB00)**
  We use this flow type for the buy side of an FX transaction, and it will use the Principal Increase flow category.
- **Sell Foreign Exchange (PS00)**
  We use this flow type for the sell side of the FX transaction, and it will also use the Principal Increase flow category.
- **Cash Settlement (PC00)**
  We use this flow type when there's a cash settlement.

The settings for the flow types should be as shown in Figure 4.69. The buy and sell sides of the transaction should be relevant to posting, position, and valuation since they are all relevant to the FX transaction. If a payment is required using the treasury payment program, we should set up the payment request settings accordingly. Our example indicates that we can use the flow type for both incoming and outgoing payments. Here, we're just highlighting the flows required for FX transactions; additional details on creating flow types can be found in Chapter 3, Section 3.13.1.

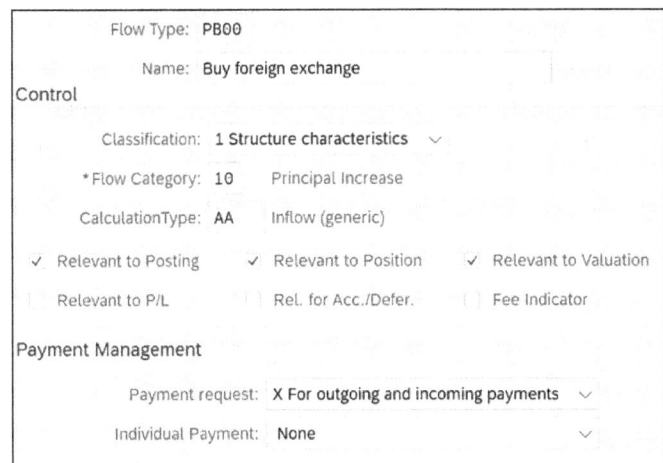

Figure 4.69 FX Flow Types

Figure 4.70 shows the required settings for the cash settlement. The primary difference in the settlement is how the flow category is indicated as **31: Cash Settlement**.

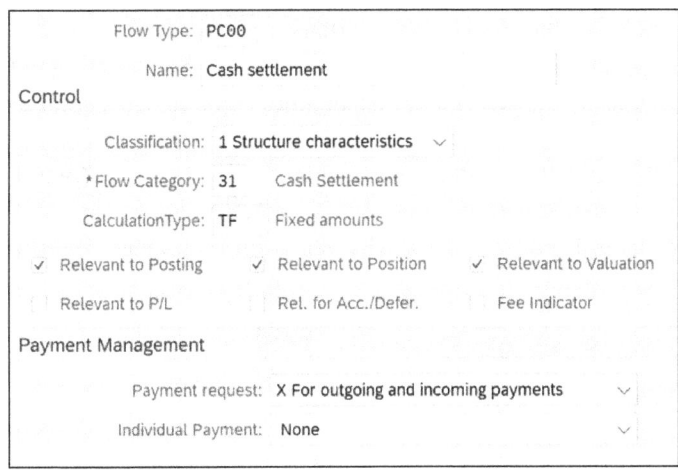

Figure 4.70 Flow Category 31: Cash Settlement

After we create the flow types, we should assign them to the correct product types and transaction types in **Assign Flow Types to Transaction Type – FX Transactions**. The full menu path is **Financial Supply Chain Management • Treasury and Risk Management • Transaction Manager • Foreign Exchange • Transaction Management • Flow Types • Assign Flow Types to Transaction Type – FX Transactions**. This is mapped the same way as we described in Chapter 3, Section 3.13.2.

4 Contract-Specific Configuration

4.2.6 Assigning Flow Types to Update Types

As with other transactions in treasury and risk management, we must create flow types for the defined flow types in the earlier sections and assign them to the **Transaction Management** update usage type. Once that's complete, we can directly link the update types to the flow types in the **Foreign Exchange • Transaction Management • Update Types • Assign Flow Types to Update Types** menu path. In Figure 4.71, we can see an example of the update types we've assigned to our previously created flow types. We should assign all of these to the **Foreign Exchange** contract type, as shown in Figure 4.71.

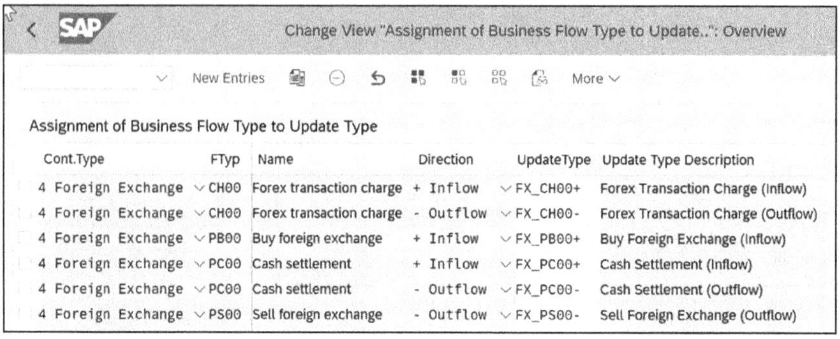

![Assignment of Business Flow Type to Update Type table showing Foreign Exchange contract type with flow types CH00, PB00, PC00, PS00 mapped to update types FX_CH00+, FX_CH00-, FX_PB00+, FX_PC00+, FX_PC00-, FX_PS00-]

Figure 4.71 Assigning Flow Types to Update Types

4.2.7 Position Management Procedures

In addition to these contract-specific configurations, there are some basic things we need to set up for the FX transactions. A position management procedure specific to FX is required, and specific details on the position management procedure configuration are covered in Chapter 3, Section 3.6.2. The main requirement is to assign a position management procedure that uses the **Position Management Cat.** of **5 Forex Transactions**. We use the key date valuation area if we need to execute valuations for the FX transactions. In the example in Figure 4.72, a security valuation is assigned, and that will drive how we run the Run Valuation app or Transaction TPM1.

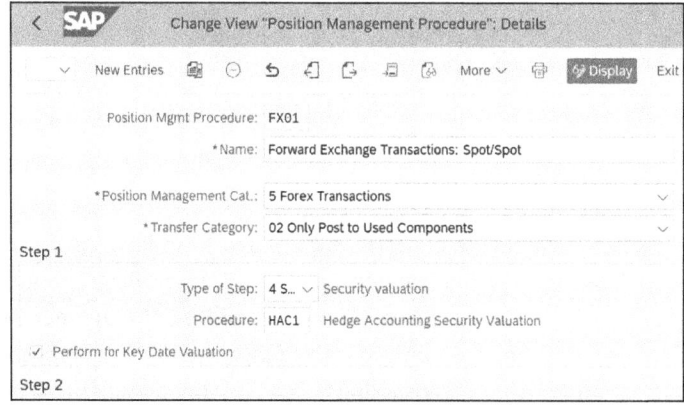

Figure 4.72 Position Management Procedure

4.2.8 Set Effects of Update Types on Position Components

To ensure that the valuations of the FX transaction are calculated correctly, we need to detail the position-relevant update types in the **Financial Supply Chain Management • Treasury and Risk Management • Transaction Manager • General Settings • Accounting • Settings for Position Management • Set Effects of Update Types on Position Components** menu path. We assign the required update types to the relevant position change category (**PCC**) in this customizing. To all the flows we create for the FX process, we can assign the position change category (**PCC**) of "1006" (**Indirect Position Change**), as shown in Figure 4.73. We add this by creating a line in this configuration, assigning an update type, and mapping to the position change category.

UType	Update Type Text	VA	ACC	PCC	Position Change Cat.
FX_PB00+	Buy Foreign Exchange (Inflow)			1006	Indirect Position Change
FX_PC00+	Cash Settlement (Inflow)			1006	Indirect Position Change
FX_PC00-	Cash Settlement (Outflow)			1006	Indirect Position Change
FX_PS00-	Sell Foreign Exchange (Outflow)			1006	Indirect Position Change

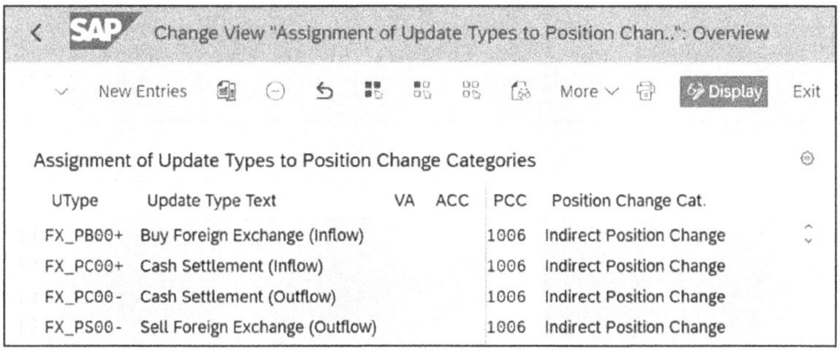

Figure 4.73 Set Effects of Update Types on Position Components

4.2.9 Assigning Update Types for Valuation

When we're running valuations for FX transactions, a series of update types are required for the valuation postings. We should create the update types with an update type usage of **Key Date Valuation**. We then assign these in customizing in the following menu path: **Financial Supply Chain Management • Treasury and Risk Management • Transaction Manager • General Settings • Accounting • Key Date Valuation • Update Types • Assign Update Types for Valuation**. This area assigns the update types by position management procedure, so we first need to make sure that our position management procedure is reflected in the initial screen. To assign the update types, we can click into the position management procedure that is assigned to the FX product type that needs update types assigned, as shown in Figure 4.74.

SAP S/4HANA provides a series of update types that are already assigned to the valuations, and generally, we can just select these. In this area, the update type reflects the label on the right side of the update type entry. In the **Security Write-up** field shown in Figure 4.75, the value is V200, and the update type that we can assign is also V200. The update types we must fill in on this screen will depend on the settings we previously made in the position management procedure configuration.

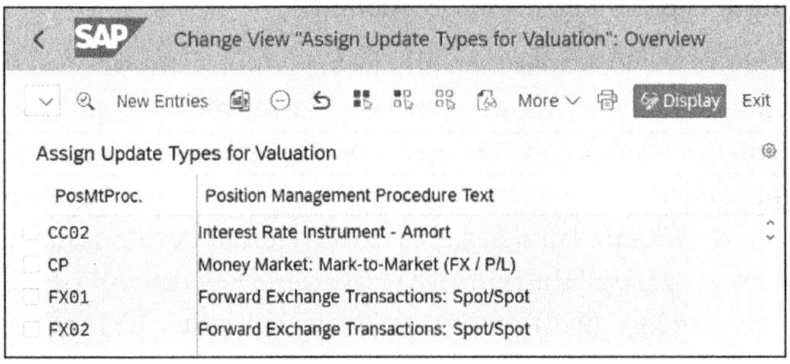

4 Contract-Specific Configuration

Figure 4.74 Assigning Update Types for Valuation

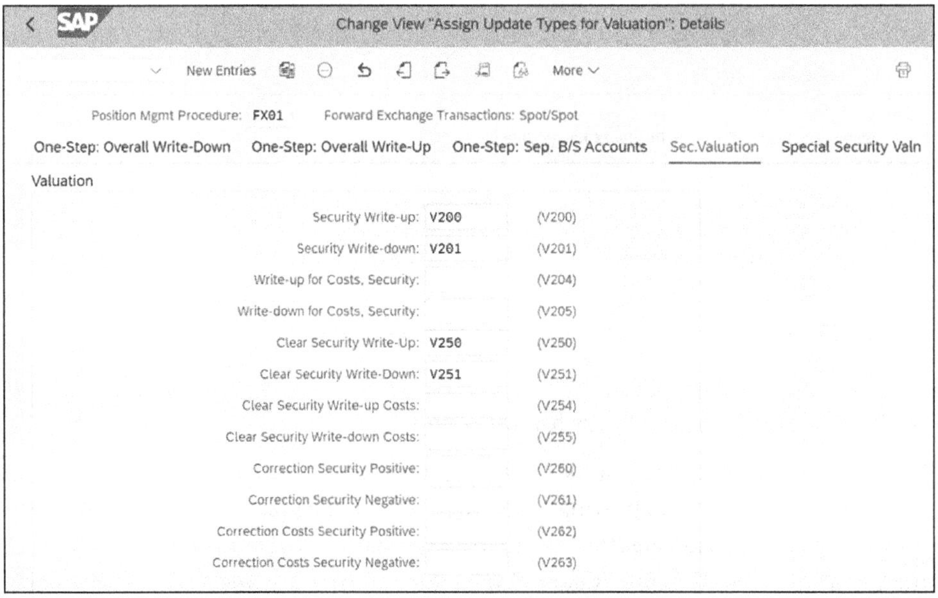

Figure 4.75 Assigning Update Types for Security Valuation

4.2.10 Assigning Update Types for Derived Business Transactions

A series of derived business transactions are required for both the calculation and the posting of FX transactions. We should create the update types with an update type usage of **Derived Business Transaction**, as shown in Figure 4.76. The relevant activities that have derived business transactions are in each of the folders on the left side of the screen. To assign an update type, we double-click into the folder that requires the update. The example in Figure 4.76 shows that we're looking into the **Position Outflows** folder. In this area, we add the update type to reflect the activity that's triggering the posting.

4.2 Foreign Exchange

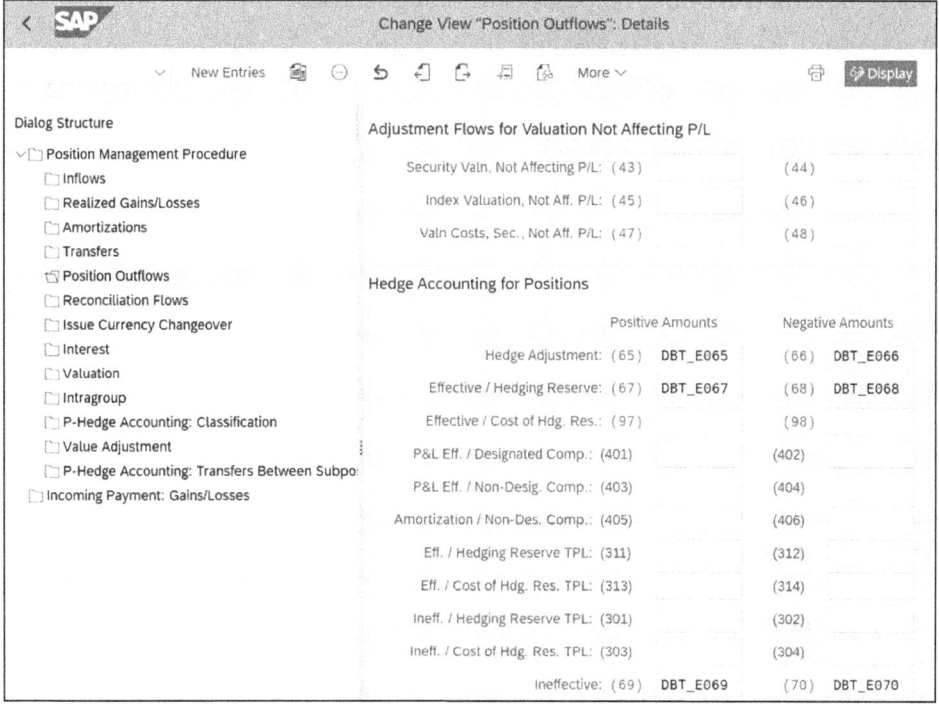

Figure 4.76 Assigning Update Types for Derived Business Transactions

Update Types: Naming Derived Business Transactions

The update types for derived business transaction come delivered from SAP S/4HANA with a standard naming convention. The naming follows a standard convention: the first three characters represent the purpose of the update type, they are followed by an underscore, and the last four characters represent which folder and field the update type should populate. Let's look at this in more detail:

- The first three characters of the update types start with **DBT**. Some update types in this configuration are specific to hedge accounting, and those update types start with **HAC** for hedge accounting.
- The fifth character determines to which folder the update type is assigned. The folders go in alphabetical order, so if we look at the folders, the first folder is **Inflows**, so all update types in that folder would have an *A* as the fifth character. The next folder is **Realized Gains/Losses**, and all update types would use a *B* as the fifth character.
- Each box in this configuration is assigned a number. The last three characters of the update type represent the same numbered box that the update type should be placed in.
- As we can see in Figure 4.76, the first update type that is populated in this configuration is **Hedge Adjustment**. Accordingly, the update type assigned is DBT_E065 since it's a derived business transaction, the **Position Outflows** folder is the fifth folder down, and the field for this activity is **65**.

267

4.2.11 Alternative Update Types for Position Outflows

The update types in a transaction ultimately can drive the accounting postings that we both debit and credit when running the various posting programs. Generally, an update type is set for a specific activity, and we always use that when running the posting programs. For example, in Section 4.2.9 , we showed how to use V200 and V201 for the security write-up and the security write-down, respectively, as shown in Figure 4.77.

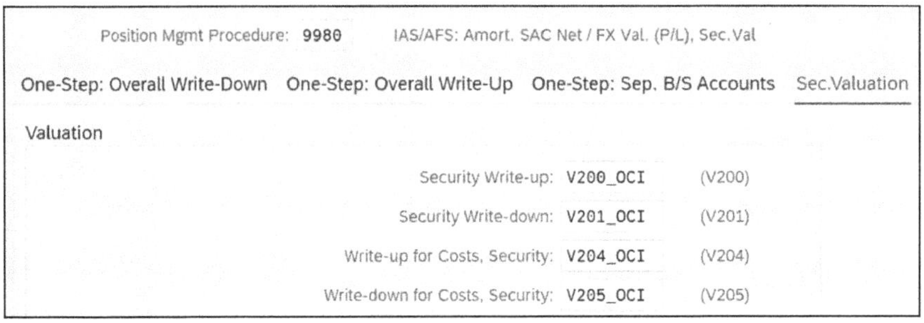
Position Mgmt Procedure: 9980 IAS/AFS: Amort. SAC Net / FX Val. (P/L), Sec.Val
One-Step: Overall Write-Down
One-Step: Overall Write-Up
One-Step: Sep. B/S Accounts
Sec.Valuation
Valuation
Security Write-up: V200_OCI (V200)
Security Write-down: V201_OCI (V201)
Write-up for Costs, Security: V204_OCI (V204)
Write-down for Costs, Security: V205_OCI (V205)

Figure 4.77 Alternative Update Types for Position Outflows

There may be scenarios in which the accounting treatment needs to differ for FX transactions, and if we want to post to different accounts for regular FX transactions than we do when posting hedging transactions, we can handled it in the configuration located in the **Financial Supply Chain Management • Treasury and Risk Management • Transaction Manager • General Settings • Accounting • Derived Business Transactions • Update Types • Alternative Update Types for Derived Business Transactions** menu path.

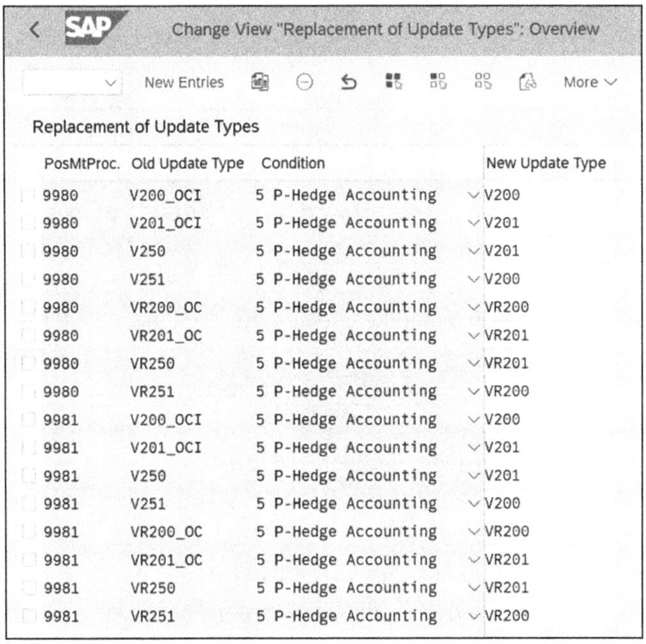
Change View "Replacement of Update Types": Overview
New Entries
More
Replacement of Update Types
PosMtProc. Old Update Type Condition New Update Type
9980 V200_OCI 5 P-Hedge Accounting V200
9980 V201_OCI 5 P-Hedge Accounting V201
9980 V250 5 P-Hedge Accounting V201
9980 V251 5 P-Hedge Accounting V200
9980 VR200_OC 5 P-Hedge Accounting VR200
9980 VR201_OC 5 P-Hedge Accounting VR201
9980 VR250 5 P-Hedge Accounting VR201
9980 VR251 5 P-Hedge Accounting VR200
9981 V200_OCI 5 P-Hedge Accounting V200
9981 V201_OCI 5 P-Hedge Accounting V201
9981 V250 5 P-Hedge Accounting V201
9981 V251 5 P-Hedge Accounting V200
9981 VR200_OC 5 P-Hedge Accounting VR200
9981 VR201_OC 5 P-Hedge Accounting VR201
9981 VR250 5 P-Hedge Accounting VR201
9981 VR251 5 P-Hedge Accounting VR200

Figure 4.78 Configuration Assigning New Update Types When Condition Is Met

We do this by determining what condition we want to use to drive a new update type in certain scenarios. In Figure 4.78, we can see the configuration of the first two entries. In position management procedure 9980, the update types V200_OCI and V201_OCI are assigned in the configuration for the security valuation key date valuation. When using hedge accounting, we need to create different posting specs, so we want to assign an alternative set of update types. We will change the update types in this scenario to "V200" and "V201" without needing to create a completely different position management procedure for hedge accounting.

4.2.12 Defining Nondeliverable Currencies

We typically use NDFs for currencies that either are not frequently traded or are currencies for which we don't wish to handle physical settlement. These forwards allow us to manage currency risk without the need to actually deliver or receive the underlying currency. To facilitate this, SAP S/4HANA provides a customization activity called **Define Non-Deliverable Currencies**, which allows us to specify which currencies should be classified as nondeliverable.

We can assess this configuration through the following path: **Financial Supply Chain Management • Treasury and Risk Management • Currencies • Define Non-Deliverable Currencies.** In this section, we can list the currencies we wish to designate as nondeliverable. We do this by adding the three-character International Organization for Standardization code for the currency in this configuration.

When we're creating an NDF using the Create Financial Transaction app, the system references the table of nondeliverable currencies that we've set up. This ensures that during transaction entry, nondeliverable currencies will not appear as options for settlement. Instead, the system will automatically default to the deliverable currency as the settlement currency. This process simplifies the transaction setup and prevents errors related to currency selection, ensuring that only valid, deliverable currencies are used for settlement purposes. A sample list of nondeliverable currencies can be found in Figure 4.79.

As illustrated in Figure 4.80, we can observe that the settlement currency (**Settlement Crcy**) has been automatically set to USD. This default setting occurs because the Philippine peso (PHP) has been designated as a nondeliverable currency according to the configuration in the system. The configuration specifies that PHP is not eligible for physical settlement, so the system defaults to using a deliverable currency (in this case, USD) for the settlement process. This automated adjustment ensures compliance with the nondeliverable currency rules and simplifies the transaction setup by eliminating the need for manual intervention in selecting a suitable settlement currency.

4 Contract-Specific Configuration

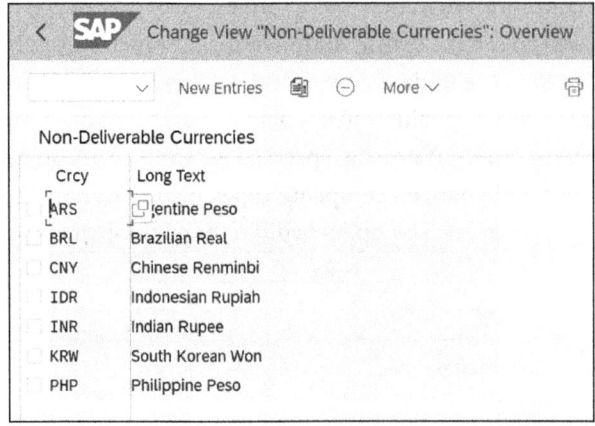

Figure 4.79 Defining Nondeliverable Currencies

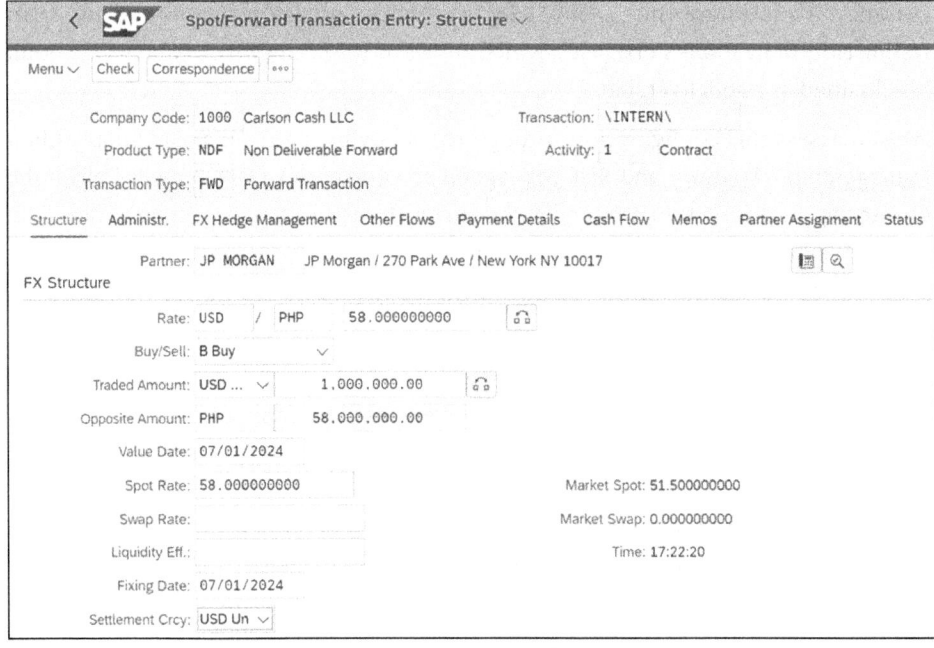

Figure 4.80 How Settlement Currency is Automatically Determined

4.2.13 Foreign Exchange Mirroring

As with the configuration of intercompany loans, it's common for organizations to engage in trading FX contracts among their internal entities. This facilitates currency exchange between two entities within the same company. When such a transaction occurs, an FX contract is recorded under one company code and linked to an intercompany business partner that represents the other participating entity.

4.2 Foreign Exchange

To ensure accurate financial reporting, it's important for the other entity to record a corresponding FX transaction that mirrors the initial entry. This mirrored transaction guarantees that both sides of the trade are accurately reflected in each entity's financial records.

To streamline this process and avoid the need for manual entry of the opposite transaction, we can configure specific nodes in SAP S/4HANA customizing. Once we've set up these nodes, the system can automatically generate the corresponding mirror transactions, thereby ensuring precise and efficient recording of intercompany FX trades.

Before going into the configuration of the FX mirroring, we'll go through an example of the mirror transactions. In Figure 4.81, we have an example of a trade that was created in company code 1000 with intercompany business partner ICL-2000. We've added the proper settings in customizing to create a mirror transaction, and as a result of this, we can see that the transaction in the originating company code sells the Canadian dollar amount and buys the US dollar amount. The opposite happens in the opposite company code: the Canadian dollar amount of $1,000,000 is purchased, and the US dollar amount is sold.

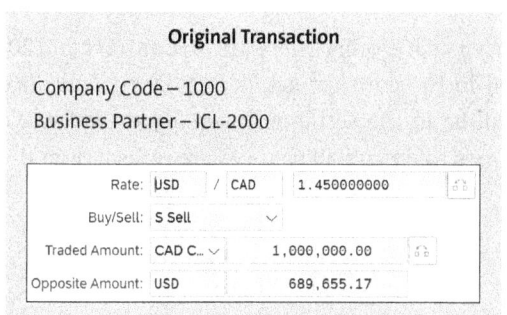
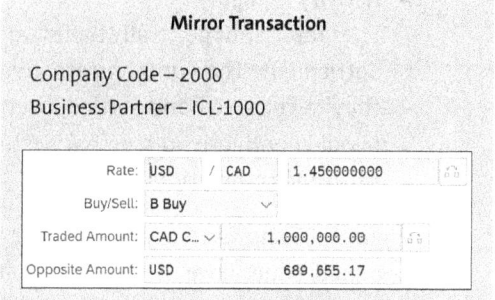

Figure 4.81 Example Output of Mirrored FX Transaction

To set up this functionality, we can find the customizing nodes in the following menu path: **Financial Supply Chain Management • Treasury and Risk Management • Transaction Manager • General Settings • Organization • Transaction Management • Intercompany Trading: Distribution of Mirror Transactions**. All five of the following sections are in this menu path location.

The following five sections detail how to configure the FX mirroring to create a mirrored FX transaction. We need to complete the following steps to make this happen:

- Determine the product types and transaction types that should be mirrored.
- Create a mapping between the product type and transaction type that are created with the transaction to the product type and transaction type that are created in the mirrored transaction.
- Map which activities are relevant to transaction mirroring.

4 Contract-Specific Configuration

- Assign the company code to the intercompany business partners.
- Determine whether to create hedging classifications in mirrored transactions.

Maintaining Relevant Product Types and Transaction Types

The first step of this configuration is to define the relevant scenarios to trigger the mirroring for the intercompany FX trades. Defining the scenario includes defining when mirroring should trigger for a company code, product type, transaction type business partner, activity category, and activity type. Let's look at those last two in more detail:

- **Activity Type**
  The activity type is the function that is performed on the contract. The most common activity types that are performed for these transactions are **01: Add or create** and **02: Change**. When a new financial transaction is created, that activity will be classified as 01, and when a transaction is changed, SAP S/4HANA will identify that activity as 02. We should assign these activities to ensure that any creation or change activities on the transactions are successfully mirrored to the opposite transactions.

- **Activity Category**
  FX transactions generally have two activity categories. These are **10: Contract** and **20: Settlement**. The contracts are created in the contract activity category, and once they're reviewed and settled, they will be in the settlement activity category. We should configure both activity categories to ensure all changes are reflected on the mirrored transaction.

Figure 4.82 shows an example where we should create mirrored transactions for both company codes if we're creating an intercompany FX trade. When company code 1000 creates an FX trade with the internal business partner ICL-2000, then a mirror transaction is generated. This also happens in the opposite scenario. When company code 2000 creates an FX trade with the ICL-1000 internal business partner, mirroring will also be triggered.

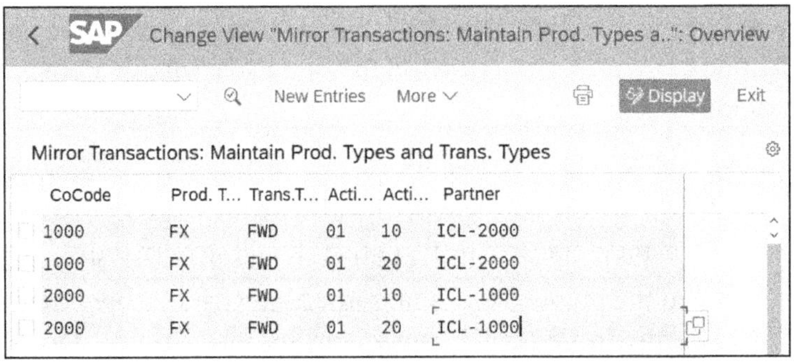

CoCode	Prod. T...	Trans.T...	Acti...	Acti...	Partner
1000	FX	FWD	01	10	ICL-2000
1000	FX	FWD	01	20	ICL-2000
2000	FX	FWD	01	10	ICL-1000
2000	FX	FWD	01	20	ICL-1000

Figure 4.82 Determining Relevant Product Types and Transaction Types for Mirroring

4.2 Foreign Exchange

Mapping Product Types and Transaction Types

The next step of this customizing is to further define the scenario for the mirror transaction. This is where we determine the creation transaction and the mirrored transaction. To properly tie these transactions together, we need to determine the following settings:

- **Company Code, Product Type, Transaction Type**
  Both sides of the transaction need to be reflected in these fields. This includes the creation side and the mirror side of the transaction.

- **Direction**
  In this setting, the outgoing direction determines the sending/creation company code, and the incoming direction determines the mirror side of the transaction.

- **Function**
  In this setting, **01 Mirror Image** is determined to be the function of all line items for the mirroring.

- **MetaText**
  The **MetaText** field is a free-form field, but we use it to link the two sides of the transaction together. When the **MetaText** is the same for an outgoing transaction and an incoming transaction, SAP S/4HANA has enough information to create the mirror image. Since we may need to create many scenarios for mirroring in this table, it's important to come up with a unique naming convention. To highlight this, we'll cover the naming convention being used in Figure 4.83 and specifically the **MetaText** "2000/1000 – 1000 FWD."

In Figure 4.83, we determine the company code for both sides of the transaction in the first part of the **MetaText:** "2000/1000," which determines that this is a mirroring transaction between company code 2000 and 1000.

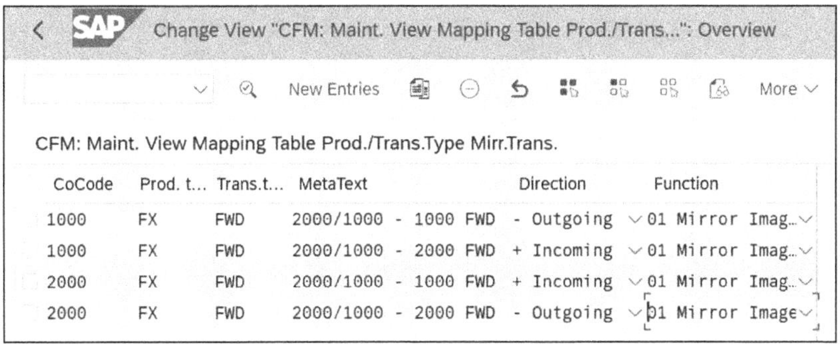

CoCode	Prod. t...	Trans.t...	MetaText	Direction	Function
1000	FX	FWD	2000/1000 - 1000 FWD	- Outgoing	01 Mirror Imag...
1000	FX	FWD	2000/1000 - 2000 FWD	+ Incoming	01 Mirror Imag...
2000	FX	FWD	2000/1000 - 1000 FWD	+ Incoming	01 Mirror Imag...
2000	FX	FWD	2000/1000 - 2000 FWD	- Outgoing	01 Mirror Image

Figure 4.83 MetaText is Key to Mapping Between Created Transaction and Mirrored Transaction

The second portion of the **MetaText** determines the outgoing side of the FX transaction. For example, "1000 FWD" highlights that this is an FX forward being created in company code 1000. This occurs from 1000 since this line is applicable to the outgoing direction.

4 Contract-Specific Configuration

The offsetting transaction copies this **MetaText** to link the lines together, and it will use the attributes for company code, product type, and transaction type to create the mirror image of this FX trade.

Processing Incoming Data

The next step is to maintain the **Process Incoming Data** node. This configuration is created from the receiving or mirrored side of the transaction. When we're expecting a mirrored transaction to be created, this node creates the details of the transaction. We need to create configuration for both sides of the FX transaction, as shown in Figure 4.84. Mapping the **CoCode**, **Prod. Type**, **Trans.Type**, **Activity Category**, and **Counterparty** in this configuration defines that when this specific scenario is triggered in the mirroring, the system can successfully create the "incoming" mirrored transaction.

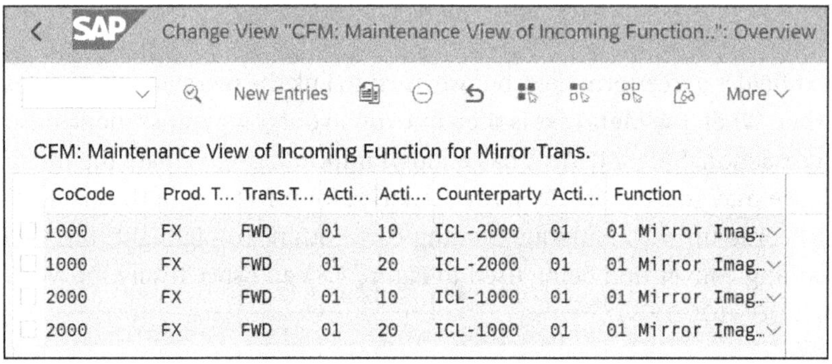

CoCode	Prod. T...	Trans.T...	Acti...	Acti...	Counterparty	Acti...	Function
1000	FX	FWD	01	10	ICL-2000	01	01 Mirror Imag...
1000	FX	FWD	01	20	ICL-2000	01	01 Mirror Imag...
2000	FX	FWD	01	10	ICL-1000	01	01 Mirror Imag...
2000	FX	FWD	01	20	ICL-1000	01	01 Mirror Imag...

Figure 4.84 ICT: Processing Incoming Data

Assigning a Company Code to a Partner

The next configuration node maps the company code to the corresponding business partner. In Figure 4.85, we show how company code 1000 is mapped to business partner ICL-1000 and company code 2000 is mapped to business partner ICL-2000. Once we've set up this final step for the mirroring, the FX trades will mirror for the intercompany transactions.

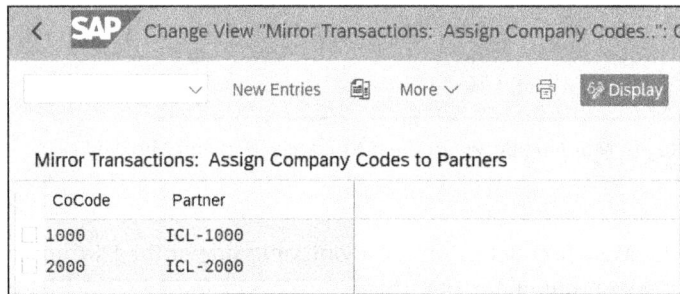

CoCode	Partner
1000	ICL-1000
2000	ICL-2000

Figure 4.85 Assigning Company Codes to Partners

Copy Hedging Classifications and Hedge Request IDs in Mirror Transactions

Hedging transactions have more details that are added to the contracts to facilitate the hedging process. Two of these fields that are entered in the **Administr.** tab in the Create Financial Transaction app are **Hedging Classification** and **Hedge Request ID**. We use these fields to tie the hedge request to the contract so we can create the hedge relationship. This configuration node enables these two fields to be copied to the mirror transaction. To activate the fields to be copied, we need to check the corresponding box as shown in Figure 4.86. We check the **Copy Hedging Classification** box to copy the hedging classification to the mirrored transaction, and we check the **Copy Hedge Request ID** to copy the hedge request ID to the mirrored transaction.

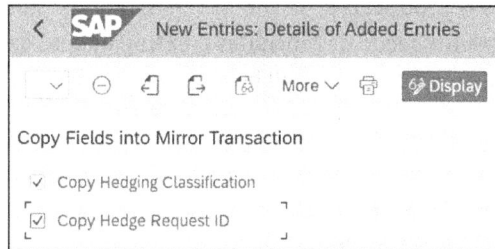

Figure 4.86 Configuration for Copy Hedging Classification and Hedge Request ID in Mirror Transaction

4.3 Securities

Now, we move on to the process of configuring securities within treasury and risk management. Securities include stocks, bonds, investment funds, and other investment instruments, and they are essential elements of a corporate treasury's portfolio in which the company can invest excess cash and optimize returns.

In this section, you'll learn how to set up and configure various securities, focusing on the key securities contract types that are most commonly used by corporate treasury departments. The securities we'll focus on in this section are bonds and investment funds such as money market funds. These two types of instruments are widely used within treasury, and we thought they would serve as a logical introduction to the securities setup within SAP S/4HANA. While the setup of money market instruments is relatively simple, securities configuration introduces added layers of complexity due to the diverse nature of securities and the need to track market values, calculate yields, and manage interest payments or dividends, as well as the complexities involved in the settlement process, which often requires coordination among multiple counterparties.

Although securities differ in many ways from other financial instruments, we'll follow many of the same fundamental principles we used in previous configurations. Unlike other contracts, where we start the configuration with the product types, securities involve some additional configuration elements that we must address before jumping into the product type setup. Condition types are generally configurations that are completed after the creation of the product type and transaction types. We'll start with the setup of the condition types, and then we'll move on to the setup of the product type and transaction type.

4.3.1 General Configuration

The first section of configuration to set up for securities is the base settings needed for the transactions. We start by creating the condition types to create the types of flows related to securities. Next, we define the product type to create different types of securities contracts, including the bond purchase product type covered later in this chapter. Configuring the rest of these base settings for securities includes creating additional assignments for the product types and defining data that will be assigned when creating the securities class.

Condition Types

The first step in customizing securities within treasury and risk management is to configure the condition types. While this may seem out of sequence compared to the setup of all other instrument type configurations—where product types and transaction types are created before condition types—in the case of securities, this order is necessary.

For securities, we establish condition types first because they are important to defining the financial characteristics of the product types. Securities often involve various conditions—such as interest payments, dividends, and other financial obligations—that need to be accurately represented in the system. Without these condition types in place, we would not be able to assign the necessary conditions to the securities product and transaction types.

By starting with condition types, we ensure that all relevant financial conditions are properly defined and available for assignment when creating securities product types. This approach is essential for accurately reflecting the diverse nature of securities, and we can find this configuration in the **Financial Supply Chain Management • Treasury and Risk Management • Transaction Manager • Securities • Master Data • Product Types • Condition Types** menu path. In the following sections, we'll use all three configuration nodes in this area:

- **Define Condition Types**
- **Define Groups for Securities**
- **Assign Condition Types to Condition Groups**

4.3 Securities

Defining Condition Types

The first definition of the condition type involves defining each type of condition flow for the security transactions. This customizing defines a three-character identifier for the **Condition Type** and a calculation category for the cash flow calculator (**FIMA Calc. Category**). Each of these categories defines different rules for the conditions.

Instead of covering all sixty or more calculation types, we'll go through a couple of examples. In Figure 4.87, condition types 501, 502, and 503 will be assigned to a product type for bond purchases. These are assigned to a series of calculation categories for the following purposes:

- **501 assigned to TZ**

  This is for nominal interest, and we normally assign a fixed interest rate to this calculation category. If we require variable interest, we could create an additional condition type with ZA to allow for the usage of a variable interest rate with interest rate adjustments.

- **502 assigned to UT**

  This calculation category allows us to define the repayment rate for the bond.

- **503 assigned to TTEN**

  We assign this to define the final repayment date and amount.

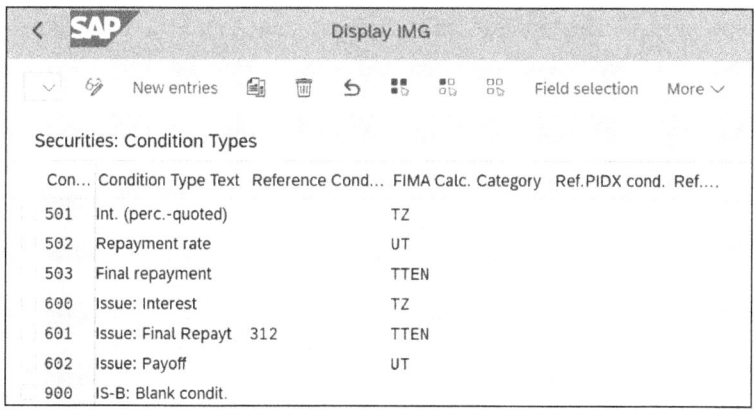

Figure 4.87 Defining Condition Types: Securities

Defining Groups for Securities

The next step involves grouping the conditions so we can assign them to a product type. In the **Define Groups for Securities** customization, we create a three-character condition group identifier (**CGr**) and give it a **Name**, as shown in Figure 4.88. Following the conditions we configured earlier, we now have condition group 50 available for fixed-interest bonds.

4 Contract-Specific Configuration

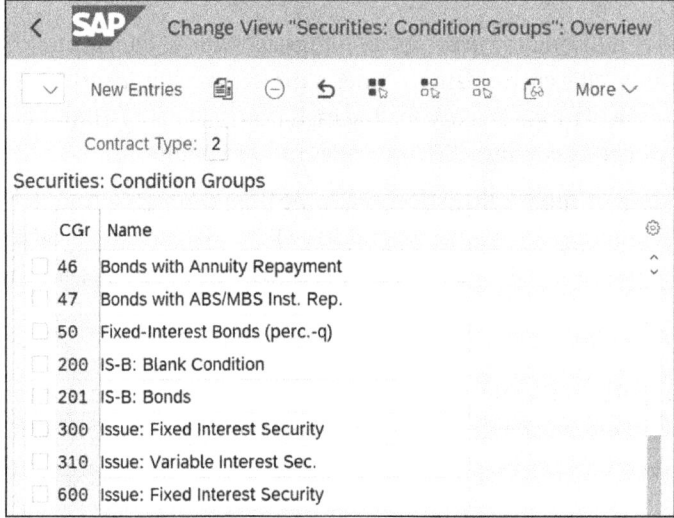

Figure 4.88 Creating Condition Group in Configuration; Sets of Conditions Will Be Assigned to Condition Group

Assigning Condition Types to Condition Groups

The final step in configuring the condition types involves assigning them to their respective condition groups. This process links multiple condition types to a single condition group, which will later be associated with the relevant product type. The assignment in this configuration includes adding a new line and assigning the condition group. Then, we need to add a new line to assign the different condition types for the group. As shown in Figure 4.89, condition types **501**, **502**, and **503** have been assigned to condition group **50**, which was specifically created for fixed-interest bonds.

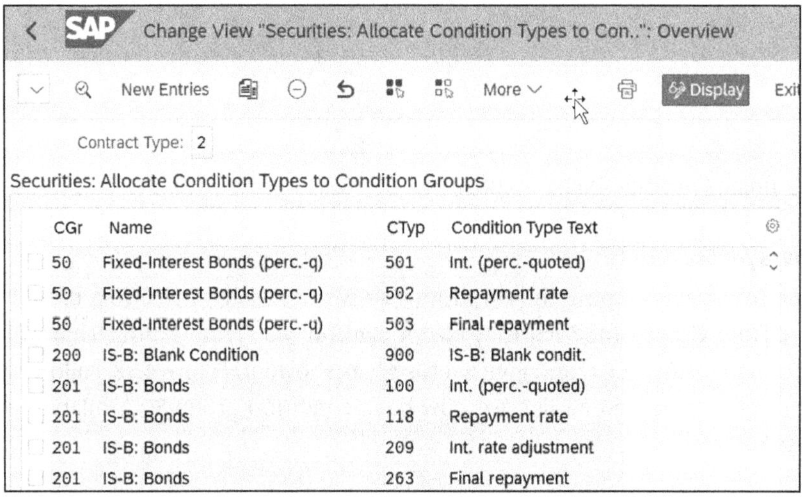

CGr	Name	CTyp	Condition Type Text
50	Fixed-Interest Bonds (perc.-q)	501	Int. (perc.-quoted)
50	Fixed-Interest Bonds (perc.-q)	502	Repayment rate
50	Fixed-Interest Bonds (perc.-q)	503	Final repayment
200	IS-B: Blank Condition	900	IS-B: Blank condit.
201	IS-B: Bonds	100	Int. (perc.-quoted)
201	IS-B: Bonds	118	Repayment rate
201	IS-B: Bonds	209	Int. rate adjustment
201	IS-B: Bonds	263	Final repayment

Figure 4.89 Allocating Condition Types to Condition Groups

This assignment is important because it consolidates the condition types under a single group, ensuring that all relevant financial conditions are correctly applied when we're setting up the product type. Additionally, by double-clicking on any of these condition types, we can set default values for the contract entry screen. These defaults can include percentage rates, interest calculation methods, working day adjustments, and specifications for which days should be considered inclusive in interest calculations.

Product Types

When we're creating the product types for securities, we'll see why we started with the condition type and group configuration. The configuration to create the product types is located in the **Financial Supply Chain Management • Treasury and Risk Management • Transaction Manager • Securities • Master Data • Product Types • Define Product Types – Securities** menu path. We'll first assign the product category to define what type of security will use this product type. A full list of the available product categories is as follows:

- 010: Stock
- 020: Investment Fund
- 030: Subscription Right
- 040: Bond
- 042: Installment Bond
- 060: Warrant Bond
- 070: Convertible Bond
- 111: Index Warrant
- 112: Equity Warrant
- 113: Currency Warrant
- 114: Bond Warrant
- 160: Shareholding

The calculation method section defines defaults for interest calculations, and the effective interest method (**Eff.Int. Method**) field drives how the system calculates the effective interest rate for the transaction. The interest calculation method (**Int.calc.method**) determines the default interest calculation method that is assigned when we're creating the securities class. Both of these settings are defaults, but we can change the selection for the effective interest method and interest calculation method when creating the transactions. Next, the **Condition group** that was previously created is assigned to the **Product Type**, as shown in Figure 4.90.

4   Contract-Specific Configuration

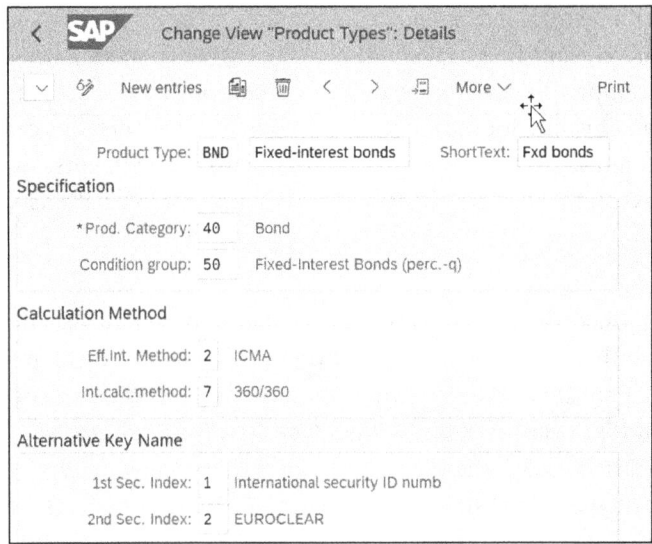

Figure 4.90 Product Type Creation

> **Parallel Interest Conditions: Bonds**
>
> Entering parallel interest conditions is covered in detail in Chapter 6, Section 6.1.3, on debt and investments transactions. So that we can enter bonds into the parallel interest conditions for compound interest calculations, like the calculations used for daily SOFR contracts, we must set the **Cash Flow Generation** section of the product type configuration to **Parallel Interest Conditions** instead of **Single Interest Conditions**.

Assigning Repayment Types to Product Types

Another difference for securities is that we need to determine the repayment types of the product types in a separate configuration. We complete this in **Assign Repayment Types to Product Types**. The full menu path for this is **Financial Supply Chain Management • Treasury and Risk Management • Transaction Manager • Securities • Master Data • Product Types • Assign Repayment Types to Product Types**. In this configuration, we can assign a separate repayment type (**ReT**) to each product type (**PTyp**), as shown in Figure 4.91. The options are as follows:

- 1: Full Repayment on Maturity
- 2: Installment Repayment
- 3: Annuity
- 4: Perpetual Bonds
- 5: Others
- 6: Annuity (imputed)

4.3 Securities

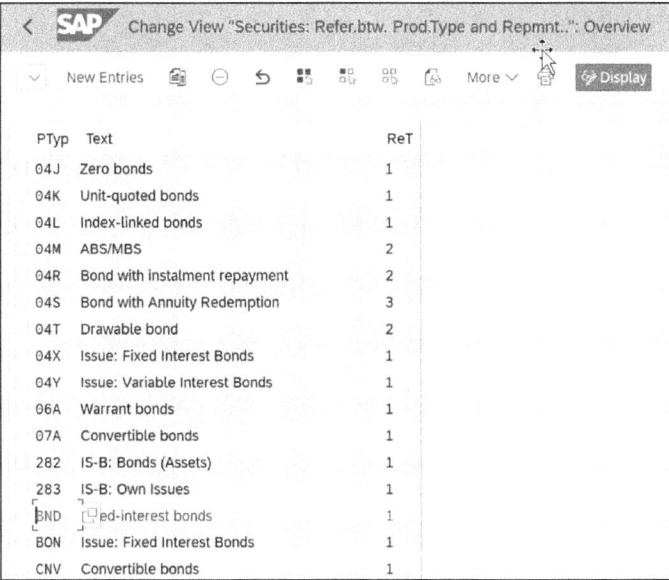

Figure 4.91 Mapping of Each Securities Product Type to Its Repayment Type

Defining Company Code-Dependent Settings for the Product Type

We can make the specific settings for each securities product type in the customizing at **Financial Supply Chain Management • Treasury and Risk Management • Transaction Manager • Securities • Master Data • Product Types • Define Company Code-Dependent Settings for the Product Type**. There are five main settings, as shown in the list below and Figure 4.92:

- **Settings for Cash Management**
  We can determine the planning level that will be generated for any cash flows in the **Memo Record Type** field. In **CM Period**, we determine how many years out the system will generate cash management records for the product type.

- **Settings for Accounting**
  With these settings, we determine whether financial postings are generated from the product type. If postings are to be generated, we can determine whether the transactions should post to the accounts receivable subledger.

- **Settings for Automatic Postings**
  With these settings, we determine whether a contract can be automatically posted using the **Automatic Debit Position and Posting** app or Transaction FWSO. If we don't set the contract to post with this app or transaction, we'll need to post it with the Manual Debit Position app or Transaction FWZE.

- **Settings for Generating Incoming Payment Flows**
  In this area, we assign the settings determined in the previous configuration step. This setting will control whether an incoming payment posting will be immediately generated.

281

4 Contract-Specific Configuration

- **Settings for Tax Generation**
  If we check the **Generate Taxes** box, tax flows will be generated for the product type.

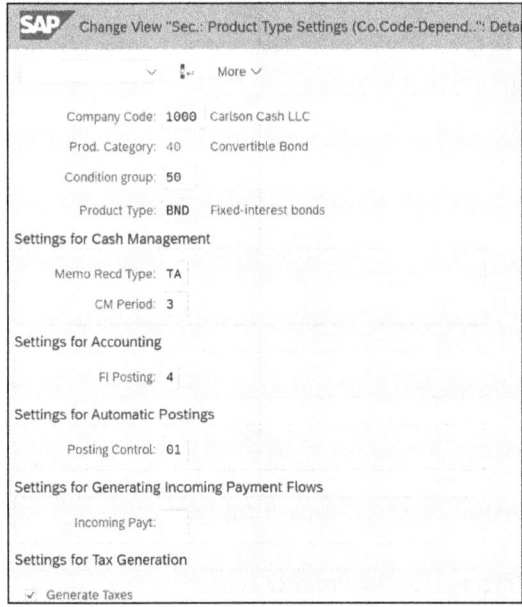

Figure 4.92 Additional Settings We Can Define in Each Company Code

Defining General Classification

We can classify securities into general groupings, and we can customize them, but SAP S/4HANA provides a number of classifications that are shown in Figure 4.93. We can assign these when creating the security class, and we'll discuss that in Chapter 6, Section 6.4.3.

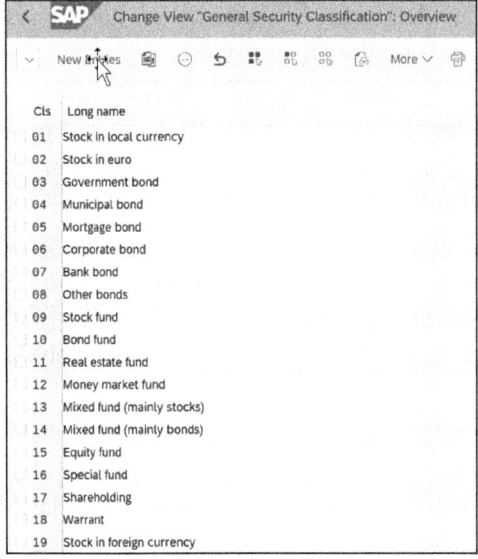

Figure 4.93 Using General Security Classification to Categorize Transactions

282

4.3 Securities

4.3.2 Transaction Management

This next section of the configuration covers the key details in the transaction types and flow types section of the securities configuration. These settings cover creating transaction types, assigning attributes to transaction types, creating flow types, and assigning those flow types to the transaction types. We need to do this to generate the correct cash flows within the various securities.

Defining Transaction Types

As always, we need to create transaction types and assign them to the corresponding product types. We configure this in the **Financial Supply Chain Management • Treasury and Risk Management • Transaction Manager • Securities • Transaction Management • Transaction Types • Define Transaction Types – Securities** menu path. The available transaction types will be determined based on the product category of the product type, and the securities-specific settings for the transaction type can be found in the **Automatic Determination** section.

The first setting in this area is the **Date Rule**, which we can see in Figure 4.94. This determines standard settings for how to determine the position value date, and it subsequently creates rules for how to determine the calculation date and payment date as it relates to the position value date.

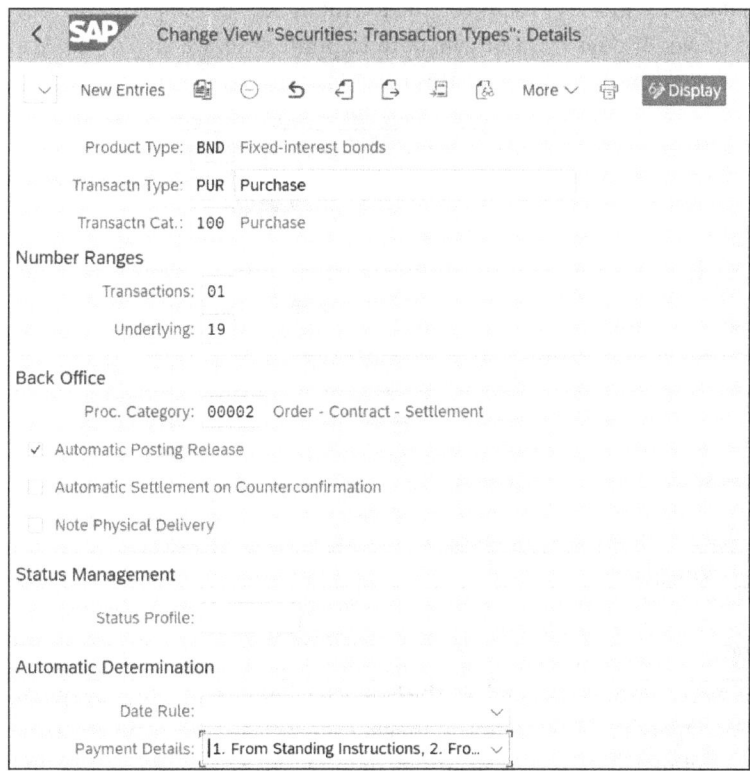

Figure 4.94 Settings We Can Make When Creating Securities Transaction Types

283

4 Contract-Specific Configuration

The payment details configuration determines how the transaction determines the payment details. We can derive this from the standing instructions from the business partner or the payment details on the securities account, or we can opt to pull the payment details from neither of these sources. In the latter case, we must enter the payment details individually into each financial transaction.

Finally, if we check the **Without Accrued Interest Calculation** box, the system won't calculate and generate a flow for the accrued interest on the bond. If we don't check this box, we can always select a similar option when creating the bond.

Defining Date Rules

We can generate date rules and assign them to the transaction types. The configuration for this is in the **Financial Supply Chain Management • Treasury and Risk Management • Transaction Manager • Securities • Transaction Management • Transaction Types • Define Date Rules** menu path. We can define three separate dates in this configuration, as shown in Figure 4.95. First, we note the system date and determine the position value date from it in the **No. of days from system date to pos.val.dat** area. In our example, we will not be changing the date, so the position value date will equal the system date and we'll leave the field blank. Now that we have determined the position value date, we can use that to determine the calculation date and the payment date. In the example, the calculation date is one day before the position value date, so in **No. of days from pos.val.date to cal.dat,** we enter "–1." However, the payment date is a day after the position value date, so in **No. of days from pos.val.date to pmnt date.** we enter "+1." After completing the configuration, we can assign this rule to a transaction type.

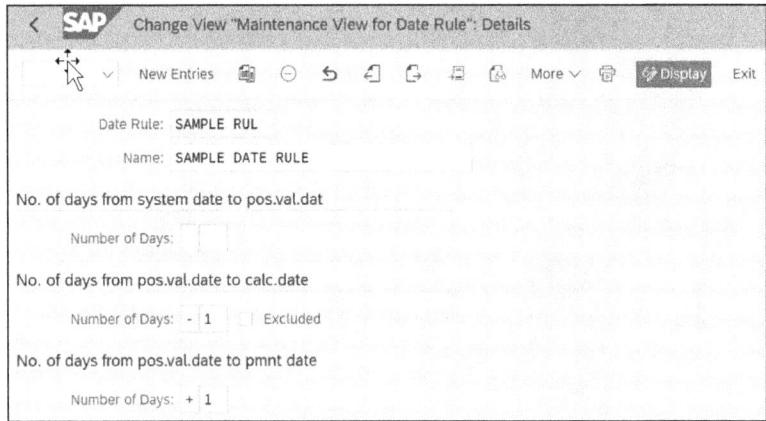

Figure 4.95 Creating Date Rule

Defining Flow Types

The definition of flow types for securities is in the **Financial Supply Chain Management • Treasury and Risk Management • Transaction Manager • Securities • Transaction Management • Flow Types • Define Flow Types – Securities** menu path. Flow types in securities work similar to the flow types in the other areas of treasury and risk management, and the main difference when it comes to securities is the inclusion of the **Include Flow Type in Net Payment Amount Calculation** checkbox, as shown in Figure 4.96. Checking this box allows us to net payments on the same day for the securities transaction.

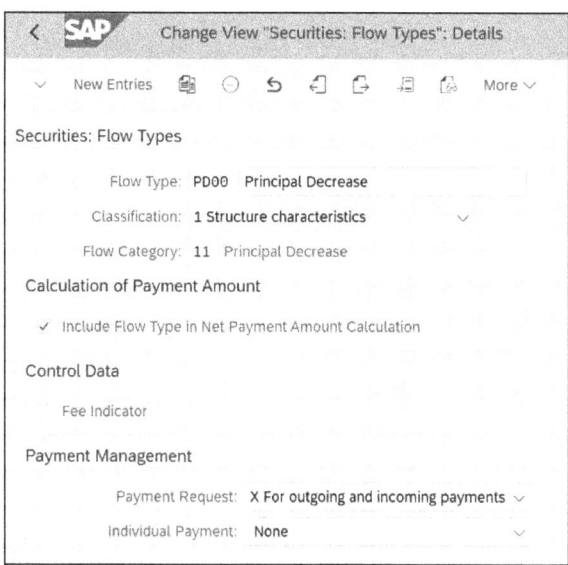

Figure 4.96 Include Flow Type in Net Payment Amount Calculation Checkbox

> **Net Payment Amount**
>
> When multiple flows have the flow type set up to be a net payment, only one payment request will be created for the net amount of the flows when they are on the same day. For example, let's say we're purchasing a bond with a face value of $1,000,000 and we purchased it at a premium of 105%. The purchase amount is $1,050,000. Let's also say we're purchasing the bond in the middle of an interest period, meaning an accrued interest flow of $20,000 is also incurred. This will appear in SAP S/4HANA as follows:
>
> - Purchase: $1,050,000
> - Accrued interest: $20,000
>
> The net payment amount will appear as $1,070,000.

Assigning Flow Types to Transaction Types

Next, we assign the flow types in the **Financial Supply Chain Management • Treasury and Risk Management • Transaction Manager • Securities • Transaction Management • Flow Types • Assign Flow Types to Transaction Types – Securities** menu path. Assigning flow types to transaction types works the same as in the other areas in treasury and risk management. We assign the relevant flow types (**FTyp**) to each product type (**PTyp**) and transaction type (**TTyp**). An example of this is shown in Figure 4.97.

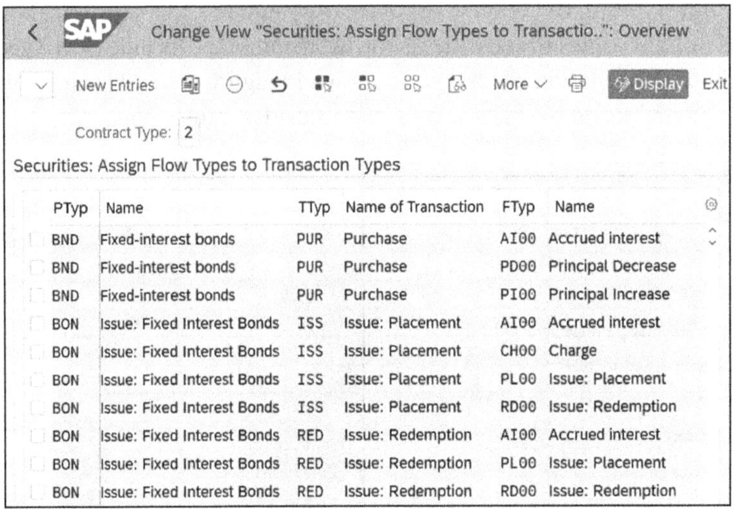

PTyp	Name	TTyp	Name of Transaction	FTyp	Name
BND	Fixed-interest bonds	PUR	Purchase	AI00	Accrued interest
BND	Fixed-interest bonds	PUR	Purchase	PD00	Principal Decrease
BND	Fixed-interest bonds	PUR	Purchase	PI00	Principal Increase
BON	Issue: Fixed Interest Bonds	ISS	Issue: Placement	AI00	Accrued interest
BON	Issue: Fixed Interest Bonds	ISS	Issue: Placement	CH00	Charge
BON	Issue: Fixed Interest Bonds	ISS	Issue: Placement	PL00	Issue: Placement
BON	Issue: Fixed Interest Bonds	ISS	Issue: Placement	RD00	Issue: Redemption
BON	Issue: Fixed Interest Bonds	RED	Issue: Redemption	AI00	Accrued interest
BON	Issue: Fixed Interest Bonds	RED	Issue: Redemption	PL00	Issue: Placement
BON	Issue: Fixed Interest Bonds	RED	Issue: Redemption	RD00	Issue: Redemption

Figure 4.97 Assigning Flow Types to Transaction Types

4.3.3 Position Management

Proper position management of securities and specifically bonds is integral to the carrying value of the bond in SAP S/4HANA and for generating correct accounting entries. This section is focused on the creation of the position management procedure and the subsequent assignment of this procedure to the update types that drive accounting in securities.

Position Management Procedure

Securities also have their own settings available within the position management procedure configuration. We configure this in the **Financial Supply Chain Management • Treasury and Risk Management • Transaction Manager • General Settings • Accounting • Define Position Management Procedure** menu path. We should assign the correct position management category to each of the securities products. The following categories are available to be assigned in the **Position Management Cat.** field (see Figure 4.98):

- Sec./Loans/M.Mkt/List. Opts Norm. Style. (w/o index-link. Bonds)
- Index-Linked Bonds
- Forwards/Repos
- Securities/Loans w/ Install. Repayt (w/o index-linked bonds)
- Index-Linked Bonds with Installment Repayment

The other main setting that we should review in the position management procedure for bonds is adding an amortization step. In Figure 4.98, the **Type of Step** has been set to **Amortization** and the **Procedure** has been assigned under it. This drives the calculation of the amortization of the bond's premium or discount.

4.3 Securities

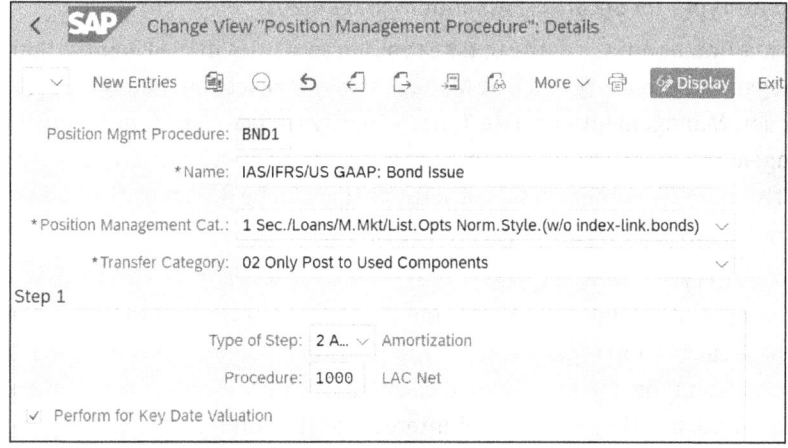

Figure 4.98 Creating Position Management Procedure for Bonds

Defining Update Types and Assigning Usages

As with other contract types, we need to create and assign the update types that are relevant to securities account management, and we do it in the **Define Update Types and Assign Usages**. The full menu path for this is **Financial Supply Chain Management • Treasury and Risk Management • Transaction Manager • Securities • Transaction Management • Update Types • Define Update Types and Assign Usages** menu path. There are a few differences in the settings and assignment of these update types in securities, and we cover this in the following section. As shown in Figure 4.99, the key detail in this configuration is to note that each of the securities update types is assigned to the **Securities Account Management** update type usage. This allows us to assign these update types in the subsequent configuration steps.

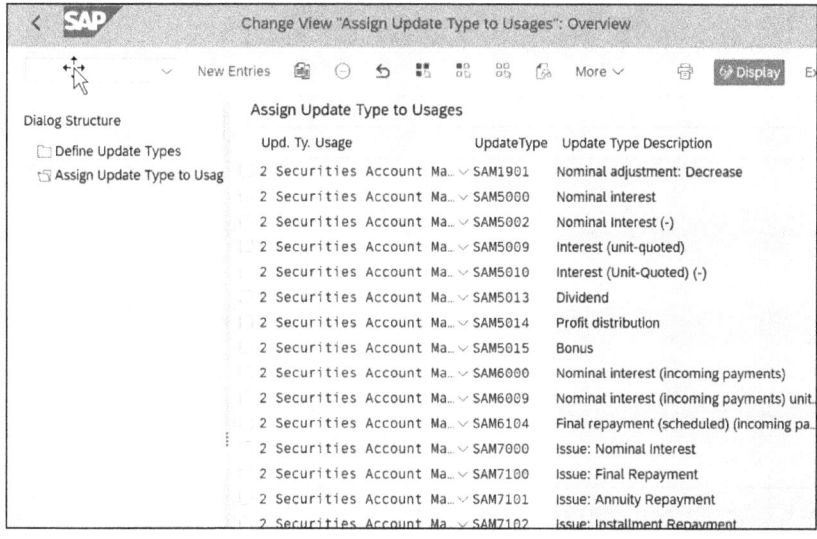

Figure 4.99 Assigning Update Types to Securities Account Management Update Type Usage

287

4 Contract-Specific Configuration

Specifying Update Types for Securities Account Management

We can assign additional settings to the update types in securities in the **Financial Supply Chain Management • Treasury and Risk Management • Transaction Manager • Securities • Transaction Management • Update Types • Specify Update Types for Securities Account Management** menu path. This configuration adds more attributes to the update types. There are five different sets of settings that can be driven by this configuration, as shown in the list below and Figure 4.100:

- **Calculation Category**
  To correctly process the transactions, we should assign them with a correct calculation type. These calculation types are similar to the ones assigned previously in the **Securities** condition type. Frequently used calculation categories can be **AA-Inflow** for a purchase of a security, **TZ-Nominal Interest** for the interest calculation, and **TTEN-Final Repayment** for the repayment of the security at maturity.

- **Effective Interest Calculation**
  If any flows related to the contract need to use the effective interest calculation, we need to ensure the correct flows are used as a basis for the calculation. We can use the checkbox and the positive/negative indicator to correctly calculate the effective interest calculation.

- **Payments**
  If we need to generate a payment request for this update type, we should specify it as an incoming or outgoing payment.

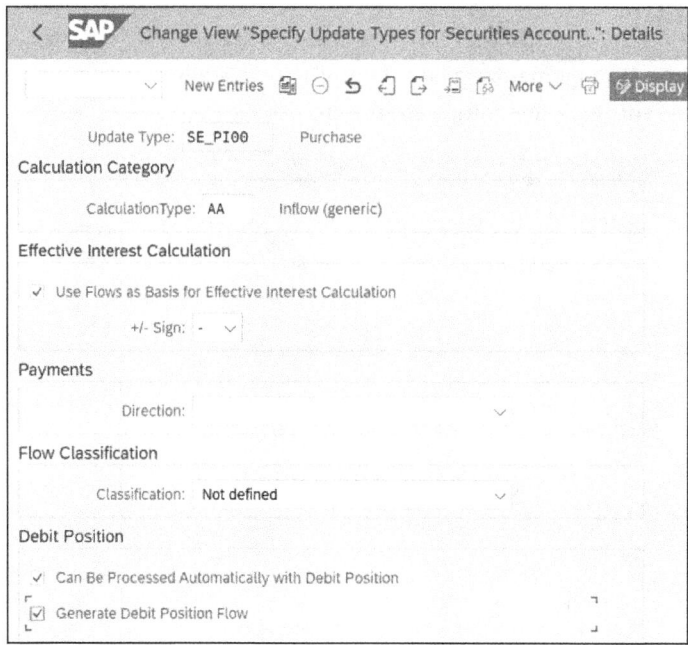

Figure 4.100 Assigning Additional Attributes to Update Types for Securities Account Management

- **Flow Classification**
  By default, the flows are determined with a classification of **Not Defined**, but if we are required to classify the flow in a different category, then we can define them here. We can determine the flows as tax, refundable tax, charges, revenue entered manually, nominal value repayment, and period end flows. This will drive whether we can determine these flows in their respective configurations in the following steps.

- **Debit Position**
  As with the product type settings for a company code, we can check the box for **Can Be Processed Automatically with Debit Position** to determine whether flows can be processed with the automatic debit position posting program. Additionally, if we want to generate a second debit position flow, then we'll check the **Generate Debit Position Flow** box.

Assigning Update Types to Condition Types

When specific condition types are called for a transaction, there need to be associated update types to drive the accounting postings. We configure this in the **Financial Supply Chain Management • Treasury and Risk Management • Transaction Manager • Securities • Transaction Management • Update Types • Assign Update Types to Condition Types – Securities** menu path. In this configuration, we'll notice that there are two separate folders for assignment. These are defined as the active position and the passive position. In general, we use the active position conditions when the financial transaction is an asset and for all the conditions related to the asset. We primarily use the passive position conditions for liability positions. In Figure 4.101, we can see how the conditions and update types have been assigned. The key details that we need to assign in this area are the payment direction (**Payment Directn**) and the **Incoming** and **Outgoing** update types. The payment direction drives the cash flow calculation, so we need to determine whether the flow is an incoming or outgoing flow. Once we do that, we assign the update types for when the defined condition creates the incoming or outgoing flow.

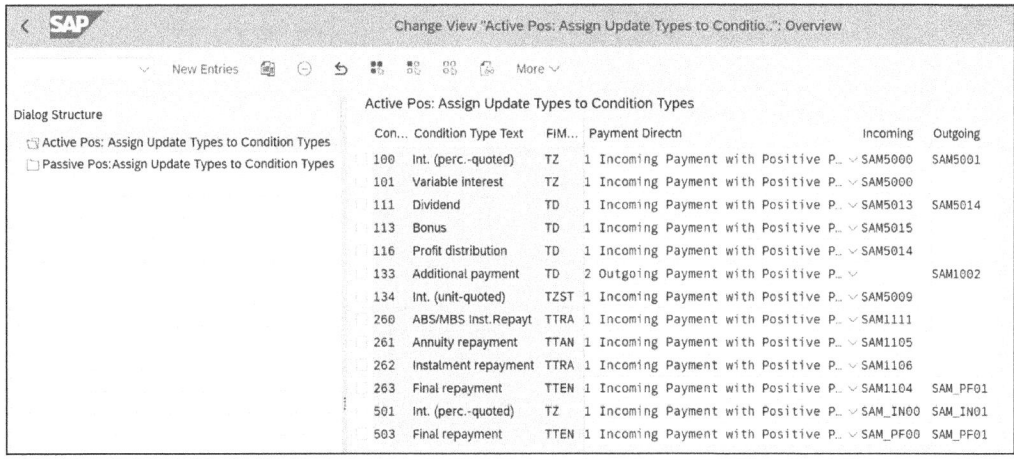

Figure 4.101 Update Types Assigned to Condition Types

4 Contract-Specific Configuration

Assigning Update Types to Functions

In the securities area, we need to define further functions of the update types. We can't simply assign update types as relevant for securities account management; we need to also assign them in the **Financial Supply Chain Management • Treasury and Risk Management • Transaction Manager • Securities • Transaction Management • Update Types • Assign Update Types to the Functions of Security Account Management** menu path. The functions that we can assign in this customizing are as follows (see Figure 4.102):

- **Update Types for Debit Position**
  We post the initial flow in securities with the Post Flows app or Transaction TBB1, but the subsequent cash flows are generated by the manual and automatic debit position programs. We need to designate them for the debit position.

- **Update Types for Manual Posting**
  If a flow is an other flow and we need to post it as a one-off entry, we need to designate the update type here.

- **Assign Update Types for Debit Position Generation**
  Update types have secondary update types that need to be generated in certain scenarios. In those scenarios, we need to tie the related update types to the main debit position update type.

- **Update Types for Capitalization of Div./Prof. Dist.**
  When there's a case when a dividend or profit distribution is given, the flow can be capitalized in the position. (This can occur with stocks and investment certificates.) We can assign this by product type in this area.

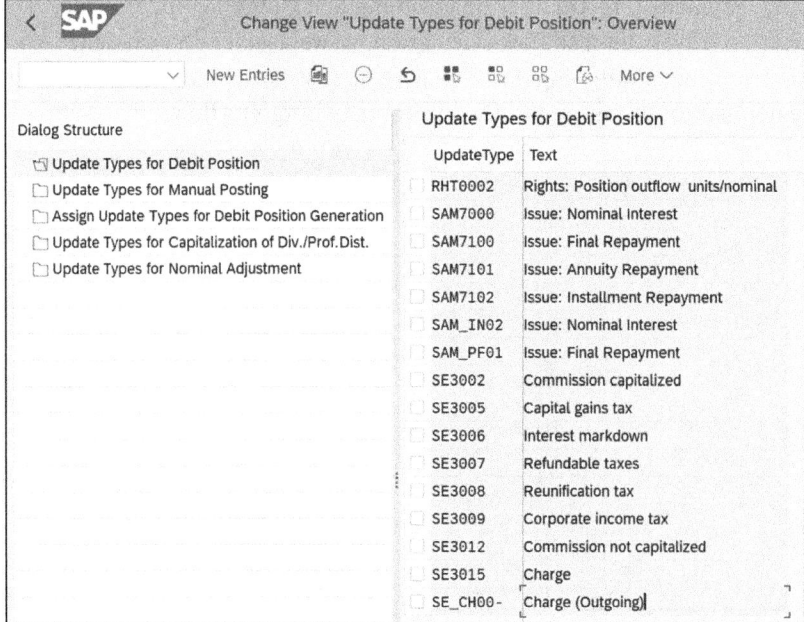

Figure 4.102 Assigning Update Types to Debit Position Functions

- **Update Types for Nominal Adjustment**
  When we need to adjust the nominal amount of a security positively or negatively, we need to determine the update types of the adjustment. We can assign different update types by product type.

Defining Derivation Rules for Tax Flows

When we need to apply taxes to any of the securities flows, we can assign the tax rate in this customizing accordingly. The menu path for this configuration is **Financial Supply Chain Management • Treasury and Risk Management • Transaction Manager • Securities • Position Management • Securities Account Management • Define Derivation Rules for Tax Flows**. The configuration in this area can be very specific, and we can determine settings by company code, product category, product type, country, securities account, and update types, as shown in Figure 4.103. There are also validity dates, so we can adjust the rates when appropriate.

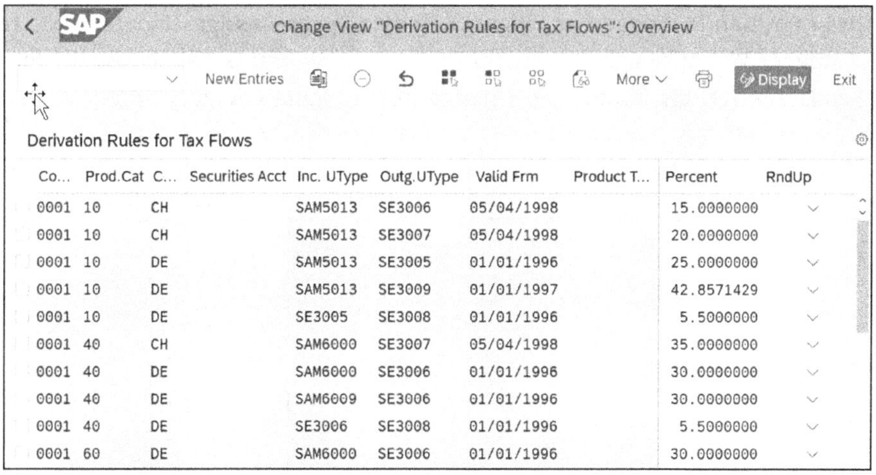

Co...	Prod.Cat	C...	Securities Acct	Inc. UType	Outg.UType	Valid Frm	Product T...	Percent	RndUp
0001	10	CH		SAM5013	SE3006	05/04/1998		15.0000000	
0001	10	CH		SAM5013	SE3007	05/04/1998		20.0000000	
0001	10	DE		SAM5013	SE3005	01/01/1996		25.0000000	
0001	10	DE		SAM5013	SE3009	01/01/1997		42.8571429	
0001	10	DE		SE3005	SE3008	01/01/1996		5.5000000	
0001	40	CH		SAM6000	SE3007	05/04/1998		35.0000000	
0001	40	DE		SAM6000	SE3006	01/01/1996		30.0000000	
0001	40	DE		SAM6009	SE3006	01/01/1996		30.0000000	
0001	40	DE		SE3006	SE3008	01/01/1996		5.5000000	
0001	60	DE		SAM6000	SE3006	01/01/1996		30.0000000	

Figure 4.103 Assigning Tax Percentage to Update Types

Assigning Update Types for Securities Account Transfer

This configuration is located in the **Financial Supply Chain Management • Treasury and Risk Management • Transaction Manager • Securities • Position Management • Securities Account Transfer • Update Types • Assign Update Types for Securities Account Transfer** menu path. The securities account transfer functionality in SAP S/4HANA allows a contract in a company code to transfer the book value from one securities account to another. To enable this, we need to assign update types for the move from one securities account to the other. The first screen in this configuration will list all securities product types, and to assign the update types for the account transfer, we highlight the line for a **Product Type** and click into the **Update Type** folder. Once in this folder, we need to assign an **UpdateType** for the **Inflow** and the **Outflow** to correspond to the transfer from one securities account to the next (see Figure 4.104).

4   Contract-Specific Configuration

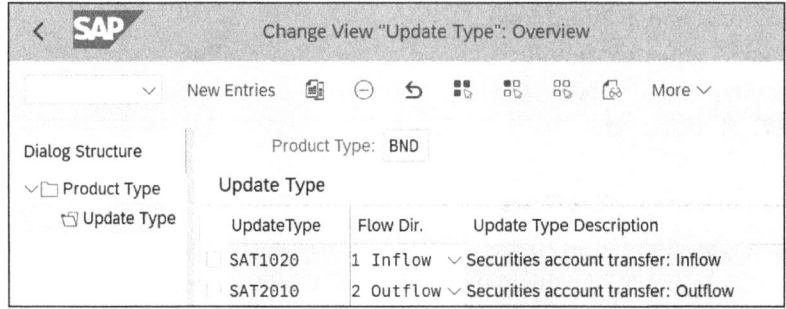

Figure 4.104 Assigning Update Types for Transfers Between Securities Accounts

Assigning Update Types to the Rights Category

There are a series of actions in a securities transaction that are defined as rights in treasury and risk management. The update types for the rights are assigned in **Financial Supply Chain Management • Treasury and Risk Management • Transaction Manager • Securities • Position Management • Rights • Update Types • Assign Update Types to Rights Category**. We can assign these to transactions for **Warrants, Callable Bonds, Puttable Bonds, Convertible Bonds,** and **Options** in the **Rights Category** dropdown list. Each of these rights can have its own set of update types that we assign, but as shown in Figure 4.105, a list of update types starting with RHT are provided for this configuration.

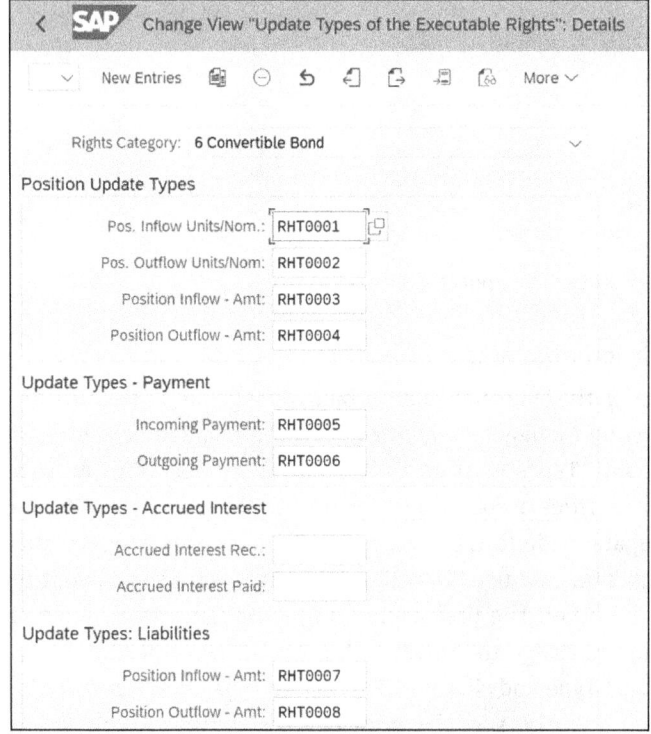

Figure 4.105 Assigning Update Types for Rights

292

4.4 Derivatives

We now turn our focus to the configuration of derivatives within treasury and risk management. Derivatives are essentials tools used by corporate treasuries to manage risk, hedge against market fluctuations, and speculate on financial outcomes. These instruments are essential for executing complex financial strategies and achieving precise risk management objectives.

In this section, we'll dive into the setup and configuration of various derivatives, emphasizing the key components necessary for effective management within the treasury and risk management framework. Unlike more straightforward financial instruments, derivatives introduce additional layers of complexity due to their intricate structures and diverse applications. Configuring derivatives requires a thorough understanding of their unique attributes, including pricing models, contract specifications, and the calculation of gains or losses.

We'll explore the process of setting up derivative product types, ensuring that all relevant parameters—such as underlying assets, strike prices, and expiration dates—are accurately defined. This process also involves managing the various contract terms and conditions, which can include specific clauses related to settlement procedures, margin requirements, and adjustment mechanisms. The first step when creating a derivative contract is to start with the product type creation, and we'll then move on to the configuration of the transaction types. The setup of the flow types and update types within a derivative contract contain some nuances related to the posting of the cash flows, and we'll cover these nuances and why they are important to keep in mind during the configuration process.

4.4.1 Product Types

Interest rate swap configuration is located within the following menu path: **Financial Supply Chain Management • Treasury and Risk Management • Transaction Manager • OTC Derivatives**. The configuration of an interest rate swap follows the same general pattern as that of any other treasury and risk management functionality. To begin the configuration, we must establish product types. The menu path to create a product type in this area is **Financial Supply Chain Management • Treasury and Risk Management • Transaction Manager • OTC Derivatives • Transaction Management • Product Types • Define Product Types – OTC Derivatives**.

Similar to other product types covered previously in this chapter, the purpose of the product type is to differentiate between different types of financial products and, in this case, different types of interest rate swaps. Figure 4.106 shows the product type creation screen. We configure interest rate swaps in the **OTC Swaps** folder, as shown in Figure 4.107.

4   Contract-Specific Configuration

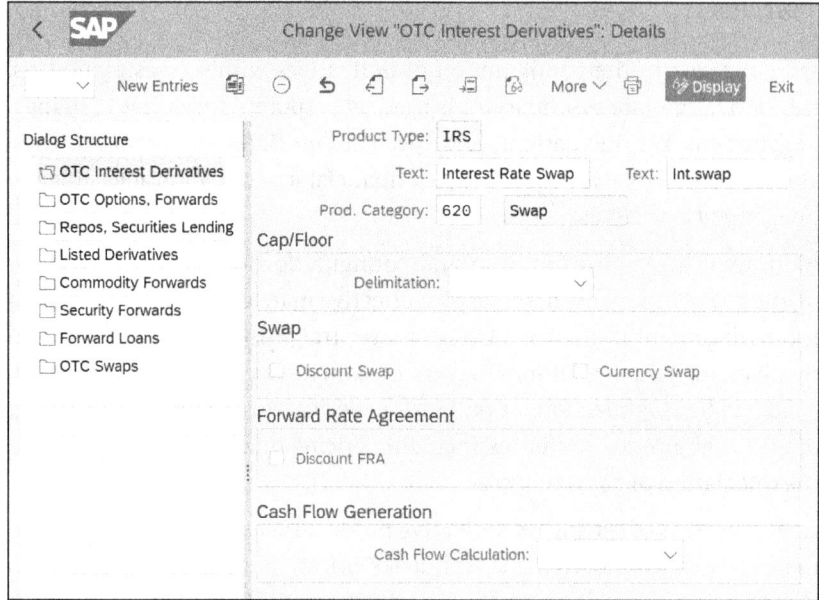

Figure 4.106 Creating Interest Rate Swap Product Types

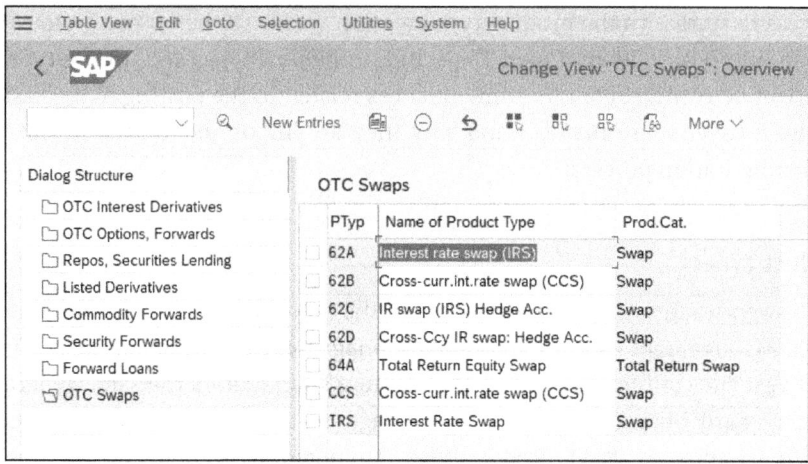

Figure 4.107 Creating Interest Rate Swap Product Types

During configuration, the **Product Type** is assigned a three-character value, **Text, Short Text,** and product category. The product category (**Prod. Cat.**) that we must use for interest rate swap transactions is 620. The **Prod. Cat.** drives other types of functionalities that are available and used by product categories of the same type. There are additional optional settings available during the product type configuration, as follows:

- **Cap/Floor Delimitation**
  With this setting, we can determine whether the configured product type has an upper or lower limit.

- **Cap with Upper Limit**
  With this setting, we can establish that the product has an upper limit.
- **Floor with Lower Limit**
  With this setting, we can establish that the product has a lower limit.
- **Discount Swap**
  With this setting, we can indicate to the system that these products should be created as discount swaps.
- **Currency Swap**
  With this setting, we can indicate to the system that these products should be created as currency swaps, meaning that the nominal amounts on the incoming and outgoing sides of the contract must have different currencies.
- **Cash Flow Generation**
  With this setting, we can influence the operation of the generated cash flows and allow for parallel conditions.

4.4.2 Transaction Types

Now that we've set up the product type, we can move on to transaction type creation for the interest rate swap. The menu path for interest rate swap transaction type configuration is located at **Financial Supply Chain Management • Treasury and Risk Management • Transaction Manager • OTC Derivatives • Transaction Management • Transaction Types • Define Transaction Types – OTC Derivatives**. Specifically for interest rate swaps, we may need to create separate transaction types, depending upon the direction of the swap being recorded.

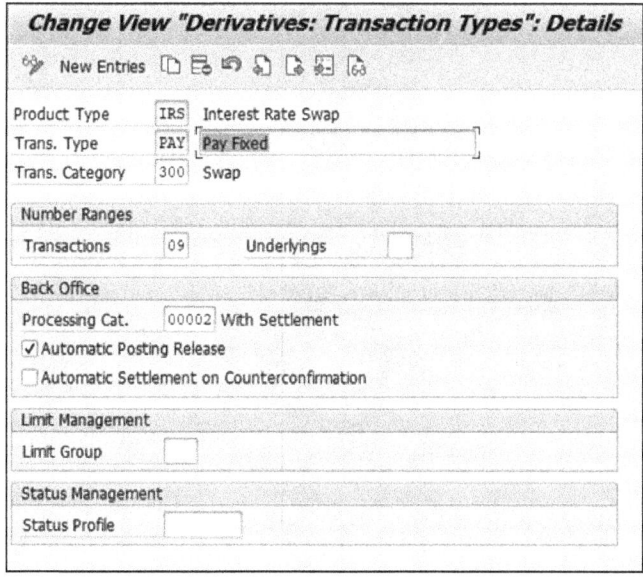

Figure 4.108 Configuring Interest Rate Swap Transaction Types

4 Contract-Specific Configuration

We'd create one transaction type for a payer-based swap and another transaction type for a recipient-based swap. We'll assign all interest rate swap transaction types the "300" **Trans. Category**, as shown in Figure 4.108. Detailed instructions for defining transaction types can be found in Chapter 3, Section 3.11.

4.4.3 Defining Flow Types

Next, we'll define the applicable flow types required for an interest rate swap. The configuration for the flow types is located in the following menu path: **Financial Supply Chain Management • Treasury and Risk Management • Transaction Manager • OTC Derivatives • Transaction Management • Transaction Types • Define Flow Types – OTC Derivatives**. We should establish flow types for all transactional flows dependent upon the actual contract structure between the entity and the counterparty.

That being said, interest rate swaps are primarily settled in cash, and therefore, the nominal contract flows for an interest rate swap are not posted to the general ledger and are used solely for the calculation of the associated interest payments. Since the nature of an interest rate swap includes both an inbound and an outbound leg for the interest payments, there are four primary flows that we should use for interest rate swaps:

- **Interest Nominal (IN00)**
  This flow represents the interest payment on both legs of the transaction, as shown in Figure 4.109.

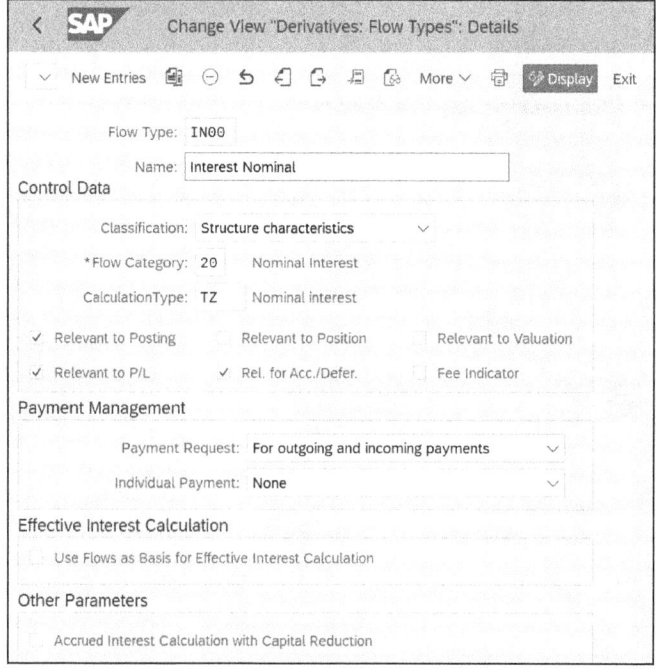

Figure 4.109 Interest Rate Swap Flow Type IN00

4.4 Derivatives

The interest payments for an interest rate swap are posted to the general ledger and are therefore relevant to posting. The interest payments should also include a cash movement request and a create a payment request for both incoming and outgoing payments; therefore, this is indicated in the **Payment Request** dropdown list.

- **Principal Increase – No Post (PIN5)**
  This flow represents the principal increase flow that is only notional and is not posted, as shown in Figure 4.110. Therefore, we don't check the **Relevant To Posting** box.

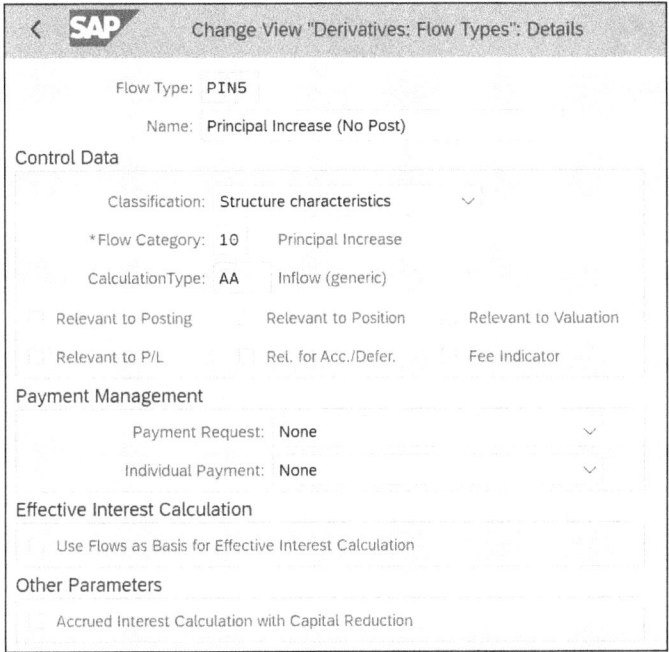

Figure 4.110 Interest Rate Swap Flow Type PIN5

- **Principal Decrease – No Post (PDN5)**
  This flow represents the principal decrease flow that is only notional and is not posted, as shown in Figure 4.111. Therefore, we don't check the **Relevant To Posting** box.

- **Final Repayment – No Post (PFN5)**
  This flow represents the final repayment that is only notional and is not posted, as shown in Figure 4.112. Therefore, we do not check the **Relevant To Posting** box.

4 Contract-Specific Configuration

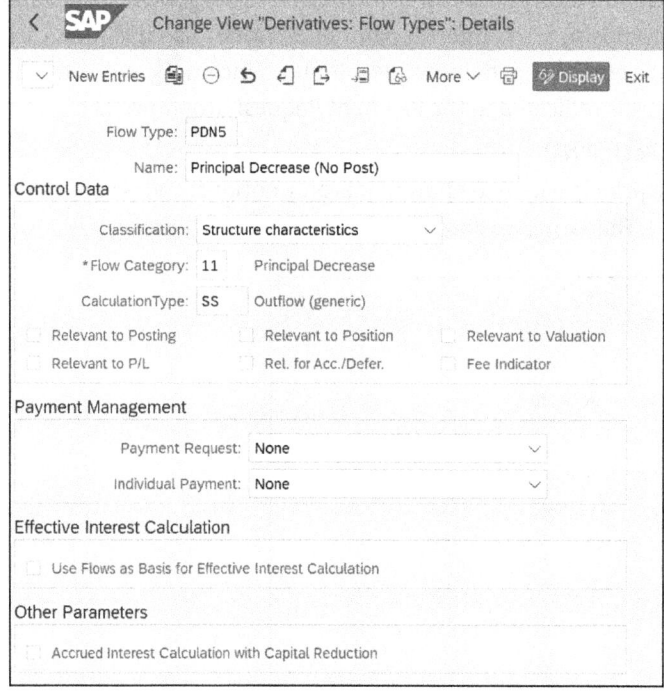

Figure 4.111 Interest Rate Swap Flow Type PDN5

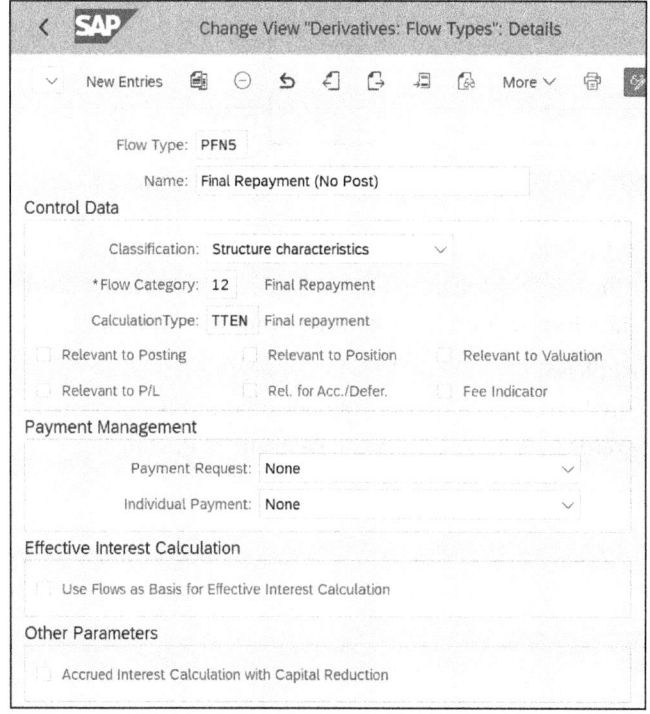

Figure 4.112 Interest Rate Swap Flow Type PFN5

4.4.4 Assigning Flow Types to Update Types

As with other product types in treasury and risk management, after the necessary flow types have been created, we'll need to assign them to an appropriate update type. Update types are used by the SAP S/4HANA system to further derive the necessary accounting settings for the direction of each flow type. We can complete this configuration in the following menu path: **Financial Supply Chain Management • Treasury and Risk Management • Transaction Manager • OTC Derivatives • Transaction Management • Update Types • Assign Flow Types to Update Types**. There are two necessary assignments in this section. First, we must assign the **UpdateType** to a contract type. For interest rate swaps, we should assign all update types to the **Derivatives** contract category (**Cont.Type**). Lastly, we'll need to assign each flow type (**FType**) to an update type. In Figure 4.113, we can see an example of the update types that are assigned to our previously created flow types.

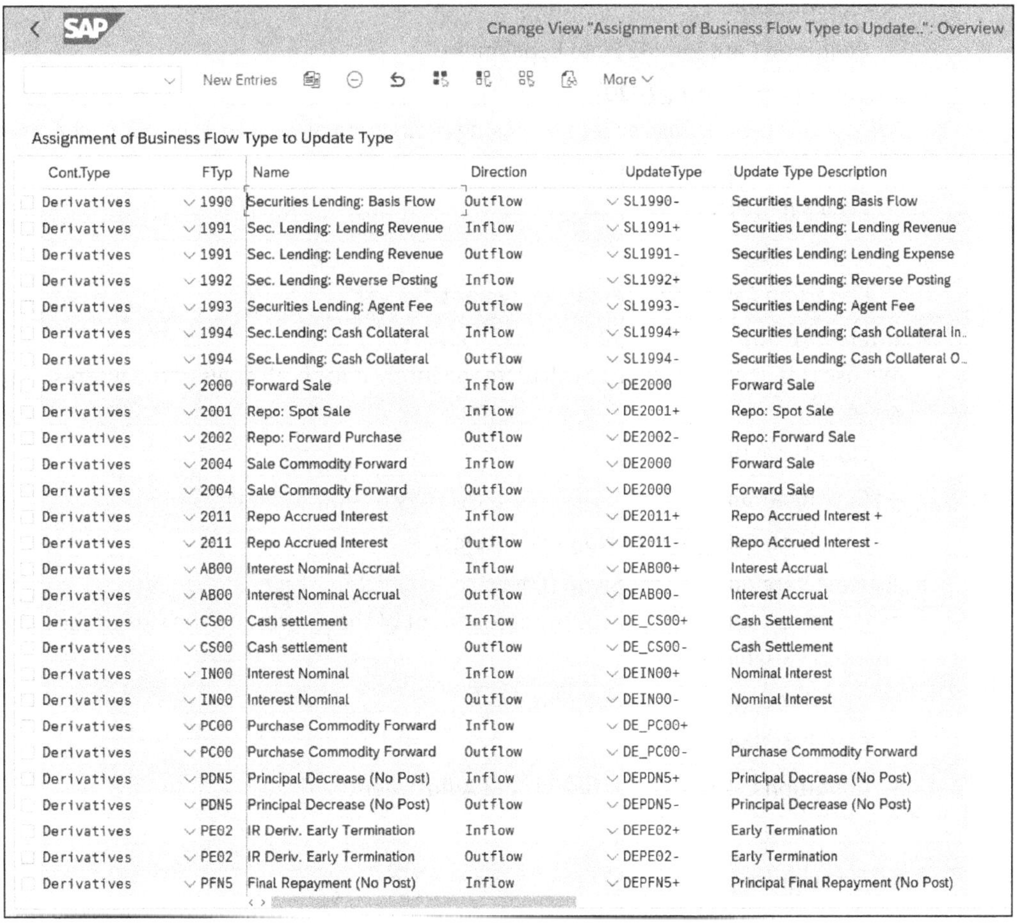

Cont.Type	FTyp	Name	Direction	UpdateType	Update Type Description
Derivatives	1990	Securities Lending: Basis Flow	Outflow	SL1990-	Securities Lending: Basis Flow
Derivatives	1991	Sec. Lending: Lending Revenue	Inflow	SL1991+	Securities Lending: Lending Revenue
Derivatives	1991	Sec. Lending: Lending Revenue	Outflow	SL1991-	Securities Lending: Lending Expense
Derivatives	1992	Sec. Lending: Reverse Posting	Inflow	SL1992+	Securities Lending: Reverse Posting
Derivatives	1993	Securities Lending: Agent Fee	Outflow	SL1993-	Securities Lending: Agent Fee
Derivatives	1994	Sec.Lending: Cash Collateral	Inflow	SL1994+	Securities Lending: Cash Collateral In..
Derivatives	1994	Sec.Lending: Cash Collateral	Outflow	SL1994-	Securities Lending: Cash Collateral O..
Derivatives	2000	Forward Sale	Inflow	DE2000	Forward Sale
Derivatives	2001	Repo: Spot Sale	Inflow	DE2001+	Repo: Spot Sale
Derivatives	2002	Repo: Forward Purchase	Outflow	DE2002-	Repo: Forward Sale
Derivatives	2004	Sale Commodity Forward	Inflow	DE2000	Forward Sale
Derivatives	2004	Sale Commodity Forward	Outflow	DE2000	Forward Sale
Derivatives	2011	Repo Accrued Interest	Inflow	DE2011+	Repo Accrued Interest +
Derivatives	2011	Repo Accrued Interest	Outflow	DE2011-	Repo Accrued Interest -
Derivatives	AB00	Interest Nominal Accrual	Inflow	DEAB00+	Interest Accrual
Derivatives	AB00	Interest Nominal Accrual	Outflow	DEAB00-	Interest Accrual
Derivatives	CS00	Cash settlement	Inflow	DE_CS00+	Cash Settlement
Derivatives	CS00	Cash settlement	Outflow	DE_CS00-	Cash Settlement
Derivatives	IN00	Interest Nominal	Inflow	DEIN00+	Nominal Interest
Derivatives	IN00	Interest Nominal	Outflow	DEIN00-	Nominal Interest
Derivatives	PC00	Purchase Commodity Forward	Inflow	DE_PC00+	
Derivatives	PC00	Purchase Commodity Forward	Outflow	DE_PC00-	Purchase Commodity Forward
Derivatives	PDN5	Principal Decrease (No Post)	Inflow	DEPDN5+	Principal Decrease (No Post)
Derivatives	PDN5	Principal Decrease (No Post)	Outflow	DEPDN5-	Principal Decrease (No Post)
Derivatives	PE02	IR Deriv. Early Termination	Inflow	DEPE02+	Early Termination
Derivatives	PE02	IR Deriv. Early Termination	Outflow	DEPE02-	Early Termination
Derivatives	PFN5	Final Repayment (No Post)	Inflow	DEPFN5+	Principal Final Repayment (No Post)

Figure 4.113 Assigning Update Types to Flow Types for Interest Rate Swaps

4.4.5 Defining Condition Types

Since a condition type is required for flow types that rely on a calculation to generate the value of a flow, interest rate swap–related transactions also require the assignment of a condition type. The menu path for the configuration settings to define condition types for an interest rate swap is located here: **Financial Supply Chain Management • Treasury and Risk Management • Transaction Manager • OTC Derivatives • Transaction Management • Condition Types • Define Condition Types – OTC Derivatives**. Condition types for interest rate swaps are similar in construction to condition types for other treasury and risk management instruments, and the necessary settings will depend upon specific requirements. In our example, we'll have the following condition types:

- **Final Repayment (1005)**
  We use this condition type for the calculation of the final repayment:
  - Flow Type: PFN5
  - Classification: 1
  - Condition Category: 12 Final Repayment
- **Closing Rel. to Posting (1120)**
  We use this condition type for the closing or final repayment in a situation where we want to settle the contract in cash:
  - Flow Type: PFN5
  - Classification: 1
  - Condition Category: 12 Final Repayment
- **Interest (1200)**
  We use this condition type to calculate the interest for both sides of the interest rate swap:
  - Flow Type: IN00
  - Classification: 1
  - Condition Category: 20 Nominal Interest
- **Interest Rate Adjustment Swap (1215)**
  We use this condition type for the adjustment of the interest on the side of the swap that has variable interest:
  - Flow Type: No flow type required
  - Classification: 1
  - Condition Category: 21 Interest Rate Adjustment

4.4.6 Assigning Condition Types to Transaction Types

While the definition of the condition types may be similar in multiple treasury instruments, we'll need to assign several condition types to each defined interest rate swap transaction type. Depending upon the nature of the interest rate swap, there are a minimum of

4.4 Derivatives

three condition types that we'll need to assign: a condition type for the closing of the swap contract, a condition type for the calculated interest of the swap, and a condition type for interest rate adjustment. In many cases, the condition type for the closing of an interest rate swap is nonposting; however, if the swap contracts are being settled in cash, then an additional condition type for the closing of the swap with a cash settlement posting may also be required. Figure 4.114 shows the four primary condition types being assigned to each interest rate swap transaction type. For detailed instructions on how to assign condition types to transaction types, see Chapter 3, Section 3.15.2.

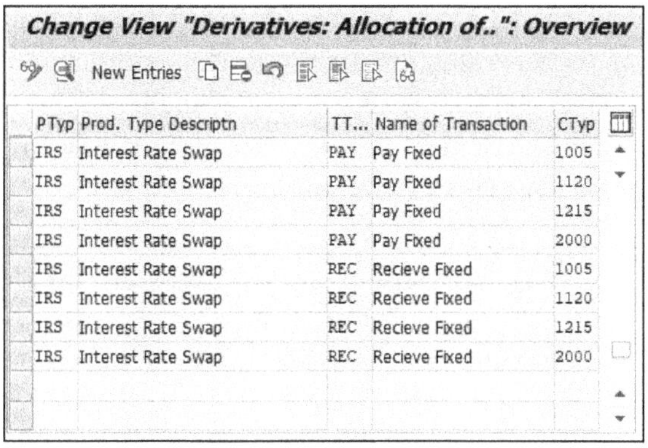

PTyp	Prod. Type Descriptn	TT...	Name of Transaction	CTyp
IRS	Interest Rate Swap	PAY	Pay Fixed	1005
IRS	Interest Rate Swap	PAY	Pay Fixed	1120
IRS	Interest Rate Swap	PAY	Pay Fixed	1215
IRS	Interest Rate Swap	PAY	Pay Fixed	2000
IRS	Interest Rate Swap	REC	Recieve Fixed	1005
IRS	Interest Rate Swap	REC	Recieve Fixed	1120
IRS	Interest Rate Swap	REC	Recieve Fixed	1215
IRS	Interest Rate Swap	REC	Recieve Fixed	2000

Figure 4.114 Assigning Product and Transaction Types to Condition Types

4.4.7 Position Management Procedure

After we've completed the product configuration, there are additional settings that are required to handle the position management and accounting aspects of interest rate swaps in treasury and risk management. First, a position management procedure specific to interest rate swaps is required, and additional details on the configuration of this are covered in Chapter 3, Section 3.6. The primary requirement pertaining to interest rate swaps in relation to the position management procedure is to assign a position management procedure that uses a **Position Management Cat.** of **OTC Derivatives**, as shown in Figure 4.115.

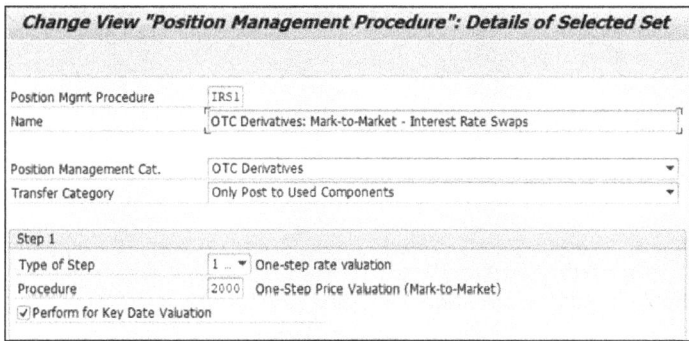

Figure 4.115 Interest Rate Swap Position Management Procedure

4.4.8 Assigning Update Types for Valuation

If interest rate contracts require a valuation, we must assign a series of update types to generate the valuation postings. This functionality is commonly utilized in mark-to-market interest rate swap contracts during a month-end closing cycle. To allow for an update type to be used during valuation postings, we must create the update type with an update type usage of **Key Date Valuation** on the **Define Update Types and Assign Usages** screen. SAP S/4HANA provides a series of preconfigured valuation update types, and in many cases, these will suffice. However, the option remains to create custom valuation update types should that need arise. After we've created the update types, we'll need to assign them to the various valuation functions. This assignment is located in the following menu path: **Financial Supply Chain Management • Treasury and Risk Management • Transaction Manager • General Settings • Accounting • Key Date Valuation • Update Types • Assign Update Types for Valuation**.

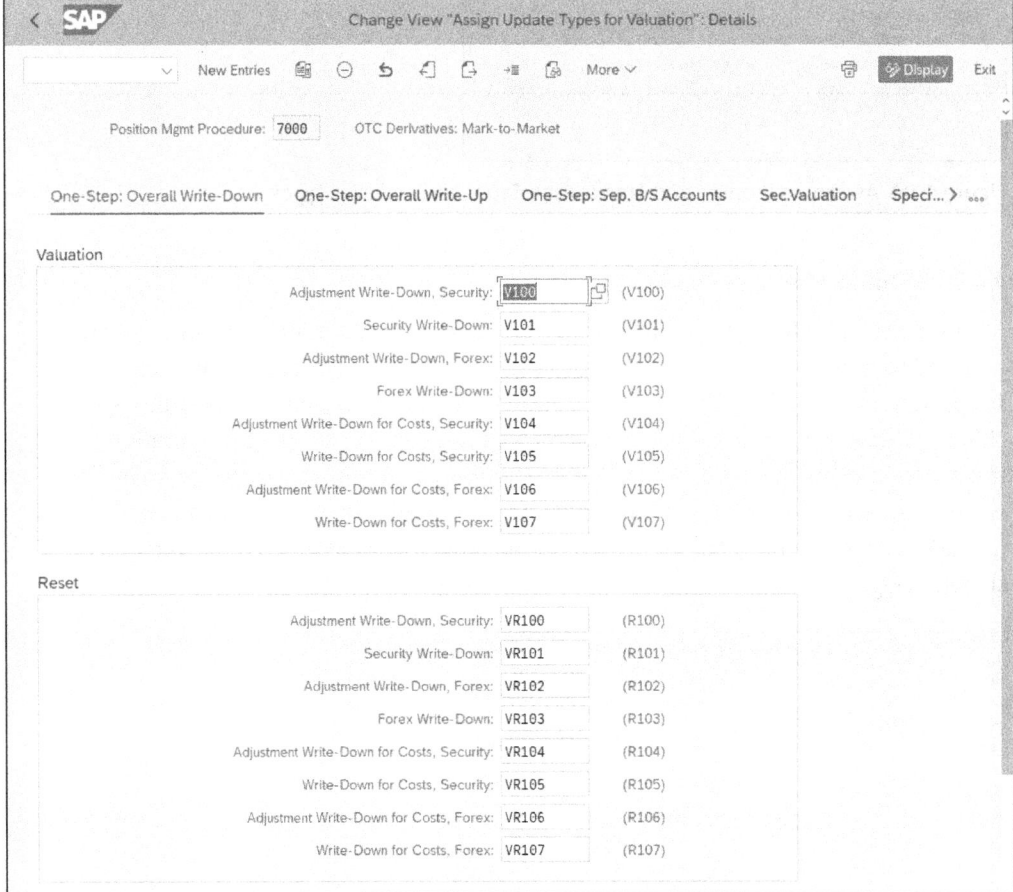

Figure 4.116 Assigning Update Types for Valuation of Interest Rate Swaps

We initially assign the valuation update types by position management procedure. We locate the proper position management procedure and double-click the entry to enter the assignment screen. Once in the assignment screen, we'll see different valuation functions and have the opportunity to assign an update type to different activities. As we can see in Figure 4.116, SAP S/4HANA has conveniently named predelivered update types that correspond to the different valuation activities. For example, the **Security Write-Down** activity in Figure 4.116 is reference V101. There's a predelivered update type named V101 that is intended to reflect that activity from an accounting perspective in most cases. While the predelivered update types may be adequate, we can also modify these settings and add custom update types.

4.4.9 Assigning Update Types for Derived Business Transactions

As with the update types for valuation, the execution of interest rate swap contracts can potentially generate other derived business transactions. An example of a derived business transaction for an interest rate swap could be the calculation of a gain or loss on the contract after it has been closed within the system. Akin to the valuation activities, we can assign derived business transactions specific update types to capture this activity in financial accounting. In related fashion to the valuation update types, SAP S/4HANA provides a series of predelivered derived business transaction update types that cover a broad range of activities. In many cases, we can adequately use and assign these update types, but we can also use custom update types if necessary.

So that we can use an update type during derived business transaction postings, we should create the update type with an update type usage of **Derived Business Transaction** on the **Define Update Types and Assign Usages** screen. To assign update types for derived business transactions, we can use the following menu path: **Financial Supply Chain Management • Treasury and Risk Management • Transaction Manager • General Settings • Accounting • Derived Business Transactions • Update Types • Assign Update Types for Derived Business Transactions**.

Once again, we must select the proper position management procedure as this drives which set of derived business transaction update types we should use for a particular contract. After selecting the proper **Position Mgmt Procedure**, we double-click on the left-hand folder for the activity to which derived business transaction update types should be assigned. Once in the assignment screen, we can assign update types to the different activities as needed. Additionally, like the valuation update types, the predelivered derived business transactions have a naming convention that can assist with their assignment. For example, in Figure 4.117, the **Security Price Gain** activity is labeled as activity **(1)**, and the derived business transaction update type that corresponds to this activity is **DBT_B001**. The same applies to activities **(2)**, **(3)**, **(4)**, and so on. We can use the predelivered update types in this scenario, while the option to utilize custom update types is also available.

4   Contract-Specific Configuration

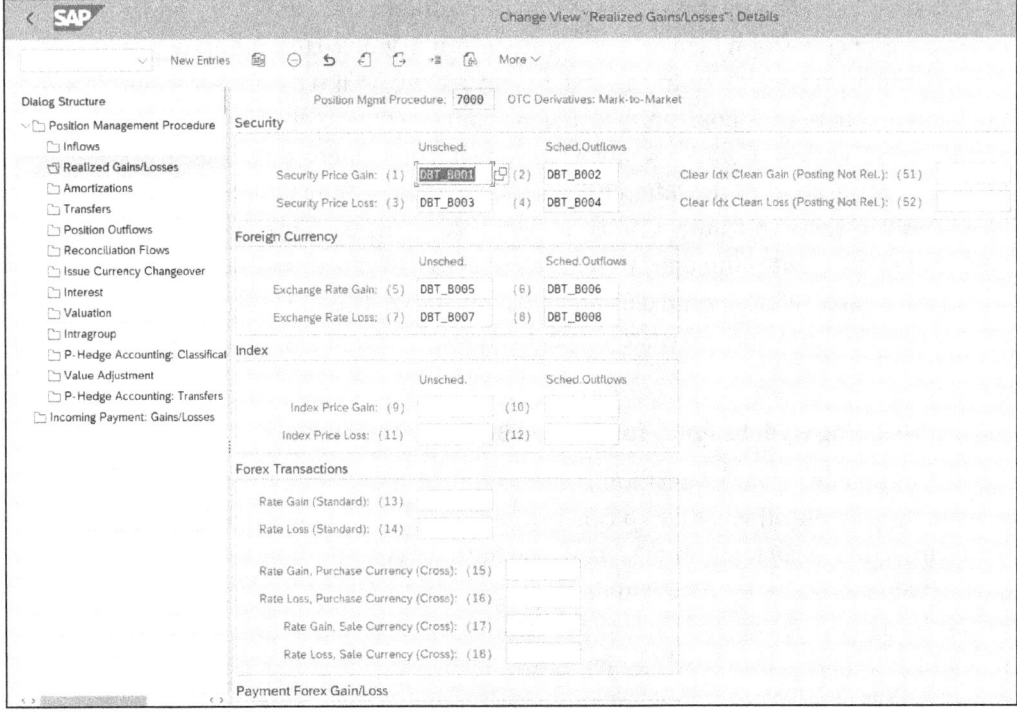

Figure 4.117 Details of Realized Gains and Losses

> **Update Types: Naming Derived Business Transactions**
>
> The update types for derived business transaction come delivered from SAP S/4HANA with a standard naming convention. The naming follows a standard convention in which the first three characters represent the purpose of the update type and are followed by an underscore. The last four characters of the update type represent which folder and field the update type should populate.
>
> - The first three characters of most update types are **DBT**. Other update types in this configuration are specific to hedge accounting, and those update types start with **HAC**.
> - The fifth character determines what folder the update type is assigned to. The folders go in alphabetical order, so the first folder is **Inflows**, and all update types in that folder would have an A as the fifth character. The next folder is **Realized Gains/Losses**, and all update types would use a B as the fifth character.
> - Each box in this configuration is assigned a number, and the last three characters of the update type represent the same numbered box that the update type should be placed in.

4.5 Commodity Derivatives

We'll now shift gears and focus on the creation of commodity derivative contracts. These contracts are essential for managing price risks associated with commodities, making them crucial tools for corporate treasury departments to manage and hedge risk. Gaining an understanding of the setup of these contracts is important because they all function somewhat similar to one another. SAP S/4HANA supports the following commodity derivatives:

- Futures
- Forwards
- Swaps
- Options

The setup of commodity derivative contracts is important because they all share certain operational processes. Whether we're dealing with forwards, futures, options, or swaps, the same principles and processes remain consistent among these instruments.

In the following sections, we'll dive into the detailed steps involved in setting up commodity derivative contracts within SAP S/4HANA. We'll cover the essential configuration steps and nuances we need to understand to effectively create and manage these contracts in treasury and risk management.

4.5.1 Commodity Types

The first requirement to configure a commodity contract in treasury and risk management is to have master data on the underlying commodity. As with a financial contract, we must classify a commodity by defining a commodity type. To do so, we follow the menu path at **Financial Supply Chain Management • Treasury and Risk Management • Basic Functions • Market Data Management • Master Data • Commodities • Define Commodity Types**.

We can specify the required commodity types (**CommTyp**) for classifying our commodities. We enter the name (**Commodity Type Text**) and description (**Commodity Category Description**) of the commodity type and assign it to a commodity category (**CommCat**), which indicates the group to which the commodity belongs. The commodity categories are predefined in the system, and the following values are available for selection:

- 1: Metals
- 2: Base Metals
- 3: Minor Metals
- 4: Grains
- 5: Oil Seeds
- 6: Livestock

4 Contract-Specific Configuration

- 7: Softs
- 8: Energy
- 9: Power
- 10: Gas
- 11: Oil
- 12: Weather

As shown in Figure 4.118, we've created a new commodity type for **ALUMINUM** and assigned it to commodity category 1: **Metals**.

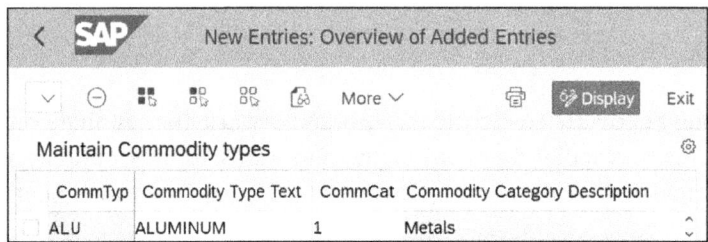

Figure 4.118 Creating Aluminum Commodity Type

4.5.2 Specifying Commodities

The next step involves defining the commodities that will be traded. We do this by navigating to **Financial Supply Chain Management • Treasury and Risk Management • Basic Functions • Market Data Management • Master Data • Commodities • Specify Commodities**. This is an important step, as it ensures that all of the relevant commodities are configured and that we can use them when creating a commodity derivatives contract. To begin, we create each commodity that falls within the scope of our trading activities, as shown in Figure 4.119.

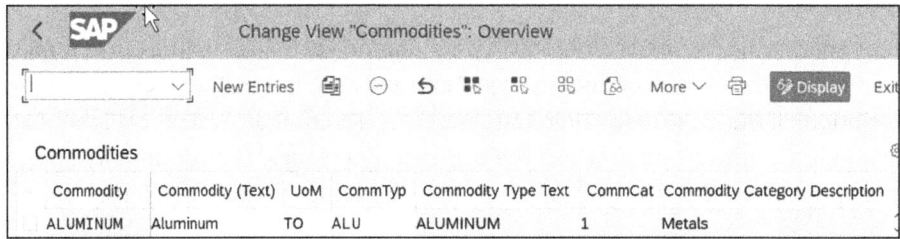

Figure 4.119 Specifying Aluminum Commodity and Assigning To Unit of Measure

Next, we assign each **Commodity** to a specific unit of measure (**UoM**), which standardizes how quantities of the commodity will be recorded and managed. The unit of measure we use for each commodity should match the unit of measure used by the exchange. This is an important thing to keep in mind for maintaining consistency and accuracy when performing valuation of the contract positions throughout the life of the contract.

Finally, we link each commodity to a commodity type (**CommTyp**) that we previously created. This classification helps in organizing and categorizing various commodities, facilitating more accurate tracking and reporting.

4.5.3 Specifying Market Identifier Codes

The next step involves configuring the market identifier codes (MICs) for all the exchanges where the specified commodities in scope are traded. A MIC is a four-character alphanumeric code that uniquely identifies each exchange, and we assign it to each exchange and use it globally to standardize and simplify the identification of the various exchanges. During trading, the MIC helps ensure accuracy by specifying the exact venue where a transaction took place, and it is especially important for commodities that trade on more than one exchange.

To configure the MIC, we navigate the following menu path: **Financial Supply Chain Management • Treasury and Risk Management • Basic Functions • Market Data Management • Master Data • Specify Market Identifier Codes**. We begin by specifying the four-character **MIC** that represents the exchange, along with the **MIC Description**, the **Country/Reg.**, the **City** of the exchange, the relevant **Calendar**, and the exchange's **MIC URL**.

Next, we make sure the box indicating the MIC's usability is checked, confirming that the **MIC Can Be Used**, as shown in Figure 4.120. Additionally, we update the **Status** of the MIC to **Active** to enable its use after creation. While this step is simple and straightforward, it's crucial in the configuration process. The MIC is a core component referenced in several subsequent steps, making this an important part of the setup.

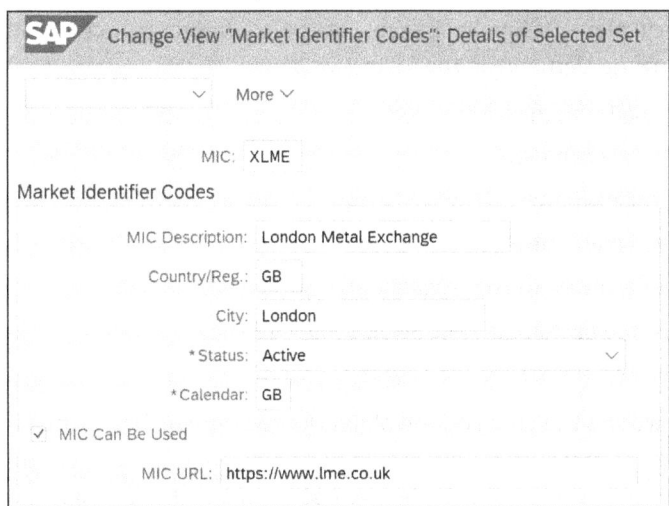

Figure 4.120 Creating MIC for XLME

4   Contract-Specific Configuration

4.5.4   Defining the Exchange

After we complete the MIC configuration, we can define the physical exchange where the commodities contracts are purchased. This step is important because it defines the exchange that will be tied to the MIC in the next step. As shown in Figure 4.121, we've defined a number of exchanges that we plan to trade on. The configuration is a two-step process: we enter the **Exchange** and then enter the **Long name** of the exchange.

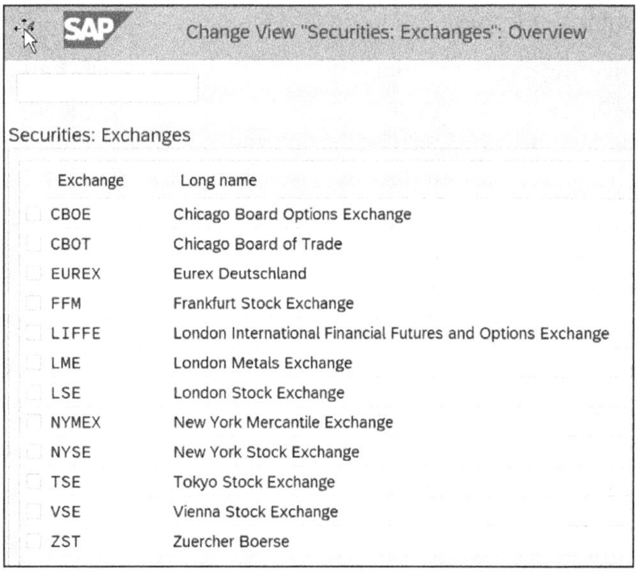

Exchange	Long name
CBOE	Chicago Board Options Exchange
CBOT	Chicago Board of Trade
EUREX	Eurex Deutschland
FFM	Frankfurt Stock Exchange
LIFFE	London International Financial Futures and Options Exchange
LME	London Metals Exchange
LSE	London Stock Exchange
NYMEX	New York Mercantile Exchange
NYSE	New York Stock Exchange
TSE	Tokyo Stock Exchange
VSE	Vienna Stock Exchange
ZST	Zuercher Boerse

Figure 4.121   Creating Exchanges Used for Commodity Trading

To create a new exchange, we navigate to this menu path: **Financial Supply Chain Management • Treasury and Risk Management • Basic Functions • Market Data Management • Master Data • Securities • Define Exchange**. On the configuration screen, we should configure the following elements in the following ways:

- Define an up-to ten-character name for the **Exchange.**
- Create a **Long Name** for the exchange.
- Create a **ShortNme** for the exchange.
- Specify the **Country/Reg** where the exchange is located.
- Choose a **Calendar** that corresponds to the calendar of the exchange.
- Specify the **Currency** of the exchange, noting that this currency is not always the same as the currency of the country.

As shown in Figure 4.122, we've added the London Metals Exchange, which is located in Great Britain and conducts all trades in US dollars. At this point, we have fully configured the exchange, and we can move on to the next step, in which we assign the exchange to the market identifier code.

4.5 Commodity Derivatives

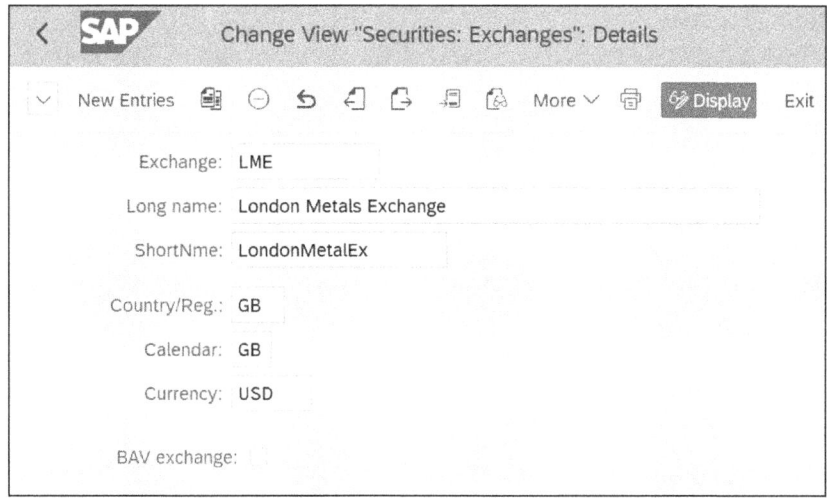

Figure 4.122 Details of Securities Exchange

4.5.5 Assigning Exchanges to Market Identifier Codes

Now that we've created the MIC as well as the exchange, the next step is to assign the MIC to the exchange. We start by navigating to this menu path: **Financial Supply Chain Management • Treasury and Risk Management • Basic Functions • Market Data Management • Master Data • Commodities • Derivative Contract Specification • Assign Exchanges to Market Identifier Codes.**

This configuration step is important for ensuring that all trading activities are correctly identified and recorded within the system under the correct exchange. In this step, we map the previously created MIC to the identifier representing the exchange. By doing this, we establish a link between the MIC and the exchange, which allows for accurate tracking and reporting of trading activities.

To illustrate, let's look at our example in Figure 4.123. We've created the **MIC** "XLME" for the London Metals Exchange (LME). In the configuration screen, we'll map XLME to LME, and this mapping links the MIC to the actual exchange and ensures that trades conducted on the London Metals Exchange are correctly identified by the XLME MIC in our system.

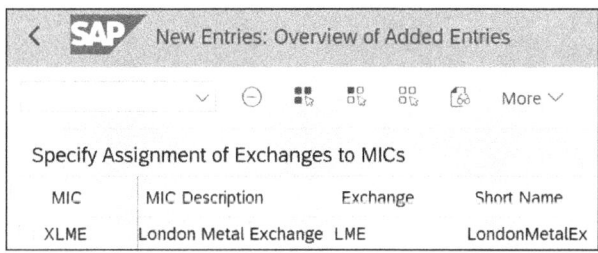

Figure 4.123 Assigning MIC XLME to Exchange

309

4.5.6 Defining Product Types

As with the previously configured product types, the first step when setting up a commodity derivative contract within treasury and risk management is to create the product type. This initial step is essential as the product type serves as the basis for structuring the entire commodity contract. Once we've defined the product type, we assign it to a specific product category, which further refines and organizes the contract's framework. Treasury and risk management supports several product categories for commodities, and each is designed to handle different types of commodity transactions. Configuring the correct product category is essential for ensuring that the commodity derivative contract operates as expected within the organization's broader financial and risk management strategies. The available product categories that we can be configure for commodities include the following:

- 700: Commodity Future
- 760: OTC Options
- 800: Commodity Forwards

A common commodity derivative that many companies utilize is a commodity forward. In this example, we'll walk through the creation of a commodity forward, as it serves as a solid foundation for setting up additional commodity derivative contracts. We start by navigating to the **Financial Supply Chain Management • Treasury and Risk Management • Transaction Manager • OTC Derivatives • Transaction Management • Product Types • Define Product Type – OTC Derivatives** menu path. We then click on the **Commodity Forwards** folder because we're creating a commodity forward contract. As shown in Figure 4.124, we've assigned a **Product Type** of "CF" and also entered a **Text** as well as a **ShortTxt** for the product type. We've assigned **Prod. Category** "800," which is used for commodity forwards.

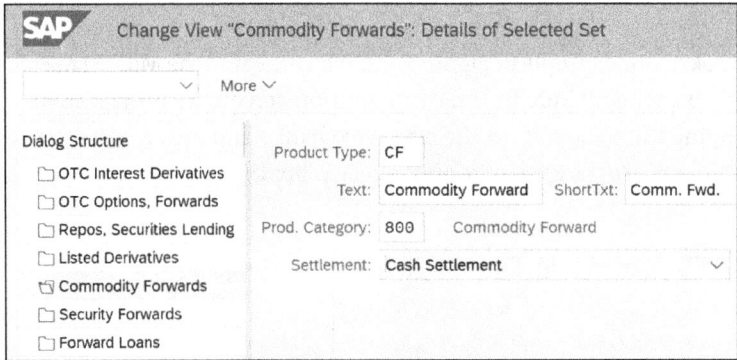

Figure 4.124 Creating Commodity Forward Product Type

An important configuration element to consider when creating the product type is the method of **Settlement** at contract maturity. This is an important aspect as it dictates how the obligations of the contract will be fulfilled once the contract reaches its end date. The **Settlement** dropdown list in the product type allows us to specify whether the transaction will involve physical delivery of the underlying commodity or if it will be settled through a cash payment.

Physical delivery means that the actual commodity will be exchanged between the parties involved, and we typically use it when we intend to use the commodity in operations. On the other hand, cash settlement involves calculating the difference between the agreed-upon price in the contract and the current market price at maturity, with the resulting amount being paid in cash. **Cash Settlement** is very common and is used primarily when purchasing over the counter. In our example, we'll opt for cash settlement of the contracts at maturity. This choice simplifies the settlement process and avoids complexities associated with physical delivery.

4.5.7 Defining Transaction Types

Now that we've successfully created the product type, we can move on to the creation of the transaction types for our commodity forwards. To start this configuration, we navigate to **Financial Supply Chain Management • Treasury and Risk Management • Transaction Manager • OTC Derivatives • Transaction Management • Transaction Numbers • Define Transaction Types – OTC Derivatives**. Within product category 800, there are two possible transaction categories:

- **Transaction Category 100: Buy**
- **Transaction Category 200: Sell**

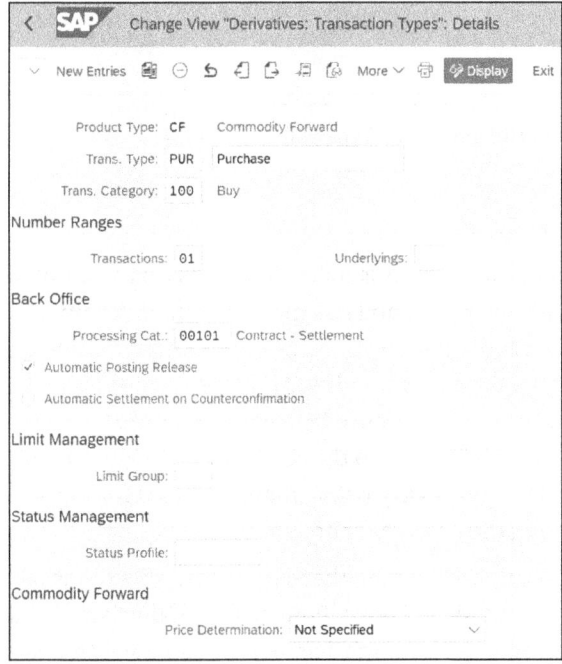

Figure 4.125 Creating Commodity Forward Transaction Type Buy

As demonstrated in Figure 4.125 and Figure 4.126, we've created two distinct transaction types to cover both the buying and the selling sides of the transaction. On the buying side, we've assigned **Trans. Category** "100" to indicate a purchase. Conversely, on the selling side, we've assigned **Trans. Category** "200" to represent a sale. For detailed instructions on how to define transaction types, see Chapter 3, Section 3.11.

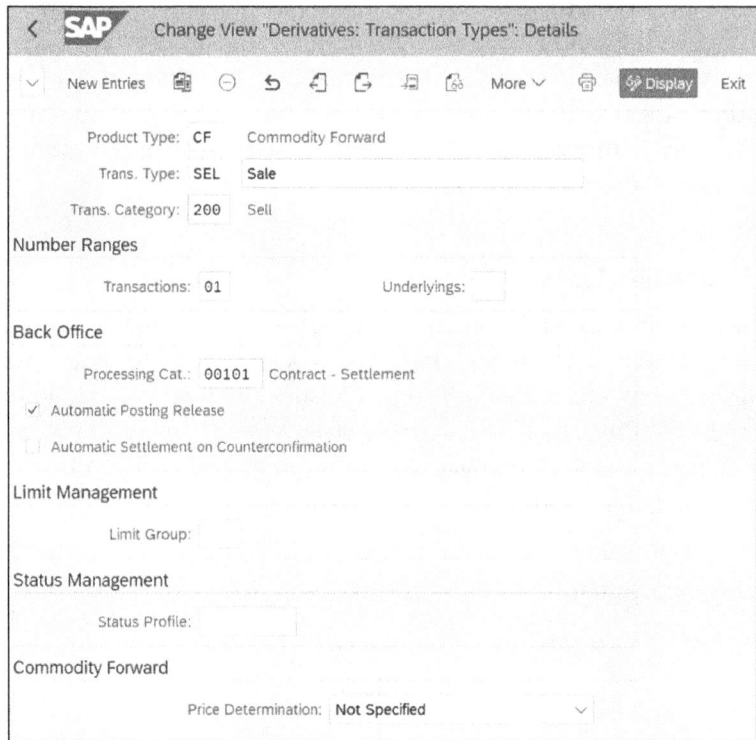

Figure 4.126 Creating Commodity Forward Transaction Type Sell

4.5.8 Defining and Assigning Flow Types

The setup and configuration of commodity forwards follow a process that's comparable to those of other instruments within treasury and risk management. A commodity forward contract starts with the initial agreement on the terms and conditions, including the commodity, price, and quantity of the commodity. At maturity, the contract is settled though a cash settlement, where the difference between the agreed-upon forward price and the market price at maturity is paid to the counterparty. This structure is convenient as a company can hedge commodity risk without the need for physical delivery of the commodity, simplifying the process for both parties.

It's at this stage of the configuration that we'll define the various cash flows that take place through the life of the contract and assign the cash flows to the product and transaction types. The flows related to a commodity forward are relatively straightforward

4.5 Commodity Derivatives

and involve either a purchase or a sale of the commodity, as well as either an outgoing or an incoming cash flow at the end of the contract, representing the cash settlement. The cash settlement is the difference between the agreed-upon forward price and the market price of the commodity at time of contract maturity. Depending on whether the contract is in a gain or a loss position, we will either send or receive funds when the contract matures. The settlement cash flow is the only cash flow that will result in a physical posting to the general ledger because the flows related to the purchase and sale of the commodity are not true cash flows and are only used in the contract to calculate the cash settlement amount. We can find configuration at **Financial Supply Chain Management • Treasury and Risk Management • Transaction Manager • OTC Derivatives • Transaction Management • Flow Types • Define and Assign Flow Types**. Refer back to Chapter 3, Section 3.13, for detailed instructions on flow types.

To account for the cash flows within a commodity forward contract, we must configure the following three flow types:

- **Purchase Commodity Forward (PC00)**
  As shown in Figure 4.127, this flow type represents the purchase of a commodity forward contract that is an outgoing flow. We only use this flow type to track the nominal cash value of the flow, and the flow will appear in the cash flow tab of the contract but will be set up to not create a posting to the general ledger.

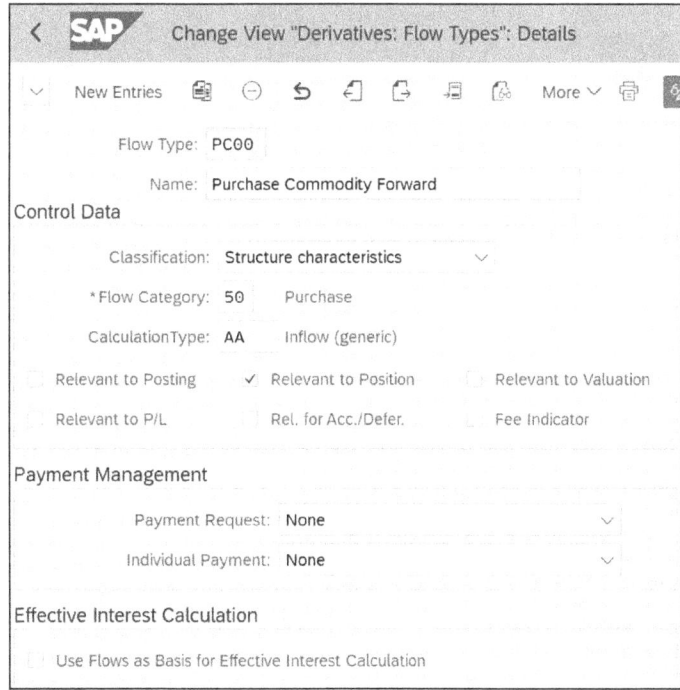

Figure 4.127 Purchase Commodity Forward Flow Type

313

4   Contract-Specific Configuration

- **Sale Commodity Forward (SC00)**
  As shown in Figure 4.128, this flow type represents the sale of a commodity forward contract that is an incoming flow. As with **Purchase Commodity Forward (PC00)**, we only use this flow type to track the nominal cash value of the flow. The flow will appear in the cash flow tab of the contract upon creation, but it will be set up to not create a posting to the general ledger.

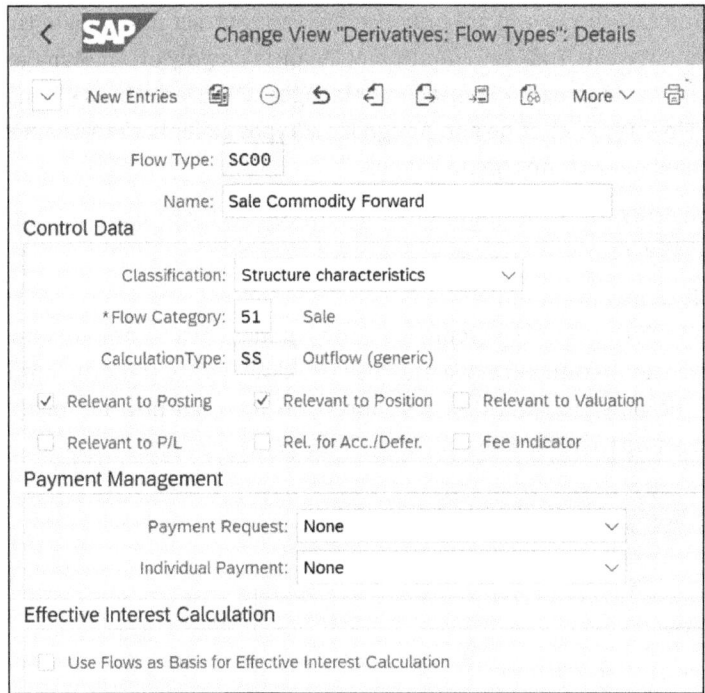

Figure 4.128 Sale Commodity Forward Flow Type

- **Cash Settlement (CS00)**
  As shown in Figure 4.129, this flow type represents the cash settlement that takes place at the maturity of the contract. The flow type could be either an incoming or an outgoing flow type, depending on whether the contract is in a gain or a loss position. It will be set up to post to the general ledger and will create a payment request that is used to settle the incoming or outgoing funds.

Now that we've created the three applicable flow types for a commodity forward, we can proceed to assign them to the two transaction types. As illustrated in Figure 4.130, the PC00 flow type (**FTyp**) is only necessary for the purchase ("PUR") transaction type (**TTyp**), while SC00 is only needed for the sale ("SEL") transaction type. Since both transaction types will involve cash settlement at contract maturity, we'll assign the cash settlement CS00 flow type to both the purchase and the sale transaction types.

4.5 Commodity Derivatives

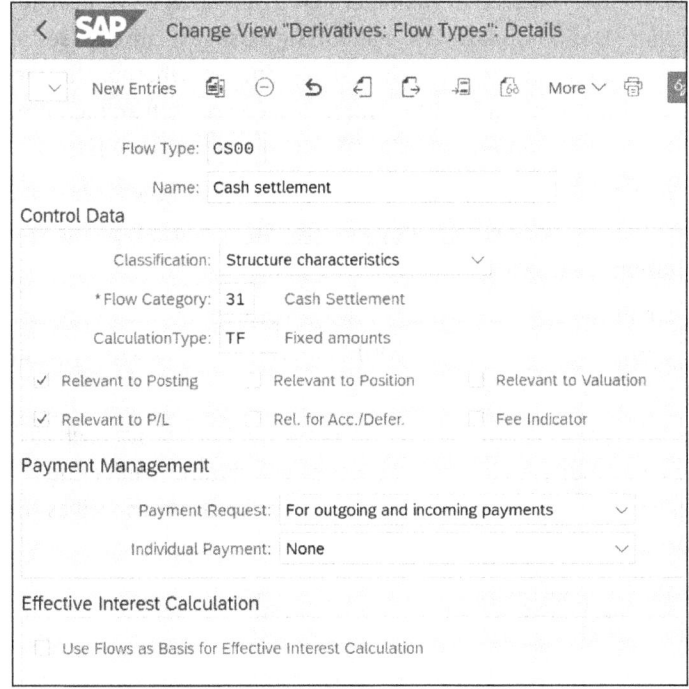

Figure 4.129 Cash Settlement Commodity Forward Flow Type

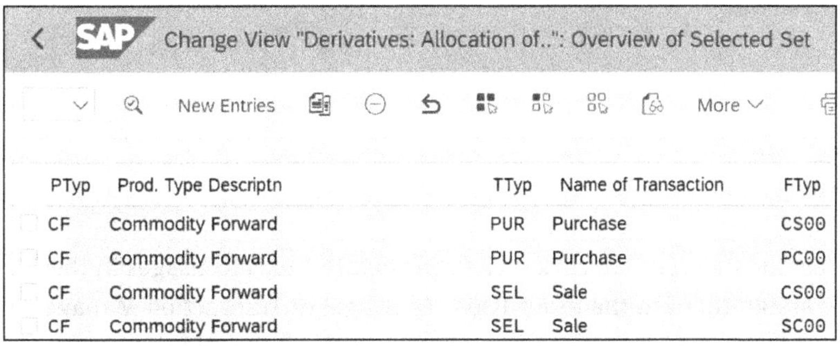

PTyp	Prod. Type Descriptn	TTyp	Name of Transaction	FTyp
CF	Commodity Forward	PUR	Purchase	CS00
CF	Commodity Forward	PUR	Purchase	PC00
CF	Commodity Forward	SEL	Sale	CS00
CF	Commodity Forward	SEL	Sale	SC00

Figure 4.130 Assigning Commodity Forwards Product Type to Flow Types

We've created the necessary flow types and assigned them to the product and transaction type combinations, so we can now move on to the creation of update types. Notice that we're skipping past the setup of any condition types because there are no calculations that take place in any of the flows, and therefore, a condition type is not required for any of the flow types on a commodity forward.

4.5.9 Defining Update Types and Assigning Usages

Next, we can proceed to creating the applicable update types for each of the flow types required for a commodity forward. We navigate to **Financial Supply Chain Management** •

4 Contract-Specific Configuration

Treasury and Risk Management • Transaction Manager • OTC Derivatives • Transaction Management • Update Types • Define Update Types and Assign Usages. The cash settlement flow type can be either an inflow or an outflow, depending on whether the contract is in a gain or a loss position at maturity. Therefore, we'll create both an incoming and an outgoing flow type for "Cash Settlement" (CS00), as shown in Figure 4.131.

The commodity purchase flow type will always be an outgoing cash flow since we're purchasing the commodity. Thus, we'll set up only an outgoing update type for flow type **PC00: Purchase Commodity Forward**.

Conversely, the sale of a commodity forward will always involve an incoming cash flow, so we'll map "Sale Commodity Forward" (SC00) to an incoming update type; an outgoing update type is not required for this flow type.

Having created these update types, we now have a configuration in place for each cash flow associated with the commodity forward, and we're now accurately reflecting the financial movements resulting from the purchase and sale transactions as well as the cash settlement at maturity.

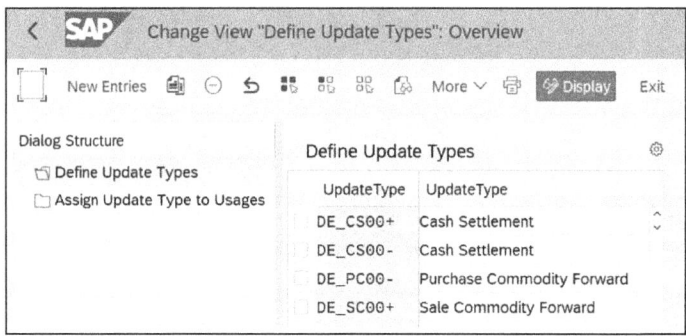

Figure 4.131 Creating Commodity Forward Update Types

After creating the update types, we click on **Assign Update Types to Usages** in the **Dialog Structure** to assign them to the usage (**Upd. Ty Usage**) of **Transaction Manager**, as shown in Figure 4.132. This is an important step, and we don't complete it, the update types will not be triggered when we post the contract.

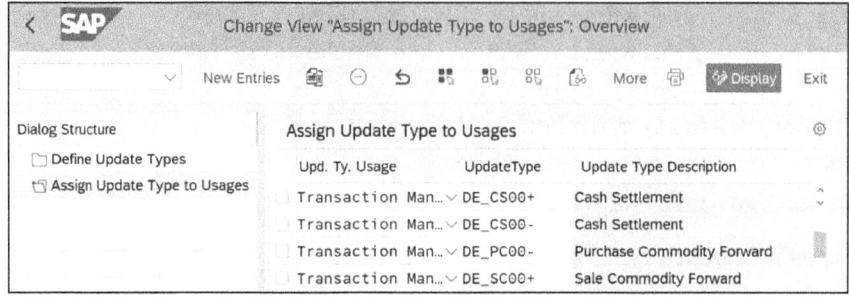

Figure 4.132 Assigning Update Types to Usages for Commodity Forward

4.5 Commodity Derivatives

The final step that completes the update type configuration is to assign the four update types to the correct contract type as well as their corresponding flow types. We navigate to **Financial Supply Chain Management • Treasury and Risk Management • Transaction Manager • OTC Derivatives • Transaction Management • Update Types • Assign Flow Types to Update Types**. We should assign all the update types to the **Derivatives** contract type (**Cont.Type**), as shown in Figure 4.133. The next step is to assign each flow type to the corresponding update type in the **UpdateType** column. It's important to note that both listed and OTC derivatives are represented by contract type derivatives when we're assigning the contract type in this node.

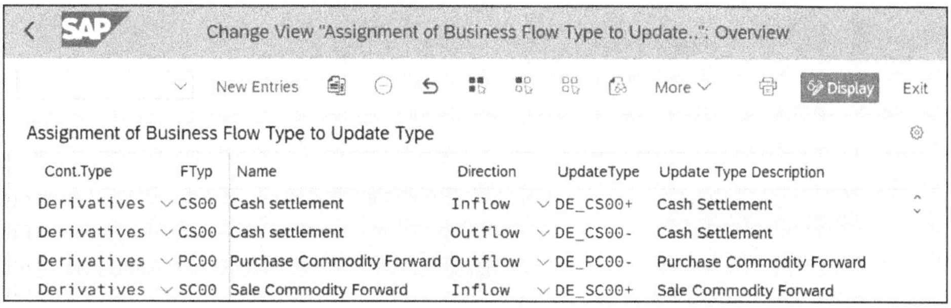

Cont.Type	FTyp	Name	Direction	UpdateType	Update Type Description
Derivatives	CS00	Cash settlement	Inflow	DE_CS00+	Cash Settlement
Derivatives	CS00	Cash settlement	Outflow	DE_CS00-	Cash Settlement
Derivatives	PC00	Purchase Commodity Forward	Outflow	DE_PC00-	Purchase Commodity Forward
Derivatives	SC00	Sale Commodity Forward	Inflow	DE_SC00+	Sale Commodity Forward

Figure 4.133 Assigning Flow Types to Update Types for Commodity Forwards

4.5.10 Updating Types for Position Updates

An important configuration element that is unique to derivative configuration is the assignment of update types for position updates. To start the configuration, we access node **Financial Supply Chain Management • Treasury and Risk Management • Transaction Manager • OTC Derivatives • Transaction Management • Flow Types • Assign Update Types for Position Update**.

Within this configuration node, we determine which update types should be triggered when opening and closing contracts. These update types will not post to the general ledger; they are solely used to manage the opening and closing of derivative positions. We do the configuration by product and transaction type so that we can use different update types for various products and transaction types.

In this setup, we assign one update type for opening a position (**Update Type for Open**) and another for closing it (**Update Type for Close**). SAP S/4HANA provides standard update types for this purpose: OTC001 and OTC002. We should assign these preconfigured update types to both purchase and sale transaction types, as illustrated in Figure 4.134 and Figure 4.135.

4 Contract-Specific Configuration

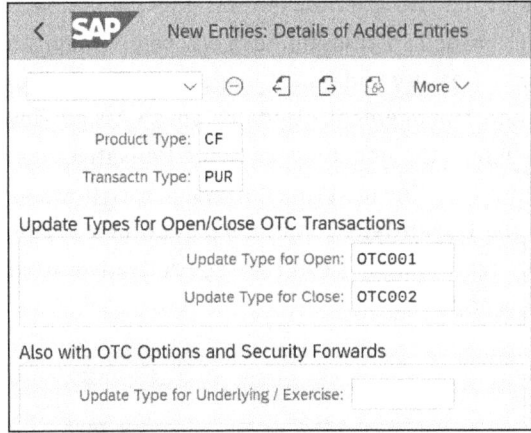

Figure 4.134 Update Types for Position Updates for Purchasing

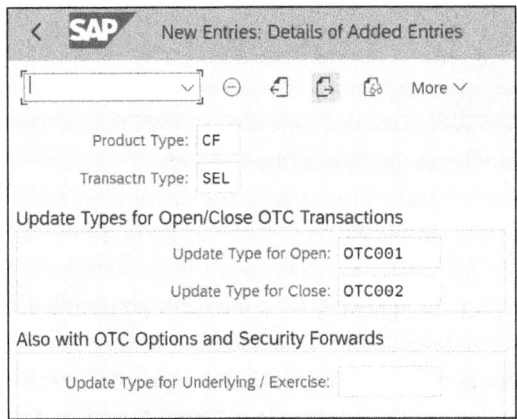

Figure 4.135 Update Types for Position Updates for Selling

4.5.11 Indicating that Update Types Are Relevant to Posting

An important step in setting up a commodity forward is determining which update types should be relevant for posting to the general ledger. This configuration ensures that only the appropriate financial transactions impact the accounting records and post to the general ledger.

To start the configuration, we navigate to **Financial Supply Chain Management • Treasury and Risk Management • Transaction Manager • General Settings • Accounting • Link to other Accounting Components • Indicate Update Types Relevant to Posting**. This path leads us to the node where we'll specify the update types that should be posted to the general ledger.

For commodity forwards, the key aspect to consider is that only the cash settlement update types should have an accounting impact. The update types associated with cash settlements represent the financial outcome and cash movement of the forward contract at maturity. These update types need to reflect the actual gain or loss resulting from the contract, so we'll set up the **UpdateType** for cash settlement (DE_CS00+ and DE_CS00–) as relevant for posting by checking the **Rel** box. This configuration ensures that the actual cash movements resulting from the settlement are accurately recorded in the general ledger.

On the other hand, the update types for the purchase and sale of the commodity forward are nonposting update types. These update types are required, and we use them to calculate the difference between the purchase price and the price at settlement, which ultimately determines the financial result of the forward contract. However, they don't represent actual cash movements and therefore, they should not impact the general ledger. The update types for commodity purchase (DE_PC00–) and sale (DE_SC00+) will not have an accounting impact, so we won't check the **Rel.** box for these types. In this table, we can still include the update types that are not relevant for posting as long we don't check the **Rel.** box. However, it's not mandatory to maintain them in this table.

By accurately configuring the relevant update types, we ensure that resulting postings accurately reflect the true cash events of the commodity forward transactions. Figure 4.136 shows the setup for reference.

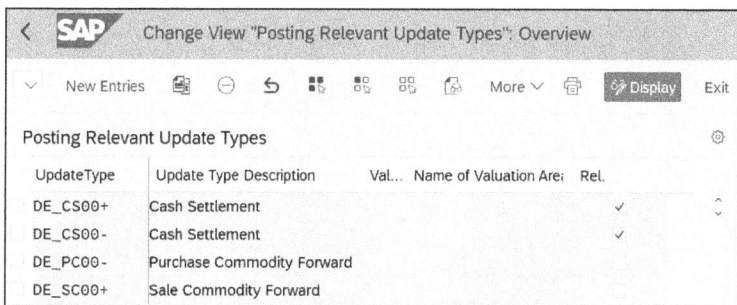

Figure 4.136 Indicating Commodity Forward Update Types That Are Relevant to Posting

4.6 Summary

In this chapter, we explored the contract-specific configuration elements for the most commonly used financial instruments in treasury management. Since contracts can vary significantly in structure, scope, and purpose, our goal was to offer a broad sampling by focusing on the most commonly used contracts used by corporate treasury departments: debt, investment, intercompany loans, FX, and derivative contracts. Each of these instruments plays a key role in a company's treasury operations, and understanding their individual configuration requirements will prepare you to create derivations of the contracts covered in this section.

While it would be impossible to cover every possible contract type used by treasury departments across different industries, the diverse contract types we examined serve as foundational examples. With the concepts and configuration techniques outlined in this chapter, you should be well equipped to expand on this knowledge and tailor the treasury and risk management system to handle a variety of other contract types that your organization may require. Whether you're dealing with more specialized financial instruments or evolving business needs, the principles you've learned here will provide the flexibility you need to customize the system as necessary.

We'll now shift our attention to the user side of contract processing. We'll dive into how end users can efficiently manage these contracts in day-to-day operations, from entering transactions to executing payments and posting to the general ledger.

Chapter 5
General Contract Processes

This chapter details the best-practice treasury management process, detailing how contracts are captured and entered into treasury and risk management. It also describes the end-to-end process for each of the primary product types within SAP S/4HANA, from creation, approval and settlement, posting, and month-end activities to payment processing.

The primary role of SAP S/4HANA Finance for treasury and risk management is managing financial transactions within SAP S/4HANA. These transactions reflect agreements between a company and its business partners. Recorded in SAP S/4HANA, these agreements are processed in accordance with the transaction terms. The data generated from these transactions not only supports reporting but also contributes to accounting, ultimately helping treasury and risk management form a cohesive treasury solution.

This chapter serves as an introduction to the general transaction process within treasury and risk management. This process includes several key steps that create a complete solution. The process starts with creating the financial transaction, and each detail of the transaction is recorded to ensure all cash flows and calculations are tracked correctly. Section 5.1 covers entry of the agreements and assignment to business partners. Section 5.2 covers the review and approval process after the transaction is created. The necessary reviewers need to ensure the transaction was entered correctly before further processing can be executed. Section 5.3 covers the processing of the transaction after the details are validated, and this processing includes posting accounting entries and calculating valuations. Section 5.4 covers the conclusion of the process with processing the payment flows for the transactions. SAP S/4HANA automates payment execution to ensure timely processing and recording of the incoming and outgoing payments.

5.1 Create Financial Transaction App

SAP S/4HANA Finance for treasury and risk management is a powerful tool that companies can use to track all aspects of their financial transactions. It creates a centralized view for the structure of transactions, reviews the position details, updates the transactions, creates accounting postings, and initiates payments related to the incoming and outgoing flows from the transactions. This introductory section will detail the important features that are executed when we enter transactions in SAP S/4HANA, and it will show the end-to-end process all the way to the final settlement and payment of the transaction.

We'll begin by looking at a high-level process flow that describes how transaction are entered, and we'll then delve into the Create Financial Transaction app itself, looking at its initial screen, header information, the all-important **Structure** tab, and the remaining tabs.

5.1.1 High-Level Process Flow

Let's look at the generic process flow for each of the types of transactions that are entered in treasury and risk management. Even though each type of transaction has its own nuances, each transaction at a minimum follows a process of create, settle, post, and pay:

❶ **Create**
Financial transactions are created with all their relevant details to create the structure of the transaction. There are many varying options in this process, and we'll detail how each of the transactions need to be entered.

❷ **Settle**
An optional step in the process is to create a settlement or approval process for entering the transactions. If we need to create a segregation of duties or a dual-control principle for transactions before someone can process the transaction further, we can ensure that the transaction has the proper approvals before they can be fully processed.

❸ **Post**
As the transactions are processed, there are accounting entries that need to be posted. Common postings for the transactions include posting cash-related entries, accruals, deferrals, valuations, and interest.

❹ **Pay**
Each of the treasury transactions includes payments. This could be outgoing payments to a counterparty or recording an incoming payment. Each of these payment flows can be automated and run through the treasury payment run. When the treasury payment functionality is configured, these payments can generate payment instruction files that are sent to banking partners.

The process flow in Figure 5.1 details each of the key steps in the process.

5.1 Create Financial Transaction App

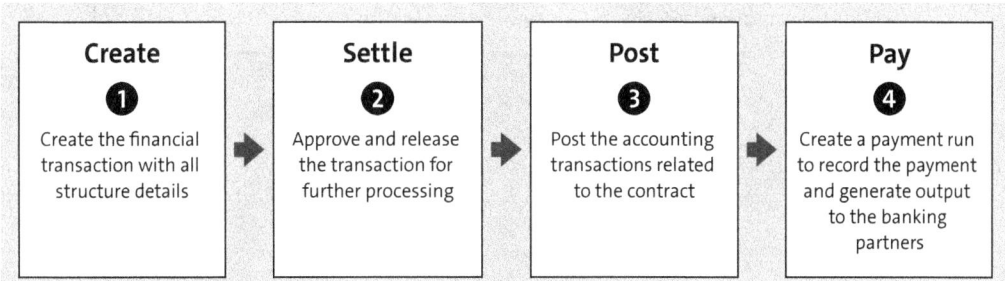

Figure 5.1 Generic Process Flow for Contract Lifecycle

5.1.2 Initial Screen

The key aspect of any financial transaction we enter in treasury and risk management is the structure of the contract. The details we enter into the structure will drive all of the following processes when we create postings or generate reporting, so it's extremely important to understand the entry screen and what selections to make when creating transactions. We'll therefore look at the entry screen and detail all the fields. Part of the entry screen includes a number of tabs that we'll also cover and that we can use for various functions. We can use the tabs to add information to the transaction, create payment details, and review related cash flows. After that, we can save the financial transaction, and it will be ready for further processing.

To start with the entry of the contract, we'll look at the Create Financial Transaction app (see Figure 5.2). The equivalent via SAP GUI is Transaction FTR_CREATE. When we enter this transaction, we need to enter some base information to move on to the next step.

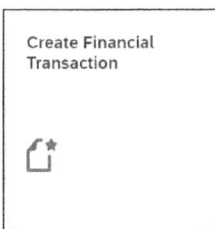

Figure 5.2 Create Financial Transaction App

On the initial screen of the Create Financial Transaction app, the minimum fields that need to be populated are **Company Code, Product Type, Transaction Type,** and business **Partner**. See Figure 5.3 and the following list for details:

5   General Contract Processes

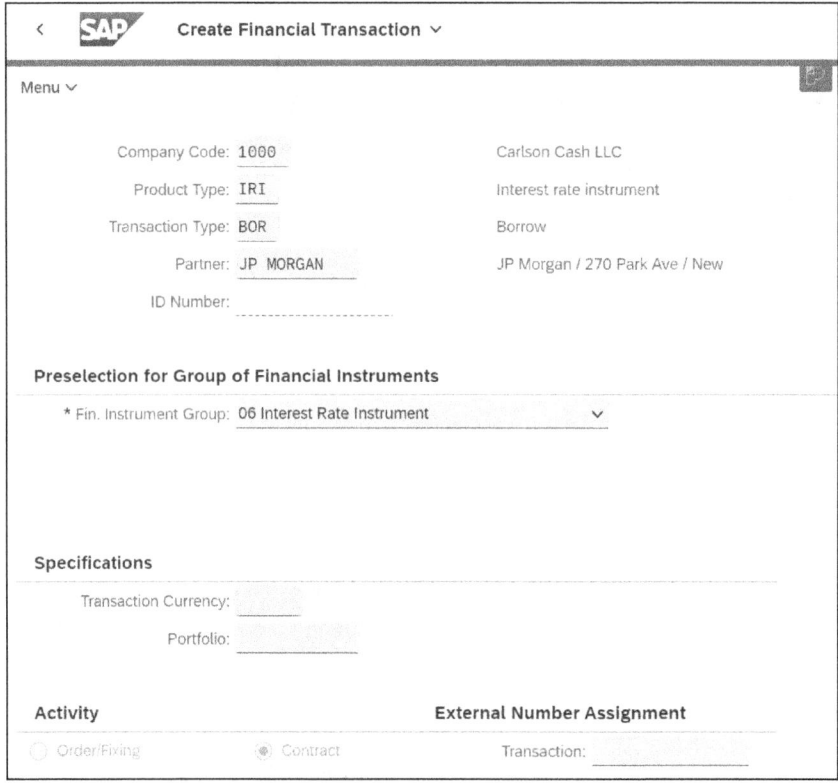

Figure 5.3 Create Financial Transaction App Initial Screen

- **Company Code**
  The *company code* is the entity where the financial transaction will be created. The accounting postings will be posted to this entity when they are generated.

- **Product Type**
  The *product type* defines the type of financial transaction we are creating. There are separate product types for FX, debt, investments, trade finance, and securities transactions. Depending on processing and posting requirements, these are broken out into different product types to drive the flows of the contracts.

- **Transaction Type**
  The *transaction type* is a subcategory of the product type. One product type can have different transaction types based on their attributes. For example, in FX, we can have an FX product type, and we can assign the spot and forward transaction types to that product type. These transaction types fall into the same general category, but we can define them differently when entering the transaction. Another example is for an interest rate instrument. We can have a product type to define the interest rate instrument, and we can assign a separate transaction type of borrowing or investment to that product type.

- **Business Partner**
  The business partner defines the counterparty that the transaction was executed with, and it also has settings to drive the sending and receiving payment details of the transaction. We discussed the settings of the business partner in detail in Chapter 2, Section 2.5.

Additional fields are also available on this screen, depending on our requirements. These fields are as follows:

- **ID Number**
  For securities transactions, we add the securities ID in this field.

- **Financial Instrument Group**
  We can change this field to drive what entry fields are available in this tab, depending on the type of transaction. For example, only securities transactions will use the **ID Number** field. If we select a **Financial Instrument Group of Foreign Exchange**, then the **ID Number** field will be grayed out.

- **Transaction Currency**
  When we create a financial transaction, the transaction will assume that we are creating it in the company code currency. If the transaction needs to be in a different currency, then we populate this field with the applicable currency.

- **Portfolio**
  We can add a portfolio to this field to group the transaction with other transactions. If we don't populate this field, then we can always add the portfolio in the **Administration** tab.

- **Activity**
  This drives whether the transaction is an order for a transaction or a completed financial transaction. The entry screen will default to **Contract** in this selection.

- **Transaction**
  SAP S/4HANA will automatically assign a transaction number based on the number ranges assigned to the transaction type.

After we have populated all the fields, we can click the **Create** button to move to the transaction entry screens.

5.1.3 Header Information

Once we are in the transaction, the entry screen will vary, depending on the specific product type and transaction type. The entry screen is shown in Figure 5.4, and there are a series of common fields that are displayed at the top of the screen that reflect the **Company Code**, **Product Type**, **Transaction Type**, **Business Partner**, and **Activity** that were entered on the previous screen.

We use the series of tabs below this header information to drive the structure of the contract and additional details. These tabs will vary, depending on the product type we

5 General Contract Processes

are using and whether the tabs are required for that financial transaction. For example, the **Other Flows** tab will only appear if flow types designated as **Other Flows** have been assigned to the product type.

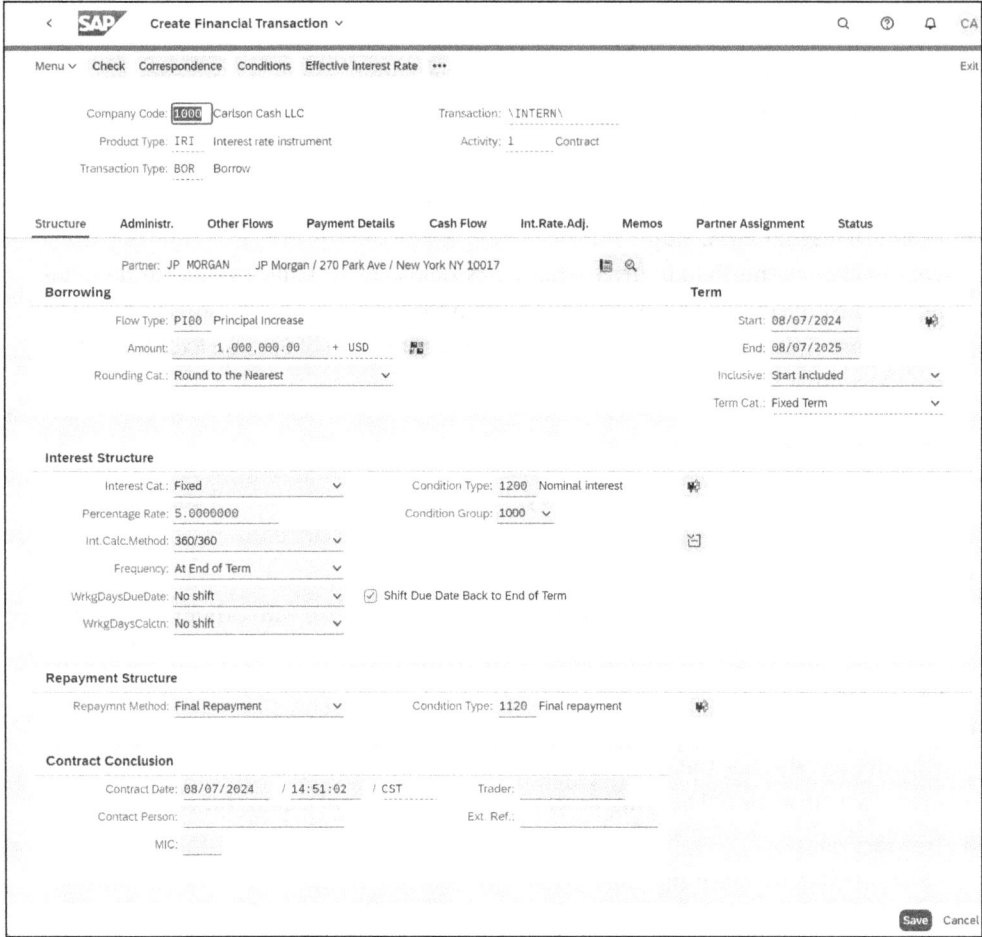

Figure 5.4 Contract Entry Screen Example

Additionally, if there are tabs or fields that are not required, we can define them in the **Financial Supply Chain Management • Treasury and Risk Management • Transaction Manager • General Settings • Transaction Management • Define Field Selection** menu path, which we covered in Chapter 3, Section 3.12. Once we have made the field selections, we can assign them by company code, product type, and transaction type.

5.1.4 Structure Tab

The **Structure** tab is the most important tab to cover in the Create Financial Transaction app. We use it to define the core structure elements of financial instruments. This includes everything from the structure of a foreign exchange contract, the nominal amount of a transaction, term dates, and the interest structure to the repayment structure. At the very bottom of the page, there are additional fields in which we can enter the contract date, bank contact information, internal traders, and the external reference. This section will dive deeper into each of these fields and how entering data into them affects the created contract and subsequent cash flows.

One thing to consider about the **Structure** tab is that the options in it will vary, depending on the type of contract we are creating. For example, the structure of an FX transaction is very different from the structure of an interest rate instrument. To cover the **Structure** tab, we'll present a generic example of the entry of a contract into this tab. To look more closely into the different types of transactions and how to enter them into the **Structure** tab, please refer to the various sections of Chapter 6:

- Section 6.1: Debt and investments
- Section 6.2: Facilities
- Section 6.3: Intercompany loans
- Section 6.4: Securities
- Section 6.5: Foreign exchange
- Section 6.6: Trade finance

To explore a standard transaction's **Structure** tab, we'll look at an interest rate instrument in the **Borrowing** screen of the tab. The fields will vary between this example and the additional examples in Chapter 6. In this entry screen, we have sections for borrowing, term, interest structure, repayment structure, and contract conclusion.

Borrowing

Depending on whether we use a debt or investment transaction type, the first section will be **Borrowing** or **Invest**. This section lets us determine the nominal amount of the contract and other amount details of the structure. As shown in Figure 5.5, we can enter additional details in each of the fields, as follows:

- **Flow Type**
  This will be defaulted based on the flow type that has been assigned to the product types in the **Financial Supply Chain Management • Treasury and Risk Management • Transaction Manager • Money Market • Transaction Management • Flow Types • Assign Flow Types to Transaction Type – MM Transactions** menu path. We can select an alternate flow type if we have assigned multiple principal flows in the customizing section.

Multiple Flow Types Assigned

We can assign multiple flow types while customizing a specific type of category. In this example, the category is either principal increase or principal decrease. When we've

5   General Contract Processes

assigned these, SAP S/4HANA needs to come up with the default flow type to populate when this transaction is called. This will be populated in alphabetical order, so if we have a principal increase with a flow type name of OI00 and another with a name of PI00, then OI00 would automatically be populated when this transaction is entered.

- **Amount**
  We use the main flow of the financial instrument to represent the nominal amount borrowed or the amount of the investment.

- **Other Changes in Capital Structure**
  When there are changes to the nominal amount of the instrument, we can enter them by clicking on the **Other Changes in Capital Structure** button . **Flow types** are available to both increase and decrease the borrowing along with adding changes to the amount, payment, and calculation date. We use the calculation date to determine the interest calculation. For example, to decrease the amount borrowed by $100,000 on 09/07/2023, we'd make the entry shown in Figure 5.6. To save this entry, we'd click the **Copy** button.

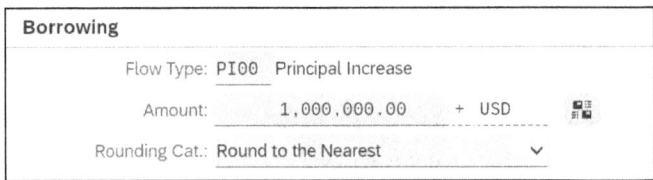

Figure 5.5 Borrowing Section of the Structure Tab

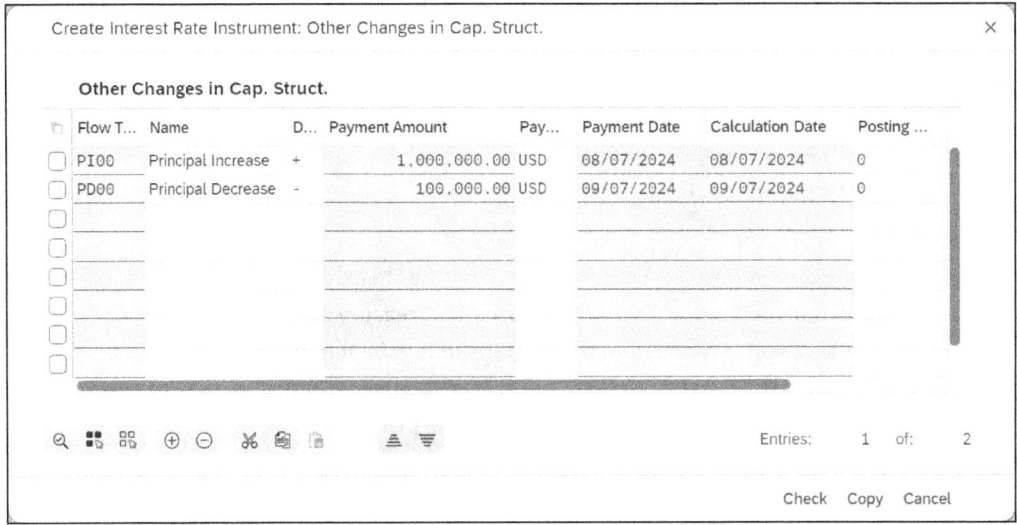

Figure 5.6 Other Changes in Capital Structure: Updating Principal Decrease

5.1 Create Financial Transaction App

- **Rounding Category**
  When generating cash flows for the contract, we use this area to determine how the system will round numbers to a tens digit. If we select **Round to the Nearest**, the system will round up numbers of 5 or more and round down numbers of 4 or less. **Round Down if 5** works the same way, except that it rounds down 5 instead of rounding it up. The other two options are **Round Down,** which rounds down all numbers, and **Round Up,** which rounds up all numbers.

Term

The **Term** section defines the general structure of the key dates of the transaction. It covers the starting and ending dates of the transaction, and it includes the fields in Figure 5.7:

- **Start**
  This field lets us determine the start date of the transaction.

- **End**
  This field lets us determine the end date of the transaction.

- **Detail View: Term**
  The **Detail View: Term** button provides an option for viewing the transaction's term. Clicking the button brings up the popup in Figure 5.8.

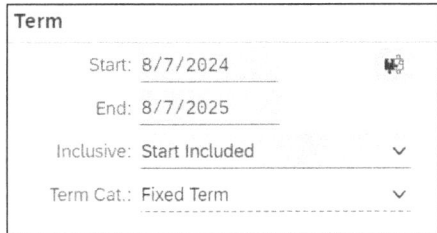

Figure 5.7 Term Section Displayed to Determine Start and End Date of Transaction

Figure 5.8 Overview of Start and End Dates of Transaction

5   General Contract Processes

- **Inclusive**

  We use this field to determine whether the start and end dates are included in the interest calculations. The options are **Start Included**, **End Included**, **Start and End Included,** and **Start and End Excluded**. If a date is included, then it is used in the interest calculation. In our example, we're using the **Start Included** option, so since the start date of the contract is 08/07/2024, the interest calculation would start on that date. Since the end date is not included in this example, the interest calculation would end 08/06/2025 (one day before the **End of Term** in the example). No interest would be calculated on 08/07/2025 since that date is not included.

- **Term Cat.**

  Two options are available for the term category, as follows:

  - **Fixed Term**

    We use this option when the contract has a predefined start and end date. Since we determine the term during contract creation, we can generate the repayment flows.

  - **At Notice**

    The end date of the term of this kind of interest rate instrument is not defined during contract creation, so the repayment flows are not generated automatically. For these instruments, we need to use the **Give Notice** utility to determine the notice date and generate the repayment flow.

Interest Structure

The **Interest Structure** section shown in Figure 5.9 is specific to interest rate instruments. This section lets us determine how the transaction calculates the interest and also lets us determine the key dates that are used to determine the calculation and payment. There are many options for entering the interest rate details into transactions, and we cover them in Chapter 6, Section 6.1.

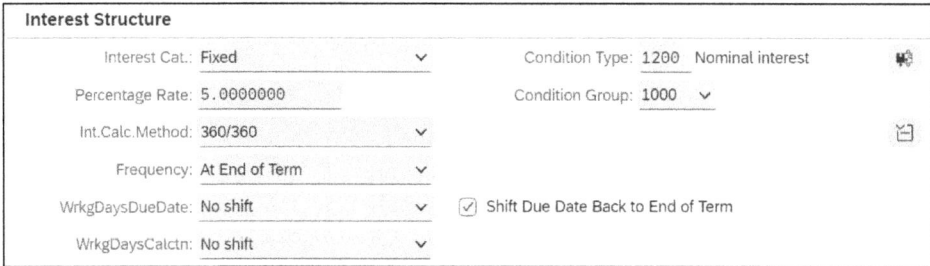

Figure 5.9 Interest Structure Information on Structure Tab

Repayment Structure

The **Repayment Structure** section shown in Figure 5.10 is also used for specific transactions. It's relevant to this transaction since the transaction is an interest rate instrument that the borrower needs repay at the end of the contract. The current selections in this area determine that the full amount of the contract needs to be repaid at term end. There are additional options for the repayment, and these are also covered in detail in Chapter 6, Section 6.1.

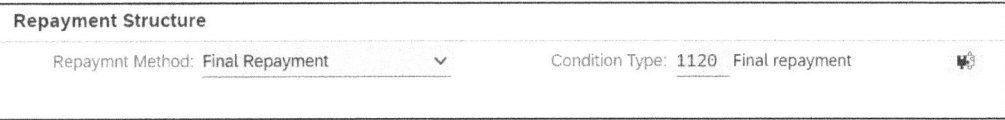

Figure 5.10 Repayment Structure Information on Structure Tab

Contract Conclusion

Every transaction in SAP S/4HANA includes a **Contract Conclusion** section, as shown in Figure 5.11. This includes some additional data points on the contract, plus the following fields:

- **Contract Date**
  This is a date and time stamp for the contract creation. The date and time will default to the system date and time. If we're testing contracts or creating a contract that has a term start date that's in the past, then we also need to update the contract date. If the term start date comes before the contract date, then an error will appear and the contract won't be saved. The error will state **Contract date is after start of term**.

- **Contact Person**
  We can enter the contact person for the transaction in this field.

- **MIC**
  The market identifier code (MIC) is a four-character code that defines exchanges and trading platforms. This follows the list of MIC codes defined under the ISO 10383 international standard.

- **Trader**
  The internal trader can have an ID that is populated in this field. We cover the full functionality of trader creation, assignment, and usage in the master data section.

- **Ext. Ref.**
  We can populate the external reference field with an external reference number. Generally, if we have a separate reference number at the counterparty, we can place it in this field.

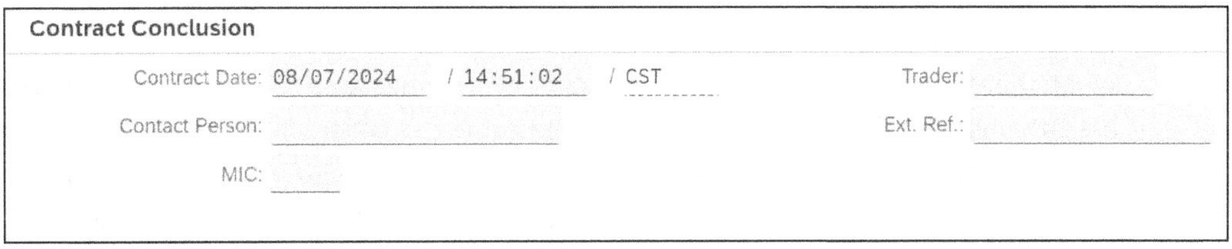

Figure 5.11 Contract Conclusion Section on Structure Tab

5   General Contract Processes

5.1.5   Additional Tabs

There are additional tabs in which we can enter and review data in the Create Financial Transaction app. This section will cover all the relevant shared tabs that we'll see in all transactions. Some specific financial transactions have their own tabs that only appear when we create those transactions. In those scenarios, we'll cover those special tabs in their respective areas.

Administration Tab

We use the **Administr.** tab in Figure 5.12 to further categorize the instrument for reporting. We use the **Position Assignment** section to group contracts together and to determine a general valuation class (**Gen. Valn Class**) to group similar contracts into short-term and long-term debt or investments. We can use the **Additional Fields** section to generate further references to the transactions and for client-specific purposes. We can also determine an **Authorization Group** and assign a level of security that only gives access to certain individuals who are authorized to process their transactions. The **Rating** section can define the credit rating for the counterparty business partner.

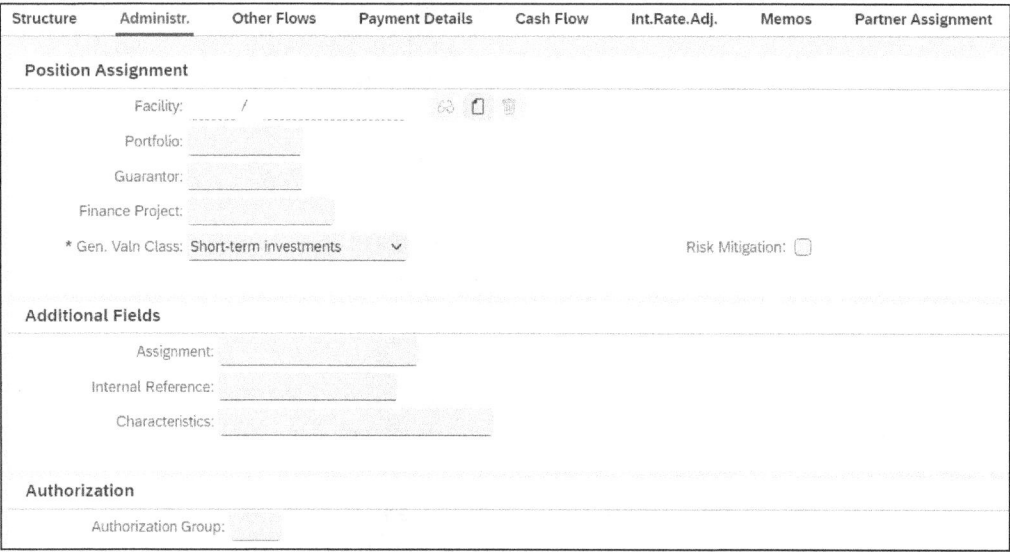

Figure 5.12 Administration Tab in Create Financial Transaction App

Other Flows Tab

When the standard contract flows don't cover all postings for a contract, we should use the **Other Flows** tab. When other fees are applicable to a contract, we can create a separate flow type in the customizing and designate it as an *other flow*. In Figure 5.13, there's an example of an other flow that has been added to a contract. This flow is an additional charge that needs to be paid on 08/09/2024. Since this is a separate flow type, it will be assigned its own update type. As a result, we can use the update type to drive the

5.1 Create Financial Transaction App

accounting for this flow. Frequently, it makes sense to create multiple other flows if we want them to be posted to different general ledger accounts.

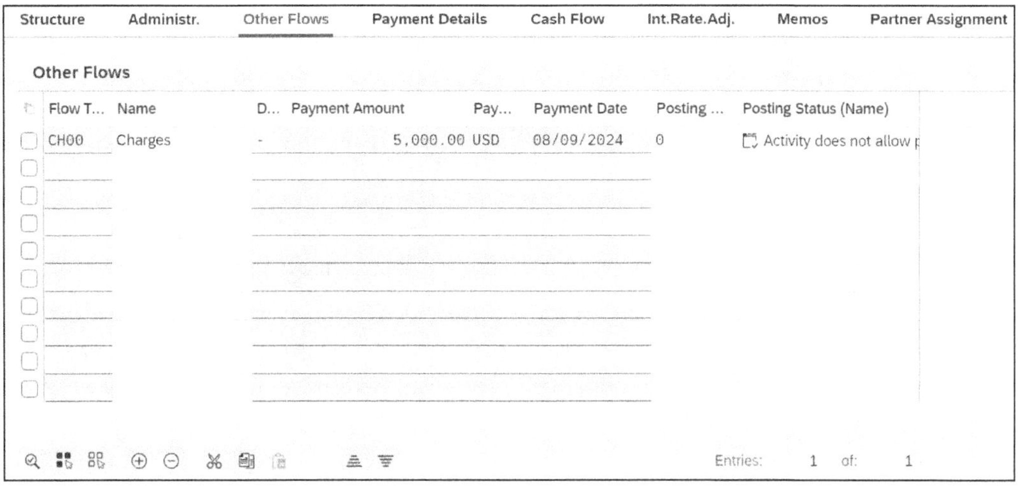

Flow T...	Name	D...	Payment Amount	Pay...	Payment Date	Posting ...	Posting Status (Name)
CH00	Charges	-	5,000.00	USD	08/09/2024	0	Activity does not allow p

Figure 5.13 Other Flows Tab Showing How to Add Charge to Contract

We can add details to the flow if we double-click into the line item. This screen is displayed in Figure 5.14.

Charges - Change

Payment
- *Payment Date: 08/09/2024
- Pmnt Amount: 5,000.00 - USD
- PaytReason:

Calculation Base
- Calculatn From: 08/07/2024 ☑ Inclusive ☐ Month-End
- Calculation To: 08/07/2025 ☐ Inclusive ☐ Month-End
- Int.Calc.Method: 360/360   Linear Interest Calculation
- Number of Days: 360
- No. Base Days: 360
- Base Amount:
- Percentage Rate:

Document
- Assignment:

✓ Check   Copy   Percentage   Pro Rata Temporis   ✕

Figure 5.14 Popup Showing Additional Details Where We Can Add Information for Accrual Period

5 General Contract Processes

If the other flow has been set up in the customizing to be relevant to accrual and deferral, the **Calculation Base** dates will be important. In the case of an accrual, the **Calculatn From** date will reflect the first relevant date in the accrual period, and the **Calculation To** date will reflect the last relevant accrual date and payment date.

In the case of a deferral, the **Calculatn From** date will be the first relevant date in the accrual period, and it will also be the payment date of the charge. The **Calculation To** date will reflect the last relevant date in the deferral period.

Payment Details Tab

The **Payment Details** tab (shown in Figure 5.15) drives all payment-related data for the instrument. We can set up this information on each business partner, and it will default into the contract automatically. We'll need to assign separate instructions for both the incoming and the outgoing payments. This will determine the bank account at the counterparty that will be paid and issue payments, and it will also determine the accounts we are paying out of for the contract.

Structure	Administr.	Other Flows	Payment Details	Cash Flow	Int.Rate.Adj.	Memos	Partner Assignment
Payer/Payee of Transaction							
		Payer/Payee: JP MORGAN	JP Morgan / 270 Park Ave / New York NY 10017				

	D...	Crcy ...	Valid From	FType ...	Name	House Bank	Account ID	Payment	Payment Req...
☐	+	USD				JPM01	PAY02	☐	☑
☐	-	USD				JPM01	PAY02	☐	☑
☐								☐	☐
☐								☐	☐
☐								☐	☐

Figure 5.15 Payment Details Tab

We can manually enter information into each of the lines in the payment details tab, or they can be populated with default data from the standing instructions on the business partner. We covered the assignment of this in Chapter 2. Also, once we've entered the information, we can drill down into each line of the **Payment Details** tab to view additional information on payments, and we can add more specific payment information on this page as well. We'll cover each section of the payment details separately because we can enter many different kinds of information in this area. We detail all these entry options in the following list, and they are also shown in Figure 5.16:

- Control
  - Posting
    Here, we click one of the radio buttons to indicate whether the payments should be posted to a customer account or general ledger accounts. If the payment should go to a customer, then the customer will be populated in the **Payment** section in the **Payer/Payee** field. If the payment will be posted to general ledger

accounts, then the posting will follow the posting rules determined in the configuration.

- **Payment Request**
  This indicator lets us determine how we want to post this transaction. When we run the Post Flows app, we can either post directly to a general ledger or customer account, or we can create a payment request. If we click the **With** radio button to tell the program to create a payment request, then the payment will be run through the treasury payment process. The **Without** radio button tells the program to make a posting to a general ledger account without creating a payment request.

- **House Bank** and **Account ID**
  These fields let us determine the house bank and account ID that the payment should be paid from or to. This will also drive the bank side of the posting. If we've opted to create a payment request, the payment settings will post to this account's defined clearing accounts upon payment. If we've opted not to create a payment request, then the account entered int the account ID field will be used for the bank side posting.

- **Repetitive Code**
  On this tab, we can assign repetitive codes that have been created in the treasury payments functionality. We use this in cases where specific reference text or intermediary banking information is required for the payments.

- **Payment**
  - **Payer/Payee**
    This is the business partner we are making a payment to or receiving a payment from, and it's also generally the same business partner that we've defined as the counterparty of the transaction. We can also determine a customer can in this field, and the payment details will be driven from the customer in this case.
  - **Partner Bank**
    The partner bank type is a field that is determined on business partners. It's a four-character field that determines the payment details on the counterparty side of the transaction.
  - **Pmt Meth.Suppl.**
    If a payment method supplement is applicable to the payment, we enter it in this field.
  - **Group Determin**.
    We use this selection to determine how to group payments together when executing the treasury payment run. We use this to help reconcile cash flows and make sure that we net payments correctly if required by the counterparty. For example, if a principal repayment and the interest on it need to be sent in one payment, then we can group them together by using the group determination. Additionally, we can group all flows together by product type or product category,

or we can group together everything within treasury and risk management into one payment. These will only be grouped together if the sending and receiving accounts are the same, so we cannot group payments together if they are set up to pay different counterparties.

- **Payment Methods**
  This field lets us determine the payment methods for the incoming and outgoing sides of the payment request.

- **Individual Payt.**
  By default, SAP S/4HANA tries to group payments together. If we leave this box unchecked, the payment program will be able to group similar payments as defined in the **Group Determination** section. The same details need to be on the payments for them to be grouped together, and this means that the sending bank account and the receiving bank accounts need to be the same to group the payments together. If we check this box, this payment won't be grouped with other payments.

- **Same Direction**
  When payments are grouped together, the incoming and outgoing payments can be grouped together to create a net payment. If we only want to group together payments that go in the same direction, then we should check this box.

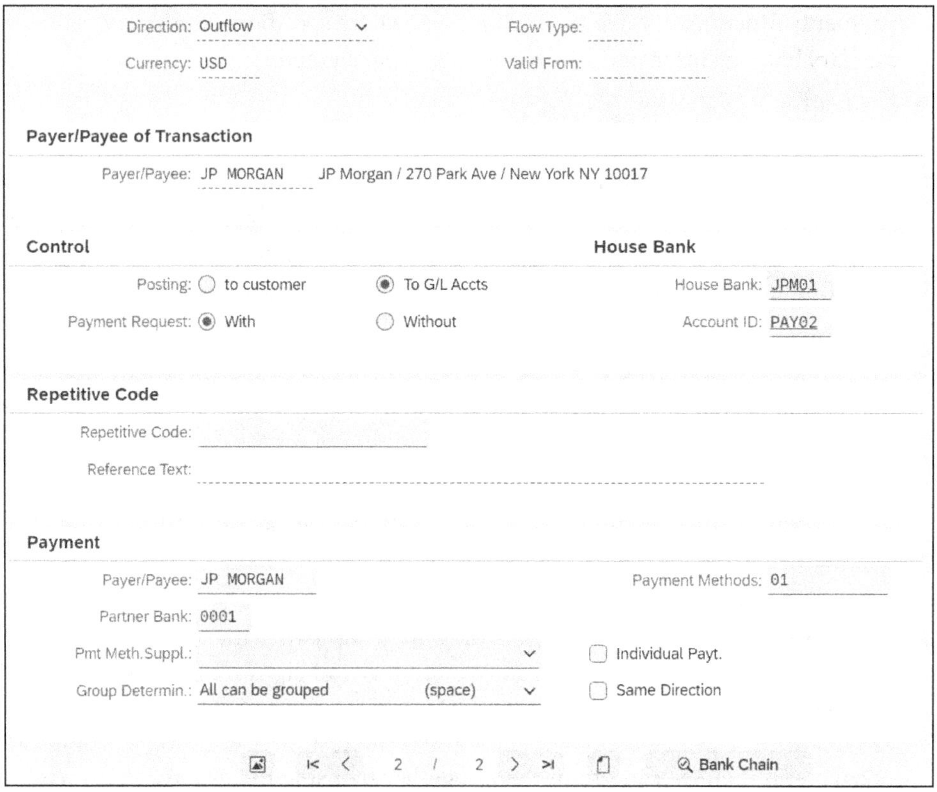

Figure 5.16 Double-Clicking into Payment Detail Allows Us to Enter Additional Details

> **Payment Methods on the Payment Details Tab**
>
> We can assign multiple payment methods in the **Payment Details** tab as long as they pay in different directions. Since cash flows go both in and out, we use multiple payment methods to ensure all bank-related cash flows are reflected correctly.

Cash Flow Tab

The **Cash Flow** tab provides a detailed view of the cash flows for the contract. This page is useful when we're validating whether the contracts have been entered correctly, because it not only details the cash flows but also shows how the interest flows have been calculated. Standard views are available, and we can edit them as required. Some useful views are as follows:

- **Basic View**
  This view (shown in Figure 5.17) provides initial information, including the flows with the payment date, payment amount, and cash flow direction.

- **Calculation View**
  This view details the interest calculation with the calculation method, calculation dates, and number of days relevant to each interest calculation period.

- **Payment View**
  This view includes all payment-related details, including the paying account, payee bank details, payment methods to be used, and payment request number (once it's available).

- **Posting View**
  This view highlights the posting status of the cash flows. Before the contract has been settled, the flows will show as not available for posting. After settlement, the flows will appear as **Flagged for Posting**, and after the posting program has been run, they'll show as **Posted**.

- **Interest Rate Adjustment View**
  For variable-rate contracts, this view is available to detail the interest rate fixing date along with whether the rate has actually been fixed and posted.

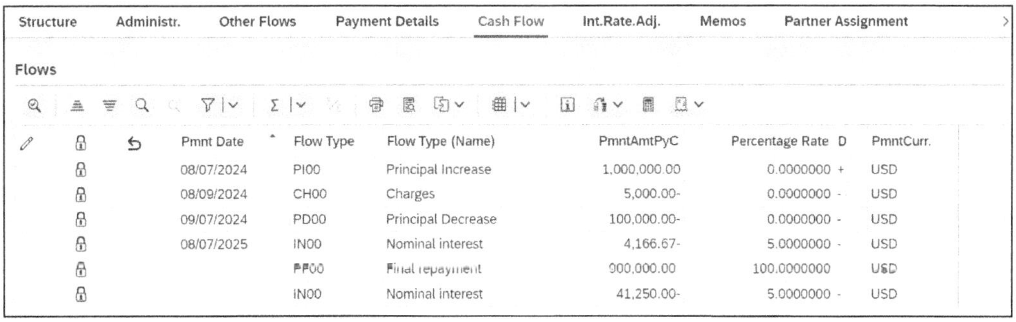

Pmnt Date	Flow Type	Flow Type (Name)	PmntAmtPyC	Percentage Rate	D	PmntCurr.
08/07/2024	PI00	Principal Increase	1,000,000.00	0.0000000	+	USD
08/09/2024	CH00	Charges	5,000.00-	0.0000000	-	USD
09/07/2024	PD00	Principal Decrease	100,000.00-	0.0000000	-	USD
08/07/2025	IN00	Nominal interest	4,166.67-	5.0000000	-	USD
	PP00	Final repayment	900,000.00	100.0000000		USD
	IN00	Nominal interest	41,250.00-	5.0000000	-	USD

Figure 5.17 Cash Flow Tab Basic View Showing Future Contract Cash Flows

5   General Contract Processes

As mentioned, the **Cash Flow** tab can show the calculations for the interest flows. One thing to note in Figure 5.17 is that there are two interest flows on the final repayment date. The cause of this is the principal decrease that was added on 09/07/2024. In this case, SAP S/4HANA is calculating an interest amount using the borrowed amount of $1,000,000 for the first month of the transaction and a different interest amount for the remainder of the transaction, when only $900,000 of debt was outstanding. We can see this by double-clicking on one of the nominal interest flows, which will lead us to the view in Figure 5.18.

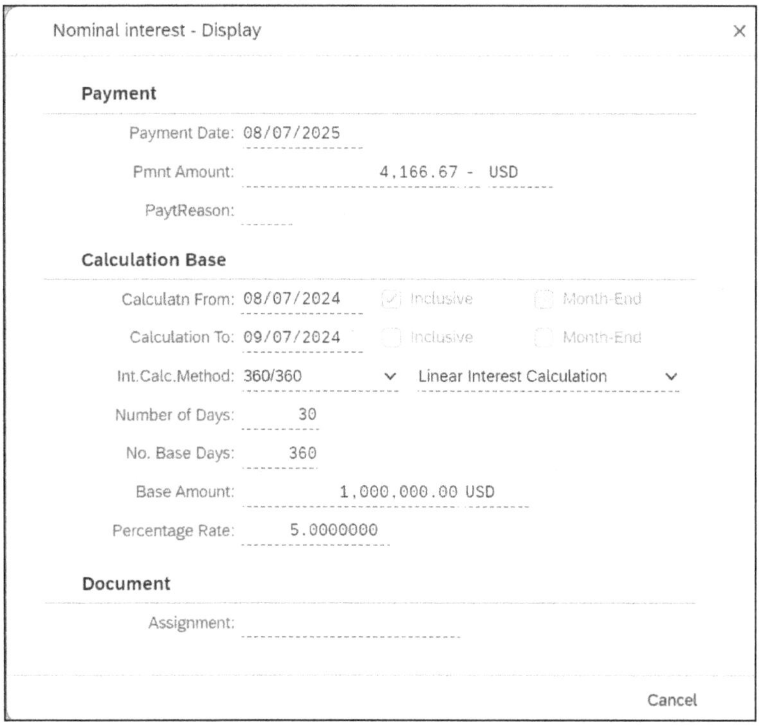

Nominal interest - Display			×
Payment			
Payment Date:	08/07/2025		
Pmnt Amount:	4,166.67 - USD		
PaytReason:			
Calculation Base			
Calculatn From:	08/07/2024	☑ Inclusive	☐ Month-End
Calculation To:	09/07/2024	☐ Inclusive	☐ Month-End
Int.Calc.Method:	360/360 ⌄	Linear Interest Calculation ⌄	
Number of Days:	30		
No. Base Days:	360		
Base Amount:	1,000,000.00 USD		
Percentage Rate:	5.0000000		
Document			
Assignment:			
			Cancel

Figure 5.18 Interest Calculation for First Month of Transaction

There are other useful features in the **Cash Flow** tab. In addition to the interest flows, we can double-click any line item to see additional details of cash flow calculations and payment dates. Sometimes, there are rounding differences between this tab and the counterparty, and we need to edit the cash flow. If a flow doesn't have the correct amount, then we can edit it by clicking the **Editing Mode** button. Once in editing mode, we click on the **Flows** button and select **Edit Flow**. This will open the change screen (shown in Figure 5.19), where we can change the cash flow. If a flow has been edited, an icon will be displayed on the screen to indicate that the flow has been changed (as in Figure 5.20).

We can follow the same process to reverse the cash flows. We need to select the **Editing Mode** dropdown list and to click the **Reverse** option in this area, as shown in Figure 5.21.

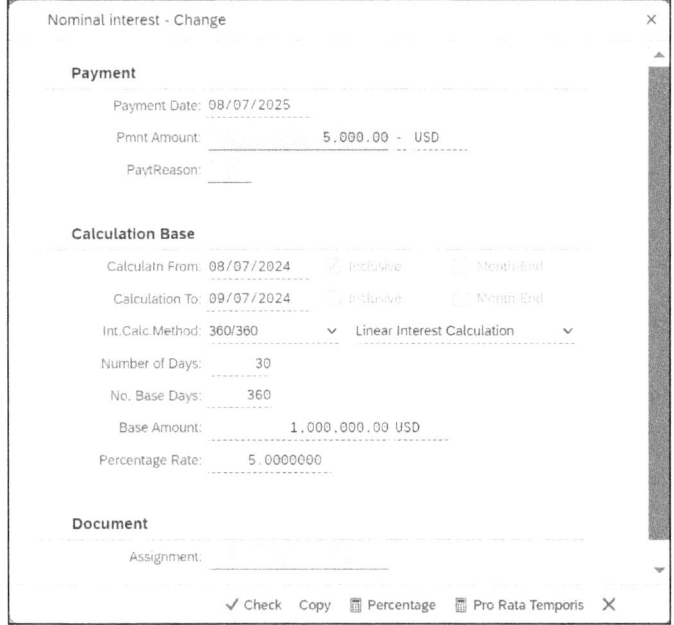

Figure 5.19 Edit Flows Screen with Flow Updated to Have Payment Amount of $5,000

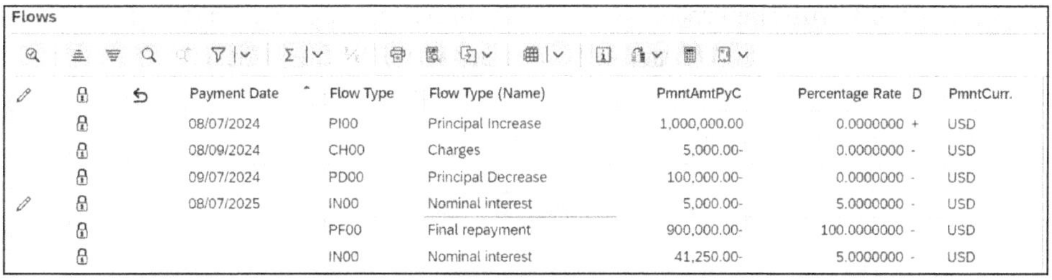

			Payment Date	Flow Type	Flow Type (Name)	PmntAmtPyC	Percentage Rate D	PmntCurr.
	🔒		08/07/2024	PI00	Principal Increase	1,000,000.00	0.0000000 +	USD
	🔒		08/09/2024	CH00	Charges	5,000.00-	0.0000000 -	USD
	🔒		09/07/2024	PD00	Principal Decrease	100,000.00-	0.0000000 -	USD
✏	🔒		08/07/2025	IN00	Nominal interest	5,000.00-	5.0000000 -	USD
	🔒			PF00	Final repayment	900,000.00-	100.0000000 -	USD
	🔒			IN00	Nominal interest	41,250.00-	5.0000000 -	USD

Figure 5.20 Modified Indicator to Show Flow Has Been Updated

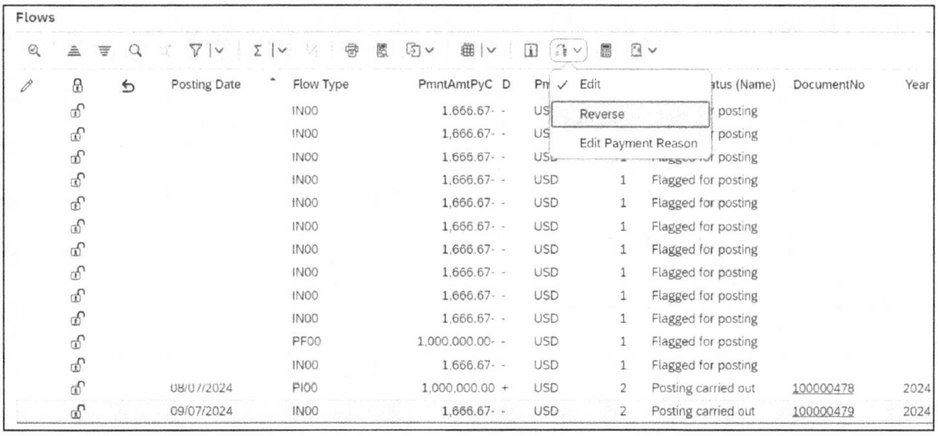

			Posting Date	Flow Type	PmntAmtPyC	D	Pm			tus (Name)	DocumentNo	Year
	🔓			IN00	1,666.67	-	US	✓	Edit	r posting		
	🔓			IN00	1,666.67	-	US		Reverse	posting		
	🔓			IN00	1,666.67	-	US		Edit Payment Reason	posting		
	🔓			IN00	1,666.67	-	USD	1		Flagged for posting		
	🔓			IN00	1,666.67	-	USD	1		Flagged for posting		
	🔓			IN00	1,666.67	-	USD	1		Flagged for posting		
	🔓			IN00	1,666.67	-	USD	1		Flagged for posting		
	🔓			IN00	1,666.67	-	USD	1		Flagged for posting		
	🔓			IN00	1,666.67	-	USD	1		Flagged for posting		
	🔓			PF00	1,000,000.00	-	USD	1		Flagged for posting		
	🔓			IN00	1,666.67	-	USD	1		Flagged for posting		
	🔓		08/07/2024	PI00	1,000,000.00	+	USD	2		Posting carried out	100000478	2024
	🔓		09/07/2024	IN00	1,666.67	-	USD	2		Posting carried out	100000479	2024

Figure 5.21 Select Highlighted Option to Enter Reverse Mode for Line Item

5   General Contract Processes

Next, we can use the **Flows** button to select the option to **Reverse Flow** (see Figure 5.22).

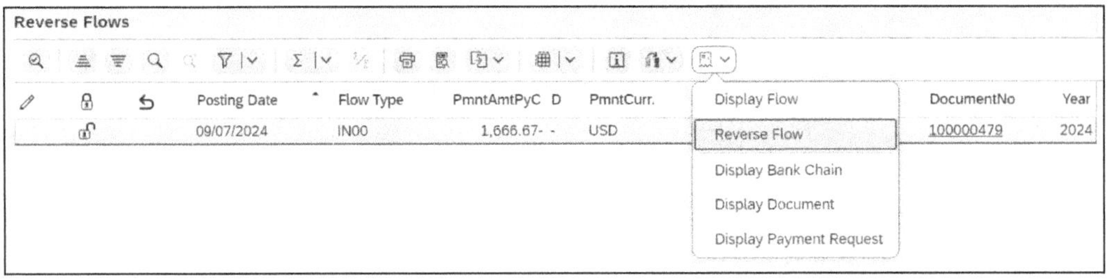

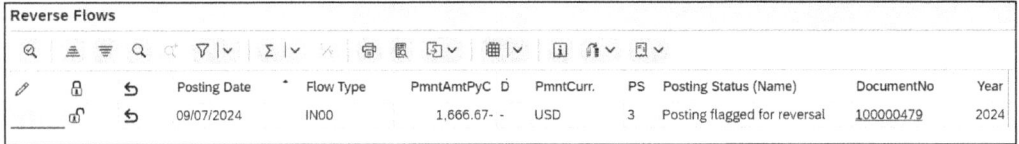

Figure 5.22 Selecting Option to Reverse Cash Flow

After we complete this, the flow has been flagged for reversal, as shown in Figure 5.23. To fully save this change, we need to select the **Save** option for the transaction.

Figure 5.23 Flow That Has Been Successfully Flagged for Reversal

Next, we need to run the Process Business Transactions app or Transaction TPM10 to fully reverse the cash flow. We can filter this transaction by the transaction ID, and once we've run the initial screen for this transaction, we'll see the transactions that are available to be processed (as shown in Figure 5.24).

Figure 5.24 Selecting the Flow for Reversal in Process Financial Transactions

We select the **Execute** button to reverse the flow. Figure 5.25 shows that the reversal has been successfully executed and the flow has been fully reversed.

5.1 Create Financial Transaction App

Figure 5.25 Successful Reversal of Cash Flow for Financial Transaction

Memos Tab

We use the **Memos** tab to store additional information on a financial transaction. Whether we use this tab is dependent on our setting up the memo types in configuration. We create the memo types in the **Treasury and Risk Management • Transaction Manager • General Settings • Transaction Management • Define Memo Book** menu path, and the available notes in Figure 5.26 show the standard delivered memo types.

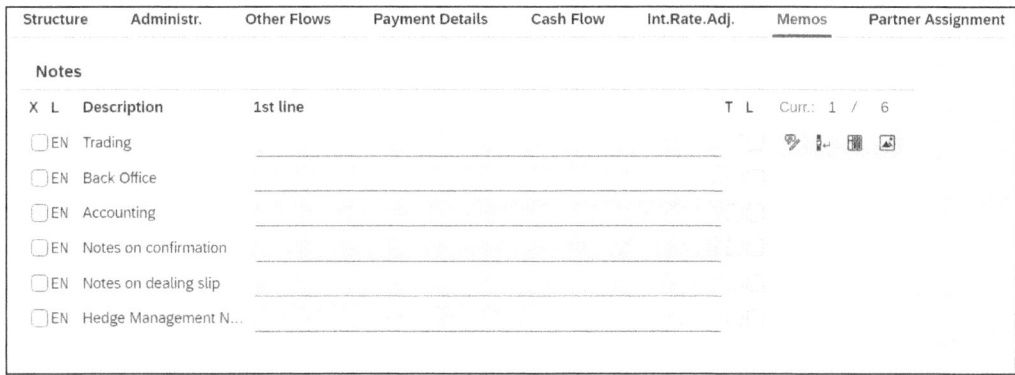

Figure 5.26 Memos Tab in Create Financial Transaction App

Status Tab

The **Status** tab provides a view that shows various available statuses for the transaction, as shown in Figure 5.27. It includes the following areas:

- **Correspondence**
  This area defines the status of a transaction if it's set up to send and receive correspondence messages. In this example, both fields are determining that correspondence is not required, but this area tracks the messages if the confirmation and/or counterconfirmation are required for the financial transaction.
- **Activity**
  This section details the activities that have been completed on the contract, and it also tracks whether the contract has just been created, whether it's been settled, etc.

341

5 General Contract Processes

Additionally, this area highlights the user who made the most recent changes to the transaction and when the changes were made.

- **Transaction**

    This area shows the processing category (**Processing Cat.**) of the transaction. This is a setting that was determined in the transaction type configuration, and the processing category shows the steps required for each transaction. This area also details whether the transaction needs to be approved in the workflow. The transaction in Figure 5.27 is currently being reviewed in the workflow since the **Release Status** says that a release is required.

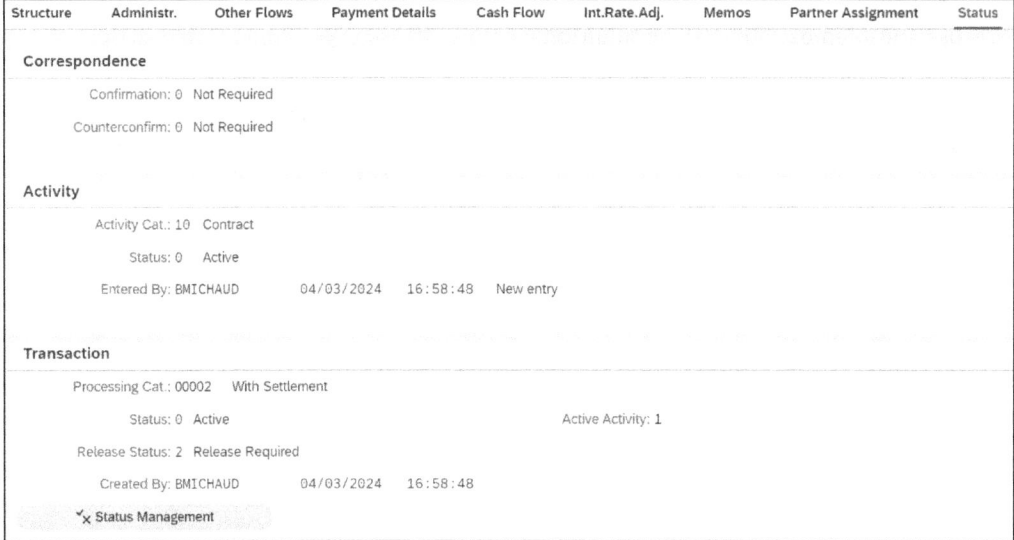

Figure 5.27 Display of Status Tab

5.2 Contract Settlement

Once we have created the contract, we should review the details. The review process for contracts is customizable to fit a company's specific requirements. We can handle this by allowing users to settle their own contracts and having stricter controls over the release of payments, ensuring a dual-control process is set up in security, or using SAP S/4HANA's workflow to ensure that multiple people review the contracts, as follows:

- **Dual control**

    We can implement security measures on the activity types of a contract so that only certain users can create, settle, or approve them. The downside of this approach is that some users will always be designated as the creators and a different group of users will always be the approvers. Since a user can't both create contracts and approve other users' contracts, this may not be the best solution for smaller treasury departments.

5.2 Contract Settlement

To run the settlement portion of a transaction, we use the Process Financial Transaction app or Transaction FTR_EDIT. The app for this transaction is shown in Figure 5.28.

Figure 5.28 Process Financial Transaction App

To settle the transaction, we enter the **Company Code** and **Transaction** ID and then click on the **Settle** button (see Figure 5.29). The next screen will show the transactions details and structure that we saw in the Create Financial Transaction app. Once we have reviewed everything, we can save the transaction. This completes the settlement process, and the **Activity** for the transaction will be updated from **Contract** to **Contract Settlement**. The exact flow and statuses of the contracts are dependent on the configuration for the transaction type.

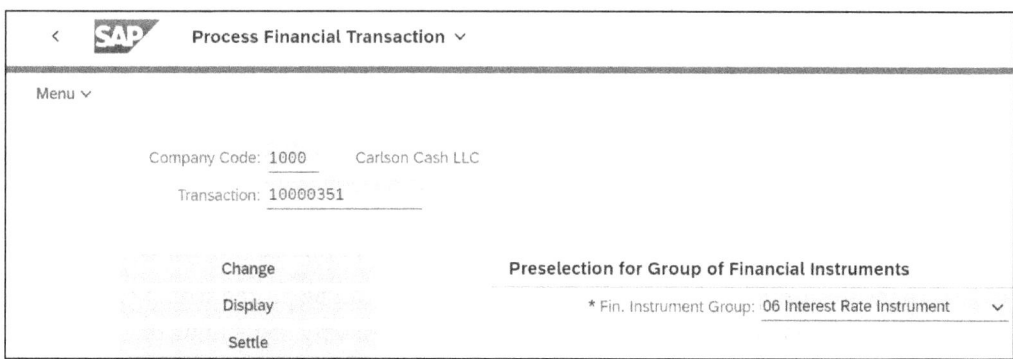

Figure 5.29 Processing Financial Transaction Settlement

- **Workflow**
  Workflows can define the release procedures for each step of the contract creation and editing process. Creation of and changes to the contract can kick off a workflow that is sent to the approver's worklist and has to be released before the creation of or edits to the contract are active. The customizing of the treasury and risk management workflow is detailed in Chapter 3, Section 3.19. When a transaction is in a workflow, we can approve the transaction in the My Inbox app or Transaction SBWP (see Figure 5.30).

343

5 General Contract Processes

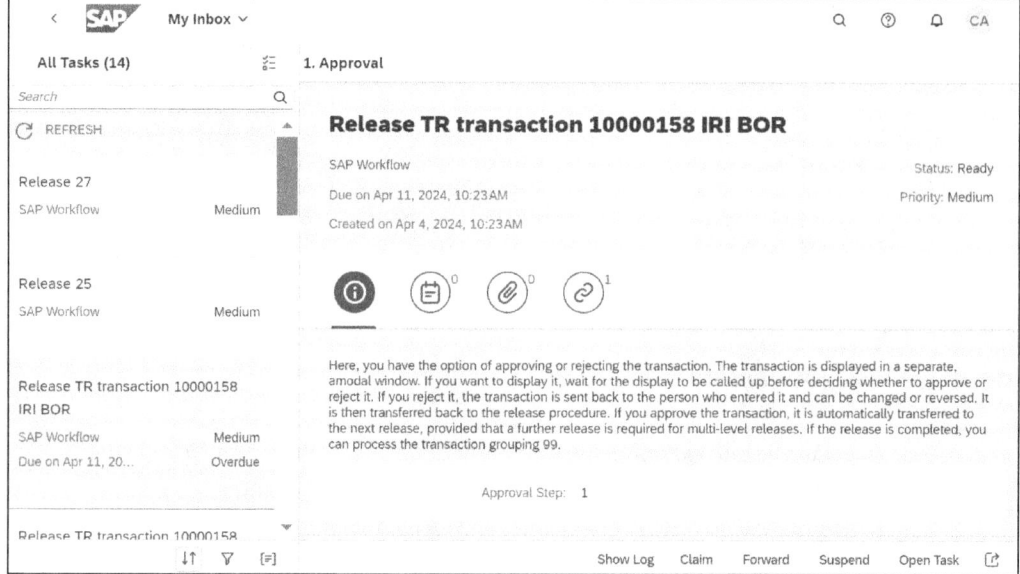

Figure 5.30 Workflow Approval Screen for Treasury and Risk Management Transactions

5.3 Transaction Posting

One of the key features of treasury and risk management is how transactions can all have automated accounting postings that are both calculated and posted. This simplifies the process since many companies have separate systems or spreadsheets that calculate the flows of a contract and are required to create separate accounting entries. This information can be generated and processed automatically, which frees us from the redundant and error-prone process of tracking this information on spreadsheets.

There are a couple ways that each type of contract can be posted within treasury and risk management. The transactions that will be used vary, depending on the types of contracts and the flows that need to be posted. The postings can be separated into the following categories:

- **Structure flows**
  These are flows associated with the creation, changes to, and repayment of the nominal amount of the contract.
- **Interest flows**
  These flows are calculated, posted, and paid according to the schedule defined in the interest conditions.
- **Accrual and deferral flows**
  These flows are dependent on the interest flows and are generally posted monthly. If an interest flow is accrual relevant, the postings for the accrual will work alongside the interest posting to correctly reflect the interest expense in the correct period.

5.3 Transaction Posting

- **Valuation flows**
  When a contract is relevant to a mark-to-market adjustment, we need to either calculate the valuation or manually input it. These flows will post the unrealized gain or loss.

- **Premium/discount flows**
  These are applicable if there's a premium or a discount associated with the contract and the premium or discount needs to be amortized through the life of the contract.

- **Other flows**
  If a cash flow doesn't fall into the above categories, an other flow can be created and assigned to the product type in customizing. Frequently, these are used for other fees that need to be tracked and posted on the contract.

In the following sections, we'll walk through topics relevant to posting transactions. We'll start with an introduction to account assignment reference and then move on to the Post Flows app, running accrual and deferral, running a key date valuation, and transferring a contract from long term to short term.

5.3.1 Account Assignment Reference

Before we cover the programs that create the postings in treasury and risk management, we need to cover the *account assignment reference*, which is a tag that is added to all financial transactions. Its main purpose is to assign the general ledger accounts when the transaction is posted. Even if transactions have the same product type and transaction type, the account assignment reference can assign different general ledger accounts to the posting transactions. This tag is generally added automatically, using derivation rules that are covered in Chapter 3, Section 3.22.2. This section will also cover the creation of the posting rules and the subsequent assignment to the general ledger accounts based on the account assignment reference.

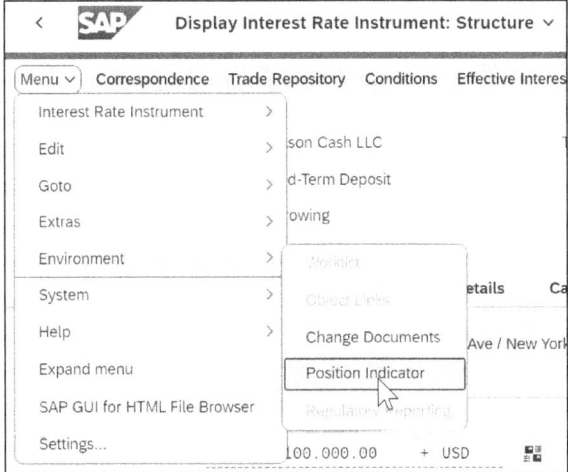

Figure 5.31 Navigation to Display Position Indicator

5 General Contract Processes

If we want to check the account assignment reference on any transaction, we can look in the Process Financial Transaction app. When we're viewing the transaction screen, we can navigate to **Menu • Environment • Position Indicator, as** in Figure 5.31.

In this area, we can view the field account assignment reference (**AcctAssRef**). In Figure 5.32 we've assigned "IRI," which assigns the general accounting treatment for the interest rate instruments in this system.

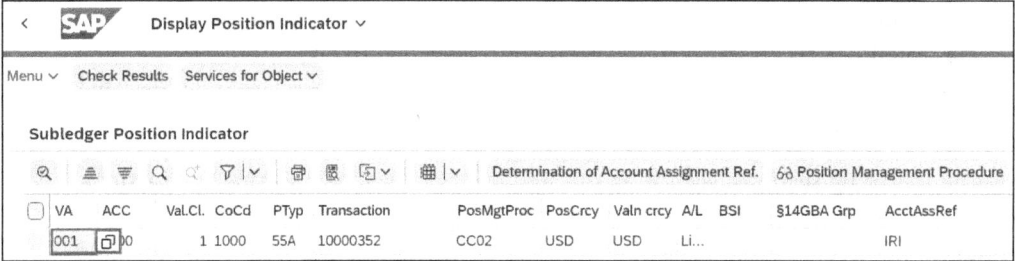

Figure 5.32 Viewing the Account Assignment Reference in Transaction

5.3.2 Post Flows

The first transaction that we'll cover is used for all types of contracts. When any cash-related flows need to be posted, we use the Post Flows app (shown in Figure 5.33) or Transaction TBB1. The key flows that are posted include structure flows that will include any principal increases and decreases from transactions, plus the incoming and outgoing interest flows. If the transactions have been set up in customizing to initiate payments through the treasury payment functionality, then this SAP Fiori app will create the payment requests for these transactions.

Figure 5.33 SAP Fiori App for Post Flows

There are three sections that we need to populate before we can run the Post Flows app. The **Application** section is a high-level filter that lets us select the types of transactions that should be pulled into the Post Flows app. For general debt and investment product types, we'll select **Money Market**. We can also filter other types of transactions like **Foreign Exchange, Derivatives, Securities,** and **Trade Finance** (as in Figure 5.34).

In the **General Selections** area, we can filter the types of transactions to be posted even further. These selections are displayed in Figure 5.35. We can filter by specific product types and business partners, and we can even filter by the transaction ID assigned to the contract upon creation. In our example, to demonstrate the posting, we're going to filter by a specific **Transaction** ID. This screen also lets us determine which flows to select by due date (**Up to and Including Due Date**) and posting date (**Up to and Incl. Posting Date**). All flows will be selected up to the date that we define in this area.

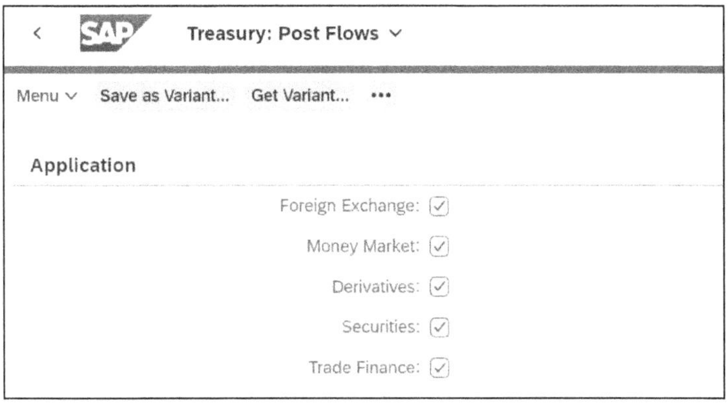

Figure 5.34 Selection of Applications in Post Flows

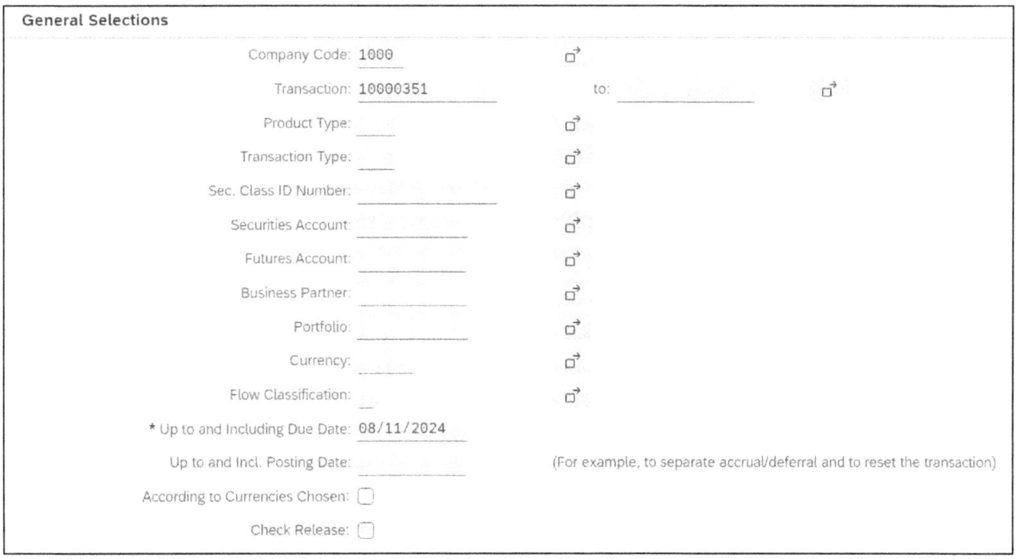

Figure 5.35 General Selections Area of Post Flows

> **Check Release in Post Flows**
>
> When we set up a workflow process in treasury and risk management, the transactions have unreleased and released statuses. We use the checkbox for **Check Release** to tell this posting program to only check for cash flows that have been fully released in the workflow.
>
> If any flows haven't been released, the Post Flows app won't be able to post them.

5 General Contract Processes

> **Filtering in Post Flows**
>
> During the testing phase of a project, it's useful to filter by specific transaction IDs. This ensures that only specific transactions are pulled into the transaction and that it's easy to trace the flows that are being posted.
>
> During more comprehensive testing or when we're in a live system, we'll generally use the more generic filtering options to create all of the flows for similar product categories or product types.

The final area, **Posting Control**, lets us determine how to post the flows. This is shown in Figure 5.36. By default, all flows will post on the defined due date, but we can change the **Posting Date** and **Document Date**. If we want to view the proposed postings before actually posting the documents, we can select the **Test Run** option.

Figure 5.36 Posting Control Area of Post Flows

Once we have entered all the selections, we can execute the transaction by clicking the **Execute** button. This will output messages to show the transaction has been successfully run in test mode or production mode. Once this is run, we'll see a **Logs and Messages** screen detailing information on the Post Flows app run. The successful output of this report can be seen in Figure 5.37.

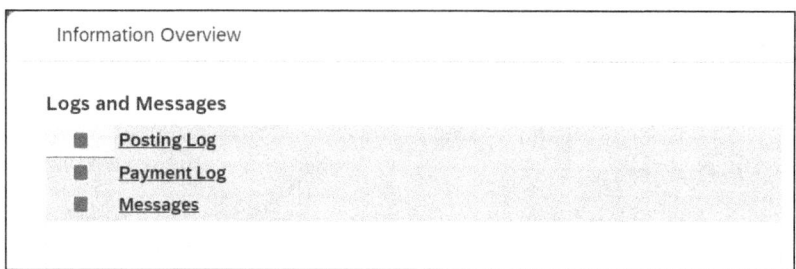

Figure 5.37 Logs and Messages from Post Flows

We can review each of these messages by clicking on the hyperlink. The first log available is the **Posting Log,** which we can see in Figure 5.38. It shows the accounting postings that have been generated from the Post Flows app.

Status	Transaction	CoCode	Key	G/L Account	G/L Acct Long Text	Update Type	Update Type Description	Amount	TC	Posting Date
	10000351	1000	40	109999	PAYMENT REQUEST CLRG	MM_PI00+	Principal Increase	1,000,000.00	USD	08/07/2024
	10000351	1000	50	113310	I/C NOTES RECEIVABLE	MM_PI00-	Principal Increase	1,000,000.00-	USD	08/07/2024
	10000351	1000	40	109999	PAYMENT REQUEST CLRG	MM_CH00+	Charges	5,000.00	USD	08/09/2024
	10000351	1000	50	361099	BANK FEES MISCELLANEOUS	MM_CH00-	Charges	5,000.00-	USD	08/09/2024
	10000351	1000	40	113310	I/C NOTES RECEIVABLE	MM_PD00-	Principal Decrease	100,000.00	USD	09/07/2024
	10000351	1000	50	109999	PAYMENT REQUEST CLRG	MM_PD00-	Principal Decrease	100,000.00-	USD	09/07/2024
	10000351	1000	40	113310	I/C NOTES RECEIVABLE	MM_PF00-	Final Repayment	900,000.00	USD	08/07/2025
	10000351	1000	50	109999	PAYMENT REQUEST CLRG	MM_PF00-	Final Repayment	900,000.00-	USD	08/07/2025
	10000351	1000	40	221010	ACCRUED INT EXP	MM_IN00-	Interest	5,000.00	USD	08/07/2025
	10000351	1000	50	109999	PAYMENT REQUEST CLRG	MM_IN00-	Interest	5,000.00-	USD	08/07/2025
	10000351	1000	40	221010	ACCRUED INT EXP	MM_IN00-	Interest	41,250.00	USD	08/07/2025
	10000351	1000	50	109999	PAYMENT REQUEST CLRG	MM_IN00-	Interest	41,250.00-	USD	08/07/2025

Figure 5.38 List of Accounting Entries Generated Using Post Flows Transaction

> **Incorrect Transactions**
>
> If there are any transactions that haven't been fully set up, we'll see them in the **Incorrect Transactions** area of the accounting **Hierarchy** on the left side of the screen. This generally occurs if posting specifications haven't been fully assigned to an update type or if the posting specifications haven't been assigned general ledger accounts in the configuration.
>
> The message that appears in the **Incorrect Transactions** area and the output messages will give us the information we need to fix any incorrect transactions.

The next option to review is the **Payment Log** (see Figure 5.39). This shows the payments that were generated from the transaction. For each flow that needs to create a payment, a payment key or payment request ID will be displayed in the payment log. We can run this number through the treasury payment process to initiate the payment.

CoCd	Transactn	K...	Payment Amount	PmtCurrAmt	PCrcy	Payt Date	HBank	Account	Payer/Payee	PBank	PaymStatus	Payment Status (Text)	Year	Line Item Text
1000	10000351	237	1,000,000.00	1,000,000.00	USD	08/07/2024	JPM01	PAY02	JP MORGAN	0001	2	Created	2024	*MM_PI00+ Principal Increase 0000010000351
		238	5,000.00	5,000.00-	USD	08/09/2024	JPM01	PAY02	JP MORGAN	0001	2	Created	2024	*MM_CH00- Charges 0000010000351
		239	100,000.00	100,000.00-	USD	09/07/2024	JPM01	PAY02	JP MORGAN	0001	2	Created	2024	*MM_PD00- Principal Decrease 0000010000351
		240	900,000.00	900,000.00-	USD	08/07/2025	JPM01	PAY02	JP MORGAN	0001	2	Created	2025	*MM_PF00- Final Repayment 0000010000351
		241	5,000.00	5,000.00-	USD	08/07/2025	JPM01	PAY02	JP MORGAN	0001	2	Created	2025	*MM_IN00- Interest 0000010000351
		242	41,250.00	41,250.00-	USD	08/07/2025	JPM01	PAY02	JP MORGAN	0001	2	Created	2025	*MM_IN00- Interest 0000010000351

Figure 5.39 Payment Log in Post Flows

The **Messages** area will show any messages that are applicable to the Post Flows app run. If everything is successful, the message will look like Figure 5.40. If there are any

5 General Contract Processes

issues with the postings, the messages in this area will display an error indicator. This log will also describe the error to explain why the run was not successful.

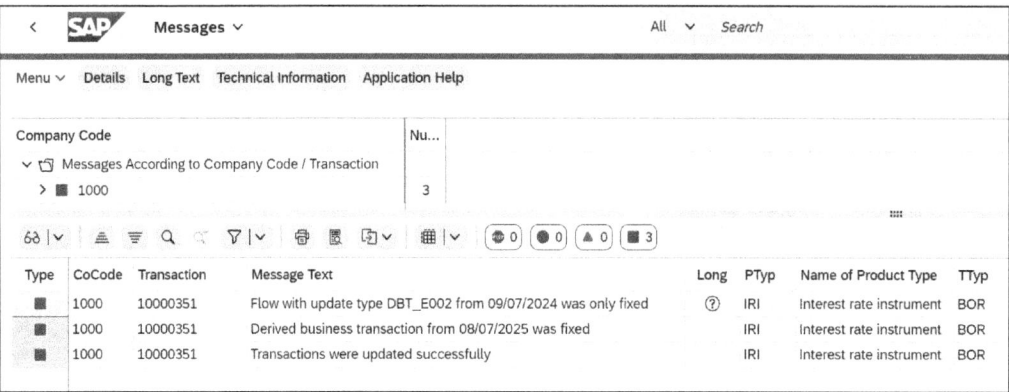

Figure 5.40 General Messages Log for Post Flows

5.3.3 Run Accrual/Deferral

There are a couple of ways to post and process interest-relevant transactions. When we post the full amount of an interest payment or receipt to an expense or income account, then the full interest amount is recognized in the profit and loss statement at that time. Generally, we'll want to recognize the interest income or expense in the period it's relevant for and not necessarily when the interest is paid or received. When this is a requirement, we can configure the interest flows to be relevant for the following:

- **Accrual**
  An accrual of interest occurs when we allocate the interest to the profit and loss each period. The interest payment can occur at a future date.

- **Deferral**
  A deferral uses a similar concept, but this interest flow is prepaid. The cash is exchanged first, and the profit and loss postings will occur in the subsequent periods.

When cash flows are set up in customizing to require an accrual or deferral, we use the Run Accrual/Deferral app or Transaction TPM44. There are two different ways we can run the accrual and deferral. The *incremental* or *difference procedure* in SAP S/4HANA will calculate the accrual/deferral amount and will incrementally post the change for the specified period. It will record a running total and only post the difference when this transaction is run. The other posting method is called *book and reset*, in which the full amount of accrual or deferral is posted on the specified date and the full entry is reset in the following period. This ensures that the correct accrual or deferral totals are reflected in the general ledger accounts at month end. Both of these procedures will result in the same balances in the accrual and prepaid general ledger accounts, but the postings will appear different for each procedure.

It's easier to understand these concepts with examples of the related postings. In Figure 5.41, there's an example of the t-accounts for interest accruals and interest payment postings for the difference procedure. The interest is accrued and will post to the expense account each month of the accrual, and we can see this posting in the three months that are reflected. The total amount of the accrued interest goes up each month, and it's also reflected in the interest expense account. When the interest is ultimately paid to the counterparty, we won't post to the interest expense account since that was recorded already. This entry will post to the bank account for the outgoing payment, and the offsetting entry will clear out the accrued interest account.

Activity	Transaction	Accrued Interest		Interest Expense		Bank Account	
Post 1st Month Accrual	Run Accrual/Deferral		100	100			
Post 2nd Month Accrual	Run Accrual/Deferral		100	100			
Post 3rd Month Accrual	Run Accrual/Deferral		100	100			
Post Interest Payment	Post Flows	300					300

Figure 5.41 Example Postings for Accrual that Posts Using Difference Procedure

The postings for the book and reset procedure will look slightly different. In the t-accounts in Figure 5.42, each month's accrual resets the following day. Due to this, the full accrual amount for the interest period is posted to make sure the accrued interest account and the interest expense account both reflect the correct balances at the end of each period. When the interest is finally paid, the full interest amount will be posted to the interest expense account. As we can see with these two procedures, the outcome is the same on each key date, but the postings have some differences when they are recorded.

Activity	Transaction	Accrued Interest		Interest Expense		Bank Account	
Post 1st Month Accrual	Run Accrual/Deferral		100	100			
Reset 1st Month Accrual		100			100		
Post 2nd Month Accrual	Run Accrual/Deferral		100	100			
Reset 2nd Month Accrual		100			100		
Post 3rd Month Accrual	Run Accrual/Deferral		100	100			
Reset 3rd Month Accrual		100			100		
Post Interest Payment	Post Flows			300			300

Figure 5.42 Example Postings for Financial Transaction that Uses Book and Reset for Accruals

Now that we understand the concept of what we are posting with this transaction, we can look at how we run the app. Figure 5.43 shows the SAP Fiori app that runs accruals and deferrals.

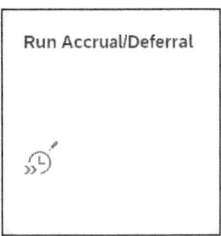

Figure 5.43 Run Accrual/Deferral App

To run this SAP Fiori app, we can use the selection screen to filter the transactions. We use the **Product Groups**, **General Selections**, **Securities**, **Listed Options/Futures**, and **OTC Transactions** sections to further refine which transactions will be selected in the accrual/deferral run. Which fields we can enter information into on this screen will depend on the **Product Groups** that we select at the top of the screen. For example, Figure 5.44 shows a section labeled **OTC Transactions** since that group has been selected. If we also selected the **Securities** product group, then fields that are specific to filtering securities transactions would also appear.

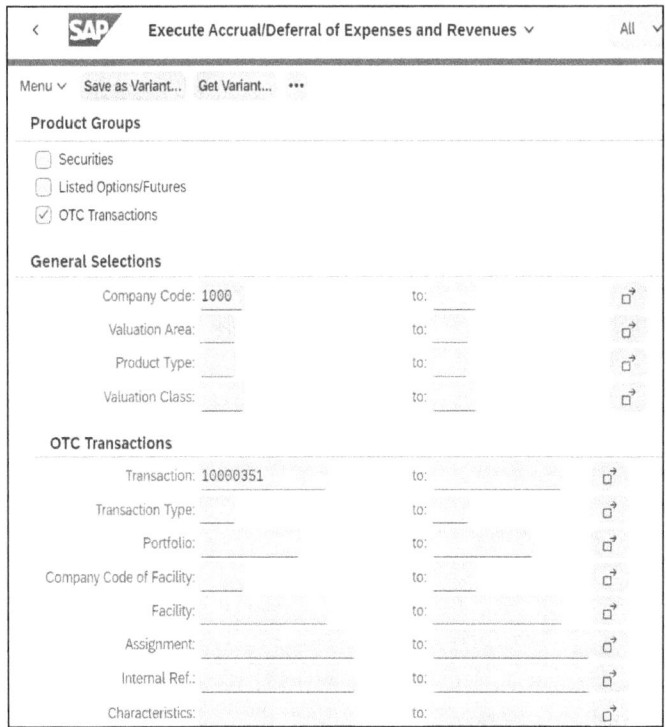

Figure 5.44 Execute Accrual/Deferral App Entry Screen

Next, the **Selection Parameters** section defines more about the transactions we are selecting. The most important selection in this section is the **Accrual/Deferral Key Date**. This determines the date until which we are accruing interest. If we want to accrue all the way up to that date, then we also need to check the **Including Key Date** box, as shown in Figure 5.45.

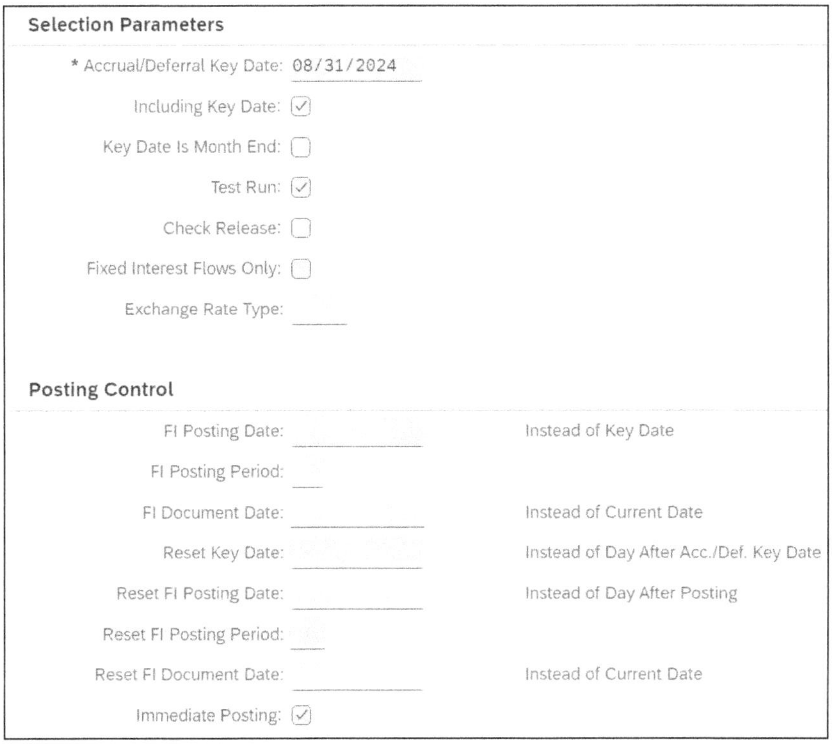

Selection Parameters		
* Accrual/Deferral Key Date:	08/31/2024	
Including Key Date:	✓	
Key Date Is Month End:	☐	
Test Run:	✓	
Check Release:	☐	
Fixed Interest Flows Only:	☐	
Exchange Rate Type:		

Posting Control		
FI Posting Date:		Instead of Key Date
FI Posting Period:		
FI Document Date:		Instead of Current Date
Reset Key Date:		Instead of Day After Acc./Def. Key Date
Reset FI Posting Date:		Instead of Day After Posting
Reset FI Posting Period:		
Reset FI Document Date:		Instead of Current Date
Immediate Posting:	✓	

Figure 5.45 Selection Parameters and Posting Control Sections of Run Accrual/Deferral

Finally, we use the **Posting Control** section to alter any of the posting dates if required. If we use the book and reset procedure, the **Reset Key Date** will automatically be set to the day after the defined accrual/deferral key date. However, we can update it in this section.

When we're ready to execute the posting of this transaction, we click the **Execute** button. The output screen shown in Figure 5.46 will then display details of the accrual/deferral run. Some of the key numbers in the output of this report are as follows:

- ❶ The amount in this field shows the total amount of interest that will be posted for the defined interest period.
- ❷ This is the total amount that will be posted for the current accrual/deferral posting.
- ❸ This is the total number of calculation days in the defined interest period.
- ❹ This shows the number of days used to calculate the interest amount in the current accrual/deferral posting.

5   General Contract Processes

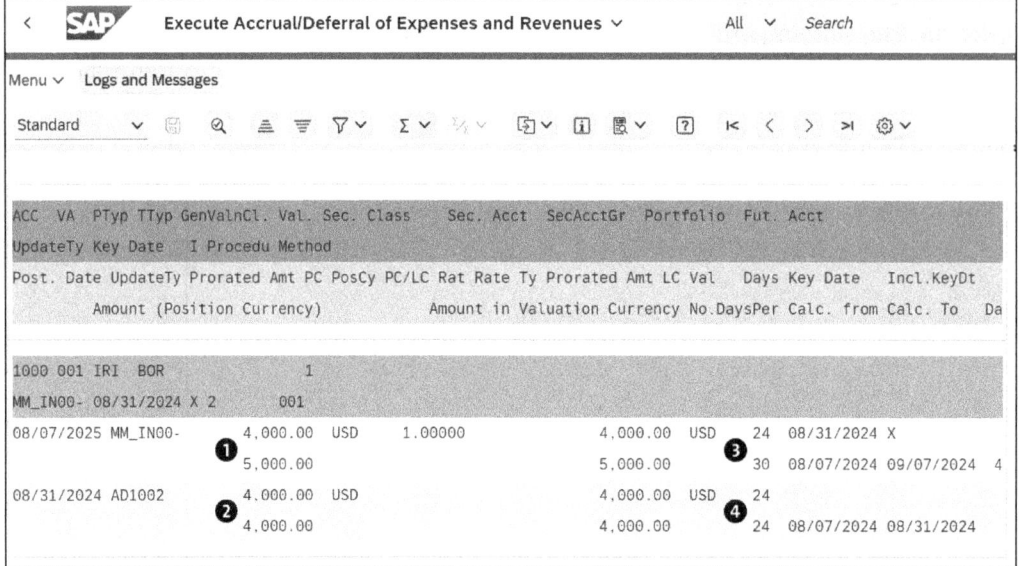

Figure 5.46 Output of Run Accrual/Deferral That Shows Date Calculation and Accrual Amounts

5.3.4 Run a Key Date Valuation

There are various valuations that we can run for a transaction. The valuations are set up in each transaction's position management procedure, and each of the individual types of valuations is covered in detail in that section. Some of the more common valuation procedures include the security valuation procedure to create a mark-to-market valuation and an amortization procedure that we use to post the amortization of a financial transaction's premium or discount. We'll cover the security valuation procedure in more detail when we cover the FX process in Chapter 6, Section 6.5.

For the purposes of this example, we'll run a key date valuation for an amortization. In this example, a debt/investment instrument has been issued at a premium or a discount, and we can amortize the difference between this amount and the book value through the life of the deal. When a discount or premium is relevant to a transaction, we need to indicate it in the transaction type in customizing via the **Financial Supply Chain Management • Treasury and Risk Management • Transaction Manager • Money Market • Transaction Types • Define Transaction Types – MM Transactions** menu path. This will make an additional field appear in the **Structure** tab. The discount or premium can be reflected in the **Amount** and **Nominal Amount** fields as shown in Figure 5.47. In this example, the contract has a $10,000 discount that will be amortized throughout the life of the contract.

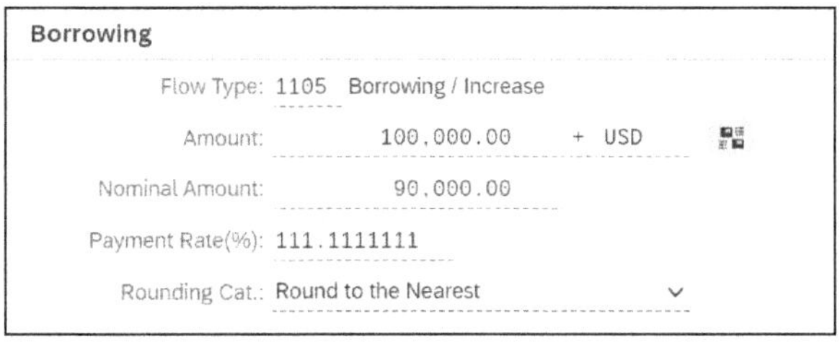

```
Borrowing
           Flow Type:  1105   Borrowing / Increase
              Amount:           100,000.00    +  USD
      Nominal Amount:            90,000.00
    Payment Rate(%):  111.1111111
       Rounding Cat.:  Round to the Nearest
```

Figure 5.47 Example of Entry of Discount on Structure Tab of Debt Contract

When a debt or investment has a premium or discount, we post the amortization using the Run Valuation app or Transaction TPM1. We use this app to run key date valuations and amortizations or premiums and discounts. To run this transaction, we use the **Product Groups** and **General** selections to filter the transactions and thus make sure we're running the valuation for the correct transactions. These sections work the same in this transaction as they do in the Run Accrual/Deferral app we described previously. Figure 5.48 shows the top half of the entry screen in the Run Valuation app.

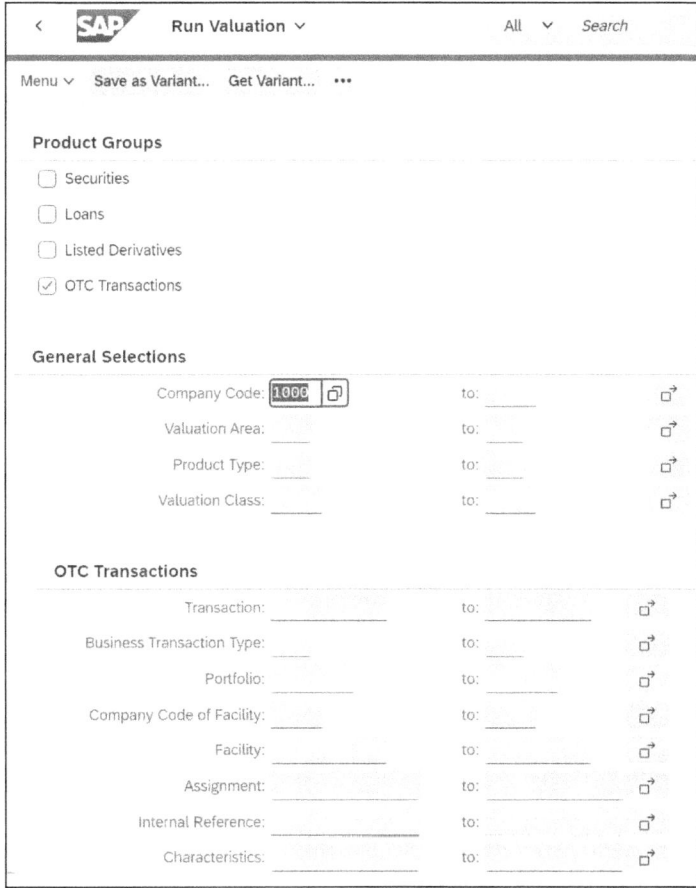

Figure 5.48 Entry Screen of Run Valuation App

The **Valuation Parameters** area (see Figure 5.49) lets us determine when the valuation run is posted and the type of posting that will be executed. Similar to the accrual/deferral transaction, the valuation can be run with a book and reset procedure or a difference procedure when posting. We use the difference procedure when the **Valuation Category** is a valuation without reset. The full list of valuation categories is as follows:

- **Year-End Valuation**
  This selection runs a valuation at the end of the year in accordance with the assigned position management procedure.

- **Mid-Year Valuation with Reset**
  This creates a valuation posting using the book and reset procedure.

- **Mid-Year Valuation without Reset**
  This creates a valuation posting using the difference procedure.

- **Manual Valuation with Reset**
  This selection allows us to write up or down to a book value that is entered manually in the system. This posting uses the book and reset procedure. The available valuation steps that can use a manual valuation are the security, foreign currency, one-step price, and index valuations.

- **Manual Valuation without Reset**
  This option works the same as the previous one, except that the postings follow the book and reset procedure.

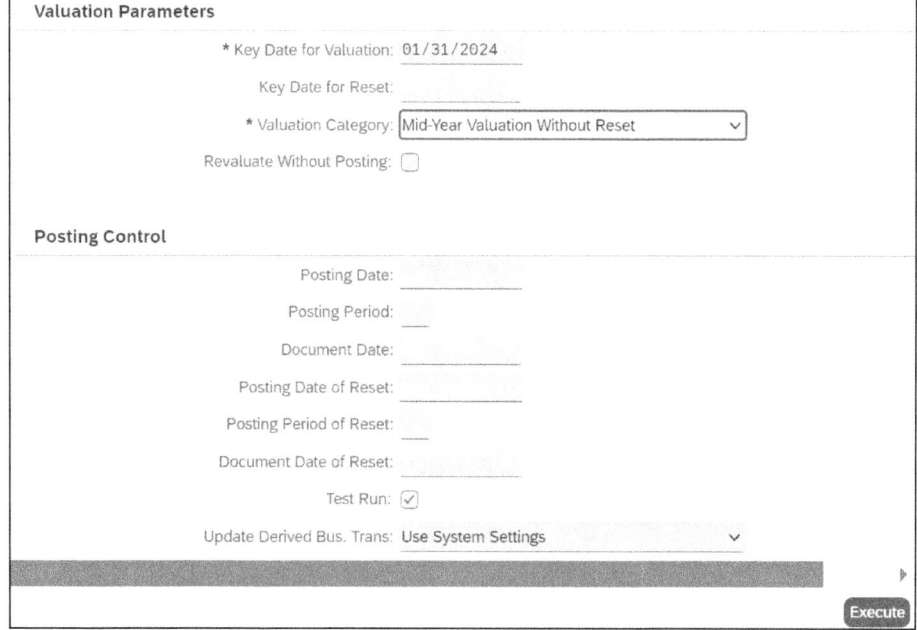

Figure 5.49 Bottom Section of Entry Screen for Run Valuation

5.3 Transaction Posting

Once we have finalized the selection screen, we can execute the transaction. There are two steps involved in running this transaction. First, we review the initial output screen that displays a summary of the transactions that we selected in the valuation run. Then, once we have validated these transactions, we click the **Run Valuation** button (see Figure 5.50).

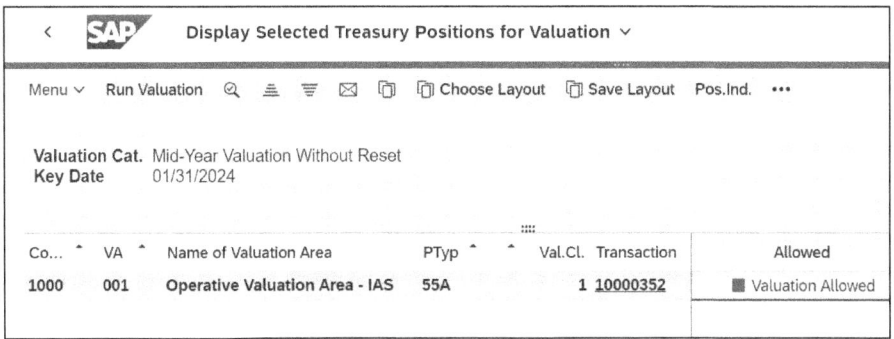

Figure 5.50 Initial Output Screen of Run Valuation

The next screen (shown in Figure 5.51) outputs the details of the amortization. The key details to note in this screenshot are that the book value started at $100,000, there was a posting to amortize $899.02, and the new book value after the amortization is $99,100.98.

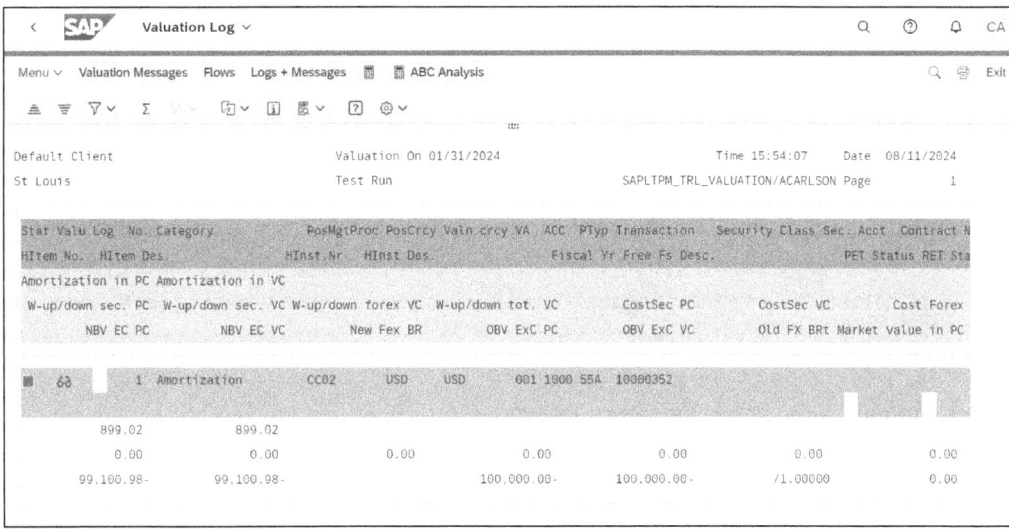

Figure 5.51 Run Valuation Output Showing Amortization Amount That Has Been Posted

By clicking on the **Logs + Messages** button, we can view the amortization log and posting log. The amortization log is shown in Figure 5.52, and the posting log is shown in Figure 5.53. To view the amortization log, we need to check the **Amortization Log** box in the **Output Selection** portion of the selection screen in the Run Valuation app.

5   General Contract Processes

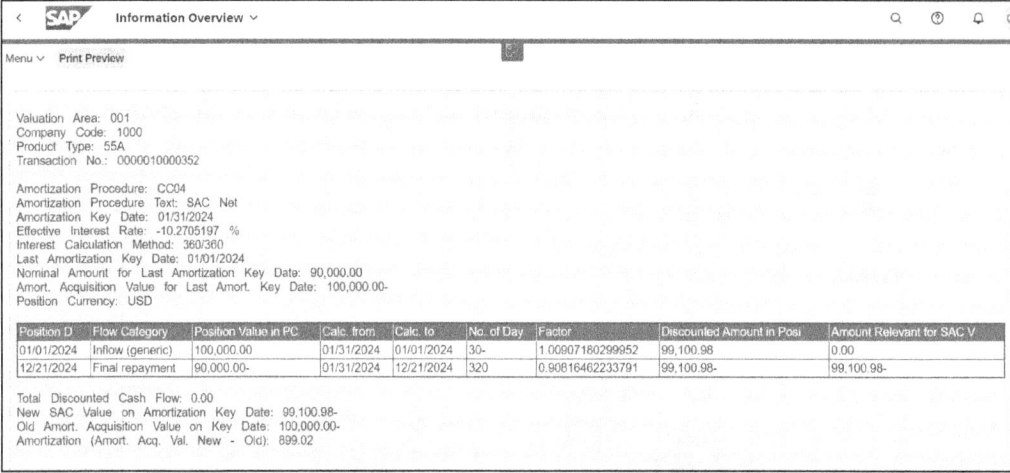

Figure 5.52 Amortization Log That Displays Calculation of New Value on Key Date

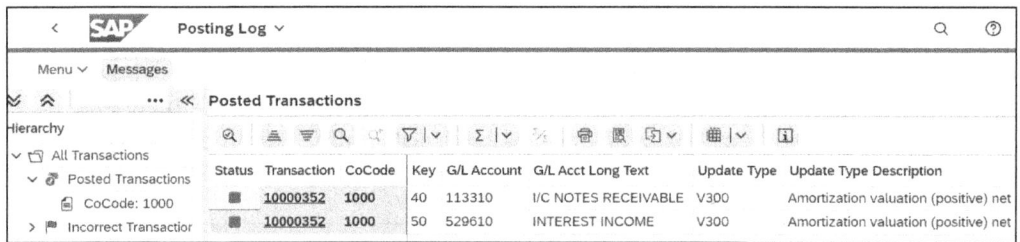

Figure 5.53 Posting Log Detailing All Postings That Occurred During Valuation Run

5.3.5   Transfer from Long Term to Short Term

When a contract is due to mature within a year, we can transfer the book value of an instrument from a long-term account to a short-term account. We can do this by using one of two processes in SAP S/4HANA: by running an account assignment reference transfer or with a valuation class transfer. These processes can vary, so we'll cover each one in detail in the following sections.

Account Assignment Reference Transfer

We covered the account assignment reference earlier in this chapter. Since the main factor we want to change when reclassifying a transaction from long-term to short-term is the accounting treatment, it makes sense that we would want to update the account assignment reference. The account assignment reference transfer allows for the transfer of the ledger position of the transaction, and the account assignment reference is the main driver for the accounts that are posted to the general ledger when we're running any of the posting programs. When we run the transfer, the position account is cleared out, passes through a technical clearing account, and is posted to a new position account. This is demonstrated in the t-accounts in Figure 5.54.

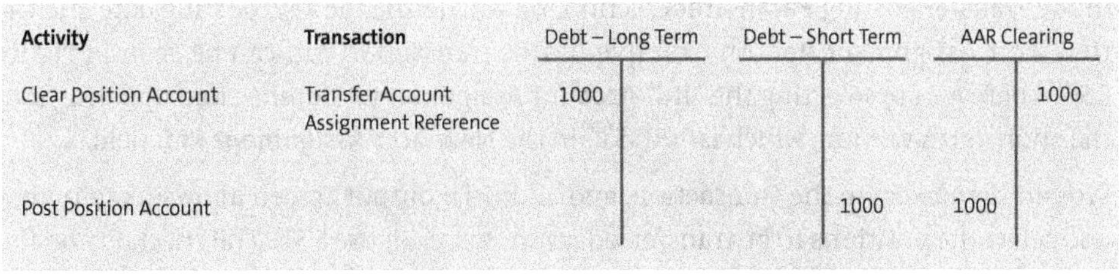

Activity	Transaction	Debt – Long Term	Debt – Short Term	AAR Clearing
Clear Position Account	Transfer Account Assignment Reference	1000		1000
Post Position Account			1000	1000

Figure 5.54 Sample T-Accounts Showing How Account Assignment Reference Transfers Balance from Long Term to Short Term

To access this transaction, we can go to the Transfer Account Assignment Reference app (shown in Figure 5.55) or Transaction TPM28.

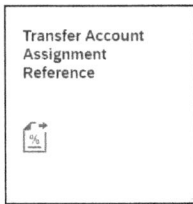

Figure 5.55 Transfer Account Assignment Reference App

Once we access this app, we'll see an entry screen that's similar to the other screens we've seen in this section. First, we need to select the applicable transactions in the **Product Groups** and **General Selections** sections. Then, in the **Special Selections** area, we need to select the originating **Account Assignment Reference** for the transfer.

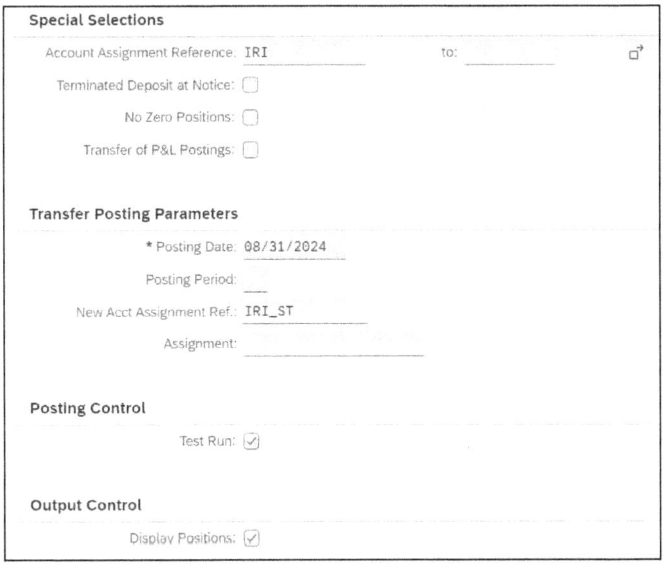

Special Selections

- Account Assignment Reference: IRI   to:
- Terminated Deposit at Notice: ☐
- No Zero Positions: ☐
- Transfer of P&L Postings: ☐

Transfer Posting Parameters

- *Posting Date: 08/31/2024
- Posting Period:
- New Acct Assignment Ref.: IRI_ST
- Assignment:

Posting Control

- Test Run: ☑

Output Control

- Display Positions: ☑

Figure 5.56 Entry Screen Showing Required Fields for Transferring Account Assignment Reference

5 General Contract Processes

In the **Transfer Posting Parameters** section, we will define the key/posting date and the **New Acct Assignment Ref**. An example of this transaction run can be seen in Figure 5.56, where we are selecting the "IRI" **Account Assignment Reference** and updating it to the short-term version, which is "IRI_ST" in the **New Acct Assignment Ref.** field.

We can then execute the transaction, and an initial output screen allows us to review and select the positions to be transferred (as shown in Figure 5.57). The checkbox on the left side of the screen allows us to unselect positions that we don't wish to transfer.

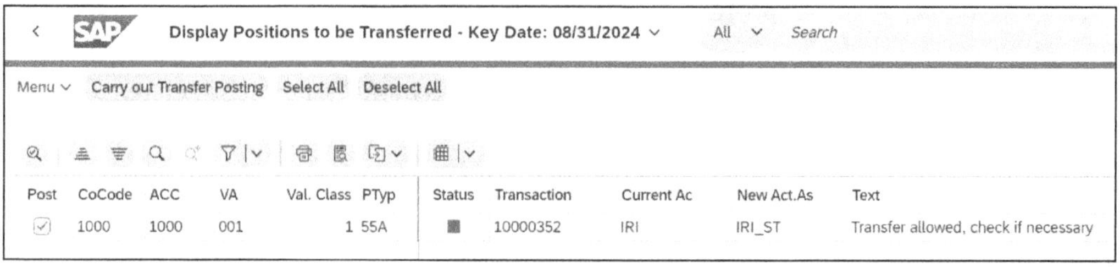

![Figure 5.57]

Figure 5.57 Initial Display of Account Assignment Reference Transfer Proposal

Once we have completed the review, we can carry out the transfer by clicking the **Carry out Transfer Posting** button. This will result in the successful transfer of the account assignment reference and postings, as shown in Figure 5.58.

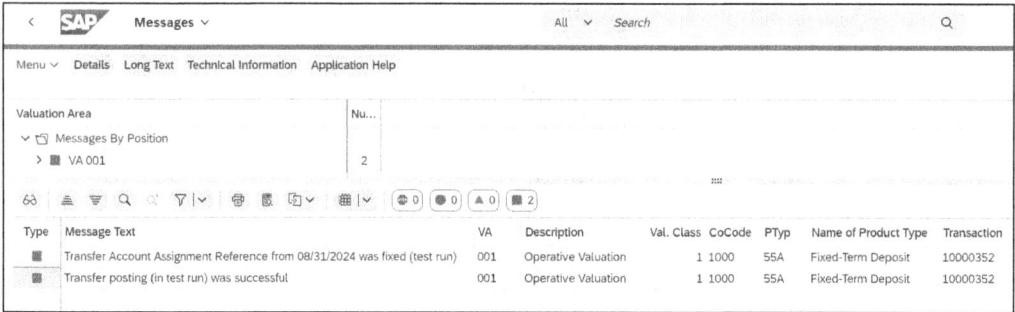

Figure 5.58 Final Output of Account Assignment Reference Transfer

There are a few consistency checks we need to keep in mind when running the account assignment reference transfer. This transaction changes the posting logic, so there are some restrictions when running the transfer:

- First, the transfer is run on a specific key date to transfer the position, so there must not be any postings that have been carried out after the key date.
- All planned business transactions must also be posted before the key date of the transfer to avoid any inconsistencies in the postings.
- Additionally, once the transfer has been run, the system will restrict any new business transactions from being entered before the key date of the transfer.

Valuation Class Transfer

The other process for transferring the book value of an instrument is the *valuation class transfer*, in which we change the *valuation class*, which defines the transactions by type. The valuation class is initially set when the instrument is created, but if we require a valuation class change, then we can update the accounting treatment of a contract that needs to change. The transfer can update the instrument's position management procedure and update the accounting from long-term position accounts to short-term position accounts. We can run valuation class transfers for both securities and over the counter (OTC) transactions using the Execute Valuation Class Transfer app or Transaction TPM15M, as shown in Figure 5.59.

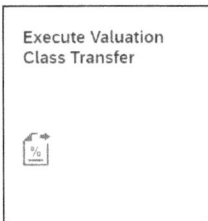

Figure 5.59 Execute Valuation Class Transfer App

There are two main ways to run this transaction. When we need to transfer the full value of a contract, we can update the valuation class for the whole contract. To do this, we use the **Product Groups** and **General Selections** areas of the selection screen to filter the contracts to transfer. In the **Specific Selections for Outgoing Position** area, we select which valuation classes to transfer. Finally, in the **Transfer Posting Parameters** section, we define the updated or target valuation class from the valuation class transfer along with the date when the positions will transfer. Figure 5.60 shows the details of the selection screen.

We can also update the account assignment reference while running the valuation class transfer. To do this, we have to set up the account assignment move in the customizing in the **Financial Supply Chain Management • Treasury and Risk Management • Transaction Manager • General Settings • Accounting • Link to Other Accounting Components • Define Account Assignment Reference Determination (OTC Transactions)** menu path. We also need to set the account assignment determination as a **Move** determination rule. In the **Definition** tab, we should set the destination account assignment reference as a **Constant** and set the **Target Field** as "AA_REF."

Let's look at an example of how to transfer the account assignment reference along with the valuation class. Figure 5.61 shows how we need to define the valuation class transfer with the destination account assignment reference. After we complete this, we can design the **Condition** to move the account assignment reference when running this transfer transaction (as shown in Figure 5.62). We use the **Condition** tab to determine what actually triggers the transfer. When an IRI product type is assigned a valuation class of 11, then this condition is applicable.

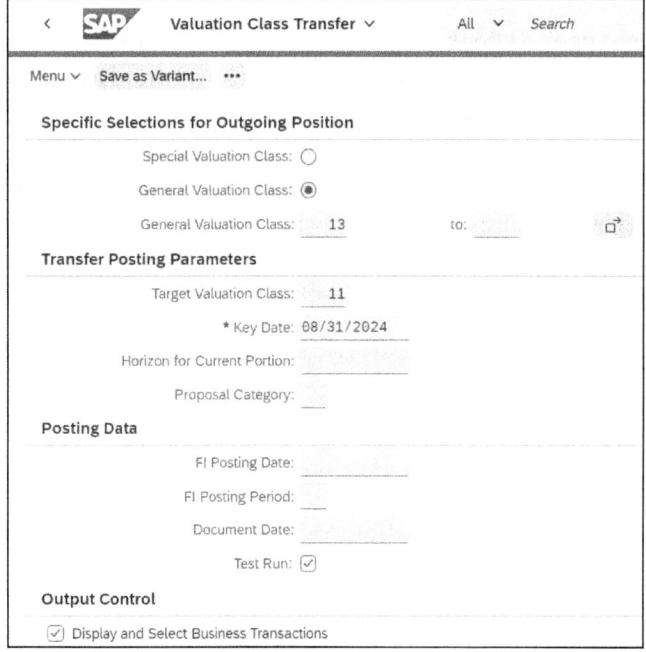

Figure 5.60 Valuation Class Transfer Entry Screen Showing Transfer from General Valuation Class 0013 to 0011

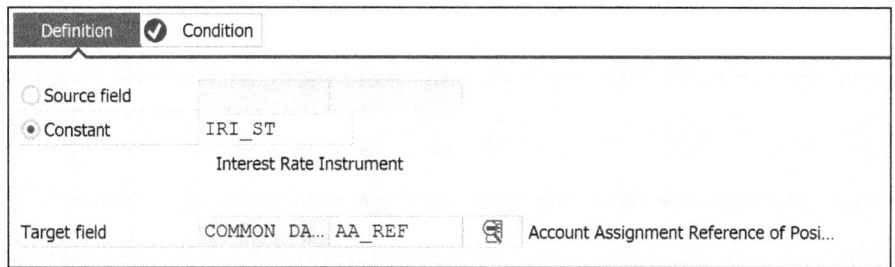

Figure 5.61 Screen Showing How to Determine New Account Assignment Reference in Mapping

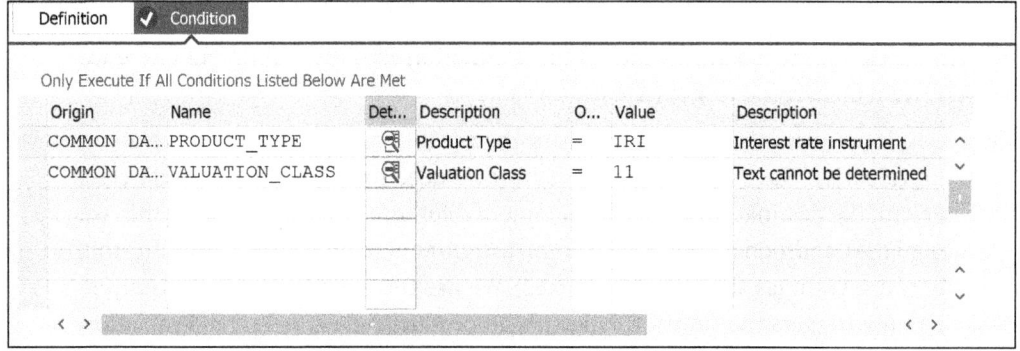

Figure 5.62 Condition Tab

Another function of the valuation class transfer occurs when we need to transfer the current portion of long-term debt. We frequently use this when a debt contract has installment or annuity payments. We need to classify the payments that are due within a year in a short-term general ledger account instead of a long-term account, and to handle this scenario, we need to follow these steps:

1. Classify the target valuation class as a current portion of long-term debt in the customizing in the **Financial Supply Chain Management • Treasury and Risk Management • Transaction Manager • General Settings • Accounting • Settings for Position Management • Define and Assign Valuation Classes** menu path. (Also see the discussion of the **Current Portion Handling** column in Chapter 3, Section 3.5.1.)

2. Set the flag for **OTC Transactions**. The current portion of long-term debt functionality is only available for OTC transactions and not for securities.

3. Select and define the source **General Valuation Class** and assign a **Target Valuation Class** that was previously defined in customizing. When this is done, a new **Horizon for Current Portion** field will appear on the selection screen. Any cash flow that occurs before the horizon date will be transferred as a current portion of long-term debt.

To demonstrate this, we'll look at an example. Generally, we reclassify all cash flows within a year to the short-term account, but in this example, we'll look at a shorter-term horizon date to demonstrate the functionality. In Figure 5.63, we have a transaction that includes monthly installment repayments of $50,000.

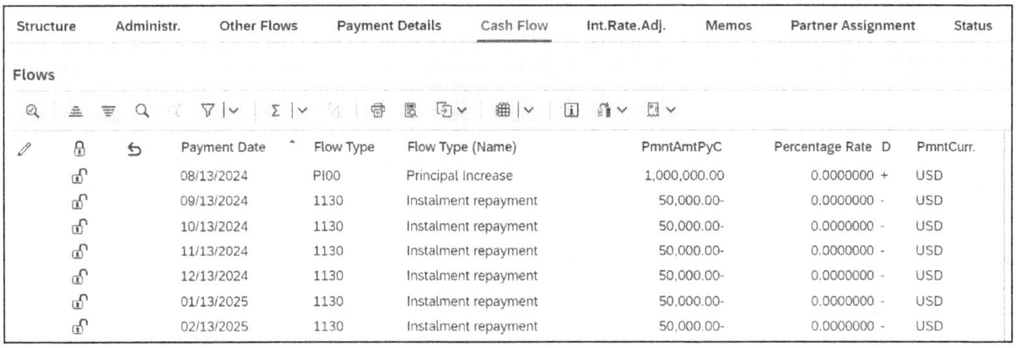

Payment Date	Flow Type	Flow Type (Name)	PmntAmtPyC	Percentage Rate	D	PmntCurr.
08/13/2024	P100	Principal Increase	1,000,000.00	0.0000000	+	USD
09/13/2024	1130	Instalment repayment	50,000.00-	0.0000000	-	USD
10/13/2024	1130	Instalment repayment	50,000.00-	0.0000000	-	USD
11/13/2024	1130	Instalment repayment	50,000.00-	0.0000000	-	USD
12/13/2024	1130	Instalment repayment	50,000.00-	0.0000000	-	USD
01/13/2025	1130	Instalment repayment	50,000.00-	0.0000000	-	USD
02/13/2025	1130	Instalment repayment	50,000.00-	0.0000000	-	USD

Figure 5.63 Sample of Transaction with Installment Repayments

This transaction has already had its initial principal increase flow posted. In Figure 5.64, we run the valuation class transfer on 08/31/2024 and add a **Horizon for Current Portion** in the selection screen for 12/31/2024. This means that the transfer will post on 08/31/2024 and any repayment flows that happen on or before 12/31/2024 will be reclassified from the long-term account to a short-term account. In this case, we have four payments of $50,000 that happen within this timeframe.

5 General Contract Processes

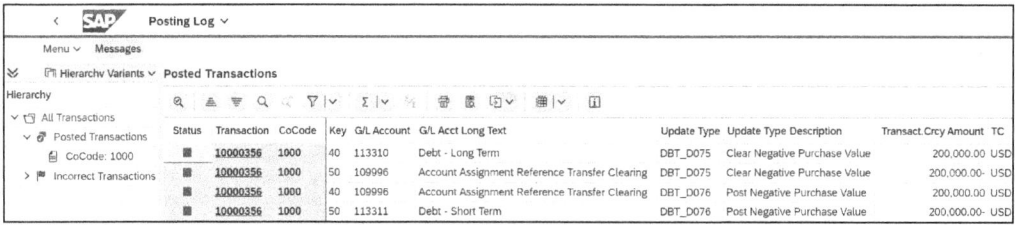

Figure 5.64 Example of How to Populate Entry Screen for Current Portion of Long-Term Debt Transfer

Once we've selected the positions and executed the transaction, the valuation class will be updated and the desired positions will be transferred to their respective new accounts. An example of the successful posting of this transaction is shown in Figure 5.65.

Figure 5.65 Postings Showing Successful Transfer of Current Portion of Long-Term Debt

5.4 Payment Processing

There are cash flows that are relevant to the treasury payment run in treasury and risk management processes. We can generate both incoming and outgoing payments, and we can generate payment requests for both that will be run through the Automatic Payment Transactions for Payment Requests app or Transaction F111. Generally, we configure outgoing payments to generate payment instruction files using the payment medium workbench. Incoming payments frequently are initiated by our business partners, so no payment instructions are needed for them. The exception to this is if we are directly debiting an account to pull the funds. To enable any of the flows for the treasury payment request, we need to make a few configuration and master data settings. These settings are detailed in Chapter 2, Section 2.5, and Chapter 12, Section 12.3.

We'll cover the exact types of payment transactions for each type of contract in the following sections, which detail contract-specific considerations. In general, we can create payment transactions for all types of incoming and outgoing payments. These include principal increases, principal decreases, interest, FX settlements, and other flows on contracts.

> **Additional Details on Outgoing Payments**
>
> Since the treasury payment functionality is automated based on the details entered into the business partner and the individual contracts, there are scenarios in which we may need to add details to the payment file when sending payment instructions to the bank. This is where the repetitive code is useful. We can add the following details to the treasury payments if we add a repetitive code:
>
> - **Bank Chain ID**
>   The bank chain can add payment instructions for intermediary bank requirements.
> - **Reference Text**
>   This is text that is sent with the payment. If the counterparty requires specific text to identify a payment, we can enter it in this field.
> - **Payment Method Supplement**, **Payment Reference**, and **Instruction Keys**
>   These are other available fields that we can add using repetitive codes.

Once the payment request has been generated by the Post Flows app, we can pay it using the Automatic Payment Transactions for Payment Requests app or Transaction F111. The app for this is shown in Figure 5.66.

Figure 5.66 Automatic Payment Transactions for Payment Requests App

The payment run will work exactly the same as the payment run for free-form and repetitive payments. The initial payment run screen is shown in Figure 5.67, and we can enter the **Run Date** as the current date. The **Identification** field is a free-form alphanumeric field that we use to enter the payment run ID.

365

5   General Contract Processes

Figure 5.67 Initial Entry Screen of Automatic Payment Transactions for Payment Requests App

We click the **Parameters** button to navigate to the selection screen, where we will need to populate the **Posting Date, Next Payment Run On, Company Code,** and **Payment Methods** fields (see Figure 5.68). If the incoming and outgoing payments are being netted together, we need to populate both of the necessary payment methods in a single payment run. The payments will have a unique **Origin**, which is "TR-TM". This will separate these payments from other payments in the treasury payment run since those payments will have an origin of TR-CM-BT for bank-to-bank transfers and FI-BL for free-form and counterparty repetitive payments.

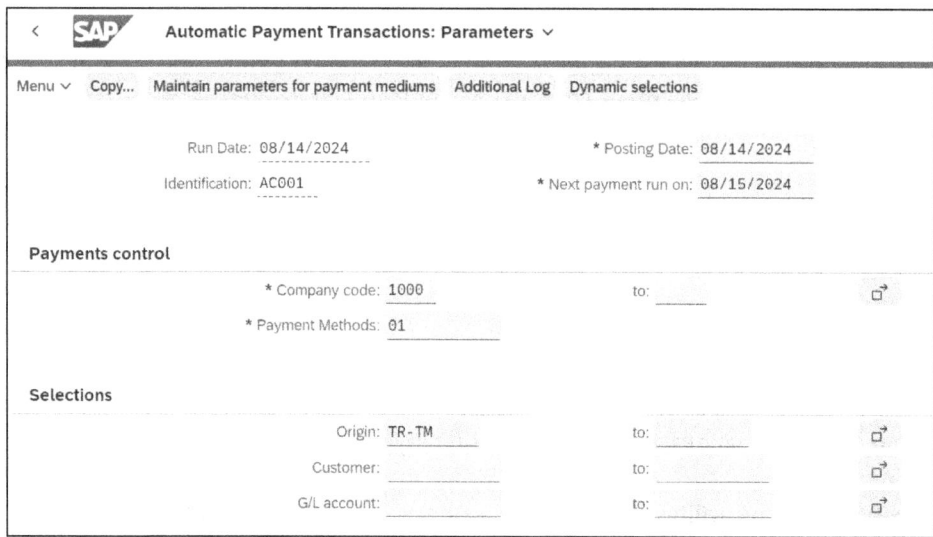

Figure 5.68 Entry Screen for Automatic Payment Transactions for Payment Requests

The **Dynamic selections** button is useful to us if we need to filter the payments further. Executing the payment run with the company code, origin, and payment method will pay off all payment requests that are due at that time. In the **Dynamic Selections** area, we can filter the payments to only pay off the desired payment requests. As displayed in Figure 5.69, any attribute of the payment is available for filtering, including the payment request **Key number**.

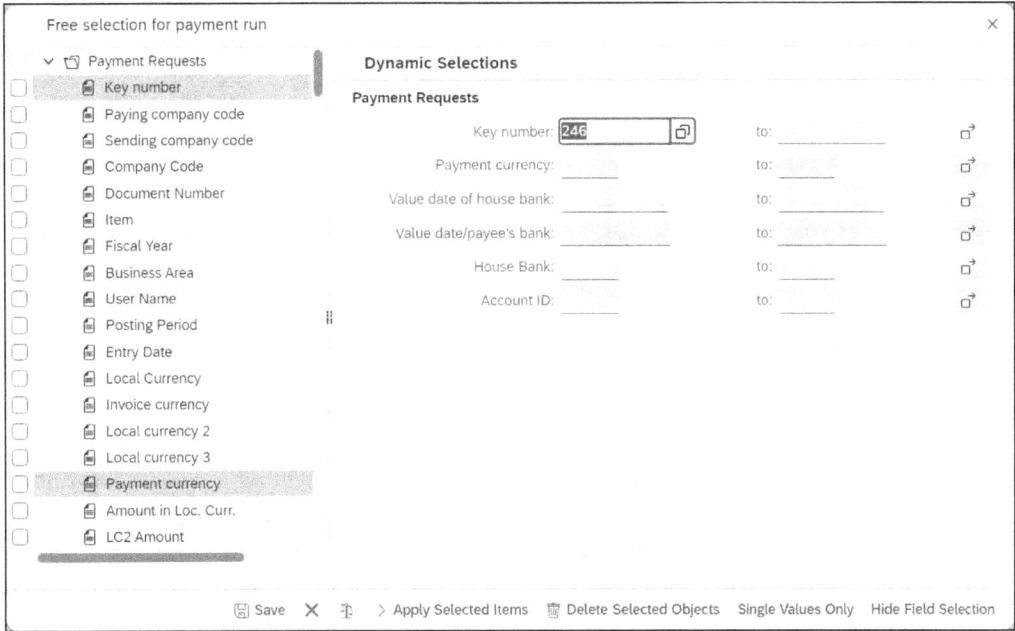

Figure 5.69 Options in Dynamic Selections

The **Additional Log** section is useful for generating outputs for both the payment proposal and the payment run so that we can see which payments are being selected, which accounts are making the payments, the accounting details behind the payments, and which account is receiving the funds. We can also use these logs to investigate whether a payment is not being picked up in the proposal for the payment run.

Once the payments are successfully made in this transaction, we can move on to generating a payment file or implementing the payment approval process in SAP Bank Communication Management. To see the full accounting postings for the interest payment, we can view the documents in the Display Journal Entries – In T-Account View app. In the example in Figure 5.70, the long-term account is credited, the transaction passes through the payment request clearing account, and the bank cash clearing account is debited from the payment run since this is an incoming payment.

5 General Contract Processes

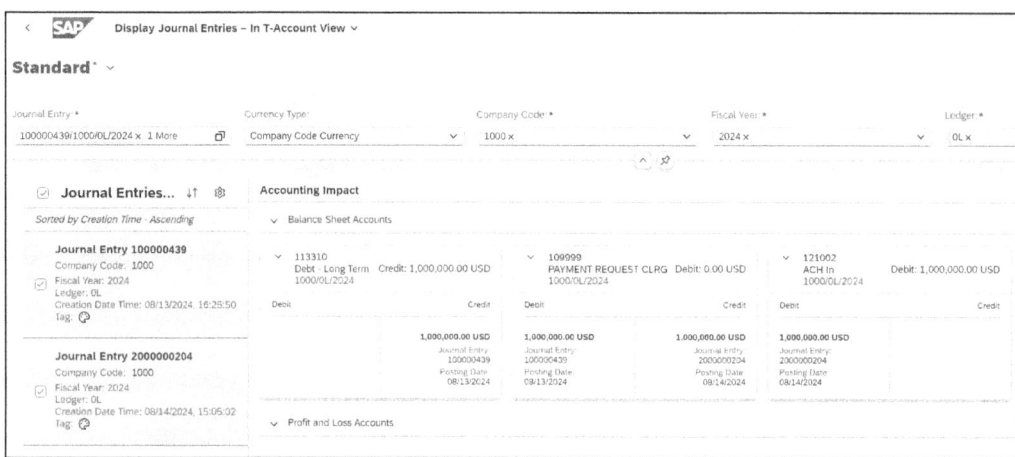

Figure 5.70 Journal Entry Impacts of Payment Request Creation and Payment Run

5.5 Summary

This chapter gives us an overview of the general treasury and risk management contract process. This process is pivotal in treasury and risk management, and it streamlines the management of financial transactions. By providing a central location to enter transactions, produce valuations, and create accounting for transactions, these functionalities help enhance organizations' decision-making and compliance efforts. Leveraging the topics we've already covered in the general contract process, the next chapter will dive deeper into these processes and highlight many of the common types of transactions that we enter and track by using treasury and risk management.

Chapter 6
Treasury Management–Specific Processes

Each financial contracts has its own nuances, and SAP S/4HANA has its own specific functions for various contract types. Depending on the transaction, the contract structure may vary, requiring different processes and apps to ensure accurate transaction flows and accounting. This chapter covers many of the common financial transactions typically encountered by treasury and risk management users.

Different financial instruments function differently, so each requires distinct processes to manage its lifecycle within SAP S/4HANA Finance for treasury and risk management. This chapter provides an in-depth exploration of the most commonly used contracts we see time and time again in most treasury and risk management implementations. These include contracts such as FX, money market instruments, debt management, derivatives, and intercompany loans. Each type of contract has its own unique requirements for setup, processing, and accounting, so we have to tailor workflows to each one to ensure accuracy and efficiency.

As part of its examination of the key stages of contract processing, this chapter provides an in-depth understanding of how we manage these transactions in treasury and risk management. It highlights the specific steps and processes necessary to manage contracts effectively. Whether we're creating a simple debt contract, an intercompany loan, or an FX contract, this chapter will help streamline our understanding of the contract entry process and the entry screens involved.

6.1 Debt and Investments

We use debt and investment management in SAP S/4HANA to manage investments and borrowings. Debt and investment management includes many varieties of contracts, including fixed-term deposits, deposits at notice, commercial paper, and interest rate instruments. Once we enter the instruments into treasury and risk management, we can process all subsequent flows, including principal flows, accruals, interest payments, and repayments. In the following section, we'll dive into the specific requirements when we enter each of the debt and investment contracts in SAP S/4HANA.

6 Treasury Management–Specific Processes

Figure 6.1 provides a list of the SAP-delivered product types for debt and investment product types—and as we learned in the previous chapter, we can create additional product types in configuration. In this section, we'll cover general debt and investments. We'll cover facilities in Section 6.2.

PTyp	Name of product type	Product Category
51A	Fixed-Term Deposit	Fixed-Term Deposit
52A	Deposit at Notice	Deposit at Notice
53A	Commercial paper	Commercial Paper
54A	Cash flow transaction	Cash Flow Transaction
55A	Fixed-Term Deposit	Interest Rate Instrument
55B	Int. Rate Instr Hdg. Acc.	Interest Rate Instrument
55C	Int. Rate Instr Hdg. Inst. HAC	Interest Rate Instrument
56A	Facility	Facility
56B	Syndicated Facility	Facility
57A	Fiduciary Deposit	Fiduciary Deposit
58A	Current Acct-Style Instrument	Current Acct-Style Instrument

Figure 6.1 List of Product Types Available for Debt and Investments

This first section will dive into debt and investments. In Section 6.1.1, we'll look at the general process flow. Many of the steps in this area will mirror what we already covered in Chapter 5, so we'll focus more on specific topics that are unique to debt and investments.

In Section 6.1.2, we'll cover the many different ways to calculate interest and repayment, which are some of the main features for these transactions. Section 6.1.3 details how compounded interest rates affect the interest calculation, and Section 6.1.4 shows how varying interest rates are fixed for the transactions. Finally, since interest rates can vary for debt and investments and planned flows can change over time, Section 6.1.5 shows how to manage planned records to show a more reliable forecast of future interest flows.

6.1.1 Process Flow

Before looking into each entry screen within debt and investments, we'll cover the high-level process flow shown in Figure 6.2. Each type of contract can have some unique nuances, and the process will vary depending on whether there are valuations, accruals, and any other flows.

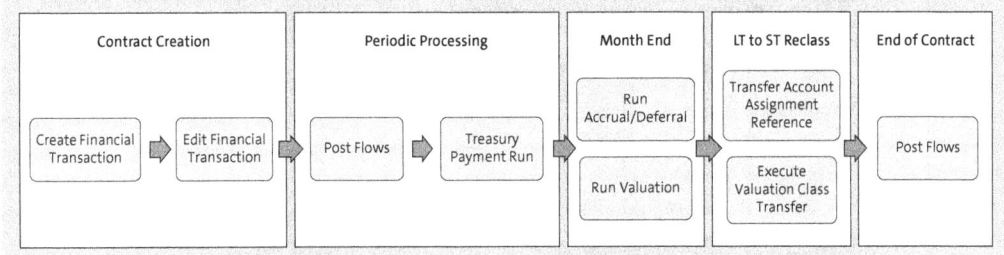

Figure 6.2 Debt Contract High-Level Process

There are similarities and differences between this process and the general contract process that we covered in Chapter 5. The key differences that we'll cover are as follows:

- **Create financial transaction**
  - **Interest**
    These transactions have both standard and variable interest calculations. We'll cover how to enter and review these calculations.
  - **Repayment**
    Debt and investment transactions have repayment schedules that should be reviewed. Some transactions may have a very simple repayment schedule in which the principal is repaid in full at the maturity date, but others have annuity and installment repayment schedules.
- **Interest rate adjustments**
  Transactions with variable interest rates leverage the market data in SAP to determine the correct calculation of interest. Transactions have to be run to fix the interest rate for the calculation.
- **Cash flow updates**
  By default, SAP S/4HANA can create forecasted cash flows in both the cash flows tab and in cash management. Certain changes to the transactions can update the planned records, and we'll cover those scenarios.

6.1.2 Borrowing or Invest

Depending on whether we use a debt or investment transaction type in Transaction FTR_CREATE or the Create Financial Transaction app, the first section will be **Borrowing** or **Invest**. In the top section of the screen, we'll determine the initial nominal amount of the transaction as well as the relevant dates. For debt and investment transactions, the top of the screen shows the starting and ending dates of the transaction. More specific to these types of transactions are the **Interest Structure** and **Repayment Structure** sections, shown in Figure 6.3.

6 Treasury Management–Specific Processes

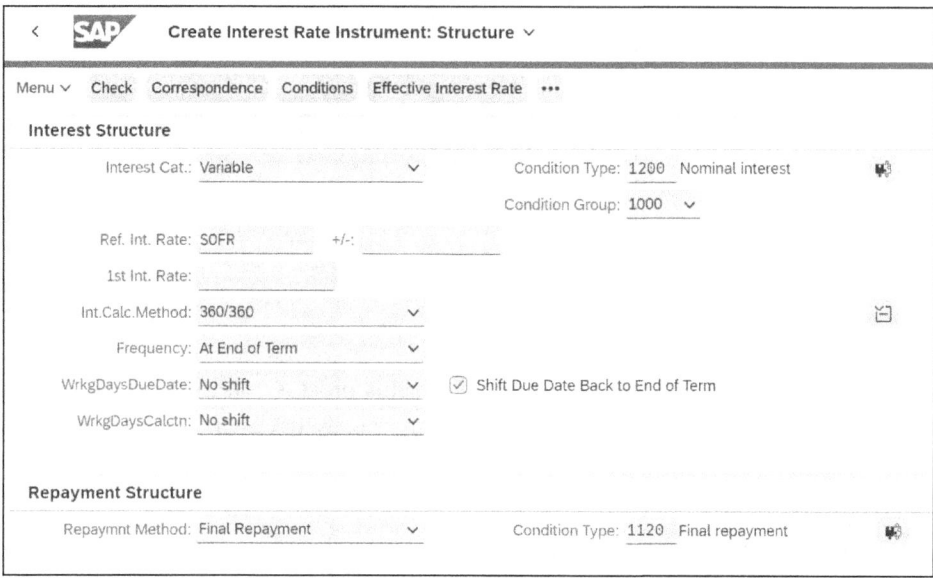

Figure 6.3 Interest Structure and Repayment Structure Sections in Create Financial Transaction

The following sections will cover the detailed instructions for how to create the interest structure and modify it to fit the requirements of any debt or investment contract. Additionally, this section will cover the various types of repayments that we can define for the transactions.

Interest Structure

The **Interest Structure** section shown in Figure 6.3 drives all details of how the interest is calculated, which rates are used, and on which dates the interest is paid. The options in the interest structure area are driven by the conditions that are assigned to the product types and transaction types. We can use the available conditions to calculate standard interest, variable interest, and capitalized interest.

The **Interest Cat.** in Figure 6.4 determines the base structure for how to calculate the interest.

Our options in this dropdown list are as follows:

- **Fixed Interest**
  We use this to assign a set interest rate percentage to the instrument. The interest will be calculated across the year, depending on the interest calculation method that we determine in this section.

- **Variable Interest**
  Here, we assign a reference interest rate along with an interest spread. The interest calculation will be based on the market rates available and will use a different interest rate for each calculation period.

- **Amount**
  We use this to assign a single amount that is applicable for each interest period.
- **Scaled (Interval)**
  We can define incremental nominal balances with their own corresponding interest rates. The interest is calculated in each scale.
- **Scaled (Incremental)**
  This category uses the same setup as **Scaled (Interval)**, but whichever scale the balance falls in, the full amount of interest is calculated with the assigned interest rate. This differs from the previous calculation since the full balance is calculated at that scaled level instead of a portion of the interest.

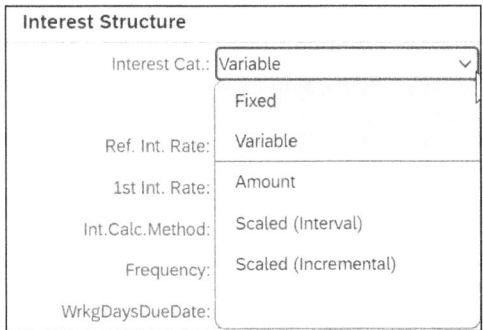

Figure 6.4 Interest Categories Available for Interest Structure

---

Variable Interest Rate

For the variable-interest option to appear in the dropdown list, we need to create a condition type using condition category **21-Interest Rate Adjustment** and assign it to the product type using the following menu paths:

- **Condition Creation: Treasury and Risk Management • Transaction Manager • Money Market • Transaction Management • Condition Types • Define Condition Types – MM Transactions**
- **Condition Assignment: Treasury and Risk Management • Transaction Manager • Money Market • Transaction Management • Condition Types • Assign Condition Types to Transaction Type – MM Transactions**

The interest **Condition Type** is an additional identifier we use to determine how to treat the interest flows. There are two main interest condition types used for interest on debt and investment transactions, as follows:

- **Nominal Interest: 1200**
  The nominal interest is calculated based on the book value of the debt or investment transaction. The interest is calculated based on the additional attributes determined in the interest structure and is paid in accordance with the defined date structure.

- **Interest Capitalization: 1150**
  We use the interest capitalization condition type to add the calculated interest back into the principal balance of the underlying contract. The capitalized interest condition operates in a manner similar to the concept of compounding interest, whereas calculated interest is added back into the principal balance and that new balance is used for ongoing interest calculations. Like nominal interest, this interest condition is based on the book value of the debt or investment transaction, is calculated based on the underlying contract attributes, and is paid in accordance with a defined date structure.

The following fields determine additional details of how and when the interest is calculated:

- **Percentage Rate**
  This determines the yearly interest percentage calculated for the contract. If we select a variable rate, fields will appear for us to enter the variable interest rate and applicable interest spread.

- **Interest Calculation Method**
  The calculation of the interest varies depending on the number of days used for the calculation. This method determines how the interest days are calculated. For variable-rate contracts, we assign a standard interest calculation method to each reference interest rate. We can assign a different calculation method at the contract level, but SAP S/4HANA will give us a warning to make sure we assign the correct method before we can save the instrument.

- **Frequency**
  The frequency determines when the interest is calculated and paid. We can set this to only calculate at the end of the contract or at regular intervals. Generally, we'll calculate the interest on a monthly or quarterly basis. To achieve this, we would set the frequency to be monthly. After we select this, another field will appear for us to determine how many months to include in an interest period. We can select "1" to calculate the interest monthly or "3" to calculate the interest quarterly.

- **Working Days Due Date**
  The due date is the date when the interest payment needs to be paid. If the payment date has been assigned to a nonworking day like a weekend or a holiday, we can determine a rule for how to handle the payment. The date can stay the same, or we can shift it to the previous working day or the following working day. The working day shifts are defined as follows:
  - **No Shift**
    The date is not changed, even if it falls on a weekend or holiday.
  - **Next Working Day**
    The date is shifted to the next working day, as on the assigned calendar.

- **Next Working Day Modified**
  The date is shifted to the next working day, unless that date would fall in the next calendar month. In that case, the previous working day is assigned.
- **Previous Working Day**
  The date is shifted to the previous working day, as on the assigned calendar.
- **Previous Working Day Modified**
  The date is shifted to the previous working day unless that date would fall in the previous calendar month. In that case, the next working day is assigned.
- **Next Working Day in Same Calendar Week**
  The date is shifted to the next working day, as long as it's in the same calendar week. If the next day is not in the same calendar week, then the date will shift to the previous working day.
- **Next Working Day in Same Calendar Year**
  The date is shifted to the next working day, as long as it's in the same calendar year. If the next day is not in the same calendar year, then the date will shift to the previous working day.
- **Previous Working Day in Same Calendar Year**
  The date is shifted to the previous working day, as long as it's in the same calendar year. If the previous day is not in the same calendar year, then the date will shift to the next working day.

- **Working Days Calculation**
  We use this calculation to determine when the interest is calculated. We can shift this as required using the same concept as the **Working Days Due Date**.

> **Calendar Assignment**
> When we assign any working-day shift to the interest period or due date, a calendar assignment will become available for entry on this screen. The working day shift will reference this calendar assignment to determine the working days. The calendar assignment includes a definition of weekend days and the rules to determine the banking holidays. We finish creating the factory calendar in Transaction SCAL.

Interest Structure Condition Screen

If filling out these fields is not sufficient for us to fully determine the interest structure of the instrument, then we can use the **Conditions** button to further define the interest structure. We can use many of the fields from the **Structure** tab to populate the **Interest Condition** page. The conditions entry page is shown in Figure 6.5, and there are additional criteria that we can add in the **Amounts** and **Dates** tabs of this conditions area. The key differences that we'll cover lie in how we can drive the dates of the interest payments, as follows:

6 Treasury Management–Specific Processes

- In the **Calculation date** section, we can further define the frequency of the interest calculation. We can define standard calculations to calculate the interest with a set period amount. We can also determine special rules to generate interest on variable dates, single dates, or manually input dates. In the example in Figure 6.5, we created the calculation date to be relative to another field. In this case, we're saying look at the due date to determine the calculation date.

- The **Due date** drives when the interest will be paid, and it can use the same working day shifts that we previously defined. In this example, we're defining an interest due date for each month. The first due date will be on 09/14/2024, and the subsequent due dates will fall on the 14th of each month.

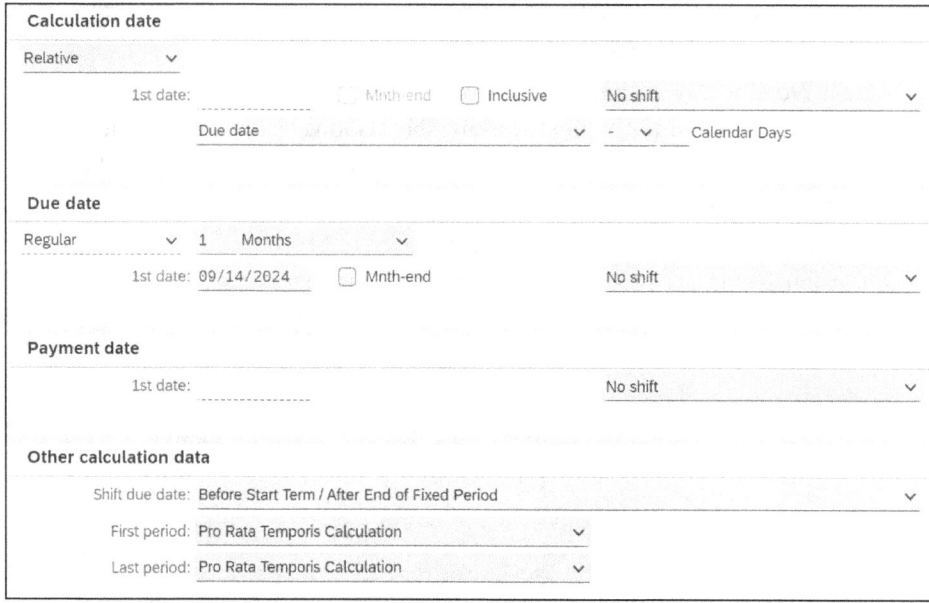

Figure 6.5 Dates Screen in Interest Condition

Interest Rate Adjustment Condition

When we assign a reference interest rate to a variable-rate contract, the interest rate adjustment condition becomes required for the instrument. This condition drives when the interest amount will be adjusted and fixed. We can edit this condition by clicking the **Conditions** button at the top of the screen. Once we're in the **Overview of Conditions** page, we can select the interest rate adjustment **Condition Type** by double-clicking the line item on this screen (see Figure 6.6).

The **Interest rate adjustment** condition has two date areas that we can define. The interest rate adjustment determines when a new interest rate affects the interest calculation, and we can set this at the beginning of an interest period. We can enter a regular frequency or individual dates. The **Interest rate fixing date** determines which day's interest rate is assigned to the interest rate adjustment, and it can follow a regular defined schedule or be set to be relative to the interest rate adjustment.

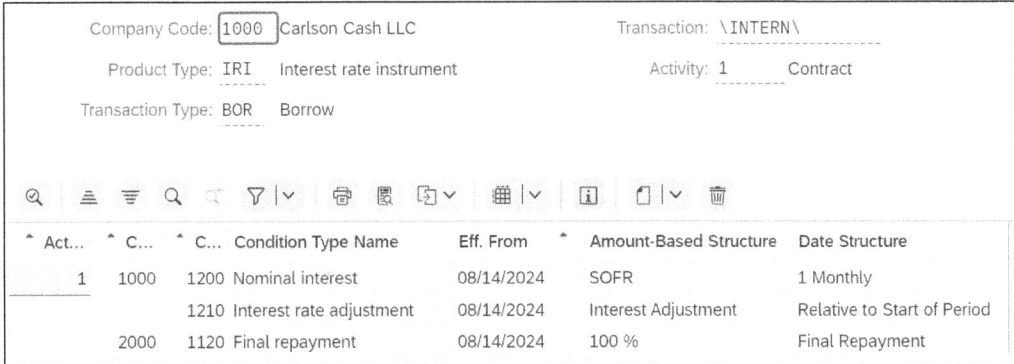

Figure 6.6 Transaction Conditions Page with Interest Rate Adjustment Condition

Let's look at a sample interest rate adjustment in Figure 6.7. The interest rate would be adjusted each month, on the last day of the month. The **Interest rate fixing date** is **Relative** to the **Interest Rate Adjustment Date** and would be fixed two working days before the adjustment date. In this case, we would use the start of the interest period to create the interest rate adjustment. Then, the interest rate fixing would occur two business days before the start of the interest period.

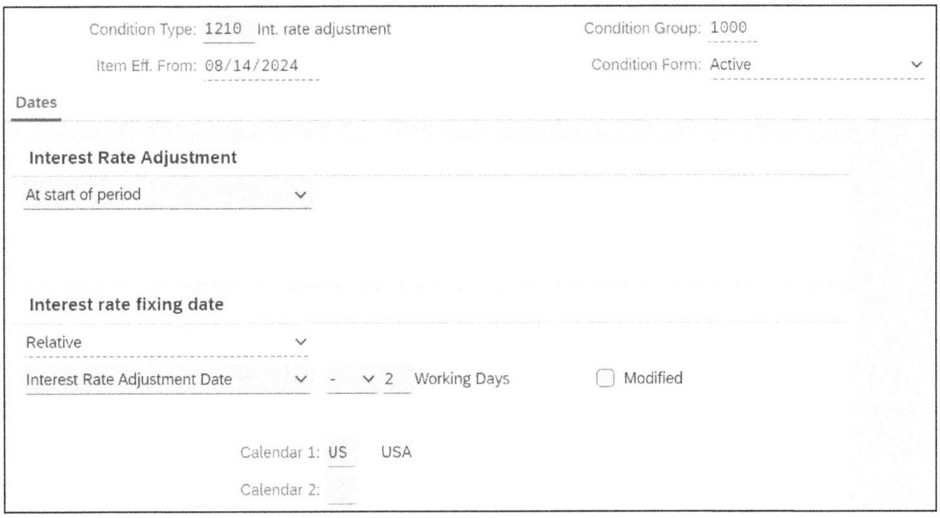

Figure 6.7 Interest Rate Adjustment Condition Entry

Repayment Structure

We use the **Repayment Structure** to determine how the transaction is repaid: all at once or using other methods. There are three different repayment conditions that we can

6   Treasury Management–Specific Processes

assign by either updating the repayment method or by selecting the applicable **Condition Type**:

- **Final repayment** uses a simple structure in which the full amount of capital is paid at term end.
- **Annuity repayments** use a fixed payment amount, but the actual principal and interest amounts paid vary throughout the contract. The interest amount decreases over the life of the deal while the principal amount increases.
- **Installment repayments** use a fixed amount of principal that is repaid at defined intervals.

In any of these repayment conditions (and particularly in annuity repayments and installment repayments), there may be working-day shifts or repayment amounts to define. The detail view button in the repayment condition's area will navigate to these settings as required. Figure 6.8 shows an example of the conditions screen for an installment repayment condition. This installment repayment has a monthly repayment amount of $65,000, and if we need to further define any dates or details of this condition, clicking the **Detailed View Repayment Condition** button will bring us to a screen where we can edit the details. In the example below, these are the key fields that we would fill in:

- **Repaymnt Method**
  In this field, we determine whether the repayment will be a final repayment, installment repayment, or an annuity repayment.
- **Condition Type**
  This is the four-character identifier that we use to determine how the transaction is repaid.
- **Amount**
  The amount of an installment repayment determines the fixed amount of principal that is repaid per period.
- **Frequency**
  The frequency is how often payments are made. In this example, the $65,000 repayment is made every month for the transaction.

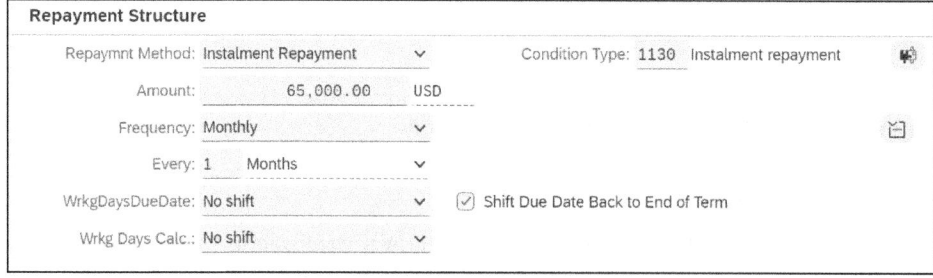

Figure 6.8 Condition Entry for Installment Repayment

6.1.3 Risk-Free Interest Rates

Starting in 2022, there was a soft transition to move many benchmark interest rates away from the London Interbank Offered Rate (LIBOR) and to risk-free rates. This transition was due to the risk that LIBOR could be manipulated, so banks transitioned to the Secured Overnight Financing Rate (SOFR) and similar rates. LIBOR was based on an average of interest rates reported by banks, but SOFR is based on actual transactions in the treasury repurchase market. SOFR and other related rates aren't susceptible to the same manipulation as LIBOR since they're based on real transactions instead of borrowing-rate estimates.

With contracts moving to use SOFR instead of LIBOR, it's important to understand SOFR and how it is used to calculate interest obligations. There are two types of SOFR rates, as follows:

- **Term SOFR**
  This is a proactive rate that is used to calculate the interest in a way similar to that of the old LIBOR contracts. The interest rate is determined and fixed at the beginning of the interest period, and it applies for the full interest period.

- **Overnight SOFR**
  This is a retroactive rate that is adjusted at the end of each day. The two calculations for overnight SOFR are average linear interest and average compounded interest. Both of these calculations look at the daily interest rate to calculate the average rate. The calculation period to determine the effective interest rate will be used for the calculation each day.

Term SOFR contracts are relatively easy to enter into SAP S/4HANA, and there are no differences from the contract entry that we described in Chapter 3, Section 3.9. Overnight SOFR contracts are more complex, and there are more entry considerations to keep in mind when entering them into treasury and risk management. Let's look at these additional complexities.

First, before entering in the contract, we have to activate this functionality in the customizing area of the product type. In the product type, we need to set the **Cash Flow Calculation** to **1 Parallel Conditions**, as in Figure 6.9. This is located in the **Financial Supply Chain Management • Treasury and Risk Management • Transaction Manager • Money Market • Transaction Management • Define Product Types – MM Transactions** menu path. This step is not required for securities contracts because we can set up the parallel interest conditions in the **Securities Class ID**.

When we're entering the contract, the **Structure** tab will look the same as in any other variable interest contract. The only difference is that the overnight SOFR rate will be assigned to the **Reference Interest Rate**. Further differences will become apparent in the conditions area of the contract for both the nominal interest and the interest rate adjustment.

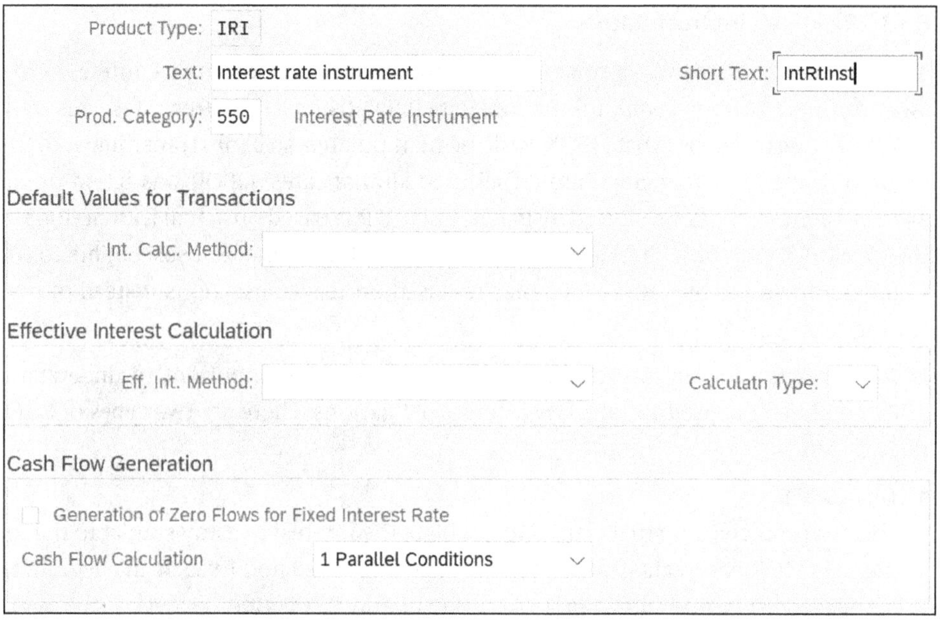

6 Treasury Management–Specific Processes

Product Type:	IRI			
Text:	Interest rate instrument		Short Text:	IntRtInst
Prod. Category:	550	Interest Rate Instrument		

Default Values for Transactions

Int. Calc. Method:

Effective Interest Calculation

Eff. Int. Method: Calculatn Type:

Cash Flow Generation

☐ Generation of Zero Flows for Fixed Interest Rate

Cash Flow Calculation   1 Parallel Conditions

Figure 6.9 Required Entry in Product Type to Enable Average Interest Rate Calculations

When we're entering the nominal interest condition, we'll see changes to the **Amounts** tab for the average interest calculations. Entering data into the key fields in Figure 6.10 is required, depending on the details of the interest calculation:

❶ **Interest Calculation Type**
The **Compound Interest Calculation** or **Average Compound Interest Calculation** interest calculation type will be required for the SOFR calculation.

❷ **Weighting**
We can calculate the average to weight the interest based on the interest rate adjustment date or the interest-fixing date.

❸ **Rounding**
This drives how many decimal points are relevant to the average interest rate.

❹ **Spread**
This applies when an additional interest amount is applicable on top of the SOFR rate.

❺ **Upper Limit and Lower Limit**
This is for setting a cap or floor on the calculated interest rates.

The **Interest Rate Adjustment** condition screen looks very similar to screens for other contracts that use reference interest rates, but the most significant change is how the interest rate is adjusted daily to calculate the average rate. Along with the adjustment, the interest rate is also fixed daily. In Figure 6.11, we cover the main entry considerations for entering the SOFR contracts, as follows:

6.1 Debt and Investments

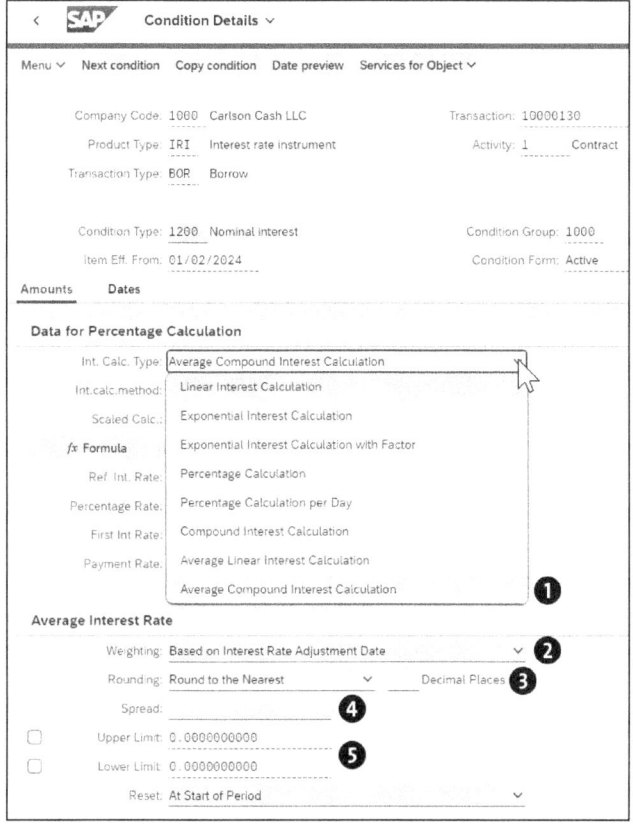

Figure 6.10 Entering Average Interest Calculation Methods

Figure 6.11 Interest Rate Adjustment Condition Screen

6 Treasury Management–Specific Processes

❶ **Frequency/Unit for Frequency**
The interest rate should be adjusted daily, and we can use the following fields to alter working day shifts.

❷ **Rule for Date Update**
The interest fixing date should also be fixed daily. In Figure 6.11, the fixing date is set to be relative to the interest rate adjustment date, but there's a day shift of -1 day. Every interest rate adjustment will use a fixed rate from the previous working day.

❸ **Lockout Period**
The lockout period is relevant to the last days in an interest period. There are no new interest rate–fixing dates in the lockout period. This is demonstrated in Figure 6.12, in which there's a three-working-day lockout period. Since 01/27/24 and 01/28/24 fall on a weekend, the last interest rate adjustment is on 01/26/24, with an interest rate–fixing date of 01/25/24. Due to this, the 5.32% interest rate is applied to all interest rate fixings that fall within the lockout period, and the rates after 01/25/24 are not considered for the average interest rate calculation.

No Lockout Period			3 Working Day Lockout Period			
Date	Int Fixing Date	SOFR Rate	Date	Int Fixing Date	SOFR Rate	Calc Rate
1/23/2024	1/22/2024	5.31	1/23/2024	1/22/2024	5.31	5.31
1/24/2024	1/23/2024	5.31	1/24/2024	1/23/2024	5.31	5.31
1/25/2024	1/24/2024	5.31	1/25/2024	1/24/2024	5.31	5.31
1/26/2024	1/25/2024	5.32	1/26/2024	1/25/2024	5.32	5.32
1/27/2024	1/26/2024	5.32	1/27/2024	1/25/2024	5.32	5.32
1/28/2024	1/26/2024	5.32	1/28/2024	1/25/2024	5.32	5.32
1/29/2024	1/26/2024	5.32	1/29/2024	1/25/2024	5.32	5.32
1/30/2024	1/29/2024	5.31	1/30/2024	1/25/2024	5.31	5.32
1/31/2024	1/30/2024	5.31	1/31/2024	1/25/2024	5.31	5.32

Figure 6.12 Example of SOFR Rate Determination with and without Lockout Period

Once we have successfully entered the contract, we can validate the interest calculations on the **Cash Flows** tab. Since there are new columns we can review in this tab, let's take a detailed look at the fields in Figure 6.13:

❶ **Payment Amount**
The interest amount calculated for each day is in this column. Note that this is calculated each day but that this contract pays interest monthly, so the first payment date isn't until 01/31/2024.

❷ **Number of Days/Days**
This displays the number of days that are considered in each line of the interest calculation.

❸ **Weighting of IR/WC**
This is the weighting of the interest rate in the interest calculation, and it is affected by the observation shift of the weighting as defined in the nominal interest condition.

❹ Cumulative Weighting of IR/CWC

This adds the number of days cumulatively that have been cumulatively weighted in the interest calculation for the current interest period.

❺ Days for Int Calc with AIR/Days

This is the total number of days calculated for the current interest period.

❻ Interest Rate Fixing Date

This determines the interest rate fixing date. The example contract has an interest rate fixing date of -1 working days, and interest is calculated for weekend days with the most recent interest rate available. An example of this is how there are three days in Figure 6.13 that have interest rate fixing dates of 01/5/2024, 01/12/2024, and 01/19/2024.

❼ Percentage Rate

This displays the interest rate that is fixed for that day. This rate is referenced in the average interest rate calculation.

❽ Average Interest Rate

This is the calculated average interest rate that looks at all variables in how the nominal interest condition was set up. We apply this calculated average interest rate to calculate the daily nominal interest amount.

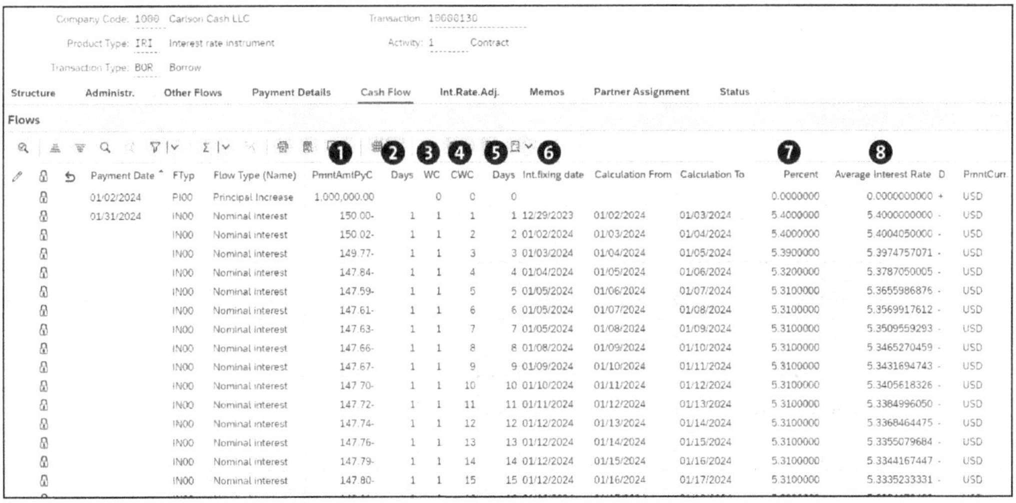
Company Code: 1000 Carlson Cash LLC   Transaction: 10000130
Product Type: IRI Interest rate instrument   Activity: 1 Contract
Transaction Type: BOR Borrow
Structure | Administr. | Other Flows | Payment Details | Cash Flow | Int.Rate.Adj. | Memos | Partner Assignment | Status
Flows

Payment Date	FTyp	Flow Type (Name)	PmntAmtPyC ❶	Days ❷	WC ❸	CWC ❹	Days ❺	Int.fixing date ❻	Calculation From	Calculation To	Percent ❼	Average Interest Rate ❽	D	PmntCurr.
01/02/2024	PI00	Principal Increase	1,000,000.00		0	0	0				0.0000000	0.0000000000	+	USD
01/31/2024	IN00	Nominal interest	150.00-	1	1	1	1	12/29/2023	01/02/2024	01/03/2024	5.4000000	5.4000000000	-	USD
	IN00	Nominal interest	150.02-	1	1	2	2	01/02/2024	01/03/2024	01/04/2024	5.4000000	5.4004050000	-	USD
	IN00	Nominal interest	149.77-	1	1	3	3	01/03/2024	01/04/2024	01/05/2024	5.3900000	5.3974757071	-	USD
	IN00	Nominal interest	147.84-	1	1	4	4	01/04/2024	01/05/2024	01/06/2024	5.3200000	5.3787050005	-	USD
	IN00	Nominal interest	147.59-	1	1	5	5	01/05/2024	01/06/2024	01/07/2024	5.3100000	5.3655986876	-	USD
	IN00	Nominal interest	147.61-	1	1	6	6	01/05/2024	01/07/2024	01/08/2024	5.3100000	5.3569917612	-	USD
	IN00	Nominal interest	147.63-	1	1	7	7	01/05/2024	01/08/2024	01/09/2024	5.3100000	5.3509559293	-	USD
	IN00	Nominal interest	147.66-	1	1	8	8	01/08/2024	01/09/2024	01/10/2024	5.3100000	5.3465270459	-	USD
	IN00	Nominal interest	147.67-	1	1	9	9	01/09/2024	01/10/2024	01/11/2024	5.3100000	5.3431694743	-	USD
	IN00	Nominal interest	147.70-	1	1	10	10	01/10/2024	01/11/2024	01/12/2024	5.3100000	5.3405618326	-	USD
	IN00	Nominal interest	147.72-	1	1	11	11	01/11/2024	01/12/2024	01/13/2024	5.3100000	5.3384996050	-	USD
	IN00	Nominal interest	147.74-	1	1	12	12	01/12/2024	01/13/2024	01/14/2024	5.3100000	5.3368464475	-	USD
	IN00	Nominal interest	147.76-	1	1	13	13	01/12/2024	01/14/2024	01/15/2024	5.3100000	5.3355079684	-	USD
	IN00	Nominal interest	147.79-	1	1	14	14	01/12/2024	01/15/2024	01/16/2024	5.3100000	5.3344167447	-	USD
	IN00	Nominal interest	147.80-	1	1	15	15	01/12/2024	01/16/2024	01/17/2024	5.3100000	5.3335233331	-	USD

Figure 6.13 Display of SOFR-Related Fields on Cash Flow Tab

6.1.4 Interest Rate Adjustments

When we assign a variable interest rate to a contract, the periodic interest payments are not known. The interest is calculated based on the assigned reference interest rate and relevant spread, and the interest payments of the contract will be displayed on the **Cash Flow** tab—but until the interest rate is fixed, the true payment amount is not known. These flows appear on the **Cash Flow** tab and in cash management as a planned record, based on the current available data in SAP S/4HANA. Before we can finalize an interest amount for a flow, we need to adjust or fix the interest rate for the cash flow. The fixed reference interest rate will depend on the reference interest rate, the spread, and the fixing dates assigned in the interest rate adjustment condition. Fixing the rates can be done automatically or manually.

Manual Interest Rate Adjustment

When market rates are not available, we can use a manual interest rate adjustment. We can use this if a market rate interface is not set up in our system or if there's a market rate that's not available from the market data interface. We can use the Create Adjustment – Rates/Prices app or Transaction TI10 for manual interest rate adjustment, as shown in Figure 6.14.

Figure 6.14 Create Adjustment – Rates/Prices App

There are additional apps and transactions that can support this process, as follows:

- **Change Adjustment – Rates/Prices app or Transaction TI11**
  We use this app or transaction to adjust rate and price adjustments as needed. We can only use these for rates that have been fixed and not posted.

- **Display Adjustment – Rates/Prices app or Transaction TI12**
  This is the display app or transaction in which we view the rate and price adjustments.

- **Reverse Adjustment – Rates/Prices app or Transaction TI37**
  We use this app or transaction to reverse rate and price adjustments.

First, when we go into Transaction TI10, we need to select the **Company Code** and **Transaction** ID that we wish to manually adjust, as in Figure 6.15.

By executing this transaction, we can enter the interest rate by double-clicking into the next interest flow. As shown in Figure 6.16, we enter the rate in the **Interest Rate** field, and then, we can save the transaction.

In our example, this process has successfully adjusted the interest rate on this contract. Now, we can follow the subsequent process to post the interest for this transaction using the Post Flows app or Transaction TBB1 on the due date. The execution of this transaction was covered in Chapter 5, Section 5.3.2.

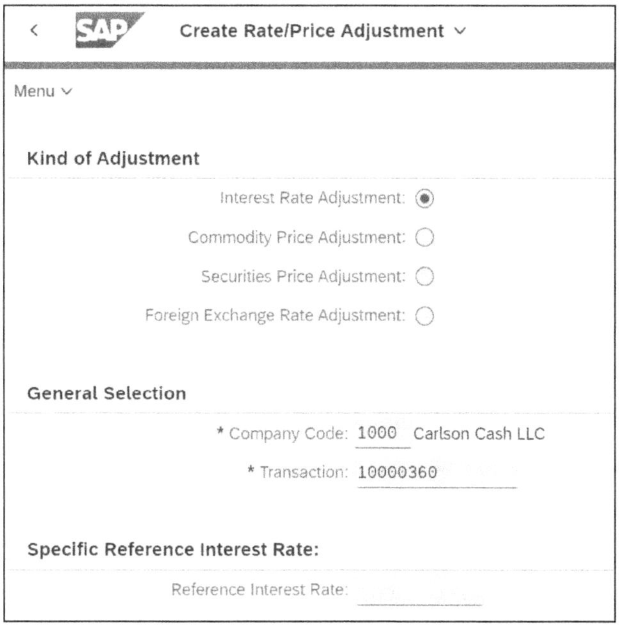

6.1 Debt and Investments

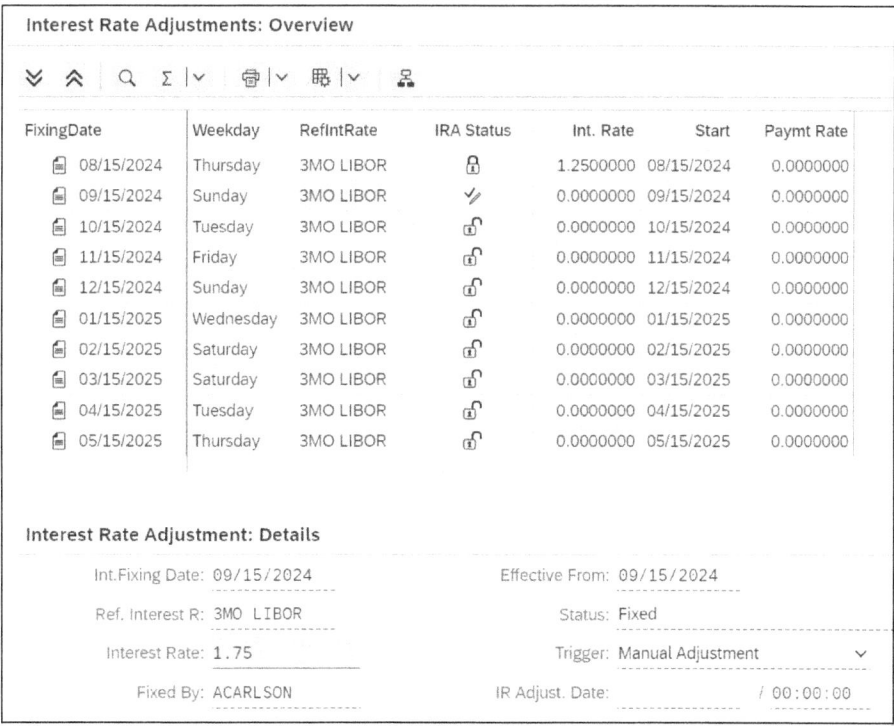

Figure 6.15 Selecting Transaction ID for Manual Interest Rate Adjustment

FixingDate	Weekday	RefIntRate	IRA Status	Int. Rate	Start	Paymt Rate
08/15/2024	Thursday	3MO LIBOR	🔒	1.2500000	08/15/2024	0.0000000
09/15/2024	Sunday	3MO LIBOR	✓✓	0.0000000	09/15/2024	0.0000000
10/15/2024	Tuesday	3MO LIBOR	🔓	0.0000000	10/15/2024	0.0000000
11/15/2024	Friday	3MO LIBOR	🔓	0.0000000	11/15/2024	0.0000000
12/15/2024	Sunday	3MO LIBOR	🔓	0.0000000	12/15/2024	0.0000000
01/15/2025	Wednesday	3MO LIBOR	🔓	0.0000000	01/15/2025	0.0000000
02/15/2025	Saturday	3MO LIBOR	🔓	0.0000000	02/15/2025	0.0000000
03/15/2025	Saturday	3MO LIBOR	🔓	0.0000000	03/15/2025	0.0000000
04/15/2025	Tuesday	3MO LIBOR	🔓	0.0000000	04/15/2025	0.0000000
05/15/2025	Thursday	3MO LIBOR	🔓	0.0000000	05/15/2025	0.0000000

Interest Rate Adjustment: Details

Int. Fixing Date: 09/15/2024   Effective From: 09/15/2024
Ref. Interest R: 3MO LIBOR   Status: Fixed
Interest Rate: 1.75   Trigger: Manual Adjustment
Fixed By: ACARLSON   IR Adjust. Date: / 00:00:00

Figure 6.16 Adding Manual Interest Rate for Adjustment

Automatic Interest Rate Adjustment

When an automated interest rate feed brings in the reference interest rates or rates are uploaded to the system, we use the automatic interest rate adjustment. The Run Automatic Adjustments – Rates/Prices app (see Figure 6.17) or Transaction TJ05 will fix the interest rates for all transactions. The fixing locks in the interest rate for that day in the transaction, and once we complete this, we can post the interest.

Figure 6.17 Run Automatic Adjustments – Rates/Prices App

An interest rate is required for the exact fixing date for the interest flow. Figure 6.18 shows this transaction being run for a specific date and a specific transaction. It's also possible to fix the interest rates for all contracts or for all contracts that use a similar **Reference Interest Rate**. To update the interest rate, we start by selecting the **Interest Rate Adjustment** indicator.

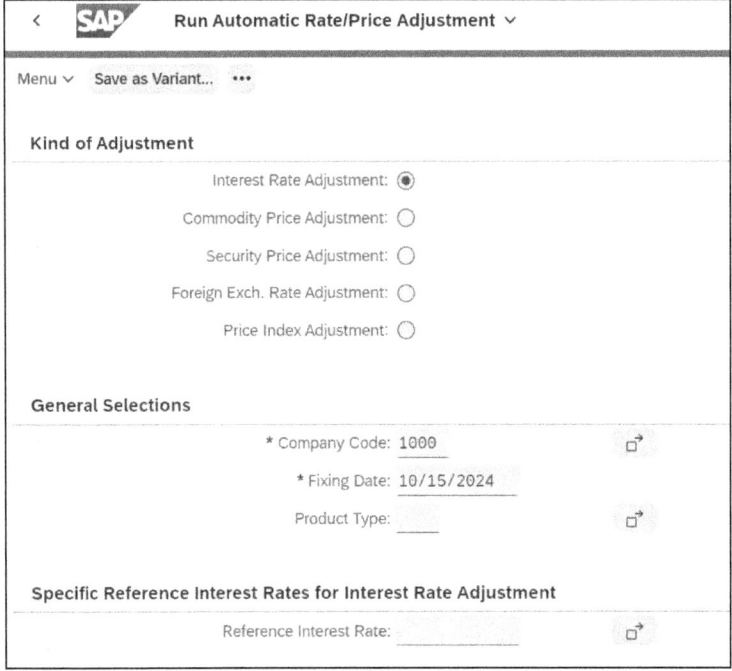

Figure 6.18 Run Automatic Adjustments – Rates/Prices Transaction Screen

Next, we update the company code and product type to select the specific types of contracts that need to be adjusted. We also need to determine the specific fixing date, and the rate needs to be available on the date indicated. We can also select the specific reference interest rates individually for the update, so only transactions using the selected rates will be fixed.

If we need to reverse any of the interest rate fixings, the Reverse Automatic Adjustments – Rates/Prices app is available. This will reverse all selected automatic interest rate fixings.

6.1.5 Planned Records

We can view the planned records for the interest payments in both the **Cash Flows** tab and the cash management tab. Depending on the customizing settings we've defined, the system will create the relevant planned interest rate payments. Once the interest rate is fixed for a transaction, it will no longer appear as a planned record and will be a fixed payment that is ready to post and pay with the necessary transactions. To direct treasury and risk management to look at the correct data for the planned records, we should configure the following customizing by company code in the **Treasury and Risk Management • Transaction Manager • General Settings • Organization • Define Company Code Additional Data** menu path. The setting is the **Planned Record Update** in the **Settings for Variable Interest Rates** section. The options are as follows:

- **Zero update**
  In this option, planned records for interest rate flows are projected with a value of 0, but we'll still see the records on the **Cash Flows** tab.
- **Update with automatically maintained interest rates**
  This option uses the most recent automatic interest rate adjustment for the planned records.
- **Update with manually maintained interest rates**
  This option uses the most recent manual interest rate adjustment for the planned records.
- **Update with current interest rates**
  This option uses the most recent interest rate adjustment to project the future interest payments. This rate can be either manually or automatically adjusted.
- **Update with automatically/manually maintained interest rates**
  This option uses the last automatic interest rate adjustment for the planned records, but if there are no automatic interest rate adjustments, then it uses the most recent manual interest rate adjustment.

An example of the interest rate projection is shown in Figure 6.19. A reference interest rate is assigned, and the most recent rate available is 1.5%, so all future flows are projected using that rate.

6 Treasury Management–Specific Processes

		Payment Date	Flow Type	Flow Type (Name)	PmntAmtPyC	Percentage Rate	D	PmntCurr.
		08/15/2024	PI00	Principal Increase	1,500,000.00	0.0000000	+	USD
		09/15/2024	IN00	Nominal interest	1,875.00-	1.5000000	-	USD
		10/15/2024	IN00	Nominal interest	1,875.00-	1.5000000	-	USD
		11/15/2024	IN00	Nominal interest	1,875.00-	1.5000000	-	USD
		12/15/2024	IN00	Nominal interest	1,875.00-	1.5000000	-	USD
		01/15/2025	IN00	Nominal interest	1,875.00-	1.5000000	-	USD
		02/15/2025	IN00	Nominal interest	1,875.00-	1.5000000	-	USD
		03/15/2025	IN00	Nominal interest	1,875.00-	1.5000000	-	USD

Figure 6.19 Cash Flow Screen Showing Projected Interest Flows Using Rate of 1.5% Starting 09/27/2024

When there are changes in the market data, the planned records don't automatically refresh. Since market data is constantly changing, we should regularly refresh the planned records to ensure the projected cash flows are accurately forecasted. We can refresh the records by individual transactions or for multiple transactions. Individual transactions are refreshed if we open a transaction in edit mode and save it. Multiple transactions are refreshed if we use the Update Planned Records app or Transaction TJ09, as shown in Figure 6.20.

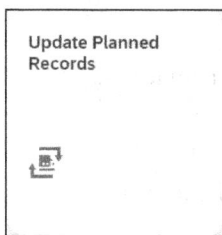

Figure 6.20 Update Planned Records App

We can update all records in one transaction run, or we can filter the update by company code, product type, and transaction number. The entry screen is shown in Figure 6.21. To update all records, we select the **Interest Rate Adjustment** option, and to update all relevant records, we select the **Update All Planned Records** indicator. There are also options within this transaction to only select specific product types, reference interest rates, and transaction numbers. Once we've updated the selections on this selection screen, we can execute the transaction.

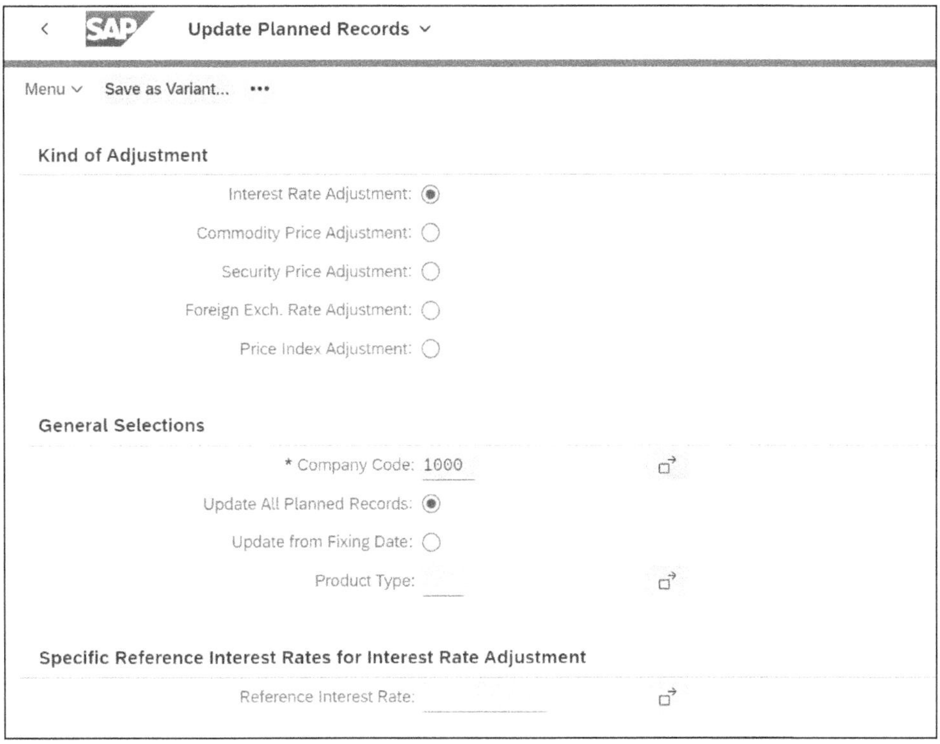

6.1 Debt and Investments

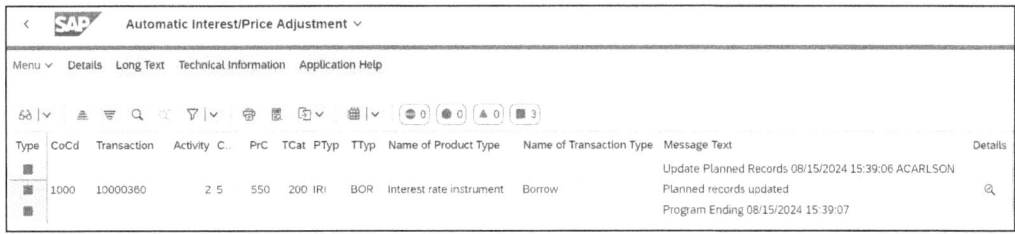

Figure 6.21 Entry Screen of Update Planned Records

When we've successfully run the transaction, the output will display the details of which transactions have been updated. In Figure 6.22, we can see that the market data has been updated for the interest rate instrument, so the planned records have been updated in Figure 6.23. This transaction previously was using a percentage rate of 1.5%, and this update run changed the rate to 1.25%.

Figure 6.22 Successful Run of Update Planned Records

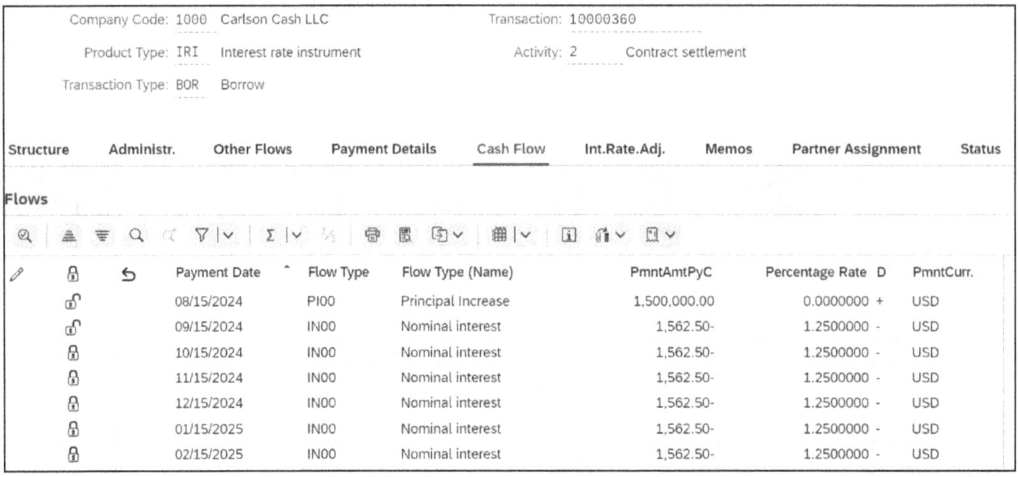

Company Code:	1000	Carlson Cash LLC		Transaction:	10000360		
Product Type:	IRI	Interest rate instrument		Activity:	2	Contract settlement	
Transaction Type:	BOR	Borrow					

Structure	Administr.	Other Flows	Payment Details	Cash Flow	Int.Rate.Adj.	Memos	Partner Assignment	Status

Flows

		Payment Date	Flow Type	Flow Type (Name)	PmntAmtPyC	Percentage Rate D	PmntCurr.
		08/15/2024	PI00	Principal Increase	1,500,000.00	0.0000000 +	USD
		09/15/2024	IN00	Nominal interest	1,562.50-	1.2500000 -	USD
		10/15/2024	IN00	Nominal interest	1,562.50-	1.2500000 -	USD
		11/15/2024	IN00	Nominal interest	1,562.50-	1.2500000 -	USD
		12/15/2024	IN00	Nominal interest	1,562.50-	1.2500000 -	USD
		01/15/2025	IN00	Nominal interest	1,562.50-	1.2500000 -	USD
		02/15/2025	IN00	Nominal interest	1,562.50-	1.2500000 -	USD

Figure 6.23 View of Financial Transaction After Planned Records Have Been Updated

6.2 Facilities

Facilities also fall within the money market portion of treasury and risk management, so they are in the same area as general debt and investments. However, there are some differences in how they operate in treasury and risk management. This is because the general structure of a facility is very different from that of a standard debt or investment contract. A *facility* is an agreement between a borrower and a lender that gives the option of drawing funds from a counterparty as a loan. This agreement has a specific structure that details how much can be borrowed, and it includes a fee structure that can depend on the total credit line and drawing amounts. These differences necessitate changes to the entry screen and the processing of facilities.

The following sections will cover the process for creating facilities, creating drawings from facilities, and the processing steps involved in the lifecycle of the transaction. In Section 6.2.1, we'll detail the high-level process flow to serve as a baseline for the facility process in SAP. In Section 6.2.2, we'll look into the entry of a facility. In Section 6.2.3 and Section 6.2.4, we'll look into the differences between bilateral facilities and syndicated facilities. Finally, in Section 6.2.5, we'll look into drawings from facilities and how the utilized balances and interest calculations are affected by these drawings.

6.2.1 Process Flow

When we're entering a facility, the first difference from entering general debt and investments is that we need to establish the structure of the credit line before we can assign drawings. To fully set this up, we establish the total drawing amount of the credit line and then create the fee structure. We can establish different fee types for the full amount of the credit line, the utilized portion, the unutilized portion, and any overdrawn amount.

Two different types of facilities are available for us to select from: bilateral facilities and syndicated facilities. In the bilateral facility product category, there's a single lender and a single borrower, and in the syndicated facility product category, there are multiple lenders and the portion drawn from each lender will be tracked. Before diving into each step of how to create and track a facility, let's look at Figure 6.24, which gives us a high-level view of the end-to-end process flow for facilities in treasury and risk management.

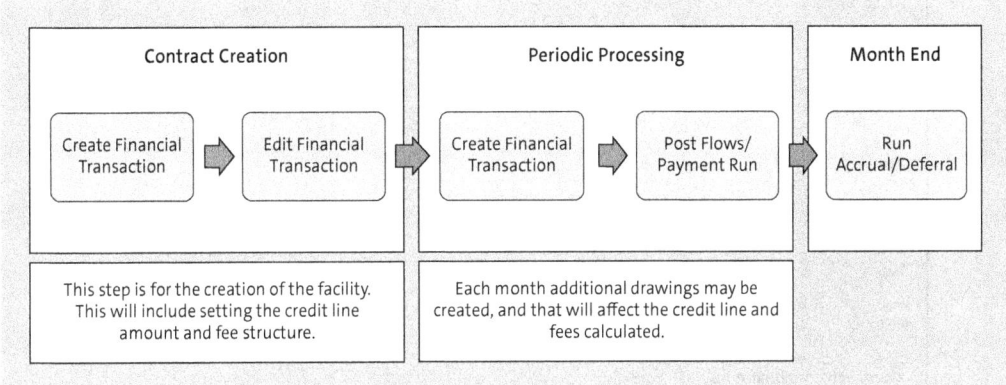

Figure 6.24 High-level Process Flow for Facilities in Treasury and Risk Management

The process for facilities has similarities to the process for other contracts. We still need to generate interest calculations and payment flows. However, there are also the following key differences:

- **Create financial transaction**
    - The entry screen for creating the structure of the facility is different.
    - Additional tabs in the facility entry screen let us add the fee structure, profile, and facility rules.
- **Drawings on the facility**
  When there are drawings on the facility, we need to tie them to the facility to update the tracking on the credit profile and the fee calculation.

6.2.2 Facility Entry Screen

We can also enter different types of facilities in this area. The structure of facilities is quite different from that of other types of debt, so the **Structure** tab for facilities has different options. The facilities entry screen is shown in Figure 6.25. In the **General Details** section, we determine the **Start of Term** and **End of Term** of the facility. The new field in

6 Treasury Management–Specific Processes

this tab is **Exceedance of End of Term**. When drawings are assigned to the facility, the system will only allow transactions to be assigned if their end date occurs before the end of the term of the facility. If we tag this as **Allowed**, then the transactions can have an end date that exceeds the facility end of term.

Figure 6.25 Facilities Entry Screen: Structure Tab

Since facilities have a fee structure and tracking of the utilization, there are a few additional tabs and entry options that exist for these transactions. First, in the fee calculation area of the **Charges** tab, we create the fee structure to determine the fees for the utilized and unutilized portions of the facility. Then, there's the **Profiles,** tab where we can create and change both the total credit line for the facility. There are also options in this tab to review the drawings against the facility. There's also the **Rules** tab, where we can add structure to the facility to limit the types of transactions that can create drawings against the facility.

Charges Tab

We use charges on facilities to define how the fees are calculated. These fees can be paid on a regular basis, and there are a few attributes we can assign to the fees. The fees are structured to look at the drawings against the facility, and that can change the fee amount that is calculated. Following are the calculation scenarios that we can assign to the facility fees:

- **Not Utilized**
  This creates a fee on the unutilized portion of the facility.

- **Utilized**
  This creates a fee on the utilized portion of the facility.

- **Total Credit Line**
  This creates a fee on the total credit line of the facility. Drawings against the facility don't change the calculation of this fee.

- **Overdrawn**
  If drawings exceed the total credit line, this fee applies to the overdrawn amount.

There are two different ways in which we can enter facilities charges. In SAP S/4HANA 2022 and earlier, there's a **Charges** tab in which we can enter the facility fee structure (see Figure 6.26).

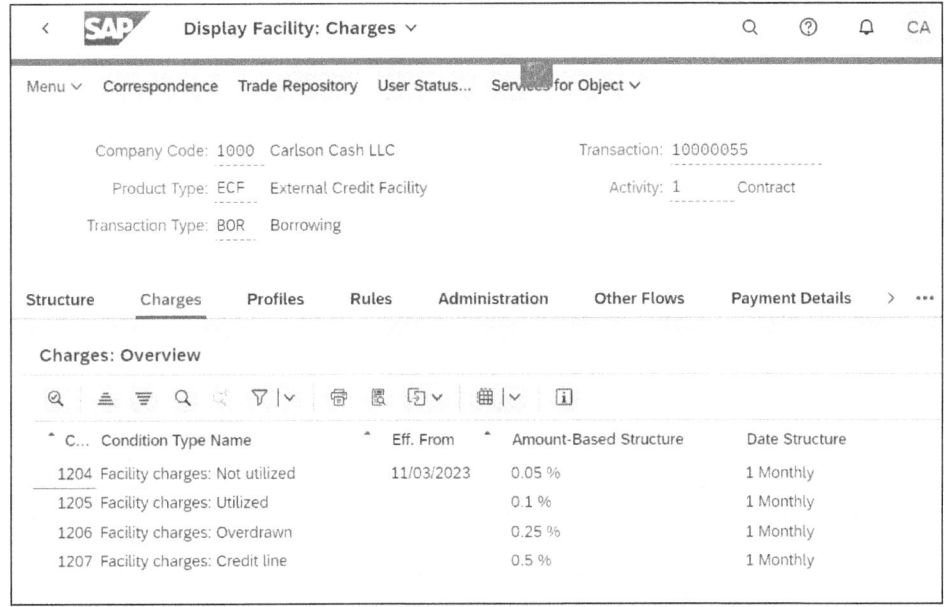

Figure 6.26 Example of Entry of Facility Charges in SAP S/4HANA 2022 and Earlier Versions

This shows that there's a different interest rate depending on the type of charge against the facility. Starting with SAP S/4HANA 2023, the facility fees have changed location, and there's no longer a **Charges** tab on the entry screen. We now enter the facility fee structure by clicking the **Conditions** button at the top of the screen in the Create Financial Transaction app.

Entering the charges in this area adds functionalities that drive the due date and calculation date (see Figure 6.27).

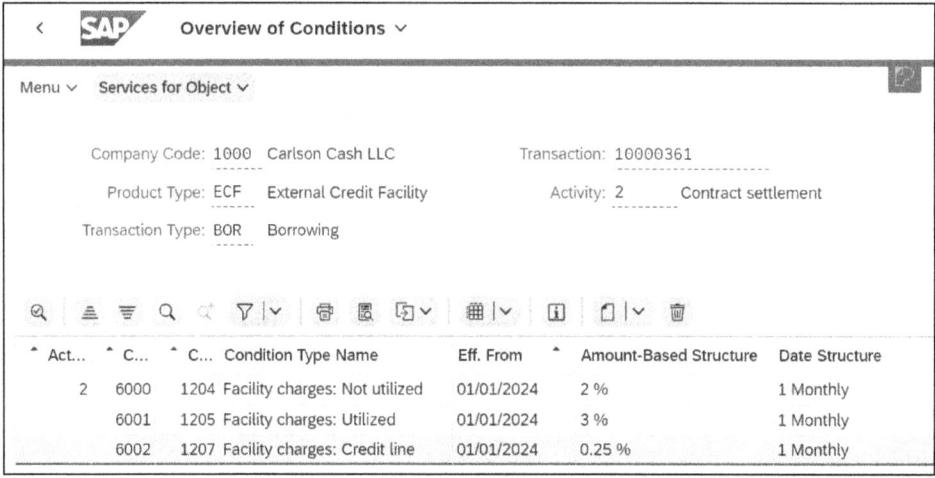

Figure 6.27 Example of Entry of Facility Charges Starting in SAP S/4HANA 2023

To update and change the conditions for charges, we double-click into one of these line items. Figure 6.28 shows what the **Facility Charges: Not utilized** conditions look like when we drill down into the line item. The interest calculation type, interest calculation method, reference interest rate, and percentage rate all drive the interest calculation.

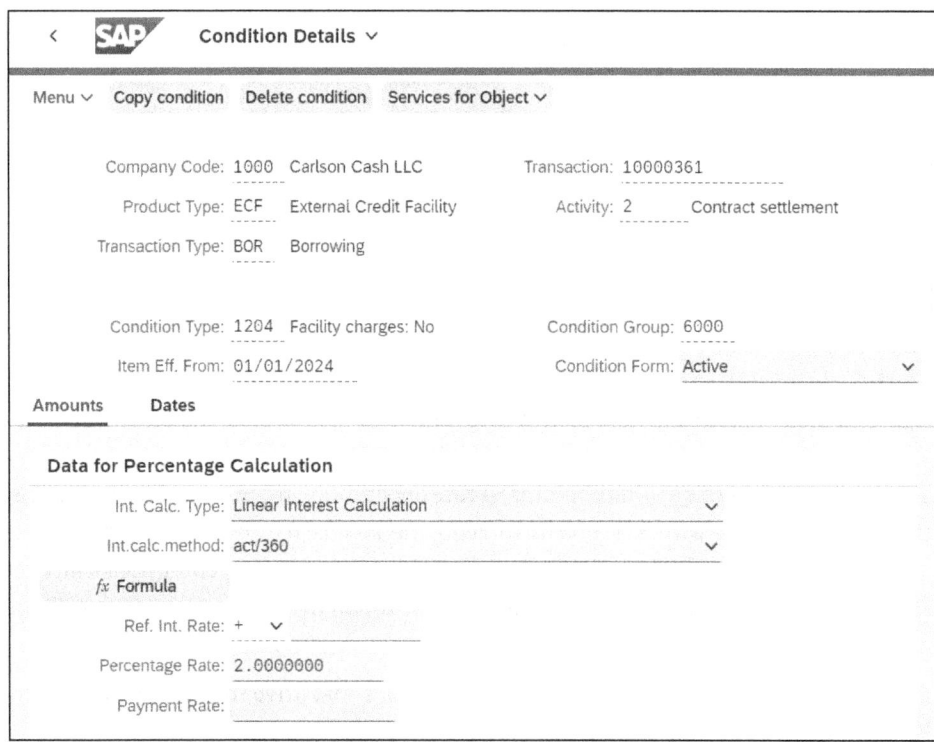

Figure 6.28 Updating Charges Condition for Credit Facility

6.2 Facilities

Next, we can go to the **Dates** tab to update when the interest is calculated for this fee condition. This screen is shown in Figure 6.29, and our example shows an interest calculation that is generated every month, starting on 02/01/2024.

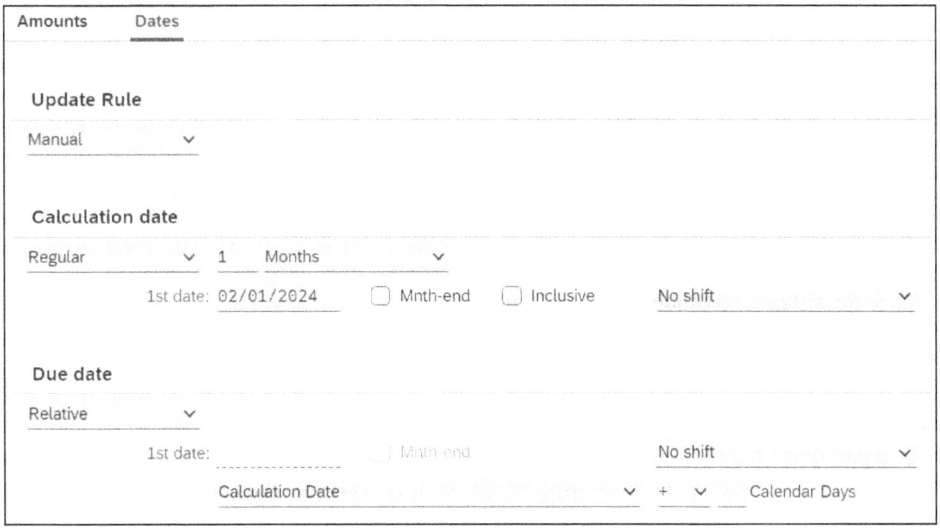

Amounts	Dates						
Update Rule							
Manual ⌄							
Calculation date							
Regular ⌄	1	Months ⌄					
	1st date: 02/01/2024	☐ Mnth-end	☐ Inclusive		No shift		⌄
Due date							
Relative ⌄							
	1st date:		☐ Mnth-end		No shift		⌄
		Calculation Date		⌄	+ ⌄	Calendar Days	

Figure 6.29 Dates Entry Screen for Interest Condition

Profiles Tab

First, we use the **Profiles** to create and change the total credit line. Then, we can use this tab to view the details of the drawings and how they affect the utilized, unutilized, and overdraft amounts of the facility. The following list and Figure 6.30 detail the full array of views we can select from the **Overview** dropdown list:

- **List: Drawing Objects**
  This details the changes in drawing amounts by contract number.
- **List: Drawings**
  This details each individual drawing change.
- **Profile: Total Credit Line**
  This view is where we can create and change the total credit line for the facility.
- **Profile: Amount Utilized**
  This view details the changes to the utilization of the credit lines and the dates of the changes.
- **Profile: Overdraft**
  This selection shows details of whether the credit line is overdrawn.
- **Profile: Credit Line Not Utilized**
  This view shows the opposite of the utilized view: the unutilized amount and the dates of the changes.

6 Treasury Management–Specific Processes

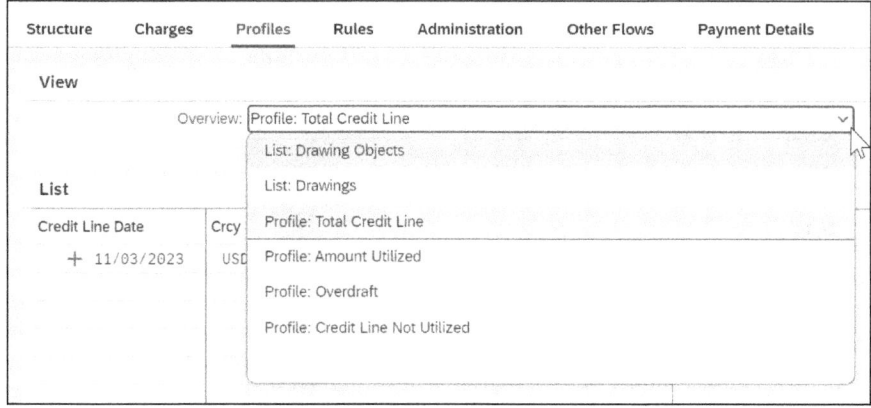

Figure 6.30 View of Available Facility Profiles

Rules Tab

The **Rules** tab determines specific restrictions of the facility. Here, we can limit the business partners, currencies, company codes, and types of transactions that can be assigned to the facility. For example, Figure 6.31 shows that this facility can only have drawing objects assigned that are in USD.

Figure 6.31 Rules Tab on Facility Creation Screen Showing It's Restricted to US Dollar Drawings

Other currencies will be restricted from assigning drawings to this facility. The following is a full list of variables that we can limit in the facility rules in the **Rule Category** dropdown list:

- **Company Code**
  Here, we can limit the entities that can create drawings.
- **Business Partner**
  Here, we can limits the business partners that can make a draw against the facility.
- **Transaction Type**
  Here, we can limit the product types and transaction types that are used as drawings against the facility. Fields are also available for us to set a minimum and maximum number of days for the contract terms for the drawings.
- **Transaction Currency**
  We've covered this already; here, we can limit the currency of the drawings against the facility.

6.2.3 Bilateral Facility Entry

In previous sections, we covered the unique tabs of the facility entry screen, but this section will run us through the complete process of creating the facility and adding drawings against the facility. The first thing we'll cover is entering a bilateral facility. The bilateral facility is an agreement between a lender and a borrower, and we'll assign it to one counterparty in SAP S/4HANA. First, we have to create the facility in the **Structure** tab, and we follow the same instructions as in Section 6.2.2 to define the start and end dates of the credit line (shown in Figure 6.32).

Figure 6.32 Initial Facility Entry Screen Determining Start and End of Term

6 Treasury Management–Specific Processes

Next, we'll create the details of the credit line of the facility. To do this, we'll navigate to the **Profiles** tab and select **Profile: Total Credit Line**. We must also establish the credit line for the facility to allow drawings against the facility. To create the credit line, we need to click the **Create** button and then enter the **Credit Line Amount** and the validity date in the **Credit Line Date** field (as shown in Figure 6.33). We can also edit the credit line amount, and to update the credit line at a future date, we'll click the **Create** button again and enter the new total credit line along with a validity date.

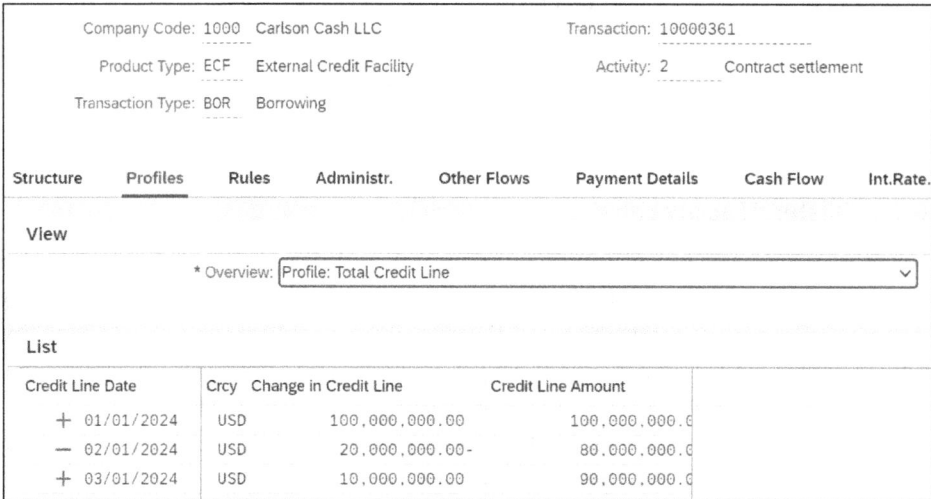

Figure 6.33 Creation of and Changes to Credit Line in Profiles Tab

Then, we can assign the facility charges for each of the types of facility fees. Each of these charges and how it is assigned is detailed in Section 6.2.2.

6.2.4 Syndicated Facility Entry

Since syndicated facilities can contain multiple lines of credit and have multiple counterparties, there are some differences between how we enter a syndicated facility and how we enter bilateral facility. The biggest difference on the **Structure** tab is the bottom section, which determines the syndicate partners for the syndicated facility. We can assign a different **Rank** to categorize the banks correctly. In Figure 6.34, we can see that JP Morgan is the assigned **Syndicate Manager** and the other two business partners are **Syndicate Banks**. This will allow us to add credit lines to each of the banks in the subsequent step.

Next, in the **Profiles** tab, we assign the total credit line in the same way as with the bilateral facility, except that we also use the **Syndication** button at the bottom of the screen. After we assign the total credit line, we click the **Syndication** button and assign a portion of the credit line to each of the syndication banks. We then populate the portion of the credit line that we assign to each partner in the **New Commitment** column. For example, in Figure 6.35, the total credit line of $100,000,000 has been distributed among the three counterparties.

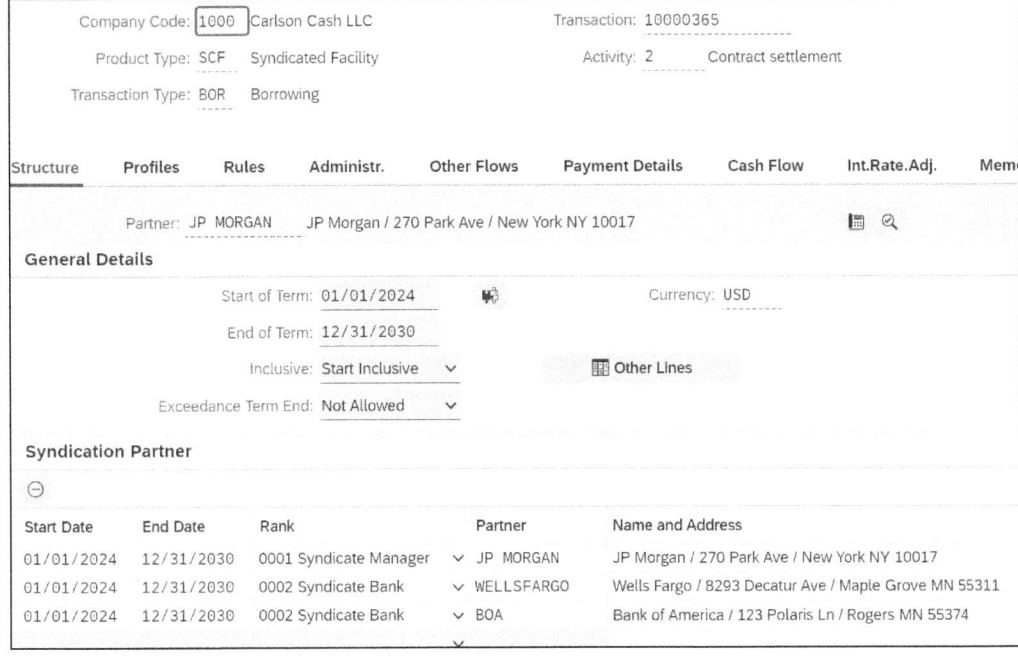

Figure 6.34 Syndicated Facility Entry Screen Showing Multiple Syndicate Banks

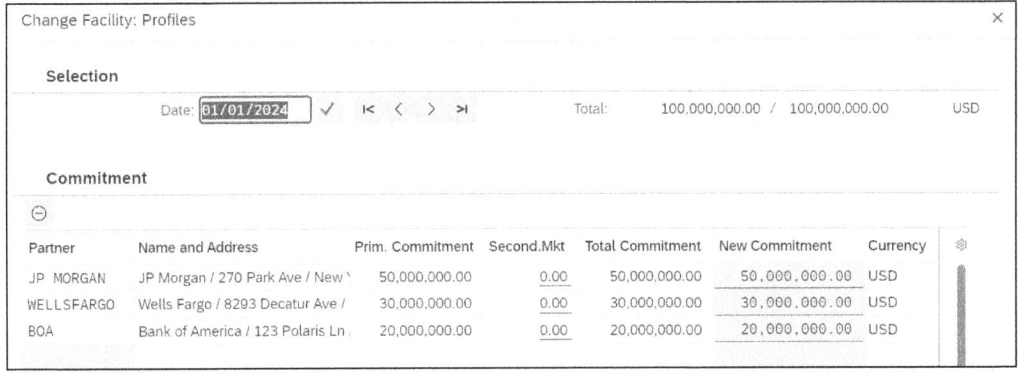

Figure 6.35 Syndicate Banks Are Assigned Different Portions of Total Credit Line

6.2.5 Facility Drawings

Now that we've gone through the creation of both types of facility, we can go through drawing against the facilities. We can assign different types of debt transactions to a facility, or we can create a generic credit facility drawing. We enter this using Create Financial Transaction app or Transaction FTR_CREATE. There's no difference in the structure entry screen for drawings against the facility. To create drawings against the facility, we can enter drawings as "other product type," or we can set up **CFD: Credit Facility Draw** as the **Product Type**. Once we have entered the base details on the transaction entry screen, we can move to the next step. The main difference in entering the draws comes in the **Administration** tab. To assign a drawing against a facility, we click the **Create** button in the **Position Assignment** section as in Figure 6.36.

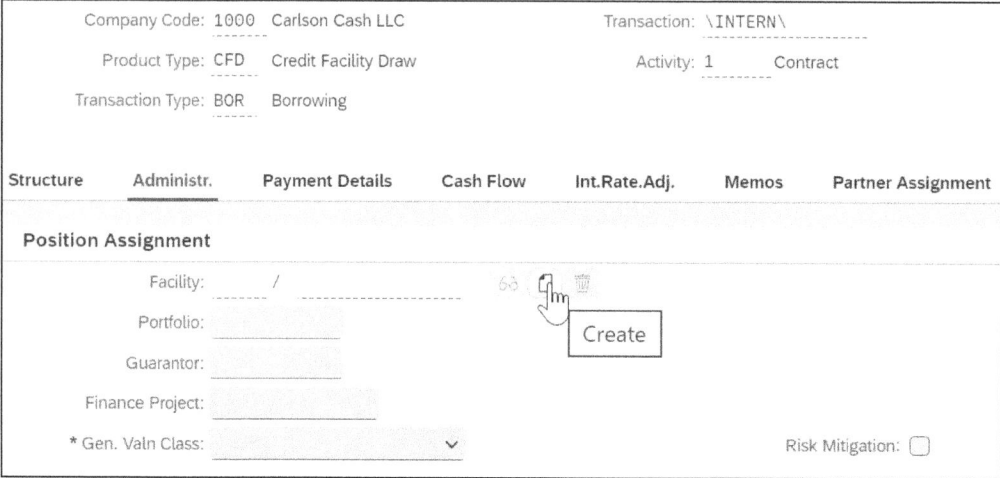

Figure 6.36 Display of Facility Assignment Section

When we click on this button, a popup screen will appear. We can select the facility on this screen by populating the facility's **Company Code** and **Transaction** ID. This screen varies slightly depending on whether we're assigning the drawing against a bilateral or a syndicated facility. We also need to assign the syndicate bank to the syndicated facility on this screen, and we select the bank by choosing the applicable **Partner**. The differences are shown in an example of a bilateral facility assignment in Figure 6.37 and an example of a syndicated facility assignment in Figure 6.38.

Figure 6.37 Assignment of Drawing against Bilateral Facility

```
Create Interest Rate Instrument: Admin.                                    ×

  Financial Transaction
       Company Code:  1000              Transaction:  \INTERN\
        Trans. Crcy:  USD

  Facility
       Company Code:  1000              Transaction:  10000103
            Partner:  JP Morgan / 270 Park Ave / New York NY 10017    ⌄
       Utilize From:  08/16/2024        Utilize To:   02/05/2030

                                                      Choose   Cancel
```

Figure 6.38 Assignment of Drawing against Syndicated Facility

Now that there are drawings against the credit facility, we can review the **Profiles** tab to see all details of the facility. In this tab, we can review the changes in the total credit line, changes to the utilization, and a list of all drawings. In the case of a syndicated facility, we can review each of these categories for each of the syndicate banks. Figure 6.39 shows an example in this **Profiles** tab of a facility with a drawing. We can find additional reports to review the facility and its drawings by using Transactions TM_60 and TM60A, which are detailed in Chapter 13, Section 13.3.6 and Section 13.3.7.

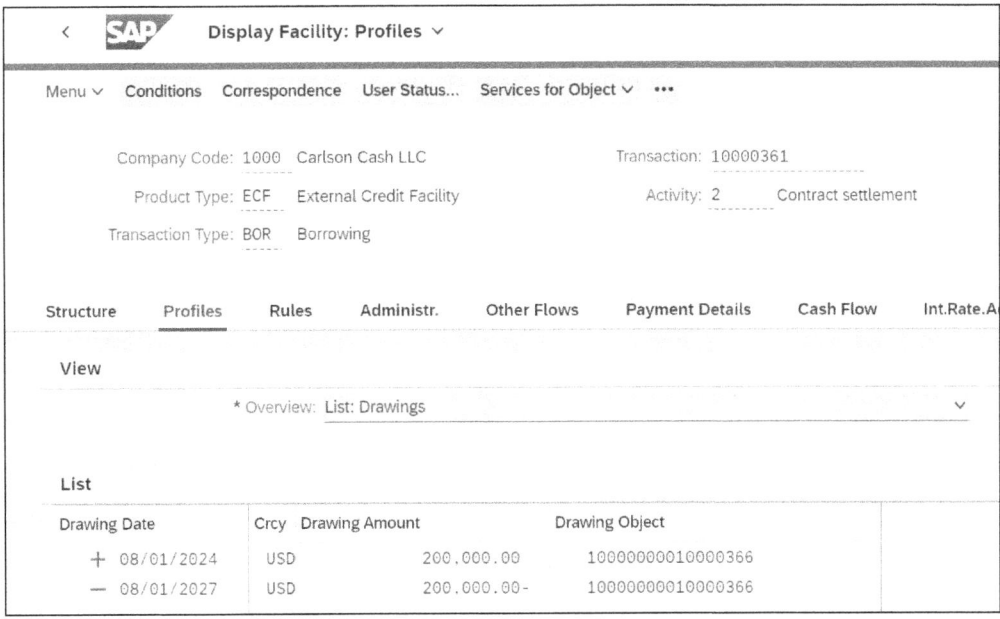

Figure 6.39 Drawings Against Facility Detailed in Profiles Tab

Based on the utilization and changes each month, the interest accruals and interest postings will be posted using the standard Run Accrual/Deferral app or Transaction TPM44 and the Post Flows app or Transaction TBB1, as required.

6.3 Intercompany Loans

Managing intercompany loans within treasury and risk management closely mirrors the handling of traditional debt and investment contracts. The procedures for entering contracts related to intercompany loans largely resemble those used for other debt or investment agreements. The steps involved in contract entry for intercompany loans are, in many respects, almost identical to the processes applied when dealing with conventional debt or investment instruments.

Given that we thoroughly covered the fundamentals of contract entry in Chapter 5, this section won't dive into those basic processes again. Instead, our focus will shift toward exploring the distinctive features and differences that separate intercompany loans from other forms of debt and investment contracts. We'll examine specific aspects such as loan mirroring, the handling of intercompany loan terms and conditions, regulatory considerations unique to intercompany transactions, and any specific contract entry requirements that may apply. Understanding these nuances is important for understanding how intercompany loans are managed and how they deviate from standard debt or investment contracts within treasury and risk management.

Section 6.3.1 will cover the overall process flow for intercompany loans, and Section 6.3.2 will cover how to create intercompany loans and define the specific details of such loans. Next, Section 6.3.3 will cover the concept of mirroring in intercompany loans. An intercompany loan assumes that there's a corresponding loan in the opposite company code, and this functionality will demonstrate how SAP S/4HANA can automatically create the opposite or mirror transaction. Then, Section 6.3.4 will cover the settlement function, and it will show how the loan is reviewed and released for further processing. Finally, Section 6.3.5 will cover what happens if any changes to the loans are required. Changes to the principal of the loan will also mirror and impact the linked intercompany loan in the opposite company code.

6.3.1 Process Flow

The primary difference between intercompany loans and other contracts managed in treasury and risk management lies in the internal nature of the contract. An intercompany loan is established between different company codes within the same organization, rather than between a company code and an external counterparty. We can facilitate this internal contract in SAP S/4HANA by creating the contract between a company code and an internal business partner representing another company code.

This internal contract setup fundamentally affects how intercompany loans are structured in SAP S/4HANA. Instead of a single unified contract to represent the intercompany loan, the system requires the creation of two distinct contracts. Each contract reflects one side of the transaction: one for the borrowing entity and the other for the lending entity. These two contracts are interconnected through a unique identifier, allowing for simultaneous management of both sides while maintaining clarity that they are separate contracts within the system.

It is important for us to understand and keep in mind this dual-contract approach when setting up intercompany loan agreements in SAP S/4HANA, as it drives the specific process for contract creation and downstream contract management. It ensures that both the borrowing and the lending aspects are accurately captured and maintained within treasury and risk management. In Figure 6.40, we can see how the overall process from contract creation to contract conclusion varies very little when compared to the process for a debt or investment contract. The key differentiating factor is that during the contract creation phase, the contracts are simultaneously created, settled, and maintained through the lifecycle of the contract.

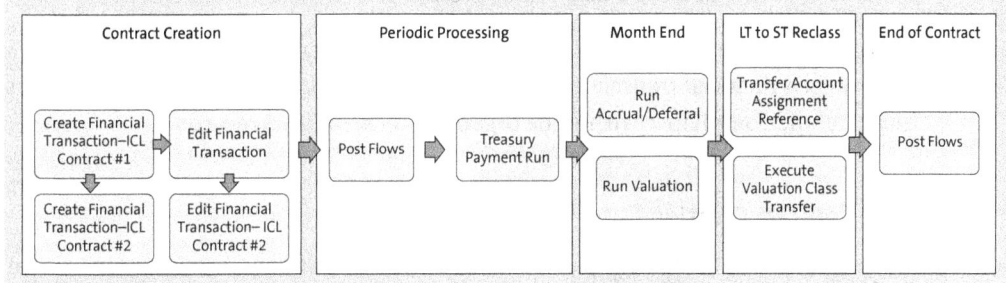

Figure 6.40 Process Flow for Intercompany Loans

This process ensures that every aspect of an intercompany loan is accurately and automatically reflected in the financial records of both the lending entity and the borrowing entity, thus maintaining consistency across both loans and reducing the risk of errors associated with maintaining both sides of the loan independently. Let's take a look at an example:

1. **Loan initiation**
   At the outset, let's consider a scenario in which company code 1000, acting as the lending entity, initiates a principal increase, allocating, for instance, $1,000,000 to the intercompany loan. Correspondingly, treasury and risk management automatically generates a mirrored transaction for company code 2000, the borrowing entity. This mirrors a principal increase entry, thereby reflecting a liability of $1,000,000 in company code 2000's financial records.

2. **Interest calculation**

   As the loan progresses, the system proceeds to calculate interest receivable for company code 1000 based on the predefined loan terms. Simultaneously, treasury and risk management mirrors this computation for company code 2000, calculating the interest payable using the same parameters. This synchronization ensures that both entities accurately reflect the accrued interest in their respective financial statements.

3. **Repayment**

   Subsequent to interest calculations, repayments are made toward the intercompany loan. Company code 1000 records principal repayment entries, reducing its outstanding liability. In tandem, treasury and risk management automatically generates corresponding repayment entries for company code 2000, reflecting a reduction in its liability. This synchronized process maintains balance and transparency across both entities' financial records.

4. **Final repayment**

   As the loan nears its conclusion, company code 1000 records the final repayment of the principal amount, thereby settling the intercompany loan. Mirroring this action, treasury and risk management automatically clears the liability in company code 2000's records, ensuring a harmonized resolution to the loan agreement.

Through this mirroring process, treasury and risk management facilitates accurate and synchronized financial transactions between internal entities, fostering transparency, efficiency, and compliance within the organization's financial records.

6.3.2 Contract Creation

In a process similar to setting up a debt or investment contract, the initial step in creating an intercompany loan involves navigating to the Create Financial Transaction app. This serves as the platform where we can establish the framework of the intercompany. Assuming that intercompany mirroring has been set up and configured within the system, the loan creation process can begin from either the borrowing or the lending perspective. As illustrated in Figure 6.41, we input essential details such as the **Company Code**, **Product Type**, **Transaction Type**, and business **Partner**, and then we click the **Create** button. It's noteworthy that the business partner utilized for an intercompany loan is not an external entity but rather an internal counterparty that corresponds directly to the company code of the opposite side of the loan agreement.

When we save the contract, we are transitioned to the primary contract entry screens, where we can input the essential structural elements defining the loan. This is very similar to creating other debt or investment contracts under the same product category. The fields available for input are very similar to those found in other contracts, and navigating through the **Structure** tab poses no deviation from standard procedures. However, the distinctions between intercompany loans and other contracts become apparent after contract creation. Within this interface, the mandatory fields to be maintained encompass crucial details such as the loan amount, the commencement and maturity dates, and the stipulated interest rate. As with other debt and investment contracts, the contract's condition type and flow type are automatically populated within the contract. A full entry screen for intercompany loans is shown in Figure 6.42, and we can review the **Cash Flows** tab in Figure 6.43 to ensure the correct cash flows are being determined for the intercompany loan.

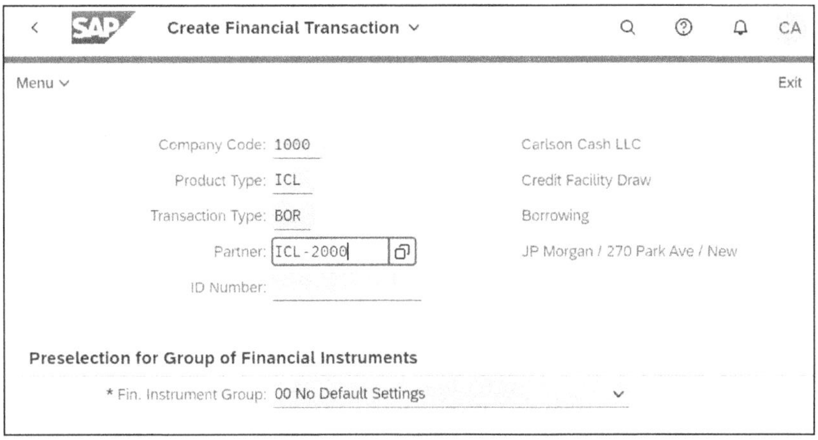

Figure 6.41 Initial Entry Screen for Creating Intercompany Loan between Two Company Codes

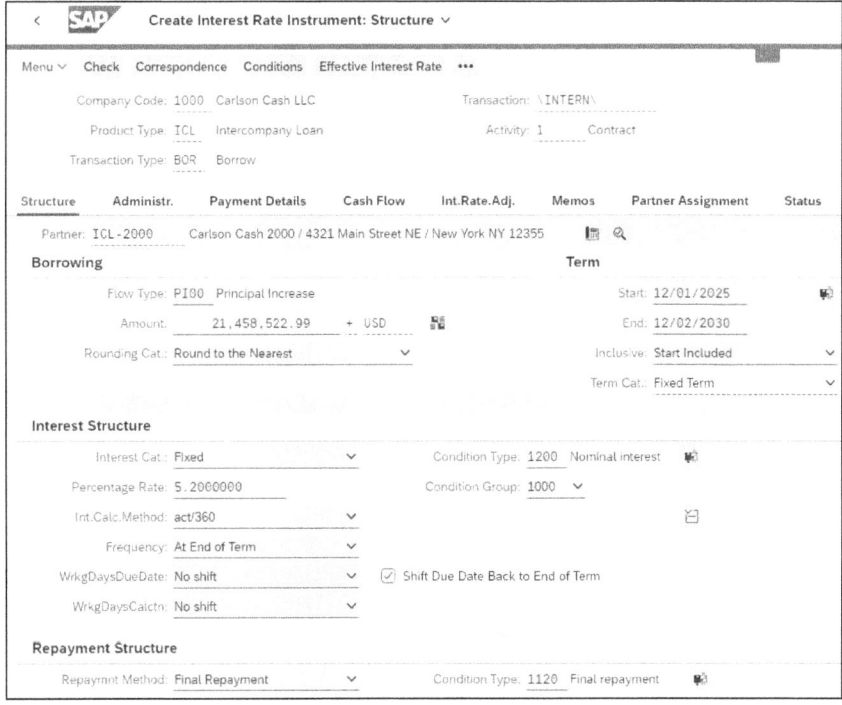

Figure 6.42 Structure Tab of Intercompany Loan Entry Screen

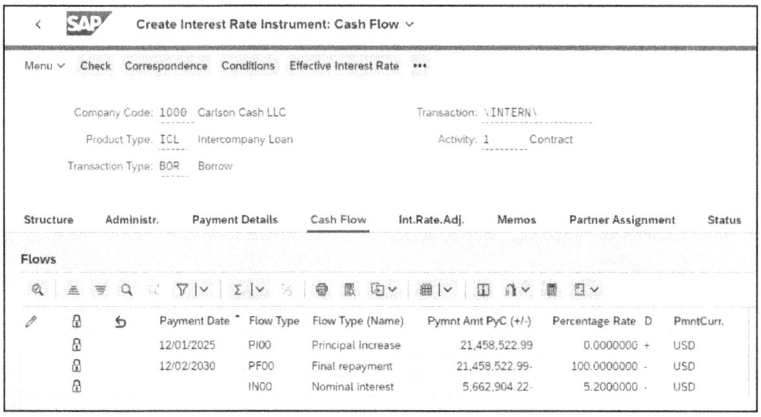

6 Treasury Management–Specific Processes

Figure 6.43 Related Cash Flows for Intercompany Loan

When we initiate the **Save** command, the contract undergoes creation within the foreground interface while concurrently, in the background, another contract is generated to mirror the opposite side of the intercompany loan. Notice that the system-generated message at the bottom of the screen in Figure 6.44 solely references the contract number generated within the foreground, omitting any indication of the creation of its mirrored counterpart. This is expected behavior; in subsequent sections, we'll dive into the various strategies for accessing and viewing both sides of the intercompany loan.

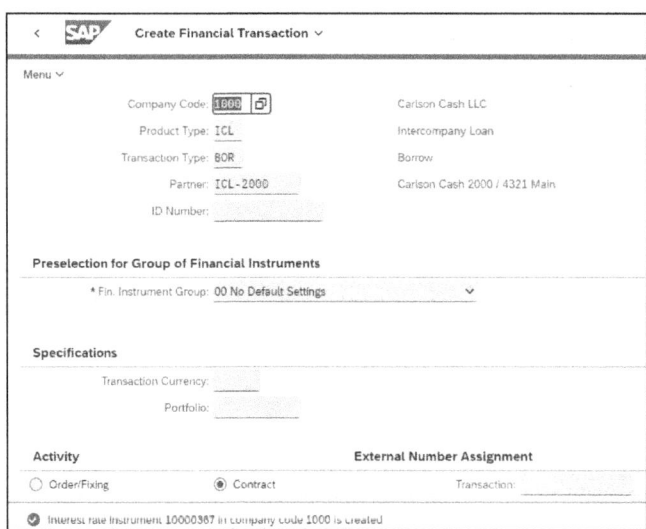

Figure 6.44 Output Reflecting Contract Just Created and Not Mirror Contract

6.3.3 Mirroring

When we save the contract in SAP S/4HANA, both sides of the intercompany loan are generated concurrently. However, it's essential to note that the system's messaging, displayed at the bottom of the screen, solely acknowledges the contract initiated in the

6.3 Intercompany Loans

foreground, without any reference to its mirrored counterpart. This discrepancy is a bit confusing and can lead us to question the automatic creation of the mirrored side of the loan. It's important to understand that if the configuration has been completed correctly, the other side of the loan has been created but the messaging doesn't signify that. Using the Process Financial Transaction app offers us a straightforward solution. By leveraging this app, we can retrieve the mirrored side of the loan by inputting the relevant details, including the **Company Code** and **Transaction** number associated with the contract initiated in the foreground. In Figure 6.45, the initial loan contract with transaction ID 10000367 is being selected.

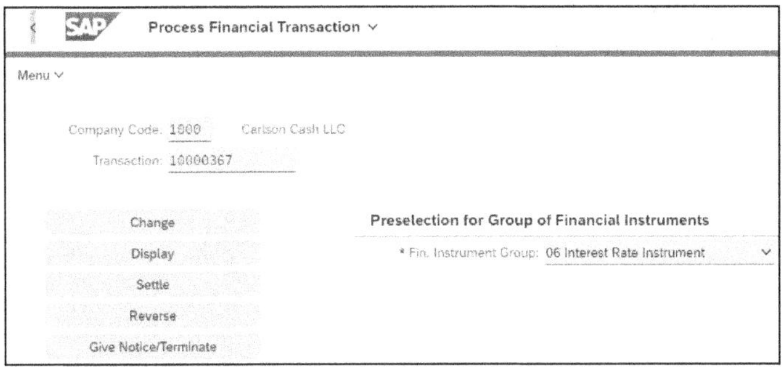

Figure 6.45 Selecting Intercompany Loan in Process Financial Transaction App

After we navigate to the contract, a notification will immediately appear at the bottom of the screen, indicating the presence of the mirrored transaction, as illustrated in Figure 6.46. This notification will appear on both sides of the contract any time we're displaying, changing, or settling a contract.

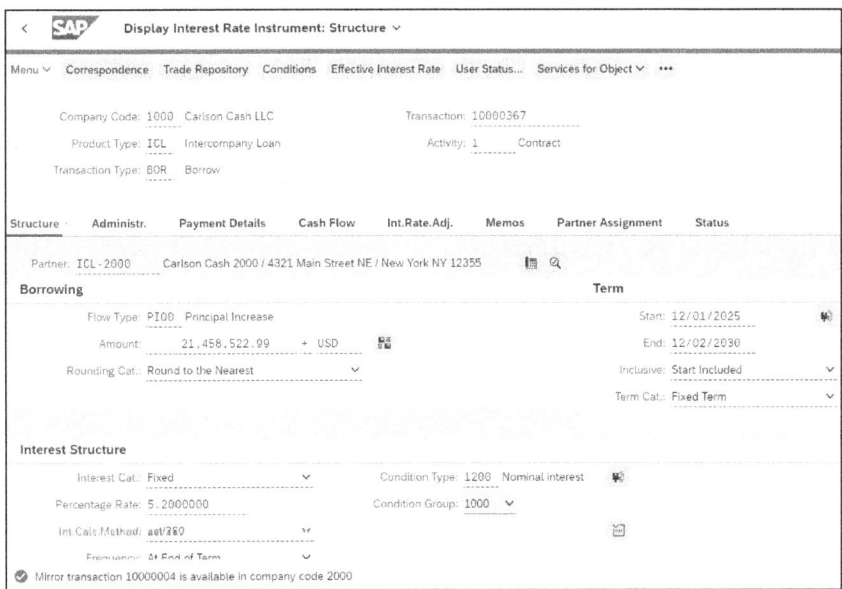

Figure 6.46 View of Intercompany Loan that Has Mirror Transaction

407

The message that appears within the transaction references the contract number or the mirrored contract along with the company code of the contract.

We can view the contract using various native SAP Fiori applications designed for contract viewing. However, we also have the option to access the contract directly from either side of the loan. To do this, we utilize the menu bar at the top of the screen as shown in Figure 6.47. We navigate through the following path: **Menu** • **Environment** • **Object Links**. This method allows us to access related objects and have a comprehensive view of the contract and its associated elements within the transaction context. This approach not only offers us convenience but also integrates the contract viewing process into the broader functionality of the system, enhancing our ability to manage and analyze contracts effectively from multiple perspectives.

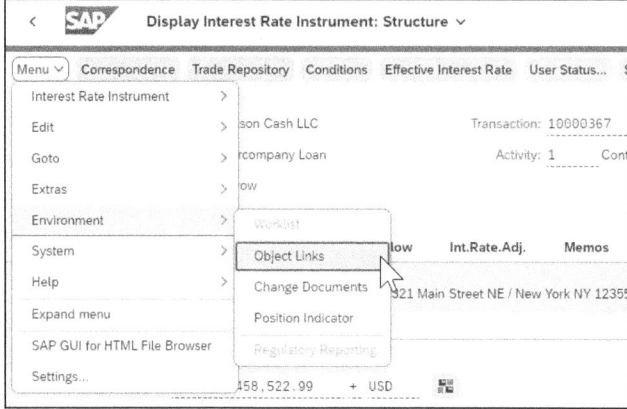

Figure 6.47 Selecting Object Links Option to View Other Side of Mirror Transaction

When we click **Object Links**, we'll be automatically navigated to a new screen, revealing the linkage between the two sides of the intercompany loan. Diving into the details, we'll notice a unique identification that's denoted as the **Reference** field in Figure 6.48 and that serves as the connective link binding both sides of the loan together. Beyond this, we'll find additional information pertaining to the contract. We should take a moment to notice the **Active** status of the contract displayed, and we should notice the details of who initiated the contract creation process, along with timestamps indicating when the contract came into existence or underwent modifications.

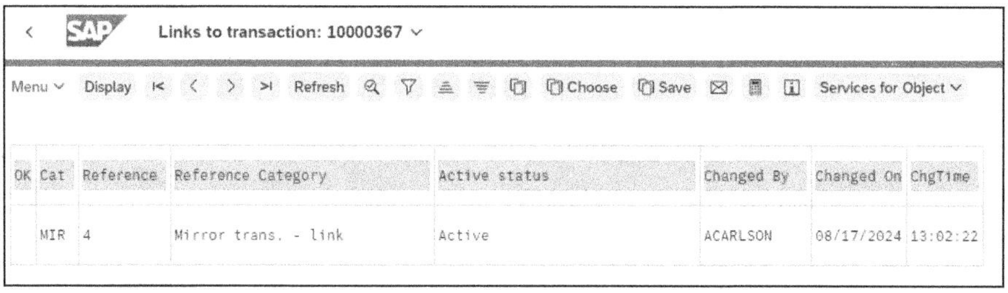

Figure 6.48 Displaying Mirror Link between Intercompany Loans

When we double-click the displayed object link, we're redirected to an additional screen, as depicted in Figure 6.49. This screen provides a detailed view, enabling us to examine both sides of the intercompany loan in a comprehensive manner. Here, we can observe that each side of the loan is distinctly represented within its respective company codes, with the associated contract numbers clearly displayed.

This screen offers a dual-perspective view of the intercompany loan, displaying how each company code is involved in the transaction. This view not only facilitates better oversight and reconciliation of intercompany loans but also enhances transparency by allowing us to verify that both sides of the transaction are accurately recorded and aligned with their respective company codes and contract details.

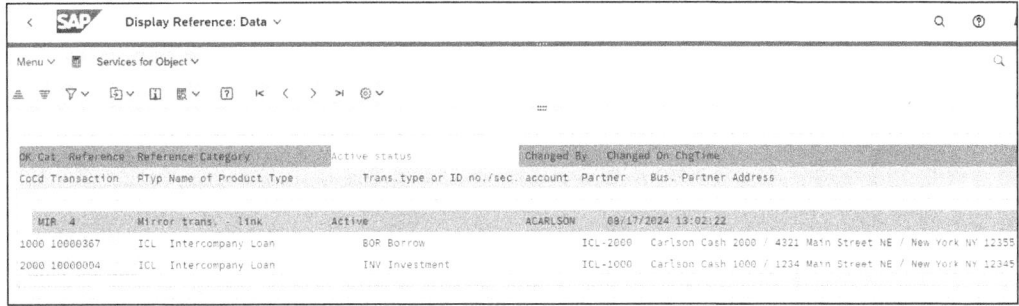

Figure 6.49 Selecting Object Link Lets Us View Both Contracts in One Place

When we click the mirrored side of the loan in company code 2000, the system navigates us further into the contract, allowing us to explore all the details of the loan, as illustrated in Figure 6.50. This deeper dive into the contract reveals that the mirrored transaction aligns perfectly with its counterparty in company code 1000.

Note that the contract details, including the start date, end date, and interest rate, are identical to those on the originating side of the contract. The only significant difference lies in the financial values: while the borrowing side displays a positive amount, the lending or investment side shows a corresponding negative value. This negative value reflects the financial obligation of the lending side, maintaining the integrity and balance of the intercompany transaction.

Additionally, the business partner details on the mirrored side reference the originating entity, noted as ICL-1000. This ensures clarity in the intercompany relationship, indicating that the mirrored transaction in company code 2000 is tied to the originating transaction in company code 1000. All other aspects, such as dates and interest rates, remain consistent across both sides of the loan, providing a full view of the contract terms.

It's also important to highlight that a notification appears at the bottom of the screen, referencing the original contract in company code 1000. This message serves as a reminder that we are viewing a mirrored transaction, ensuring our awareness of the intercompany link regardless of which side of the loan we're examining.

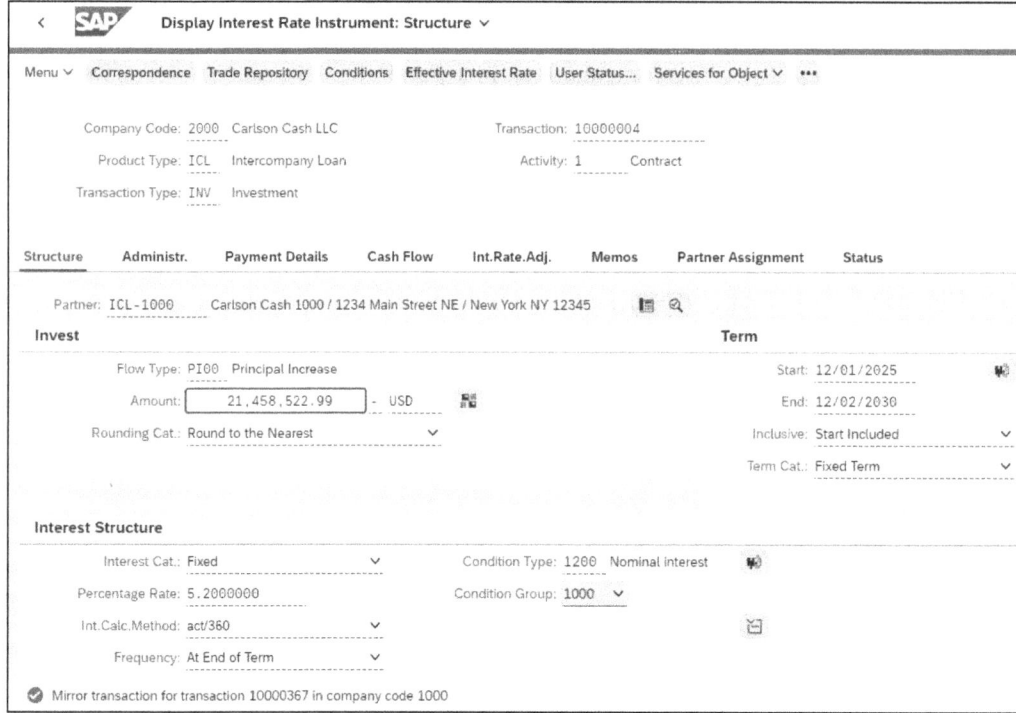

Figure 6.50 Other Side of Mirrored Intercompany Loan

Accessing the **Cash Flow** tab within the loan provides us with a detailed view of the transaction's financial movements. This tab highlights that the cash flow details on the mirrored side of the loan closely mirror those of the originating loan, with the key distinction being the direction of the cash flows. The cash flows for the mirrored loan are in Figure 6.51.

In the mirrored transaction, the cash flows are represented inversely compared to the originating side. While the originating loan might show cash inflows as positive values, indicating funds received, the mirrored side will display these same cash flows as negative values, signifying funds disbursed, and vice versa. This inverse relationship accurately reflects the financial dynamics between the lending and borrowing entities involved in the intercompany loan.

Each scheduled payment, interest accrual, and principal repayment is recorded identically on both sides, but with opposite financial implications. For instance, a cash inflow on the borrowing side, where the entity receives loan funds, will correspond to a cash outflow on the lending side, where the entity disburses funds. This inverse setup ensures that both sides of the intercompany transaction are balanced and accurately represent the financial exchanges between the two company codes.

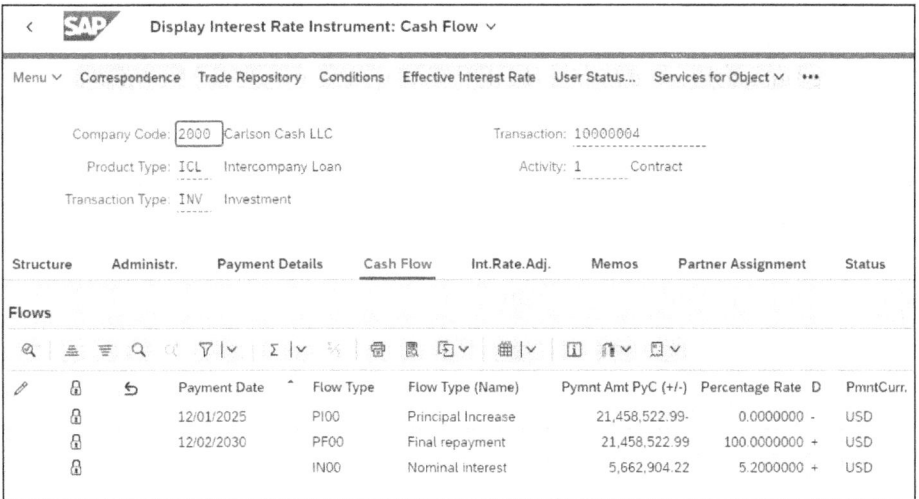

Figure 6.51 Cash Flow Tab of Mirrored Intercompany Loan

6.3.4 Settlement

With the creation of the intercompany loan complete and both sides successfully verified, we're now ready to proceed to the next phase: approving or settling the intercompany loan. This settlement process is a key step in the lifecycle of the loan, ensuring that the transaction is formally recognized and recorded within the financial system.

We can perform the settlement from either side of the loan, giving us flexibility in how to manage the process. Instead of only maintaining the loan from either the borrowing or lending side of the trade, assuming the configuration has been completed for both sides, we can manage the loans from either side. To initiate this step, we navigate to the Process Financial Transactions app. As shown in Figure 6.52, upon accessing the loan details, we'll observe that the loan's status has progressed to an activity code of **2**, signifying **Contract Settlement**. This status update indicates that the loan is now in the settlement phase and thus ready for final approval and processing.

During the settlement process, we'll be able to confirm and finalize the terms of the intercompany loan, ensuring that all conditions are met and the transaction is accurately documented. This includes validating the financial terms, payment schedules, and any other relevant contractual details. Settlement is a key step as it formalizes the financial obligations and rights of both entities involved in the intercompany loan, aligning the internal records and preparing the loan for subsequent financial activities related to the loan, such as interest accruals and repayments.

The system provides a message at the bottom of the screen (see Figure 6.52), referencing the mirrored side of the intercompany loan. This notification is a key feature because it maintains transparency and reminds us that the loan has a mirrored counterpart. This ensures that any actions taken on one side are consistently reflected on the other, thus preserving the integrity of the intercompany relationship and supporting effective reconciliation. Completing the settlement accurately is vital for maintaining precise financial records across both loans and ensuring that the intercompany loan is accurately reflected on the general ledger of both company codes. If both sides of the loan are not settled, it impacts subsequent processes because the posting step can't take place unless the loan has been settled and approved. If only one side of the loan has been settled, only the settled side of the intercompany loan will be posted to the general ledger. That's why it's extremely important to settle both the lending and the borrowing side of the intercompany loan.

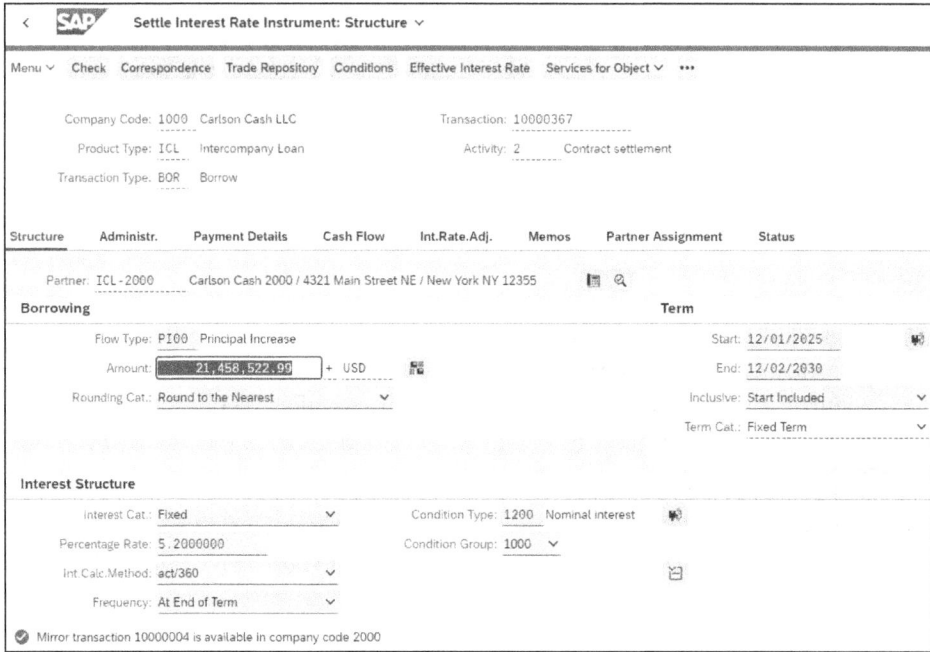

Figure 6.52 Confirming Intercompany Loan Details before Settling Transaction

To proceed with settling the intercompany loan, we click the **Save** button located in the lower right-hand corner of the screen. This action simultaneously saves and settles the loan, ensuring that both sides of the transaction are updated concurrently. By saving the intercompany loan, we're not only confirming the details of the transaction but also changing the overall status of the loan and allowing for subsequent processes to take place.

When we press the **Save** button, the system executes the settlement process for both the lending and borrowing entities involved in the intercompany loan. This mirrored approach means that any changes or confirmations made to one side of the loan are immediately reflected on the other side, maintaining consistency and accuracy across the transaction.

As depicted in Figure 6.53, when we click **Save**, a notification will appear on the screen. This message explicitly references the mirrored side of the intercompany loan, letting us know that it has been settled as well. This notification serves as a confirmation that both sides of the loan have been successfully updated and settled within the system.

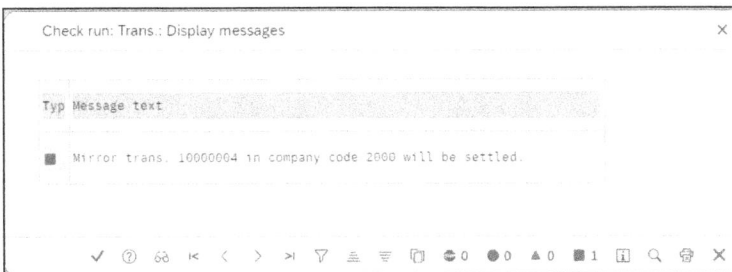

Figure 6.53 Confirmation Message Highlighting that Mirror Transaction Was Automatically Settled

6.3.5 Changes

We have the option to configure intercompany loan mirroring so that any modifications made to the contract after the settlement process are automatically reflected on both sides of the transaction. Implementing this mirroring for post-settlement changes is highly recommended and considered a best practice for managing intercompany loans in SAP S/4HANA. This approach ensures that the loan remains accurately maintained throughout its lifecycle, reducing the risk of discrepancies and simplifying ongoing contract management.

By enabling mirroring for subsequent changes, we eliminate the need for manual updates on each side of the loan. This automation enhances the efficiency and accuracy of contract maintenance, as any amendments made—such as adjusting interest rates, changing repayment schedules, and updating principal amounts—are instantly mirrored across both entities involved in the loan. This setup not only helps us maintain consistent records but also minimizes the administrative burden associated with managing intercompany loans.

To initiate a change to the loan, we navigate to the Process Financial Transaction app. We can make changes from either side of the loan, and that provides us with flexibility in how we manage updates. For example, we might adjust the loan's principal or alter the interest terms. Regardless of which side we modify, the system will automatically apply the changes to both sides of the contract, ensuring synchronization.

In Figure 6.54, we illustrate an example with our intercompany loan in company code 1000. By selecting the **Principal Increase/Decrease** option using the **Other Changes in Capital Structure** button, we can decreased the principal of the loan. This adjustment will be simultaneously updated on the mirrored side of the loan and be reflected in the corresponding records of the other company code involved in the transaction.

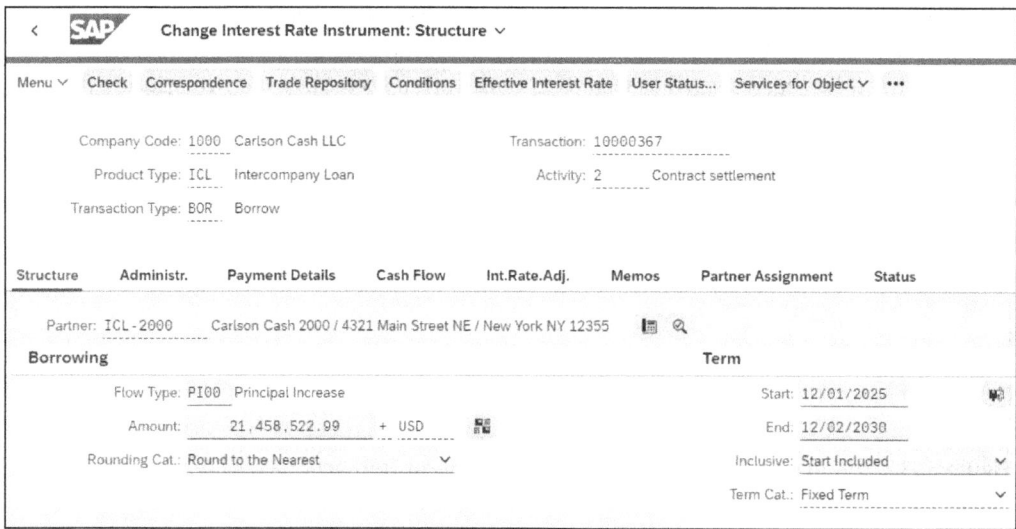

Figure 6.54 Intercompany Loan Borrowing Amount Is Adjustable by Using Change Function in Process Financial Transaction

As depicted in Figure 6.55, the initial transaction indicates a **Principal Increase** for the intercompany loan. However, should there be a need to adjust the loan's principal amount, introducing a **Principal Decrease** would enable a reduction in the overall principal balance of the loan.

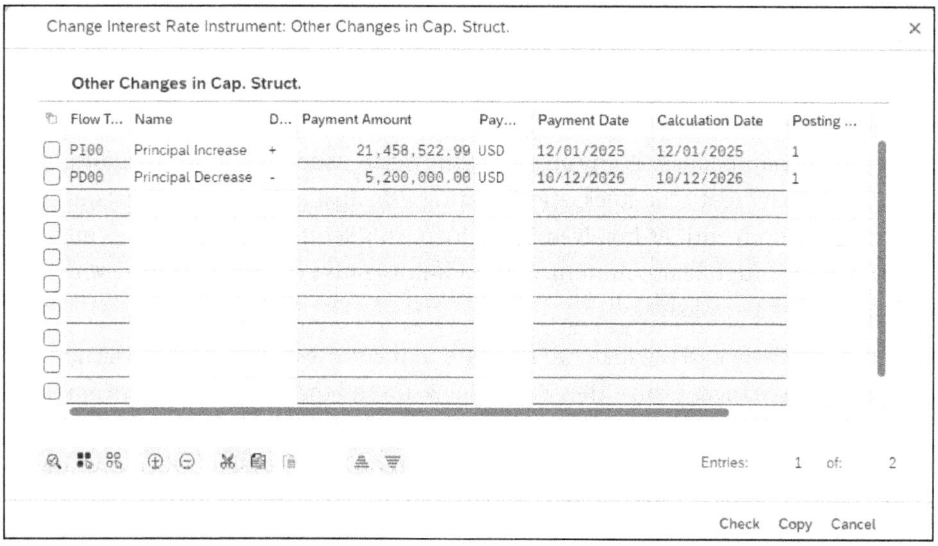

Figure 6.55 Updating Intercompany Loan Borrowing Mirrored on Other Side of Loan

After we input the reduction in the principal amount of the intercompany loan in the **Payment Amount** field, we save the changes to the loan record and a notification

6.4 Securities

appears, indicating that the mirrored counterpart of the intercompany loan has been updated accordingly. The message in Figure 6.56 confirms that both the primary and the mirrored versions of the intercompany loan have been successfully modified to reflect the principal decrease. At this point, we can be sure that the adjustments have been synchronized across the entire loan, ensuring that both sides of the intercompany loan now accurately incorporate the updated principal reduction.

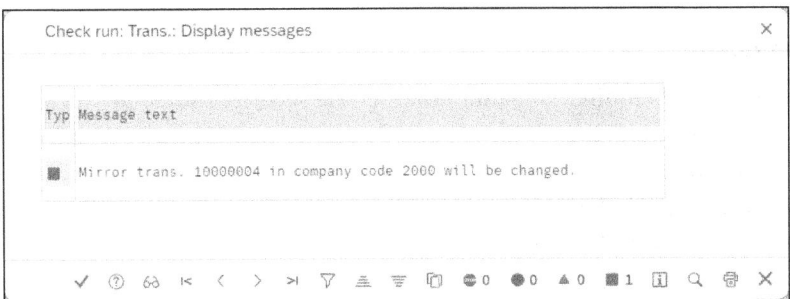

Figure 6.56 Confirmation that Changes to Intercompany Loan Were Mirrored in Corresponding Intercompany Loan in Company Code 2000

In summary, the intercompany loan process involves three main phases: creation, settlement, and changes. During the creation phase, we initiate the loan with accurate details and terms, ensuring proper documentation and alignment with financial requirements. Once we've created the loan, the settlement phase formalizes the loan, ensuring that financial obligations are recognized and recorded across both entities involved. Any subsequent changes to the loan, such as adjustments to the principal amount, are facilitated seamlessly, ensuring that both sides of the loan accurately reflect the modifications. Throughout each phase, the system's notifications and synchronization mechanisms ensure transparency and consistency, enabling effective management and monitoring of intercompany loans.

6.4 Securities

Securities transactions in treasury and risk management follow a flow that is different from that of the other general contracts. Generally, the other transactions in treasury and risk management are agreements between a company and a counterparty, but the securities are publicly traded on a market, so market data is readily available for the securities transactions. Since we base this information on the public pricing of securities in public markets, we have additional master data that is entered in SAP S/4HANA for the tracking of the value and processing of the transaction.

Securities transactions require two extra steps before we can create a contract: the securities account and the securities class. Section 6.4.1 starts by detailing the process flow required to create securities transactions. Section 6.4.2 and Section 6.4.3 detail creating the securities account and securities class, and Section 6.4.4 details the next step, which is creating the securities contract. Section 6.4.5 details the new apps and transaction codes that we need to run to generate the payments and accounting entries for these transactions. Finally, Section 6.4.6 details the other flows and exercising of rights that are involved in the one-off transactions in some of the securities contracts.

6.4.1 Process Flow

The securities account is required to define whether the security should be carried in SAP S/4HANA as an asset or liability. We can add details to the securities account, like defining payment details for when there are incoming or outgoing payments. The creation of the securities class follows the creation of the securities account, and it creates the structure of the security. The securities class also includes details of the different interest and repayment conditions. After we create the securities account and class, we create the transaction in a way that's similar to the creation of other contract types. We enter it using the Create Financial Transaction app.

To summarize what the process looks like for each of the securities transactions, we provide a process flow in Figure 6.57 that details the key steps in the process.

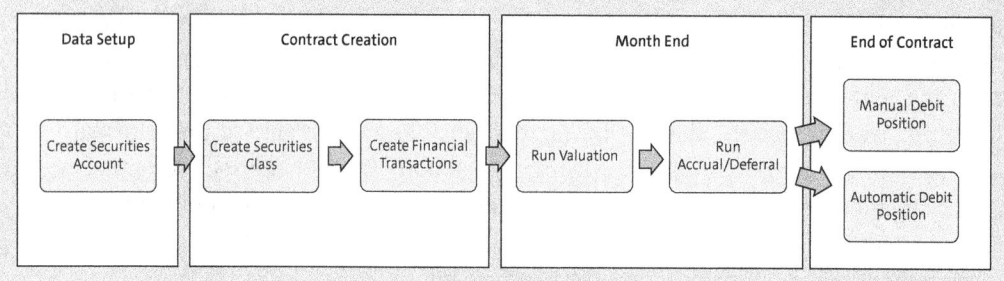

Figure 6.57 Standard Process Flow for Securities

The securities process varies from the standard process that we see in the debt, investment, and foreign exchange processes. In those processes, we enter the contract all in one screen, but due to the nature of how securities are publicly traded and how the prices can change daily, there are a few additional steps that we must followed to set up the securities transactions. First, we need to create a securities account to determine whether the securities will be assets or liabilities. There are a few other details of the securities account that are required if the cash flows will progress through the treasury payment process. Next, the securities class is required to create the overall structure of the security to determine the cash flows and how they are calculated. Finally, we can create the contract to determine the nominal amount and review the cash flows that will be created by the financial transaction.

6.4.2 Accounts

The first step of creating a securities contract is the creation of a securities account. This account is required for some base information in securities. We enter the securities account in the Manage Securities Accounts app (shown in Figure 6.58) or Transaction TRS_SEC_ACC, and we specify whether a security will be an asset, a liability, or for lending.

Figure 6.58 Manage Securities Accounts App

Each securities account is required to at least have a securities account type (to determine the type of securities account) and a depository bank. The depository bank is a business partner role that we must assign to any business partner that we will use for securities transactions. The other main function of entering a securities account is to determine payment details for any transaction that uses the assigned securities account. The payment details here are just like the standing instructions that are assigned in the treasury business partners. They determine the details of the paying bank and determine whether a payment request will be created for the associated cash flows for the securities transactions.

The details of the securities account setup are as follows (see Figure 6.59 for where we enter this data when creating the securities account):

❶ **Company Code**
Securities accounts are set up by company code, so the same account may need to exist in additional company codes.

❷ **Securities Account**
This is the field where we determine the securities account name.

❸ **Securities Account Type**
This is the field where we determine the type of securities account, which in turn determines the types of transactions we can create for the security. For example, we can use the **AKT/Asset Securities Account** for bond purchases and the **PAS/Issuance Securities Account** for bond issuances.

❹ **Depository Bank**
We enter the counterparty for the securities transactions in this field, and we must assign the **TR0152/Depository Bank** business partner role in order to populate the information in this field.

6 Treasury Management–Specific Processes

❺ **Default**

We enter payment details on this line, and we can create multiple lines by payment currency and/or update type to drive the specific payment details needed for each payment.

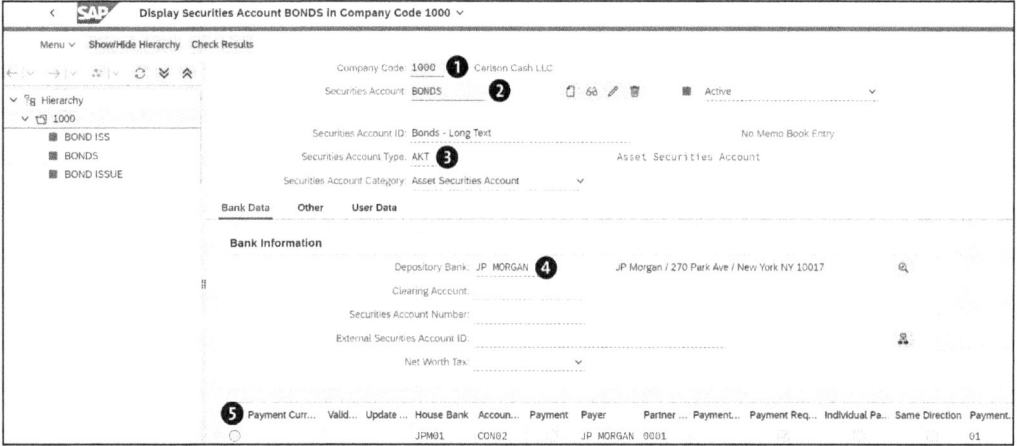

Figure 6.59 Securities Account Creation Example

> **Default Payment Details**
>
> The payment details for securities can originate from either the securities class or the business partner standing instructions. This setting is driven in customizing on the securities transaction type. This is in the following menu path: **Financial Supply Chain Management • Treasury and Risk Management • Transaction Manager • Securities • Transaction Management • Transaction Types • Define Transaction Types – Securities**. In the **Automatic Determination** area, we can derive the payment details by using one of three options:
>
> - We can set the payment details to first look at the standing instructions for the business partner, and if there are no payment details available, it should look at the securities account.
> - The payment details can also look only at the standing instructions and ignore any settings on the securities account.
> - The third option is to not automatically derive any of the payment details; in this case, we would need to define the payment details when posting the transaction in the Execute Debit Position – Manual Debit Position app or Transaction FWZE.

6.4.3 Classes

We use the securities class to build the basic structure of the securities. When a security is issued and traded, there are known attributes to those transactions, and those details are added to the securities class. Data included in this area details the description of the security, dates of issuance, nominal value, interest frequency, and interest payment rate, and it can even determine the periods when a bond is callable or puttable. This section will detail each of the key fields to consider when creating a securities class. To highlight these details, we'll cover each tab in the securities class entry screens. As shown in Figure 6.60, we can use the Manage Securities Classes app or Transaction FWZZ for this process.

Figure 6.60 Manage Securities Classes App

Search Terms Tab

The **Search Terms** tab holds basic information for the securities class. In the following examples, we'll use a purchased bond as our example when running through the rest of the securities class tabs and fields. In this tab, the **Short Name**, the **Long Name**, and some classification categories are available for us to use to report on different types of securities. The general security classification is meant to be a high-level classification to determine the type of security in SAP S/4HANA, while the bond classification is more specific and classifies the individual type of bond. We use both of these classifications for evaluation purposes. Figure 6.61 shows the initial screen for the securities class.

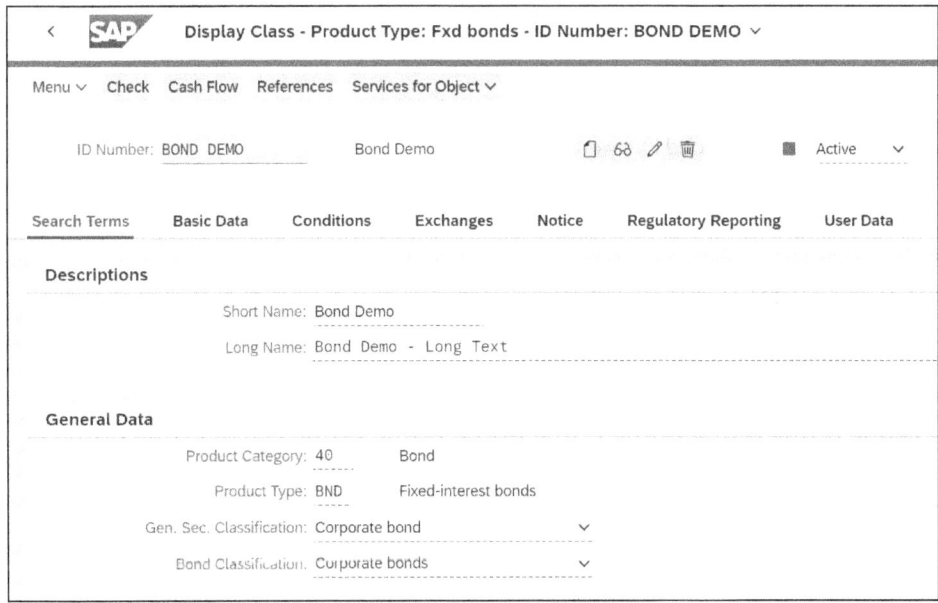

Figure 6.61 Search Terms Tab in Manage Securities Classes

6 Treasury Management–Specific Processes

Basic Data Tab

The **Basic Data** tab holds the initial data for the securities transaction. The **Issuer** will match the issuing counterparty where the security was purchased. The **Nominal Value** drives the amount we can assign to the contract, and this can equal the full value of the security issuance. We do not use the **Nominal Value** in this tab to drive the cash flows, and we only reference it to limit the contract values when we enter the security transactions. The **Issue Start** and **End of Term** for bonds represent those dates for the bond's issuance and maturity. If a bond has been purchased on the secondary market, there are different date fields that we use to ensure the cash flows follow the purchase correctly. Figure 6.62 shows an example of a bond entry.

Figure 6.62 Example Bond Entry on Basic Data Tab

Conditions Tab

The **Conditions** tab drives most of the structure for the security. Figure 6.63 shows the options available in this tab. The first two sections drive the **Interest Calculation** and which type of **Repayment** is required. Since this bond will be repaid at the end of the term, we select the **Maturity Repayment Type**.

The **Condition Items** drive the interest calculation and final repayment. Since the interest dates are slightly different for the security, let's look at them in a little more detail:

❶ We select the condition type for interest here. The standard percent calculated interest is defaulted into this screen, but other interest conditions can replace this condition as needed.

6.4 Securities

❷ Effective from (**Eff. From**) determines the start of the interest calculation period. If we're purchasing a bond on the secondary market, we enter the day the bond is acquired in this field.

❸ The calculation date (**Calc. date**) functions the same as money market transactions, so this field determines the first day we calculate the interest for the first interest payment. The due date three columns over determines when the calculated interest is paid.

❹ **MC** and **MEID** are the month-end indicators. We should check these boxes if a given date should always fall at the end of a month.

❺ Both **CR** columns determine whether the calculation date and the due date require a working-day shift.

❻ The **Frq** field is the frequency in months for the interest payment. In this case, the interest will be paid every six months.

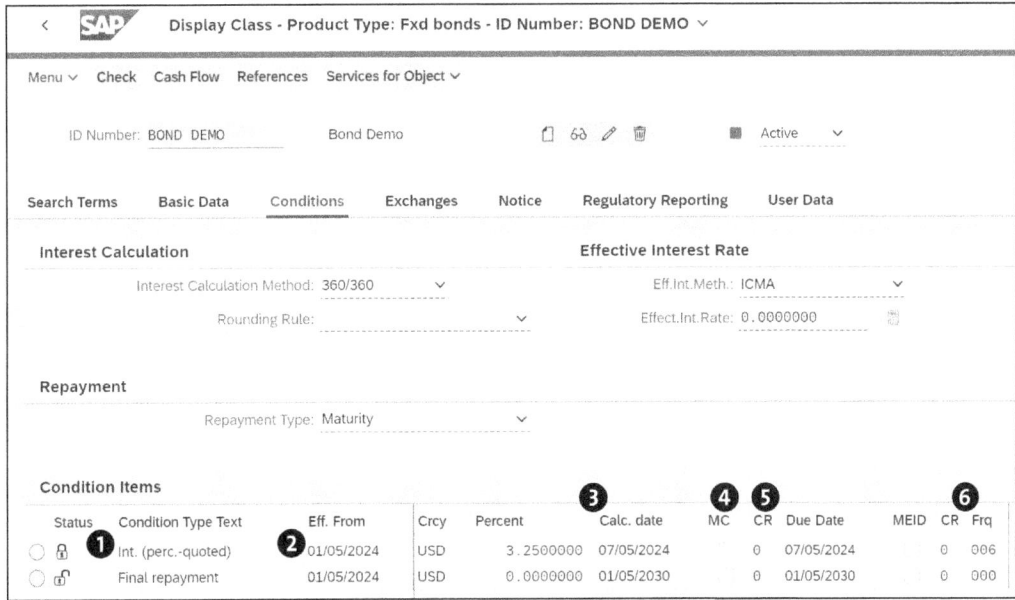

Figure 6.63 The Conditions Tab Where We Create Interest and Repayment Structures

Notice Tab

The **Notice** tab of the bond drives whether the bond is callable or puttable. The first section, labeled **Notice periods of Issuer,** will be populated if the bond is callable. We need to define the period when the bond is callable in the date fields, and these fields will restrict the dates when the bond can be called in the Exercise Rights app or Transaction FWER. If an issuer calls the bond at a premium, the premium is reflected in the **Notice rate** field.

6   Treasury Management–Specific Processes

Similarly, we use the **Notice Periods of Bondholder** section if the bond is puttable. The notice dates determine the dates when the bondholder can force the issuer to repurchase the bond. We also use the **Notice Rate** field if we need to make an adjustment to the repurchase amount. A simple example in Figure 6.64 shows the entry of both a call and a put option on the bond that spans the final year of the bond, from 01/05/2029 to 01/05/2030.

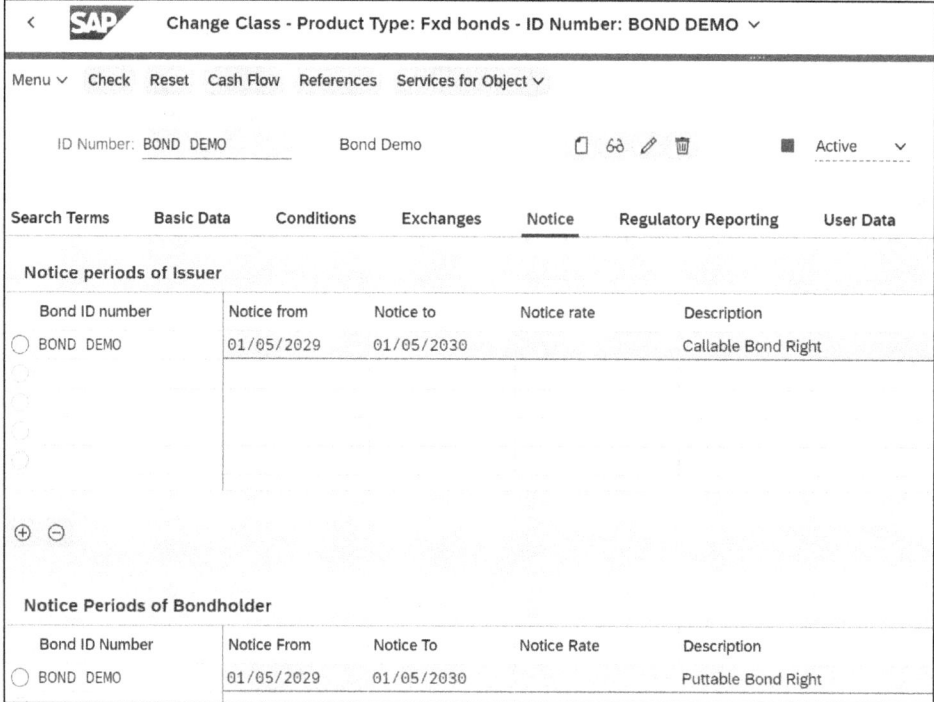

Figure 6.64 How to Add Information for Callable or Puttable Bonds on Notice Tab

6.4.4 Contracts

Now that we have created the securities account and securities class, we can enter the transaction into the Create Financial Transaction app or Transaction FTR_CREATE. In the following sections, we'll look at how to create a securities contract and then dive into two important concepts for these contracts: accrued interest and securities cash flows.

Contract Creation

The main difference between securities transactions and other transactions lies in how we need to enter the securities class into the initial screen before we can add the contract details. In Figure 6.65, this is identified by the **ID Number** field.

422

6.4 Securities

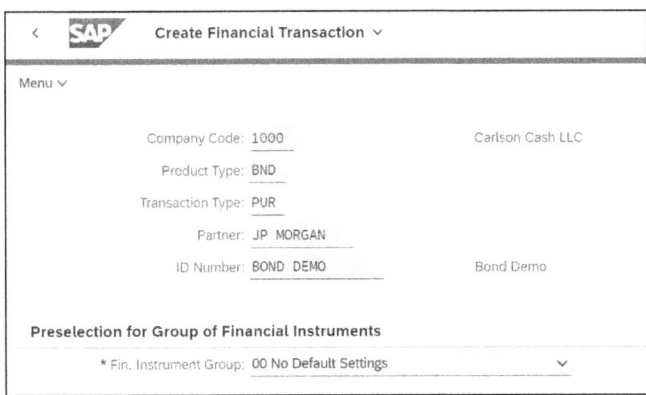

Figure 6.65 Adding Securities Class in Create Financial Transaction App for Securities Transactions

Figure 6.66 displays the structure of a bond. The key fields to enter data into are as follows:

❶ Here, we enter the applicable **Securities Account** and the correct account type. For example, the "BONDS" security account in this example is an asset securities account, so it's applicable for a purchased bond.

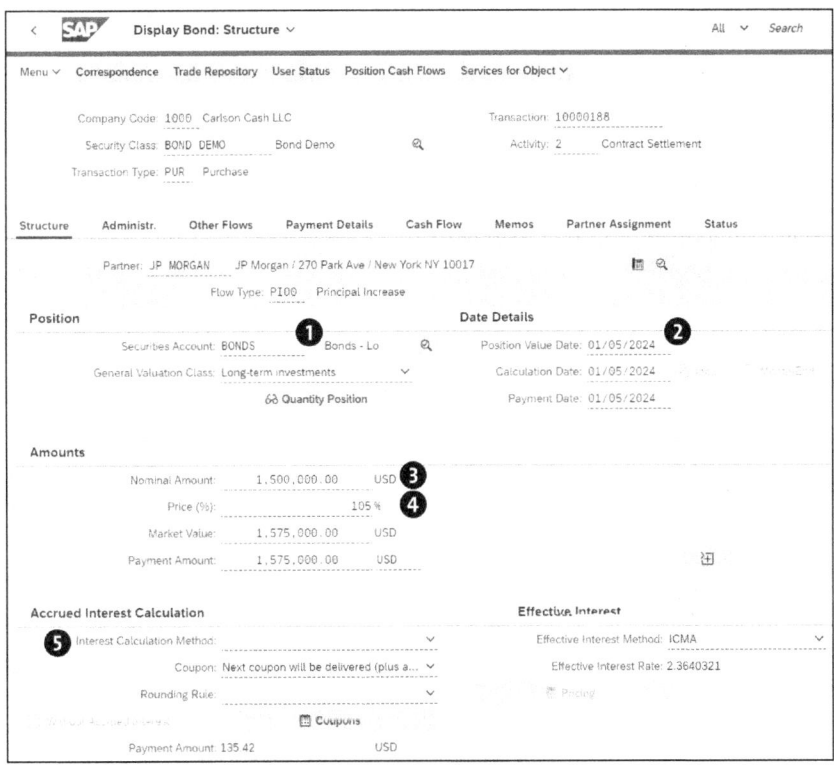

Figure 6.66 Bond Entry Screen with Important Fields Highlighted

6 Treasury Management–Specific Processes

❷ The dates we enter should reflect the purchase date of the bond and the payment date for the bond purchase.

❸ Here, we enter the **Nominal Amount** of the bond. The premium or discount is not included in this number.

❹ We use the **Price (%)** field to enter the premium or discount percentage. In our example, the bond was purchased at a premium, so the amount reflects 105%. In the subsequent fields, the market and payment amount will automatically calculate based on the nominal amount and the price percentage.

❺ In this example, the accrued interest is not applicable since we purchased the bond at issuance. Next, we'll cover how the accrued interest is calculated and represented.

Accrued Interest

When a bond is purchased on the secondary market, it's likely that it was purchased in the middle of an interest period. Interest is usually paid semiannually or annually, and the previous bondholder wants to be paid for the interest that they incurred from owning the bond for the partial period before they sold it. Typically, when a bond is purchased in the middle of the interest period, the purchaser pays the amount of the interest that the previous bondholder earned in that partial period, and this is what is covered in the **Accrued Interest Calculation** section of the entry screen.

Figure 6.67 shows an example of how this calculation works. In this example, we used some round numbers to perform an easy calculation. The bond was purchased for $1,000,000 with a 6% interest rate paid semi-annually. With this, the coupon payments are $30,000. Since we purchased the bond two months into the interest period, the accrued interest paid is $10,000.

The bond was issued on 01/05/2024 with semiannual interest payments, and we purchased the bond on 09/06/2024. The first interest period was from 01/05/2024 to 07/05/2024, but this doesn't impact our calculation. To calculate the accrued interest, we focus on the period of 07/06/2024 to 09/05/2024 since that's the period when we did not own the bond.

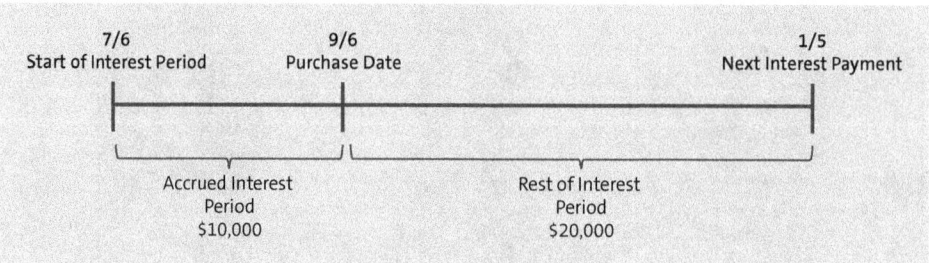

Figure 6.67 Example of Accrued Interest Period

6.4 Securities

Now that we know how the interest is calculated, let's cover how we enter it into the Manage Securities Classes and Create Financial Transactions apps. First, we need to create the securities class for this scenario. When calculating the accrued interest for a bond, we have to treat the dates differently from how we would if there were no accrued interest. In the interest condition, the effective-from date needs to start on the first day of interest calculation and not on the purchase date of the bond. These changes are detailed in the following list (see an example of the entry of this securities class in Figure 6.68):

- **Effective from (Eff. From)**
  We should set this to be the first day when the interest should be calculated. Even though this bond was purchased on 09/06/2024, the interest period started on 07/06/2024, so we should enter that into this field.

- **Calculation date (Calc. date)**
  We should set this to be the date when we'll receive the first interest payment. The effective-from date and the calculation date are six months apart to represent the first full interest period.

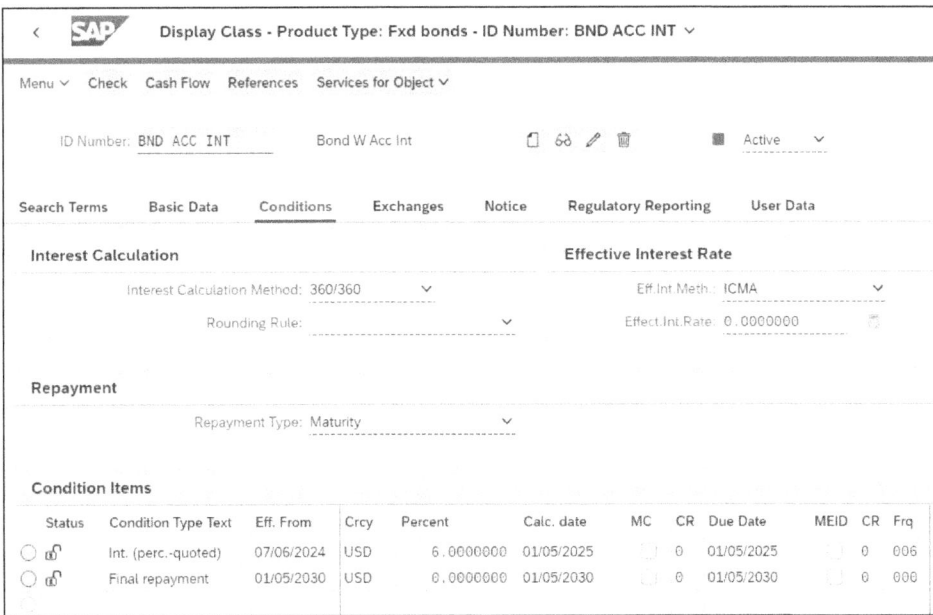

Figure 6.68 Entry of Securities Class that Will Have Accrued Interest Flow

Now we can create the transaction in the Create Financial Transactions app. The main field to focus on is the **Position Value Date**, which we set to "09/06/2024" in this example. This represents the settlement date of the bond purchase. Using this information, we can determine the accrued interest that will be paid. This is represented in Figure 6.69 with the **Payment Amount** in the **Accrued Interest Calculation** section.

425

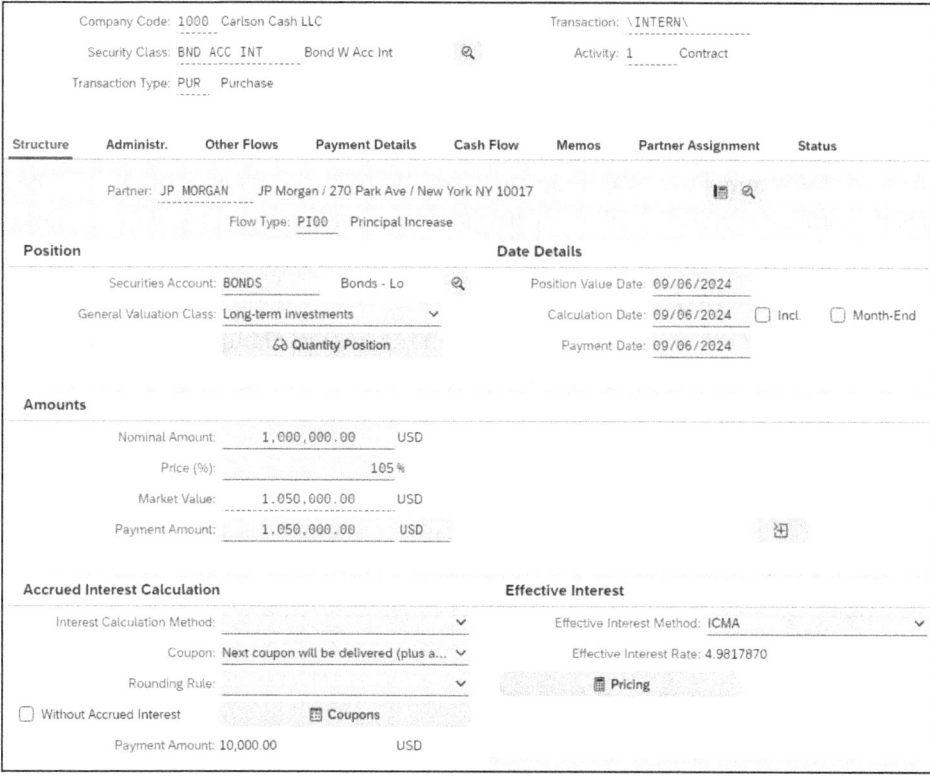

Figure 6.69 Example of Entering Bond with Accrued Interest

There are a few options on the accrued interest calculation that we should consider. There are situations where accrued interest payments are not required, and the **Coupon** field is available for us to use in those situations. The options in this field and when we use them are as follows:

- **Next coupon won't be delivered (minus accrued interest)**
  In this option, the next coupon is not delivered and the accrued interest is calculated.

- **Next coupon will be delivered (plus accrued interest)**
  In this option, the next coupon will be paid including the accrued interest before the settlement date.

- **Partial right on next coupon, no accrued interest calc**
  In this option, the next coupon is paid, but the amount paid is calculated using the number of days from the settlement to the coupon payment date. Accrued interest is not calculated for this coupon method.

- **Coupon information from coupon date**
  In this option, we can enter the next coupon date by clicking the **Coupons** button. The accrued interest will be calculated according to the structure of the interest.

6.4 Securities

- **No coupon delivery**
  In this option, the coupon is not delivered but the system will generate the accrued interest flows.

Additionally, if the accrued interest is just contained in the price of the bond, we can flag the **Without Accrued Interest** indicator to remove the additional accrued-interest cash flow.

Securities Cash Flows

Another nuance of securities contracts is how the cash flows are represented. If we navigate to the **Cash Flow** tab, we'll only see the bond purchase and the accrued interest flow. An example of this is shown in Figure 6.70. These flows will be posted with the Post Flows app or Transaction TBB1.

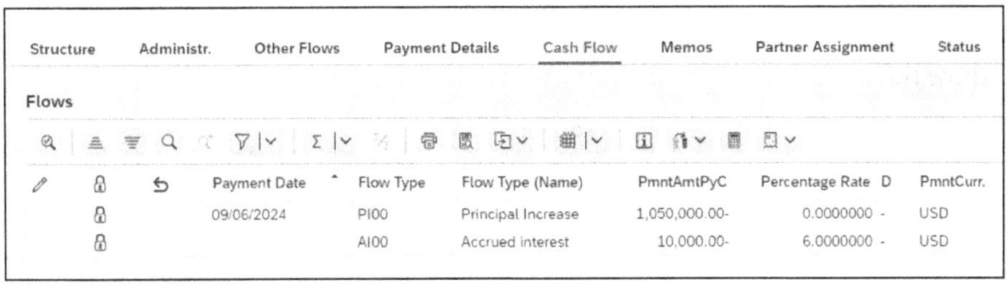

Payment Date	Flow Type	Flow Type (Name)	PmntAmtPyC	Percentage Rate D	PmntCurr.
09/06/2024	PI00	Principal Increase	1,050,000.00-	0.0000000 -	USD
	AI00	Accrued interest	10,000.00-	6.0000000 -	USD

Figure 6.70 Cash Flow Tab Showing Principal Increase and Accrued Interest Flows

We can access the rest of the cash flows by clicking the **Position Cash Flows** button at the top of the entry screen. The initial flows will also be displayed in this report, as shown in Figure 6.71.

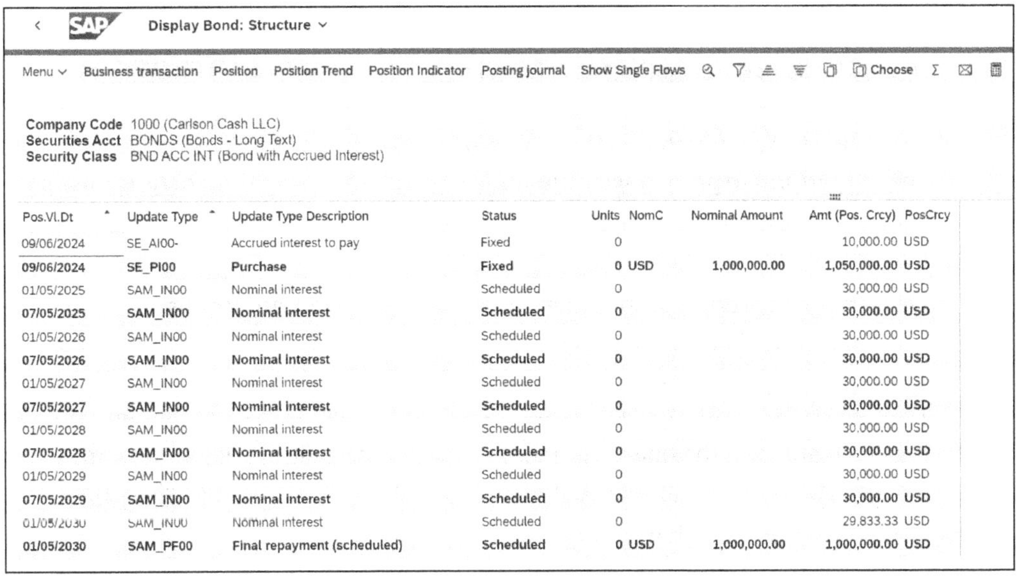

Display Bond: Structure

Company Code 1000 (Carlson Cash LLC)
Securities Acct BONDS (Bonds - Long Text)
Security Class BND ACC INT (Bond with Accrued Interest)

Pos.Vl.Dt	Update Type	Update Type Description	Status	Units NomC	Nominal Amount	Amt (Pos. Crcy)	PosCrcy
09/06/2024	SE_AI00-	Accrued interest to pay	Fixed	0		10,000.00	USD
09/06/2024	SE_PI00	Purchase	Fixed	0 USD	1,000,000.00	1,050,000.00	USD
01/05/2025	SAM_IN00	Nominal interest	Scheduled	0		30,000.00	USD
07/05/2025	SAM_IN00	Nominal interest	Scheduled	0		30,000.00	USD
01/05/2026	SAM_IN00	Nominal interest	Scheduled	0		30,000.00	USD
07/05/2026	SAM_IN00	Nominal interest	Scheduled	0		30,000.00	USD
01/05/2027	SAM_IN00	Nominal interest	Scheduled	0		30,000.00	USD
07/05/2027	SAM_IN00	Nominal interest	Scheduled	0		30,000.00	USD
01/05/2028	SAM_IN00	Nominal interest	Scheduled	0		30,000.00	USD
07/05/2028	SAM_IN00	Nominal interest	Scheduled	0		30,000.00	USD
01/05/2029	SAM_IN00	Nominal interest	Scheduled	0		30,000.00	USD
07/05/2029	SAM_IN00	Nominal interest	Scheduled	0		30,000.00	USD
01/05/2030	SAM_IN00	Nominal interest	Scheduled	0		29,833.33	USD
01/05/2030	SAM_PF00	Final repayment (scheduled)	Scheduled	0 USD	1,000,000.00	1,000,000.00	USD

Figure 6.71 Position Cash Flows View Shows All Relevant Cash Flows for Bond

6.4.5 Accounting Postings

As detailed in the process flow at the beginning of this section, there are journal entries at the purchase of the bond, at month end, periodically when interest needs to be posted, and at the end of the transaction.

We cover the execution of these postings in this section. Before we dive into each posting transaction, we'll cover a sample set of t-accounts in Figure 6.72 to show how these transactions post through the lifecycle of a bond. Note that one example of the periodic postings is reflected but would run many times. Also note the following details about our example:

- Amortization will be posted monthly, and by the end of the transaction, the entire $75,000 will be amortized.
- The accrual will be posted monthly.
- Interest will be paid according to the terms of the contract, but for simplicity's sake, we show the interest being paid after one month in the example.

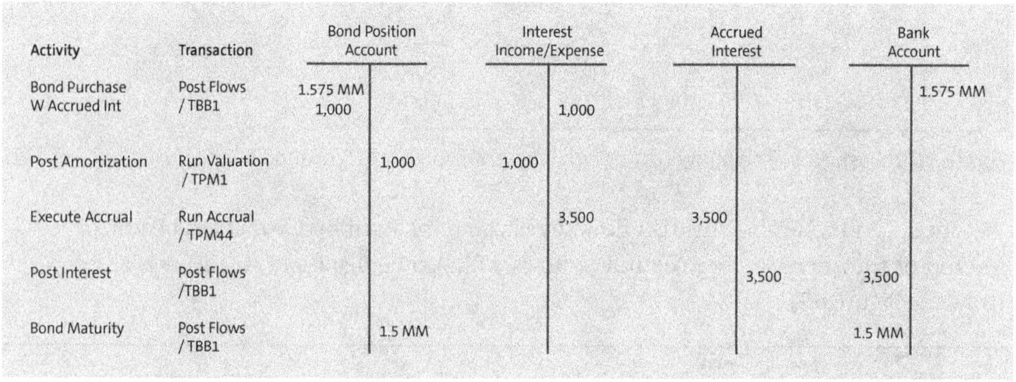

Activity	Transaction	Bond Position Account		Interest Income/Expense		Accrued Interest		Bank Account	
Bond Purchase W Accrued Int	Post Flows / TBB1	1.575 MM 1,000			1,000				1.575 MM
Post Amortization	Run Valuation / TPM1		1,000	1,000					
Execute Accrual	Run Accrual / TPM44				3,500	3,500			
Post Interest	Post Flows / TBB1						3,500	3,500	
Bond Maturity	Post Flows / TBB1		1.5 MM						1.5 MM

Figure 6.72 Sample Accounting Flows for Bond

As we've seen so far in this section, there are additional transactions and steps that we need to perform to process securities in SAP S/4HANA. Due to this, there are also additional steps that we need to cover when processing the accounting postings for these transactions.

We'll start by covering posting the initial flow for these transactions using the Post Flows app. After that, we can move on to the monthly transactions, which will post the amortization of the premium or discount of the bond as well as the interest accruals. Then, we'll cover posting the debit positions, which include the actual interest or coupon payments and the final repayment of the bond. Finally, we'll look at some of the one-off transactions that we can process, and we'll describe the securities rights and how they are initiated and posted.

6.4 Securities

Posting Flows

As we mentioned in the previous section, the accrued interest and principal flows are posted by the Post Flows app or Transaction TBB1. The difference with securities is in how we can run the posting programs with either the transaction number or the **Securities Class ID**. When we run the Post Flows app, the output will display both the posting details and the payment details if a payment request was created. Figure 6.73 shows the postings for both the purchase of the bond and the accrued interest payment.

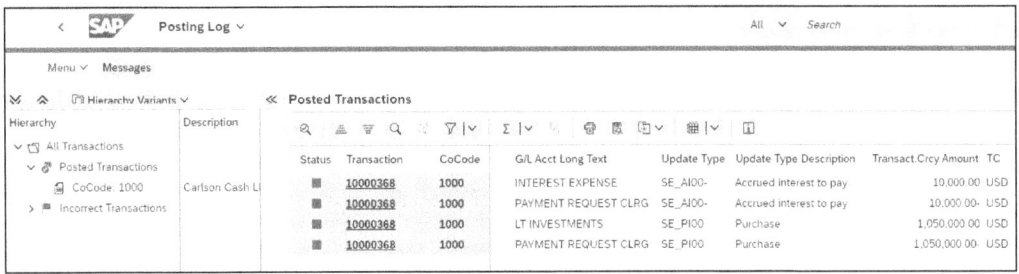

Figure 6.73 Post Flows App Will Post Both Accrued Interest and Initial Flow for Bond

Amortization

The difference between the purchase price of a bond and the nominal value of the bond is the *premium* (if the purchase price is higher than the value) or the *discount* (if the purchase price is lower than the value). That difference is amortized throughout the life of the bond, and we determine the amount and method of the amortization in customizing. The settings of the amortization procedure are tied to the position management procedure, and this results in the amortization amounts that are posted in each period.

To post the amortization, we go to the Run Valuation app or Transaction TPM1. This is the same transaction that posts the mark-to-market positions. For securities, we run this transaction by the **Securities Class ID** number and not by transaction number. Once we select the transactions, we can run them, and the amortization amounts and postings will be generated accordingly. It's important to keep in mind that this transaction needs to be run after any relevant position postings. If any interest or position flow hasn't been posted before the Run Valuation transaction is executed, an error will occur saying that there's a scheduled business transaction with a posting date before this key date valuation.

Once we run the transaction, the output screen will show the amortization details. The key figures for the amortization are as follows and are reflected in Figure 6.74:

❶ This number displays the carrying value of the bond before the amortization.

❷ This is the amortization amount that is posted as part of this valuation run.

❸ After the amortization posting, this is the new carrying value of the bond.

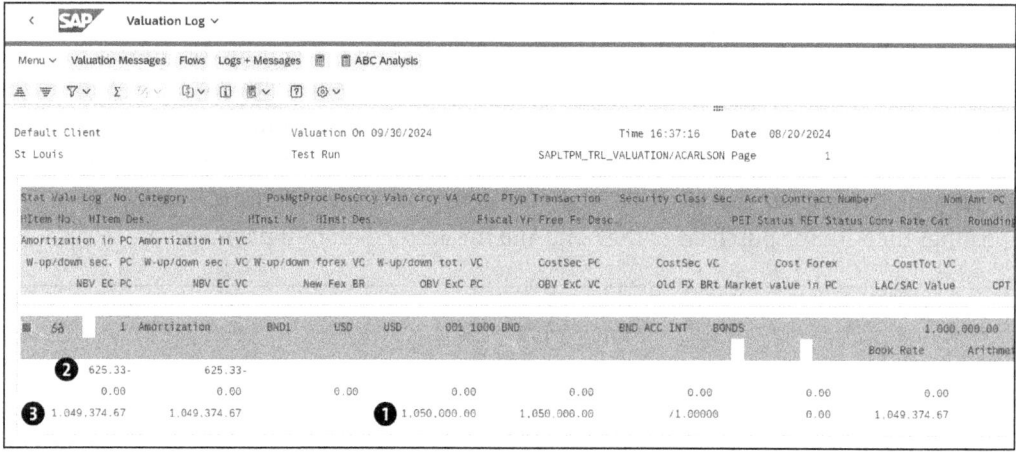

Figure 6.74 Valuation Log for Amortization of Bond Premium or Discount

Interest Accrual

Since the bond coupon payments are generally paid annually or semiannually, the interest is relevant to accrual and can be accrued monthly or quarterly. We will use the Run Accrual/Deferral app or Transaction TPM44 for the bond accruals. When we're running the program, the transaction number is not available, so the securities class ID is required for this posting. The output screen gives all details on the current accrual run and shows what will be accrued for the next interest payment. This is detailed in Figure 6.75 and as follows:

- The record labeled with the **MM_AI00+** update type shows the total amount that is accrued in the current accrual period. There are 85 days of accrual from 07/06/2024 to 09/30/2024, for a total accrued amount of $14,166.67. Each accrual will save an amount in this update type to show what was accrued in each period.

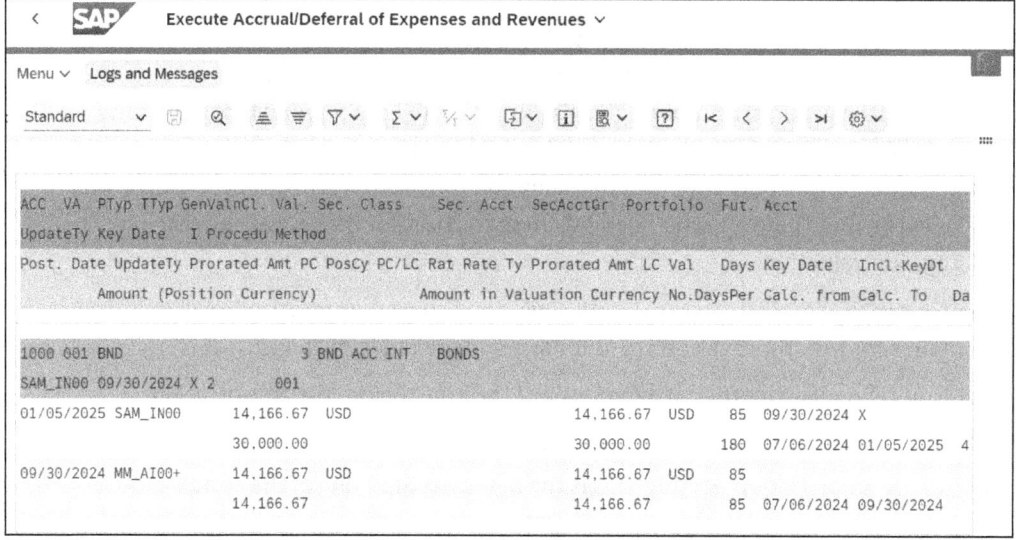

Figure 6.75 Log Screen Showing Output of Accrual/Deferral Run for Bond Interest

- The record labeled with the **SAM_IN00** update type displays the running total amount that was accrued for the six-month interest period. So far, only the 85 days' worth of interest has been accrued, but the full interest period has a total of 181 days, and the total amount of accrued interest that will be recognized over that time period is $14,166.67.

Manual Debit Position

Business transactions that we must post in the system for securities are known as *debit positions*. The Execute Debit Position – Manual Debit Position app or Transaction FWZE appear in Figure 6.76. The app can post interest and the final repayment of the bond individually.

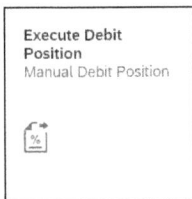

Figure 6.76 Execute Debit Position App

We run this transaction for each security class ID. After we enter the **Company Code**, **Security Class ID**, and **Securities Acct**, we can execute the program. This is shown in Figure 6.77, which displays all positions that are available for posting through the life of the transaction. To post a position, we can select a line and click the **Post Business Transaction** button. We double-click on the line item if any changes need to occur.

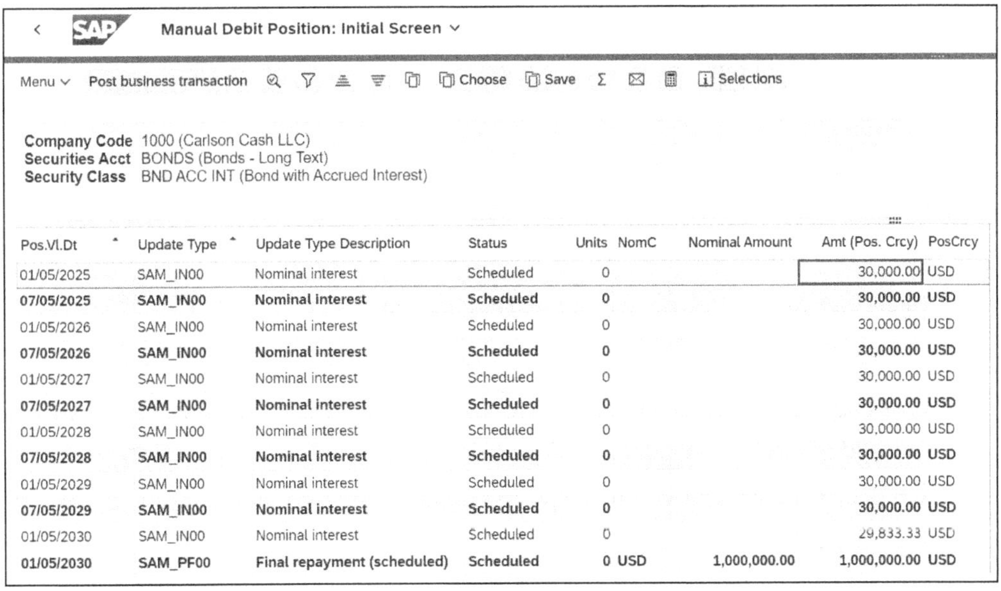

Pos.Vl.Dt	Update Type	Update Type Description	Status	Units	NomC	Nominal Amount	Amt (Pos. Crcy)	PosCrcy
01/05/2025	SAM_IN00	Nominal interest	Scheduled	0			30,000.00	USD
07/05/2025	SAM_IN00	Nominal interest	Scheduled	0			30,000.00	USD
01/05/2026	SAM_IN00	Nominal interest	Scheduled	0			30,000.00	USD
07/05/2026	SAM_IN00	Nominal interest	Scheduled	0			30,000.00	USD
01/05/2027	SAM_IN00	Nominal interest	Scheduled	0			30,000.00	USD
07/05/2027	SAM_IN00	Nominal interest	Scheduled	0			30,000.00	USD
01/05/2028	SAM_IN00	Nominal interest	Scheduled	0			30,000.00	USD
07/05/2028	SAM_IN00	Nominal interest	Scheduled	0			30,000.00	USD
01/05/2029	SAM_IN00	Nominal interest	Scheduled	0			30,000.00	USD
07/05/2029	SAM_IN00	Nominal interest	Scheduled	0			30,000.00	USD
01/05/2030	SAM_IN00	Nominal interest	Scheduled	0			29,833.33	USD
01/05/2030	SAM_PF00	Final repayment (scheduled)	Scheduled	0	USD	1,000,000.00	1,000,000.00	USD

Figure 6.77 Transaction Where We Can Review and Post Flows

We can change the **Value Date** and **FI Posting Date** on the subsequent screen. If required, we can click the **Payment Details** button to either add or update the paying house bank and account ID. If all looks good, on the output screen shown in Figure 6.78, we can select the **Post** button to post the flows.

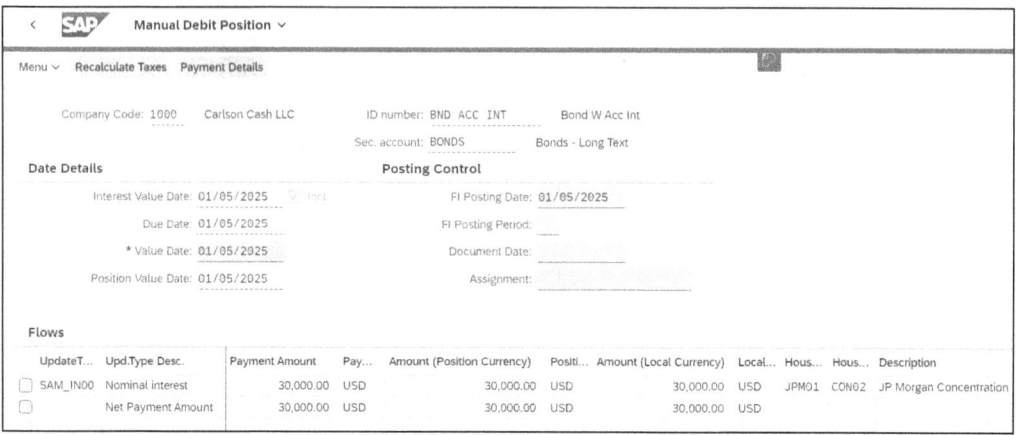

Figure 6.78 Posting Individual Flow in Manual Debit Position

Automatic Debit Position

Posting each transaction by securities class ID can be cumbersome, especially if we own a large number of securities. If this is the case, there's another transaction available that can post all cash flows up to a defined posting date, similar to how the Post Flows app or Transaction TBB1 operates for other types of transactions. To post multiple transactions in one run, we can use the Automatic Debit Position and Posting – Security Account app or Transaction FWSO, as highlighted in Figure 6.79.

Figure 6.79 Automatic Debit Position and Posting App

We can execute this transaction without filtering to pull in all securities debit position postings, or we can filter it by key attributes to ensure only the desired postings are run. The options are shown in Figure 6.80.

When we run this transaction, the postings and related payment requests will be automatically generated. The output of this transaction is shown in Figure 6.81.

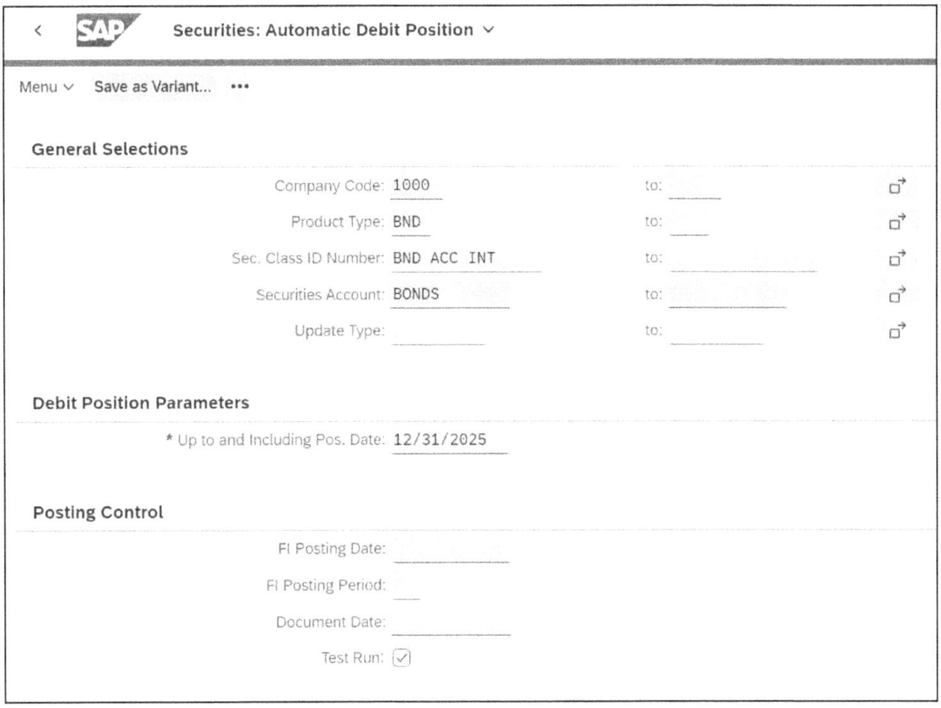

6.4 Securities

![Securities: Automatic Debit Position screen]

Figure 6.80 Selection Screen for Automatically Posting the Debit Position for Securities

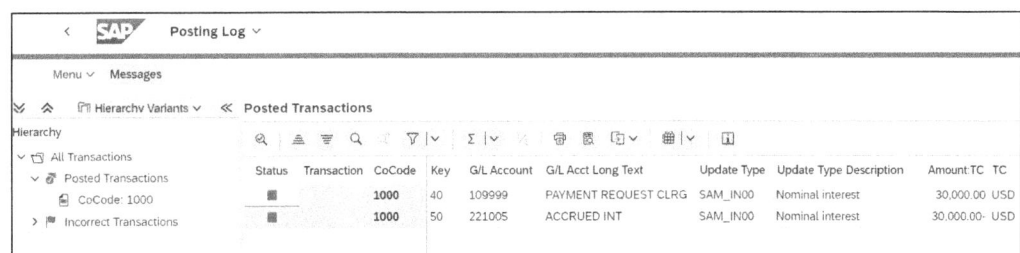

Figure 6.81 Posting Log for Automatic Debit Position Posting

Enabling Transactions for Automatic Posting

By default, we cannot post the debit position flows using the Automatic Debit Position and Posting app. We need to enable this by specifying an update type in the customizing. The menu path to the required settings is **Financial Supply Chain Management • Treasury and Risk Management • Transaction Manager • Securities • Position Management • Securities Account Management • Update Types • Specify Update Types for Securities Account Management**. To enable this functionality, we drill down into the desired update type and select the flag in the **Can Be Processed Automatically with Debit Position** setting.

433

6.4.6 One-Off Postings

Along with the regular postings for the securities transactions, there are a few types of postings that only occur in certain scenarios. One of these postings occurs when there are rights related to the securities transactions, and some of these exercisable rights apply if a bond is callable or puttable. Another one-off posting involves adding other flows to the securities contract, which occurs if an extra fee or flow needs to be added and posted for the transaction. There are also scenarios where postings from the securities transaction need to be reversed. We'll highlight the apps and transaction codes we use when we need to reverse a flow on a securities transaction.

Securities and Options: The Exercise Rights App

Certain securities have other flows that can occur throughout the life of the transaction. There are executable rights on the transactions, and they need to be recorded accordingly. Treasury and risk management supports a list of rights, known as rights categories. They are as follows:

- Equity Warrant
- Bond Warrant
- Index Warrant
- Currency Warrant
- Subscription Right
- Convertible Bond
- Warrant Bond
- Puttable Bond
- Callable Bond
- Security Swap
- Stock Option Bond Option
- Index Option
- Commodity Listed Option on Commodity Futures

When a right needs to be executed in any of these categories, the system will recognize any transaction that is relevant for exercising any of these rights in the Exercise Rights app or Transaction FWER, as shown in Figure 6.82. The applicable executable right will only appear for a security if the security has been set up to execute that right in both the configuration and the securities class. An example of this occurs when we need to populate the notice tab in the securities class if a bond is either callable or puttable. We covered the method for populating that tab in Section 6.4.3.

Figure 6.82 Exercise Rights App

As shown in Figure 6.83, there are currently transactions available to exercise a right for a convertible bond, a puttable bond, and a callable bond. If a transaction is available for rights, there will be a dropdown list option, and if we select that, we'll be able to see a list of all security class IDs under that rights category. To select any of the transactions, we double-click on the line item. Then, we need to select the key date, which needs to fall within the defined period for the selected right. In the example in Figure 6.83, we've selected a **Callable Bond** whose key date needs to fall between 01/05/2029 and 01/05/2030 (inclusive). Since the defined key date is 02/01/2029, the right can be executed. Additionally, there's an option to execute the right with a partial amount in the **Nominal Amount to Exercise** field, so if the full amount of the bond isn't called, then the correct amount can be reflected. To review the impact on the subsequent cash flows from the right execution, we can click the **Cash Flows** button. If everything looks accurate from a cash flow perspective, we can click the **Exercise Right** button to post the related documents.

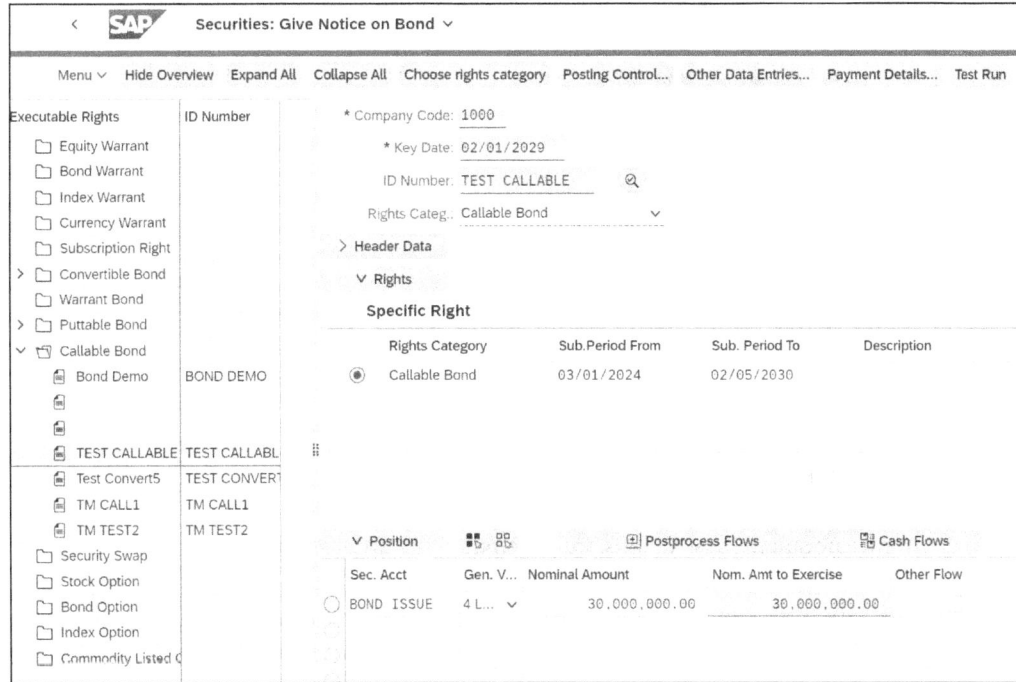

Figure 6.83 Sample Callable Bond in Exercise Rights App

Other Flows

Periodically, we may have to add other flows to a securities transaction. These can be additional fees or charges that we need to record in the transaction. After the initial date of a securities transaction, we can only add flows in a separate transaction and not directly in the Create Financial Transaction app. To add these flows, we use the Create Manual Posting app or Transaction FWBS (see Figure 6.84).

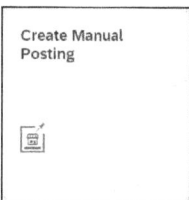

Figure 6.84 Create Manual Posting App

In this transaction, we need to designate the update type as an *other flow* in the customizing. If we do this, we can select the update type, and we also need to add transaction details (including payment amount, dates, and payment details) if the flow should create a payment. To add the flow, we start by clicking the **Insert other flow** button ☐ at the bottom of the page. The popup shown in Figure 6.85 will be displayed so we can add the flow, and we should add the other flow's **Update Type** and **Payment Amount**. After we populate the details, we click the **Copy** button to navigate back to the main screen.

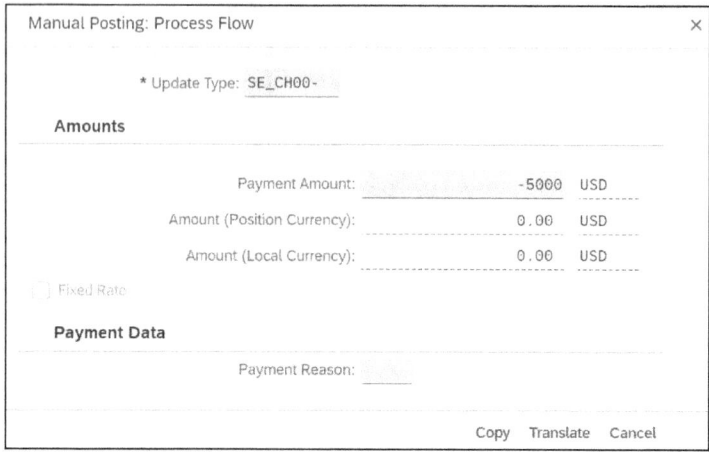

Figure 6.85 Popup with Update Type and Payment Amount

After we populate these details, we can update the different dates to drive the due date for payment, the value date, the posting date, the interest value date, and the position value date. These details can be found in Figure 6.86. We can post this flow directly in this transaction by clicking the **Post** button—or we can select **Save Without Posting**, which will create a flow that is queued up to be posted but that we will need post by using the automatic or manual debit position transaction.

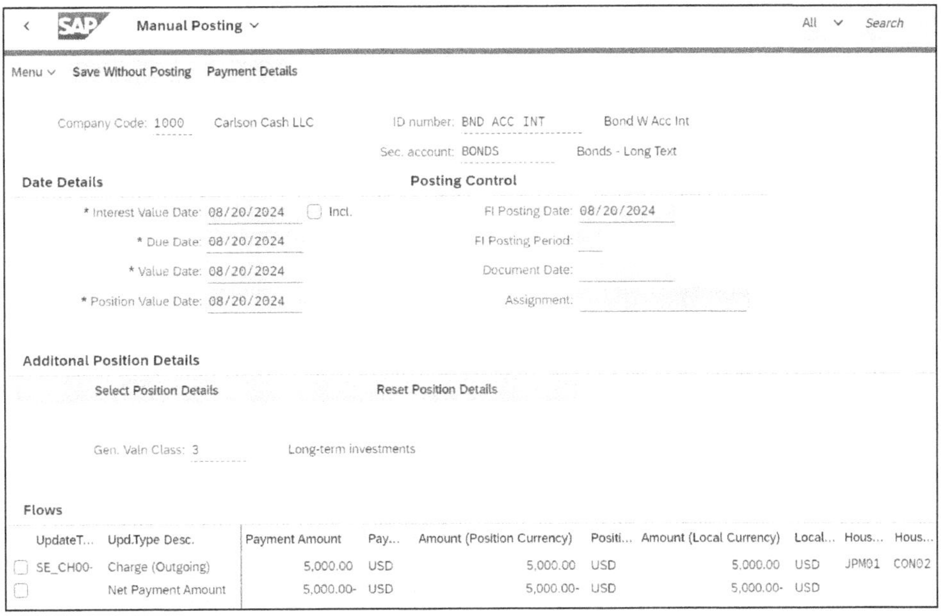

6.4 Securities

Figure 6.86 Creating Manual Posting for Securities Transaction

Defining Update Types for Manual Posting

We need to enable update types in a specific way to ensure they are relevant to the Create Manual Posting app or Transaction FWBS. We need to add them to the **Update Types for Manual Posting** tab in the following customizing menu path: **Financial Supply Chain Management • Treasury and Risk Management • Transaction Manager • Securities • Position Management • Securities Account Management • Update Types • Specify Update Types to the Functions of Security Account Management**. An example of this configuration is shown in Figure 6.87. Each of these update types added in this tab are available for manual postings.

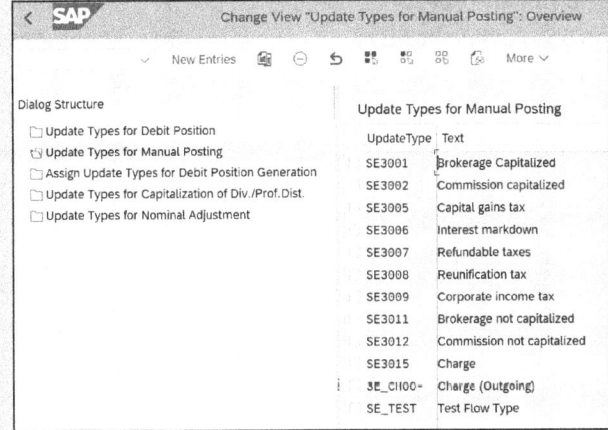

Figure 6.87 Update Types Assigned in This Tab Allow for Manual Postings for Securities Contracts

437

Securities Transaction Reversals

If we need to reverse any of the transactions within securities, we can use the transactions listed in Table 6.1.

Original SAP Fiori App/Transaction	Reversal SAP Fiori App/Transaction
Execute Debit Position – Manual Debit Position app/Transaction FWZE	Reverse Debit Position app/Transaction FWOEZ
Automatic Debit Position and Posting – Security Accounts app/Transaction FWSO	Reverse Automatic Debit Position Run – Securities Accounts app/Transaction TPM_POSTAUTREV
Exercise Rights app/Transaction FWER	Reverse Rights app/Transaction FWER_STORO_NEU
Create Manual Posting app/Transaction FWBS	Reverse Debit Position app/Transaction FWOEZ

Table 6.1 Transactions Needed to Reverse Securities Entries

6.5 Foreign Exchange

We use the FX process in SAP S/4HANA to facilitate the purchasing and selling of two currencies. We can handle these as FX spots, FX forwards, nondeliverable forwards, and FX swaps. *Spots* are FX trades on today's date, and *forwards* will be settled at a future date. To differentiate these transactions, we set them up to have different transaction types. *Non-deliverable forwards* (NDFs) involve a cash settlement of the gain or loss of the transaction. The notional amount of these transactions is not exchanged or not delivered. In Chapter 8 and Chapter 9, we cover FX financial transactions that are relevant to hedge accounting, and we also cover balance sheet and cash flow hedging.

This section will cover the end-to-end process of recording and processing foreign exchange transactions. Section 6.5.1 covers the high-level process for FX transactions in SAP to create a baseline for the following sections. Section 6.5.2 covers the differences in the Create Financial Transaction app or Transaction FTR_CREATE for FX transactions and details the additional fields that we should consider when entering contracts. Section 6.5.3 highlights some differences in FX processing, and Section 6.5.4 covers the month-end process of running valuations and postings. Section 6.5.5 looks at the settlement of FX trades and the posting of the gain/loss from each trade, and Section 6.5.6 covers some additional processes for FX trades, including processing NDFs and rolling over FX forwards.

6.5.1 Process Flow

Figure 6.88 summarizes how the process flow looks for each of the FX transactions and details the key steps in the process.

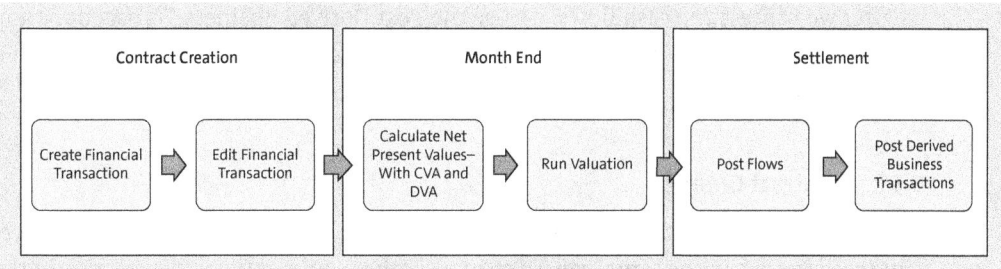

Figure 6.88 General Process Flow for FX Transactions

First, we have contract creation. The initial transactions in this process start with the creation and review of the FX trades in SAP S/4HANA. There are two aspects to this, as follows:

- **Create Financial Transaction app or Transaction FTR_CREATE**
  The general process flow for FX transactions starts with creating the trade using the Create Financial Transaction app. This is where we enter the details of the purchase and sale sides of the FX transaction, as well as the FX rates that we have agreed upon with the counterparty. The structure of the contract will drive how the transaction is processed in the following steps.

- **Process Financial Transaction app or Transaction FTR_EDIT**
  The Process Financial Transaction app allows us to review and settle the transaction. This step allows for a second set of eyes to review the trades to ensure all details have been captured correctly.

The next set of apps and transaction codes are focused on the month-end or period-end processing to calculate and post the valuation for the transaction, as follows:

- **Calculate Net Present Values – With CVA and DVA app or Transaction TPM60CVA**
  When we need to calculate a net present value (NPV) for a transaction, we execute the Calculate Net Present Values – With CVA and DVA app. This app reviews the market data available to calculate the NPV for the transactions.

- **Run Valuation app or Transaction TPM1**
  The Run Valuation app takes the NPV that was calculated in the previous processing step and generates the accounting postings.

Finally, we come to contract settlement. The final steps in this process cover the posting and paying the exchange of the currencies and recognizing the gain or loss that occurred from the transaction.

- **Post Flows app or Transaction TBB1**

  The cash flows for the FX contract occur at the end of the contract, and we process them using the Post Flows app.

- **Post Derived Business Transactions app or Transaction TPM18**

  After we process the cash flows, we use the Post Derived Business Transactions app to create postings to post and recognize the gain or loss that results from the FX transaction.

6.5.2 Contract Creation

We'll also enter FX transactions in the Create Financial Transaction app. We start by creating the contract in the same way we create all other contract types, by entering in the company code, product type, transaction type, and business partner for the transaction.

However, there are noticeable differences in the entry screen for FX transactions. The **Structure** tab now holds all the details of the buy and sell sides of the FX transaction. Details on each of the fields to enter data into in an FX forward are as follows (and as shown in Figure 6.89):

- **Rate**

  This entry has three blank fields, which are as follows, from left to right:

  - **Leading currency and following currency**

    Here, we define the leading and following currency in the currency pair.

  - **Rate of foreign exchange transaction**

    Here, we define the all-in exchange rate for the financial transaction, factoring in the swap rate for forwards.

- **Buy/Sell**

  Here, we determine whether we're buying or selling the currency amount that we determine in the next field.

- **Traded Amount**

  Here, we determine the currency and amount that we're trading.

- **Opposite Amount**

  Here, we define the opposite amount and currency for the transaction.

- **Value Date**

  Here, we define the payment and settlement date for the FX transaction.

- **Spot Rate**

  Here, we define the current spot exchange rate.

- **Swap Rate**

  Here, we define the markup or markdown from the spot rate for the forward contract.

6.5 Foreign Exchange

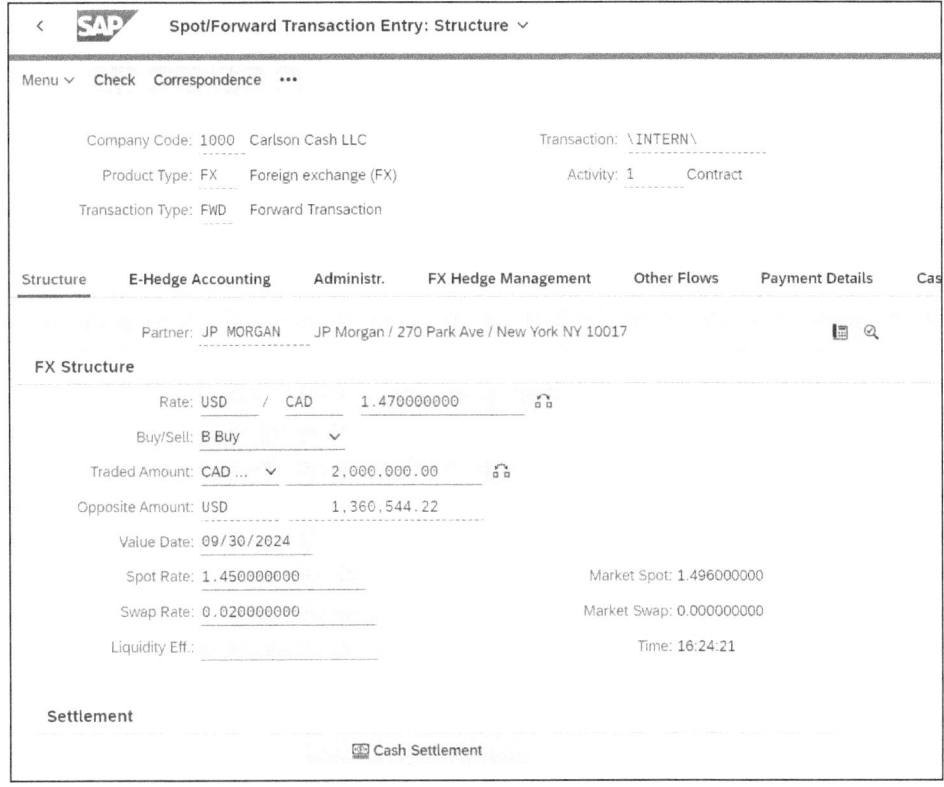

Figure 6.89 Contract Entry Screen for FX Forwards

6.5.3 Contract Posting

The flows of an FX contract differ from the flows of debt and investments since there are no cash flows until the settlement of the financial transaction. There are, however, additional calculations and postings for FX transactions due to fluctuations in FX rates and the impact those fluctuations have on the value of the transaction. Due to this, here are the common postings in FX:

- **Mark-to-market valuation**
  This posts the calculated mark-to-market value at month end.

- **Post nominal flows**
  When an FX transaction is settled, the nominal amounts are exchanged for spots and forwards. In the case of an NDF, the notional amount is not exchanged and only the difference between the notional amounts is exchanged.

- **Post gain/loss**
  Once the transaction is settled, the resulting gain or loss is posted and recognized.

6.5.4 Monthly Processing

Since the cash movements of an FX transaction happen during settlement, the first postings occur during the month-end process, which includes both calculating the NPV of the FX cash flows and posting the accounting entries for this valuation. To start examining monthly processing, we'll look into the process of calculating the NPV at month end. The following sections cover the process if the NPV was calculated in SAP S/4HANA using the market data that has been imported, or if a manual valuation was entered due to the valuation being calculated outside of SAP S/4HANA. After we calculate the valuation, we'll cover creating the accounting entries for the valuation.

Determining Net Present Values: Business Process

To determine the NPV, we can use the Calculate Net Present Values – With CVA and DVA app or Transaction TPM60CVA. This app (shown in Figure 6.90) looks at the available market data to determine the NPV.

Figure 6.90 Calculate Net Present Values App

For each currency that we're evaluating, there will need to be a currency-specific yield curve, the reference interest rate data points for that yield curve, and FX volatilities. In addition to the NPV, we can determine the following:

- **Credit value adjustment (CVA)**
  This is the value deducted from the NPV that represents the projected loss if the counterparty were to default.

- **Debit value adjustment (DVA)**
  This is the value added to the NPV that represents the projected gain if the company were to default.

We can calculate the CVA and DVA in two different ways, as follows:

- **Difference method**
  This method uses two different calculations to determine the CVA and DVA. In the evaluation type customizing area, we assign multiple yield curves, including a risk-free yield curve and a risk-based yield curve, as follows:
  - **Risk-free NPV**
    First, we use the risk-free yield curve to calculate the NPV.

- **Risk-based NPV**

    Then, we also use the risk-based yield curve to calculate an NPV. If the system is configured with credit spreads, we take those into account in the risk-based NPV calculation.

- **CVA or DVA**

    The CVA or DVA is the difference between the risk-free NPV and the risk-based NPV: Risk-free NPV – Risk-based NPV = CVA or DVA.

- **Expected exposures**

    With this method, we will also calculate the risk-free NPV, but we won't use the risk-based NPV. Instead, we calculate the CVA or DVA based on expected exposures on a key date. We calculate the expected exposures in three different ways:

    - **Constant exposure approach**

        We determine the NPV on the evaluation date, and we will assume that the NPV is constant until the end of the transaction.

    - **Variable exposure approach**

        We determine multiple maturity bands and a different NPV for each of the maturity bands. Due to this, the expected exposure varies over time.

    - **Manual entry**

        If we calculate the exposures outside of SAP S/4HANA, we can manually enter the expected exposures into SAP S/4HANA using Transaction TPMEEM.

Determining Net Present Values: Transaction Processing

To determine the NPV in the Calculate Net Present Values – With CVA and DVA app, we need to first enter some information into the entry screen to drive the execution of this program. First, the selection mode determines key details on which transactions will have their NPVs determined.

These settings will determine which transactions we can select, based on whether they are OTC transactions, what their netting groups are, and whether they are commodity or noncommodity transactions. Two selections to highlight in the selection mode are the evaluation parameter derivation and the market value decomposition. These selections, along with the additional filtering criteria, are shown in Figure 6.91.

If we check the box for the **Evaluation Parameter Derivation** indicator, many of the key details for how to run this transaction will be derived based on the evaluation parameters set in position management. The details that can be derived are the evaluation type, CVA/DVA type, price/NPV type, and valuation currency. **Market Value Decomposition** determines that market value components are calculated for the NPV.

6 Treasury Management–Specific Processes

Figure 6.91 Selection Screen for Determining NPVs

Once we have successfully selected the transactions, we need to set a few key attributes to successfully run the transaction. They are as follows (and also see Figure 6.92):

- **Evaluation Date**
  This is the key date in the valuation.

- **CVA/DVA Type**
  This determines how the CVA or DVA is calculated.

- **Evaluation Type**
  This determines the data used to run the valuation. We assign the relevant yield curves and market data to the evaluation type in customizing.

- **Price/NPV Type**
  This drives the type of NPV that is saved.

Once we have finished making these settings, we can execute the transaction. The system will display the output of the transaction, and a green indicator will appear to confirm that the NPV was successfully determined and saved. In Figure 6.93, we can see that the NPV was calculated successfully.

For any transaction, we can review the details of how the NPV was determined and what data was used to calculate the NPV. We select the **Detailed Log** button to navigate to the detailed log in Figure 6.94, which shows how the NPV was calculated and the key figures that were used to calculate the NPV.

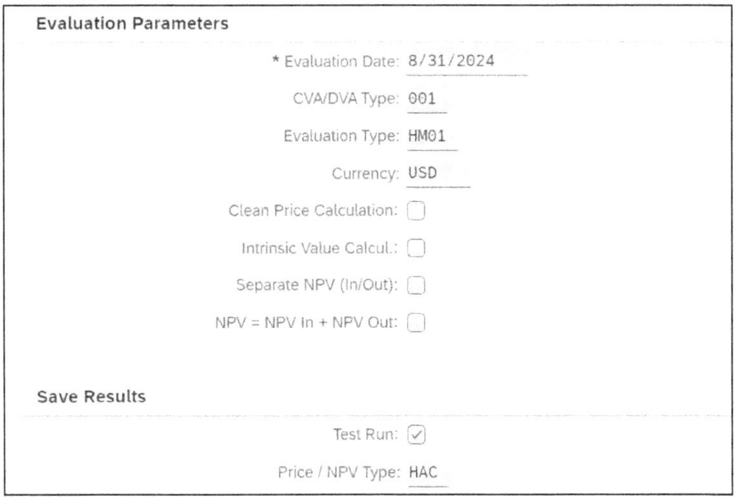

6.5 Foreign Exchange

Evaluation Parameters

* Evaluation Date: 8/31/2024
CVA/DVA Type: 001
Evaluation Type: HM01
Currency: USD
Clean Price Calculation: ☐
Intrinsic Value Calcul.: ☐
Separate NPV (In/Out): ☐
NPV = NPV In + NPV Out: ☐

Save Results

Test Run: ☑
Price / NPV Type: HAC

Figure 6.92 Evaluation Parameters for NPV Calculation

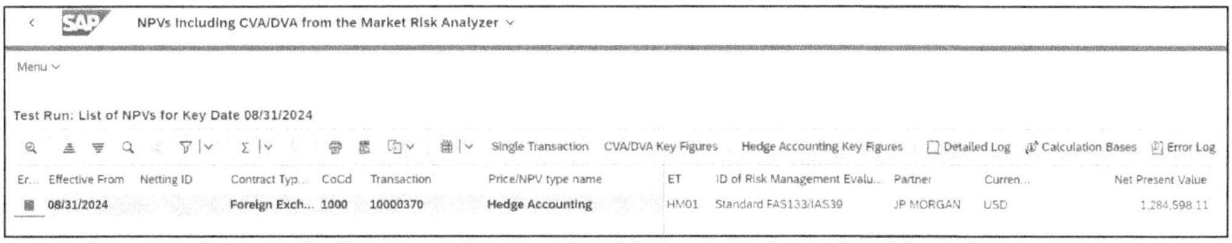

Figure 6.93 NPV for This Transaction Was Successfully Calculated on 08/31/2024

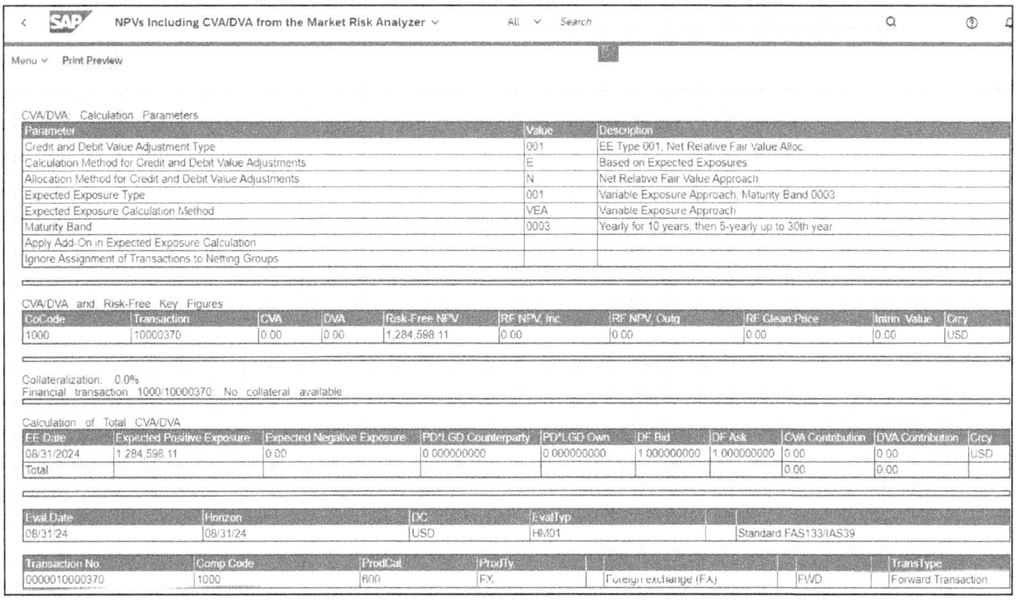

Figure 6.94 Detailed Log Showing All Details of How NPV Was Calculated

445

6 Treasury Management–Specific Processes

Finally, clicking the **Calculation Bases** button will navigate us to all the details of the market data that went into the calculation of the NPV. Figure 6.95 shows the details behind the CAD and USD yield curve along with the current FX spot rate.

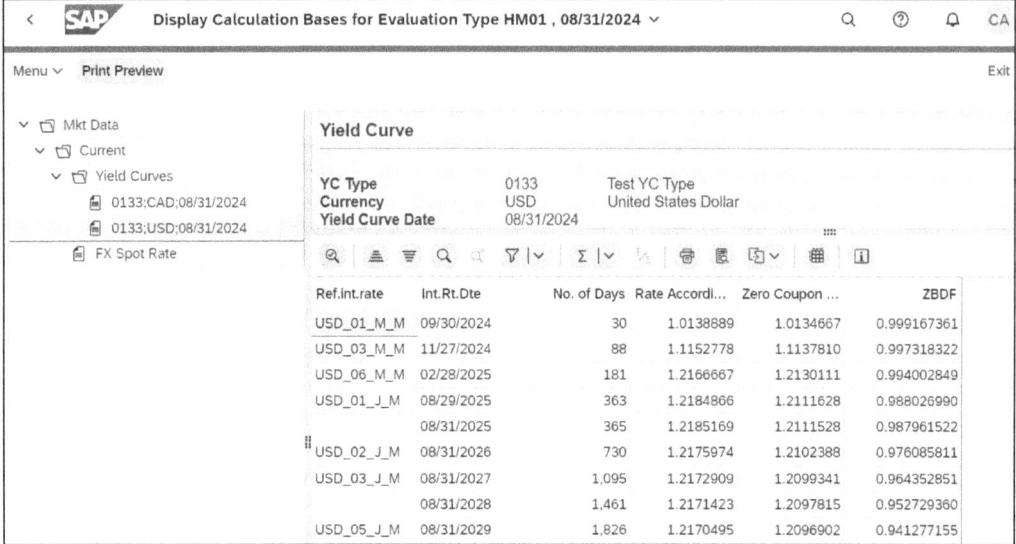

Figure 6.95 Details of Data Yield Curve Data

After we've successfully calculated the NPVs, the mark-to-market valuation will be ready for posting.

Manual Valuation

If we don't have the relevant market data in SAP S/4HANA for an NPV calculation, or if the calculations are provided by another system, then we can enter the valuations manually. The manual valuation function lets us manually enter valuations into SAP S/4HANA if the valuations are being calculated outside of SAP S/4HANA or provided by a third party. If we need to enter an NPV, we can use it to post the valuation just as if the NPV were calculated directly. Before we add the manual valuation into a transaction, we should determine the updated book value, which we'll then enter into the Enter Book Values for Manual Valuation app or Transaction TPM74 (shown in Figure 6.96).

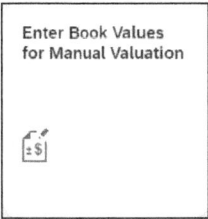

Figure 6.96 Enter Book Values for Manual Valuation App

6.5 Foreign Exchange

In this app, we can assign a valuation for a given **Key Date**. We can identify the transaction by company code (**CoCd**), **Valuation Area**, **Valuation Class**, and **Transaction ID**. **Security Class** and **Securities Account** are also available if a transaction is a security. We need to enter the book value in both the position currency and the valuation currency in this app. An example of an entered manual valuation in this transaction is shown in Figure 6.97.

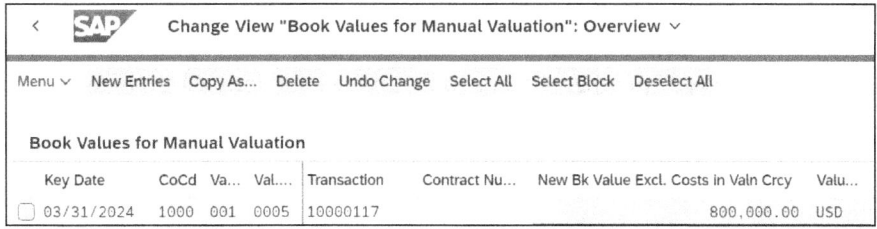

Figure 6.97 Assignment of Manual Valuation to Transaction ID 10000117

We also need to confirm that the transaction has been enabled so we can use the manual valuation. In customizing, the assigned security valuation procedure has to have the **Enable Special Write-Ups/Write-Downs for Securities** box checked. We find the screen where we check this box (as shown in Figure 6.98) in the following menu path: **Financial Supply Chain Management** • **Treasury and Risk Management** • **Transaction Manager** • **General Settings** • **Accounting** • **Settings for Position Management** • **Key Date Valuation** • **Define Security Valuation Procedure**.

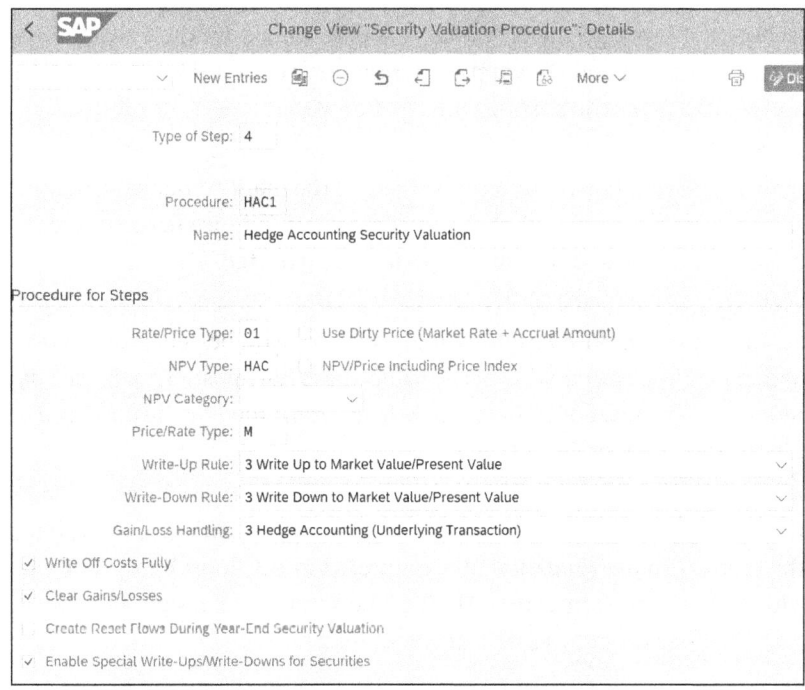

Figure 6.98 Security Valuation Procedure Enabled for Manual Valuations

447

Posting Valuations

There are many options for how to calculate and post valuations. Fortunately, once we have set up all the transactions with their position management procedure, we can run these transactions monthly without needing to review the calculations for each posting run. We successfully completed the calculation of the NPVs in the Calculate Net Present Values – With CVA and DVA app in the previous steps, and now, we're ready to post this valuation.

The Run Valuation app or Transaction TPM1 will use the previously calculated NPV and execute the posting. As with other SAP Fiori apps and transactions within treasury and risk management, making selections on the entry screen in **Product Groups** and **General Selections** helps us define the filtering to ensure that only the desired transactions are selected for this key date valuation. The two main additional selections that are required in this transaction are the **Key Date for Valuation** and **Valuation Category**. We must run the **Key Date for Valuation** on the same date when the NPV was calculated. The **Valuation Category** has a few options to drive what data this transaction is looking at and how the transaction is posted, as follows:

- Year-End Valuation

  This is meant to be run at year end and will post the change in positions according to the assigned position management procedure.

- Mid-Year Valuation

  We can run the mid-year valuation at any time, but it's generally run monthly or quarterly, depending on business needs and when we need to record the valuation. This transaction will leverage the NPV that was determined by the Calculate Net Present Values – With CVA and DVA app or Transaction TPM60CVA. When running the valuation, we can make the posting with reset or without reset, as follows:

  - Without Reset

    Running this transaction without reset will record the mark-to-market valuation each period, and the posting will be the incremental difference between the last mark-to-market posting and the current period being posted.

  - With Reset

    If we post the valuation **With Reset**, an additional flow will be generated to reset the valuation posting. The booking of the valuation is fully posted on the last day of the period, and the reset of this posting will be determined by the posting date defined in the **Key Date for Reset** field.

- Manual Valuation

  This valuation posting works the same as do the mid-year valuation options, but it looks for the manual valuation data that was entered in the Enter Book Values for Manual Valuation app or Transaction TPM74. The manual valuation also has an option to post the book differences with and without a reset.

6.5 Foreign Exchange

Once we've entered the selection screen, we can run the valuation. This transaction will display an initial screen to confirm that we want to run the valuation for all the selected transactions (see Figure 6.99).

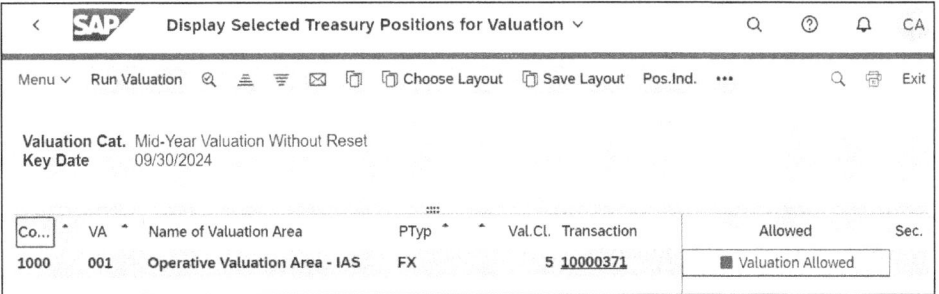

Figure 6.99 Run Valuation Transaction Showing We Can Run Valuation for This Contract

The **Valuation Log** will show all key details for the valuation posting. In Figure 6.100, a few of the key figures have been detailed with the following labels:

❶ This is the value of the previous valuation that was posted on 08/31/2024.

❷ This is the value of the current valuation calculated on 09/30/2024.

❸ Since we're posting this valuation without reset, the incremental difference between the August valuation and the September valuation is calculated and is displayed. This is the amount in the current posting.

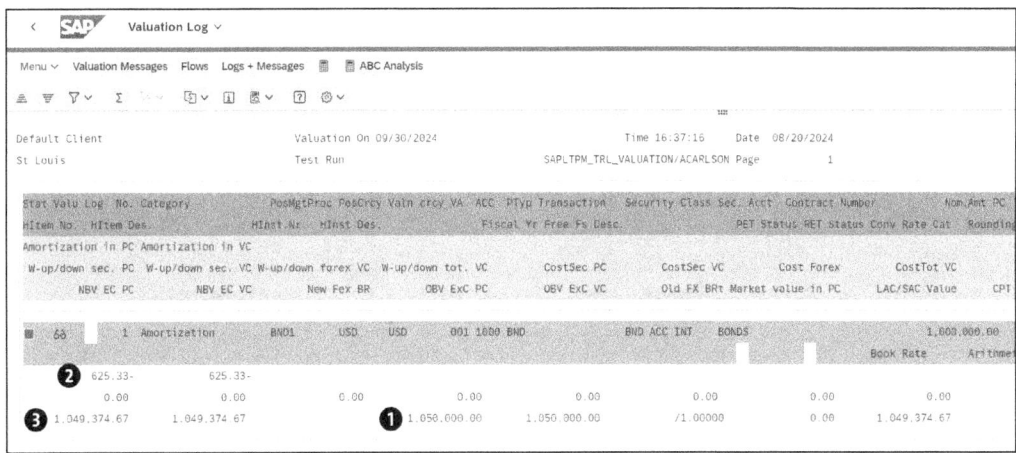

Figure 6.100 Valuation Log Detailing Current Valuation and How It Differs from Previous Valuation

The valuation will list each of the valuation steps separately for each position. The example in Figure 6.100 only has a security valuation, but there would be additional lines if there were other types of key date valuations. These could be a one step price valuation, a foreign currency valuation, a rate valuation, or an index valuation.

449

6 Treasury Management–Specific Processes

As with other FX transactions, there are other outputs available in which we can review the valuation messages and postings. The **Logs+Messages** area includes details on the **Posting Log**, **Distribution of Profit/Loss**, and **Messages**, and there's even a log on **Hedge Management** that lets us review details of how the mark-to-market postings affect the transactions related to hedging.

6.5.5 Transaction Settlement

The main postings that occur at the end of an FX transaction include the cash exchange for the currencies and the recognition of the gain or loss. These are posted with the Post Flows and Post Derived Business Transactions apps.

Posting Flows

At the conclusion of spot and forward FX transactions, we need to exchange the nominal amounts. To initiate this process, we need to post both cash flows, and we use the Post Flows app or Transaction TBB1 for this process. We can select the transactions with the filters in the **General Selections** section, and the date we enter in the **Up to and Including Due Date** field needs to be far enough out so we can select the cash flows for posting. Once we validate these, we can execute the Post Flows app.

The app will then show an output with the standard posting details, payment details, and messages. The posting log for the cash flows is shown on Figure 6.101. One thing to notice is that these flows are posted in the transaction currency. Since we're buying CAD and selling USD, we need to post these in the transaction currency to facilitate the payments.

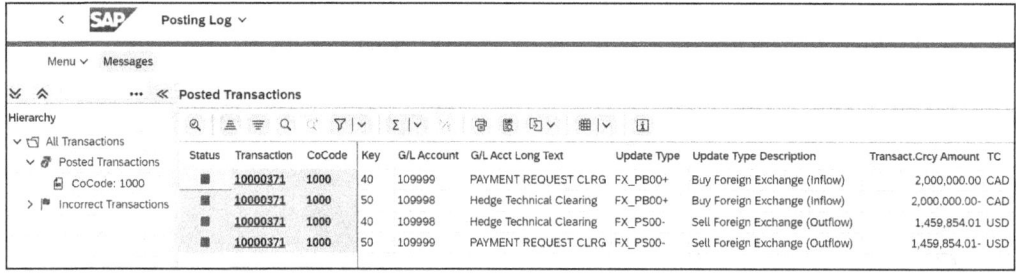

Figure 6.101 Post Flows Postings Showing Cash Impact of FX Trade

Posting Derived Business Transactions

The final step in the FX process is recognizing the gain or loss. There are a couple of different ways we can do this, and it depends on how the mark-to-market value has been posted at month end. Even though the derived business transaction posting does not use these two procedures, it needs to clear out the balances created by the previous postings that occurred in the Run Valuation app. If there are balances still sitting in the unrealized gain/loss account, we'll first need to clear out the balances. After that, we can post the gain or loss to the recognized gain/loss account. This section will cover these two scenarios and show how the postings vary based on the mark-to-market postings.

Difference Procedure

The first procedure we'll cover is the one we use if the mark-to-market postings were posted **Without Reset**. In this case, there will be a balance still sitting in an unrealized gain/loss account during settlement, and we need to clear this balance out. Figure 6.102 shows how the journal entries would work in this scenario.

To start this transaction, we're going to run through a simple example with round numbers. In our examples, the FX forward has us selling $1,000 US dollars and buying $1,300 Canadian dollars with an FX rate of 1.30. Let's get into some specifics of the transaction:

- A previous mark-to-market posting has an unrealized gain of $100.
- During settlement, the FX rate changes, and the spot rate for US dollars/Canadian dollars is now 1.40. This equals roughly $930, and that's reflected in the cash transaction. In the t-accounts, the US dollar account is credited and the Canadian dollar account is debited to reflect the cash movement. Both of these transactions pass through a technical clearing account when they are posted.
- When we run the Post Derived Business Transactions app or Transaction TPM18, the balance sitting in the unrealized gain/loss account is offset. This results in the balance being cleared, and no amount will be left in that account after this transaction is completed. Second, there's a balance remaining in the FX technical clearing account. This $70 difference will be cleared from this account and posted as a debit to recognize the resulting FX loss.

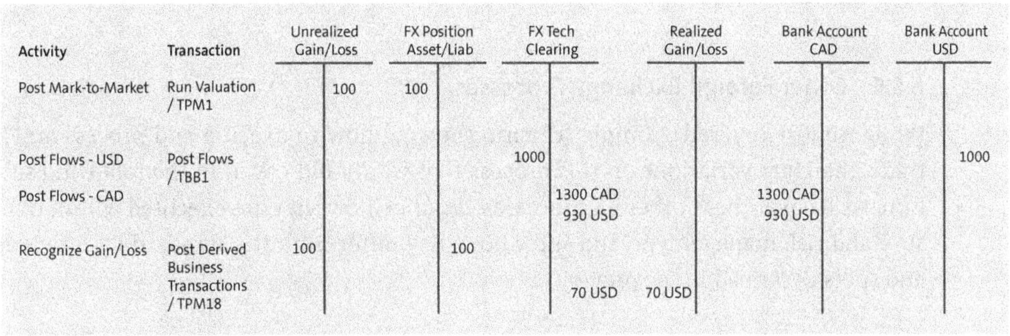

Activity	Transaction	Unrealized Gain/Loss	FX Position Asset/Liab	FX Tech Clearing	Realized Gain/Loss	Bank Account CAD	Bank Account USD
Post Mark-to-Market	Run Valuation / TPM1	100	100				
Post Flows - USD	Post Flows / TBB1			1000			1000
Post Flows - CAD				1300 CAD 930 USD		1300 CAD 930 USD	
Recognize Gain/Loss	Post Derived Business Transactions / TPM18	100	100	70 USD	70 USD		

Figure 6.102 Accounting for FX Trade Settlement Postings When Valuation Was Booked with Difference Procedure

Book and Reset Procedure

The other option is to record the mark-to-market postings **With Reset**. This means that when we post the mark-to-market valuation, there's an offsetting transaction created to reset the posting. This generally occurs at month end or quarter end. The initial journal entry is posted on the last day of the month or quarter, and the resetting entry occurs on the first day of the following period.

For this example of a transaction and how the gain or loss is posted, we'll use the same details as in the previous sample contract. The biggest difference is that the mark-to-market valuation was already posted and reset, so in this example, we don't need to reset the unrealized gain/loss account. The example of how we would handle the accounting in this scenario is shown in Figure 6.103. Here are two things to note:

- As with the previous example, the US dollars and Canadian dollars are exchanged in the bank accounts, and we should post the offsetting entry to a technical clearing account.
- We clear out the difference in the FX technical clearing account by using the Post Derived Business Transactions app, and the residual amount is posted to the realized gain/loss account.

Activity	Transaction	FX Tech Clearing	Realized Gain/Loss	Bank Account CAD	Bank Account USD
Post Flows-USD	Post Flows / TBB1	1000			1000
Post Flows-CAD		1300 CAD 930 USD		1300 CAD 930 USD	
Recognize Gain/Loss	Post Derived Business Transactions / TPM18	70 USD	70 USD		

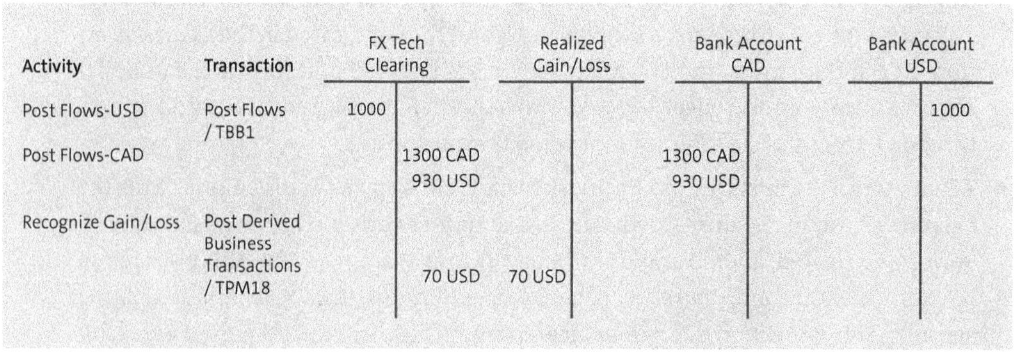

Figure 6.103 Accounting for FX Trade Settlement Postings when Valuation Was Booked with Book and Reset Procedure

6.5.6 Other Foreign Exchange Processes

While we just covered a simple scenario showing how to execute and process an FX trade, there are variations on this process that we should cover. In the following section, we'll cover these other FX processes, describe how they are executed within treasury and risk management, and show how they differ from the standard FX forwards and spots described in the previous section.

Non-Deliverable Forwards

NDFs operate slightly differently from regular FX forwards. The difference in this product type is in the settlement of the nominal cash flows. A regular FX forward will settle both sides of the FX trade in the traded currency—while the NDF instead calculates the difference in the settlement amounts of the currencies, and only the difference in these amounts is paid or received in the agreed upon settlement currency.

We generally use NDFs in currencies that are not frequently traded or when a company doesn't have a physical bank account in a foreign currency. These FX transactions are commonly used at companies that want to hedge exposures in other currencies but want to limit the cash outlay required for the settlement of the trades. In the following sections, we'll review the end-to-end process for NDFs from the entry screen to executing the FX rate fixing, and we'll then cover settlement and posting gains or losses.

Entry Screen

The entry screen for NDFs is similar to the entry screen for other FX contracts. The main difference is the inclusion of a settlement currency that drives the net settlement payment for the NDF maturity (see Figure 6.104). On the surface, the entry screen looks the same as the FX forward screen, but the key difference is the addition of a fixing date and settlement currency on the NDF. Since we need to fix a rate prior to the settlement of the NDF, we need to define the fixing date. The **Settlement Crcy** is the other addition, and we can select it manually on this entry screen. Alternatively, we can default the settlement currency based on the customizing that is detailed in Chapter 4, Section 4.2.12.

Figure 6.104 Entry Screen for NDFs

Fixing

An additional step in the NDF process is the requirement to run the fixing for the contract. We can do this in the Process Financial Transaction app (shown in Figure 6.105) by entering the **Company Code** and **Transaction** ID and selecting the **Fixing** option.

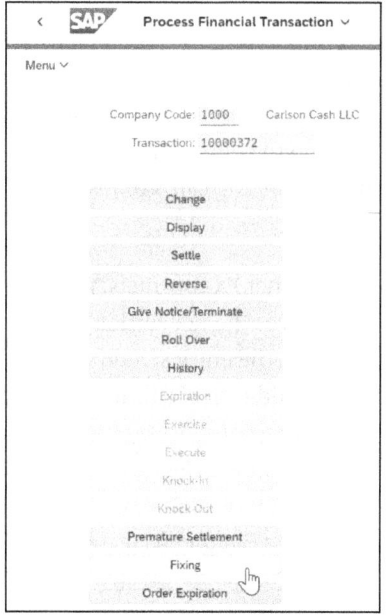

Figure 6.105 Selecting Fixing Option for NDF

We'll now see a **Settlement** section with an FX **Rate** and **Flow Type** for the cash settlement. The proposed FX rate in this area will display the current rate in the system, based on the selected currency pair. We can change this to the correct settlement rate that is provided by the business partner or trading platform. An example of the **Settlement** section is shown in Figure 6.106.

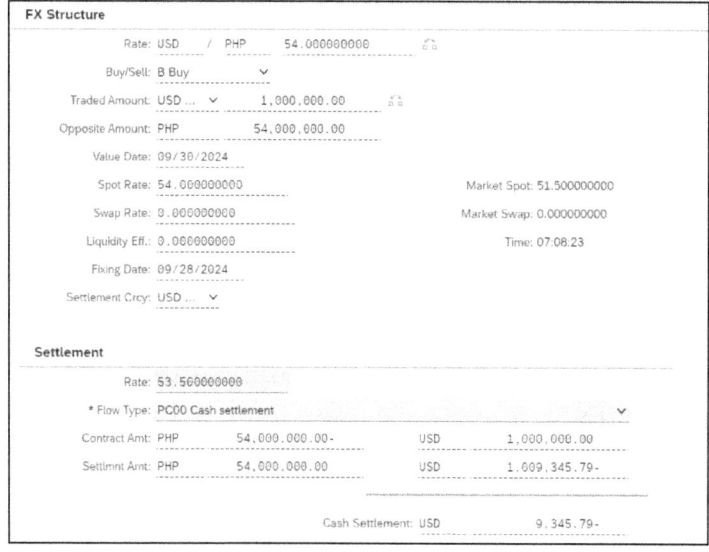

Figure 6.106 Settlement Area of Screen Allows for Determination of Flows and Rate of Cash Settlement

6.5 Foreign Exchange

Accounting

Once we have fixed the NDF rate, we can move on to the steps for the settlement. There are no differences in the month-end process for valuing NDFs, so this section's purpose is to show the flows for the settlement.

As with other processes, we use the Post Flows app to post the cash portion of the transaction. In this transaction, we pay or receive the difference between the two traded currencies in the NDF. Generally, this transaction will create either an incoming or an outgoing payment request for the cash side of the transaction, and the opposite line item will post to an FX clearing account.

The posting in Figure 6.107 is an example of the cash side of the settlement using the Post Flows app.

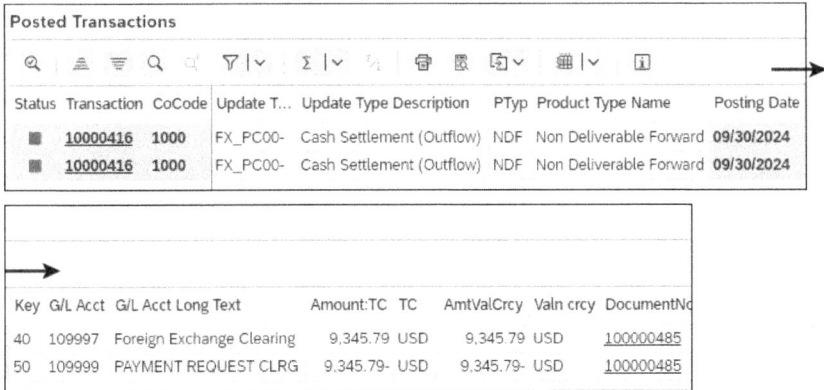

Figure 6.107 Posting Flows for NDF

Following this, we can run the Post Derived Business Transactions app or Transaction TPM18 to reflect the gain or loss from the NDF. This amount will equal the payment amount, and the log from this transaction is shown in Figure 6.108.

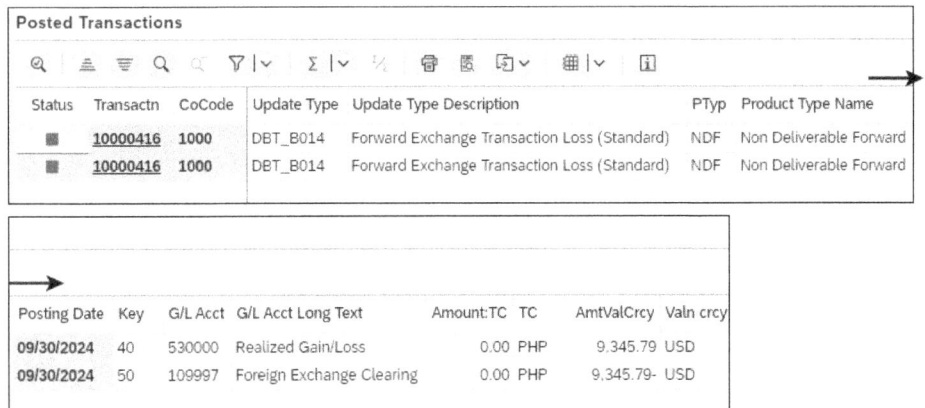

Figure 6.108 Gain/Loss Posting Using Post Derived Business Transactions App

In Figure 6.109, there's a summary of the journal entries we make when processing the postings for NDFs.

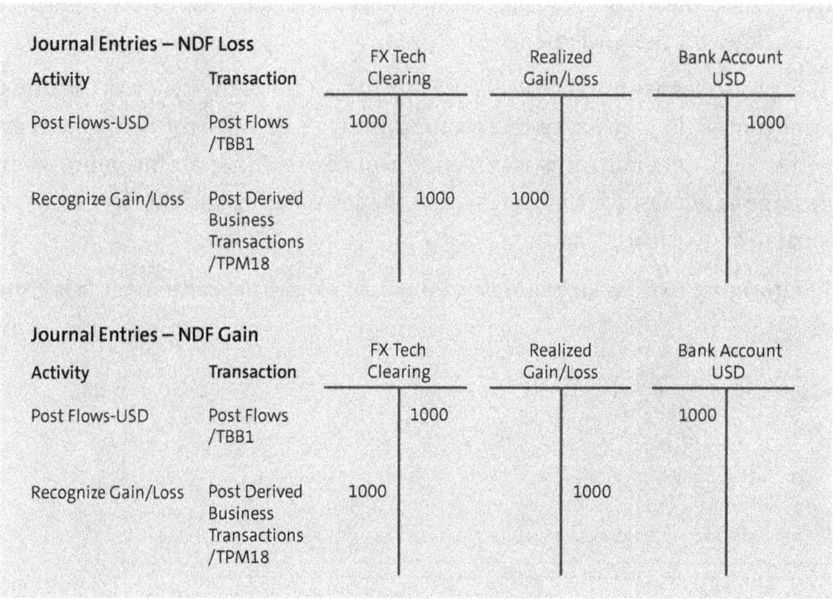

Journal Entries – NDF Loss
Activity
Transaction
FX Tech Clearing
Realized Gain/Loss
Bank Account USD
Post Flows-USD
Post Flows /TBB1
1000
1000
Recognize Gain/Loss
Post Derived Business Transactions /TPM18
1000
1000
Journal Entries – NDF Gain
Activity
Transaction
FX Tech Clearing
Realized Gain/Loss
Bank Account USD
Post Flows-USD
Post Flows /TBB1
1000
1000
Recognize Gain/Loss
Post Derived Business Transactions /TPM18
1000
1000

Figure 6.109 Accounting for NDF Settlement Flows

FX Forward: Rollovers

We need to extend some FX transactions' positions to a later settlement date, and when this scenario occurs, we don't simply change the settlement date of the transaction. Instead, we roll over the transaction to a future date. Functionally, what we need to do to initiate the rollover is in the Process Financial Transactions app. In this app, there's a **Roll Over** button that we click to initiate the change to the transaction, as shown in Figure 6.110.

Process Financial Transaction
Menu
Company Code: 1000
Carlson Cash LLC
Transaction: 10000373
Change
Display
Settle
Reverse
Give Notice/Terminate
Roll Over
History

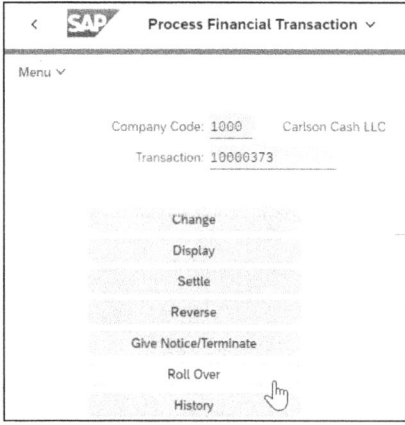

Figure 6.110 Initiating Rollover of FX Forward in Process Financial Transactions App

6.5 Foreign Exchange

When we click the **Roll Over** button, an additional screen will be displayed showing the proposed future transactions for the rollover (see Figure 6.111). The following list explains the purpose of the two transactions that we create for the rollover:

1. **Rollover transaction**
   The first transaction that is displayed is the new FX forward. For display purposes, we can see that the transaction type has been changed to **Rollover**. We can populate any of the available fields to create the next transaction. For this example, we'll update the contract's FX rate and value date to push out the maturity date. In this example, the original value date was 09/30/2024, and now the new date is being updated to 11/30/2024.

2. **Spot/netting transaction**
   The second transaction shows the FX spot that is used to offset the original FX forward. This transaction will have offsetting cash flows for the **Traded Amount** and the **Opposite Amount**. The FX rate may vary from the original rate, and if this occurs, there will be a gain/loss amount that we must settle as part of the rollover.

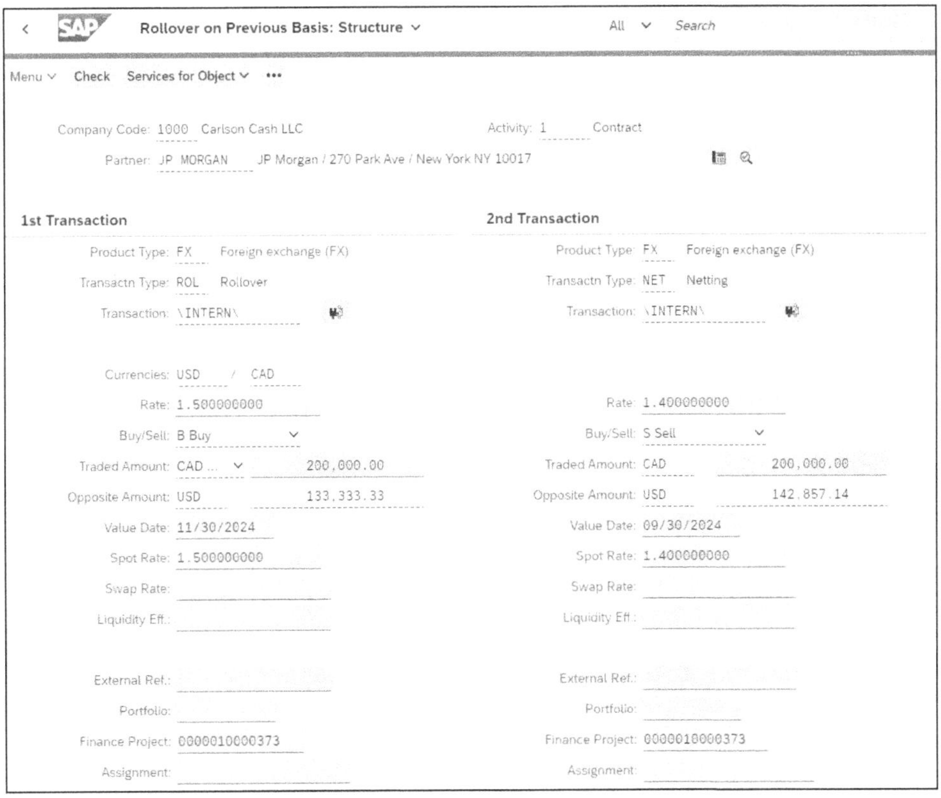

Figure 6.111 Rollover Entry Screen, Including Rollover Transaction and Netting Transaction

As a result of the rollover, two new financial transactions will be created since we created a new spot and forward. Both the spot and the forward will be tied to the original FX forward, and that can be displayed in the **Administr.** tab in the **Ref. Trans.** field, as shown in Figure 6.112.

6 Treasury Management–Specific Processes

Figure 6.112 Reference Transaction Field on Administration Tab References Original Transaction

Since rollovers have many moving parts, we'll look at the accounting that takes place as part of the rollover. The key details of the dates and transactions for the trades are as follows:

- **Original trade**
  - The original FX forward started on 08/23/2024.
  - The trade had a value date of 09/30/2024.
- **Rollover trades**
  - The offsetting spot was created with a value date of 09/30/2024.
  - The new FX forward has a value date of 11/30/2024.

When we post the settlement of the original contract and the offsetting spot trade, the amounts can offset, and only a residual amount remains for payment. In the example in Figure 6.113, the Canadian dollar amounts offset each other, and only a difference of $5.291.01 US dollars remains. We can see all the accounting entries that have been generated by the Post Flows app.

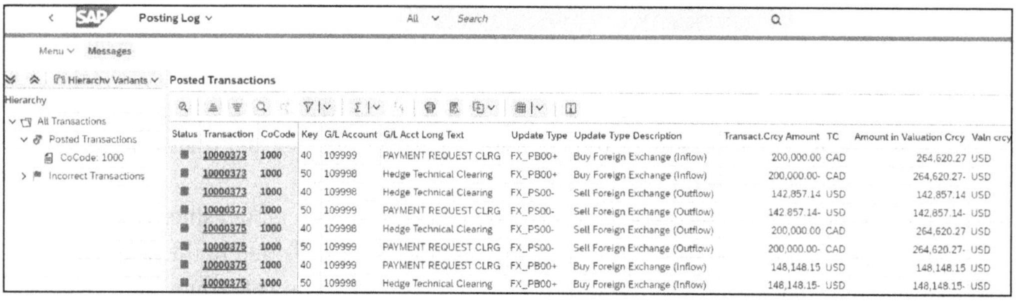

Figure 6.113 Posting Log

458

We can also see the differences in the amounts of the payments in the payment log in Figure 6.114. Each of these lines will create a payment request.

CoCd	Transactn	K...	Amount	PmtCurrAmt	PCrcy	Payt Date	HBank	Account	Payer/Payee	PBank	PaymStatus	Payment Status (Text)	Year	Customer	G/L Account	G/L acct
1000	10000373		200,000.00	200,000.00	CAD	09/30/2024	BOM01	PAY05	JP MORGAN	0002	2	Created	2024		109999	133000
			142,857.14	142,857.14-	USD	09/30/2024	JPM01	PAY02	JP MORGAN	0001	2	Created	2024		109999	121000
	10000375		200,000.00	200,000.00-	CAD	09/30/2024	BOM01	PAY05	JP MORGAN	0002	2	Created	2024		109999	133000
			148,148.15	148,148.15	USD	09/30/2024	JPM01	PAY02	JP MORGAN	0001	2	Created	2024		109999	121000

Figure 6.114 Transaction Showing Payment Log

Finally, we can post the new FX forward normally using the Post Flows app, and we can process the realized gain/loss with the Post Derived Business Transactions app.

6.6 Interest Rate Swaps

Like other types of contracts, interest rate swaps are managed within treasury and risk management. More specifically, as discussed earlier, interest rate swaps are managed within the **OTC Derivatives** subsection of treasury and risk management. OTC derivatives contain several different types of contracts: OTC swaps, OTC interest derivatives, OTC options and forwards, repos, securities lending, listed derivatives, commodity forwards, security forwards, and forward loans. Many of these contract types are outlined in detail within other sections of this book. To focus on one of the more commonly used transactions within OTC derivatives in SAP S/4HANA, this section specifically highlights the user-side processes for an interest rate swap. SAP S/4HANA uses a data element called *product category* to determine which type of contract is being configured and thus which pieces of additional functionality will be available to each configured product type. Figure 6.115 shows the initial configuration screen for OTC Derivative product types, and each folder on the left contains a set of products assigned to a specific product category. Figure 6.115 also shows the product types available in the **OTC Swaps** folder, and some of these folders also contain multiple product categories.

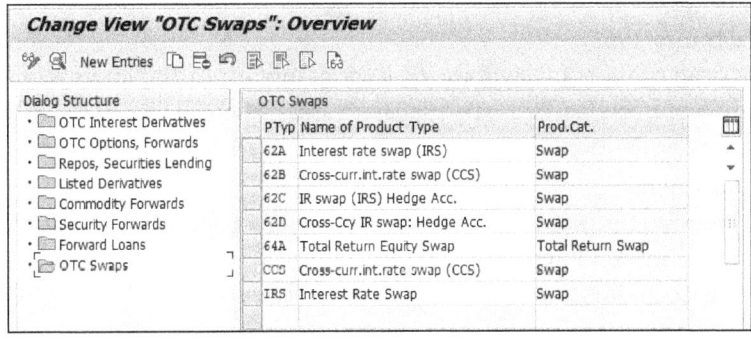

Figure 6.115 List of Product Types Available for OTC Swaps

This section focuses on interest rate swap execution in the **OTC Derivatives** area. Section 6.6.1 reviews the process flow required for the full lifecycle of these transactions, all the way from contract creation to contract maturity. Section 6.6.2 shows how the entry screen varies for these transactions due to the different transaction structure. Section 6.6.3 reviews the processing of the transaction, including fixing the interest rate adjustments, and Section 6.6.4 reviews using these adjustments to post cash flows. Additionally, interest rate swaps have the ability to run a valuation to determine how the value has changed period to period, and Section 6.6.5 explains how to run the valuation for these transactions.

6.6.1 Process Flow

Before we walk through the creation and routine processing of an interest rate swap, it's important for us to understand the process at a high level. It's also important to remember that each contract type can have varying degrees of difference from the others depending upon specific end-user requirements. However, the core processes of an interest rate swap remain largely consistent. In most cases, the overall process flow of an interest rate swap is as shown in Figure 6.116.

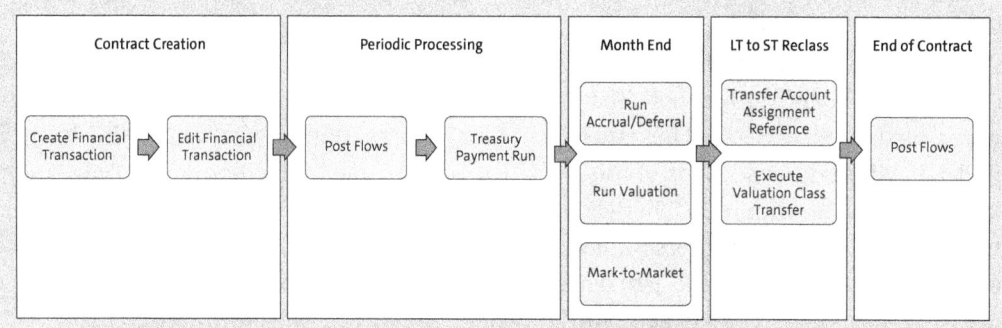

Figure 6.116 Interest Rate Swap Contract High-Level Process

This process has many similarities to the general contract process that we covered earlier. Yet, there are some key differences. In the following list, we'll highlight the primary differences for each stage:

- **Create financial transaction**
  The nature of a swap contract involves the concept of interest in and interest out. For these types of contracts, there are two separate areas where we maintain each side of the interest calculations.

- **Interest rate adjustments**
  In many cases, interest rate swaps will involve the usage of a reference interest rate to reflect a particular market interest rate. Transactions with variable interest rates leverage market data in SAP to determine the correct calculation of interest. We must run separate SAP transactions to fix the interest rate before the calculation will be updated and reflected.

6.6 Interest Rate Swaps

- **Cash flows**

  Since interest rate swaps involve independent settings for both inbound and outbound interest conditions, there's an extra tab provided in the **Cash Flows** area that allows us to view inbound and outbound cash flows separately.

- **Valuation/mark-to-market**

  A common month-end activity for an interest rate swap involves calculating a valuation of a swap contract. Since interest rate swaps typically don't involve any actual principal cash exchange, we can summarize the value of an interest rate with the value of the interest related to the swap contract.

6.6.2 Entry Screen

We can create interest rate swap contracts like other treasury and risk management contracts, by using the Create Financial Transaction app. If we select an interest rate swap product type, then the main **Structure** tab of the entry screen will look moderately different from those of other types of products. There are two columns on the main screen outlining the different settings that we can adjust for both incoming and outgoing interest. These two columns—which we call *legs*—work independently of each another, and we can adjust them separately. The leg dictates settings for outgoing interest, whereas the right leg dictates the settings for incoming interest. Figure 6.117 shows the contract creation screen **Structure** tab during creation of an interest rate swap contract.

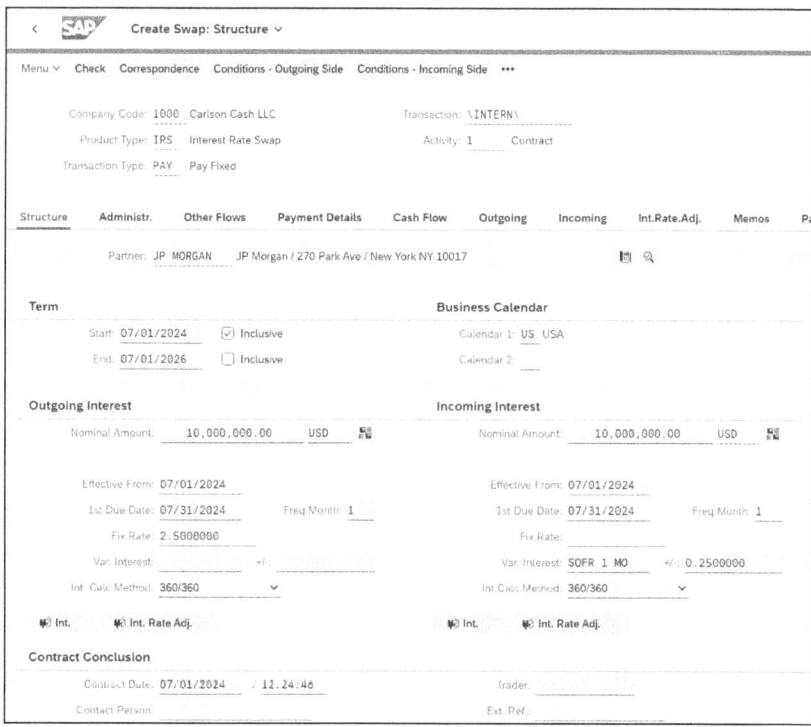

Figure 6.117 Structure Tab During Creation of Interest Rate Swap Contract

461

In the following sections, we'll walk through the interest structure and then the interest condition settings.

Interest Structure

Within the **Structure** tab on the main contract entry screen, as previously outlined, there are two separate sections where we can control the inbound and outbound interest conditions. While there are two separate sections, the available interest condition settings are consistent between the two.

The **Outgoing Interest** and **Incoming Interest** sections shown in Figure 6.117 drive all details of how the interest is calculated, which rates are used, and on which dates the interest is paid for each respective direction. The options in the interest structure area are driven by the conditions that we assign to the product types and transaction types, and we can use the available conditions to calculate standard interest, variable interest, and capitalized interest.

One common example of an interest rate swap contract structure involves paying *fix to float*. Essentially, this means that the contract is constructed so that the outbound interest (the interest being paid) is calculated at a fixed rate and inbound interest (the interest being received) is calculated using a variable interest rate. We can dictate whether a certain leg of a contract uses a fixed interest rate or a variable interest rate by selecting a reference interest rate in the **Variable Interest** field. If we don't select a reference interest rate, then the system assumes that the interest leg will use a fixed interest rate. In addition, we can apply any additional spread to the variable interest rate.

The following fields are also present on the interest structure screen, and we use them to establish different attributes of the contract interest flows:

- **Effective From**
  Here, we can determine the date from which the corresponding interest conditions should begin.

- **1st Due Date**
  Here, we enter the first date when interest is paid or received. This field can be different for each leg but is commonly the same.

- **Interest Calculation Method**
  The calculation of the interest varies depending on the number of days we use for the calculation, and in this field, we determine how the interest days are calculated. For variable rate contracts, we assign a standard interest calculation method to each reference interest rate. We can assign a different calculation method at the contract level, but SAP S/4HANA will provide a warning to make sure the correct method has been assigned before we can save the instrument.

- **Frequency (in months)**
  In this field, we determine how frequently interest is calculated, paid, and/or received. Generally, we'll calculate the interest on a monthly or quarterly basis. To

achieve this, we should input the number of months between interest payments in this field. We can input "1" to calculate the interest monthly or "3" to calculate the interest quarterly.

- **Fixed Rate**
  If we calculate the interest leg using a fixed rate, we input the necessary fixed rate here as a percentage. For example, we input 2.5% interest as 2.500.

- **Variable Interest**
  If we calculate the interest leg using a variable interest rate, we input the necessary reference interest rate here. As mentioned earlier, the reference interest rate will use SAP market data input into the system to calculate the resultant interest flows.

- **+/- (spread)**
  We can input any necessary spread on top of the assigned reference interest rate. Frequently, an interest rate will be quoted as a rate plus or minus a certain number of basis points (which we call a *spread*). We can input those basis points here to modify the overall interest rate used to calculate interest flows.

Interest Condition Settings

There are additional interest condition–related settings that we can modify for each leg. Again, the settings available are the same for each leg, but we can maintain them independently for each leg of the contract. Additional details of the interest condition settings can be found in Chapter 3, Section 3.15.

If these fields are not sufficient to fully determine the interest structure of the instrument, we can use the **Conditions** buttons to further define the interest structure. We use many of the fields from the **Structure** tab are to populate the **Interest Conditions** page. There are additional criteria that we can add in the **Amounts** and **Dates** tabs of this conditions area, and we covered the details of how to edit the interest conditions in Section 6.1.2.

When we assign a reference interest rate to a variable-rate contract, the interest rate adjustment condition becomes required for the instrument. This is no different for interest rate swaps. This condition drives when the interest amount will be adjusted and fixed, and we can edit this condition by clicking on one of the **Conditions** buttons at the top of the screen. Like other areas of contract construction, interest rate swaps can maintain independent interest conditions for both inbound and outbound interest flows. Thus, they have two available interest **Conditions** buttons at the top of the screen: one for outgoing interest conditions and one for incoming interest conditions. Once in the **Overview of Conditions** page, we can select the interest rate adjustment **Condition Type** by adding a parallel condition group. This condition is only required for the legs of the contract that use a variable interest rate to calculate interest payments. We can see this in Figure 6.118 and Figure 6.119. In Figure 6.119, notice how this variable-interest leg has the interest rate adjustment condition type assigned.

6 Treasury Management–Specific Processes

Figure 6.118 Outgoing Interest Conditions Overview

Ac...	C...	C...	Condition Type Name	Eff. From	Amount-Based Structure	Date Structure
1	1000	2000	Nominal Interest	07/01/2024	SOFR 1 MO + 0.25 %	1 Monthly
		1215	Interest Rate Adjustment Swap	07/01/2024	Interest Adjustment	Relative to Start of Period
	2000	1005	Final Repayment (No Post)	07/01/2024	100 %	Final Repayment

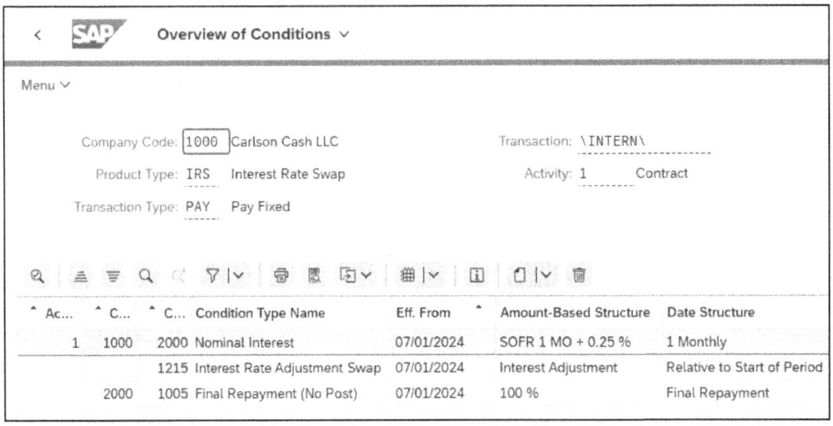

Figure 6.119 Incoming Interest Conditions Overview

6.6.3 Interest Rate Adjustments

When we assign a variable interest rate to a contract, the values of the periodic interest payments are not known at the time of contract creation. We calculate the interest based on the assigned reference interest rate and any relevant spread. The interest payments of the contract will be displayed in the **Cash Flows** tab, but until the interest rate is fixed, the actual amount of the interest payment amount is not known. Before we can finalize an interest amount for a flow, we need to adjust or fix the interest rate for the cash flow.

The fixed reference interest rate will depend on the reference interest rate, the spread, and the fixing dates we assign in the interest rate adjustment condition. Fixing the rates can be done automatically or manually, and it follows the same process that we covered in Section 6.1.4.

6.6.4 Cash Flows

Traditionally, a treasury contract has a single set of cash flows. Most product categories don't natively utilize a dual condition structure, but interest rate swaps, as we've learned, do have multiple corresponding interest conditions: a set of conditions for outgoing interest and a set of conditions for incoming interest. Because of this, we can view the cash flows for each direction of the interest rate swap independently. On the main entry screen of an interest rate swap, there are three tabs related to cash flows. The first, **Cash Flow**, shows both the outgoing and incoming cash flows on the same screen. Additionally, there are separate tabs for strictly **Outgoing** interest flows and for strictly **Incoming** interest flows. This can aid the user in removing excess noise from the cash flow area and only viewing the flows related to one leg of the contract at a time. The additional cash flow tabs can be seen in Figure 6.120.

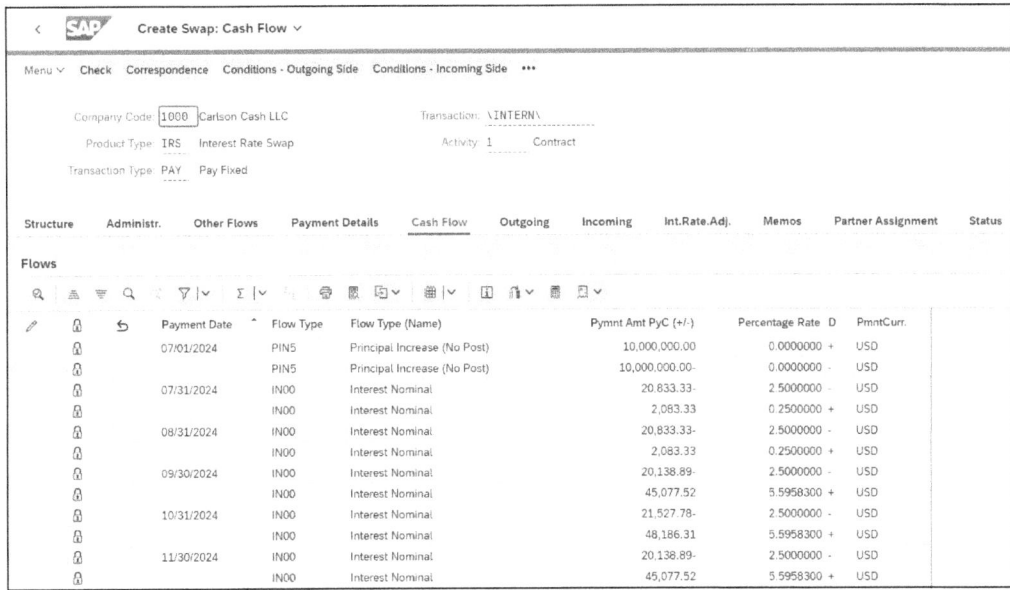

Figure 6.120 All Three Cash Flow Tabs: Cash Flow, Outgoing, and Incoming

6.6.5 Mark-to-Market Valuation

As stated earlier, traditionally, interest rate swaps don't contain an actual movement of principal cash. The principal value of an interest rate swap contract serves as a reference on which to calculate corresponding interest payments. At the same time, accounting departments often report the value of interest rate swaps within company records. To do this, SAP S/4HANA can calculate the NPV of the interest rate swap based on settings established within the position management procedure and the related configuration. The settings for a position management procedure are detailed in Chapter 3, Section 3.6.

6 Treasury Management–Specific Processes

Based on these settings and the adjoining configuration, we can calculate the value of an interest rate swap using different apps or transactions:

- **Calculate Net Present Values app or Transaction TPM60**
  We use this app or transaction to calculate the NPV of an underlying contract. The values are calculated by the Market Risk Analyzer, and the results are posted to table JBNPV for further review. We can maintain table JBNPV using Transaction JBNPV should we need to make manual adjustments.

- **Calculate Net Present Values – With CVA and DVA app or Transaction TPM60CVA**
  We use this app or transaction to calculate the NPV of an underlying contract with the added ability to include credit and debit value adjustments. These values are calculated by the Market Risk Analyzer and added to table JBNPV for further review.

The transactions function similarly, and Figure 6.121 shows the main screen of the Calculate Net Present Values app.

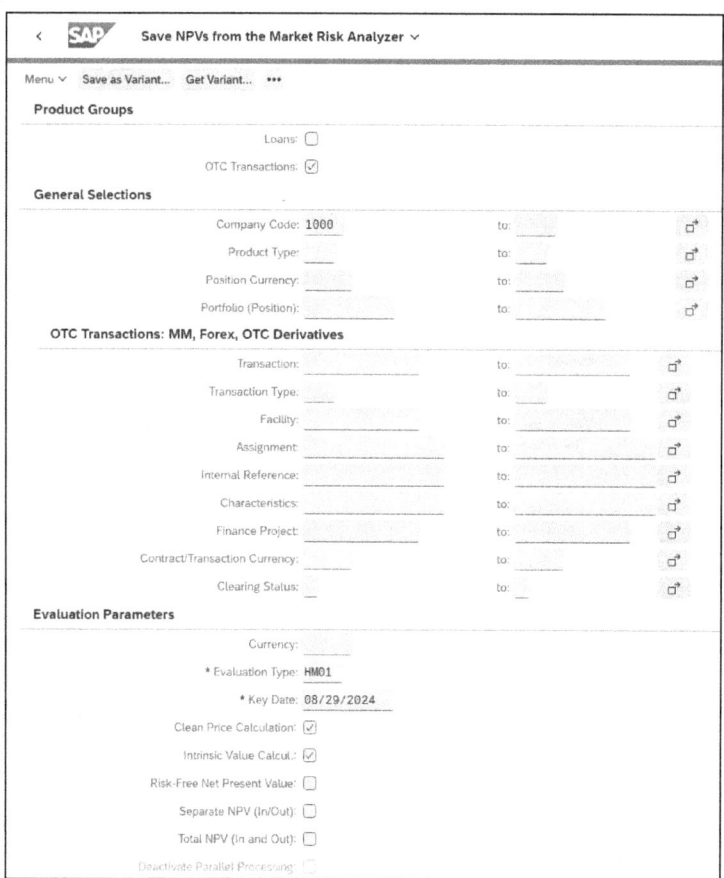

Figure 6.121 Main Entry Screen of Calculate Net Present Values App

There are many selections that we can make on this screen to derive an accurate NPV. The key data elements that we need to generate an NPV are as follows (see Figure 6.122):

- **Evaluation Type**
  This is the evaluation type the transaction should use to calculate the NPV. Evaluation types are configured market and valuation parameters that the Market Risk Analyzer uses to determine an NPV. These are oftentimes predelivered in accordance with common accounting principles (for example, FAS133 and IAS39).

- **Key Date**
  This is the date for which the valuation should be generated. Any related transactional flows up through this date will be included for valuation purposes, whereas flows that are calculated beyond this date will be excluded.

There are several other evaluation parameters (as well as important contract values) that we can use with these apps or transactions to assist with calculating an NPV. Checking each of the evaluation parameter boxes will activate a different calculation procedure. The most utilized evaluation parameters are as follows:

- **Clean Price Calculation**
  Checking this box will generate a *clean price* in addition to an NPV for any of the indicated transactions. While not especially relevant for an interest rate swap, the clean price is often used to calculate the NPV of securities contracts.

- **Intrinsic Value Calculation**
  Checking this box will make the system calculate the intrinsic value and time value of the selected contracts in addition to an NPV. In relation to an interest rate swap, the intrinsic value is generally the value of any fixed leg payments less the value of any floating leg payments. The time value calculation is not especially relevant for interest rate swaps but is more commonly associated with options contracts.

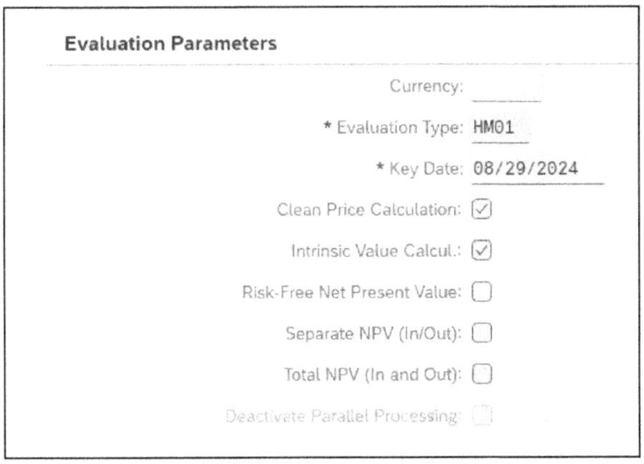

Figure 6.122 Evaluation Parameters Area of Calculate Net Present Values App

After the contract has been properly valued and any necessary valuations have been added to the NPV table `JBNPV`, we can generate postings for those values by using the Run Valuation app or Transaction TPM1. Details on how to execute this transaction are the same as those we covered for FX in Section 6.5.4.

6 Treasury Management–Specific Processes

6.7 Trade Finance

Within SAP S/4HANA's trade finance management functionality, organizations can track and oversee their trade finance activities. Trade finance encompasses transactions, including letters of credit and bank guarantees. Once these financial instruments are recorded within treasury and risk management, the system facilitates subsequent processes like issuance, amendments, payments, and settlements. In the upcoming sections, we'll dive into the specific intricacies involved in entering a letter of credit contract in SAP S/4HANA.

6.7.1 Process Flow

Before jumping into the details of each entry screen within trade finance, we need to explore the overarching process flow of the treasury and risk management process, as depicted in Figure 6.123. This high-level visual offers insights into the steps involved in managing trade finance transactions. The overall process is very similar to that for a debt or investment contract, starting with the creation of the contract, the settlement process, and the subsequent accruals, deferrals, and financial postings. The biggest difference between trade finance contracts and a money market contract is the process of creating a contract, once the contract has been created the same programs are used to calculate and post the accruals and deferrals as well as the financial postings.

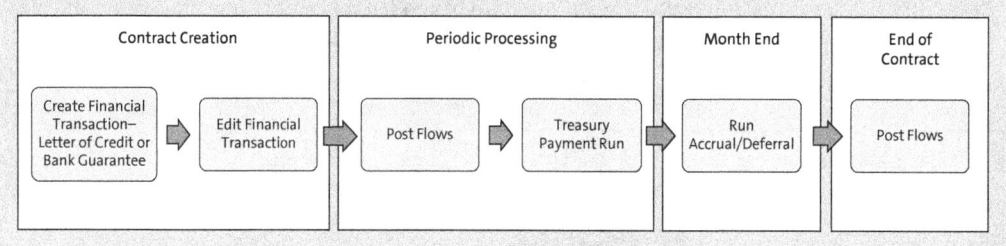

Figure 6.123 Trade Finance Process Flow

6.7.2 Letter of Credit Contract Creation

To start the setup of a letter of credit contract, we use the Create Financial Transaction app to input the initial contract details. We first enter our **Company Code**, followed by **Product Type** "LCS" and **Transaction Type** "REC." We then specify the business **Partner** that we've entered into the letter of credit with; in this example, we're using "JP Morgan," as we can see in Figure 6.124. One we've entered in the initial details of the contract, we click **Create,** which takes us to the next screen, where we enter the structural details of the contract.

This section serves as the entry screen where we can enter the structural components of the letters of credit. It lets us enter important details such as the nominal amount of the contract, term dates, fee structure, and repayment frequency. Near the bottom of the screen, there are supplementary fields unique to trade finance that we use to input additional information such as the shipping method, shipping period, and documents required as part of the agreement.

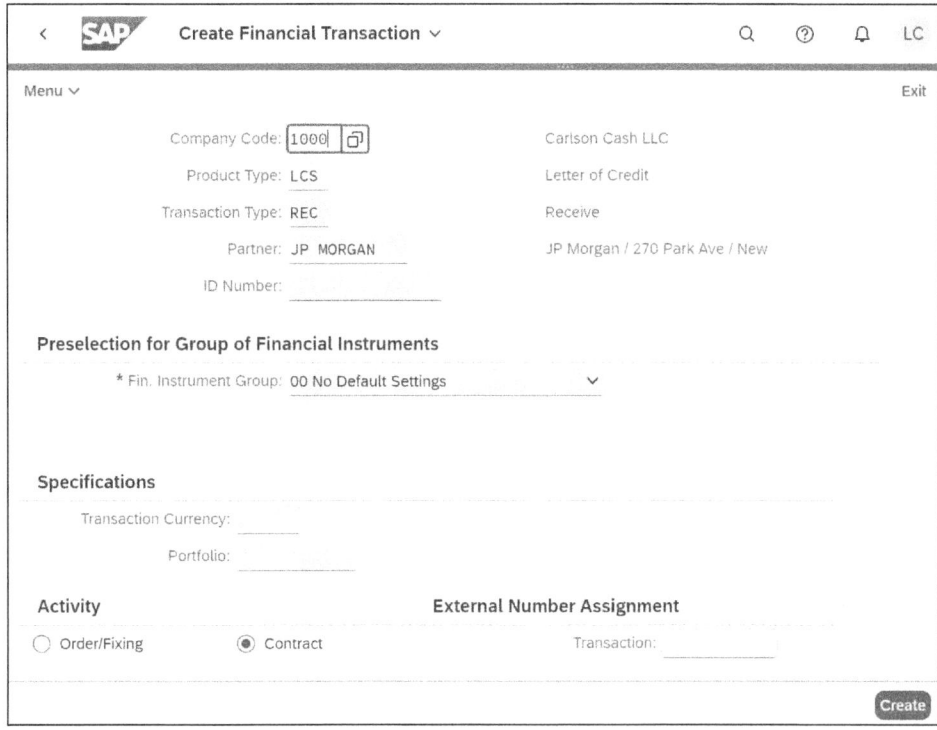

Figure 6.124 Letter of Credit Creation Using Create Financial Transaction App

Let's start with the key details required for any letter of credit contract. These are the most important elements of the contract, and they provide the overall structure of the contract. The key fields that we must enter data into in this area are as follows (see Figure 6.125):

- **Flow Type**
  Here, we enter the **Remaining Credit Amount** available on the letter of credit. The default flow type in our example is "RC00."
- **Amount**
  Here, we enter the nominal amount or value of the contract.
- **Term Start**
  Here, we enter the start date of the letter of credit.
- **Term End**
  Here, we enter the end date of the letter of credit.

6 Treasury Management–Specific Processes

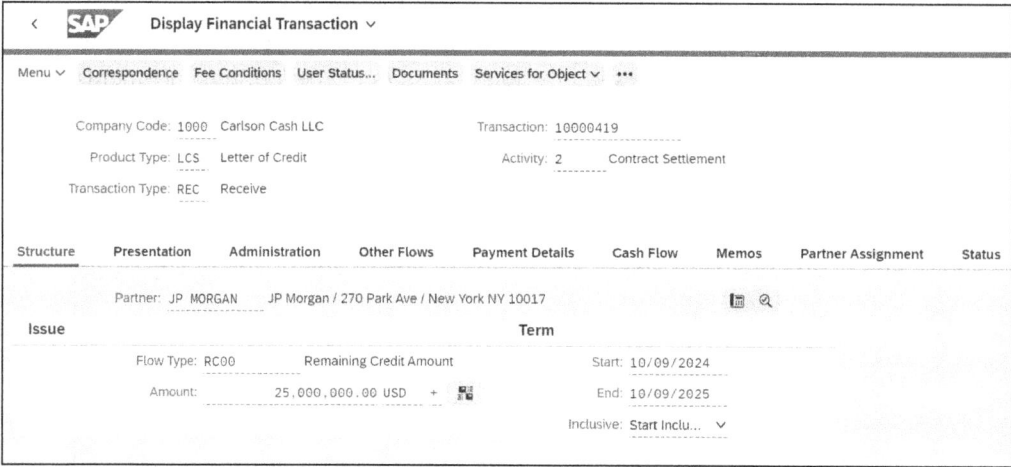

Figure 6.125 Letter of Credit Structure Tab Details

Now that we've entered the most important structural details of the contract, we can move on to the **More or Less Terms**. This is a unique section of the **Structure** tab that is not available in other contract types and is only used for letters of credit. When we check the **More or Less Terms** box, the **Tolerance** radio button appears and we can enter a positive or negative tolerance. The tolerance is based on the amount of the letter of credit and allows for a positive or negative tolerance over and above the nominal amount of the letter of credit. As we can see in our example in Figure 6.126, we have two additional considerations:

- **Tolerance**
  We click the tolerance radio button to indicate that there's a percentage tolerance either under or above the remaining credit amount of the letter of credit. If we leave the tolerance blank, there's no tolerance outside of the remaining credit amount value.

- **Maximum Amount**
  We click the maximum amount radio button to indicate that the letter of credit amount can be less than the negotiated amount of the letter of credit but can't exceed the negotiated amount.

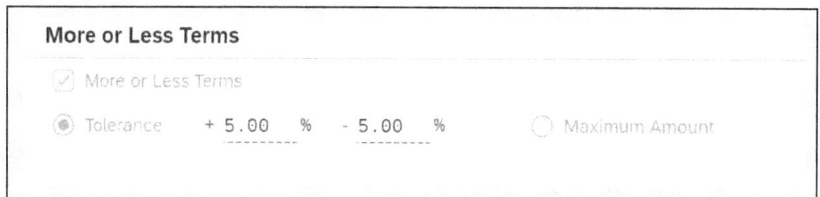

Figure 6.126 Letter of Credit More or Less Terms

We can now enter in the **Main Data** and **Manual Entry of Applicant** sections of the **Structure** tab of a letter of credit (see Figure 6.127). Similar to the previous section, these two

6.7 Trade Finance

sections of a letter of credit contract are unique to trade finance contracts. The following fields are available for entry of information within the main data section of the letter of credit contract:

- **Main Data**
  - **L/C Number**
    Here, we enter the letter of credit number or contract number.
  - **Place of Expiry**
    Here, we enter the location where the letter of credit beneficiary presents documents.
  - **Time Zone**
    Here, we enter the time zone of the place of expiration.
  - **Confirm Instructions**
    Here, we specify whether the letter of credit must be confirmed by a confirming bank.
  - **Applicant**
    Here, we enter the name of the applicant for the letter of credit. We can either select the name of a business partner or manually enter the details of the applicant.
  - **Issuing Bank**
    Here, we enter the name of the bank issuing the letter of credit. We can either select the name of a business partner or manually enter the details of the issuing bank.
- **Underlying Transaction**
  - **Sales Order**
    Here, we tie the letter of credit to the sales order in SAP S/4HANA.
  - **Incoterms Version**
    Here, we enter the Incoterms version that's published by the International Chamber of Commerce and that outlines a set of standardized international trade terms for transportation.

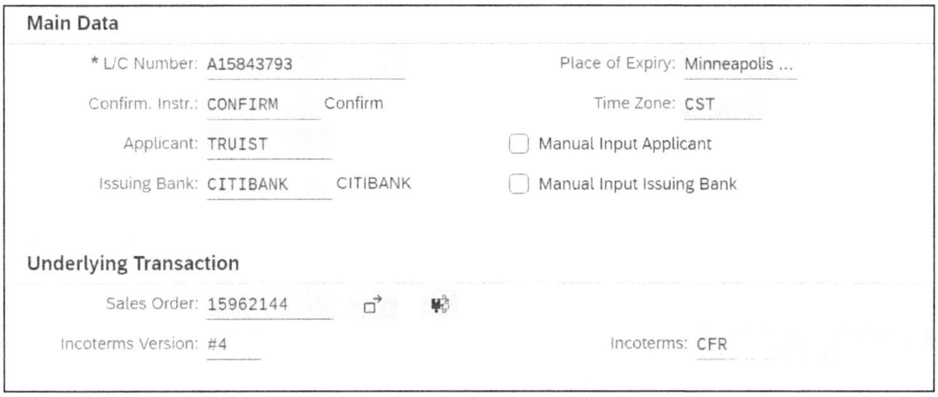

Main Data			
* L/C Number: A15843793		Place of Expiry: Minneapolis ...	
Confirm. Instr.: CONFIRM	Confirm	Time Zone: CST	
Applicant: TRUIST		☐ Manual Input Applicant	
Issuing Bank: CITIBANK	CITIBANK	☐ Manual Input Issuing Bank	
Underlying Transaction			
Sales Order: 15962144			
Incoterms Version: #4		Incoterms: CFR	

Figure 6.127 Main Data and Underlying Transaction Entry Screen for Letter of Credit

6 Treasury Management–Specific Processes

- Incoterms

    Here, we enter the Incoterms that define internationally recognized rules that both the shipper and the receiving party must follow to ensure the successful completion of a shipping transaction.

Now that we've entered the main data, we can move on to the **Goods and Shipping** section of the letter of credit entry screen. All of these fields are optional fields when maintaining a letter of credit, so not all of the fields are used in the example in Figure 6.128. We use the fields in this section to track the specifics of the transport documents. We can enter data into the following fields when maintaining the letter of credit contract:

- **Shipment Period**
    Here, we enter the dates when the goods can be shipped.

- **Shipping Method**
    Here, we specify how the goods are shipped.

- **Place of Receipt**
    Here, we specify the place of receipt, which is where the seller hands the goods over to the carrier.

- **Place of Delivery**
    Here, we specify final destination where the carrier delivers the goods to the buyer.

- **Port of Loading**
    Here, we specify the port of loading.

- **Port of Discharge**
    Here, we specify the port of discharge.

Figure 6.128 Letter of Credit Entry of Goods and Shipping Details

- **Partial Shipment**
  We click this radio button to indicate that a partial shipment is allowed.
- **Transhipment**
  We click this radio button to indicate that the shipped goods can travel to a different destination before reaching the final destination.
- **Description of Goods and Services**
  Here, we can enter free text, most commonly to provide a detailed description of the goods or services.

The next step is to populate the **Presentation** and **Payment** details within the letter of credit. This is not required, but we can use these two sections to track the presentation details of the contract. The *presentation* of the letter of credit is the process in which the beneficiary submits the required documents to the issuing bank to request payment under the terms of the letter of credit. We can use the following fields to track the presentation and associated payment details (as shown in Figure 6.129):

- **Presentation Period**
  Here, we can enter date by which the documents must be presented (expressed as a number of days after the presentation condition).
- **Additional Comments**
  Here, we can add any comments related to the presentation.
- **Payment Bank**
  Here, we can specify whether **Any Bank** or a **Nominated Bank** can present the required documents.
- **Payment at**
  Here, we can specify when the payment will take place after the documents have been submitted to the authorized bank. The payment can take place immediately at **Sight** or **Deferred** by a number of days after the **Presentation Date**.

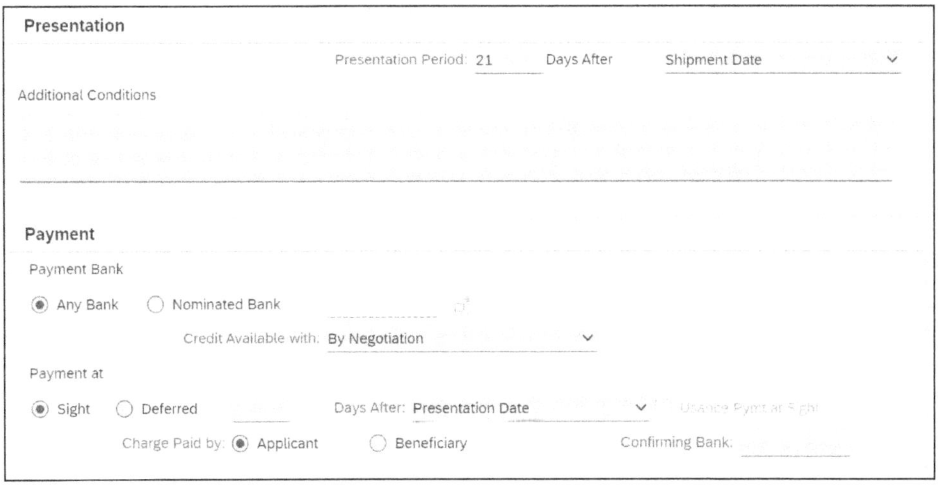

Figure 6.129 Letter of Credit Presentation and Payment Details

6 Treasury Management–Specific Processes

- **Charge Paid by**
  Here, we can select the **Applicant** radio button or the **Beneficiary** radio button to indicate who will pay the charge.
- **Confirming Bank**
  Here, we can specify the bank that will confirm the letter of credit.

Finally, we can populate the **Document** section of the letter of credit **Structure** tab with additional information. This is not required, and we only populate and store the applicable document types required for the letter of credit. Once the documents have been received, we can attach copies of them in their corresponding document types. As we can see in Figure 6.130, various document types applicable to the contract are contained as part of the contract within treasury and risk management.

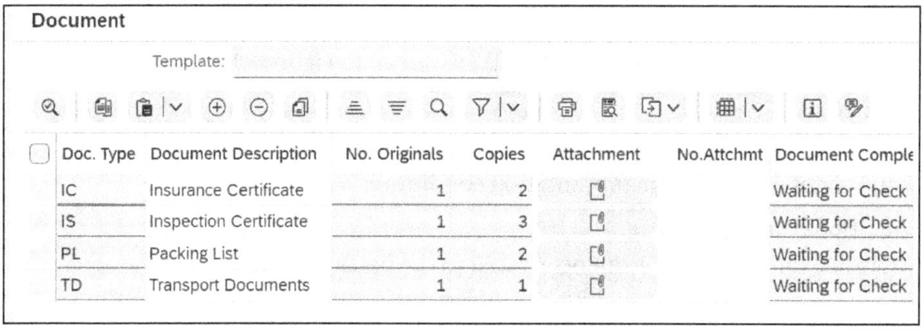

Doc. Type	Document Description	No. Originals	Copies	Attachment	No.Attchmt	Document Comple
IC	Insurance Certificate	1	2			Waiting for Check
IS	Inspection Certificate	1	3			Waiting for Check
PL	Packing List	1	2			Waiting for Check
TD	Transport Documents	1	1			Waiting for Check

Figure 6.130 Letter of Credit Document

The final step when creating a letter of credit is adding the associated fees. We click on the **Fee Conditions** button at the top of the entry screen and then click on the **Create Fee Condition** button. We then choose one of the fees from the popup selection screen as shown in Figure 6.131. When we click on one of the fee conditions, we're brought to the next screen, where we can set the start date of the fee.

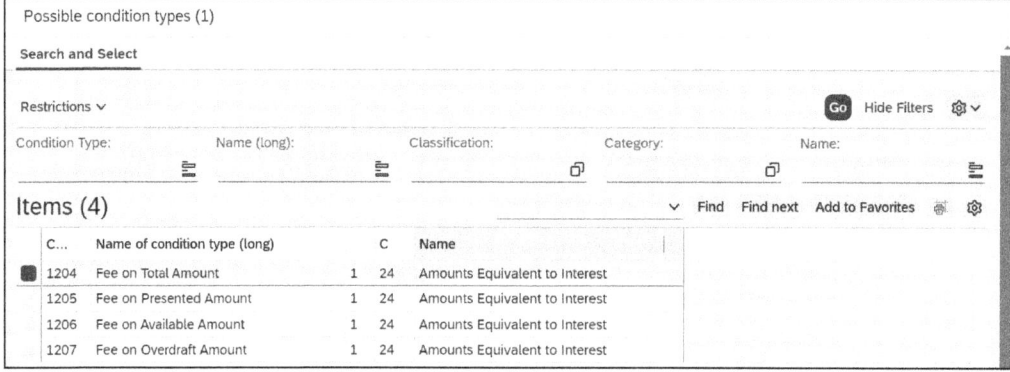

Figure 6.131 Letter of Credit Fee Conditions

When we select the fee, a popup box will appear, and we can populate the start date of the fee there by entering a date into the **Eff. From** field (see Figure 6.132). Once we've entered the start date, we press the **Copy** button, which will navigate us to the **Condition Details** screen.

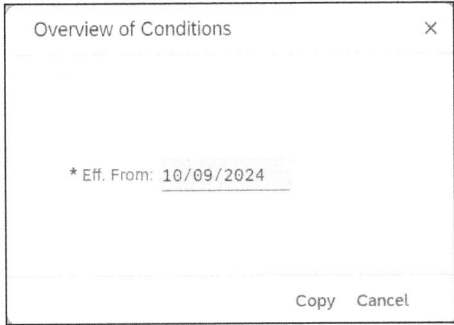

Figure 6.132 Overview of Conditions Effective-From Date

The next step is to populate the details of the fee condition. Fee conditions function very similar to the charges discussed in Chapter 5, Section 5.1.5. As shown in Figure 6.133, we'll first populate the **Int. Calc Type**; the options available are as follows:

- Linear Interest Calculation
- Exponential Interest Calculation
- Percentage Calculation
- Percentage Calculation Per Day

We can then populate the **Int. Calc. Method** for the fee, as well as the **Percentage Rate** or **Payment Rate** of the fee. The last section worth mentioning within the fee conditions is the **Round Amounts** section. Based on how the fee structure is set up and how it should be paid, we can choose to round the calculated fee to match the fee that is calculated by the issuer of the letter of credit. We can select from the following rounding categories:

- **Round to the Nearest**
  With this option, if the first digit after the decimal place is 4 or less, then the fee will be rounded down. If the first digit is greater than or equal to 5, then the fee will be rounded up.
- **Round Up**
  With this option, the fee will always be rounded up.
- **Round Down**
  With this option, the fee will always be rounded down.
- **Round Down If 5**
  This option functions like **Round to the Nearest**, except that if the number 5 is in the first dropped decimal place and there are no further nonzero decimal places after the number 5, then the system rounds the fee down.

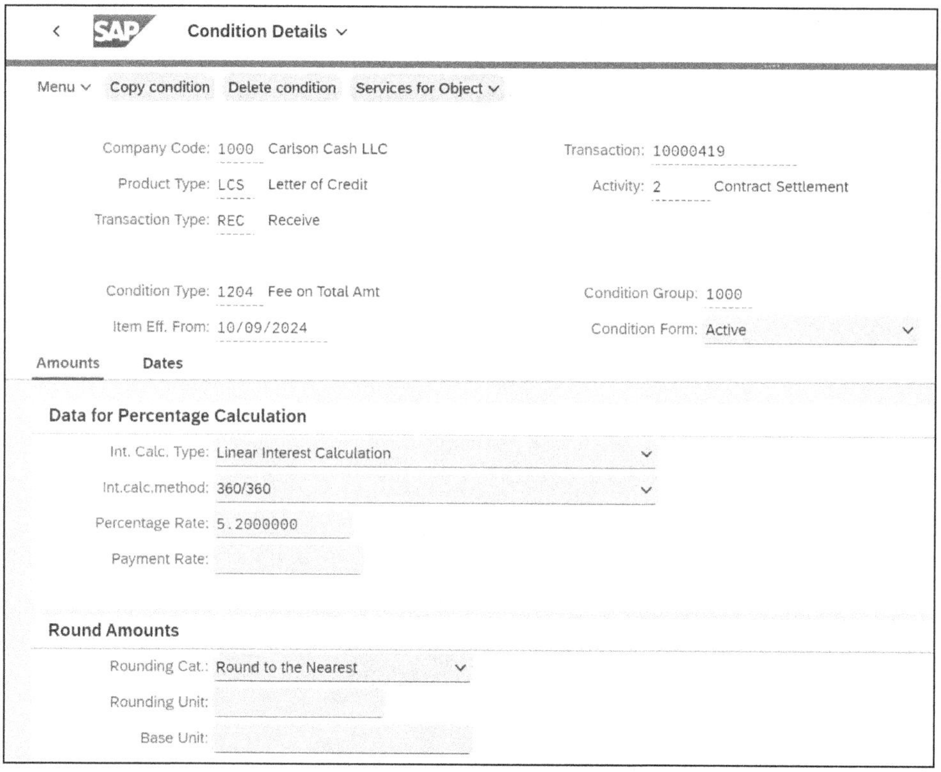

6 Treasury Management–Specific Processes

Figure 6.133 Creation of Letter of Credit Fee Conditions

6.8 Commodity Contracts

We'll now shift our attention to the creation of a commodity contract and walk through the various steps required to process a commodity derivative contract. As always, we'll start by introducing the process flow and then dive into contract creation, cash settlement, and contract settlement.

6.8.1 Process Flow

This process is similar to the creation of other product types within treasury and risk management. Figure 6.134 outlines the full lifecycle from initial contract creation to the end of the contract in SAP S/4HANA. The process of managing a commodity forward contract begins with creating the contract itself. This is done by entering all the relevant details into the Create Financial Transaction app. Once the contract is created, its value will fluctuate based on changes in the price of the underlying commodity.

The current value of the contract can be viewed and saved through either an ad-hoc or end-of-day valuation. At month-end, we use the Calculate Net Present Values app to determine the contract's present value. This is followed by using the Run Valuation app, which uses the NPV to post any unrealized gain or loss for the contract.

6.8 Commodity Contracts

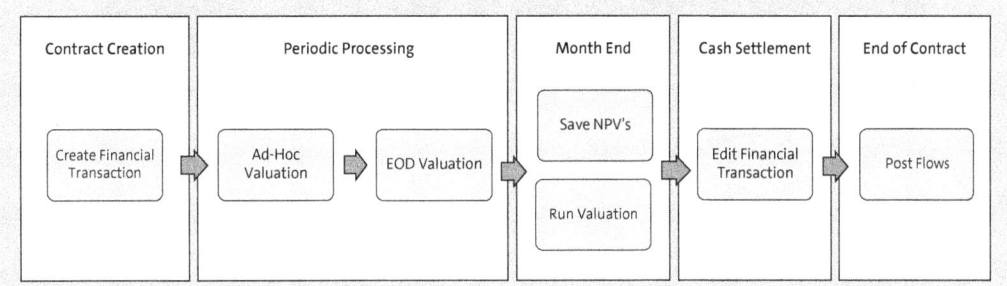

Figure 6.134 Commodity Contracts Process Flow

At the contract's maturity, where it is cash settled, we start by updating the contract with the current market price using the Edit Financial Transaction app. The system will then calculate the cash settlement value. Finally, the cash flows and any associated realized gain or loss are posted using the Post Flows app.

6.8.2 Contract Creation

The first step is to navigate to the Create Financial Transaction app. Upon entering the app, we'll be presented with the main screen, where we need to specify several key details. First, we need to specify the **Company Code**, which identifies the legal entity for which the contract is being created. Next, we choose the **Product Type**, which in this case is a commodity forward contract that we enter as "CF."

Following this, we select the **Transaction Type**, which could be either a purchase or a sale, depending on the nature of the commodity derivative contract we are creating. In our example shown in Figure 6.135, we're purchasing a commodity forward contract. This transaction type determines the flow of the commodity and the associated cash flows within the system.

Additionally, we must specify the business **Partner** involved in the trade. The business partner represents the counterparty to the transaction. Once we've entered all of these details, we review the information to ensure that it's accurate and then press the **Create** button located in the lower right-hand corner of the screen.

We're now at the main contract entry screen, where we'll establish the structural details of the commodity contract. The first and most important element to enter into the contract is the derivative contract specification of the commodity being traded. If we choose the incorrect derivative contract specification within the contract, the details of the contract won't be accurate. The derivative contract specification serves as the foundation for the commodity contract and contains essential information about the commodity's trading parameters. When we input the derivative contract specification into the contract details, the system will automatically populate various fields based on the preconfigured data stored within the derivative contract specification.

477

6 Treasury Management–Specific Processes

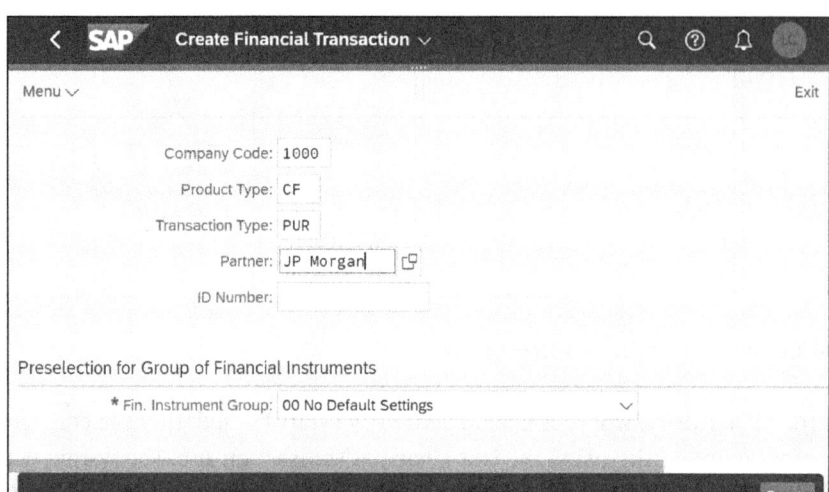

Figure 6.135 Commodity Forward Creation Using Create Financial Transaction App

Once we enter the derivative contract specification into the **DCS ID** field, the system will automatically fill in several important trade attributes in the contract. These attributes include the market identifier code, which specifies the exchange where the commodity is traded; the **Commodity** type, which identifies the specific commodity being dealt with; the **Contract Size**, which indicates the standard lot size for the commodity; the **Unit of Measure**, which defines the measurement unit for the contract; and the **Derivative Category**, which classifies the type of derivative being used.

As demonstrated in Figure 6.136, the details contained within the derivative contract specification are automatically integrated into the contract. This automation streamlines the contract creation process, ensuring consistency and accuracy by reducing the need for manual entry of repetitive trade details. This integration also helps to minimize errors and makes sure that all contract details align with the preconfigured contract specifications stored in the derivative contract specification.

The next step involves entering the quantity and price details associated with the contract. Recall that when we created the derivative contract specification, we specified the lot size or contract size within it. Specifying this in the derivative contract specification gives us some flexibility in how to enter the contract into the system. There are a few options that we can use to enter the size of the contact into SAP S/4HANA:

- **Number of Contracts**
  We can specify the number of lots or contracts being traded. Based on the number of contracts we enter, the system will automatically calculate the total quantity field. This is possible because the system references the contract size specified in the derivative contract specification to determine the quantity per contract.

- **Quantity**
  Alternatively, we can specify the total quantity being traded, and the system will then calculate the number of contracts required for the trade. The system uses the lot sizes defined in the derivative contract specification to perform this calculation.

6.8 Commodity Contracts

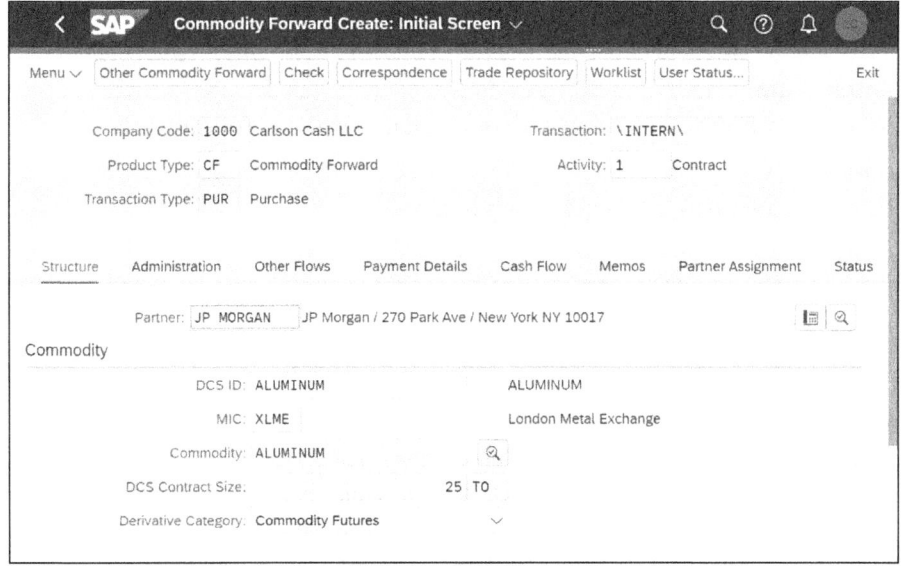

Figure 6.136 Creation of Aluminum Commodity Forward Contract

Now that we've entered the contract quantity, we can enter the price, currency, and unit of measure of the contract. Then, the system will calculate the price of the total contract and populate the **Amount** field. Keep in mind that the price we enter into the screen represents the price of one unit and not the entire lot; the total amount of the contract is calculated as the price times the number of units within the lot. This is depicted in Figure 6.137, where $2,350 × 25 = $58,750.

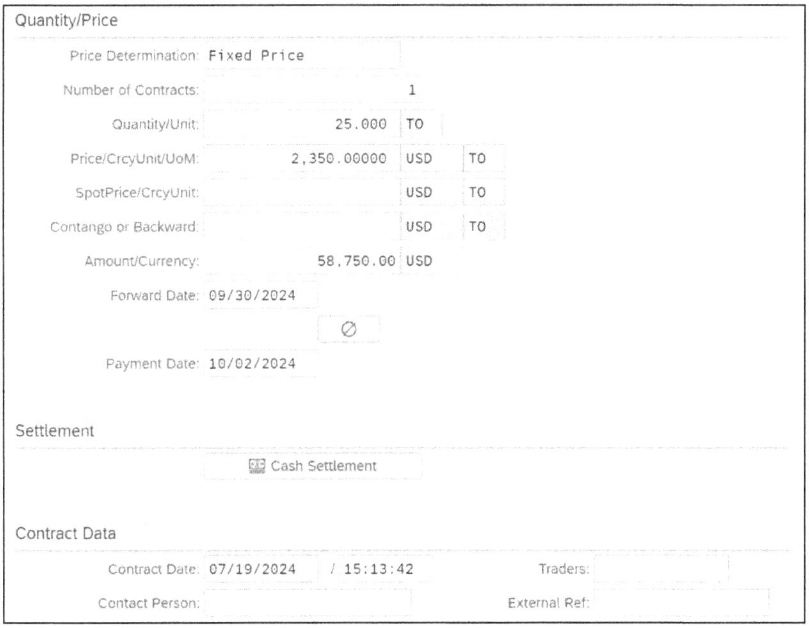

Figure 6.137 Commodity Forward Contract Price and Quantity

479

Finally, we must enter the various dates related to the contract. The following dates are available for us to populate:

- **Contract Date**
  This is the date when the contract was entered into with the counterparty.

- **Forward Date**
  This is the date when the contract ends or matures.

- **Payment Date**
  This is the date when the settlement payment takes place. Often, the payment date falls a few days after the forward date.

At this stage, we've entered all the essential information into the contract. It's always a good idea to review the **Cash Flow** tab before saving the contract, as it outlines when the various cash flows associated with the contract will occur. Figure 6.138 depicts the purchase of a commodity forward as a cash flow. While this entry doesn't produce an actual cash flow and won't result in any accounting postings, having it visible in the **Cash Flow** tab is important for tracking and clarity.

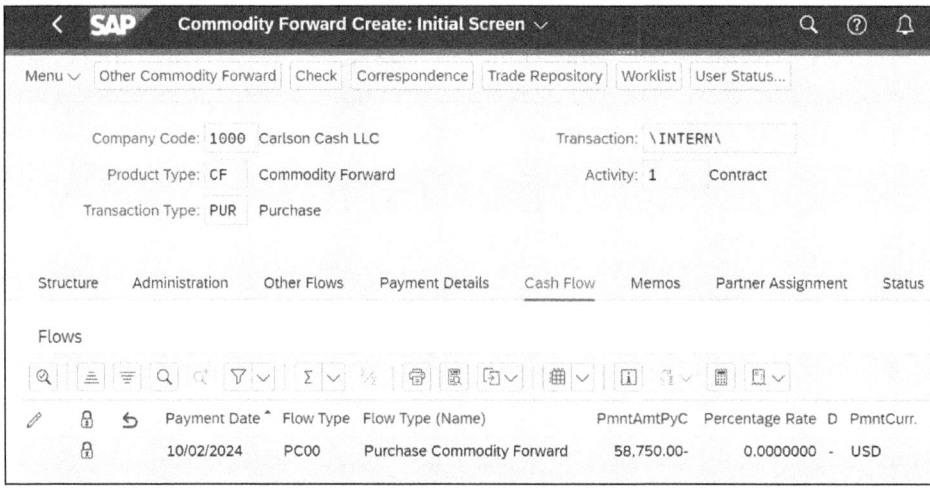

Figure 6.138 Review of Commodity Forward Cash Flows Using Cash Flow Tab

At this point, we've entered all the details into the contract and verified the cash flow information—so we can save the contract and move to the next step, which is performing cash settlement of the contract.

6.8.3 Cash Settlement

We can now fast-forward to the end of the contract. We're settling this commodity forward contract at its maturity by cash settlement, which means that we're settling it through a cash payment rather than through physical delivery of the underlying

6.8 Commodity Contracts

commodity. To do this, we navigate to the Process Financial Transaction app, enter the **Company Code** and contract number of the contract (**Transaction**), and press the **Change** button, which directs us to the contract details, as shown in Figure 6.139.

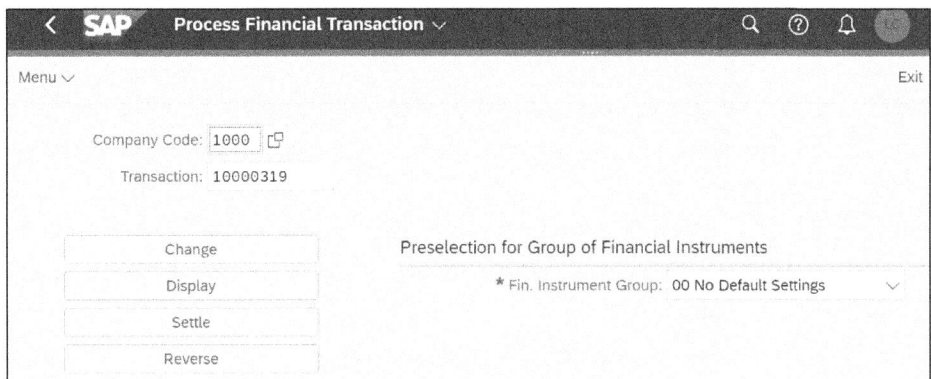

Figure 6.139 Initial Step of Cash Settlement of Commodity Forward Contract

Near the bottom of the screen, there's a **Cash Settlement** button, as shown in Figure 6.140. Clicking this button is the key to triggering the cash settlement process for the contract.

Figure 6.140 Cash Settlement Button in Commodity Forward Contract

After we click the button, the screen will expand and display a new **Cash Settlement** section within the contract. At this time, if required, we can change the cash settlement or payment date on the contract. All of the other contract elements—such as **CurrencyUnit**, **Currency,** and **UoM**—are prepopulated, as shown in Figure 6.141. At this point, all that we need to do is to enter the current spot price of the commodity derivative, which is also the price that the counterparty uses to determine the gain or loss on the contract.

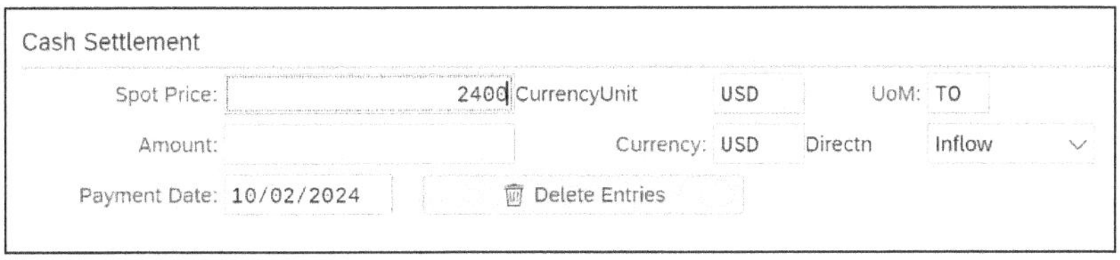

Cash Settlement						
Spot Price:	2400	CurrencyUnit	USD		UoM:	TO
Amount:		Currency:	USD	Directn	Inflow	
Payment Date:	10/02/2024	Delete Entries				

Figure 6.141 Cash Settlement Process for Commodity Forward Contract

After we enter the spot price, SAP S/4HANA will calculate the difference between the spot price and the contract price and populate the amount of the cash settlement. Also, the direction of the dropdown field will change based on whether the contract is in a gain or a loss position.

The system derives cash settlement amount using the following steps:

1. It determines the agreed-upon forward price within the commodity forward contract.
2. It determines the spot or market price that we manually entered into the contract.
3. It calculate the price difference by subtracting the forward price from the current market price.
4. It multiplies the price difference by the number of contracts purchased:
   - If the price has now gone up and the difference is positive, it indicates a gain and represents an inflow of cash.
   - If the price of the commodity has gone down and the difference is negative, it indicates a loss and represents a cash payment that must be made to the counterparty.

Let's look at an example calculation:

- Forward contract price: $2,350
- Price at contract maturity: $2,400
- Price difference: $50
- Quantity: 25 Metric Tons
- Cash settlement amount: $50 x 25 = $1,250

As we can see in Figure 6.142, the cash settlement amount is an inflow of $1,250.

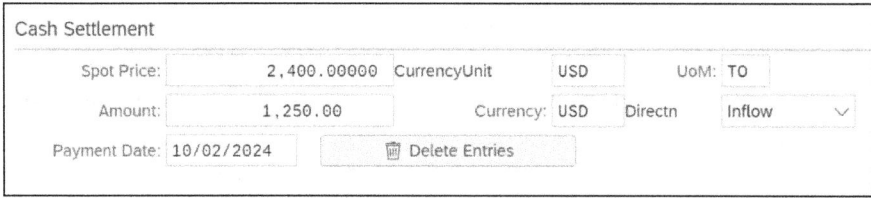

Figure 6.142 Cash Settlement Screen with Final Cash Settlement Price Details

As we did when we initially created the contract, let's review the **Cash Flow** tab. As we can see in the **Cash Flow** tab in Figure 6.143, the $1,250 gain is represented as an incoming cash flow with flow type CS00 – Cash Settlement. This cash flow is a true cash movement and will produce a cash flow as well as journal entries when the contract is posted in a subsequent step. If the contract were in a loss position and this were an outgoing cash flow, we would also create a payment request when this is posted during the post flows process to make a cash settlement with our counterparty.

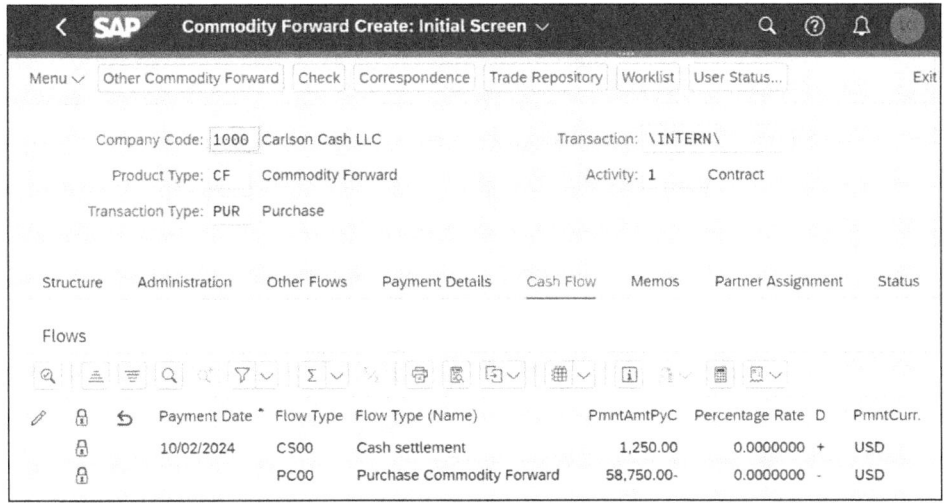

Figure 6.143 Review of Cash Flows Related to Cash Settlement of Commodity Forward

At this point, we've created the contract and performed cash settlement at contract maturity. If we've configured the system to require a second set of eyes or settlement of the contract, we'll now move to that step in the process.

6.8.4 Contract Settlement

Moving on to the final step in the contract creation process, we have the option of including a contract settlement step. This step is integral to the overall process and serves a similar purpose for commodity forwards as it does for any other instrument type within treasury and risk management. The contract settlement acts as an additional verification point after the contract has been initially created.

For commodity forwards, especially those that are settled in cash, this step becomes even more important. The cash settlement phase is where the financial outcome of the contract is crystallized, determining whether the contract has resulted in a gain or loss. This step generates either an incoming or an outgoing cash flow, reflecting the financial result of the forward contract at maturity.

The best practice in this scenario is for us to configure the product type to require that a settlement contract can't be processed further without the contract settlement step, as shown in Figure 6.144. By doing this, we introduce a mandatory checkpoint that verifies the accuracy of the cash settlement calculations and confirms that all relevant details have been entered into the contract before any further processing.

To summarize, incorporating a contract settlement step as the last stage in the contract creation process for commodity forwards is an important practice. It acts as a safeguard, ensuring the accuracy of cash settlements and the integrity of the contract prior to the cash posting of the contract to the general ledger.

6 Treasury Management–Specific Processes

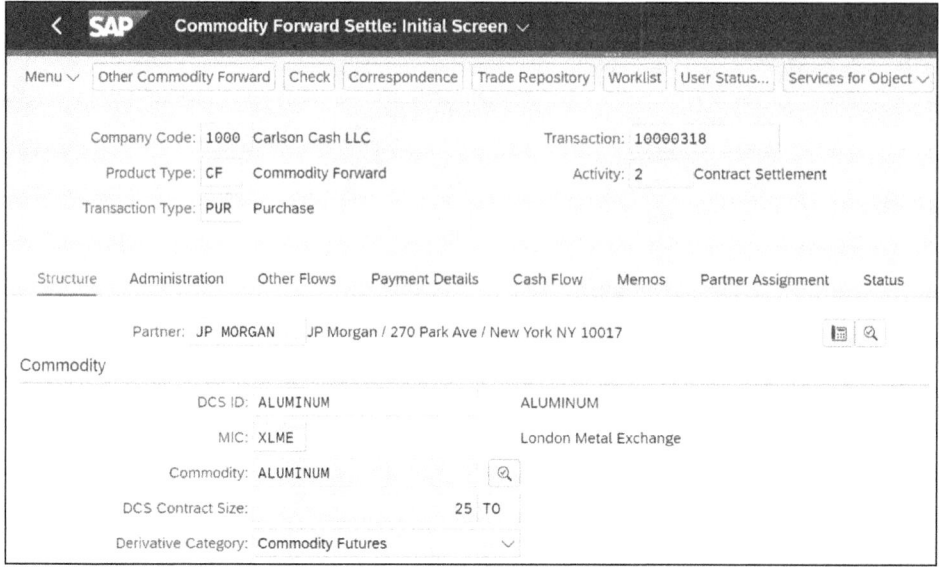

Figure 6.144 Settlement Step of Commodity Forward Contract

6.9 Summary

This chapter has provided a comprehensive overview of the various contract-processing flows within treasury and risk management. It has focused on the most commonly used financial instruments, such as FX contracts, money market instruments, debt management, and intercompany loans. Each contract type has unique processing requirements, from setup and configuration to managing cash flows, settlements, and accounting entries. By understanding and mastering these processes, treasury professionals can ensure accurate financial reporting, effective risk management, and optimized treasury operations. This chapter serves as a guide to help users efficiently manage the lifecycle of financial contracts within treasury and risk management.

Chapter 7
Exposure Management

In this chapter, we'll focus on the creation of exposures across treasury and risk management and the configuration that supports it. Throughout this chapter, we'll talk about where exposures will come from and how we can manually create them or upload them into the SAP S/4HANA system. These exposures will then integrate with additional capabilities available for hedging purposes.

Exposure management 2.0 is a key component of SAP S/4HANA Finance for treasury and risk management. Exposure management allows us to capture inflows and outflows that are associated with currency risk or commodity price risk. These can be either committed cash flows or projected cash flows. Here, we'll identify the risks in the payment flows that can link into the hedge management capabilities, which we'll cover in Chapter 8 and Chapter 9.

Key uses for exposure management include the following:

- Doing business in a country or in countries where the currency is different from the company's functional currency, which means cash flows could have fluctuations in underlying currency
- Buying or selling commodities where prices are continually changing

Before we dig into the use of exposure management 2.0, we'll first define a few key terms, as follows:

- **Raw exposure**
  This is a type of risk position that is either entered manually or imported from another system in the enterprise.

- **Subexposure**
  This is a further breakdown of a raw exposure into individual characteristics. For example, if a company that uses US dollars company is trading a steel commodity in euros, then the overall commodity exposure would be defined as a raw exposure, while two subexposures would be created: one for the commodity and one or the EUR component

- **Exposure position**
  Once a subexposure is created and released with derivation rules defined in configuration, it's converted into an exposure position that can then be processed through hedging or risk analyzers (which we will discuss in Chapter 11).

7 Exposure Management

In this chapter, we'll explore how to create cash flow exposures, review all open exposures across the organization, and review the configuration that supports the exposure management process.

7.1 Raw Exposures

Creating an exposure is similar to creating any other deal types discussed in earlier in the book. When we have identified that we have an exposure, we'll either manually create or upload multiple exposures by using a spreadsheet. With either method, this will create a raw exposure. Once we have created and saved that exposure based on our system settings, that exposure can be automatically released or it can go through a workflow review process that will validate the information that has been provided. Figure 7.1 shows the typical process for creating and releasing an exposure.

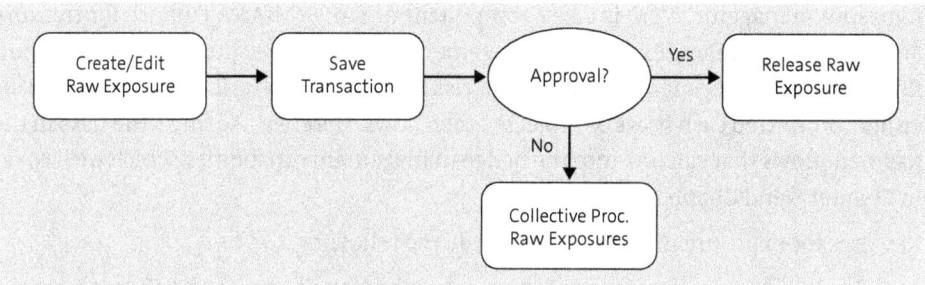

Figure 7.1 Exposure Management Overview

To create a raw exposure, we utilize the Process Raw Exposure app. We can also use this app to change, release, or delete raw exposures. To generate a raw exposure, we first identify the type of exposure we're going to capture.

We'll have at least two types of exposures: incoming and outgoing. However, we can additionally segment those incoming and outgoing exposures into more defined categories, as determined by our requirements.

We've identified that we have an incoming FX exposure, so we've selected the "YFXI" **Exposure Activity Type**. Since this is a new exposure that is all that is required for entry on this screen, we then select the **Create** option as shown in Figure 7.2.

As we can see in Figure 7.3, there are three key tabs associated with each exposure:

- Header Data
- Line Item Data
- User Data

We'll now look into how to input all necessary data into our exposure.

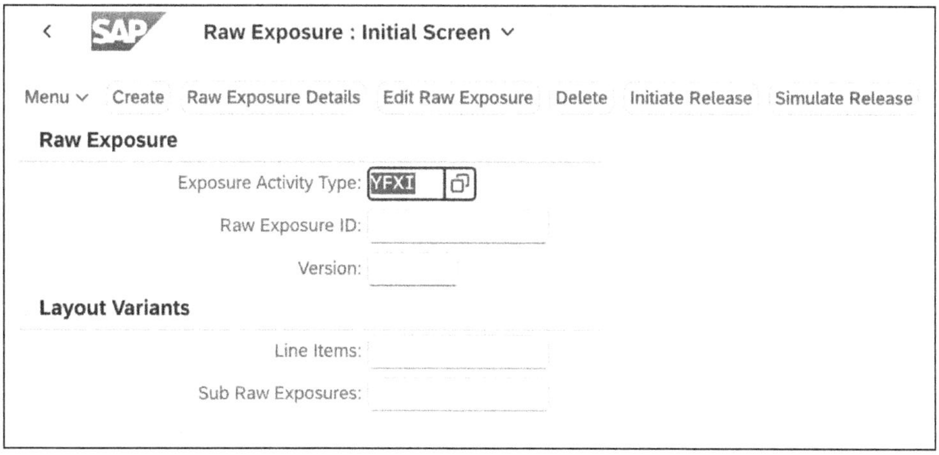

7.1 Raw Exposures

Figure 7.2 Raw Exposure Entry Screen

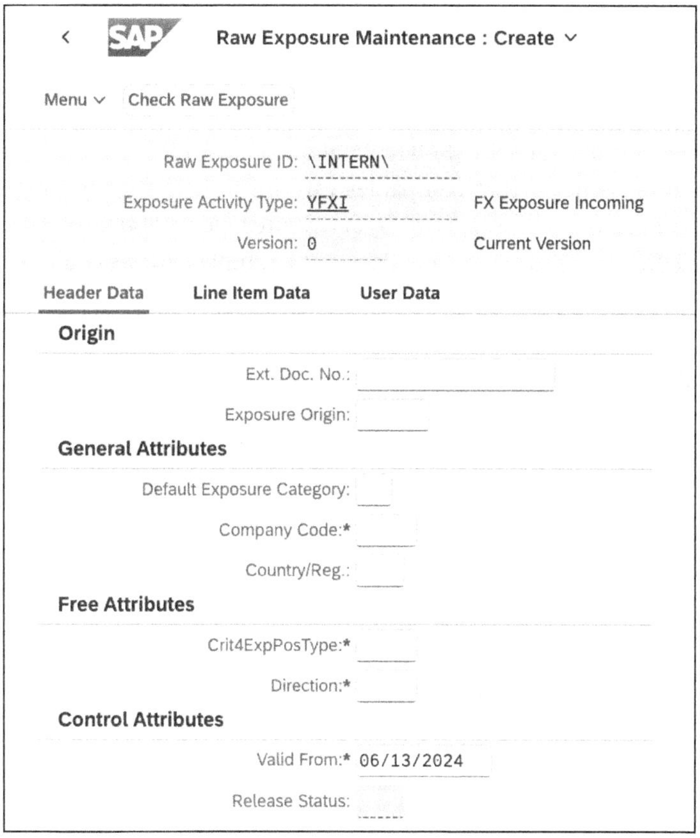

Figure 7.3 Process Raw Exposure

As we can see in Figure 7.4, there are four major sections of the **Header Data** tab, as follows:

- **Origin**

    This section helps us define where the exposure comes from. We can use the external document number and exposure origin to delineate this. The external document number (**Ext. Doc. No.**) is a free-form field that could match to an underlying cash flow transaction either inside or outside of SAP S/4HANA. The **Exposure Origin** is a configured field, and for this example, we have selected "AMBU," which is an asset transaction that we will define later in the exposure origins configuration discussed in Section 7.4.10.

- **General Attributes**

    This section allows us to further define our exposures for the **Default Exposure Category,** which are predefined as **Firm Commitment**, **Asset/Liability Transaction**, and **Forecasted Transaction**. For this example, we have selected "03," **Asset / Liability**. We then identify the **Company Code** (1710) and the **Country/Reg.** (US). This will help with the integration of the exposure to hedge accounting, which we will discuss in future chapters.

- **Free Attributes**

    This is an optional section in which SAP S/4HANA allows companies to customize their screens by adding elements that companies can use to help segment their exposures. In our example, we have added two free attribute fields: **Crit4ExpPosType** (critical exposure position type) and **Direction**. These are examples, but we could create as many additional elements as we need.

- **Control Attributes**

    Here, we set the validity date (**Valid From**) of the exposure, the default value for which is today's date. Additionally, there's a **Release Status** field. Note that since this is still in creation status, the release status is blank but will be populated when we save the exposure. This could be displayed as one of the following statuses:

    - Editable/Unreleased
    - Release Initiated
    - Released
    - Archivable

> **Note**
>
> All required fields are marked with an asterisk (*). This is controlled in the configuration but helps the user know what is required on the input screens. Additionally, if any information can be derived from other data, those settings can be configured in the derivation rules. For example, if we have a US company code, we can make the United States the default country rather than having end users key that in each time.

7.1 Raw Exposures

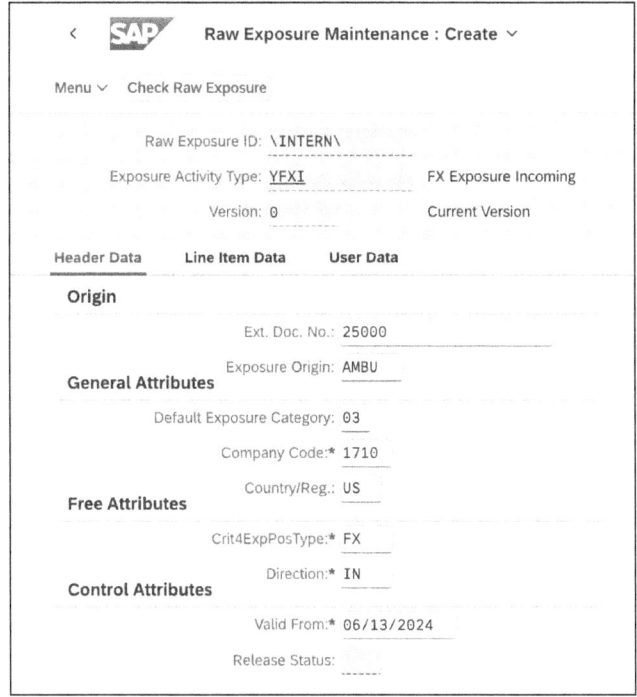

Figure 7.4 Process Raw Exposure Header Data

Next, we'll click on the **Line Item Data** tab, as shown in Figure 7.5. This contains the transactional data of the exposure. We could enter the exposure as a single line item, or if needed, we could have one raw exposure with multiple exposure line items over varying months for the future asset exposure. To do this, we select the **+** icon, which will populate a new raw exposure line. Since we're planning our exposure for three months, we click the icon three times so that all lines are available for input.

Now, we'll populate our exposures month by month. Notice that the **LI ID** (line item ID) is already populated sequentially for us. First, we enter the external item number, which is free-form text identifying that monthly exposure. Next, we'll enter the period, M06, which relates to June. Then, we enter the planned year of the exposure (e.g., 2024 and an exposure date). Next, we key in the **Exposure Amounts** and the exposure currency (**ExpAmtCrcy**). The target currency field will already be populated, based on logic from the company code; in this case, 1710 is set up in US dollars. We'll also notice additional fields like **Profit Center**, **Portfolio**, etc., that we can populate as needed. We'll repeat this process for as many exposures as needed, in our case, three months. In the example in Figure 7.5, we have a forecasted asset for which we expect to have FX exposures over the next three months.

For further details on our exposure, we can select our raw exposure line items, and the sub raw exposure will populate in the lower section of the screen. Here, we've selected our M07 exposure, and we see that the sub raw exposure has been populated.

489

7 Exposure Management

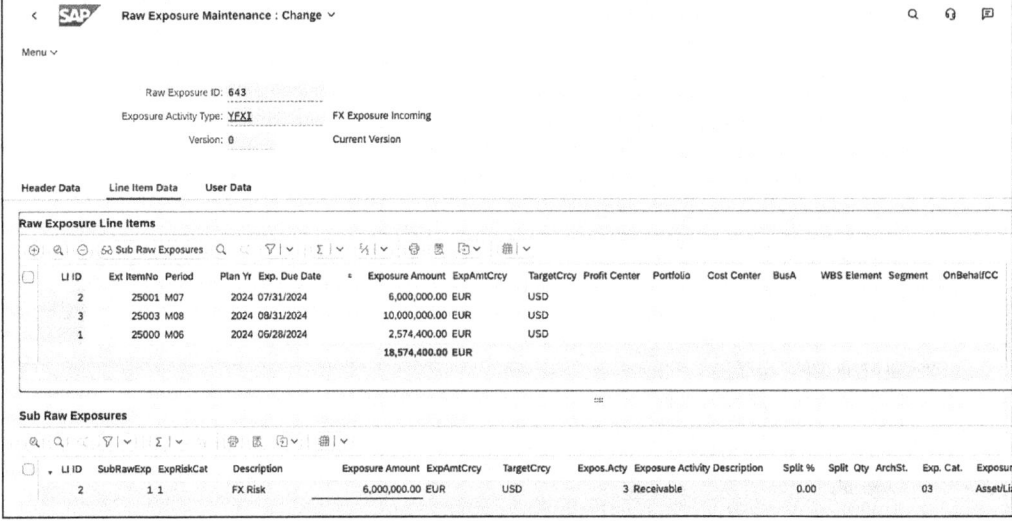

Figure 7.5 Process Raw Exposure Line Item Detail

The final tab available on the raw exposure screen is the **User Data** tab, shown in Figure 7.6. This is a display-only tab, but it provides important information as to who created the exposure and when it was first entered. Additionally, as we make updates to the exposure, an additional **Change Data** section will populate and provide information on who made the last change and the time of that change, as in Figure 7.6.

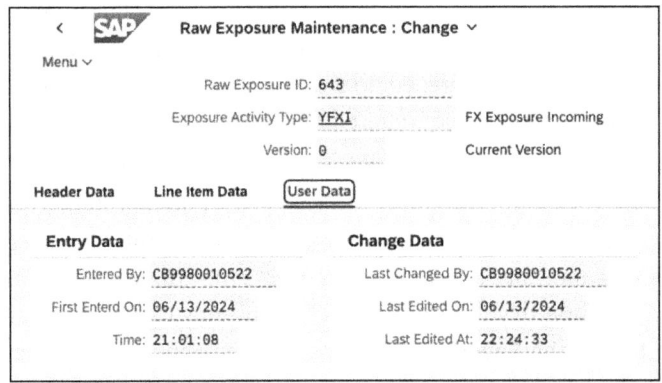

Figure 7.6 Process Raw Exposures User Data

> **Note**
> The current version of the exposure is always noted with "000," and other version numbers will indicate older versions of the exposure that we can use for historical reporting.

Once we're satisfied with our entry of the raw exposure, we can click **Save** in the bottom right corner of the screen and our exposure number will be represented in a popup. In our example in Figure 7.7, we can see that the raw exposure has been successfully saved and released. If any workflow release is required, it will be triggered and routed appropriately.

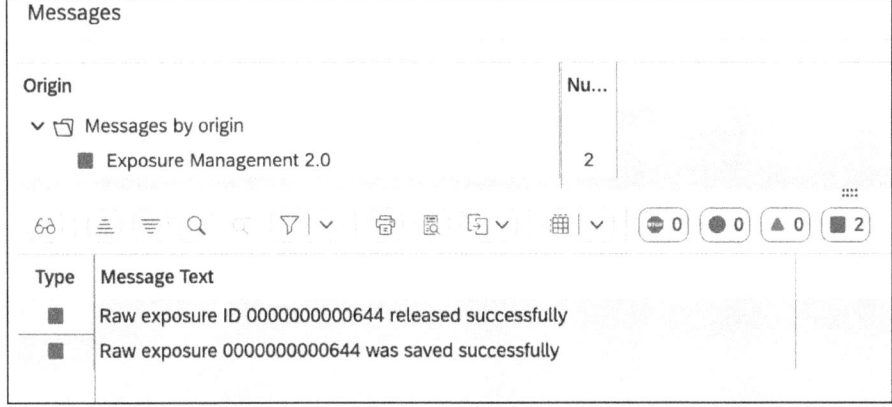

Figure 7.7 Process Raw Exposure Saved and Released

7.2 Raw Exposures Apps

Now that we've successfully created and released our exposure, we'll want to manage that exposure. In this section, we'll share apps that help us do so: the Process Raw Exposures – Collective Processing app and the Import Raw Exposures from Spreadsheets app.

7.2.1 Process Raw Exposures – Collective Processing App

As exposures are now created in the system, we can utilize the collective reporting functions in SAP S/4HANA to display, change, release, and delete any exposure in the system. We accomplish this using the Process Raw Exposures – Collective Processing app.

Upon entry into the Process Raw Exposures – Collective Processing app, we can see that the report is very robust and that we can run it by selecting any attribute that has been set up for the exposure. In Figure 7.8, we've simply entered the previously created **Raw Exposure ID**. Once we've entered our parameters, we click **Run** at the bottom right-hand corner (not shown).

When we run the report, the screens shown in Figure 7.9 will be displayed. Since this is the raw exposure collective processing report, we can see the list of raw exposures we've selected. Since we've only selected one raw exposure, we see that, but we also see all of the sub raw exposures we created for the three months.

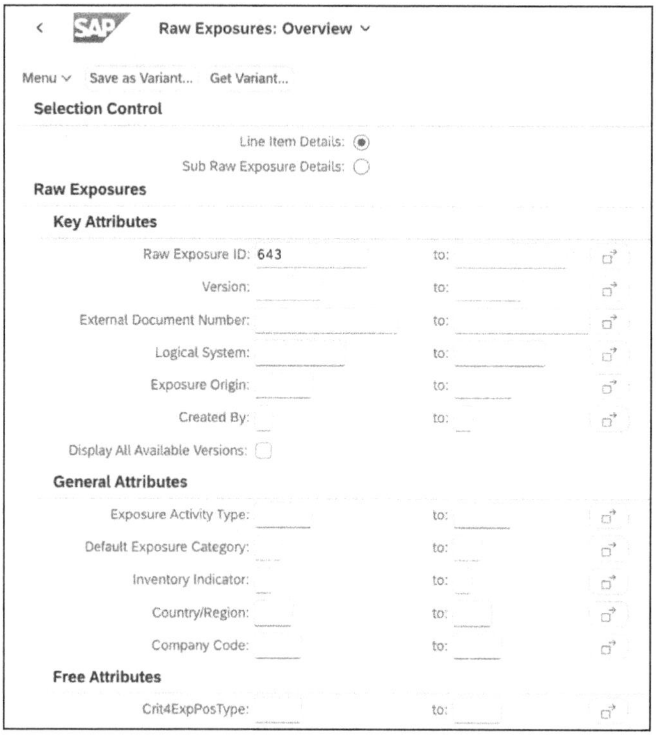

7   Exposure Management

Figure 7.8 Overview of Raw Exposures

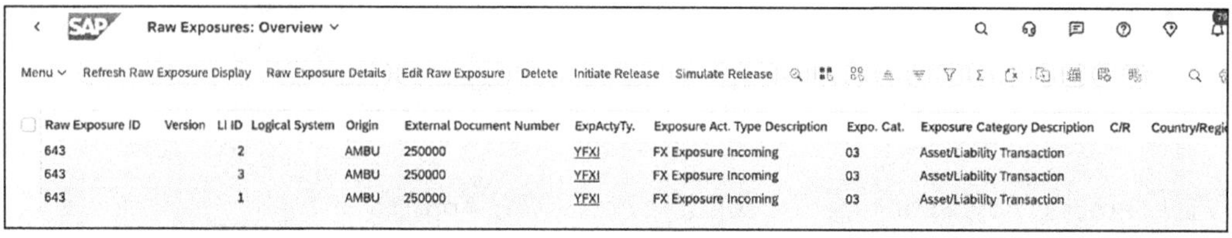

Figure 7.9 Display of Raw Exposures

Additionally, we can use this report as a dashboard for managing our exposures. There are several options for executing additional functions:

- **Refresh Raw Data Exposure Display**
  This option allows us to refresh our raw exposure list, and it will update output if any changes occur.
- **Raw Exposure Details**
  This option allows us access the Process Raw Exposures – Collective Processing app in display mode, as shown in Figure 7.10.
- **Edit Raw Exposure**
  This option allows us to access the Process Raw Exposure app in edit mode. Here, we'll adjust our M08 exposure from 10 Million to 15 Million, as shown in Figure 7.11.

7.2 Raw Exposures Apps

- **Delete**
  This option allows us to delete the raw exposure, but note that to delete a raw exposure that has been released, we need to undo the release.

- **Initiate Release**
  If an exposure requires our approval, we can release it from this dashboard.

- **Simulate Release**
  If an exposure requires our approval, we can simulate the release from this dashboard, which allows us to check whether any errors would occur upon release.

Figure 7.10 Raw Exposure Details

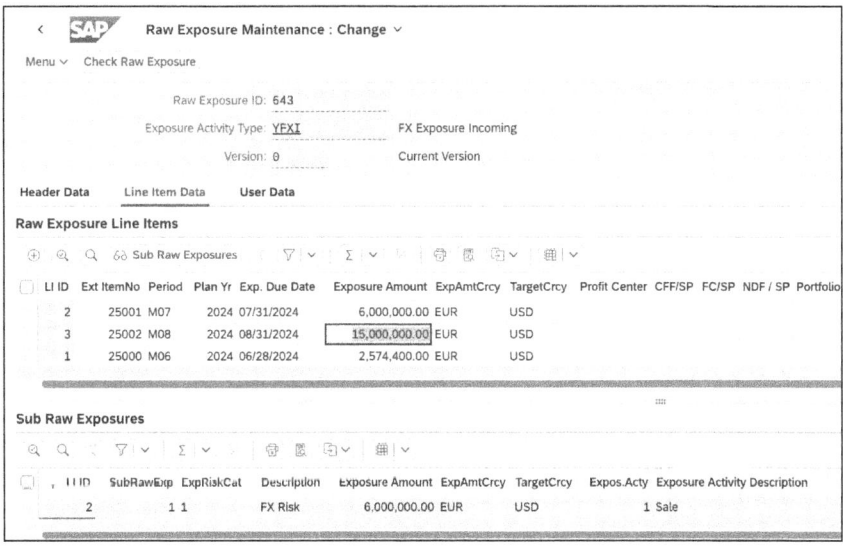

Figure 7.11 Editing Raw Exposure

7   Exposure Management

7.2.2 Import Raw Exposures – Spreadsheet App

While adding manual exposures is useful, oftentimes, companies have many exposures, and manual input is not scalable. To address this, the Import Raw Exposures – Spreadsheet app is available. This allows us to collect all of our raw exposures and subexposures into a spreadsheet and upload them en masse. Once we have populated our Excel spreadsheet, we can navigate to the app and select the file name. We can pull the file from local or shared locations, and then, we can select the processing options (see Figure 7.12):

- **Display Data Before Processing**
  If we select this option, as the file is loaded, we'll see an import screen with all the exposures formatted before the exposures are created.

- **Test Run**
  If we select this option, we can review the data to make sure it's what we expected before we create the exposures.

> **Note**
> As of the current version, the only standard way to upload exposures en masse is through the Import Raw Exposures – Spreadsheet app. Exposures haven't yet been added to the migration cockpit.

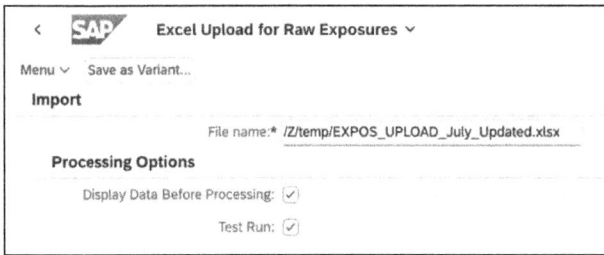

Figure 7.12 Excel Upload for Raw Exposures

Once we've made the proper settings, we'll click **Run** in the bottom right corner, and the output shown in Figure 7.13 will be displayed. This lists all the exposures that were part of our upload.

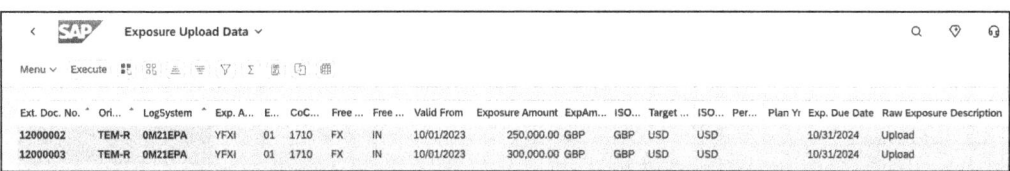

Figure 7.13 Uploading Exposure Data

If we're satisfied with the results, we can then click the **Execute** button. This will create the exposures shown in Figure 7.14, where the **Raw Exposure ID** has been populated and the status is shown in green.

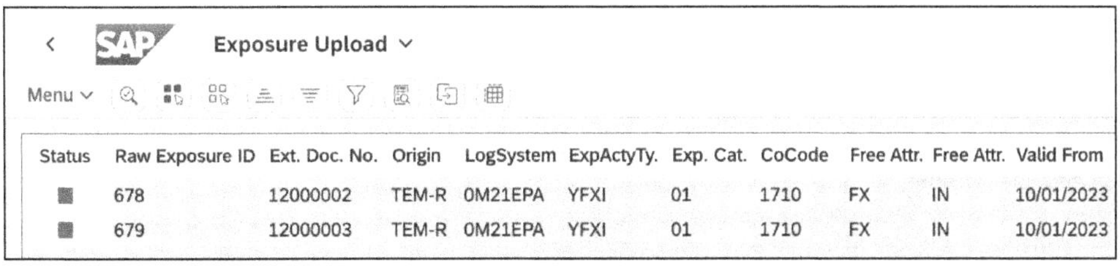

Status	Raw Exposure ID	Ext. Doc. No.	Origin	LogSystem	ExpActyTy.	Exp. Cat.	CoCode	Free Attr.	Free Attr.	Valid From
■	678	12000002	TEM-R	OM21EPA	YFXI	01	1710	FX	IN	10/01/2023
■	679	12000003	TEM-R	OM21EPA	YFXI	01	1710	FX	IN	10/01/2023

Figure 7.14 Exposure Upload

> **Note**
>
> In addition to manually inputting and uploading exposures, we can use a Business Application Programming Interface (BAPI) to assist us with integrating FX exposure positions from materials management and SAP S/4HANA Sales. With the BUS5990 BAPI, we can create exposures, update versions, and have those exposures integrate automatically with the processes outlined in this chapter.

7.3 Releasing Raw Exposures

We've discussed creating and changing raw exposures; however, once we have set up an exposure, the exposure gets added to other exposures and allows a user to see all risks across the organization and take action to hedge those positions. If desired, we may elect to turn on the release of exposures to help us control what is being created as an exposure. These exposures can be released manually or automatically.

In Section 7.1, we discussed how each exposure activity is set up and then noted how the release category is set up for each exposure activity. If the exposure activity is set up to be manually released, then there are a few options for how to release the exposure:

- **Using the Process Raw Exposure app**
  As shown in Figure 7.15, upon entering the app, we'll need to fill out the **Exposure Activity Type** and **Raw Exposure ID** and then select **Initiate Release** at the top of the screen.

  When we select **Initiate Release**, the system runs validation checks on the exposure. If the validations are passed, the exposure is released, and we'll get success messages like the ones shown in Figure 7.16. If any errors are noted, we can repair them.

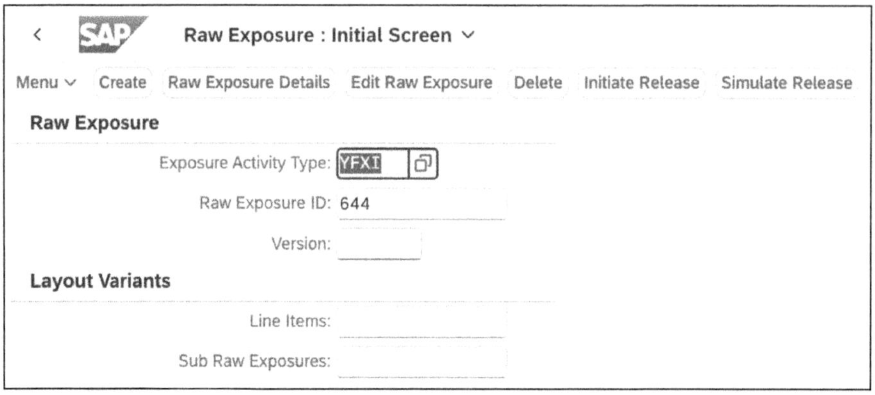

7   Exposure Management

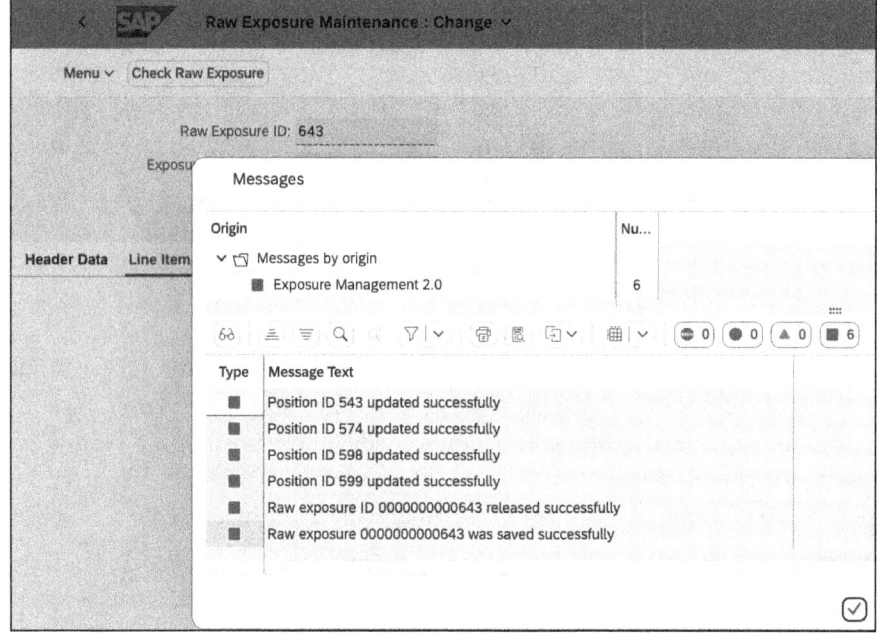

Figure 7.15 Process Raw Exposure: Initiate Release

Figure 7.16 Initiate Release

- **Using the Process Raw Exposures – Collective Processing app**
  Alternatively, we can use the Process Raw Exposures – Collective Processing app to release our exposures. To do this, we navigate into the app and look for the **Release Status** section. From there, we can select the status—for example, we selected **Release Initiated** as shown in Figure 7.17.

  When we select this, a list of all unreleased exposures will be listed as shown in Figure 7.18. Then, we can click **Initiate Release,** which will release the transaction.

7.4 Exposure Configuration

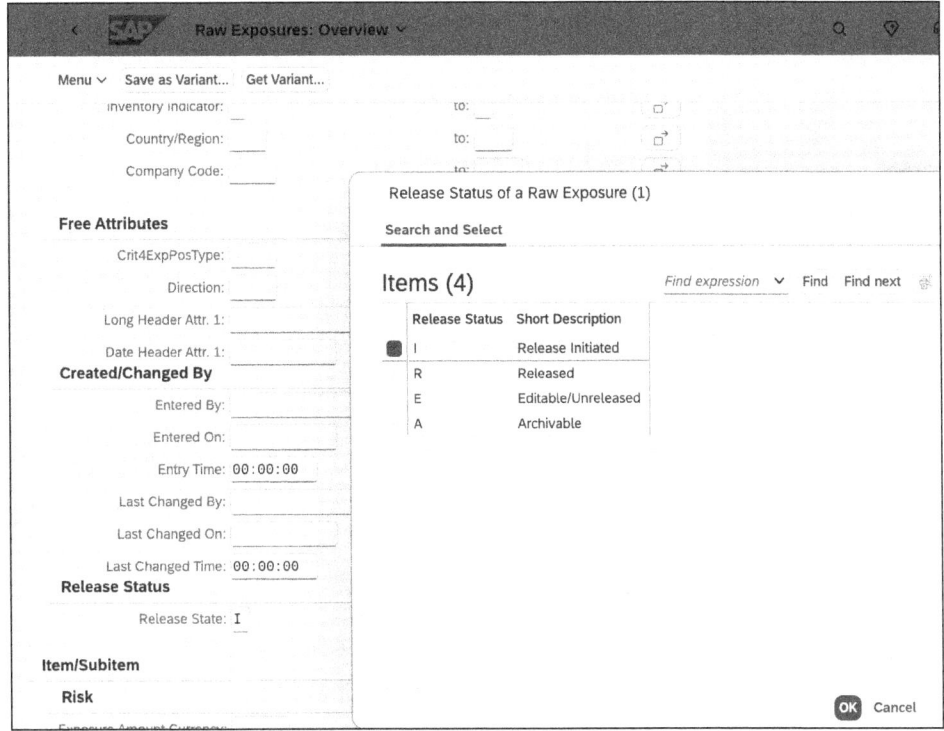

Figure 7.17 Release Initiated

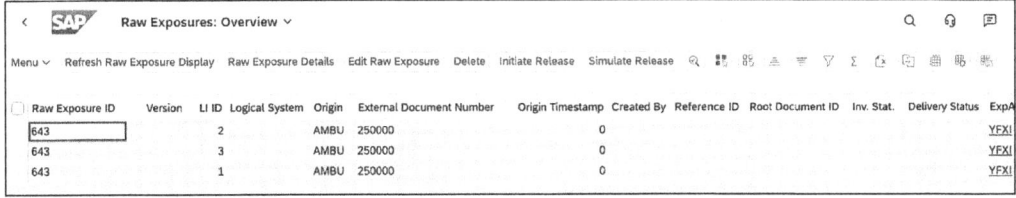

Figure 7.18 Process Raw Exposures – Collective Processing App: Initiate Release

Note

If we have many exposures, we may not want to initiate a release of all exposures manually, so if we have the automatic exposure release indicated in the exposure activity, using it can be very helpful. We discuss this in more detail in Section 7.4.3.

7.4 Exposure Configuration

We now understand how to capture exposures in the system, so we'll transition into the system configuration settings that will drive the previously discussed actions for

7 Exposure Management

the users. We'll define global settings to make exposure management active, we'll configure the exposure types available to users, and we'll discuss derivation and workflow process that help with exposure capture and release.

7.4.1 Defining Global Settings

First, we'll define the global settings for exposures. We start by navigating to the following menu path: **Financial Supply Chain Management • Treasury and Risk Management • Transaction Manager • General Settings • Exposure Management 2.0 • Define Global Settings**. In this configuration node, we select one of the following settings, as shown in Figure 7.19:

- **Automatic Position Matching Allowed**
  If we want to select this setting, a new raw exposure is required. For example, if an exposure has been changed from a planned transaction to a firm commitment, then we need to create a new raw exposure for the firm commitment. We need to do additional configuration to define exposure position types to facilitate the automatching of exposures, and we'll cover that in Section 7.4.8.

- **Transaction Category Change Allowed**
  When selecting the **Transaction Category Change Allowed** setting, we don't have to create a new raw exposure if the exposure category changes. For example, if an exposure has been changed from a planned transaction to a firm commitment, when the exposure is released, the planned transaction goes to 0 and the firm commitment position is created.

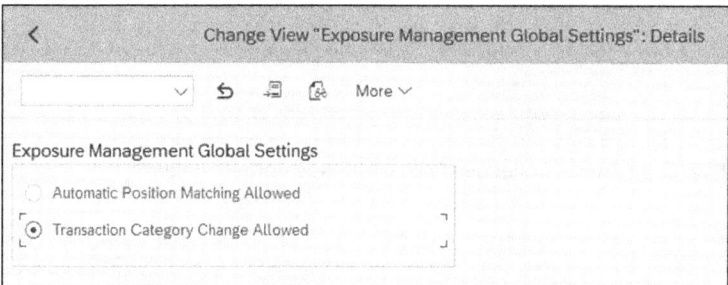

Figure 7.19 Defining Global Settings for Exposures

7.4.2 Defining Periods

Next, we'll define the periods to be used for exposure planning. We can elect to create our own periods, or we can leverage our setup in financial accounting. We start by navigating to the following menu path: **Financial Supply Chain Management • Treasury and Risk Management • Transaction Manager • General Settings • Exposure Management 2.0 • Define Periods**. We then define one month (for example, January). Here, we enter a three-digit unique period in the **Per** field (for example, M01), and then, we enter **Period**

Text that correlates with the month of January. Next, we define the planning year variants in the **PV** field; we have chosen K4 (a calendar year with four special periods). Next, we indicate the **Period, which** in this case is a month, so we input "1." We'll repeat this for all the periods that are necessary; in Figure 7.20, we've done this for all twelve months.

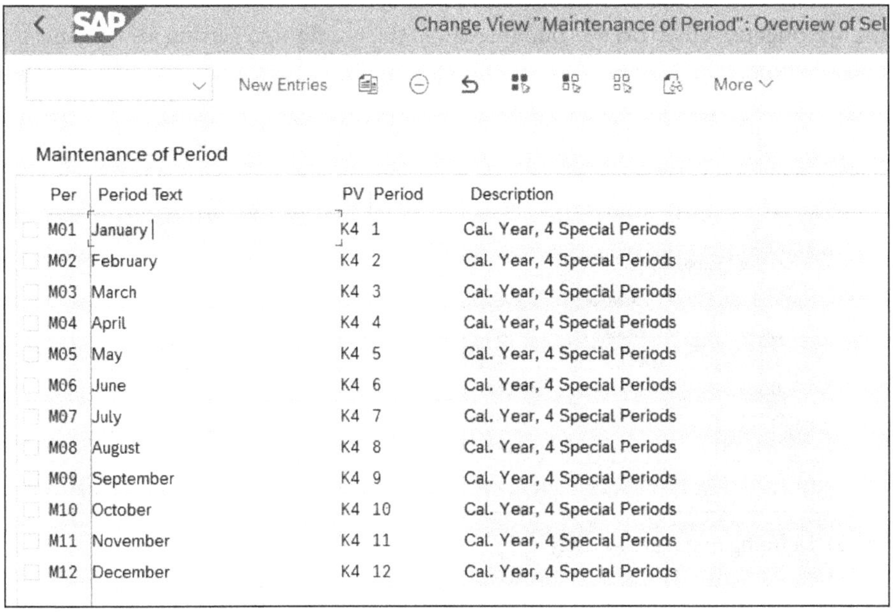

Per	Period Text	PV	Period	Description
M01	January	K4	1	Cal. Year, 4 Special Periods
M02	February	K4	2	Cal. Year, 4 Special Periods
M03	March	K4	3	Cal. Year, 4 Special Periods
M04	April	K4	4	Cal. Year, 4 Special Periods
M05	May	K4	5	Cal. Year, 4 Special Periods
M06	June	K4	6	Cal. Year, 4 Special Periods
M07	July	K4	7	Cal. Year, 4 Special Periods
M08	August	K4	8	Cal. Year, 4 Special Periods
M09	September	K4	9	Cal. Year, 4 Special Periods
M10	October	K4	10	Cal. Year, 4 Special Periods
M11	November	K4	11	Cal. Year, 4 Special Periods
M12	December	K4	12	Cal. Year, 4 Special Periods

Figure 7.20 Define Periods

7.4.3 Defining Exposure Types

We'll now define the types of exposures we want to capture in the system. We start by navigating to the following menu path: **Financial Supply Chain Management • Treasury and Risk Management • Transaction Manager • General Settings • Exposure Management 2.0 • Define Exposure Activity**.

As shown in Figure 7.21, we'll want to indicate the destination of any FX-based exposures in the **Exp Cat.** (exposure category) field as either **Incoming** or **Outgoing**. Here, we set up a unique five-character exposure type in the **ExpActyTy.** field. This will help us segregate the different types of raw exposures we plan to have in the system. We can also determine how to release our exposures, either manually or automatically. If we choose **Automatic**, then once we save the raw exposure, the release automatically starts. If we choose **Manual**, then we'll have to follow the procedure described in Section 7.3 to release the exposure. The final step is to confirm how we want to have the exposure activity defined. Remember that back in Section 7.4.1, we indicated either **Automatic Position Matching Allowed** or **Transaction Category Change Allowed**. In the **Change GS** field, we can choose one of the following options:

7   Exposure Management

- **Default of Global Settings**
  This option lets us use the setting that was previously configured in Section 7.4.1.
- **Automatic Position Matching**
  This option lets us use this method instead of the configured setting for this specific exposure type.
- **Transaction Category Change Allowed**
  This option lets us use this method instead of the configured setting for this specific exposure type.

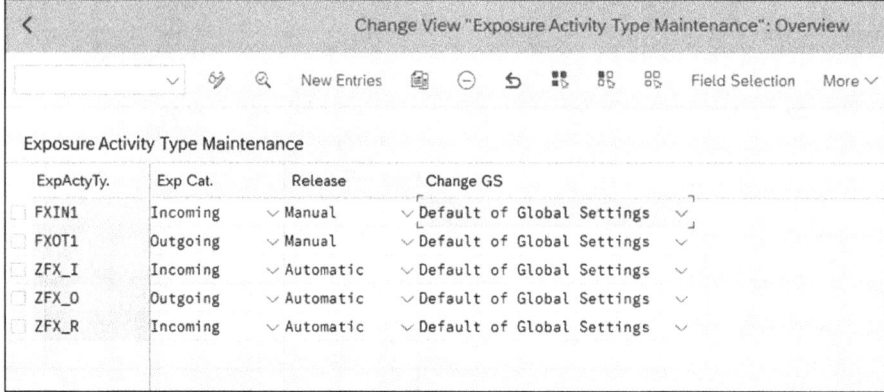

Figure 7.21 Defining Exposure Activity

Additionally, we can drill down into the overall maintenance view if desired, as shown in Figure 7.22.

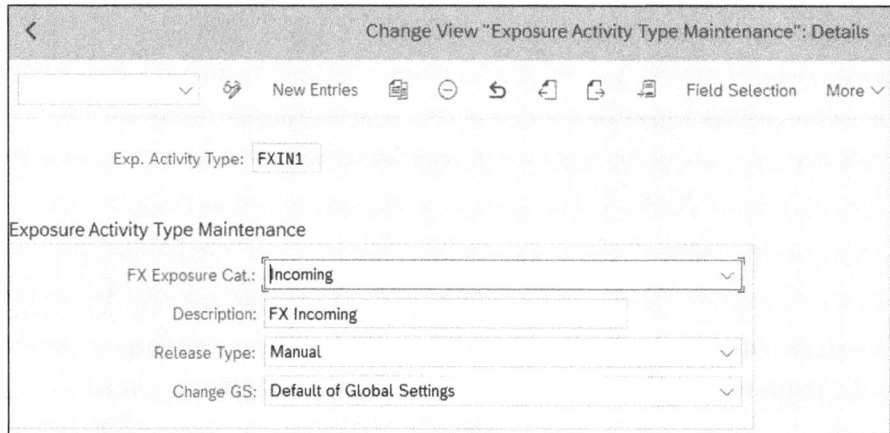

Figure 7.22 Details of Defining Exposure Activity

Once we have defined our **FX Exposure Cat.**, **Description**, **Release Type**, and **Change GS**, we may also want to control the field selection of each exposure. To do this, we select the **Field Selection** in the maintenance view.

7.4 Exposure Configuration

There are three sections that we can control, as we saw in the Process Raw Exposures – Collective Processing app in Figure 7.4 for the **General Attributes**, **Free Attributes**, and **Line Item Attributes**. Here, we select the attributes we wish to update, as shown in Figure 7.23.

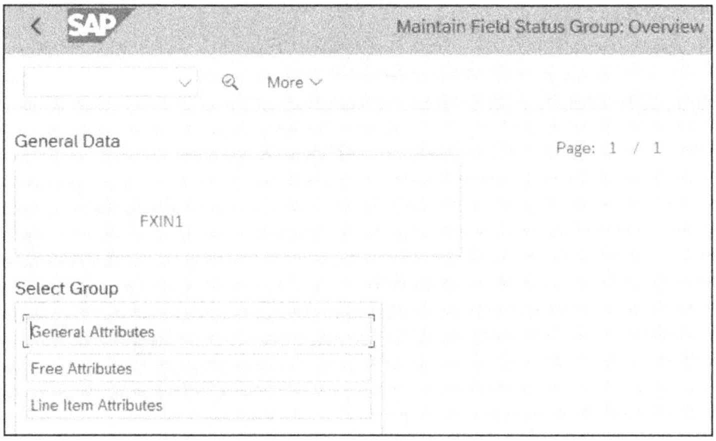

Figure 7.23 Select Attributes

We can drill down to each of the attributes to control what will be displayed on the entry of each raw exposure. We can also choose to suppress, require entry, optional entry, or display only. These options will help us streamline the entry of exposures so that they can be reported consistently and alert users if something required is missing. This is shown in Figure 7.24.

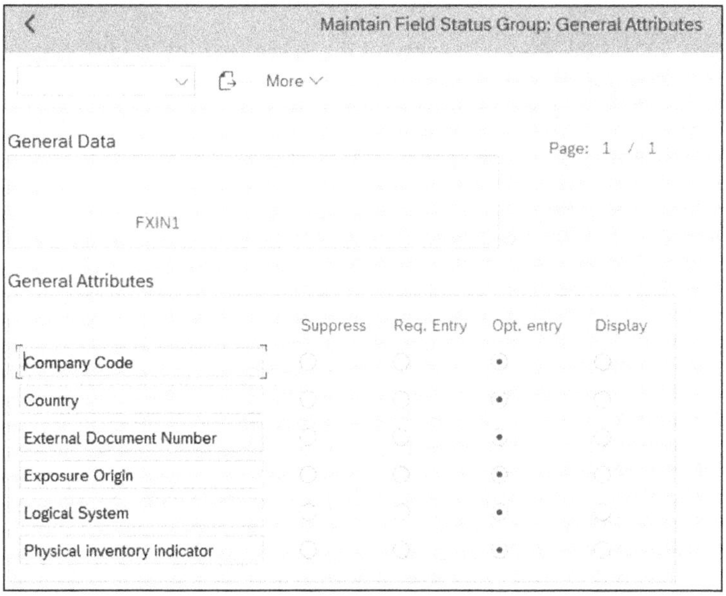

Figure 7.24 Defining Key Fields for Exposures

501

7.4.4 Settings for Free Attributes

While SAP S/4HANA provides several standard dimensions that we can use to segment exposures, there may also be client-specific attributes that we would like to add. We can use the settings for free attributes to set up any client-specific attributes. There are three types of free attributes, as follows:

- Short attributes, which we define with four characters or less
- Medium attributes, which we define with five to ten characters
- Long attributes, which we define with eleven to twenty characters

> **Note**
>
> Characteristics like short text, description, and long name are displayed in the exposure creation screens. If we support multiple languages, we can set up multiple language attributes there. If the text can't be translated into another language, it will remain in English.

To access the free attributes, we navigate to the following menu path: **Financial Supply Chain Management • Treasury and Risk Management • Transaction Manager • General Settings • Exposure Management 2.0 • Settings for Free Attributes • Define Headings, Values, and Texts for Short Attributes**.

To add a new short, medium, or long attribute, we first need to select an **Attribute ID**. There will be several options, but it's key to note the embedded naming convention—for example, SH01 versus SI01. SH01 is an attribute on the header level (as indicated by the H in the second position), while SI01 is an attribute on the line-item level (as indicated by the I in the second position).

Once we've selected the **Attribute ID**, we add free-form **Short Text** (up to ten characters), a **Description** (up to twenty characters) and a **Long Name** (up to forty characters). Then, we can save as shown in Figure 7.25.

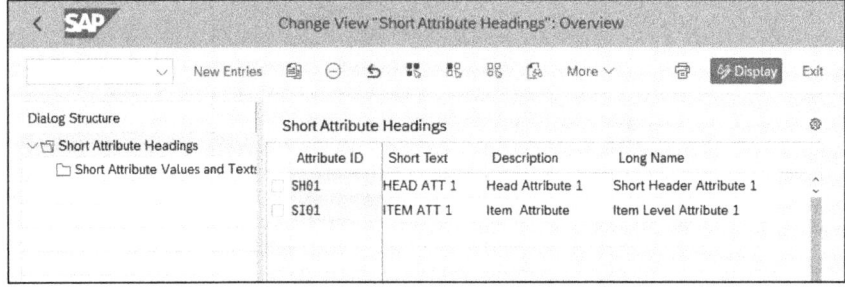

Figure 7.25 Short Attribute Headings

Once we have saved our short attribute heading, we can assign values and text by selecting the **Short Attribute Values and Texts** subfolder. Here, we assign a **Free Attr.**

(a free attribute of four characters) and the associated text for the attribute as shown in Figure 7.26.

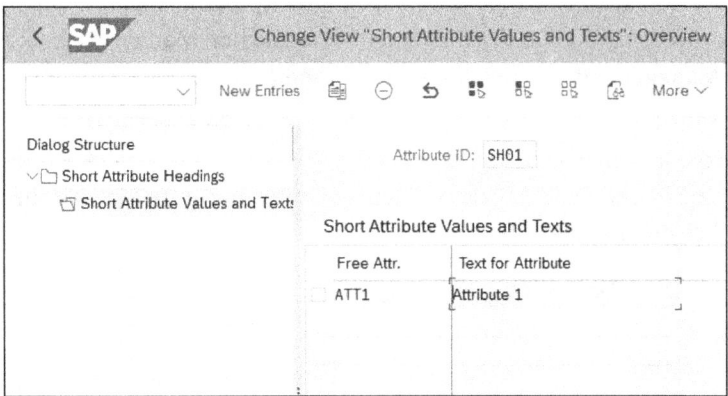

Figure 7.26 Short Attribute Values and Texts

7.4.5 Maintaining Release Procedures

Next, we'll look into the maintenance of the release procedure. We can access this via the following menu path: **Financial Supply Chain Management • Treasury and Risk Management • Transaction Manager • General Settings • Exposure Management 2.0 • Release • Assign Release Procedure to Release Object**.

Here, we'll define the release object for TRM_EM (Treasury – Exposure Management). We'll also define the number of release steps: dual control or more to meet our requirements. As shown in Figure 7.27, we first indicate the number of **Required Release Steps**, and since we want dual control, we select **01**. Additionally, we want our workflow to always be running, so under **Run Release Workflow**, we select **Always**. Then, we click **Save**.

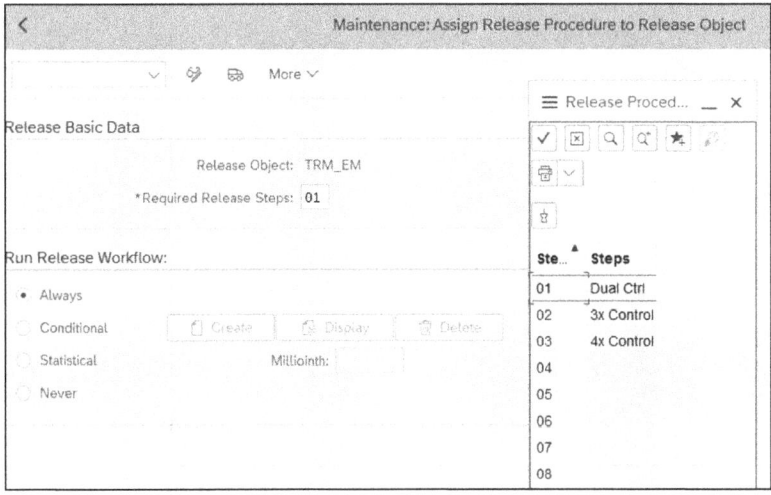

Figure 7.27 Maintenance: Assign Release Procedure to Release Object

7.4.6 Assigning Users and Roles to Release Steps

Once we have saved the release procedure, we can begin to assign users and roles to the release procedure. We can access this via the following menu path: **Financial Supply Chain Management • Treasury and Risk Management • Transaction Manager • General Settings • Exposure Management 2.0 • Release • Assign Users/Roles to Release Steps**.

This will be the first rule that we'll create, so we want to input the **Release Object** "TRM_EM," which is for exposure management, and then, we want to input our **Required Release Step**(s) and the workflow (**WF**) **Release Step**, as shown in Figure 7.28. We then select the row and click **Create Rule**.

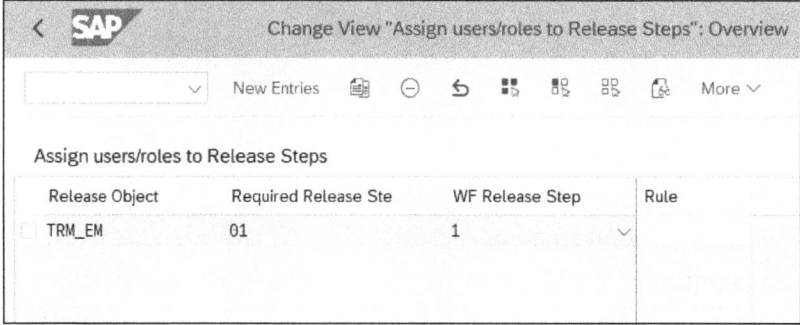

Figure 7.28 Assigning Users and Roles to Release Steps

After we select **Create Rule**, Figure 7.29 will be displayed with the available rule parameters. We'll create a simple rule-based workflow based on exposure activity type, but note that we could select multiple options during workflow rule creation. We'll select **EXP_FLOW_TYPE** and then click **Create Rule Now**.

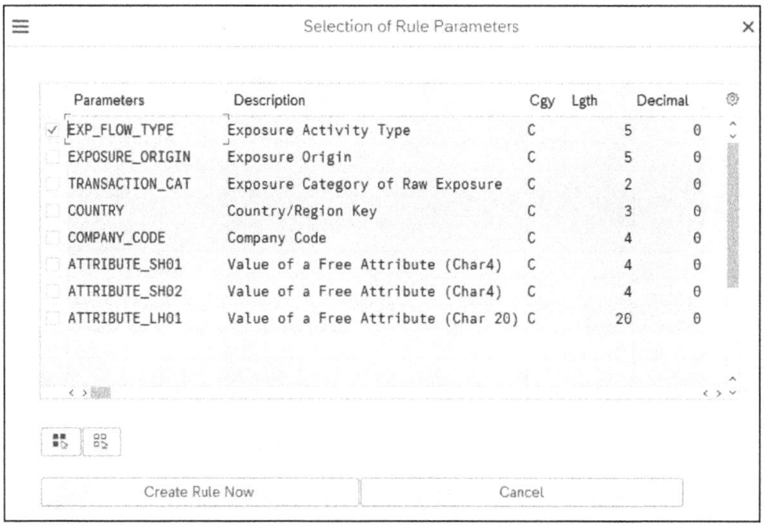

Figure 7.29 Selection of Rule Parameters

7.4 Exposure Configuration

The **Change Object Directory Entry** prompt is prepopulated with the ABAP **Package** and user ID (**Person Responsible**). We could change these if required, but we'll use the default values as shown in Figure 7.30. Then, we click **Save**.

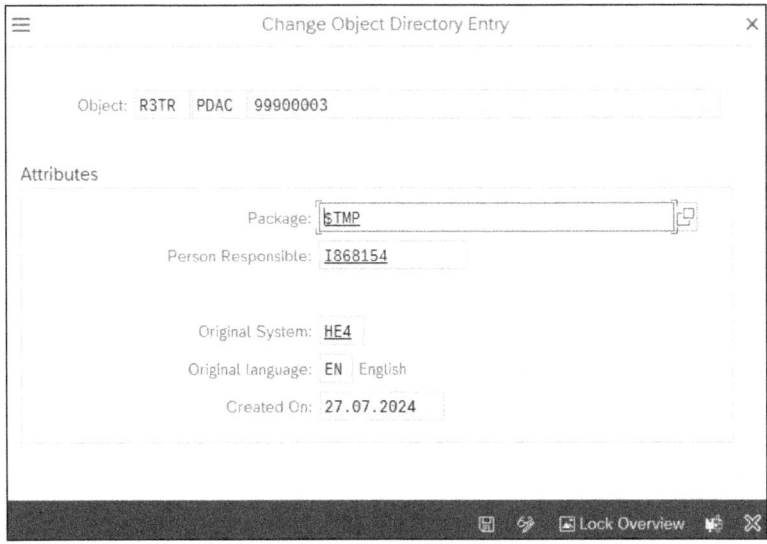

Figure 7.30 Prompt after Generation of Rule

We'll save this and then select **Apply Generated Rule**, as shown in Figure 7.31. This will automatically assign a generated rule number to our release procedure.

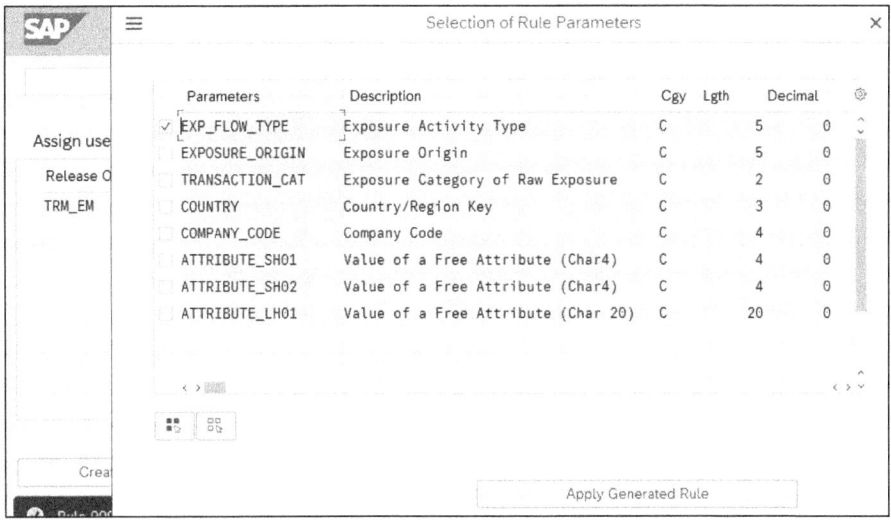

Figure 7.31 Applying Generated Rule

Upon saving, we are returned to the main screen of the workflow responsibilities, where we now can see that our rule has been populated, as shown in Figure 7.32.

505

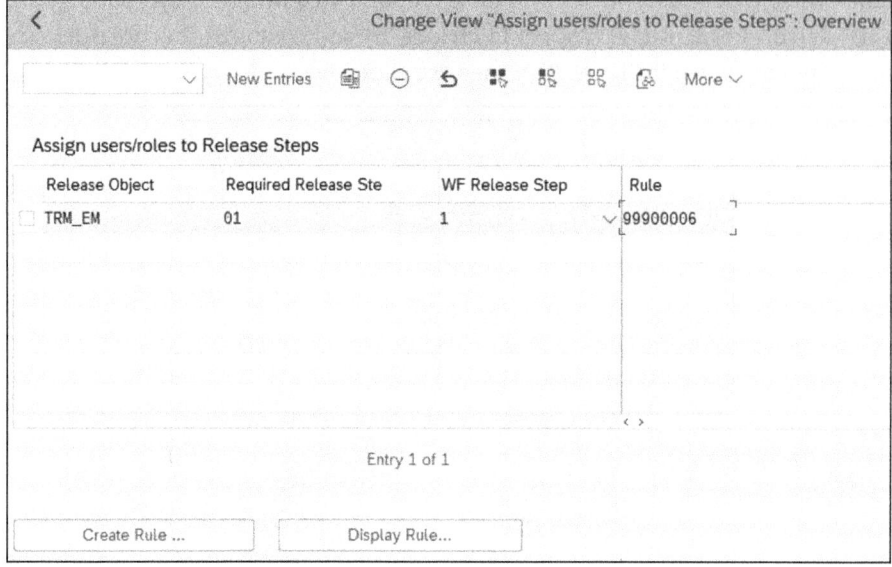

Figure 7.32 Rule Has Been Populated

Now that we have defined the rule, we can set the responsibilities. We select the item and click **Display Rule,** as shown in Figure 7.32. When we do that, the **Responsibilities for Rule** screen will be displayed. We then click the **Create Responsibility** button, which will take us to the **Create responsibility** prompt, where we'll maintain our overall rule. The **Object abbr.**, **Name**, **Start date**, and **End Date** will autopopulate but change as needed. As Figure 7.33 shows, we've updated the name to "TRM_EM Exposure Attribute." Then, we click the checkmark at the bottom right.

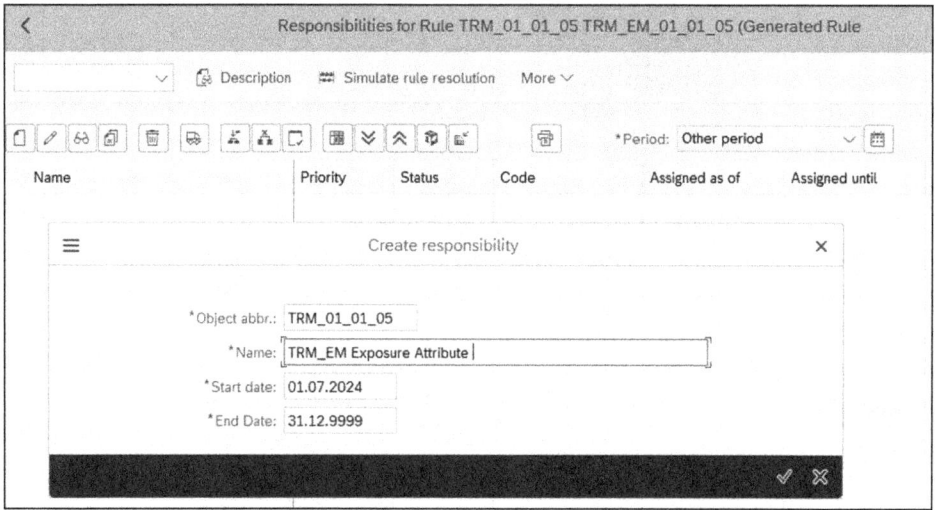

Figure 7.33 Creating Responsibility Rules

7.4 Exposure Configuration

Now, we'll specify the values of the underlying exposure activity types that we want to initiate the workflow, and we'll populate the **of** and **to** values with our exposure types. In Figure 7.34, we've included the full range of exposure types available. If we don't want to input a range but want to input individual values instead, we can input each exposure type individually by using the **+** icon on this screen.

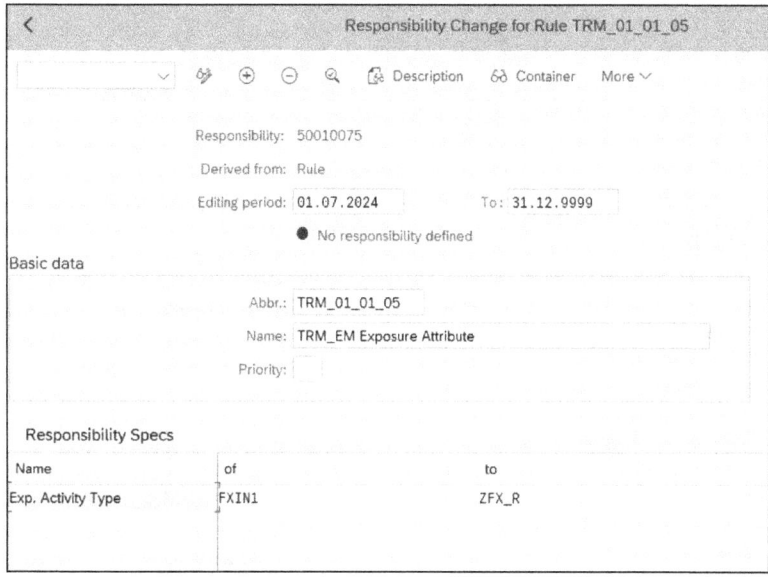

Figure 7.34 Defining Rules for Responsibilities

Once we've entered all values, we click **Save**. Note that once saved, the screen will update and show **Responsibility complete,** as shown in Figure 7.35.

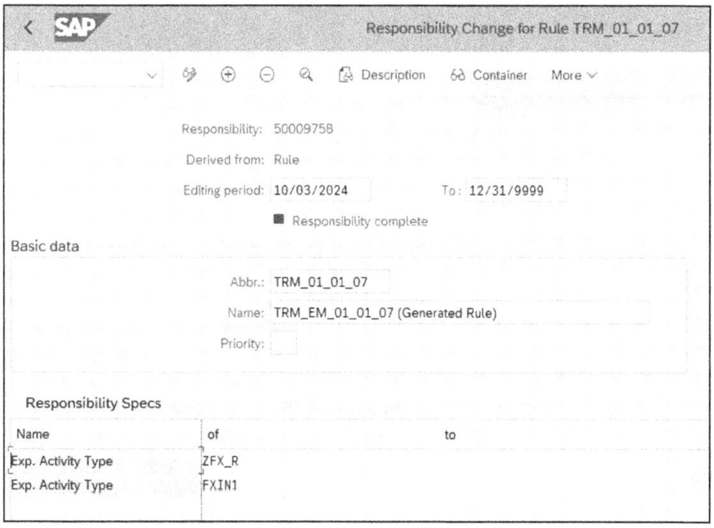

Figure 7.35 Responsibility Complete

507

7   Exposure Management

After we create our rule, it will be available on the **Responsibilities** page. To add responsibilities, we select the icon (the seventh icon from the left in Figure 7.36), which allows us to add key workflow attributes (e.g., person, position, job).

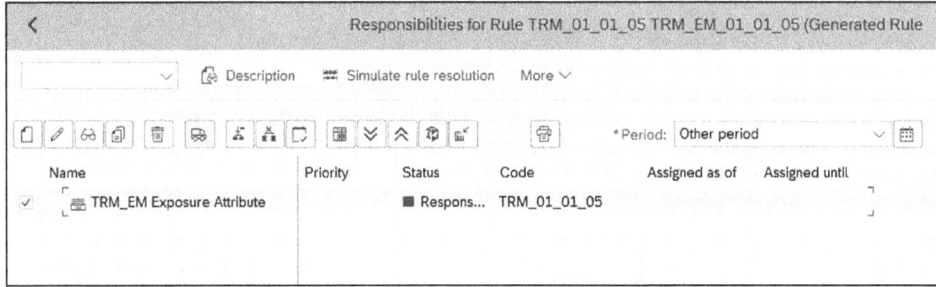

Figure 7.36 Assigning Users and Roles: Defining Responsibilities

Once we select the agent, we'll fill in the **Chose User** box with an appropriate ID. Once we've selected the user ID, we click the checkmark to save our values as shown in Figure 7.37.

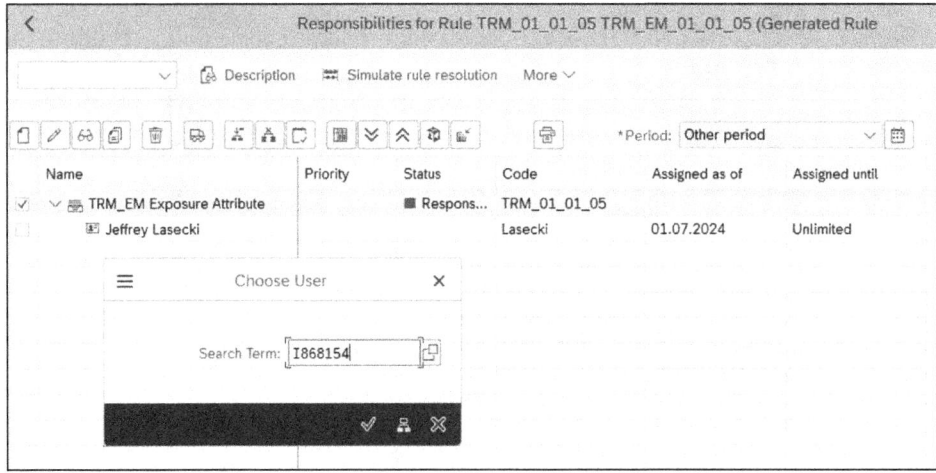

Figure 7.37 Assigning Users: Choose User

Now that we have identified a user, they are now part of the workflow process, and any time an exposure is created or changed, it will flow to this user for approval.

> **Note**
> There's a useful simulation tool (**Simulate Rule Resolution**) that is part of this workflow setup and that we can select from the **Responsibilities for Rule** screen. This helps us validate the workflow routing that we have set up. This functionality is shown in Figure 7.38.

7.4 Exposure Configuration

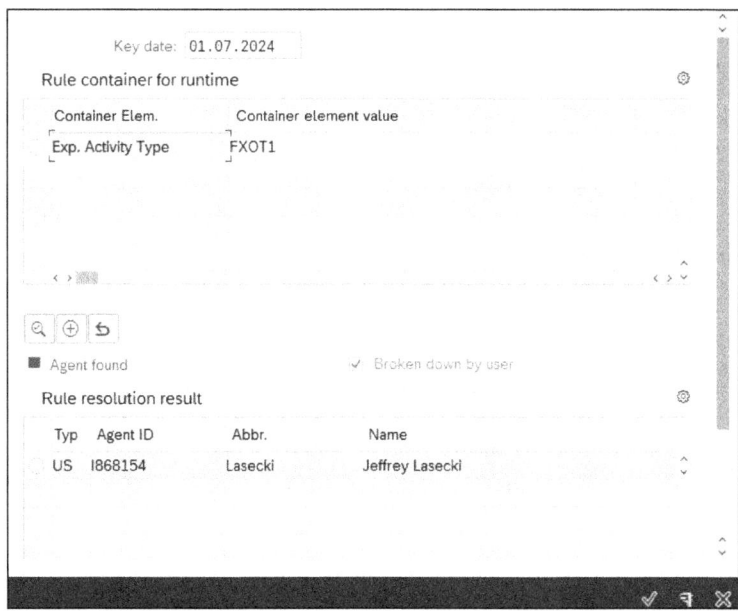

Figure 7.38 Exposure Workflow Routing Tool

7.4.7 Defining Product Types for Exposures

Next, we'll define the product types (much like those we use in debt, FX, and derivatives) for exposures. These exposure product types will be connected to the exposure position types. To define the exposure product type, we enter a unique three-digit **Product Type**, **Text** describing the product type, and the **Prod. Category** "990" **Exposure**, as shown in Figure 7.39. We can find this setup in the following menu path: **Financial Supply Chain Management • Treasury and Risk Management • Transaction Manager • General Settings • Exposure Management 2.0 • Define Product Types for Exposures**.

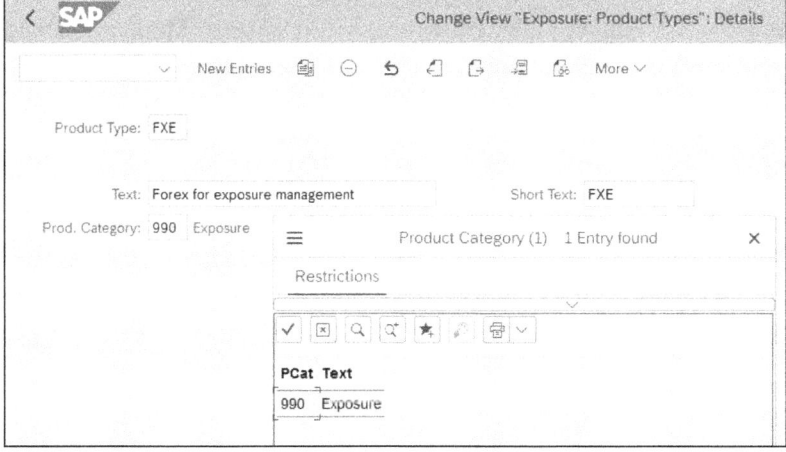

Figure 7.39 Defining Product Types for Exposures

509

7 Exposure Management

7.4.8 Defining Exposure Position Types

Next, we'll define the exposure position types in the following menu path: **Financial Supply Chain Management • Treasury and Risk Management • Transaction Manager • General Settings • Exposure Management 2.0 • Define Exposure Position Types**. We use these position types to analyze and manage exposure positions throughout the organization. The exposure position type also makes the link between the exposure type and the product type we've just defined in Section 7.4.7.

To define the exposure position types, we first enter a unique four-digit **Exposure Position Type** and an associated **Description**. Then, we use checkboxes to determine whether we want to look at the positions in **Aggregate** or whether we want to look at the **Default** position type. Then, we link the exposure position type to the **Product Type** as desired. If our global setting is set to automatically match as defined in Section 7.4.1, then we can indicate matching exposure position types that will roll up when we look at the reporting. Here, we also define the planning year variant. If we check the **No Planning Period** box, it will tell the system to not take the date into account; this should not be turned on if **Aggregate** is already selected, as shown in Figure 7.40.

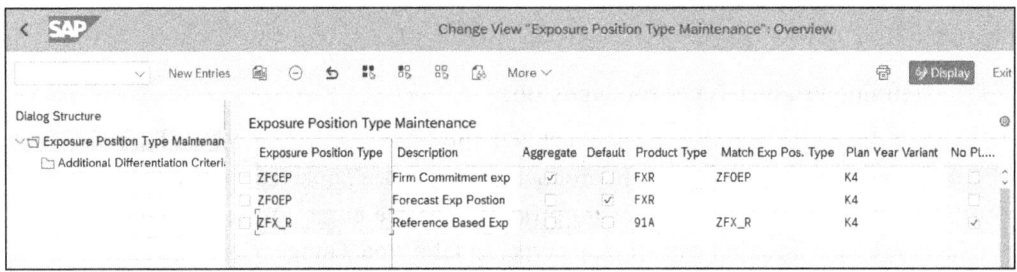

Figure 7.40 Defining Exposure Position Types

7.4.9 Defining the Derivation Strategy for Exposure Fields

Next, we'll define how data can be pulled from the raw exposure to populate the exposure position. We can find this configuration in the following menu path: **Financial Supply Chain Management • Treasury and Risk Management • Transaction Manager • General Settings • Exposure Management 2.0 • Define Derivation Strategy for Exposure Fields**.

We can use all fields that are part of the exposure to segment the position. The **Exposure Position Type** is one of the key fields of the exposure that needs to be derived from the raw exposure. Within this configuration element, we can set up one of the following steps (see Figure 7.41):

- **Derivation rule**
  We use this with multiple source fields where the information can be pulled from the raw exposure into specific target fields in the exposure position.

7.4 Exposure Configuration

- **Move**
  A move action takes a single element from the raw exposure to the exposure position.

- **Clear**
  A clear action removes items from fields in the exposure position that are not needed.

- **Enhancement**
  If any additional enhancement is needed, a BAdI is available to help us fill in the raw exposure fields.

For this example, we'll create a new derivation rule that defaults to the country when a company code is entered.

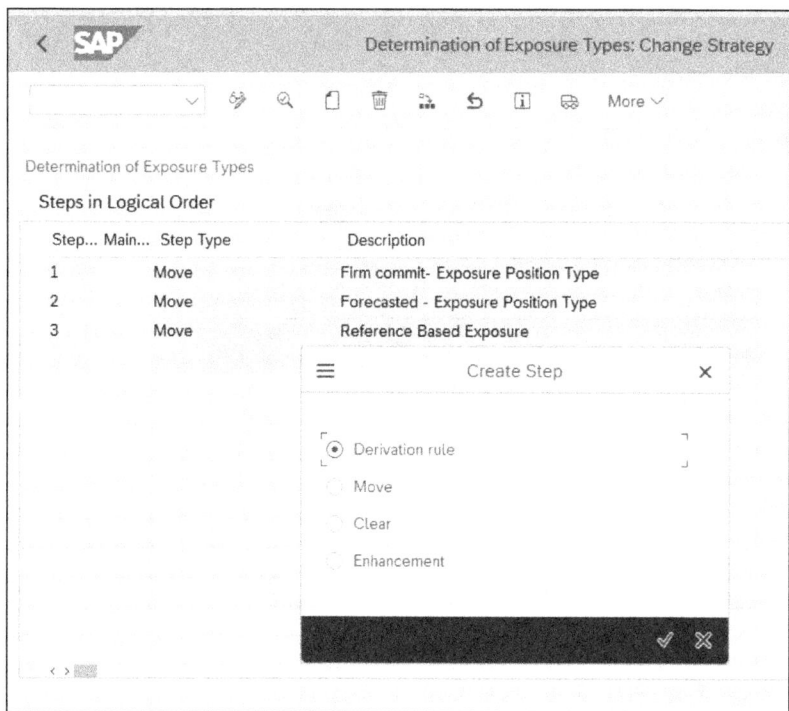

Figure 7.41 Derivation Strategy for Exposures

We enter a **Step Description** and then move on to the definition of the source field. For our example, when we have a company code, we want to default to the country automatically. We define our source field as "COMPANY_CODE" and add "COUNTRY" to the target field, as shown in Figure 7.42.

Then, we select the **Condition** tab, where we populate the conditions in which the derivation rule will take effect. Figure 7.43 shows the setup for the "COMPANY_CODE" conditions and their corresponding **Value** setups in SAP S/4HANA (e.g., 1710 and 1010).

7 Exposure Management

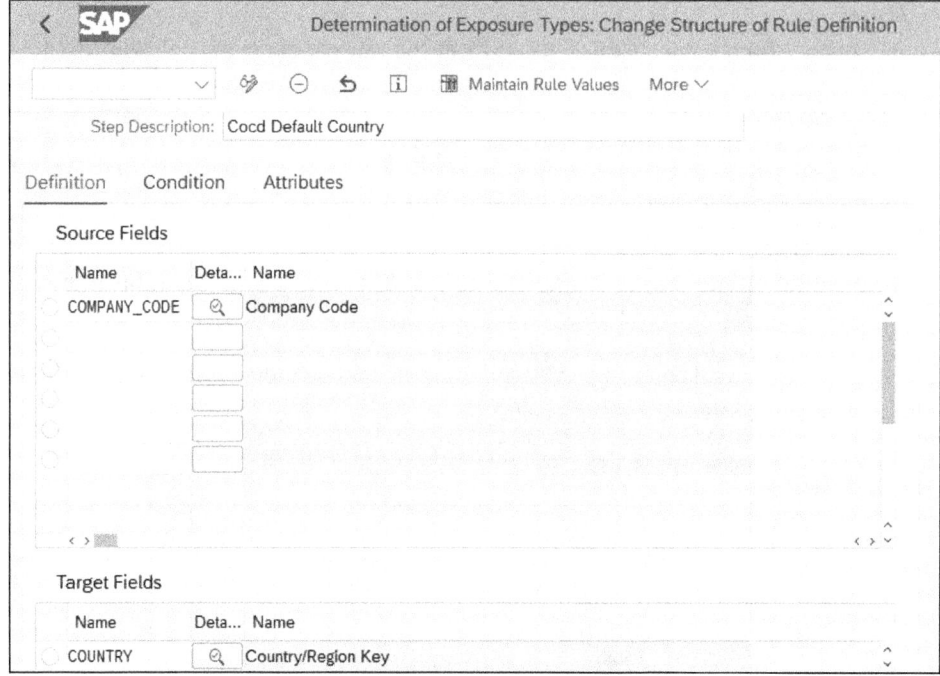

Figure 7.42 Derivation Strategy Definition

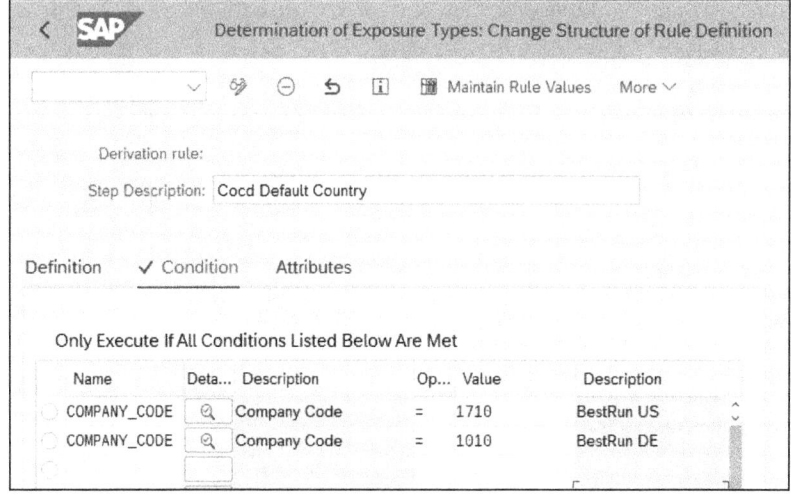

Figure 7.43 Creating Step for Exposure Derivation

Now, we select the **Maintain Rule Values** option and key in the source and target that have been previously defined. Here, we see that **Company Code 1710** will derive a US company code, as shown in Figure 7.44.

7.4 Exposure Configuration

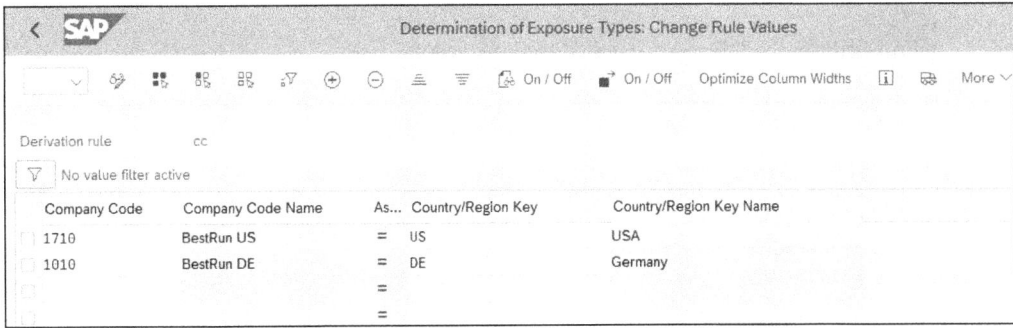

Figure 7.44 Exposure Conditions That Drive Derivation

Now that the derivation rule has been set the next time a user creates a raw exposure, when the company code is entered the country will also default. This can be used for any standard or free attribute defined throughout this section.

7.4.10 Defining Exposure Origins

Next, we'll define the exposure origin via the following menu path: **Financial Supply Chain Management • Treasury and Risk Management • Transaction Manager • General Settings • Exposure Management 2.0 • Define Exposure Origins**. This setting is designed to help further identify where the exposure originates. By having this setup, we can bring in external document numbers and identify what logical systems the exposures came from. To set this up, we need to configure a five-digit origin name (**Orgn**) and a description, as shown in Figure 7.45.

> **Note**
>
> This configuration element is contained in a cross-client table. We should always be careful to determine whether we need to make a change here.

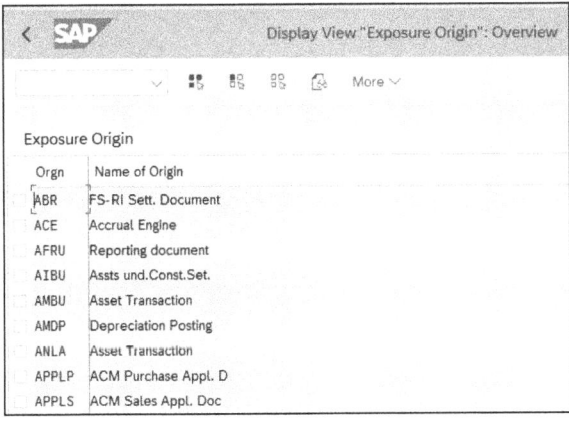

Figure 7.45 Defining Exposure Origins

513

> **Additional BAdI for Exposure Management**
>
> In addition to the standard configuration available to support exposure management 2.0, there are several BAdIs that we can implement. They include the following:
>
> - BAdI: Enhancement for Deriving Exposure Fields
> - BAdI: Enhancement for Commodity Splits
> - BAdI: Enhancement for Position Data in Other Applications

7.5 Summary

We've just completed the end-to-end process for creating cash flow exposures. We can either manually create exposures or import them using a spreadsheet. Additionally, we've learned how to manage those exposures and subexposures and ensure that data is being reviewed. We've also explored all the configurations that are available to make the exposures fit our company's requirements. The exposure is the basic building block for cash flow hedging, which we will cover in the next chapter.

Chapter 8
Cash Flow Hedging

By utilizing the functionality established for the contract types in treasury and risk management, SAP offers a robust solution for cash flow hedging. This solution lets us assess exposures and gives us the capability to hedge against them.

Since there's a vast difference in the hedging process between balance sheet hedging and cash flow hedging, there's a different process we follow for cash flow hedging. It includes completely different transactions and an overall process flow when we create our hedging rules, create hedges, and determine how the hedges are managed going forward. SAP S/4HANA Finance for treasury and risk management has established a new process for cash flow hedging that replaces Transaction THMEX. This process includes creating a hedging area with many of the necessary settings for generating and designating the hedges, details on how the documents are generated for hedge accounting, and different views to gauge the overall hedges against target hedge percentages. Before going into the details of what the process looks like in SAP S/4HANA, we'll cover a high-level overview of both the configuration elements and the end user process for cash flow hedging:

- **Hedge management configuration**
  In Section 8.1 we'll start by covering the initial configuration required for the cash flow hedging process. These initial settings will allow us to create the subsequent settings required for a fully functional solution. We'll cover creating a category known as the hedging classification, which must be assigned to any hedging area in this solution. We'll also cover the process of generating the target quota settings that allow us to assign the target hedging percentages to each hedging scenario. We'll also create the hedge request reason that serves as a classification for the hedge request we created. We'll reference these settings in the following sections, and they'll serve as a baseline for the functionality in this hedge management solution.

- **Hedge accounting for positions configuration**
  In Section 8.2, we'll cover the rest of the required settings in the hedge accounting for positions configuration area. We'll detail all settings related to setting up hedge accounting for hedges, and we'll determine what types of transactions are included in hedge accounting and how we can generate the designation for the hedging instrument to the hedged item. We'll cover settings related to the calculation for hedge accounting and settings related to setting up different types of effectiveness

testing for hedge management. We'll assign these calculations and effectiveness testing parameters to a hedging profile and all of these settings will be ultimately assigned to a hedging area in Section 8.3. The final settings we'll cover in this section will assign update types to different hedge management related business processes. Each of these update types can influence the accounting postings that are processed in hedge accounting.

- **Importing exposures**
  Cash flow hedging in SAP is dependent on exposures, which generally are created by forecasted numbers established based the forecasted foreign currency risks. Exposure management 2.0, which is covered in Chapter 7, covers how the exposures are brought into SAP S/4HANA and how they're approved. These exposures can be driven based on exposures generated in exposure management or based on records in one exposure from operations in cash management.

- **Hedging area creation**
  In Section 8.3, we cover the hedging area in SAP S/4HANA where we create the structure for the hedging policy. Here, we define at how granular a level we want to manage hedges based on available data points. Due to the amount of data available in SAP S/4HANA, filters are available for us to define the specific types of transactions the hedging area should consider for exposures and for reviewing the created hedges.

  Also included in hedging area creation are the definition of exposure currencies, target quotas, default hedging instruments, designation settings, designation splitting settings, and assignment of the profiles required to automatically generate the hedge accounting within the hedging area. Once we have entered the exposures and determined these settings, we can run a snapshot to lock in the exposure and hedge details for further processing. We do this processing in the hedge management cockpit.

- **Hedge management cockpit**
  In Section 8.4, we'll use the hedge management cockpit to view a multitude of information within our established hedging area. This includes reviewing the overall health of our hedging strategy, and the report displays totals and drilldown capabilities for all key figures in the hedging area. Some of the key data points that we can review are all incoming and outgoing exposures, the net exposure amount, details on existing hedges, current hedge percentages, and the proposal for the amount to hedge based on the target quotas established in the hedging area. This cockpit will compare the exposures to the hedges to define the current hedge quota percentage and compare them with the target quota to determine the proposed amount to hedge. From this screen, we can manage the whole hedging area by creating automated hedge requests, looking into the current hedges, and even creating swaps to roll over any hedges that we need to extend to a new settlement date.

- **Releasing and importing trades**
  When we create new hedge requests in the hedge management cockpit, we can review the requests and release them before further processing. We can also edit the request amount in this review process before we send the trade request to the

trading platform. After we confirm and release all the hedge requests, we need to execute the trades in the trading platform to confirm the details of the trades, including the counterparties and related FX rates. We can then create the confirmed trades in SAP S/4HANA either manually (in the Create Financial Transaction app or Transaction FTR_CREATE) or by using an interface to import the executed trades from a trading platform. Creating the transaction will tie each FX trade to the hedge request/exposure and allow for the automatic designation of the hedge. When the hedge is designated, the predefined rules in the hedging area will determine how to run the hedge effectiveness testing and documentation for the hedge.

- **Month/period end**
Section 8.5 covers the month-end process, which is one of the key processes of hedge management and includes integral processes that must be run at month end for the hedging relationships. First, we valuate the positions of the hedging relationship using market data to determine the mark-to-market valuation amounts. Next, we post the amounts that were determined in the valuation to reflect the changes in market value. At month end, we can also run effectiveness tests to determine the effectiveness of the hedging relationship. Finally, we can run the classification for the hedges. This step reviews the effectiveness tests of the hedging relationship and separates the postings of the valuation between the effective and ineffective accounts.

- **Hedge maturity**
Section 8.6 details additional tasks that we need to carry out during the maturity of the hedging process. At the maturity date, we need to designate the hedging instrument from the hedged item. We also need to post the cash flows and initiate payments to settle the trade with the counterparty. Once we do this, the gain or loss from the FX difference can be realized. We can also post the other comprehensive income (OCI) amount at this time and reclassify it in subsequent periods based on the defined OCI release schedule.

In this chapter, we'll cover the full cash flow hedging process, including configuration considerations and subsequent processing in the hedge management cockpit that demonstrates the end-to-end process.

8.1 Hedge Management Configuration

There are two main areas that we need to configure when looking into cash flow hedging in treasury and risk management. The first area is in the customizing in the **Financial Supply Chain Management • Treasury and Risk Management • Transaction Manager • General Settings • Hedge Management** menu path. This section will cover the base settings for hedge management in the following four configuration nodes (see Figure 8.1):

- Define Hedging Classifications
- Define Target Quota Types

8 Cash Flow Hedging

- Define Authorization Groups for Hedging Areas
- Define Hedge Request Reasons

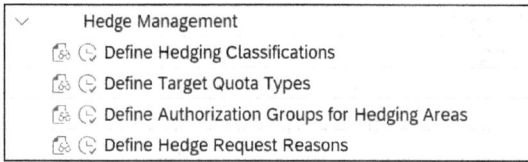

Figure 8.1 Hedge Management Configuration Area That Creates Base Settings for Cash Flow Hedging

8.1.1 Defining Hedging Classifications

The first configuration in the hedge management customizing area is **Define Hedging Classifications**. Creating the hedging classification creates a base setting that we must define when creating the hedging area. When assigning this setting to the hedging area, we also tie due date shifts and a definition to how the system searches for the correct exposure items. Additionally, this configuration determines whether the hedging classification is for a hedging area where we're creating hedge accounting or not. In Figure 8.2, all the hedging classifications are active for hedge accounting in this system.

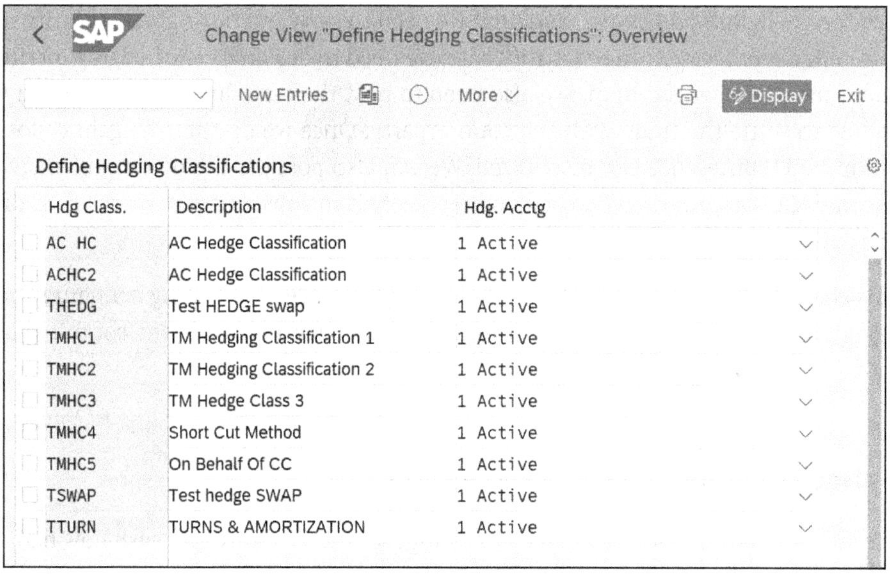

Hdg Class.	Description	Hdg. Acctg
AC HC	AC Hedge Classification	1 Active
ACHC2	AC Hedge Classification	1 Active
THEDG	Test HEDGE swap	1 Active
TMHC1	TM Hedging Classification 1	1 Active
TMHC2	TM Hedging Classification 2	1 Active
TMHC3	TM Hedge Class 3	1 Active
TMHC4	Short Cut Method	1 Active
TMHC5	On Behalf Of CC	1 Active
TSWAP	Test hedge SWAP	1 Active
TTURN	TURNS & AMORTIZATION	1 Active

Figure 8.2 Define Hedging Classification Entry Screen

If the hedging classification is active for hedge accounting, then we can only assign it to one hedging area. If the hedging classification is not active for hedge accounting, then we can assign it to multiple hedging areas. The applicable fields that are added in this configuration are as follows:

- **Hdg Class.**
  With this tag, we can create a hedging classification identifier that's up to five characters long. The hedging classification will be assigned in the hedging area.
- **Description**
  We can enter a longer description in this configuration to create a more descriptive explanation of the hedging classification.
- **Hdg. Acctg**
  This field allows us to indicate whether the hedging classification is active or inactive for hedge accounting.

8.1.2 Defining Target Quota Types

The *target quota type* is a determination of how the system calculates the amount that needs to be hedged in a certain time period. The amount that is determined depends on the specific target quota types that we create and assign. We use the target quota type to define the percentage variables for the recommended hedge amount in the hedge management cockpit. These target quota types will be assigned in the hedging area based on the key figures we determine. To set up the target quota type, we first need to determine the name for it. This name can be up to five characters long, and we assign it in the hedging area. There's also a column where we can add a longer description of the target quota type. The most important field to fill in for this configuration is the one where we assign a target quota category (**TQ Cat.**). There are three different calculation categories available in this customizing area, and they drive the options we can use to calculate a target quota:

- **Limit**
  In this option, each period has a target percentage that is used to hedge. If we determine that the target limit for a period is 80%, then the proposed hedge amount will reflect that percentage.
- **Band**
  In this option, we define a lower limit and an upper limit. This works differently depending on the scenario, as follows:
  - If the hedged percentage is inside the target quota band when the hedge requests are generated, then no hedge request will be generated for that period.
  - If the hedged percentage is outside the target quota band when we generate hedge requests, then a hedge request is created. If the hedged percentage is below the band, then SAP S/4HANA will propose a hedge amount to make sure the resulting hedged percentage equals the lower end of the target quota band. If the hedged percentage is above the target quota band, then SAP S/4HANA will propose a hedge amount to make sure the resulting hedged percentage equals the upper end of the target quota band.

- **Target Limit within Band**
  This option is a combination of the above two options. In it, we establish a target quota band and a target limit within that band. This can result in one of the following two scenarios:
  - If the hedged percentage is inside the target quota band when we generate hedge requests, then no hedge request will be generated for that period.
  - If the hedged percentage is outside of the target quota band when we generate hedge requests, then a hedge request will be created. The hedge request that is created will adjust the hedged amount to equal the target limit we defined.

This configuration allows us to define the target quota categories. The actual assignment of the percentages for the target limits and bands will occur during the creation of the hedging area. An example of the configuration of each of the target quota types is shown in Figure 8.3.

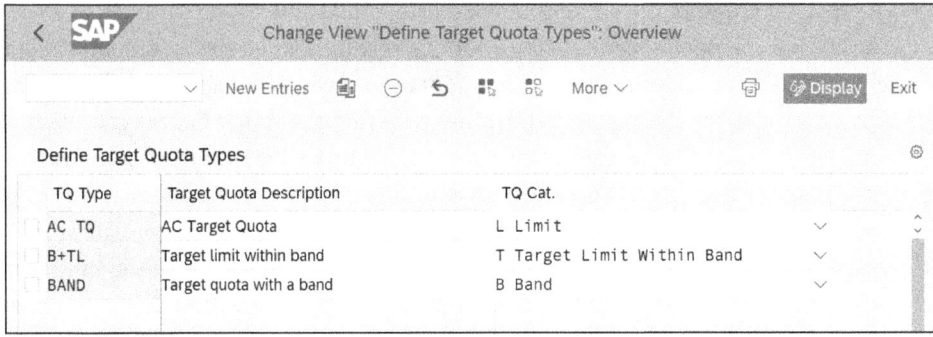

Figure 8.3 Configuration of Target Quota Types

Creating the time pattern for the target quota isn't configuration, but it's related to the setup of the target quota within a hedging area. This will be clear when we're creating the hedging area that is detailed in Section 8.3. The time pattern is the definition of the periods when we generate hedges within the hedging area, and defining the time pattern will allow us to determine whether we should generate hedges monthly, quarterly, or at a different frequency. To create a time pattern, we need to navigate to the Define Time Pattern app or Transaction TOE_TIME_PATTERN. The following are the main options that we can select when creating an entry in this table:

- **Calendar Related**
  If we have the time pattern follow a calendar, it will follow a twelve-month calendar.
- **Fiscal Year**
  When we assign the fiscal year, we also need to assign the applicable fiscal year variant to fully define the periods.
- **Number of Periods**
  In this field, we define the number of periods we can assign target quota percentages

to in the hedging area. If we designate that the time pattern has twelve periods, then we'll be limited to assigning the target quote percentages to those twelve periods.

- **Period Length**
  In this field, we define whether each period is a month, a quarter, or a year.

- **Absolute Time Pattern**
  When we're assigning an absolute time pattern, we define a finite amount of periods in the time pattern. When we start in the hedging area, we'll have this limited amount of periods, and we'll need to assign a new time pattern and target quota once the hedging area's time periods expire.

- **Add Periods Until**
  In this field, we define when the last period will be when we select the **Absolute Time Pattern** option. The options we can choose from the dropdown list are as follows:

  - **Blank**
    If we leave this field blank, then the last time period will be the last period that is defined based on the number of periods and period length settings.

  - **End of Last Quarter**
    If we choose this option, then the last period will be the end of the calculated last quarter. For example, if our time period should end in April, then periods will be added until the end of June since April is not the end of the quarter in a regular calendar year.

  - **End of Last Year**
    If we choose this option, the last period will be the end of the calculated last year. For example, if our time period should end in April, then periods will be added until the end of December since April is not the end of the year in a regular calendar year.

When we create the target quota in the hedging area, we'll see how the defined time pattern affects the settings. Figure 8.4 shows a time pattern we defined. This is a standard time pattern that follows the calendar. We defined 60 monthly periods, and we'll show how this affects the target quota percentage assignment in Section 8.4.2.

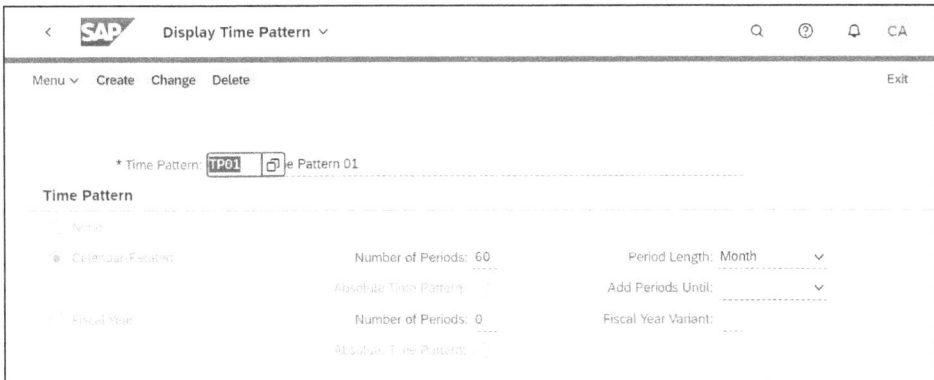

Figure 8.4 Time Pattern Creation in SAP Fiori

8 Cash Flow Hedging

8.1.3 Defining Authorization Groups for Hedging Areas

Assigning the authorization group is mandatory when creating the hedging area. Creating an entry in this configuration will allow security to limit access to hedging areas if we want to separate security for different hedging areas. There are three different authorization objects that leverage the authorization group in security, as follows:

- Hedging area: T_TOE_HA
- Hedge management snapshot: T_TOESNP
- Hedge management cockpit: T_TOE_HMC

The only assignments we need to make in the authorization group are an up-to-five-character identifier in the **Auth.Group** field and a **Description** as shown in Figure 8.5.

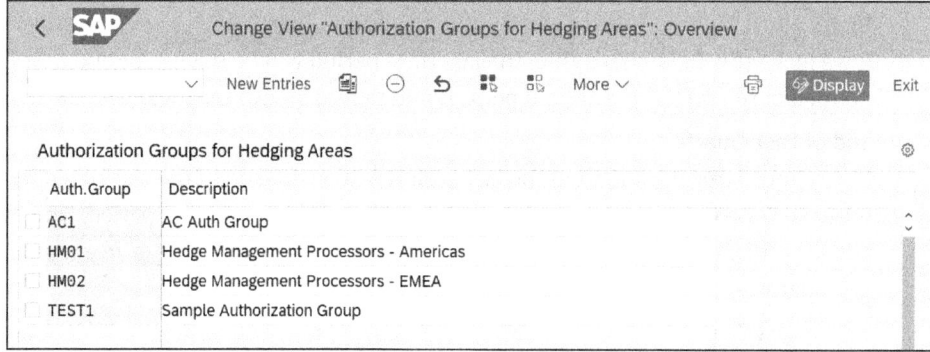

Figure 8.5 Creation of Authorization Groups for Hedging Areas

8.1.4 Defining Hedge Request Reasons

We can generate hedge requests in SAP S/4HANA for different reasons. When creating a regular hedge request, a swap request, or a dedesignation request, we need to assign a hedge request reason to the hedge. We assign these different categories and can use them to analyze and generate reporting for the hedges. For the example in Figure 8.6, we created a hedge request reason for each type of hedge request.

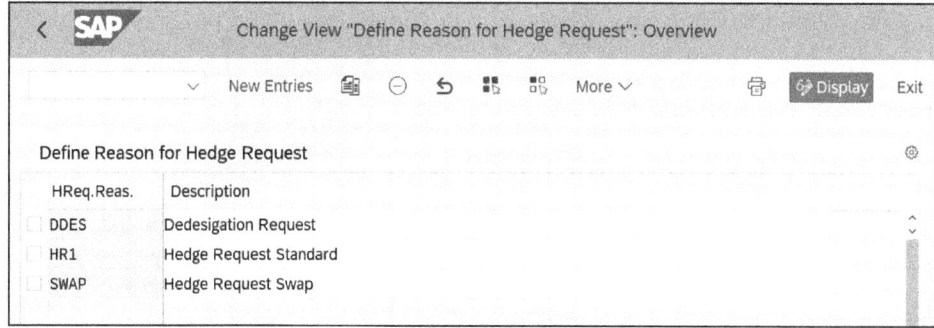

Figure 8.6 Samples of Hedge Request Reasons

This configuration is the creation of the hedge request reason (**HReq.Reas.**) identifier (which can be up to four characters long) and a long **Description** of the hedge request reason.

8.2 Hedge Accounting for Positions Configuration

The second main area that we need to configure for cash flow hedging in treasury and risk management is the definition of the details behind the hedging profiles and settings for the designation of the hedges. This is in the customizing area in the **Financial Supply Chain Management • Treasury and Risk Management • Transaction Manager • General Settings • Hedge Accounting for Positions** menu path. This section will cover the detailed settings for the designation and accounting for the cash flow hedges. We can find the first five configuration nodes in **Settings for Automated Designation of Exposure Items (FX Risk)** in Figure 8.7, as follows:

- Define Designation Types
- Define Product Types for Exposure Subitems
- Assign Update Types to Product Types for Exposure Subitems
- Assign General Valuation Class to Product Type
- Define Hedge Accounting Calculation Types

Once we've made our way through these nodes, we'll cover effectiveness tests, defining hedging profiles, defining update types and assigning usage, assigning update types to business transactions, and defining and activating groups.

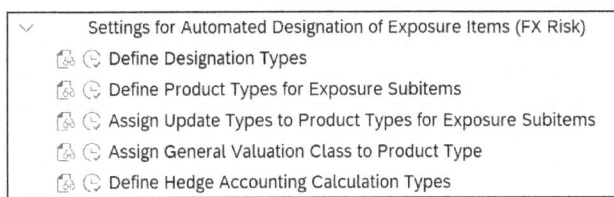

Figure 8.7 Settings for Automated Designation of Exposure Items

8.2.1 Defining Designation Types

The first configuration in this area is the definition of designation types, which is a way to define how a hedge is designated. A designation type also holds additional details on what hedging instrument we use when creating the hedge designation. The designation type is another setting that we'll assign when creating the hedging area on the user side. and after assigning a short designation type name (**Des.Type**) and **Description**, we can assign the following additional settings as shown in Figure 8.8:

8 Cash Flow Hedging

- **Designation category (Desig. Cat)**
  We assign different designation categories for different types of FX trades:
  - **One Instrument Designation Pattern**
    This is the most common category used in this pattern. We would assign this to any standard FX forwards or NDFs.
  - **Two Instruments Designation Pattern**
    If the hedge is an FX option, we assign this designation pattern.
  - **N Instruments Designation Pattern**
    We need to use this designation pattern with all FX swaps.
- **Require counterconfirmation (Req. CConf)**
  If we select this option, the designation for the hedge can only be released once a counterconfirmation is received for the applicable trade in the correspondence monitor. If we don't select this option, this requirement doesn't apply, and we can successfully designate the hedge immediately.
- **End-of-day designation (EOD Des.)**
  Generally, the hedge trade and hedged item can be automatically designated immediately when we create the transaction. If we check the **EOD Des.** box, then this doesn't happen, and the hedge is only designated once we reprocess the designation using the Reprocess Transactions – Automated Designation app or Transaction TPM104. We need to check the **EOD Des.** box when running Transaction TPM104.
- **Required release (ReqRelease)**
  If a workflow needs to be triggered to release the hedging relationship, then we should check this box. The hedging relationship will only be designated if the workflow is successfully released.

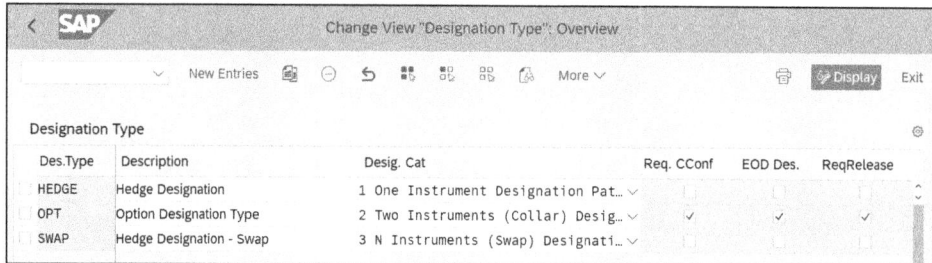

Figure 8.8 Assigning Designation Type to Transaction to Determine Requirements for When Transaction Can Be Designated

8.2.2 Defining Product Types for Exposure Subitems

During the automatic designation process, an exposure subitem is generated for the hedged item. We need to assign a product type to this exposure subitem in SAP S/4HANA, and we use this to generate and drive accounting for the exposure side of the transaction. This means that different accounting postings can be generated for either the financial transaction or the exposure subitem. We also assign the exposure subitem product type when creating the hedging area—specifically, we assign it in the **Hedge Accounting II** tab, which we'll cover in detail in Section 8.3.9. If different scenarios require us to create different sets of postings based on different criteria, we would create multiple product types in this customizing area. Figure 8.9 shows an example of this configuration that we will assign in a hedging area in Section 8.3. These exposure subitems always have a **Prod. Category** of 991, so we can't edit this field when creating this product type. The fields that we need to edit in this area the **Product Type** and the **Short Text** and long description (**Text**) fields.

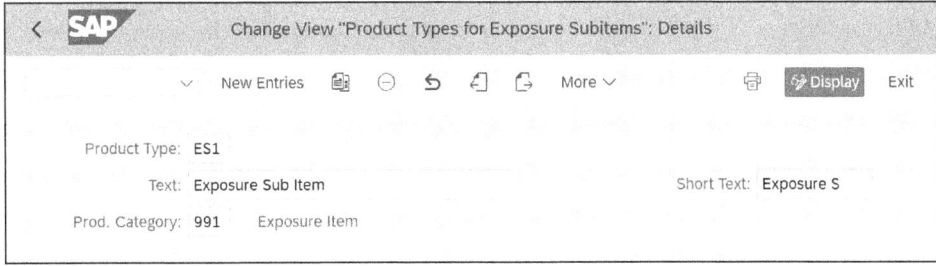

Figure 8.9 Exposure Subitem Is Required in Hedging Area for Hedge Accounting

8.2.3 Assigning Update Types to Product Types for Exposure Subitems

We use an update type to both open and close the exposure subitem when we designate and dedesignate the hedge. This customizing assigns the update types to be used when these activities occur. This is required for position management and ensures that the position amounts are correct in SAP S/4HANA for the balance and valuation calculations. Figure 8.10 shows the product types with their related update types assigned for the opening and closing of the positions, and this configuration assigns the update types that are referenced during the designation and dedesignation of the hedge. We also need to assign an open and close update to the exposure subitem product type defined in Section 8.2.2.

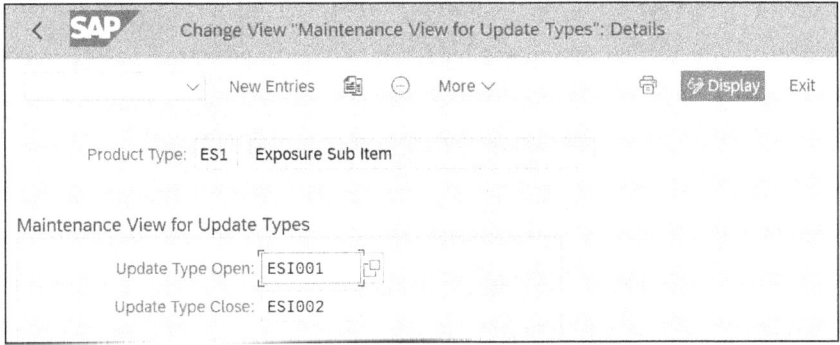

Figure 8.10 Update Types for Designation and Dedesignation of Hedge for This Exposure Subitem

8 Cash Flow Hedging

8.2.4 Assigning General Valuation Classes to Product Types

We need to assign general valuation classes to each of the treasury and risk management product types. We generally make assignments in other areas of the treasury and risk management configuration, but we use this area to assign the general valuation class to the exposure subitem. As shown in Figure 8.11, we also need to assign the general valuation class (**GenValnCl.**) to our exposure subitem **ES1** by defining the product type in the product type column and by mapping the general valuation class in the third column. We can assign this in this area for all product types or assign it for each company code if we go to the **Setting per Company Code** option.

We can navigate to this option by highlighting a product type in this view and double-clicking on the **Setting per Company Code** folder.

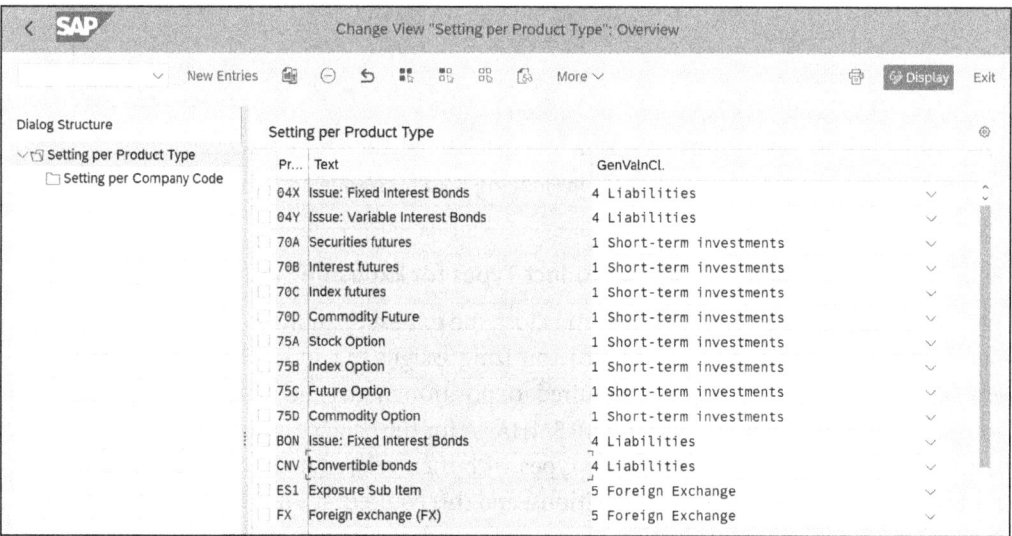

Figure 8.11 Assigning General Valuation Class to Exposure Subitem

8.2.5 Defining Hedge Accounting Calculation Types

The hedge accounting calculation types drive some base settings for the calculation of the hedge relationship. This drives when the hypothetical derivative is created and whether the credit value adjustment (CVA) or debit value adjustment (DVA) is calculated for the valuation, and it determines what market data is used for the calculations. We will assign the hedge accounting calculation type in the **Define Hedging Profiles** configuration in Section 8.2.7.

The configuration screen for the accounting calculation types is in Figure 8.12.

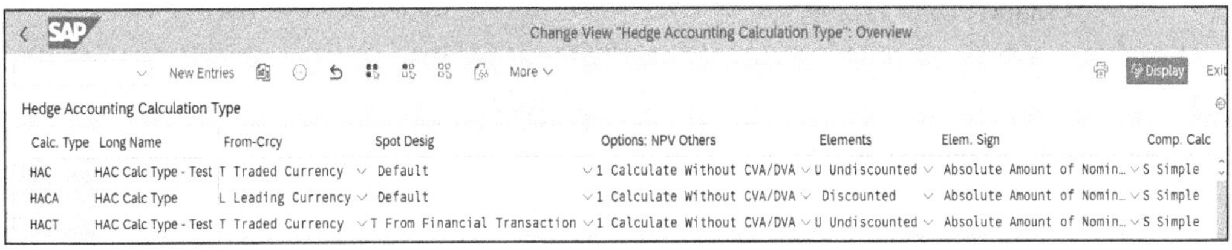

8.2 Hedge Accounting for Positions Configuration

Figure 8.12 Example Settings for Hedge Accounting Calculation Types

Now, we'll cover each of the settings of the hedge accounting calculation types in detail:

- **From-Currency**
  When we're calculating the hedge accounting key figures, we need to define the from-currency and the to-currency. To assign these, we determine which currency we will use to determine the from-currency. We can set it as the leading currency, the following currency, or the traded currency. This defined from-currency then drives the creation of the hypothetical derivative and which currency is used to calculate the hedge effectiveness.

- **Spot Designation**
  We can utilize different FX rates when releasing the designation of a hedging relationship. This setting drives which spot FX rate is used at this time. The three options for this are default, financial transaction, and market data table:
  - **Default**
    In this option, the determination of the spot rate at the release of the designation is determined as follows:
    - **Hedging Instrument**
      This uses the spot rate on the hedging instrument.
    - **Hypothetical Derivative**
      This uses the spot rate from the hedging instrument.
    - **Market Value Components**
      This uses the spot rate from the market data table on the contract date.
  - **Financial Transaction**
    In this option, the system looks at the spot component of the transaction/hedging instrument and hypothetical derivative. Since this is driven directly from the contract, the market value components are zero at the designation.
    - **Hedging Instrument**
      This uses the spot rate on the hedging instrument.
    - **Hypothetical Derivative**
      This uses the spot rate from the hedging instrument.
    - **Market Value Components**
      This uses the spot rate from the hedging instrument.

527

- **Market Data Table**
  In this option, SAP calculates the NPV the same way as the Calculate NPV – Including CVA/DVA app or Transaction TPM60CVA does.
  - **Hedging Instrument**
    This uses the spot rate from the market data table on the contract date/designation date.
  - **Hypothetical Derivative**
    This uses the spot rate from the market data table on the contract date/designation date.
  - **Market Value Components**
    This uses the spot rate from the market data table on the contract date/designation date.

- **Options: NPV Others**
  This setting is specific to options. It can ignore the market value for the NPV other component, or it can be calculated. **No Calculation** determines that there's no calculation for NPV other, and the **Calculation without CVA/DVA** option calculates NPV other as the difference between the option's premium and the calculated NPV.

- **Elements**
  When the designation of the hedging relationship is released, SAP S/4HANA calculates the elements for the hedging instrument and hypothetical derivative to determine the posting logic for the cost of the hedging reserve. These elements include the key figures of ELEM_FWD, ELEM_CCBS, and ELEM_OTHER. During this calculation, the elements can be one of the following:
  - **Discounted**
    If there's a payment term shift in the hedging area settings, the hypothetical derivatives could have different maturity dates than the contracts/hedging instruments. The discount factor between these dates will influence the posting logic.
  - **Undiscounted**
    In the same scenario with the payment term shift, the undesignated amount will relate to the amount in the cost of hedging reserve accumulation. This occurs at the end of the hedging relationship.

- **Element Sign**
  This setting is related to the above determination of the elements. The difference is that this setting determines whether the direction of the nominal amount can be used to determine the elements or can be ignored for the calculation.

- **Market Value Component Calculation**
  There are two different settings that we can determine when calculating the market value components for the hedging instruments and hypothetical derivatives. SAP has the option to determine and save various market value components, depending on whether the hedging instrument is an FX forward or an FX option. These are as follows:

- Spot
  This is determined by the difference between the spot rates at the inception and valuation dates.
- Forward
  This is determined by the difference between the forward rates at the inception and valuation dates.
- CCBS
  This shows the impact of the cross-currency basis spreads on the market value.
- Others
  This is determined by any other flows. One example of this is bank fees on the contract.
- Intrinsic Value
  This value is used for options and reflects the difference between the spot/forward rate and the strike rate.
- Time Value
  This value is used for options and reflects the risk from changes of the option value due to the underlying instrument's volatility.

Now that we've defined each of the market value components, we can cover what is included in each of the options in this configuration:

- Complete
  The treatment of this calculation is different for forwards and options, as follows:
  - FX Forwards
    The market value is calculated for four components, including spot, forward, cross-currency basis spreads, and others. The spot value is discounted in this calculation.
  - FX Options
    The market value is calculated for the intrinsic value, time value, and cross-currency basis spreads.
- Simple
  The treatment of this calculation is also different for forwards and options, as follows:
  - FX Forwards:
    The market value is calculated for two components, including spot and forward. The spot value is not discounted in this calculation.
  - FX Options
    The market value is calculated for the intrinsic value and the time value.
- Hypothetical Derivative Forward Rate
  When we're creating the hypothetical derivative, the system needs to determine the forward rate for the hedging relationships. There are four options to select for this setting.

8 Cash Flow Hedging

- Yield Curve

  The yield curve from both sides of the trade is used to create a theoretical forward rate.

- Swap Rate

  The system combines the spot rate and the swap rate according to the swap rate's curve. These two values are added together to assign the forward rate.

- Evaluation Type

  We use the settings in the evaluation type configuration to determine the forward rate. If we use the **Discounting Before Currency Conversion** valuation method, then we use the yield curve to create the forward rate. If we use the **Currency Conversion Before Discounting** valuation method, then the **Swap Rate** method determines the forward rate.

- Financial Transaction

  The forward rate is copied from the transaction/hedging instrument.

- Fixing

  This setting determines how the presence of a fixing date affects the creation of a hypothetical derivative and whether a fixing date is also in that hypothetical derivative. This setting drives the calculation and dates for the hypothetical derivative's NPV during calculations. We can select three settings in this configuration:

  - None

    If we make this selection, the hypothetical derivate won't be assigned a fixing date. It has the same treatment if the hedging instrument is an FX forward or if it's an NDF.

  - From NDFs Only

    If we make this selection, the fixing date from the settlement currency will be adopted by the hypothetical derivative only if the hedging instrument is an NDF. There's no assignment of a fixing date if the hedging instrument is an FX forward.

  - From FX Forwards and NDFs

    As with the previous option, if the hedging instrument is an NDF, then the fixing date from the settlement currency is copied to the hypothetical derivative. If the hedging instrument is an FX forward with a due date shift, then the calculated due date is set as the fixing date on the hypothetical derivative. In this case, the hypothetical derivative would look like an NDF, but the hedging instrument would still be an FX forward.

From-Currency Example

For the British pound/US dollar currency pair, the British pound is generally set up as the leading currency for this pair. We can define either the British pound or the US dollar as the from-currency using this configuration. Let's look at the effect of defining the currency as the from-currency for each of the listed currencies.

8.2 Hedge Accounting for Positions Configuration

In our example, an FX forward between the currencies is created: 100,000 British pounds are bought and 120,000 US dollars are sold, as follows:

- **From-currency defined as the leading currency: British pound**
  In this scenario, the hypothetical derivative would have a nominal amount of 100,000 British pounds, and the US dollar amount would be calculated. The rate of 1.2 would be decomposed for the calculations.
- **From-currency defined as the following currency: US dollar**
  In this scenario, the hypothetical derivative would have a nominal amount of 120,000 US dollars, and the GBP amount would be calculated. The rate of 0.83 would be decomposed for the calculations.

8.2.6 Effectiveness Test

The effectiveness test configuration determines the type of effectiveness tests that are available to assign for hedge accounting. This configuration is completed in the **Financial Supply Chain Management • Treasury and Risk Management • Transaction Manager • Hedge Accounting for Positions • Effective Test • Effectiveness Test Method** menu path. The effectiveness tests have a set of rules that are applied to determine whether the hedge is effective or not. Depending on the effectiveness test, a portion or all of the hedge may be deemed ineffective. When this occurs, there needs to be a separation in the accounting between the effective and ineffective portions of the valuation. The entry screen for the effectiveness test methods is shown in Figure 8.13.

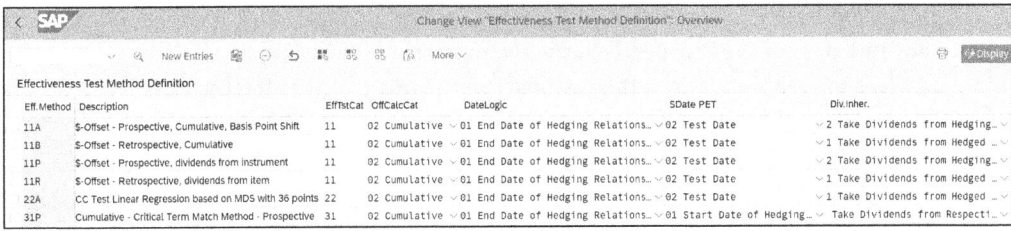

Figure 8.13 Configuration Screen for Creating Effectiveness Test Methods

The configuration options for the effectiveness test methods are as follows:

- **Eff Method**
  This is an up-to-three-character identifier for the effectiveness test method.
- **Description**
  This is a long description of the effectiveness test method.
- **Effectiveness Test Category (EffTstCat)**
  This category determines the type of effectiveness test that is used to calculate and compare the hedging instrument and hedged item. The options for this are as follows:
    - 11: Dollar Offset Ratio
    - 12: Schleifer Noise Method

- 13: Basis Point Dollar Offset
- 21: Linear Regression Time Ref.
- 22: Linear Regression with MDS
- 31: Critical Term Match Method
- 91: Upload of Dollar Offset Test Results
- 92: Upload of Linear Regression Test Results

- **Offset Calculation Category (OffCalcCat)**
  This setting determines how the change in values between the dates in the effectiveness test plan is calculated. This setting determines whether we are comparing the change in values from period to period or cumulatively.

- **Date Logic**
  The selection in this configuration determines how the end date is calculated for the prospective effectiveness testing. The options for this are as follows:
  - **End Date of the Hedging Relationship**
    This setting means that the end date of the hedging relationship is the end date of the period.
  - **According to Outstanding Days of Test Plan**
    This setting looks at the test plan frequency. We can set the frequency of effectiveness testing to be monthly, quarterly, or something else. The end date calculation for the effectiveness test is determined as the end date for each defined period.
  - **'n' Equidistant Dates in the Period Until End Date of HR**
    This setting divides the time to maturity into equidistant periods for the hedging relationship.

- **Start Date of the Prospective Effectiveness Test (SDate PET)**
  This setting determines the start date of the prospective effectiveness tests. We can set this to be equal to the testing date or the start date of the hedging relationship.

- **Dividend Inheritance Method (Div.Inher.)**
  If a dividend is applicable, this setting determines how the dividend is treated when we're calculating the value of the hedging instrument and the hedged item. The options for this are as follows:
  - Take dividends from respective objects.
  - Take dividends from hedged items.
  - Take dividends from hedging instruments.
  - Disregard dividends.

- **Slice Type**
  When a period needs to be split into slices, this indicator determines how the period is split. The options are as follows:
  - **1: Split Without Considering Calendar**
    In a period, the days are split evenly by the number of slices that have been determined without looking at holidays and date shifts.

- **2: Split and Adjust Based on Calendar**

  In this option, the dates are split evenly by the number of slices, but each date that has been determined is subject to a working day shift if the day is on a holiday or weekend.

- **3: Split Based on Working Days**

  In this split method, only working days are considered, and the period is split based on the number of working days.

- **Calendar (Cal)**

  We can set a calendar in this field to drive the working day shifts.

- **Holiday Movement Type (Holiday MT)**

  If a date is on a holiday, this setting determines whether the data point should be shifted to the next working day or the previous working day.

- **Effectv. If Points=0**

  If the linear regression test method calculates successfully but calculates the data points as zero, then by default, we can't apply the testing method. If this scenario occurs and the result of the effectiveness test should be deemed as effective, we set this indicator.

  We only use this setting for the **Linear Regression with MDS** testing method.

- **Assessment Method**

  This determines how the linear regression effectiveness test is determined. The options are as follows:

  - **Independent**

    This allows us to determine which parameters are considered. The options are **Coefficient of Determination (R^2)**, **Slope**, **Intercept**, and **T-statistics**.

  - **Slope and R^2**

  - **BAdI**

    We can design a custom condition for the hedging relationship by using the following BAdI: `BADI_THXE_LIN_REGR_EFFECTIVE`.

- **Condition Type**

  We assign the linear regression condition type in this field.

- **Check R^2**

  We check this to calculate hedge effectiveness using the R^2 of linear regression.

- **Check Slope**

  We check this to calculate hedge effectiveness using the slope of linear regression.

- **Check Intercept**

  We check this to calculate hedge effectiveness using the Intercept of linear regression.

8 Cash Flow Hedging

- **Check T-Statistic**
  We check this to calculate hedge effectiveness using the t-statistic of linear regression.

- **Prospective Market Data Calculation Logic**
  This determines how the market data is used to calculate the effectiveness test. The options are as follows:
  - MDCR
  - Basis Point Shift
  - Market Data Scenario
  - Market Data Shift
  - Not Applicable

  We use this for the critical term match effectiveness test method.

- **Effectiveness Test Critical Term Type (CT Type)**
  When we set a critical terms match effectiveness test, this setting drives the critical terms that the system compares during the effectiveness test.

- **Suppress Recalculation**
  This indicator is relevant when the NPV is calculated including the CVA/DVA. When this indicator is set, the CVA/DVA is calculated once and applied to the NPV, and the NPV is calculated for each market data calculation. If this indicator is not set, the CVA/DVA is recalculated each time the NPV is calculated.

8.2.7 Defining Hedging Profiles

The hedging profile creates the settings around the designation and dedesignation of the hedges. The hedging profile is also a setting that we assign when we create the hedging area, and we create and assign the hedging profile in the hedge accounting settings covered in Section 8.3.9. The configuration path for the hedging profiles is **Financial Supply Chain Management • Treasury and Risk Management • Transaction Manager • Hedge Accounting for Positions • Define Hedging Profiles**. The key settings determined in the hedging profile are as follows and as shown in Figure 8.14:

- **Scenario**
  The hedging relationship scenario drives when we can use a hedging relationship. Specific product categories are allowed because the hedging instrument and certain product categories are allowed as the hedged items, depending on the selected scenario. Table 8.1 contains a list of all hedging scenarios available, along with their allowed product categories. In these hedging relationship scenarios, the scenarios starting with CFH (cash flow hedge) are used for the hedge accounts for exposure items.

8.2 Hedge Accounting for Positions Configuration

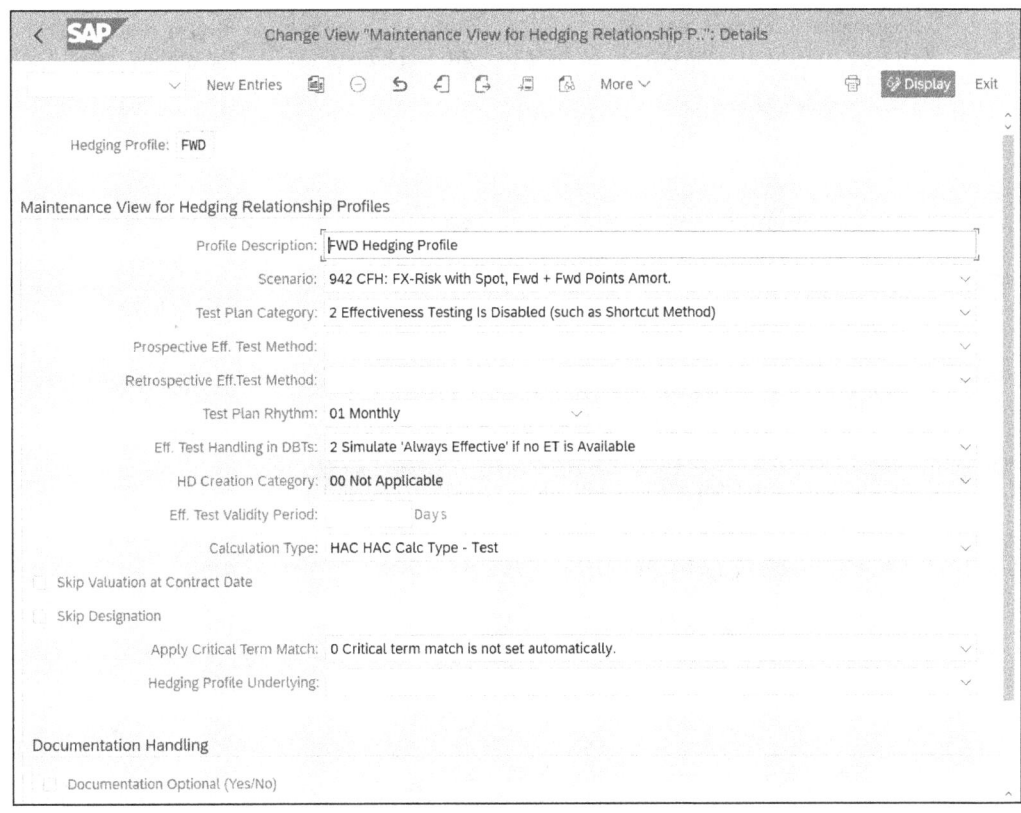

Figure 8.14 Creating Hedging Profiles

Hedging Relationship Scenario	Hedging Instrument: Product Categories	Hedged Item: Product Categories
110 – FVH: Price Risk with Adjusted Spot-Spot Value w/o FX	740	010, 020, 160
120 – FVH: Stocks Hedged with OTC Options	760	010
130 – FVH: Stock Hedged with Total Return Swap	640	010
150 – FVH: Bond (Asset) Hedged w. IRS – maturities equivalent	620	040
151 – FVH: Bond Hedges w. IRS – maturities not equivalent	620	040
160 – FVH: Bond (Liability) Hedged w. IRS – maturities equivalent	620	040, 510, 550

Table 8.1 Available Hedge Relationship Scenarios and Which Product Types and Exposures Use Them

535

8 Cash Flow Hedging

Hedging Relationship Scenario	Hedging Instrument: Product Categories	Hedged Item: Product Categories
510 – UoV: FX Forwards as Hedging Instruments	600	010, 020, 030, 040, 041, 160, 300, 310, 320, 330
520 – UoV: Swaps as Hedging Instruments	620	040, 041, 300, 310, 320, 330
521 – UoV: Futures as Hedging Instruments	700	010, 040, 041
522 – UoV: FSTs as Hedging Instruments	740	010, 020, 030
523 – UoV: Forwards as Hedging Instruments	780	010, 020, 030
530 – UoV: FX Forwards Hedged with FX Forwards	600	600
531 – UoV: Swap Hedged with Swaps	620	620
532 – UoV: FST Hedged with FSTs	740	740
710 – CFH: Bond Hedged with IRS	620	040
720 – CFH: Loans Hedged with IRS	620	300, 310, 330
910 - CFH: FX Risk with Spot, Forward + CCBS	600	991
913 – CFH: FX Risk with Spot, Forward + CCBS + Others	600	991
920 – CFH: FX Risk with Forward + Spot, CCBS	600	991
923 – CFH: FX Risk with Forward + Spot, CCBS + Others	600	991
942 – CFH: FX-Risk with Spot, Fwd + Fwd Points Amort.	600	991
943 – CFH: FX-Risk with Intrinsic, Time + Premium Amort.	760	991
980 – CFH: FX-Risk with Intrinsic, Time + CCBS	760	991
981 – CFH: FX-Risk with Intrinsic, Time + CCBS + Others	760	991
990 – NIH: FX-Risk with Spot, Forward	600	991

Table 8.1 Available Hedge Relationship Scenarios and Which Product Types and Exposures Use Them (Cont.)

- **Test Plan Category**
  This selection determines the frequency of the effectiveness testing. The options are as follows:
  - **Test Plan Acts as a Proposal – All Tests are Valid**
    With this option, we can run effectiveness tests, but they are optional. There's only a proposed schedule of dates for the effectiveness testing, and if we're using these tests for posting valuations, then the program will look at the validity dates of the effectiveness test to determine whether it's valid.
  - **Test Plan is Mandatory – Optional Additional Tests**
    When effectiveness testing needs to be executed at a defined frequency, we select this option. The effectiveness testing will be mandatory, and we can run additional tests if required.
  - **Effectiveness Testing is Disabled (such as Shortcut Method)**
    With this option, we are assuming that the hedges are always effective with the shortcut method. No effectiveness test plan will be generated, and the manual running of any effectiveness testing will be disabled.
- **Prospective and Retrospective Effectiveness Test Method**
  If the effectiveness testing is enabled, then we can assign the test methods. We can also determine the tests discussed in Section 8.2.6, and this will drive the data points used for the effectiveness testing.
- **Test Plan Rhythm**
  We can set a monthly, quarterly, yearly, or manual frequency to drive when to run the effectiveness tests.
- **Effectiveness Test Handing in DBTs**
  This option determines how the system looks at the effectiveness tests. With this, there are the following two options:
  - **Strict Reading of Effectiveness Tests**
    The valuation will look at the settings for the test plan category to ensure the effectiveness tests are reviewed when generating the derived business transactions.
  - **Simulate 'Always Effective' if no ET is Available**
    With this setting, the system will ignore whether an effectiveness test is not available and will generate the derived business transactions assuming the test was effective.
- **HD Creation Category**
  There are three options for the category used for the hypothetical derivative upon creation, as follows:
  - **Blank**
    Not applicable. We use this if the hedging relationship won't use a hypothetical derivative.

- **Create Hypothetical Derivative with NPV = 0**
  In this option, the NPV of the hypothetical derivative is set to zero during designation.
- **Create Hypothetical Derivative with NPV = - NPV Desig Swap**
  In this option, the value is set to be the inverse of the NPV of the designated swap.

- **Effectiveness Test Validity Period**
  When we've determined that the effectiveness test is optional, we need to set the validity period for the tests. If we run the valuation during the effectiveness test validity period, then we can use it to post the valuation entries.
- **Calculation Type**
  We assign the hedge accounting calculation type created in Section 8.2.5 to the hedging profile. This drives the creation of the hypothetical derivative.
- **Skip Valuation at Contract Date**
  By default, a valuation is run at the designation date of a hedge resulting in a gain or loss. If we want to skip this step and not run this valuation, we check this box.
- **Apply Critical Term Match**
  This setting only applies when the hedging relationship is using the **Critical Term Match** method. If we check this box, the contract is assumed to be effective under this method.
- **Hedging Profile Underlying**
  This is where we assign the underlying instrument's hedging profile.
- **Documentation Optional**
  We check this box if we don't want to create the PDF documentation for the hedging relationship.
- **Semi-Automatic Documentation**
  We select this option if we need to generate documentation for the hedging relationships. When we release the designation for the relationship, the hedge documentation is automatically generated and can be viewed in the **Documentation** tab in the Manage Hedging Relationships app or Transaction TPM100.
- **PDF-Based Forms: Form Name**
  We assign the PDF form for the hedge documentation. The standard form available is TR_F_THS_NOTE_HREL_FXRISK_FX, and if we need to modify the form to meet specific needs, we can do it in the Form Builder app or Transaction SFP.

8.2.8 Defining Update Types and Assigning Usages

We've already covered creating update types and assigning usages in Chapter 3, Section 3.14, so we'll just reference this setting in this section. The main difference is that we're assigning update types so they can be assigned in the following configuration activity in Section 8.2.9. We need to assign to each of the relevant update types the usage of **9011 Hedge Accounting for Positions** like in Figure 8.15. This configuration is located in the

8.2 Hedge Accounting for Positions Configuration

Financial Supply Chain Management • Treasury and Risk Management • Transaction Manager • General Settings • Hedge Accounting for Positions • Update Types • Define Update Types and Assign Usages menu path.

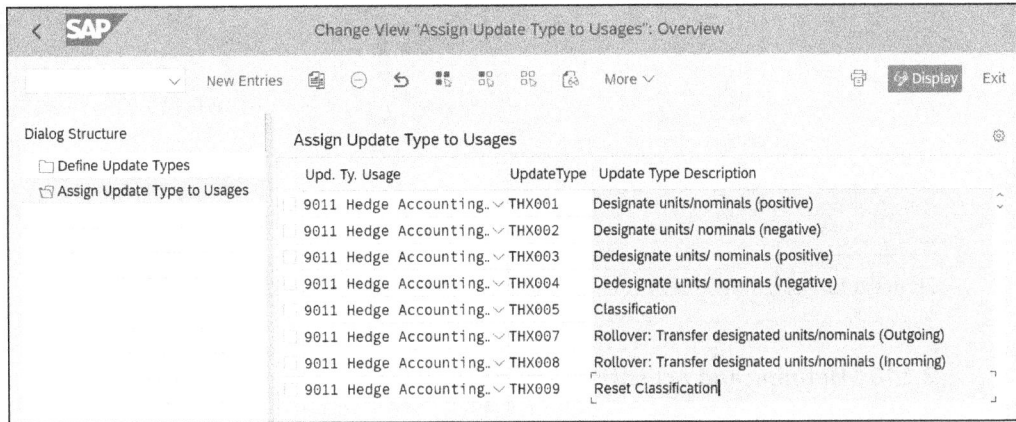

Figure 8.15 Assigning Update Types to Update Type Usages

8.2.9 Assigning Update Types to Hedging Business Transactions per Product Type

Hedging in SAP S/4HANA includes transactions executing different activities than they would if they were not being hedged. We need to assign update types in this configuration so we can post each of these activities, and we can find this configuration is located in the Financial Supply Chain Management • Treasury and Risk Management • Transaction Manager • General Settings • Hedge Accounting for Positions • Update Types • Assign Update Types to Hedging Business Transactions per Product Type menu path. When we run these different activities for the transactions in SAP S/4HANA, additional product types are assigned to ensure that the transactions can be processed and posted. These activities include designation, dedesignation, classification, swap transfers, and rollovers. When we run these activities, this customizing is referenced to generate the position changes and/or postings.

As shown in Figure 8.16, we need to assign different product types to many update types depending on how we hedge and process the transactions. To assign update types in this configuration, we need to assign them to the Hedge Accounting for Positions update type usage as shown in the previous section. This configuration works by assigning different update types to each product type, depending on the activity that is being processed. For example, if we're designating a hedge and the value is positive, the THX001 update type would be called. Similarly, if the designation has a negative value, TIIX002 will be called during the designation. Each of these fields works in this way, and we're assigning the specific update types for different hedging activities.

539

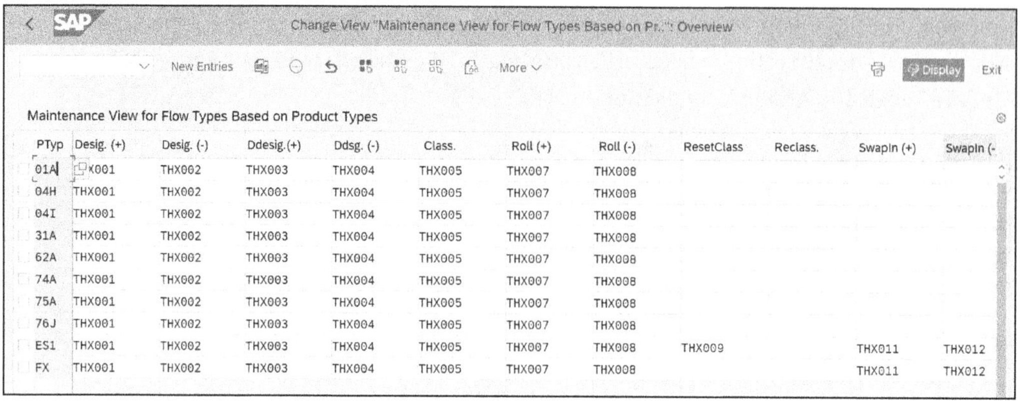

8   Cash Flow Hedging

PTyp	Desig. (+)	Desig. (-)	Ddesig.(+)	Ddsg.(-)	Class.	Roll (+)	Roll (-)	ResetClass	Reclass.	SwapIn (+)	SwapIn (-)
01A	X001	THX002	THX003	THX004	THX005	THX007	THX008				
04H	THX001	THX002	THX003	THX004	THX005	THX007	THX008				
04I	THX001	THX002	THX003	THX004	THX005	THX007	THX008				
31A	THX001	THX002	THX003	THX004	THX005	THX007	THX008				
62A	THX001	THX002	THX003	THX004	THX005	THX007	THX008				
74A	THX001	THX002	THX003	THX004	THX005	THX007	THX008				
75A	THX001	THX002	THX003	THX004	THX005	THX007	THX008				
76J	THX001	THX002	THX003	THX004	THX005	THX007	THX008				
ES1	THX001	THX002	THX003	THX004	THX005	THX007	THX008	THX009		THX011	THX012
FX	THX001	THX002	THX003	THX004	THX005	THX007	THX008			THX011	THX012

Figure 8.16 Assigning Relevant Update Types to Activities in This Configuration

8.2.10 Defining and Activating Groups

The **Define and Activate Groups** customizing activity allows for the creation and naming of grouping fields that we can assigned to the hedging relationships. This is in the **Financial Supply Chain Management • Treasury and Risk Management • Transaction Manager • General Settings • Hedge Accounting for Positions • Define and Activate Groups** configuration location. The groups that we create in this configuration can be assigned in the Manage Hedging Relationships app or Transaction TPM100 in the **Hedging Relationship Details** tab. We can create and activate three groups, as in Figure 8.17. To do this, we click **New Entries**, select the **H.Rel.Grouping**, check the **Grouping active** box, and add a free-text **Grouping Description**. The descriptions in the figure are examples; you can write whichever ones fit your requirements.

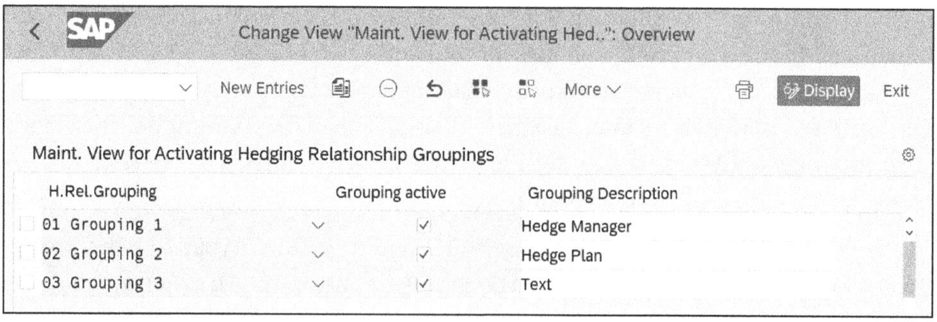

H.Rel.Grouping	Grouping active	Grouping Description
01 Grouping 1	✓	Hedge Manager
02 Grouping 2	✓	Hedge Plan
03 Grouping 3	✓	Text

Figure 8.17 Activating Hedge Relationship Grouping

Grouping 1 and **Grouping 2** have additional areas that are available in the dropdown list and that that we can set up for the options. We can create the values and assign them in the user menu in the **Accounting • Financial Supply Chain Management • Treasury and Risk Management • Hedge Management and Accounting • Hedge Accounting for Positions • Master Data • Grouping Key 1/Grouping Key 2** menu path location. The entry of the first **Grouping Key** is in Figure 8.18. This is a free-text definition that we use to create the options for **Grouping Key 1** and **Grouping Key 2**. In this example, we are assigning different names to designate the hedge manager.

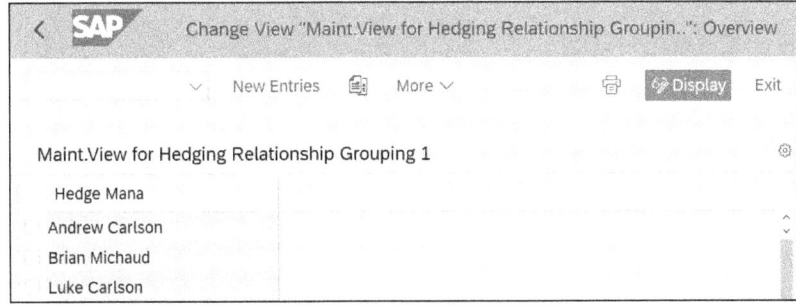

Figure 8.18 Creating Grouping Key 1

Grouping Key 3 is a free-text field, so no additional setup is required.

8.3 Hedging Area

We create the core master data and setup for cash flow hedging in the hedging area, which drives many aspects of how we manage and monitor FX risk. The hedging area has many settings that we need to define to ensure that the correct exposures are viewed and that the FX trades appear in the reporting and settings specific to the hedge accounting and designation of the trades. This section will cover each of the settings in detail to show the end-to-end process of setting up the hedging area. Before we can start processing data for this process, we need to create the hedging area in the Define Hedging Area app or Transaction TOE_HEDGING_AREA. Once we're in the hedging area, there are a series of tabs that we need to populate. First, to create a new hedging area, we must populate a name in the **Hedging Area** field and click the **Create** button (see Figure 8.19). This will bring up a popup where we can add more information for the hedging area, as follows:

- **Authorization Group**
  Here, we add the authorization group that we created in Section 8.1.3.
- **Valid From**
  Here, we establish the validity date of the hedging area.
- **Without Template** and **With Template**
  Here, we have the option of copying the hedging area from an existing hedging area and a specific version. The hedging area copy has some limitations, and the **Hedge Accounting I** tab and **Hedge Accounting II** tab settings don't get carried over.

8   Cash Flow Hedging

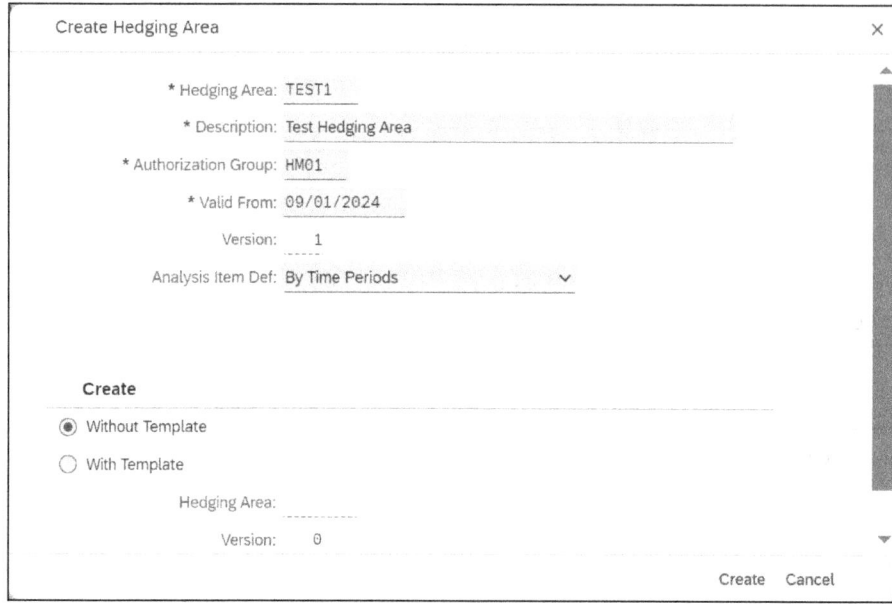

Figure 8.19 Determining Initial Settings when Creating Hedging Area

Then, we click **Create** and start filling out the tabs, each of which we'll cover in the following sections.

8.3.1   Main Data

The first tab is the **Main Data** tab, which we use to create some of the base data for the hedging area. This includes defining how the items are analyzed in the hedging area, how the risk-free currency is defined, the reporting time periods, the integration settings, and additional information on the target quotas. We cover each of these sections in detail and provide context on why each option is available.

Risk-Free Currency

The *risk-free currency* is the base currency for the hedging area, entity, or transaction. Exposures in currencies outside the risk-free currency will be available to hedge against when we're creating requests in the hedge management cockpit. Depending on the settings in the hedging area, there are three options that we can select from for the risk-free currency, as follows (see Figure 8.20):

- **Single Risk-Free Currency**
  If we need to define a single risk-free currency for the hedging area, we select this radio button. This option is only available if the hedging area is not applicable for hedge accounting.

8.3 Hedging Area

- **Local Currency**
  This option looks at each company code's local currency to define the risk-free currency. In this option, there will be multiple risk-free currencies in the hedging area since it's based on the company code's local currency.

- **Currency Defined by the Source**
  This option uses the data available in SAP S/4HANA to define the risk-free currency and looks at the information available in exposure management and cash management:
  - When exposures are created in exposure management, the exposures have a target currency and an exposure currency. The target currency in the exposure is defined as the risk-free currency.
  - For exposures sourced from cash management, the risk-free currency is defined as the local currency of the company code.

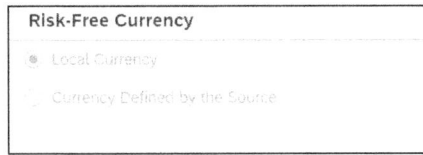

Figure 8.20 Determining Risk-Free Currency in Hedging Area

Reporting Time Pattern

The reporting time pattern creates the structure for how the hedging area reports time periods. If we're working with a reporting time pattern of the next twelve months, then we'll see twelve periods in the hedge management cockpit. Figure 8.21 shows an example of how we can assign twelve calendar months to the reporting time pattern. These fields all operate the same as the time pattern that we defined in Section 8.1.2.

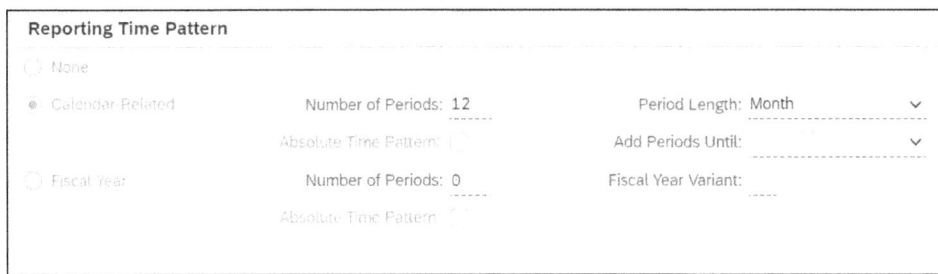

Figure 8.21 Assigning Time Pattern for Hedging Area to Determine Twelve-Month Calendar Year

Integration

This section includes the integration-based data for the hedging area, as shown in Figure 8.22:

8 Cash Flow Hedging

- **Hedge Accounting**
  This activates hedge accounting for the hedging area and will allow us to populate the **Hedge Accounting I** tab and **Hedge Accounting II** tab.
- **Product Type for Exposure Subitems**
  Here, we can assign the exposure subitem product type that we created in Section 8.2.2.
- **Activate SAP Trading Platform Integ.**
  This lets us send the trades through the SAP trading platform integration. We detail the functionality of the SAP trading platform integration in Chapter 9, Section 9.5.

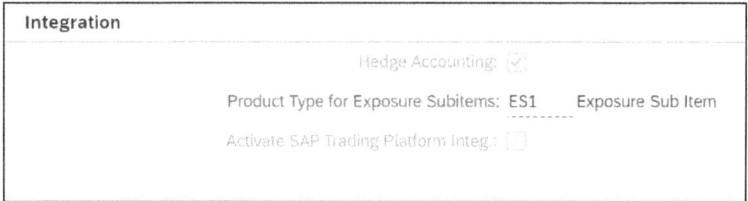

Figure 8.22 Assigning Integration Section of Hedging Area

Consideration of Financial Transactions

This area defines how the hedging area treats value dates and how those dates are assigned for hedging instruments, options, and NDFs. The following options are shown in Figure 8.23:

- **Consider Hedging Instrument Until**
  When the hedging area is looking at the hedging instruments, this filter looks at the key date entered in the report. Based on this key date, we can consider hedging instruments in the key figures in two different ways:
  - **End of the Period**
    With this option, the hedging instrument shows up in the key figures until the end of the period. This means that the hedging instrument's value date might have passed and that the instrument would still show up in the key figures.
  - **Value Date of the Hedging Instrument**
    With this option, as soon as the value date of the hedging instrument has passed, it will fall off the reporting in the hedge management cockpit.
- **Date for Determining Exposure Item for FX Option**
  This setting is specific to FX options. There are two different ways to determine the exposure item:
  - **Value Date of Underlying**
    In this method, the exposure item is based on the underlying transaction's value date.
  - **Exercise Date of FX Option**
    In this method, the exposure item is based on the exercise date of the hedging instrument FX option.

- **Date for Determining Exposure Item for Nondeliverable Instruments**
  This setting is specific to nondeliverable instruments like NDFs. We use this setting to define which date to use to determine the date for the exposure item. The options are to determine the exposure item based on either the value date or the fixing date of the nondeliverable instrument.

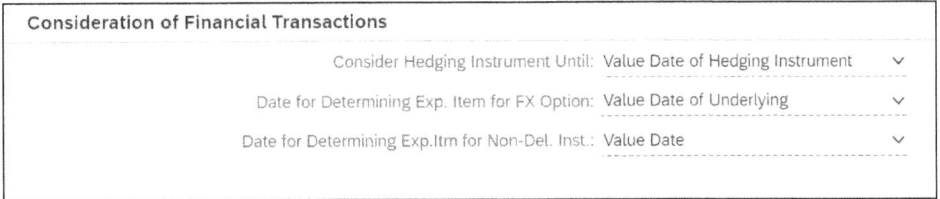

Figure 8.23 Assigning Value Dates for Hedging Instrument and Exposure Items

Target Quota

We created the settings for the target quota and time period in Section 8.1.2. We can add multiple target quotas if we have a scenario in which some currencies need to have a band-based target quota and some need to have a limit-based target quota. In Figure 8.24, we have a target quota assigned to the limit type, and we added this by clicking the **Add Target Quota Type** button ⊕ and populating the **TQ Type** and **Time Pattern ID** fields. We'll assign the actual percentages in the **Target Quotas** tab we covered in Section 8.3.6.

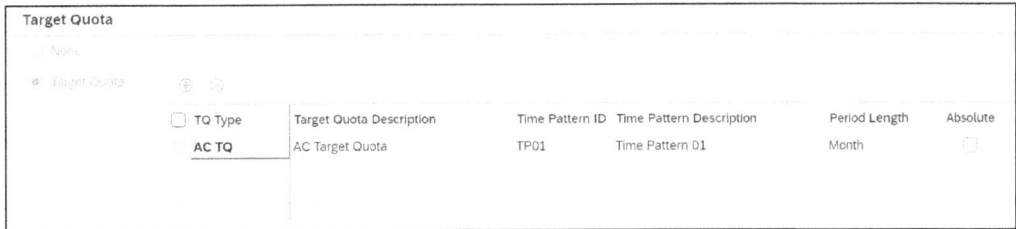

Figure 8.24 Assigning Target Quota to Hedging Area

8.3.2 General Settings

The **General Settings** tab is where we determine the key settings for the hedging area that we cover in the following sections. These settings drive the hedging company codes and criteria for how the exposures are hedged.

Company Codes

We can enter the company codes relevant to hedging in this area. If we don't enter a company code, then all company codes in SAP S/4HANA will be active for the hedging area. The assignment of the company codes can be seen in Figure 8.25.

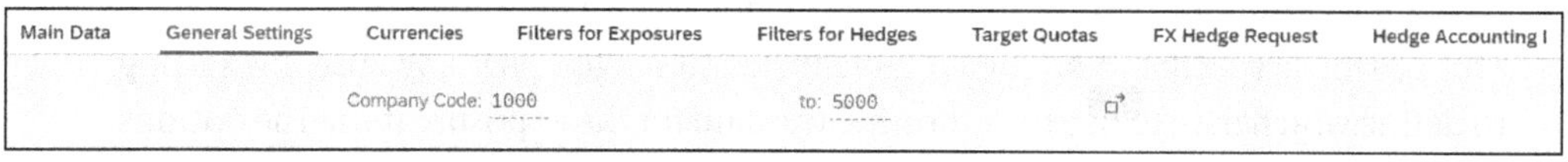

Figure 8.25 Assigning Company Codes to Hedging Area

Differentiation Criteria

When viewing the hedge management cockpit, we can view our data with different attributes. The differentiation criteria drive the level of granularity when we're viewing the data in the hedge management cockpit. We can select each of the differentiation criteria by checking the box to the left of the description. Once we've done this, we'll see that there are two additional columns we where we can check boxes. These options are as follows (see Figure 8.26):

- **Relevant for Target Quota**
  Checking this box will add this criterion to the target quota tab. This allows the target quotas to be more specific if we want to assign them in more granular detail than assigning a percentage to a currency. **Currency | Currency Group** is automatically designated as relevant for target quota.
- **Relevant for Hedge Accounting**
  When we're activating the scenarios for hedge accounting and activating different scenarios for hedge accounting, checking this box will add this criterion as a way to drive the hedge designation. We can use this if we should activate some scenarios for automated designation and if we should not activate other scenarios. **Currency | Currency Group** and **Company Code** are automatically designated as relevant for hedge accounting.

Differentiation Criteria

Selected	Differentiation Criterion	Relevant for Target Quota	Relevant for Hedge Acctg
☑	Currency | Currency Group	☑	☑
☑	Company Code	☐	☑
☐	Portfolio		
☐	Profit Center		
☐	Cost Center		
☐	WBS Element		
☐	On Behalf of Company Code		
☐	Country/Region Key		
☐	Segment		
☐	Business Area		

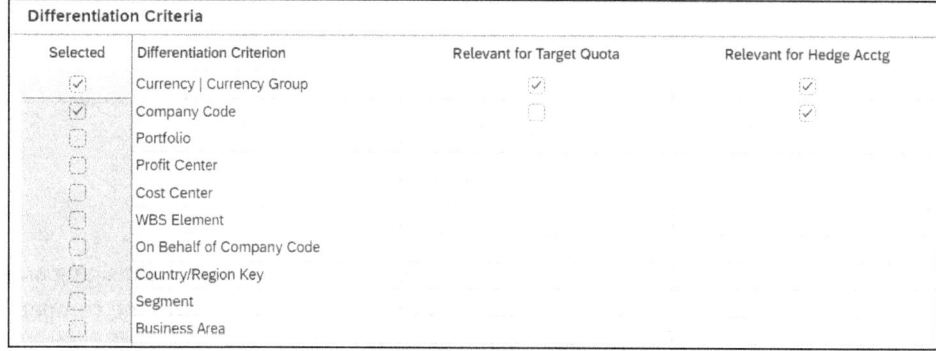

Figure 8.26 Assigning Differentiation Criteria Allows for More Granular Hedging Strategies

Hedging Classifications

We also assign hedging classifications in the **General Settings** tab. We defined the hedging classifications in Section 8.1.1. We can define multiple hedging classifications in a hedging area, and we use them to differentiate designation control in the **Hedge**

Accounting II tab. Additionally, we can add a due date shift to the hedging classification if the hedging instruments need to show up in a different month than the due date of the hedging instrument. This would allow for the hedging of exposure items in a different month. As shown in Figure 8.27, the options for **Due Date Shift** are as follows:

- **No selection**
  With this option, the due date is not shifted.

- **Beginning of Next Month**
  With this option, when a hedge is created, the end date for the hedged item is shifted to the first day of the following month.

- **End of Next Month**
  With this option, when a hedge is created, the end date for the hedged item is shifted to the last day of the following month.

- **End of Previous Month**
  With this option, the hedging instrument's value date is shifted to the last day of the previous month. We can only select this option if hedge accounting is not active for the hedging classification.

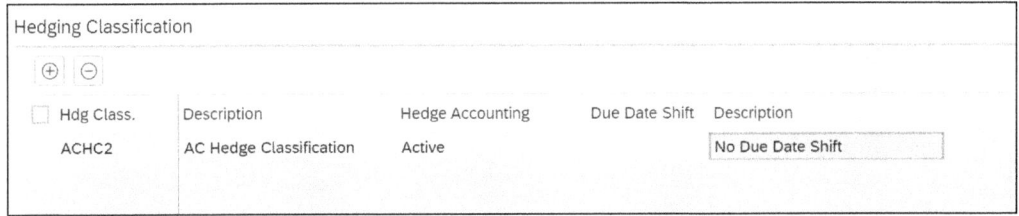

Figure 8.27 Assignment of Hedging Classification to Hedging Area

8.3.3 Currencies

The **Currencies** tab includes a couple of different settings related to currencies. First, it establishes the hedging-relevant currencies within the hedging area. Adding a currency in this tab is the first step to setting up that currency as a hedge-related currency and will allow us to progress to the additional settings related to fully establishing that currency for hedging. Second, each currency is assigned a currency group. This is a freeform field that is used to group currencies together, and this will be leveraged in the **Target Quota** tab. The **Target Quota** tab doesn't directly assign the percentage targets directly to a currency; they are assigned to the currency group. To demonstrate how to assign different currency groups, in Figure 8.28, we have a currency group used for Canadian dollars and British pounds, and we have a separate currency group for euros. To assign currencies in the hedging area, we click the **Add** button and define the three-character ISO code in the **Currency** column. Next, we assign an identifier for the **Currency Group** that can be up to three characters long. We'll define these currency groups further in Section 8.3.6 on the **Target Quota** tab.

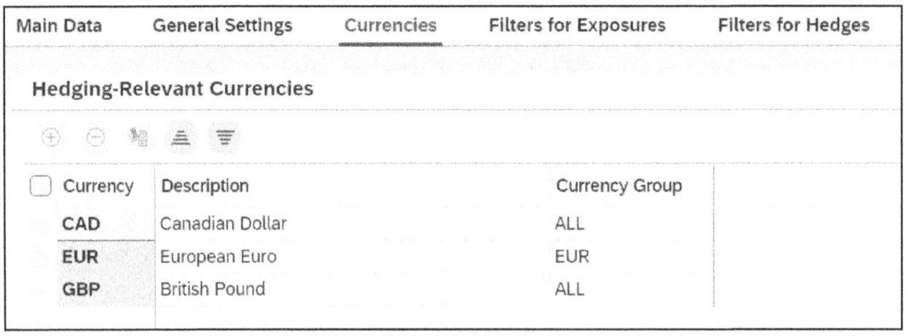

8   Cash Flow Hedging

| Main Data | General Settings | Currencies | Filters for Exposures | Filters for Hedges |

Hedging-Relevant Currencies

Currency	Description	Currency Group
CAD	Canadian Dollar	ALL
EUR	European Euro	EUR
GBP	British Pound	ALL

Figure 8.28 Assigning Different Currency Groups to Currencies

8.3.4 Filters for Exposures

The next step we need to set up in the hedging area is **Filters for Exposures**. This step determines the data we are going to pull into this report for the snapshot, and we'll ultimately use it in the hedge management cockpit for exposure reporting and hedging. First, we need to create a filter assign it to a data source. In Figure 8.29, we create a filter for both **Exposure Management 2.0** and **Cash Management** to demonstrate how we can use both of these sources to pull exposures into the hedge management cockpit.

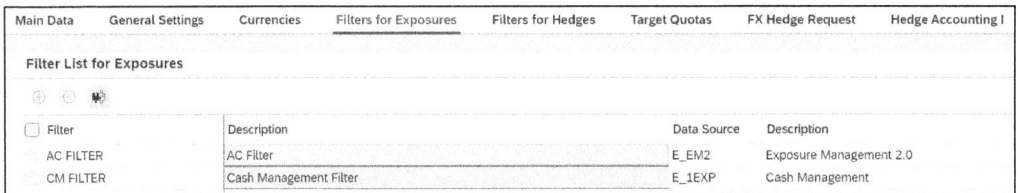

Figure 8.29 Filtering Information from Exposure Management 2.0 or Cash Management Tables

To create a new filter, click on the **Create Filter** button ⊕. A popup will appear where the filter name, description, and data source need to be determined. To create a filter for exposure management 2.0, we select **Data Source E_EM2**. After we confirm the details and click the **Create** button, the **Filter Details** section will be at the bottom of the screen, and we can define each available field for filtering to determine the exposure information that will appear in the hedge management cockpit. As we can see in Figure 8.30, some filters have been added to determine exactly which exposures we want to bring into the snapshot. In this scenario, we've filtered by company code, exposure position type, and exposure category.

> **Filters for Exposures: Company Codes**
>
> Company codes relevant to the hedging area were determined in the **General Settings** tab. If the company codes were filtered out of that tab, it wouldn't be available to bring into the filters for exposures. We can only use the company codes available in the hedging area when filtering the exposures and the hedges.

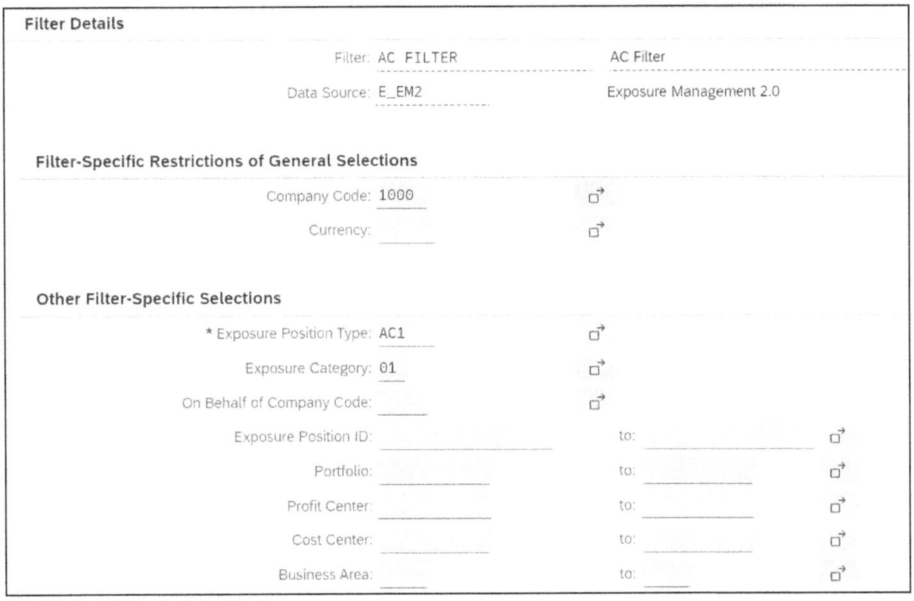

8.3 Hedging Area

Filter Details

Filter: AC_FILTER	AC Filter
Data Source: E_EM2	Exposure Management 2.0

Filter-Specific Restrictions of General Selections

- Company Code: 1000
- Currency:

Other Filter-Specific Selections

- *Exposure Position Type: AC1
- Exposure Category: 01
- On Behalf of Company Code:
- Exposure Position ID: to:
- Portfolio: to:
- Profit Center: to:
- Cost Center: to:
- Business Area: to:

Figure 8.30 How to Filter Information from Exposure Management 2.0

When we want to create a filter for cash management, we select **Data Source E_1EXP**. This allows us to select cash management records from the one exposure from operations table FQM_FLOW. Figure 8.31 shows an example of the data being filtered in cash management by planning level. Based on the settings in the cash management filter, this data will also be pulled when the snapshot is generated.

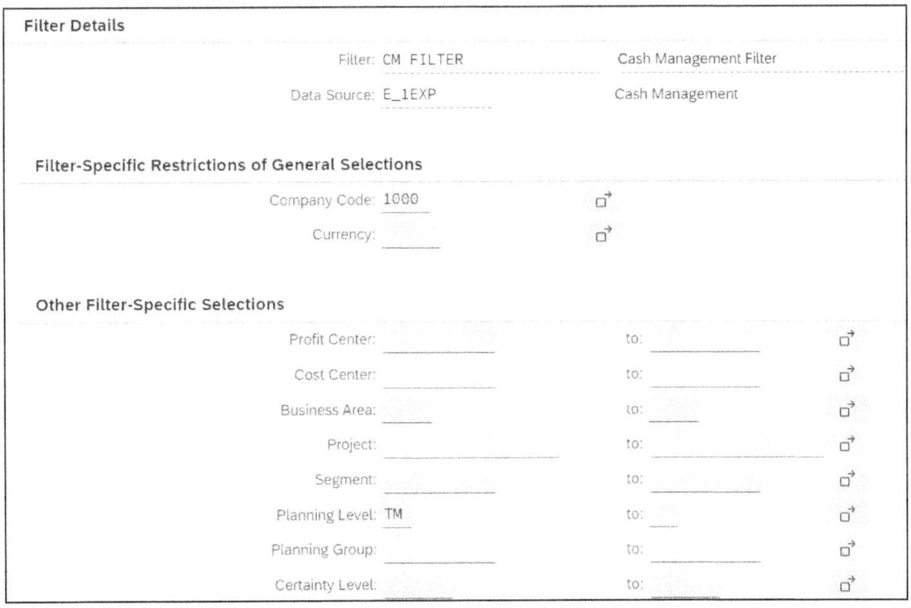

Filter Details

Filter: CM_FILTER	Cash Management Filter
Data Source: E_1EXP	Cash Management

Filter-Specific Restrictions of General Selections

- Company Code: 1000
- Currency:

Other Filter-Specific Selections

- Profit Center: to:
- Cost Center: to:
- Business Area: to:
- Project: to:
- Segment: to:
- Planning Level: TM to:
- Planning Group: to:
- Certainty Level: to:

Figure 8.31 Filtering Data Using Planning Level from Cash Management

8　Cash Flow Hedging

8.3.5　Filters for Hedges

Even though we create hedges in the hedge management cockpit, we need to determine which hedges to pull in the reporting in the **Filters for Hedges** tab. We can pull in hedges using different criteria like the filters for exposures, and commonly, we'll filter the hedges by the hedging classification or portfolio. Since the hedge classification is generally unique to a hedging area, it's a simple way to bring in all hedges relevant to the hedging area. To create the filter, we need to create the **Filter** name and **Description** as in Figure 8.32. Note that there's only one **Data Source** in the filters for hedges (**H_TM: Transaction Management**).

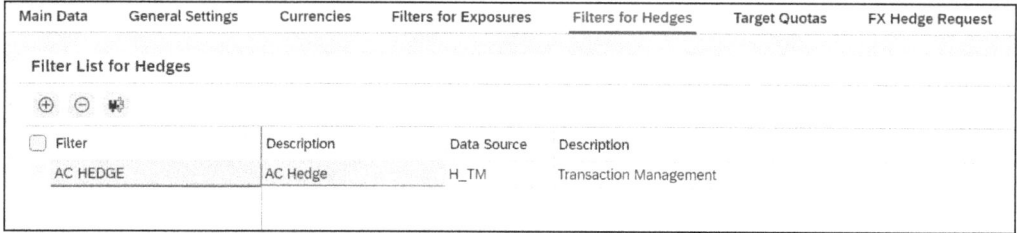

Figure 8.32 Creating Filter for Hedges in Hedging Area

Once we've created the filter, we can assign the filters just like in the filters for exposures. Different fields are available for this data source, as in Figure 8.33. The fields available in the hedge filtering will be specific to the hedging instruments. These filters are applicable to the data entered into the transaction, so we can filter these by **Hedging Classification, Product Type, Business Partner, Portfolio,** and the other key fields available.

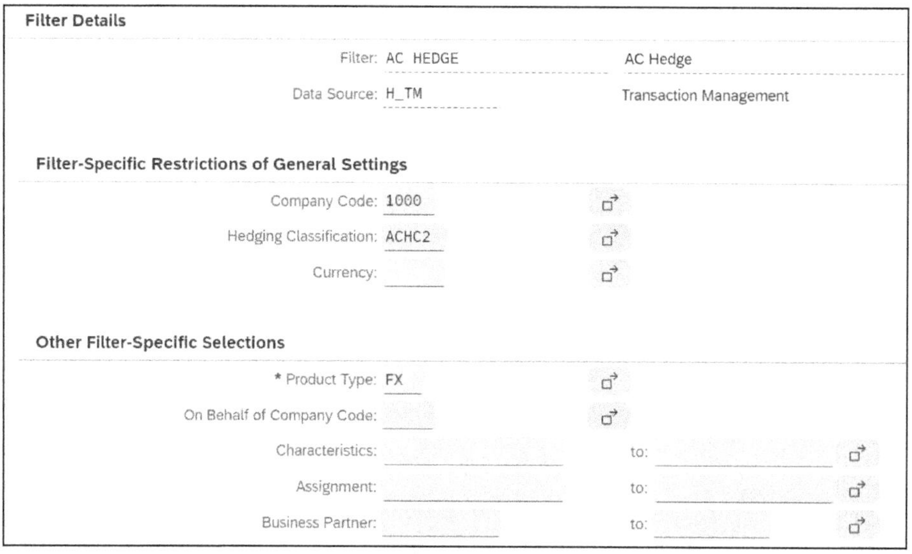

Figure 8.33 Determining Which Transactions to View as Hedges

550

8.3 Hedging Area

8.3.6 Target Quotas

Up until now, any mention of the target quota has been the initial creation and assignment of the quota. The **Target Quotas** tab finally assigns the amounts to the target quota. Based on the time period assigned in the **Main Data** tab, you can assign the different percentage target quota. This will either be a band, limit or a combination of both, depending on the customizing settings. In Figure 8.34, you can see that the limit-based target quota is being assigned with different settings based on the time periods. If multiple target quota types are assigned to the hedging area, they can be selected in the **Target Quota Type** dropdown.

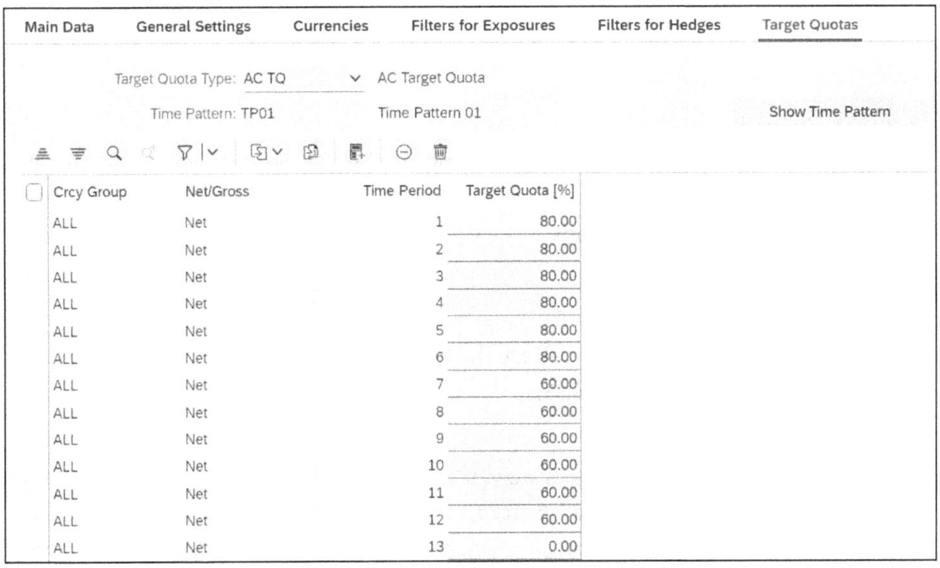

Crcy Group	Net/Gross	Time Period	Target Quota [%]
ALL	Net	1	80.00
ALL	Net	2	80.00
ALL	Net	3	80.00
ALL	Net	4	80.00
ALL	Net	5	80.00
ALL	Net	6	80.00
ALL	Net	7	60.00
ALL	Net	8	60.00
ALL	Net	9	60.00
ALL	Net	10	60.00
ALL	Net	11	60.00
ALL	Net	12	60.00
ALL	Net	13	0.00

Figure 8.34 Assigning Target Quota for Hedging Area's ALL Currency Group

The other columns in Figure 8.34 are as follows:

- **Currency Group**
  In this column, we assign the currency groups established in the **Currencies** tab to define the target quota.

- **Net/Gross**
  In this column, we define whether we're going to hedge the net or gross amount of the exposures. In the hedge management cockpit, the hedge percentages and other key figures are designed around the net amounts and don't have separate gross amounts.

- **Time Period**
  In this column, we define the month we're hedging. The number of time periods that are displayed are driven by the settings defined for the assigned time pattern.

- **Target Quota (%)**
  In this column, we set the target hedge percentage. In the following scenario, this is a rolling target quota, so if we're setting our hedge amounts in June, then time period 1 would be June, time period 2 would be July, and so forth for the twelve defined time periods in Figure 8.21. As shown in Figure 8.34, we're going to hedge 80% of the exposure amounts for the first six months and 60% for the second six months.

- **Other columns**
  We can add columns to this area if we select the **Relevant for Target Quota** option in the **General Settings** tab.

8.3.7 FX Hedge Request

In the **FX Hedge Request** tab, we determine the details of how we'll create the hedge request. This includes the process for creating the hedge request, the value date, and which financial instrument we're using to hedge the exposure. We cover the key details of this tab in the following sections.

FX Hedge Request Header

The first couple of settings in this tab are the target status and the current period exclusion, which are shown in Figure 8.35:

- **Target Status for Automation**
  When we create a hedge request, we can drive the release status of the request. We can set the status as created, submitted, or released. This will drive the initial status of the hedge request and determine whether it needs to be released before execution. These statuses will appear when we review the hedge requests in the Process Hedge Requests app:
  – **Created**
    This is the initial status of the hedge requests, and we can edit it. Once we've reviewed the request and edited it, we can submit the request for approval.
  – **Submitted**
    We can review hedge requests in this status, but we can't edit them. We can only release or reject them.
  – **Released**
    We skip the editing steps in this status, and we can't reject the hedge request.

  Any or all of these three statuses may be required, based on how many reviews and changes the hedge requests may have and depending on the business scenario.

- **Exclude Current Period**
  We check this box if we don't want to create a hedge request for the current period and only want to create a request for the next period and beyond.

8.3 Hedging Area

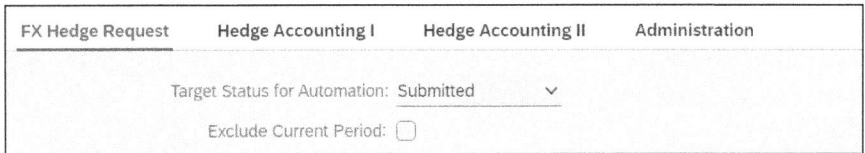

Figure 8.35 Determining Initial Settings for FX Hedge Request

Value Date

The **Value Date** section offers a comprehensive list of options for how create a default value date for the hedging area. These options are shown in Figure 8.36:

- **Value Date Definition**
  In this field, we can define the default value date for the automated FX hedge requests as being the first day in a period or the last day in a period. If the logic in the value date section is not sufficient for our business requirements, then we can select the **Custom via BAdI** option to customize the value date logic.

- **Additional Days**
  This field allows us to shift the value date forward or backward in time for the automated hedge requests. Positive amounts in this field will add days, and negative amounts will subtract days. We can shift this field by up to 365 days in either direction.

- **Working Day Shift**
  We check this box if the value date needs to be updated when it falls on a weekend or holiday.

- **Working Day**
  If we've checked the **Working Day Shift** box, in this field, we define the rule for how to shift the working day: to the previous working day or the next working day.

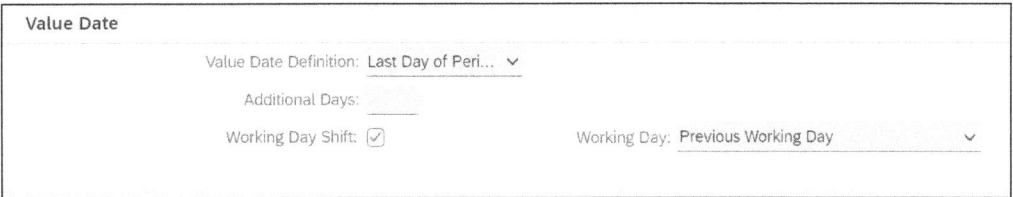

Figure 8.36 Value Date Settings Relevant to Hedge Request Determination

Hedge Request Settings

In the **Hedge Request Settings** area, we add specific settings that determine what the hedge request looks like when we chose to automatically generate the hedge requests. We can determine specific settings for different scenarios, depending on the company code and risk currency. The other columns have a few other options, which are as follows (see Figure 8.37):

553

8 Cash Flow Hedging

- **Default Instrument Category**
  In this section, we define the type of FX trade that is generated to hedge the exposure. The options in this field are as follows:
  - FXFW – FX Forward
  - FXSP – FX Spot
  - FXOP – FX Option
  - NDF – FX Non-Deliverable Forward
  - FXCO – FX Collar
- **Minimum Amount**
  In this column, we define the minimum required amount to generate a hedge. To avoid new hedge requests for small amounts, we can set a limit in this field.
- **Roundings**
  We can apply rounding to the minimum amount determination for the automatically generated requests. The values are always rounded down with this setting. In the example in Figure 8.37, 1H rounds to one hundred.

Hedge Request Settings					
Comp...	Risk C...	Default Instrument Category	Hedging Classification	Minimum Amount	Roundings
1000	CAD	FXFW	ACHC2	100.00	
1000	EUR	FXFW	ACHC2	100,000.00	1H
1000	GBP	NDF	ACHC2	1,000,000.00	

Figure 8.37 Hedge Request Settings

8.3.8 Hedge Accounting I

The hedge accounting tabs drive our ability to automatically designate the hedges we create from the hedge management cockpit. There are three areas in the **Hedge Accounting I** tab that we'll cover in the following sections.

Designation Level

In the first section of hedge accounting, we determine the **Designation Level** for the company code and valuation area. In this tab, we determine whether the exposure sub-items are generated for the net exposure amount or the gross exposures. This drives which exposure item is designated with the hedging instrument. All options for this setting are shown in Figure 8.38.

For each designation level, we can determine whether the **Splitting** designation is active for that scenario. We cover the details of designation splitting and its settings later in this section. If we enter "1" in this field, then designation splitting will be active, and if we enter "0" or leave the field blank, then designation splitting will be inactive. If designation splitting is active, then we can populate the **Designation Splitting** section for this company code. The last part of this section is the **Sequence** column, in which we determine the methodology to use when hedging relationships need to be either dedesignated or swapped. When we create a dedesignation or swap request, we need to select the hedging relationships to dedesignate or swap. This field determines the consumption sequence for this scenario. There are three options for this, as follows:

- **N (no consumption sequence)**
  When we choose this option, we need to manually select the hedging relationships when running the dedesignation request or swap.

- **F (first in, first out)**
  When we choose this option, the hedges are selected in order and the first hedge is the one that is dedesignated or swapped.

- **L (last in, first out)**
  When we choose this option, the hedges are selected in order and the first hedge is the one that is dedesignated or swapped.

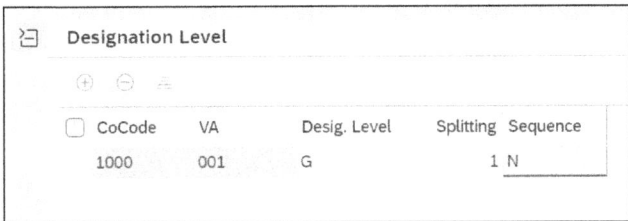

Figure 8.38 Designation Level Settings

Designation Activation

In the **Designation Activation** section, we determine the hedging scenarios that are set up to automatically designate. When we enter "1" in the **Designation** column, we activate the automatic designation as in Figure 8.39. If we enter "0" in the column, we deactivate the automatic designation. The columns that appear in this section are driven by the differentiation criteria section in the **General Settings** tab that we covered in Section 8.3.2.

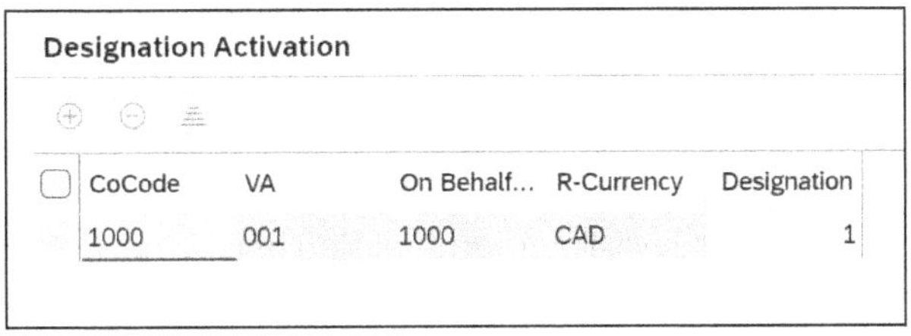

CoCode	VA	On Behalf…	R-Currency	Designation
1000	001	1000	CAD	1

Figure 8.39 Scenario Activated for Automatic Designation

Designation Splitting

In the **Designation Splitting** section, we can further define how the hedging relationships are created, and we can also control an OCI release at the end of a cash flow hedge. When a hedge is effective, the full valuation at the end of the month can be posted to the balance sheet in OCI. We need to split this between the profit and loss and the balance sheet if there's an ineffective portion of the hedge. This split is determined by the effectiveness testing at the end of the period and what effectiveness test is executed. There are a couple options for processing the hedge when the hedging relationship concludes and is dedesignated, as follows:

- **Without designation splitting**
  If designation splitting is not active, the posting programs automatically post the realized gain or loss on the maturity of the contract.

- **With designation splitting**
  When designation splitting is active, the amount in OCI can be left in that account and released to the realized gain and loss accounts at a predetermined rate. Using the settings in designation splitting, we can achieve this functionality.

To set up the designation splitting, we first need to activate splitting in the designation level area. Once it's activated, we can add lines in the designation splitting section. The first few columns are self-explanatory, and there, we can define which company code (**CoCode**), valuation area (**VA**), and risk currency (**R-Currency**) we'll use in designation splitting. The remaining columns are as follows:

- **Direction**
  Here, we define whether the exposure being hedged is incoming or outgoing (**I** or **O**). We can define different settings if the exposure is incoming or outgoing, so we can save them on separate lines with separate settings.

- **Split ID**
  This is a free-form field where we define different levels of the designation split. We can't repeat a number for the combination of the company code or on behalf of company code, valuation area, and direction of the exposure. If we need to add different settings for document splitting rules, then we need to assign a different number to each line. This is reflected in Figure 8.40. Each setting has a different line and is separated by the split ID of 1, 2, 3, and 4.

- **Ratio**
  The ratio is the percentage of OCI that is being amortized for a certain period. In the below example, the OCI amount is not going into the profit and loss in the first month. Then, 33% is amortized after 30 days, another 33% is amortized after 60 days, and the rest of the OCI is moved after 90 days. For a combination of company code, valuation area on behalf of company code, risk currency, and direction, the ratio amounts need to add up to 100% to fully reclass the OCI.

- **Extended period (ExtendPer.)**
  This selection allows us to extend of the period of the hypothetical derivative. If we set this to **Yes**, then the due date for the hypothetical derivative will be calculated as the due date of the hedging instrument plus the payment term in the next column. If we set this field to **No**, then the period will not be extended.
- **Pymt Term**
  Here, we define the number of days until invoices are paid.
- **DIO**
  This stands for "days of inventory outstanding." Here, we can determine the number of days of inventory outstanding that are relevant for the company code.

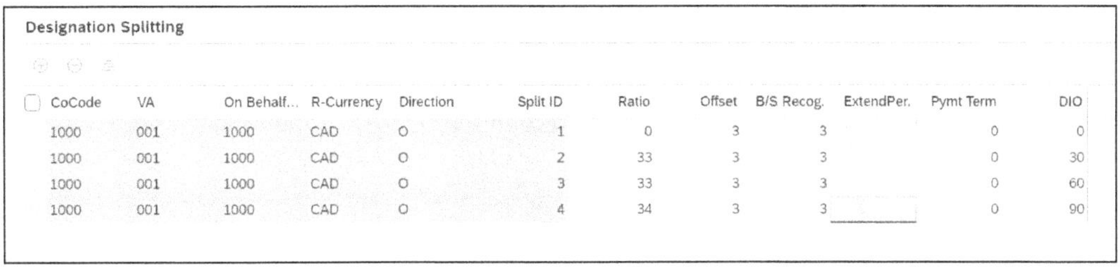

CoCode	VA	On Behalf...	R-Currency	Direction	Split ID	Ratio	Offset	B/S Recog.	ExtendPer.	Pymt Term	DIO
1000	001	1000	CAD	O	1	0	3	3		0	0
1000	001	1000	CAD	O	2	33	3	3		0	30
1000	001	1000	CAD	O	3	33	3	3		0	60
1000	001	1000	CAD	O	4	34	3	3		0	90

Figure 8.40 Splitting Designation into Multiple Periods

8.3.9 Hedge Accounting II

The second hedge accounting tab adds classification and processing details for the automatic designation of the hedge. We'll notice that many of the details that were set up in the customizing area are now being assigned in this tab. The settings for **Hedge Accounting II** are as follows (see Figure 8.41):

- **Hedging Classification**
  We can now assign the hedging classifications that we assigned in the **General Settings** tab to each scenario for hedge accounting. The addition of the classifications is covered in Section 8.3.2.
- **Designation Type**
  We can now assign the designation types that we created in Section 8.2 to each combination of company code, valuation area, and hedging classification.
- **Product Type Exposure Subitem**
  We assign the exposure subitem product type again in this section. We covered the creation of the exposure subitem in Section 8.2.2.
- **Hedging Profile**
  In the hedging profile, we assign key details of the effectiveness testing and which methods to use. We covered the background settings of the hedging profile in Section 8.2.7.

- **Market Data Set (MDS) ID**
  If an effectiveness test leverages the market data sets to calculate the effectiveness of the hedges, we can set the market data set ID in this field. The linear regression within the market data set effectiveness tests uses this data. If we need to create the market data set, we can do this in the Specify Market Data Sets app or Transaction TAN_MDS.

CoCode	VA	Hdg Class.	Desig.Type	PType ESI	HdgProfile	MDS ID
1000	001	ACHC2	HEDGE	ES1	FWD	

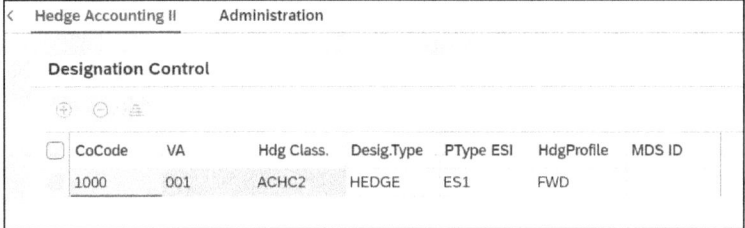

Figure 8.41 Sample of Settings in Hedge Accounting II

8.3.10 Administration

The **Administration** tab highlights the versions and changes of the hedging area. It also shows when each version of the hedging area was entered and who made the edits (see Figure 8.42).

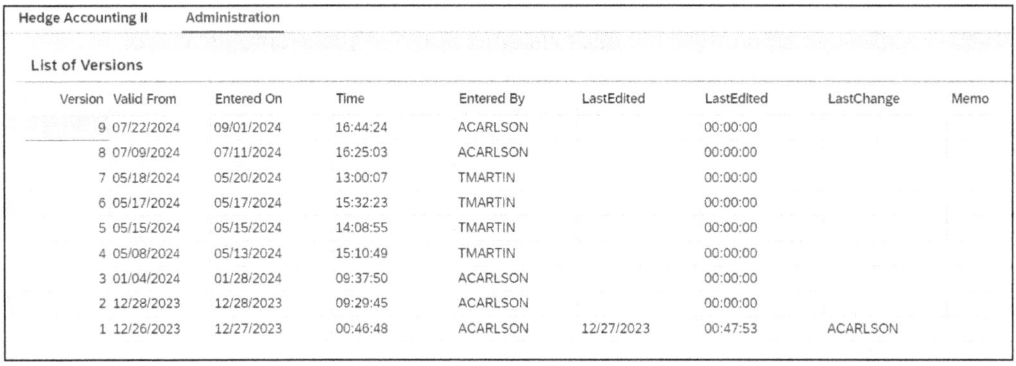

Version	Valid From	Entered On	Time	Entered By	LastEdited	LastEdited	LastChange	Memo
9	07/22/2024	09/01/2024	16:44:24	ACARLSON		00:00:00		
8	07/09/2024	07/11/2024	16:25:03	ACARLSON		00:00:00		
7	05/18/2024	05/20/2024	13:00:07	TMARTIN		00:00:00		
6	05/17/2024	05/17/2024	15:32:23	TMARTIN		00:00:00		
5	05/15/2024	05/15/2024	14:08:55	TMARTIN		00:00:00		
4	05/08/2024	05/13/2024	15:10:49	TMARTIN		00:00:00		
3	01/04/2024	01/28/2024	09:37:50	ACARLSON		00:00:00		
2	12/28/2023	12/28/2023	09:29:45	ACARLSON		00:00:00		
1	12/26/2023	12/27/2023	00:46:48	ACARLSON	12/27/2023	00:47:53	ACARLSON	

Figure 8.42 History of Changes Tracked in Administration Tab

8.4 Cash Flow Hedging Process

Now that we understand the background of the hedge management process and how it flows in SAP S/4HANA, we can go through the transactions in this process. First, we're going to assume that our system has various exposures that have been entered in exposure management 2.0 or that consist of data populated in one exposure for operations for cash management. As we discussed in the previous section, this information will be leveraged to determine the exposures for hedge management. Since the

8.4 Cash Flow Hedging Process

exposure data is all in SAP S/4HANA now, we'll cover the high-level hedge management process in the same order as shown in Figure 8.43.

In this section, we'll cover the creation portion of the cash flow hedge management process. We'll start by looking at taking snapshots, and then, we'll cover creating hedge requests in the hedge management cockpit and processing hedge requests. From there, we'll move on to creating financial transactions, managing hedging relationships, and releasing hedging business transactions. We'll cover processing trade requests in Chapter 9, Section 9.5.1.

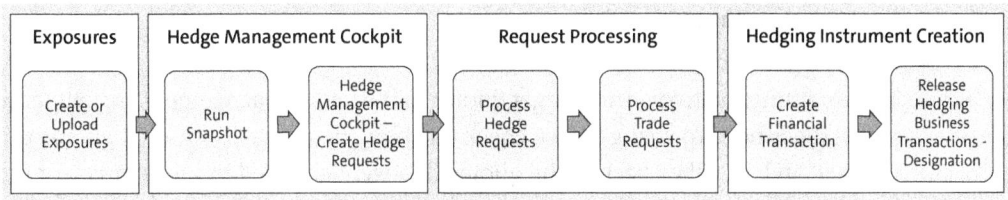

Figure 8.43 High-Level Hedge Management Process Flow for Cash Flow Hedging

8.4.1 Taking a Snapshot

Once all the relevant exposure and hedging data is reflected in the system, we can take a snapshot in the Take Snapshot app or Transaction TOESNAP. We need to run the snapshot to save the relevant information in the hedging tables. This locks in the exposure and hedging information at that time, and once we run the snapshot, we can view the saved information in the hedge management cockpit. The app for this transaction is shown in Figure 8.44.

Figure 8.44 Take Snapshot App

When we're running the snapshot, there are a few fields that are important to consider (see Figure 8.45):

- **Hedging Area**
  Here, we select the hedging area to include in the snapshot. Clicking the **Multiple Selection** button to the right will allow for multiple hedging areas to be included in one snapshot.

- **Extraction Date**
  Here, we enter the key date to extract the exposures.

- **Snapshot Date**
  The snapshot date is saved as the system date. This field only appears when the day reference box is checked.

- **Day Reference**
  This indicator is saved in the snapshot. We can save multiple snapshots per day, but the day reference indicator is required for the automated designation process for hedge accounting. The latest snapshot with the day reference indicator is used in the automated designation. It's also important to note that we can't run another snapshot if we've already run a snapshot with the day reference indicator and used it for hedge accounting on that day.

- **Reset Target Quota**
  There's a function to override target quotas in the hedge management cockpit if the percentages need to change. If we check this box, the manually entered quotas will be reset, and they'll go back to the quotas that were assigned in the settings for the hedging area.

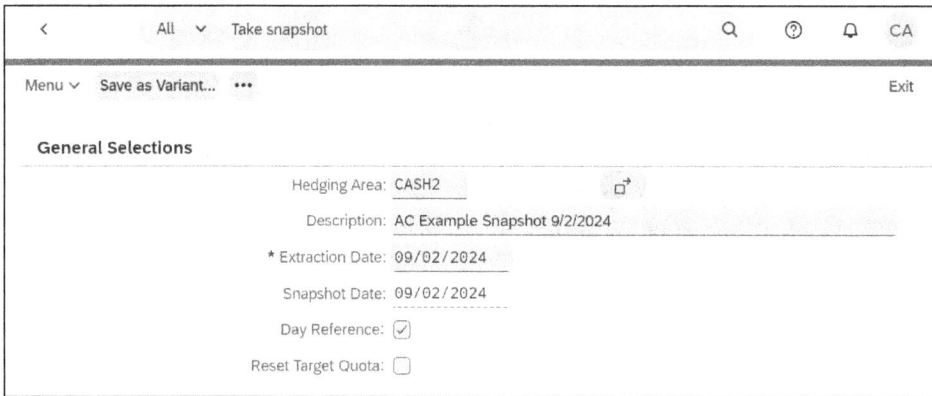

Figure 8.45 Take Snapshot App

> **Snapshot Overview**
> We can use Transaction TOESNAPO to manage the snapshots. This includes the ability to see the full list of snapshots taken in a hedging area and view the details of each snapshot, including the extraction dates, the day reference information, and the hedging area version when the snapshot was taken.

We initiate the snapshot by clicking the **Execute** button. Once we take the snapshot, an output screen will detail the status, including the number of items that were pulled in with the snapshot. This is shown in Figure 8.46.

8.4 Cash Flow Hedging Process

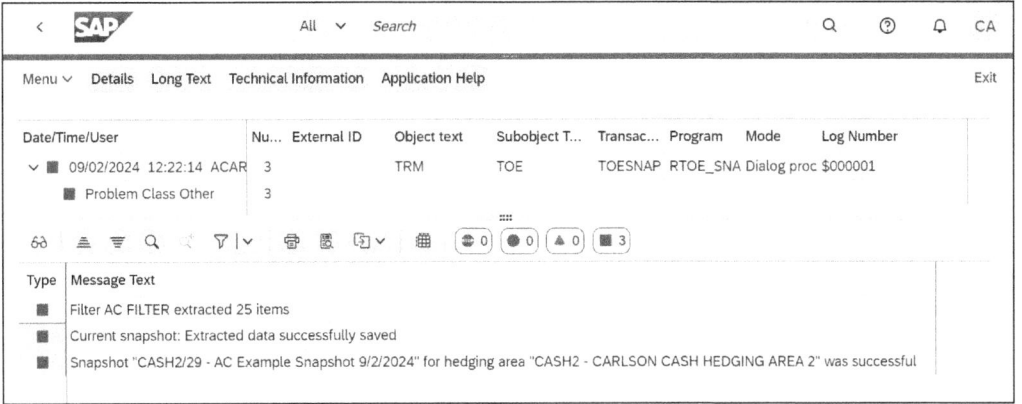

Figure 8.46 Output Screen Informing User Snapshot Was Successful and Displaying Number of Items Extracted

8.4.2 Hedge Management Cockpit

The hedge management cockpit is the central location where we'll manage the cash flow hedging process. In this report, we'll be able to generate various hedge requests, view the report of where we are in the hedging decisions, and drill down into the reporting to see what hedges and exposures are behind the numbers we see on the screen. To access the cockpit, we go to Transaction TOENE or open the Hedge Management Cockpit app, as shown in Figure 8.47.

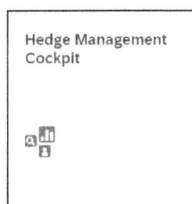

Figure 8.47 Hedge Management Cockpit App

The hedge management cockpit is very customizable, and it can present the data in many different ways to meet our requirements. In the following sections, we'll start by determining the initial fields that need to be entered to view the data in the hedging area. This is a complex report with many options, so to reflect how the screens can be edited, we'll show how to create different layouts to alter the output data. Following this, we'll detail the functions within the hedge management cockpit and demonstrate how to view the exposures and create the hedges.

Initial Settings

Before viewing any data in this report, we need to select the **Hedging Area**, the **Layout ID**, and the **Scaling** of the report. When we select the **Hedging Area**, the details of the

8 Cash Flow Hedging

most recent snapshot will auto-populate. As shown in Figure 8.48, the **Date**, **Target Quota Type**, **Snapshot ID**, and **Day Reference** fields are populated based on the hedging area that we've selected.

The scaling of this report determines whether we're viewing the actual number of exposures or hedges or whether we're going to round them to summarize the values. If we have larger amounts for our exposures and hedges, we might want to view the data rounded to thousands or millions instead of viewing the exact numbers.

We cover the layout ID in the next section since there are many options that we can select for the layout.

Figure 8.48 Selecting Hedging Area and Layout ID to View Data in Hedge Management Cockpit

Layout IDs

We'll drill down further into the layout IDs because they drive how we can view the report. The layout ID determines the report output and which fields are visible. A list of layout IDs is available as sample report layouts, but we can create new layout IDs as well. Figure 8.49 shows a combination of the standard layouts available and custom layouts. To create a new layout, we click on the **Manage Layouts** button, which will bring up all layouts that are available. We can create, edit, view, change, and delete layouts.

Within the creation of the layouts, we can alter many settings. We'll start by reviewing the **Layout ID**. Creating the **Layout ID** creates the name of the layout that we will reference when selecting the layout in the hedge management cockpit. Once this tag is created, the initial details of the layout create some of the base settings for the layout, including how exposures are viewed, how a key date is selected, and the scaling of the report. Further into the details of the layout, we'll determine the **Differentiation Criteria** to determine how granularly we view the exposure and hedge data. After that, we'll determine the **Key Figures** to drive which figures are displayed on the report. We can

8.4 Cash Flow Hedging Process

display the totals for a period, a year, and the full hedging area by making updates to the **Period** settings. Finally, the layout can drive which **Differentiation Criteria** we can use to filter data in the report. We'll review each of these settings in detail in the following sections.

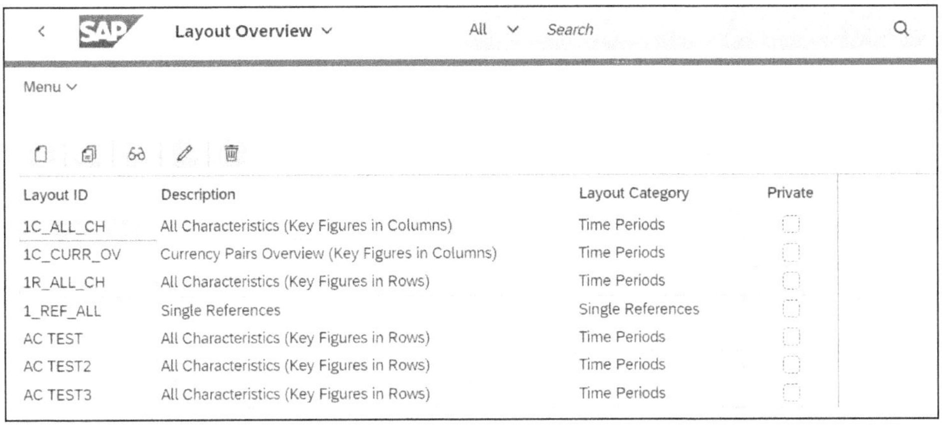

Layout ID	Description	Layout Category	Private
1C_ALL_CH	All Characteristics (Key Figures in Columns)	Time Periods	
1C_CURR_OV	Currency Pairs Overview (Key Figures in Columns)	Time Periods	
1R_ALL_CH	All Characteristics (Key Figures in Rows)	Time Periods	
1_REF_ALL	Single References	Single References	
AC TEST	All Characteristics (Key Figures in Rows)	Time Periods	
AC TEST2	All Characteristics (Key Figures in Rows)	Time Periods	
AC TEST3	All Characteristics (Key Figures in Rows)	Time Periods	

Figure 8.49 First Four Entries Are Standard Delivered Layouts, but We Can Create More

New Layout ID

If we click the **Create** button, we can assign a **Layout ID** and a **Layout Category,** and we can determine whether it will be a **Private Layout** for our user ID. These options are displayed in Figure 8.50. One thing to note is that the layout category ties back to the analysis item definition that we determined when we created the hedging area. If we've set the hedging area's analysis item definition to be by time periods, then we'll select the same information for the layout category. Checking the **Private Layout** box ensures that this layout will only be available to the user ID creating the layout. If we don't check this box, the layout will be saved as public and other users will be able to view the layout.

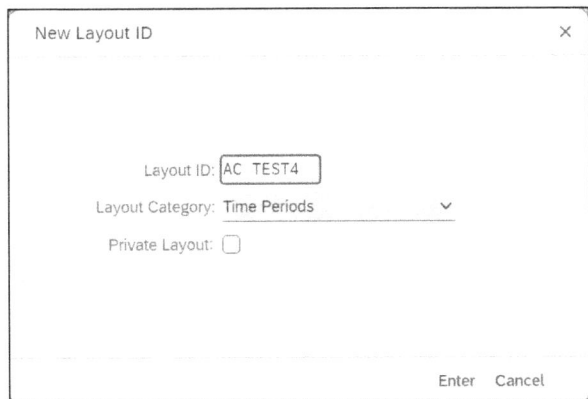

Figure 8.50 Initial Settings We Can Determine When Creating New Layout ID

563

Layout Initial Details

Next, we can determine how the data is displayed in this view. The initial details of the layout are at the top of the popup in Figure 8.51, and they include the following entries:

- **Snapshot Exposures**
  Here, we can determine which snapshot is selected. There are three options we can select:
  - **Last Snapshot**
    With this option, we select the last snapshot taken for that hedging area.
  - **Last Relevant Snapshot**
    With this option, we select the last relevant snapshot, which is the last snapshot taken with the day reference indicator selected.
  - **Snapshot ID**
    With this option, we select the snapshot directly, by snapshot ID.

- **Selected Key Date**
  Here, we can either make the current date the default key date, or we can enter the key date manually.

- **Display Mode**
  Here, we can choose to display the key figures in rows or columns. The difference between these views is shown in Figure 8.52 (rows) and Figure 8.53 (columns). In Figure 8.52, we can see that the key figures are in the rows in this view and the columns contain the periods. In Figure 8.53, we can see that the key figures are in the columns in this view and the periods are in the rows.

- **Target Quota Type**
  Here, we select the target quota for this view.

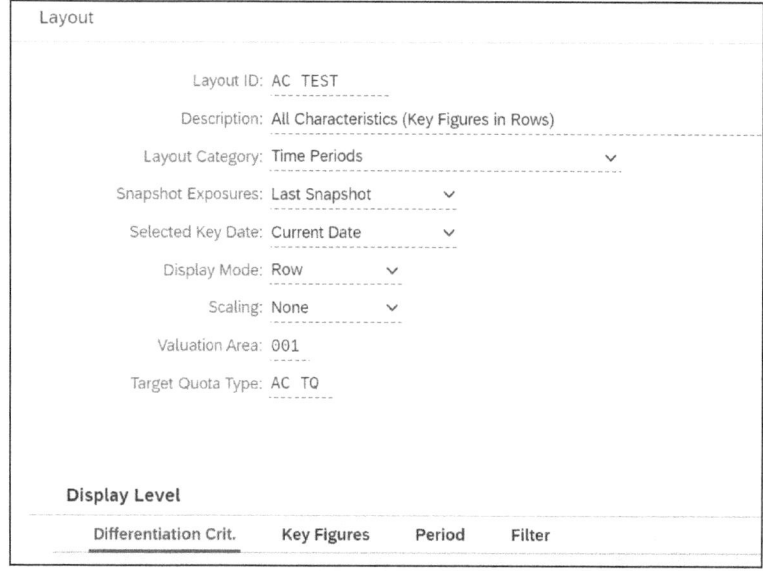

Figure 8.51 Selection Options When Editing Layout

8.4 Cash Flow Hedging Process

CoCode	Target Crcy	Risk Currency	Key Figure Name	Total	09/2024	10/2024	11/2024	12/2024	01/2025	02/2025	03/2025
1000	USD	CAD	Incoming Exposure	0.00	0.00	0.00	0.00	0.00	0.00	0.00	0.00
1000	USD	CAD	Outgoing Exposure	45,709,888.22-	26,411,111.11-	6,722,222.00-	10,210,000.00-	1,766,555.11-	100,000.00-	100,000.00-	300,000.00-
1000	USD	CAD	Amount to Hedge	6,426,577.42	5,760,000.00	0.00	0.00	626,577.42	20,000.00	20,000.00	0.00
1000	USD	CAD	Net Overhedge	0	0	0	0	0	0	0	0
1000	USD	CAD	Net Open Exposure	28,849,055.06-	22,632,222.22-	944,444.40-	4,052,500.00-	979,888.44-	40,000.00-	40,000.00-	120,000.00-
1000	USD	CAD	Net Hedges	16,860,833.16	3,778,888.89	5,777,777.60	6,157,500.00	786,666.67	60,000.00	60,000.00	180,000.00
1000	USD	CAD	Hedge Quota [%]	64.6	58.2	80.0	75.0	44.5	60.0	60.0	60.0
1000	USD	CAD	Target Quota [%]	N/A	80.00	80.00	75.00	80.00	80.00	80.00	80.00

Figure 8.52 Display Mode by Rows

CoCode	Target Crcy	Risk Currency	Period Name	Inc. Exp.	Out. Exp.	Net Exp.	NOE(HR)	HedgedRate	HdgQ [%]	TQ [%]	AmtToHedge	OpenAmtHR
1000	USD	CAD	Total	0.00	45,709,888.22-	45,709,888.22-	16,159,055.06-	1.4873	64.6	N/A	6,426,577.42	12,690,000.00
1000	USD	CAD	2024	0.00	45,109,888.22-	45,109,888.22-	15,919,055.06-	1.4892	64.7	N/A	6,386,577.42	12,690,000.00
1000	USD	CAD	09/2024	0.00	26,411,111.11-	26,411,111.11-	11,042,222.22-	1.4894	58.2	80.00	5,760,000.00	11,590,000.00
1000	USD	CAD	10/2024	0.00	6,722,222.00-	6,722,222.00-	1,344,444.40-	1.4882	80.0	80.00	0.00	400,000.00-
1000	USD	CAD	11/2024	0.00	10,210,000.00-	10,210,000.00-	2,552,500.00-	1.4892	75.0	75.00	0.00	1,500,000.00
1000	USD	CAD	12/2024	0.00	1,766,555.11-	1,766,555.11-	979,888.44-	1.4961	44.5	80.00	626,577.42	0.00
1000	USD	CAD	2025	0.00	600,000.00-	600,000.00-	240,000.00-	1.4056	60.0	N/A	40,000.00	0.00
1000	USD	CAD	01/2025	0.00	100,000.00-	100,000.00-	40,000.00-	1.4500	60.0	80.00	20,000.00	0.00
1000	USD	CAD	02/2025	0.00	100,000.00-	100,000.00-	40,000.00-	1.4500	60.0	80.00	20,000.00	0.00
1000	USD	CAD	03/2025	0.00	300,000.00-	300,000.00-	120,000.00-	1.3963	60.0	80.00	0.00	0.00
1000	USD	CAD	04/2025	0.00	100,000.00-	100,000.00-	40,000.00-	1.3500	80.0	80.00	0.00	0.00
1000	USD	CAD	05/2025	0.00	0.00	0.00	0.00	0.0000	0.0	80.00	0.00	0.00
1000	USD	CAD	06/2025	0.00	0.00	0.00	0.00	0.0000	0.0	80.00	0.00	0.00
1000	USD	CAD	07/2025	0.00	0.00	0.00	0.00	0.0000	0.0	80.00	0.00	0.00
1000	USD	CAD	08/2025	0.00	0.00	0.00	0.00	0.0000	0.0	80.00	0.00	0.00

Figure 8.53 Display Mode by Columns

Next, we'll review the four tabs in the layout screen to determine the differentiation criteria, key figures, period, and filter.

Differentiation Criteria

The differentiation criteria section drives the differentiation sections that we can view in the hedge management cockpit. This differentiation criteria section is in the lower half of the layout screen. To select one of these values and have it show up in the reporting, we need to select that value as relevant for differentiation criteria in the **General Settings** tab of the hedging area. In the example in Figure 8.54, the company code won't appear since we set it not to show by entering "0" in the **Display Level** column.

Display Level	Differentiation Criteria
0	Company Code
1	Target Crcy
1	Risk Currency
1	Portfolio
1	Profit Center
1	Cost Center
1	WBS Element
1	On Behalf of CoCode

Figure 8.54 Differentiation Criteria Drive Which Key Figures Are Shown in Report

8 Cash Flow Hedging

Key Figures

The **Key Figures** tab drives which key figures appear in the reporting and the order in which they appear. In Figure 8.55, we can see that all the key figures have the same display level, but the order goes from 1 to 8. This order will be clear when we look at this report and see these key figures appear in this defined order.

Display Level	Key Figure	Order
1	Target Quota [%]	8
1	Hedge Quota [%]	7
1	Net Hedges	6
1	Net Open Exposure	5
1	Net Overhedge	4
1	Amount to Hedge	3
1	Outgoing Exposure	2
1	Incoming Exposure	1
0	Gross Overhedge (Hedge Accounting)	0
0	Net Overhedge (Hedge Accounting)	0

Figure 8.55 Key Figures Determined on This Screen Drive Which Ones Are Displayed in Hedge Management Cockpit

Period

- In the **Period** tab, we define which periods will show up in the reporting (seen in Figure 8.56). We can just view the actual periods that were defined for this hedging area, but we can also add total columns or rows. We can select the periods, total, and total of year by entering a "1" in the display level. This can show us the aggregate of the year and the overall total for all key figures in the reporting.

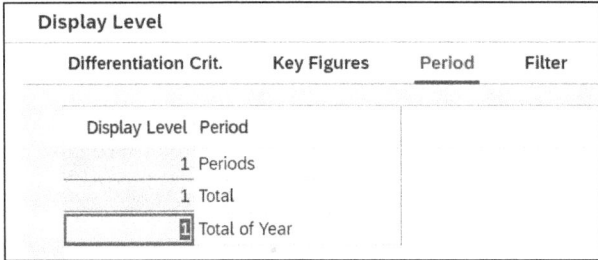

Figure 8.56 Period Tab

If we have set up all three of these periods, we can see how these totals show up in the reporting in Figure 8.57. Since we've added all three types of periods, we see the overall total, the yearly total, and the period total.

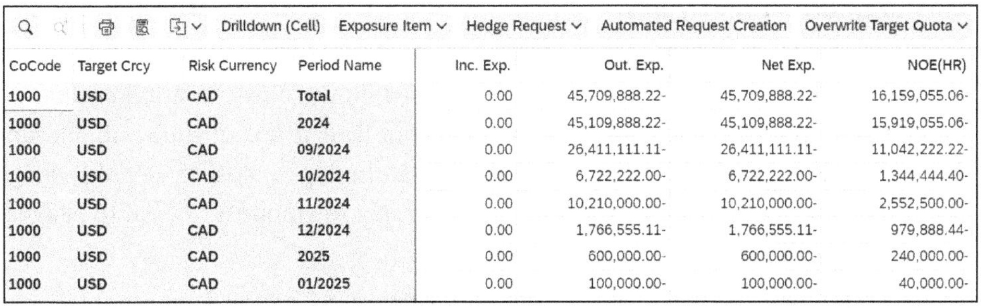

8.4 Cash Flow Hedging Process

CoCode	Target Crcy	Risk Currency	Period Name	Inc. Exp.	Out. Exp.	Net Exp.	NOE(HR)
1000	USD	CAD	Total	0.00	45,709,888.22-	45,709,888.22-	16,159,055.06-
1000	USD	CAD	2024	0.00	45,109,888.22-	45,109,888.22-	15,919,055.06-
1000	USD	CAD	09/2024	0.00	26,411,111.11-	26,411,111.11-	11,042,222.22-
1000	USD	CAD	10/2024	0.00	6,722,222.00-	6,722,222.00-	1,344,444.40-
1000	USD	CAD	11/2024	0.00	10,210,000.00-	10,210,000.00-	2,552,500.00-
1000	USD	CAD	12/2024	0.00	1,766,555.11-	1,766,555.11-	979,888.44-
1000	USD	CAD	2025	0.00	600,000.00-	600,000.00-	240,000.00-
1000	USD	CAD	01/2025	0.00	100,000.00-	100,000.00-	40,000.00-

Figure 8.57 Overall Total, Yearly Total, and Period Total

Filter

We can also filter the data in the report further if required on the **Filter** tab. To add filters, we need to check the box to the left of the key figure so we can filter it, as in Figure 8.58. We can set the filter values in this screen to default in the desired data, or we can add them at the time we're running the report.

Figure 8.58 Adding Filters to Further Refine Data in Cockpit

Now that we've added the filters, they show up in the **Filter Details** section when we're creating the initial options in this report, as shown in Figure 8.59.

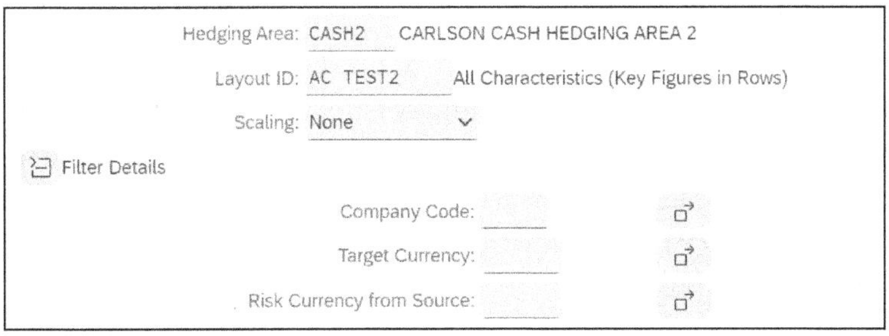

Figure 8.59 Filters Are Now on Initial Entry Screen Based on Layout ID Settings

Report Layout and Key Figures

Now that we've set up the layout, we can view the output of the hedge management cockpit report. The report will show the data based on the differentiation criteria we defined both in the hedging area and on the layout ID. The layout shows all relevant details available to make hedging decisions, including exposure details, hedge amounts, hedged rates, quota percentages, and proposed amounts of what to hedge if we automatically generate hedge requests.

Figure 8.60 shows a summary of some of the key figures we can view in this report, but we can add more by editing the layout. The relevant key figures can help us determine how to action the exposures in the hedge management cockpit.

Drilldown (Cell) | Exposure Item | Hedge Request | Automated Request Creation | Overwrite Target Quota

CoCode	Target Crcy	Risk Currency	Key Figure Name	Total	09/2024	10/2024	11/2024	12/2024	01/2025	02/2025
1000	USD	CAD	Incoming Exposure	0.00	0.00	0.00	0.00	0.00	0.00	0.00
1000	USD	CAD	Outgoing Exposure	45,709,888.22-	26,411,111.11-	6,722,222.00-	10,210,000.00-	1,766,555.11-	100,000.00-	100,000.00-
1000	USD	CAD	Net Exposure	45,709,888.22-	26,411,111.11-	6,722,222.00-	10,210,000.00-	1,766,555.11-	100,000.00-	100,000.00-
1000	USD	CAD	Net Hedges	16,860,833.16	3,778,888.89	5,777,777.60	6,157,500.00	786,666.67	60,000.00	60,000.00
1000	USD	CAD	NetOpenExp (Incl.HR)	16,159,055.06-	11,042,222.22-	1,344,444.40-	2,552,500.00-	979,888.44-	40,000.00-	40,000.00-
1000	USD	CAD	Hedged Rate	1.4873	1.4894	1.4882	1.4892	1.4961	1.4500	1.4500
1000	USD	CAD	Hedge Quota [%]	64.6	58.2	80.0	75.0	44.5	60.0	60.0
1000	USD	CAD	Target Quota [%]	N/A	80.00	80.00	75.00	80.00	80.00	80.00
1000	USD	CAD	Amount to Hedge	6,426,577.42	5,760,000.00	0.00	0.00	626,577.42	20,000.00	20,000.00
1000	USD	CAD	Open Amt Hedge Req.	12,690,000.00	11,590,000.00	400,000.00-	1,500,000.00	0.00	0.00	0.00

Figure 8.60 Key Figures Displayed in Hedge Management Cockpit

Each key figure that is displayed in the hedge management cockpit has additional details behind it. The exposures, hedges, and target amounts in this table have multiple data points that are aggregated to the totals that appear on this screen.

We can drill down into any of these data points by double-clicking on the applicable cell, and that will bring up a new screen that will provide additional details for that key figure in that period. In the following three sections, we'll drill down into a few of the key figures to show how this report can change depending on which key figure we're analyzing.

Exposure Totals

The first data that we can view in this report is the exposure totals. We can view the total incoming and outgoing exposures for the periods, and these exposures can be from both exposure management 2.0 and cash management. The incoming and outgoing exposures are totaled and shown in the net exposure amount.

At this point, we're only viewing this data as a summary amount, but we can drill down into any of these lines to view the individual exposures that add up to the totals we see on the screen. This is shown in Figure 8.61, which details the net exposure amount for period 9 in 2024. In this line, we've drilled down into the **Net Exposure** key figure, so all the exposures are displayed for that period.

8.4 Cash Flow Hedging Process

Hedging Area:	CASH2	CARLSON CASH HEDGING AREA 2
Key Date:	09/02/2024	
Valuation Area:	001	
Target Quote Type:	AC_TQ	

Key Figure

CoCode	Target Crcy	Risk Currency	Period Name	Key Figure Name	Calculated Value	Unit
1000	USD	CAD	09/2024	Net Exposure	26,411,111.11-	CAD

Exposures

Exposure Item ID (Gross)	Exposure Item ID (Net)	Filter	Sign	Amount	Date	Direction	Class	Period Start Date	Product	CoCode	Target Crcy	Risk Currency
26	28	AC FILTER	-	6,000,000.00-		O	FK0	09/01/2024	3	1000	USD	CAD
26	28	AC FILTER	-	7,000,000.00-		O	FK0	09/01/2024	3	1000	USD	CAD
26	28	AC FILTER	-	6,000,000.00-		O	FK0	09/01/2024	3	1000	USD	CAD
27	28	AC FILTER	-	0.00		I	FK0	09/01/2024	3	1000	USD	CAD
26	28	AC FILTER	-	111,111.11-		O	FK0	09/01/2024	3	1000	USD	CAD
26	28	AC FILTER	-	100,000.00-		O	FK0	09/01/2024	3	1000	USD	CAD
26	28	AC FILTER	-	100,000.00-		O	FK0	09/01/2024	3	1000	USD	CAD
26	28	AC FILTER	-	7,000,000.00-		O	FK0	09/01/2024	3	1000	USD	CAD
26	28	AC FILTER	-	100,000.00-		O	FK0	09/01/2024	3	1000	USD	CAD

Figure 8.61 Drilling Down into Line Item

Hedges

The next key figures are the hedge amounts. The **Net Hedges** key figure details the hedges that have been requested, and a hedging instrument and hedged item have been designated. Any hedge that has a value date in the period will appear. Drilling down into the net hedges key figure will show the financial transactions that add up to the total key figure, as in Figure 8.62.

Key Figure

CoCode	Target Crcy	Risk Currency	Period Name	Key Figure Name	Calculated Value	Unit
1000	USD	CAD	09/2024	Net Hedges	3,778,888.89	CAD

Hedges

CoCode	Transaction	PTyp	Text	Instrument Cat.	Instrument Cat.	Exprd/Exer	Amount in Trans. Crcy	Crcy	Pmnt Amnt in LCurr	Target Crcy	Term Start	Value Date
1000	10000095	FX	Foreign exchange (FX)	FXFW	FX Forward		3,600,000.00	CAD	2,416,107.38-	USD	01/30/2024	09/30/2024
1000	10000114	FX	Foreign exchange (FX)	FXFW	FX Forward		6,000.00	CAD	4,026.85-	USD	02/08/2024	09/30/2024
1000	10000245	FX	Foreign exchange (FX)	FXFW	FX Forward		80,000.00	CAD	55,172.41-	USD	05/22/2024	09/30/2024
1000	10000113	FX	Foreign exchange (FX)	FXFW	FX Forward		4,000.00	CAD	2,684.56-	USD	02/08/2024	09/30/2024
1000	10000234	FX	Foreign exchange (FX)	FXFW	FX Forward		88,888.89	CAD	59,259.26-	USD	05/15/2024	09/30/2024

Figure 8.62 Drilling Down into Net Hedges Shows All Hedges in Selected Period

There are also numbers in this report that show whether we have any hedge requests in transit. For example, a hedge request was created but a trade hasn't been executed, so there's not a designated hedge for that request. This can occur for many reasons. The hedge request could still be in a review status and may not have been released to create a trade with the trading platform, or the trade may have not been executed yet in the trading platform. If either of these is the case, then it can be shown in the net open exposure including the hedge requests key figure. This shows the total net exposure for the period, but it also separates the hedge amounts to show the total number of hedges that have been requested, the amounts that are open and haven't been executed and

8   Cash Flow Hedging

designated as hedge requests, and the number of hedges that are designated. In Figure 8.63, we can see the total list of contributing key figures that are used to calculate the total amount for the net open exposures.

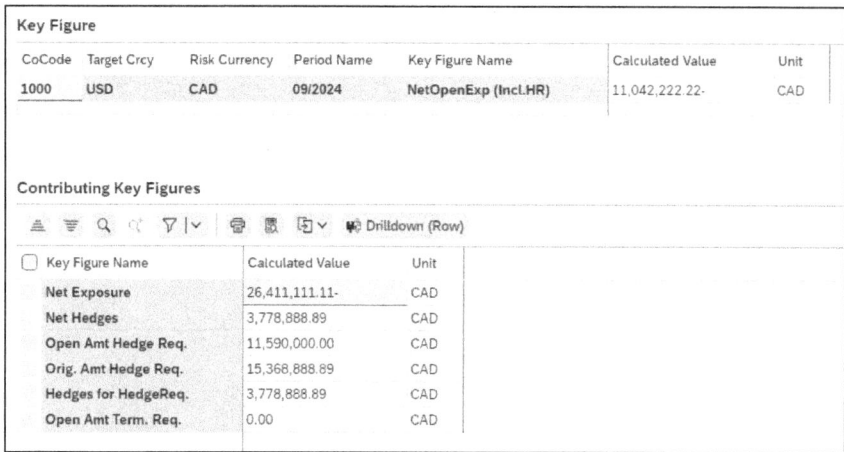

Figure 8.63 Net Open Exposure Key Figure Showing Executed Hedges versus Hedge Requests Still Open

Another hedge-related key figure is the hedged rate. Figure 8.64 details the calculated FX rate for the hedges in each period. This is calculated using the net hedge amount in the risk currency and the comparable amount in the risk-free currency.

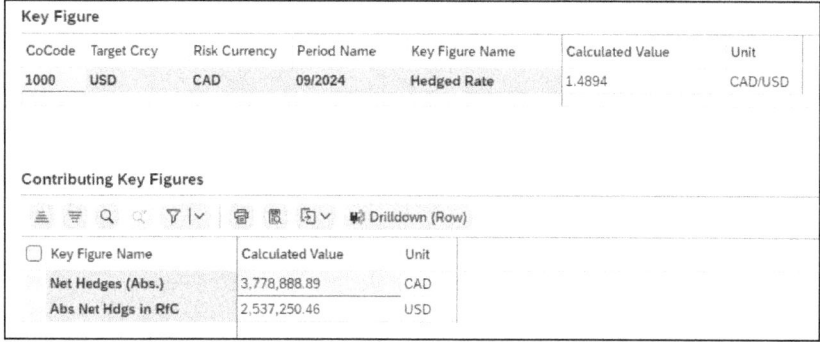

Figure 8.64 Hedged Rate Calculates Average Rate for Hedges during Selected Period

Hedge Targets

There are also key figures we use to calculate the hedge request amount. We compare the target quota that we determined in the hedging to the current hedge quota percentage. This will show if we're in an overhedged situation or if a hedge is required, and it will result in a proposed *amount to hedge*. Drilling down into this key figure will let us compare the exposures to the total requested hedge amounts. We compare the target hedge amount to the hedge request amount to determine the amount-to-hedge key figure. All data points we reviewed for this key figure are in Figure 8.65.

8.4 Cash Flow Hedging Process

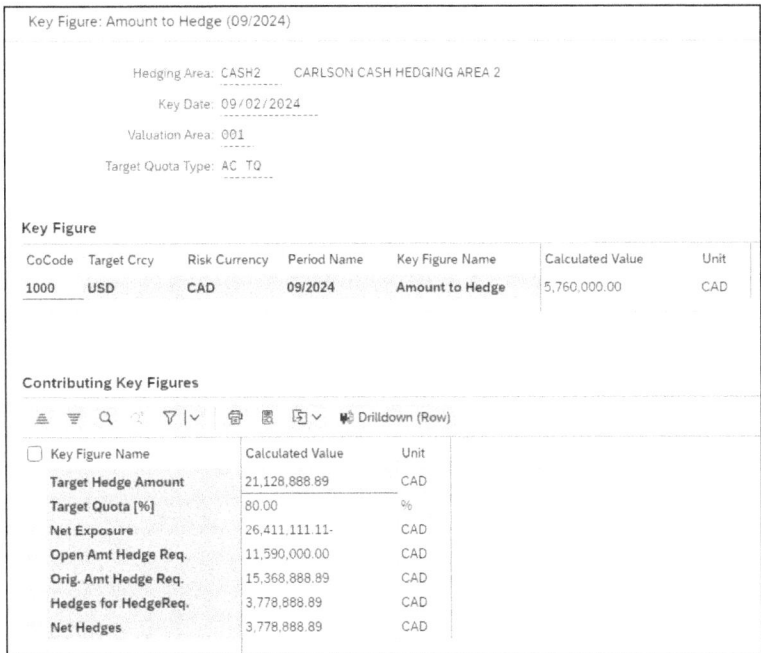

Figure 8.65 Determining Amount to Hedge

Functions in the Hedge Management Cockpit

Since we've covered all the details on how to set up the hedge management cockpit and covered the key details that we can view, let's explore how to edit the data in the cockpit and create the hedge requests. We execute this by using the series of buttons above the key data report (shown in Figure 8.66).

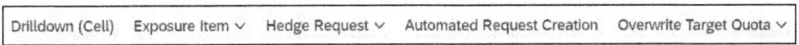

Figure 8.66 Buttons in Report that Allow for Actioning Hedges in Cockpit

In the following sections, we'll walk through the big-three buttons: **Overwrite Target Quota**, **Automated Request Creation**, and **Hedge Request** (the last of which, as you'll see, has multiple suboptions).

Overwrite Target Quota

There are situations in which we need to overwrite the target quota percentages. If we need to deviate from the target quota amounts established in the hedging area, we can do this with the **Overwrite Target Quota** function. We can run this in two different ways:

- **One Entry**
  If we need to change only one target quota, we highlight the relevant target quota field and click on **Overwrite Target Quota • One Entry**. This will bring up the popup in Figure 8.67 that allows us to change the **Target Quota [%]**, which we've updated to 75% in the figure.

571

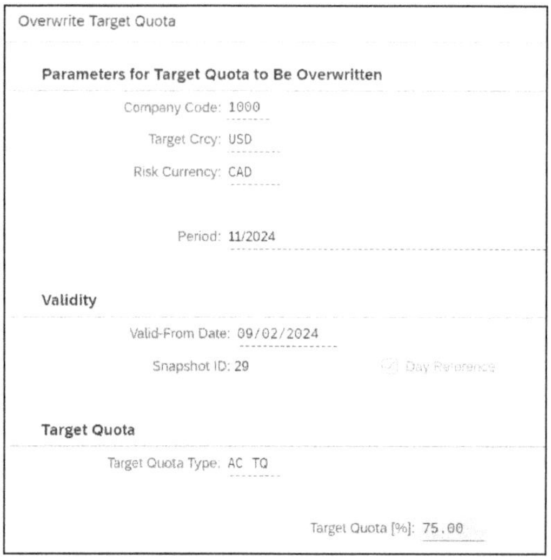

Overwrite Target Quota
Parameters for Target Quota to Be Overwritten
Company Code: 1000
Target Crcy: USD
Risk Currency: CAD
Period: 11/2024
Validity
Valid-From Date: 09/02/2024
Snapshot ID: 29
Day Reference
Target Quota
Target Quota Type: AC TQ
Target Quota [%]: 75.00

Figure 8.67 Updating Target Quota Percentage

- **Multiple Entries**
  When we need to change multiple target quota numbers, we can select the **Overwrite Target Quota • Multiple Entries** option. This will allow us to update the target quota for multiple periods and scenarios all at one time. The one limitation of this option is that we can only update the target quota percentage to the same percentage, so we'll have to execute multiple runs of this function if the target quota percentage needs to vary depending on the scenario. This entry screen requires that we determine the scenario that we need to update. In Figure 8.68, we determine that we're updating the target quotas for a certain combination of **Company Code**, **Target Crcy**, and **Risk Currency**, and we can also determine a specific **Period**.

Overwrite Target Quota
Parameters for Target Quota to Be Overwritten
Company Code: 1000
Target Crcy: USD
Risk Currency: CAD
Period:
Validity
Valid-From Date: 09/02/2024
Snapshot ID: 29
Day Reference
Target Quota
Target Quota Type: AC TQ
Target Quota [%]: 70.00

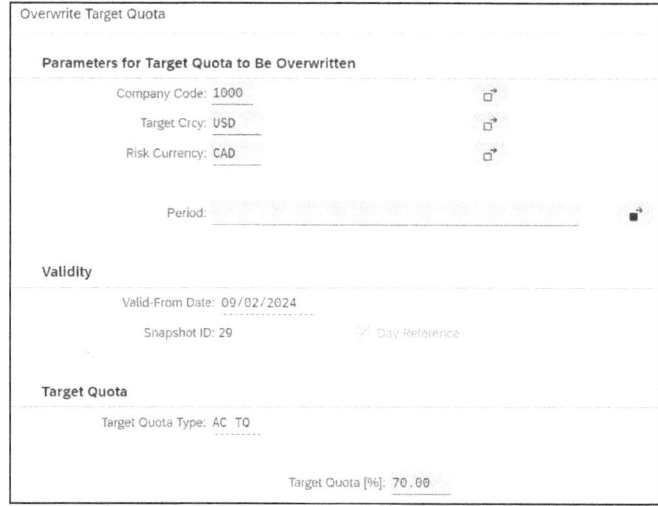

Figure 8.68 Updating Multiple Target Quotas Using Multiple Entries Option

8.4 Cash Flow Hedging Process

Once we've updated the quota numbers, the reporting will reflect the updates. Both the target quota and the amount-to-hedge key figures will update accordingly and allow for automated request creation with the new numbers.

Automated Request Creation

Hedge requests in the hedge management cockpit can be either manual requests or automated. The automated requests from the cockpit will leverage the target quota percentages and other settings determined in the hedging area to calculate and request hedges. To create a proposal for the automated requests, we click on the **Automated Request Creation** button and a popup with the proposed hedge requests will appear.

An indicator will appear on the left side of the screen in the shape of a traffic light. The green box indicates that the hedge request can be created, the yellow triangle determines that a hedge request can't be created, and the red circle highlights that the current period is in an overhedged scenario and won't create a hedge request.

You'll notice that some fields are grayed out and some are not. This can appear different based on the version of SAP S/4HANA and the theme that is selected. All hedge requests that we can create in Figure 8.69 appear with grayed-out fields for **Portfolio, Amount, Value Date, Instrument,** and **Hedging Classification**. We can edit these in this screen so before submitting the request, we can edit the portfolio, hedge request amount, value date, requested financial instrument, and hedging classification if required. We can also exclude hedges from being requested by unchecking the box in the **Selection** column. Once we want to execute the hedges, we can click the **Enter** button and the hedge requests will be saved.

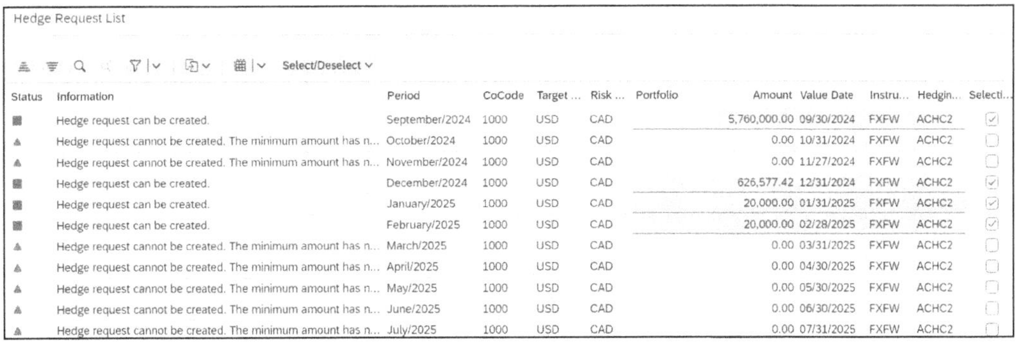

Status	Information	Period	CoCode	Target...	Risk...	Portfolio	Amount	Value Date	Instru...	Hedgin...	Selecti...
■	Hedge request can be created.	September/2024	1000	USD	CAD		5,760,000.00	09/30/2024	FXFW	ACHC2	✓
▲	Hedge request cannot be created. The minimum amount has n...	October/2024	1000	USD	CAD		0.00	10/31/2024	FXFW	ACHC2	☐
▲	Hedge request cannot be created. The minimum amount has n...	November/2024	1000	USD	CAD		0.00	11/27/2024	FXFW	ACHC2	☐
■	Hedge request can be created.	December/2024	1000	USD	CAD		626,577.42	12/31/2024	FXFW	ACHC2	✓
■	Hedge request can be created.	January/2025	1000	USD	CAD		20,000.00	01/31/2025	FXFW	ACHC2	✓
■	Hedge request can be created.	February/2025	1000	USD	CAD		20,000.00	02/28/2025	FXFW	ACHC2	✓
▲	Hedge request cannot be created. The minimum amount has n...	March/2025	1000	USD	CAD		0.00	03/31/2025	FXFW	ACHC2	☐
▲	Hedge request cannot be created. The minimum amount has n...	April/2025	1000	USD	CAD		0.00	04/30/2025	FXFW	ACHC2	☐
▲	Hedge request cannot be created. The minimum amount has n...	May/2025	1000	USD	CAD		0.00	05/30/2025	FXFW	ACHC2	☐
▲	Hedge request cannot be created. The minimum amount has n...	June/2025	1000	USD	CAD		0.00	06/30/2025	FXFW	ACHC2	☐
▲	Hedge request cannot be created. The minimum amount has n...	July/2025	1000	USD	CAD		0.00	07/31/2025	FXFW	ACHC2	☐

Figure 8.69 Clicking Automated Request Creation Button Initiates Hedge Requests

> **Settings for Automated Request Creation**
>
> The automated hedge requests will only run properly if key figures are visible in the report. We'll receive an error message (**No amount to hedge displayed for automated hedge requests**) during the automated hedge request if we haven't selected the required fields as key figures in the layout.

To ensure the correct amount is requested, we should make the following key figures available in the layout:

- Incoming Exposure
- Outgoing Exposure
- Net Exposure
- Net Hedges
- Net Open Exp (Incl. HR)
- Hedged Rate
- Hedge Quota
- Target Quota
- Amount to Hedge
- Open Amt Hedge Req.

Manual Request Creation

We can initiate additional hedge requests if we have more one-off requests we want to execute. We can initiate manual hedge requests for FX hedge requests, FX swap requests, and dedesignation requests. We do this by selecting the exposure line in the period when we want to generate the request and then selecting the hedge request type we want to generate under the **Hedge Request** button.

When we need to manually generate a hedge request (not through automated request creation), we can select the **Hedge Request • FX Hedge Request** option. When we do this, the key data will already be populated in the **Analysis Item** section that gives the details of the exposure we are hedging. Other than this information, we can select the **Instrument Category, Hedging Classification, Portfolio, Value Date,** and **Hedge Request Amount**. SAP S/4HANA will propose the same amount to hedge based on the target quota, as in Figure 8.70, but we can edit the hedge request amount before submission. Once we have validated the details, we can click the **Save** button to save the hedge request.

If we want to create an FX swap to change the value date of a hedging instrument, we can do it by making an FX Swap request. To do this, we click **Hedge Request • FX Swap Request**. To complete this request, we need to properly assign a few things:

- We should create a separate hedging classification for the swap request. This helps separate the requests and can separate the designation control in the **Hedge Accounting II** tab settings.

- We need to create a separate designation type for the swaps, and we should assign it to the proper designation category (**N Instruments Designation Pattern**), as detailed in Section 8.2.1.

8.4 Cash Flow Hedging Process

Figure 8.70 Creating Manual FX Hedge Requests

Once we've confirmed these settings, we can select a current hedge that is designated in the Net Hedges key figure and execute the FX swap request. The selection screen is slightly different for the FX swap request, and in Figure 8.71, we can see that the **Instrument Attributes** section includes an **Initial Value Date** and a **Target Value Date**. The **Initial Value Date** should reflect the payment date of the first leg of the FX swap, and the **Target Value Date** should reflect the new settlement date for the second leg of the FX swap. In this case, the swap is moving the value date from September to October.

The FX swap request has an additional **Hedging Relationships** tab that we can utilize to determine which financial transaction is being rolled over to the new settlement date. In Figure 8.72, we can see that we are requesting to roll over the full amount.

If a consumption sequence has been assigned, it can determine which trades we are rolling over by using either a first-in, first-out (FIFO) or a last-in, first-out (LIFO) method. This is relevant if there are multiple hedges in the same period and we need logic to determine which trades are being rolled over. In this example, only one trade is in this period and no consumption sequence is being assigned, so we manually select the trade by checking the **Requested** box and enter the full amount in the **Requested Amount** field. Once we've confirmed the details, we can save the swap request.

8 Cash Flow Hedging

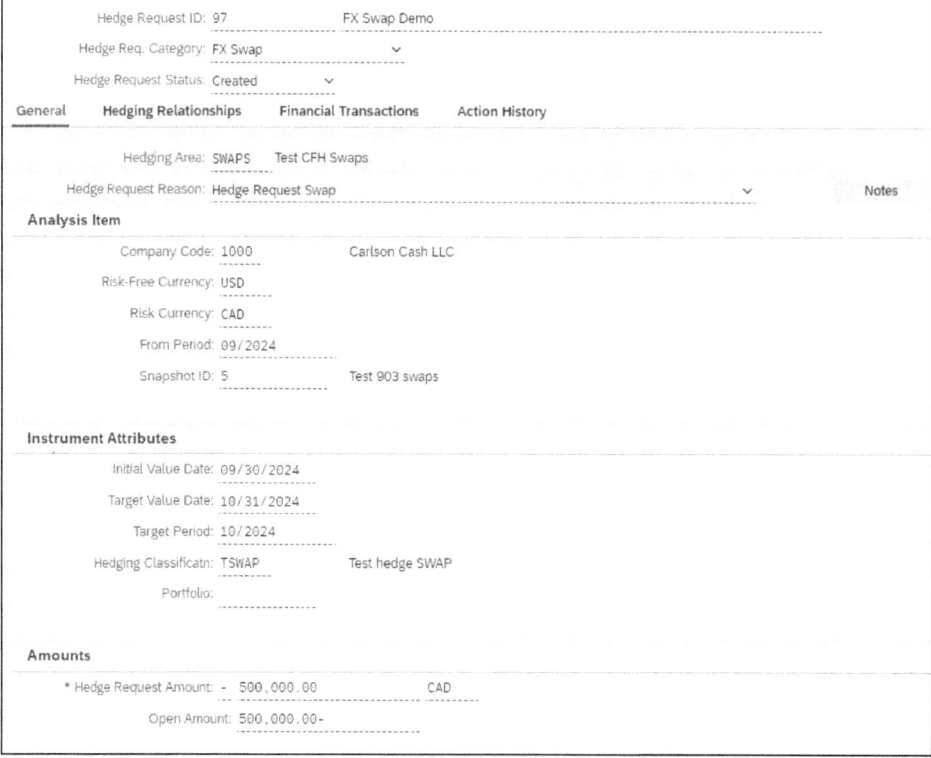

Figure 8.71 Manual FX Swap Request Rolling Transaction to Future Date

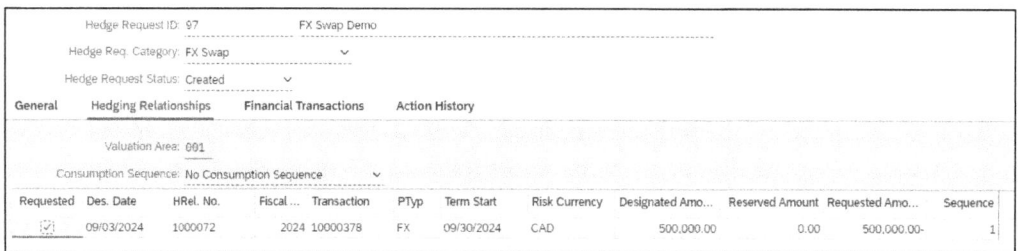

Figure 8.72 Selecting Hedging Relationships for Which Amounts to Roll Over to New Value Date

Finally, if we need to dedesignate a designated hedge for a different day than the planned dedesignation date, we can initiate this can from the hedge management cockpit. To do this, we select the **Net Hedges** key figure and click **Hedge Request • Dedesignation Request**.

Once we're in the hedge request screen, we enter the necessary details, including the **Hedge Request Reason**, **Dedesignation Date**, and **Hedge Request Amount**, as in Figure 8.73.

8.4 Cash Flow Hedging Process

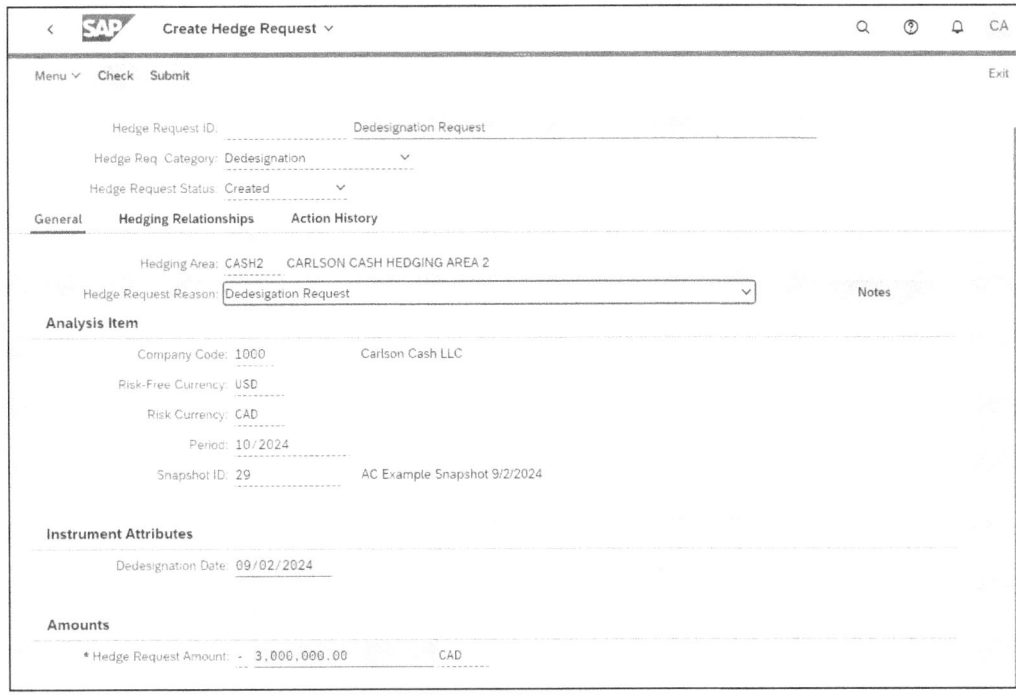

Figure 8.73 Dedesignation Request for $3,000,000 Canadian Dollars

Once we've added the details in the **General** tab, we can go to the **Hedging Relationships** tab, where we can determine which hedging relationships we need to dedesignate and the requested amount. Assigning the requested amount and sequence is the same as making FX swap requests, but we need to add a new field for the dedesignation to handle the reclassification. Generally, the reclassification of the hedging reserve and the cost of the hedging reserve occurs according to the hedging area settings, but this can vary for the dedesignation. We can determine whether we want to follow the settings in the hedging area for the reclassification as planned by selecting the **Planned Reclassification** option in the **Reclass. Handling** field. Otherwise, we can have the reclassification happen **Immediately**. Figure 8.74 shows how we can set the reclassification to occur as planned. This is also a scenario where a consumption sequence doesn't exist, so we need to manually select the items for the dedesignation. To do this, we populate three fields as follows:

- We check the **Requested** box for the line item.
- We determine the dedesignation amount in the **Requested Amount** field.
- We determine the sequence for the dedesignation in the **Sequence** column. The lowest number in the sequence will be prioritized.

8 Cash Flow Hedging

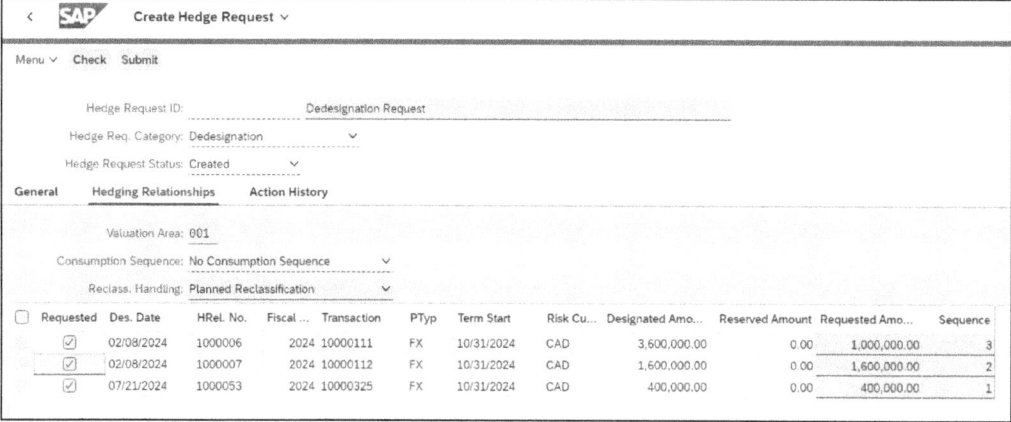

Figure 8.74 Trades in Which Dedesignation Request Will Apply to Requested Amount

8.4.3 Processing Hedge Requests

Once we've created the hedge requests, we can review and approve them in the Process Hedge Requests app or Transaction TOEHREQO. This app is shown in Figure 8.75, and once in it, we can see the list of hedge requests available for processing as shown in Figure 8.76.

Figure 8.75 Process Hedge Requests App

All hedge requests related to cash flow hedging will show up on one dashboard. This includes all requests from the hedge management cockpit, including FX hedge, FX swap, and dedesignation requests. We can update the status of a request by selecting the hedge request ID and clicking the **Process** button. This will open a dropdown list showing what the next available statuses are for the requests. Also, we can review additional details of the hedge request by double-clicking into it.

To review the outstanding hedge requests that haven't been released, we can filter the report by the **Submitted** status. We do this by unchecking all the boxes except for **Submitted,** as shown in Figure 8.77. We need to review hedge requests in this status and can release or reject them.

To further view the details in a hedge request, we can double-click any of these line items to view additional details. **Hedge Request ID 78** is shown in detail in Figure 8.78.

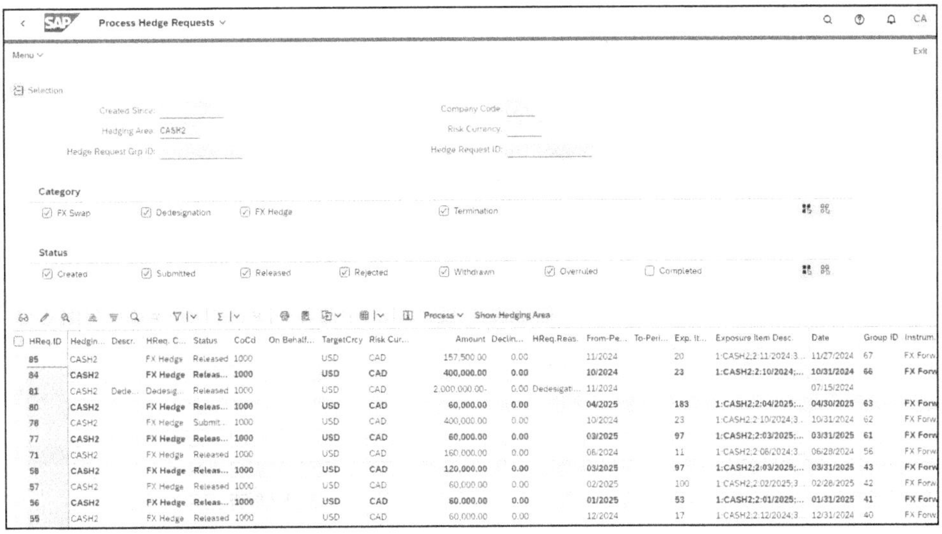

8.4 Cash Flow Hedging Process

Figure 8.76 Available Hedge Requests Displayed in Process Hedge Requests App

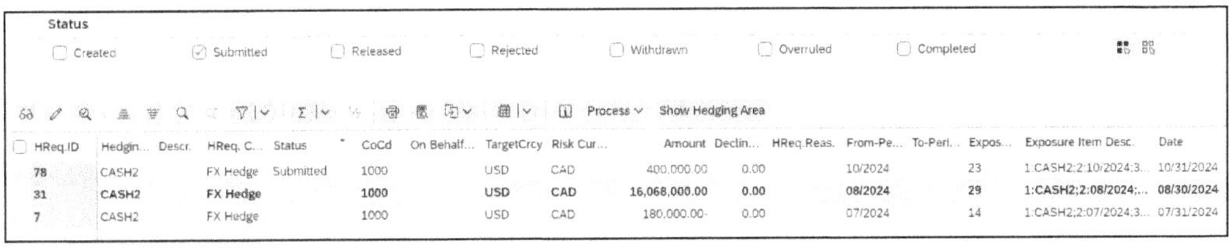

Figure 8.77 Report Filtered by Hedge Requests in Submitted Status

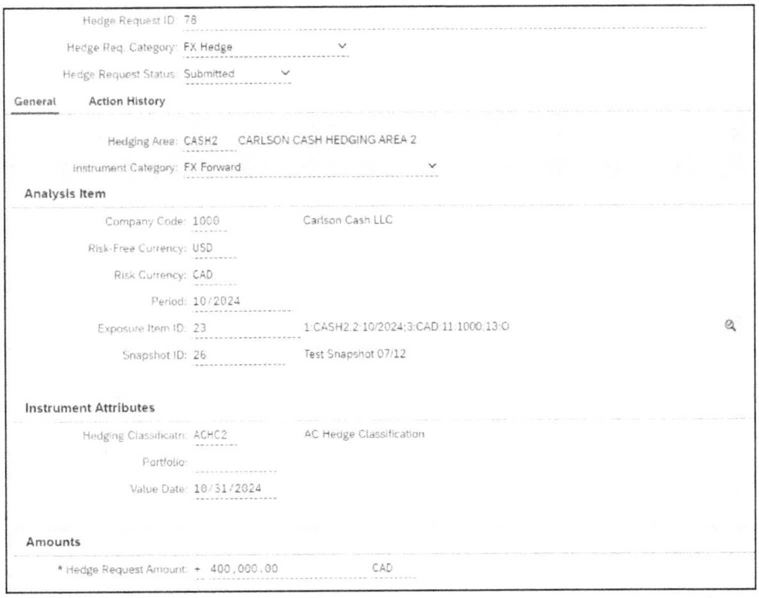

Figure 8.78 Drilling into Hedge Request to Show Additional Details

579

8 Cash Flow Hedging

After we've validated the details, we can back out to the initial screen. Highlighting the line item will allow us to click the **Process** button and select the next status for this hedge request, as shown in Figure 8.79. In this example, we have the option of approving the hedge request and clicking **Release**. If the hedge request is incorrect or we don't want to send it in for further processing, we can click **Reject**. Once we've selected either of these options, the hedge request will be updated to the next status.

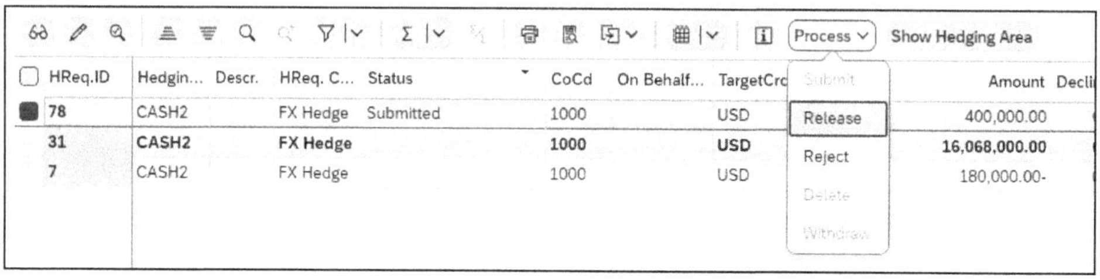

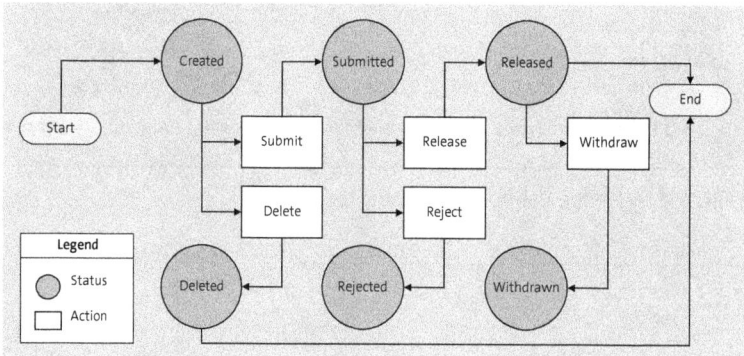

Figure 8.79 Selecting Line Item So We Can Click Process Button to Release Request

Figure 8.80 shows a graphic with each available status in this report and the process flow for how the statuses changes as the transaction is processed.

Figure 8.80 Process Diagram Showing Options for How to Process Hedge Requests

8.4.4 Creating a Financial Transaction

We covered creating a standard financial transaction in Chapter 6, Section 6.5, while running through the FX process, so this section will specifically cover the transaction creation process after a hedge request has been created in the hedge management cockpit. We can create these transactions from an interface or a trading platform, import them through SAP trading platform integration, or manually enter them using the Create Financial Transaction app. The key difference is on the **Administr.** tab and the **FX Hedge Management** tab. To properly assign the hedging instrument to the hedging item, we have to add the hedge request information to the **Administr.** tab. As shown in Figure 8.81, we'll need to add both the **Hdg. Classific.** and the **Hdg. Request ID** to this transaction in order to tie the transaction to the hedge request.

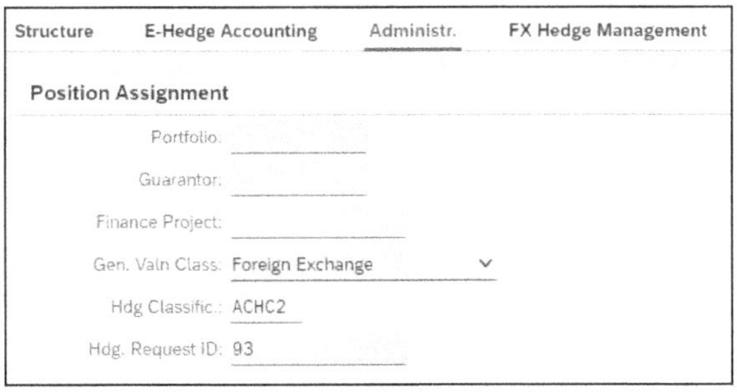

8.4 Cash Flow Hedging Process

| Structure | E-Hedge Accounting | Administr. | FX Hedge Management |

Position Assignment

Portfolio:
Guarantor:
Finance Project:
Gen. Valn Class: Foreign Exchange
Hdg Classific.: ACHC2
Hdg. Request ID: 93

Figure 8.81 Adding Hedging Classification and Hedge Request ID to Administration Tab

When we add this information, additional information will be populated in the **FX Hedge Management** tab. Figure 8.82 shows that the information in this tab includes details on the hedged exposure item. Once we've validated these tabs, we can save the trade. Assuming there are no issues in the data or customizing, the hedging relationship will automatically be created, and we can view it in the Manage Hedging Relationships app or Transaction TPM100.

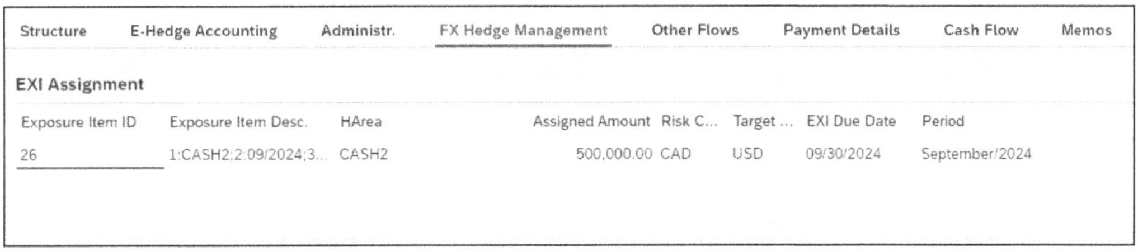

| Structure | E-Hedge Accounting | Administr. | FX Hedge Management | Other Flows | Payment Details | Cash Flow | Memos |

EXI Assignment

Exposure Item ID	Exposure Item Desc.	HArea	Assigned Amount	Risk C...	Target ...	EXI Due Date	Period
26	1:CASH2;2:09/2024;3...	CASH2	500,000.00 CAD		USD	09/30/2024	September/2024

Figure 8.82 Display of Exposure Item Details and Hedging Area in the FX Hedge Management Tab

There are scenarios in which the financial transaction might not automatically create a hedging relationship. When this occurs, we use the Reprocess Transactions – Automated Designation app or Transaction TPM104, as shown in Figure 8.83. It's available to both help diagnose the issue and create the hedging relationship when the issue is resolved. Transaction TPM104 can also designate transactions that have been configured to use an end-of-day designation (we detailed designation types in Section 8.2.1). To process these transactions, we check the **Include EOD-of-Day Designation** box and click **Execute**.

Figure 8.83 Reprocess Transactions – Automated Designation App

581

8 Cash Flow Hedging

When reprocessing failed transactions, we can run this report without filtering to view any transactions that have been assigned a hedging classification. From here, we can test transactions, reprocess them, or remove them from this report for reprocessing. The list of transactions available for reprocessing is shown in Figure 8.84.

CoCode	VA	Transaction	Hdg Area	PTyp	TTyp	Buy/Sell/C	Traded Amount	TradedCrcy	Partner	Contract Date	Term End	Entry Date	Hdg Class.
1000	001	10000192	CASH3	FX	SPT	B	4,533,332.88	CAD	JP MORGAN	04/15/2024	04/17/2024	04/15/2024	TMHC1
1000	001	10000232	CASH2	FX	FWD	B	726,666.67-	CAD	JP MORGAN	05/13/2024	12/31/2024	05/13/2024	ACHC2
1000	001	10000238	TM1	FX	FWD	S	160,000.00	CAD	JP MORGAN	05/17/2024	07/31/2024	05/17/2024	TMHC3
1000	001	10000265	TMHC4	FX	FWD	B	60,000.00-	CAD	JP MORGAN	06/05/2024	04/30/2025	06/05/2024	TMHC4

Figure 8.84 Report Displaying Each Transaction Available for Reprocessing with All Key Details

We can also highlight any transaction and click the **Reprocess (Test)** button to see why the transaction has failed. In Figure 8.85, we have an example in which this request is trying to designate a hedging instrument for an amount that is greater than the exposure. The test shows a log of why the hedging relationship hasn't been designated successfully, and we need to resolve this before we can designate the trade.

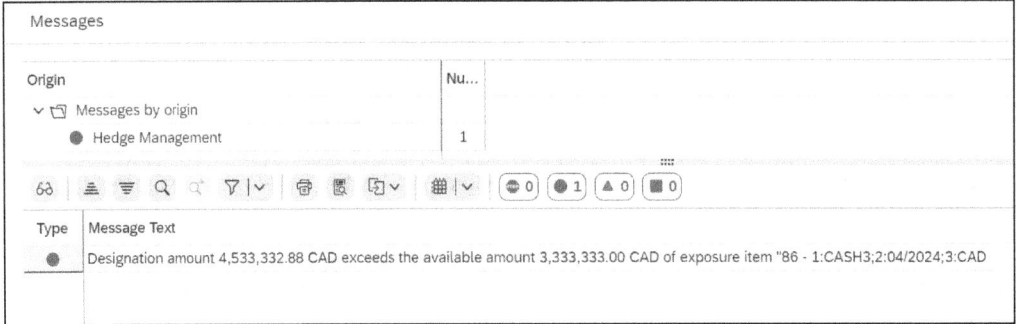

Figure 8.85 Testing Failed Transaction

8.4.5 Managing Hedging Relationships

Once we've created the financial transaction, the hedging relationship will be available. The hedging relationship ties the hedging instrument and the hedged item together, and we can view it in the Manage Hedging Relationships app or Transaction TPM100, as shown in Figure 8.86.

This app consolidates the information on the hedging relationship so we can view it all in one place. We can view the details in the corresponding tabs for the hedging relationship, hedged item, hedging instrument, documentation, effectiveness test, and user data. To view this data, we can search for the applicable hedging relationship by using the filters. Once we have identified the hedging relationship like in Figure 8.87, we can click into it to view the details.

8.4 Cash Flow Hedging Process

Figure 8.86 Manage Hedging Relationships App

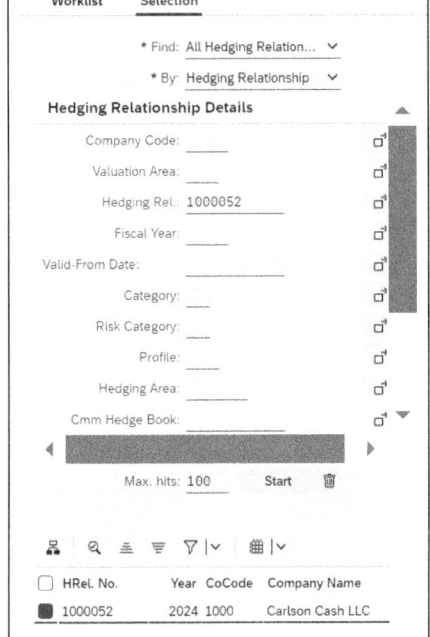

Figure 8.87 Filters Help in Search for Correct Hedging Relationship

Let's now look at the use of each of the tabs within this app, one by one.

Hedging Relationship Header

Before we go into the hedging relationship tabs, there's a top section of the hedging relationship report that gives a summary of the hedge. This area covers the overarching details of the hedge, and the status of the hedging relationship appears in the **Hedging Relationship Status** field. The list of available statuses is as follows:

- Created
- Revoked
- Designated
- Dedesignated
- Designation Planned
- Dedesignation Planned
- Designation Sent for Release
- Dedesignation Sent for Release
- All
- Rollover Sent for Release

583

There's also a field that determines the **Effectiveness Test Status** for the hedging relationship. This area can simply determine that we are using the shortcut method for the effectiveness test, or it can go into detail on whether there's a test planned and the effectiveness test is effective. The example in Figure 8.88 shows that this hedging relationship is both designated and effective, based on the executed effectiveness tests.

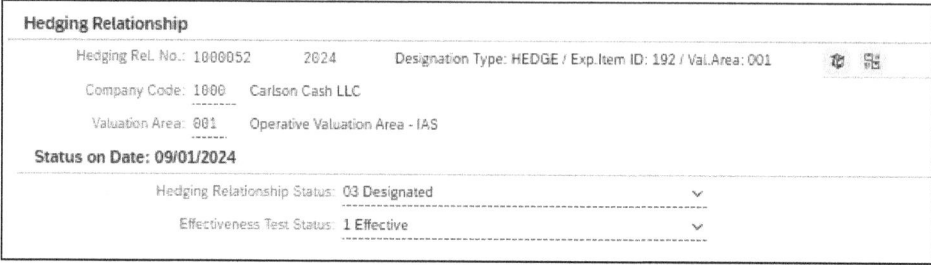

Figure 8.88 Initial Details of Hedging Relationship

Hedging Relationship Detail Tab

The **Hedging Relationship Detail** tab goes into deeper details of the overall hedging relationship and shows the **Risk Currency**, risk-free **Target Currency**, hedging **Profile**, hedge **Category**, and type of risk it's hedging. It also covers the key dates of the hedge to show when the hedge was designated and when the dedesignation needs to be executed, as seen in Figure 8.89.

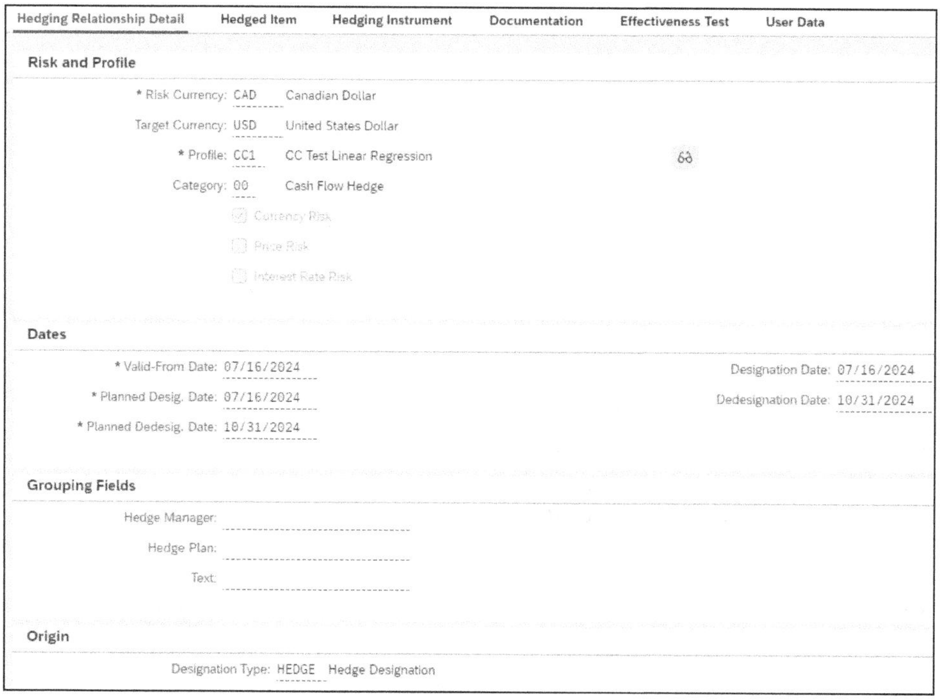

Figure 8.89 Key Details and Dates for Hedging Relationship

8.4 Cash Flow Hedging Process

If we needed to run a classification for the hedging relationship, the classification date would also appear in this section. The classification date appears in the **Dates** section when applicable.

Hedged Item Tab

The **Hedged Item** tab includes all the background information for the exposure, as seen in Figure 8.90. These details show how the hedged item ties to the actual exposure item that was in the hedge management cockpit. The **Snapshot ID** and **Hedging Area** help determine the source of this hedged item.

The **Exposure Subitem** at the bottom of the screen also ties in with the split ID in case the hedge is set up for designation splitting.

Figure 8.90 Hedged Item Tab Showing Details of Hedged Item in This Relationship

Hedging Instrument Tab

The hedging instrument for the hedge is equivalent to the financial transaction or FX forward that was created. The **Hedging Instrument** tab shows details of the hedging instrument side, and it details the transaction that was created, including the **Transaction Number** and **Start Date/End Date** details, as in Figure 8.91.

Additionally, there are buttons that are available in the **Position Details** area that navigate to the transaction, position values report, position flows report and an overview of the position management procedure.

8 Cash Flow Hedging

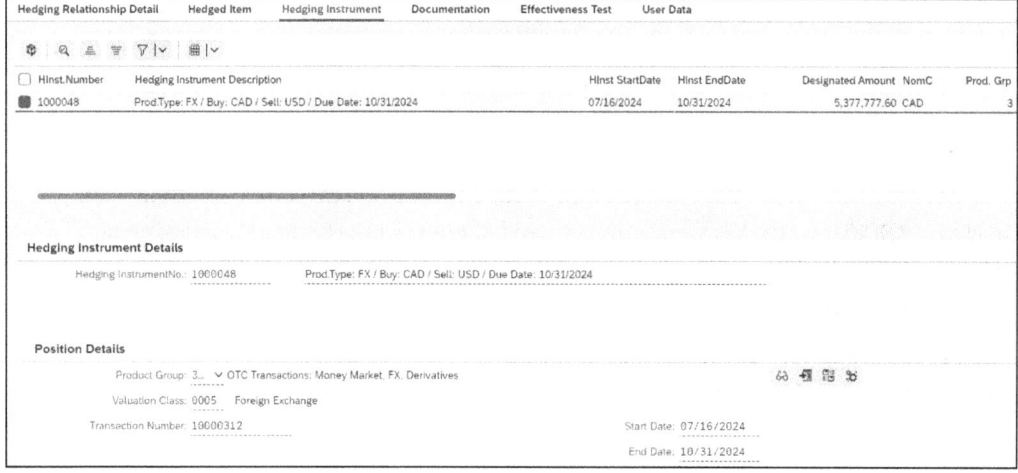

Figure 8.91 Hedging Instrument Tab Detailing Financial Instrument in Hedge

Documentation Tab

The **Documentation** tab houses the hedge documentation. This is information that can be created in a PDF form and is available in this tab. We assign the form according to the customizing detailed in the hedging profile detailed in Section 8.2.7.

When we assign the PDF forms for the hedging relationship to create documentation, we can view the information in the **Documentation** tab. At that time, it will give us information on the documentation and will show the details of where the documentation is stored. The **Documentation** tab is shown in Figure 8.92.

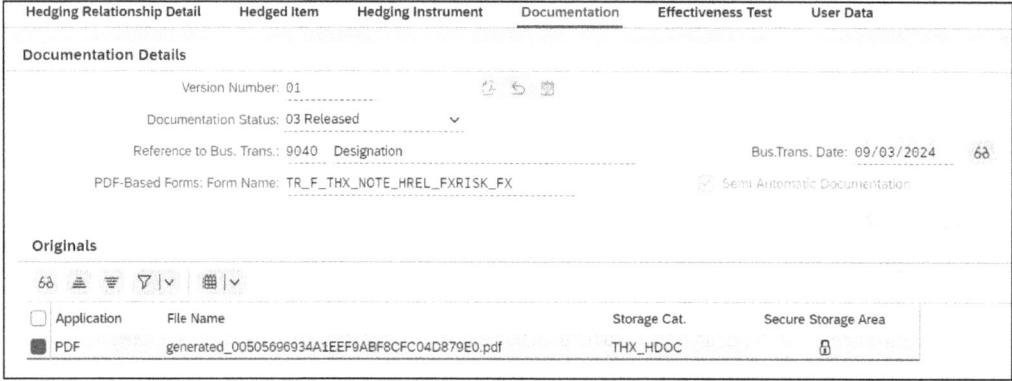

Figure 8.92 Documentation Tab Showing Header-Level Information on Hedge Documentation

To view the created PDF, we select the line item and the glasses icon will open up the PDF of the hedge documentation. The standard form creates information on the hedging documentation around the hedging relationship, hedging instrument, hedged item,

8.4 Cash Flow Hedging Process

and effectiveness assessment. The initial page of this hedge documentation is shown in Figure 8.93.

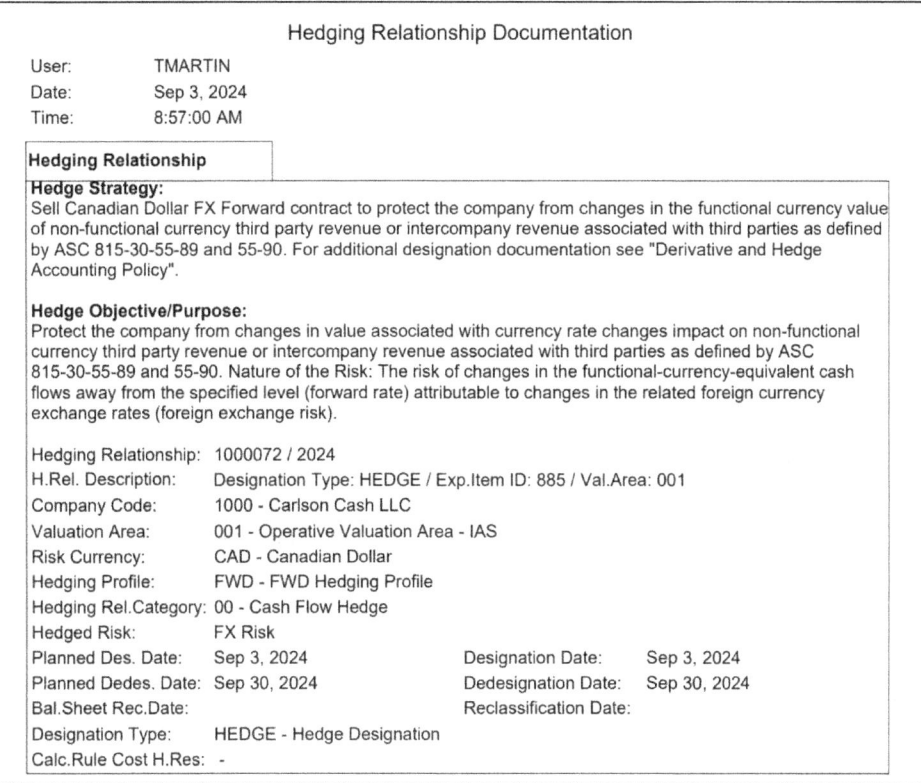

Figure 8.93 Hedging Relationship Documentation Stores All Key Details of Hedge

Effectiveness Test Tab

The effectiveness test is run periodically to compare the hedging instrument and hedged item to determine the effectiveness of the hedge. The type of effectiveness test we use for a hedging relationship is determined by two factors. First, we determine the type of effectiveness test for each hedging profile. (We covered that assignment in Section 8.2.7.) Then, we assign the hedging profile to the hedging area in the **Hedge Accounting II** tab.

An additional setting in the hedging profile is the effectiveness test plan rhythm, which determines how frequently the effectiveness test needs to be run, whether it's monthly, quarterly, yearly, or manually. Figure 8.94 shows an example of a hedge whose test plan rhythm (**Test Plan Date Logic**) is **01 Monthly**, and we can see that we need to run the effectiveness test at the end of each month.

8   Cash Flow Hedging

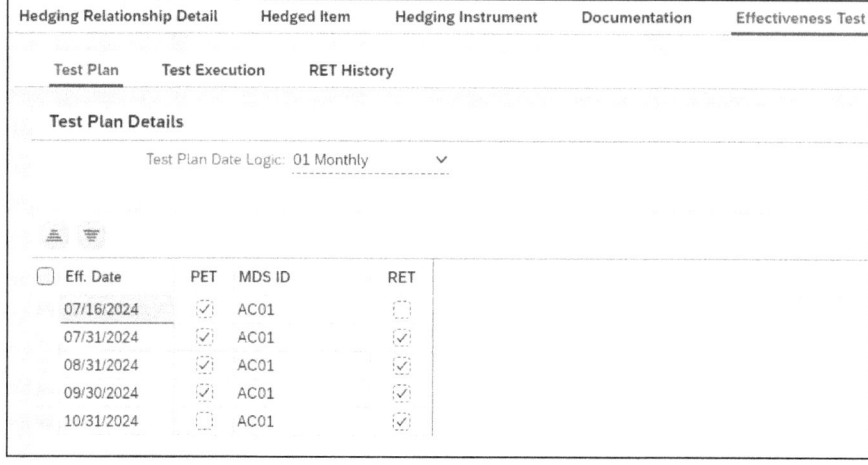

Figure 8.94 Test Plan Tab Showing When Effectiveness Tests Are Planned

There are additional functions in the effectiveness test report that we can cover in the **Test Execution** tab. This report shows all past and future effectiveness tests. If we want to view the details of any effectiveness test, we can click the applicable box in the prospective effectiveness test (**PET**) or retrospective effectiveness test (**RET**) columns. This will bring up a popup like in Figure 8.95 that displays the key details of the effectiveness test that was run.

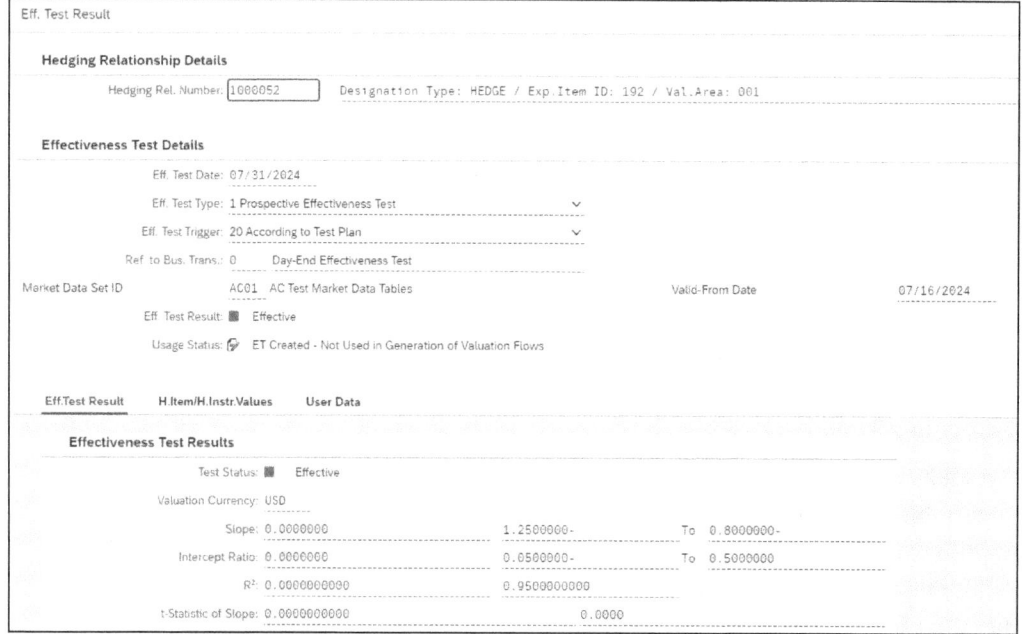

Figure 8.95 Details of Effectiveness Test

Additionally, we can run the effectiveness test directly in Transaction TPM100. To do this, we first need to make sure the transaction is in change mode by clicking on the **Display <-> Change** button. Once it's in the change mode, we can select the applicable test date by highlighting the box on the left side of the screen (as in Figure 8.96), and we can then click the **Run Effectiveness Test** button. When the effectiveness test run is completed, we can save it, and it will update the details in this **Test Execution** tab.

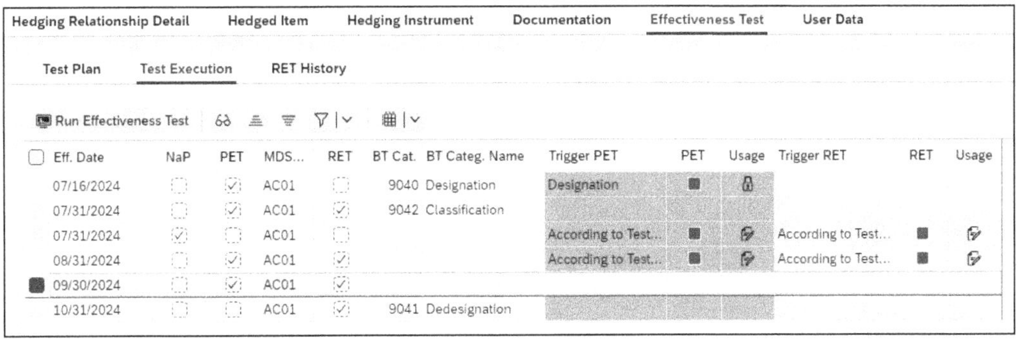

Hedging Relationship Detail		Hedged Item		Hedging Instrument		Documentation		Effectiveness Test		User Data	
Test Plan		Test Execution		RET History							
Run Effectiveness Test											
Eff. Date	NaP	PET	MDS...	RET	BT Cat.	BT Categ. Name	Trigger PET	PET	Usage	Trigger RET	RET	Usage
07/16/2024		✓	AC01		9040	Designation	Designation	■	🔒			
07/31/2024		✓	AC01	✓	9042	Classification						
07/31/2024	✓		AC01				According to Test...	■	🔏	According to Test...	■	🔏
08/31/2024		✓	AC01	✓			According to Test...	■	🔏	According to Test...	■	🔏
09/30/2024	■	✓	AC01	✓								
10/31/2024			AC01	✓	9041	Dedesignation						

Figure 8.96 Running Effectiveness Tests

The Release Hedging Business Transactions app (shown in Figure 8.97) or Transaction TPM120 can process the designation, dedesignation, and reclassification of hedging transactions.

Figure 8.97 Release Hedging Business Transactions App

To run this transaction, we need to determine the run date (**Up to Key Date**) in the **General Selection** section. Then, we can run it by selecting the radio button for either **Transaction Selection** or **Hedging Relationship Selection**. Figure 8.98 shows the key details that we can search when selecting either transactions or hedging relationships.

The second section of this screen (see Figure 8.99) determines what type of release is being run. The options are to process the designation, dedesignation, or reclassification. The standard posting control selections are also available for the transactions that we can post as part of these processes.

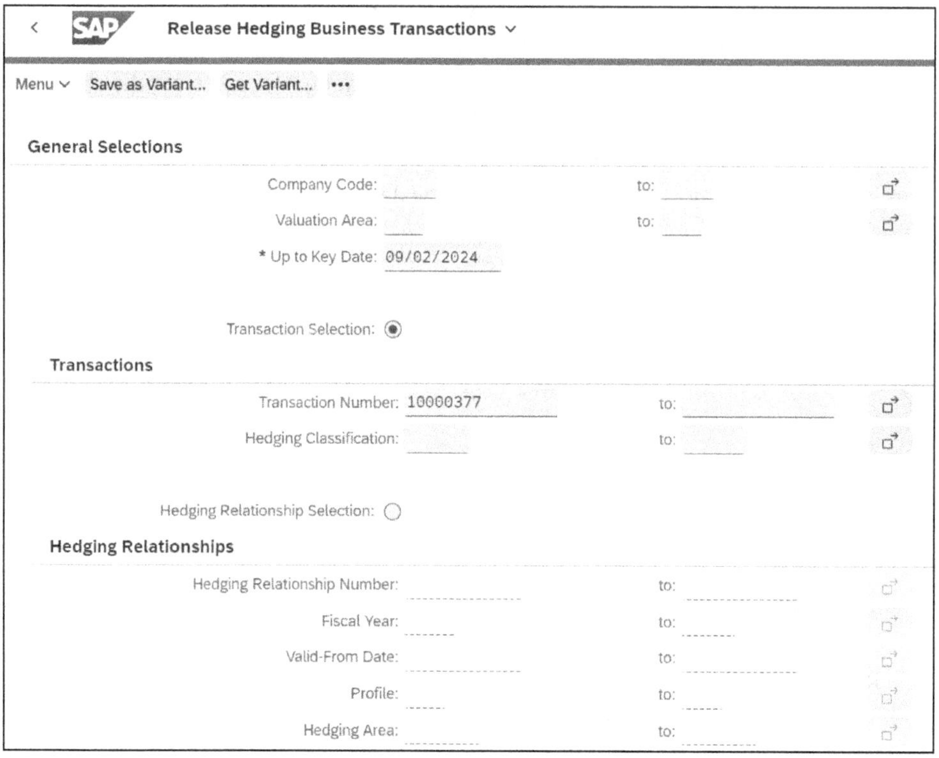

8 Cash Flow Hedging

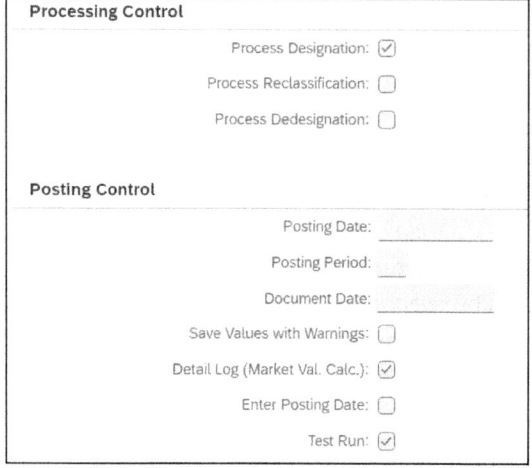

Figure 8.98 Selecting Transactions and Hedging Relationships in Release Hedging Business Transactions App

Figure 8.99 Processing Control Section Determines Type of Transactions to Process

Once we run this transaction, we can see whether the designation, dedesignation, or reclassification runs were successful (see Figure 8.100). We can also view the standard posting log and message logs by clicking the **Display Log** button.

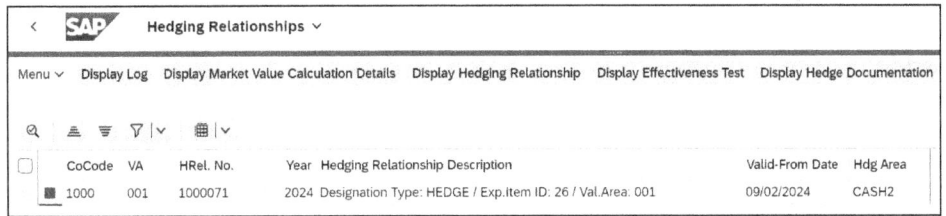

Figure 8.100 How Successful Run of This Transaction Appears when Hedging Relationship Is Successfully Designated

8.5 Period End Closing

The previous section covered the end-to-end process for creating hedges and ensuring that the subsequent hedging relationships are created. There are now periodic processing steps that we need to consider during month-end processing. Figure 8.101 shows the periodic processing for the cash flow hedges. We can set this to be at different frequencies, so it could be month-end or quarter-end processing, depending on when we want to calculate and post these transactions. The key apps or transactions to consider during period-end processing are as follows:

1. Calculate Net Present Values app or Transaction TPM60CVA
2. Run Valuation app or Transaction TPM1
3. Execute Effectiveness Test app or Transaction TPM110 or TPM100
4. Execute Classification app or Transaction TPM101

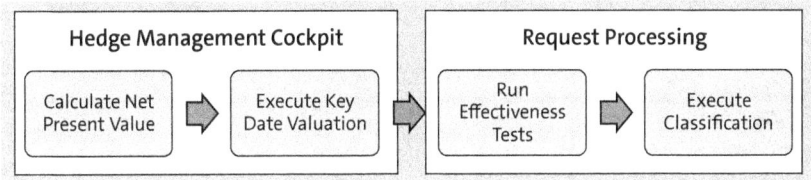

Figure 8.101 Month-End Process Flow for Hedge Management: Cash Flow Hedging

We covered running the effectiveness test in Section 8.4.5, so we'll cover the other three steps in the following sections.

8.5.1 Calculating Net Present Values

The first step of the month-end process is calculating the net present value (NPV) of the hedge. This transaction follows the same operational parameters that we discussed in Chapter 6, Section 6.5.4. Once we've generated the NPV using the Calculate Net Present Values app or Transaction TPM60CVA, we can move to the next step. The difference for the hedge management process is that there will be a calculation of the NPV of both the hedging instrument and the hedged item. Clicking the **Hedge Accounting Key Figures** button will show us the NPV details for both, as in Figure 8.102.

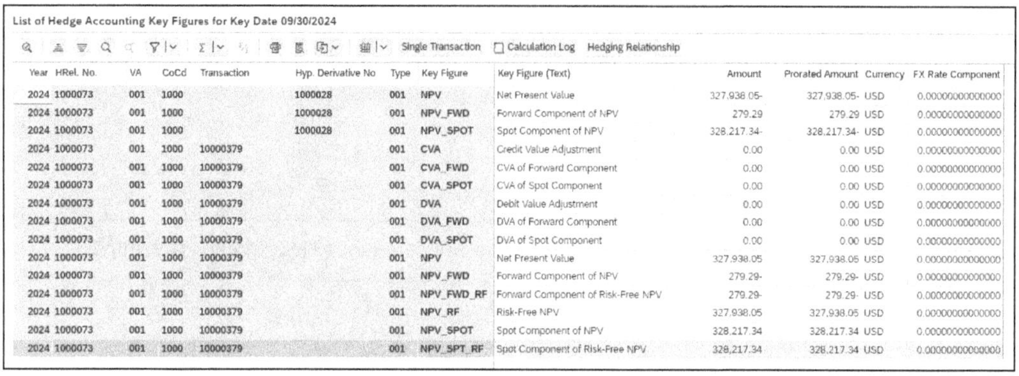

List of Hedge Accounting Key Figures for Key Date 09/30/2024

Year	HRel. No.	VA	CoCd	Transaction	Hyp. Derivative No	Type	Key Figure	Key Figure (Text)	Amount	Prorated Amount	Currency	FX Rate Component
2024	1000073	001	1000		1000028	001	NPV	Net Present Value	327,938.05-	327,938.05-	USD	0.00000000000000
2024	1000073	001	1000		1000028	001	NPV_FWD	Forward Component of NPV	279.29	279.29	USD	0.00000000000000
2024	1000073	001	1000		1000028	001	NPV_SPOT	Spot Component of NPV	328,217.34-	328,217.34-	USD	0.00000000000000
2024	1000073	001	1000	10000379		001	CVA	Credit Value Adjustment	0.00	0.00	USD	0.00000000000000
2024	1000073	001	1000	10000379		001	CVA_FWD	CVA of Forward Component	0.00	0.00	USD	0.00000000000000
2024	1000073	001	1000	10000379		001	CVA_SPOT	CVA of Spot Component	0.00	0.00	USD	0.00000000000000
2024	1000073	001	1000	10000379		001	DVA	Debit Value Adjustment	0.00	0.00	USD	0.00000000000000
2024	1000073	001	1000	10000379		001	DVA_FWD	DVA of Forward Component	0.00	0.00	USD	0.00000000000000
2024	1000073	001	1000	10000379		001	DVA_SPOT	DVA of Spot Component	0.00	0.00	USD	0.00000000000000
2024	1000073	001	1000	10000379		001	NPV	Net Present Value	327,938.05	327,938.05	USD	0.00000000000000
2024	1000073	001	1000	10000379		001	NPV_FWD	Forward Component of NPV	279.29-	279.29-	USD	0.00000000000000
2024	1000073	001	1000	10000379		001	NPV_FWD_RF	Forward Component of Risk-Free NPV	279.29-	279.29-	USD	0.00000000000000
2024	1000073	001	1000	10000379		001	NPV_RF	Risk-Free NPV	327,938.05	327,938.05	USD	0.00000000000000
2024	1000073	001	1000	10000379		001	NPV_SPOT	Spot Component of NPV	328,217.34	328,217.34	USD	0.00000000000000
2024	1000073	001	1000	10000379		001	NPV_SPT_RF	Spot Component of Risk-Free NPV	328,217.34	328,217.34	USD	0.00000000000000

Figure 8.102 Calculating NPV for Both Transaction and Hypothetical Derivative

8.5.2 Running the Valuation

Once we've calculated the net present value, we can post the changes in value by using the Run Valuation app or Transaction TPM1. We covered the entry screen of this program in Chapter 6, Section 6.5.4. This valuation run results in a writeup (displayed in Figure 8.103), and it will be posted accordingly.

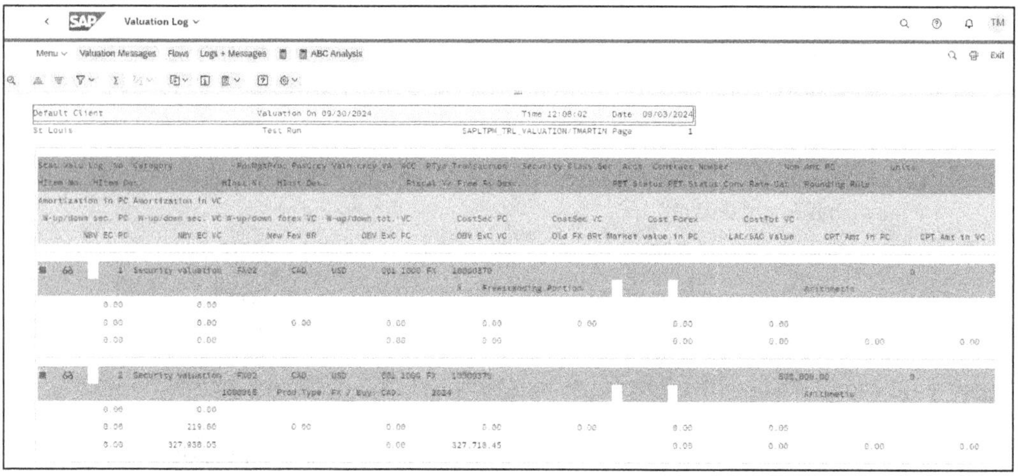

Figure 8.103 Valuation Log Posting Unrealized Gain or Loss

8.5.3 Executing Classification

The month-end step that we've not covered in another section is executed in the Execute Classification app or Transaction TPM101. Since we can apply different accounting treatments to hedges than those we apply to regular FX trades, there are additional postings that we need to generate. In the Run Classification app, we split the valuation result into different accounts, and we complete these postings in accordance with the hedge relationship scenario assigned in Section 8.2.7. Based on the defined rules and effectiveness tests, SAP will determine which amounts of the hedge were effective or ineffective and which entries should be classified into the hedging reserve or the cost of hedging reserve. Based on this, accounting will be generated for the following scenarios:

- Effective – Hedging Reserve
- Effective – Cost of Hedging Reserve
- Ineffective – Hedging Reserve
- Ineffective – Cost of Hedging Reserve
- Ineffective

The app that we use to run the classification is shown in Figure 8.104.

Figure 8.104 Run Classification App

To execute the classification run, we select either the **Transaction Number** or the **Hedging Relationship Number** on the entry screen. We also need to define a key date to ensure that the postings occur on the correct date, and we also need to input contract information. After we've verified these details as in Figure 8.105, we can execute the classification.

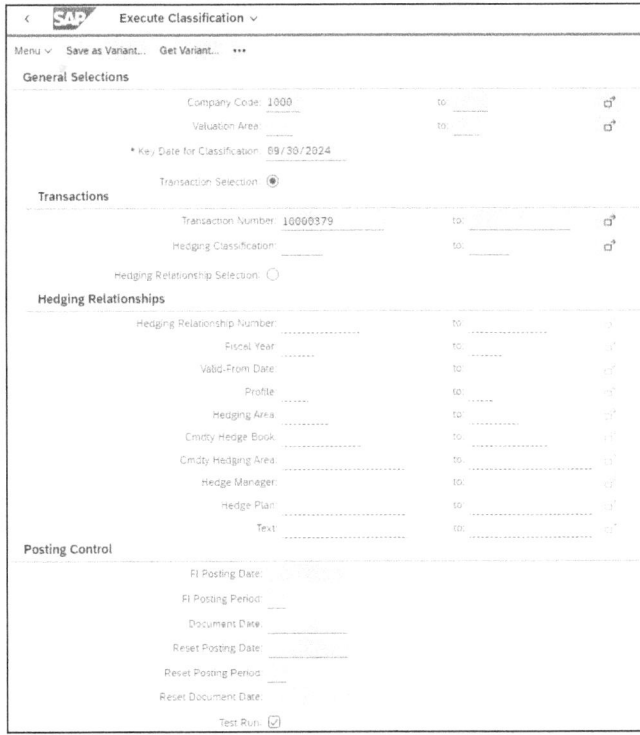

Figure 8.105 Populating Entry Screen with Key Date and Contract Information

8 Cash Flow Hedging

Once we've run this transaction, the hedging relationship will be distributed between the effective and ineffective accounts with the accounting posting. In the example in Figure 8.106, the hedging relationship was determined to be effective, so the full amount of the valuation will be posted to the effective general ledger account.

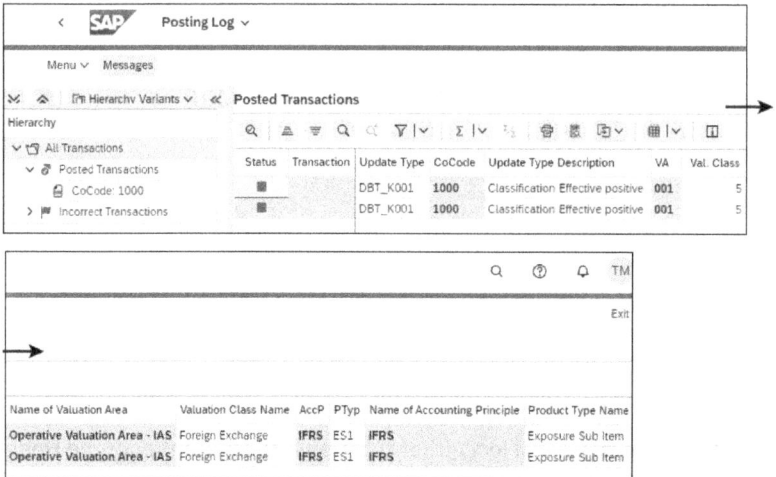

Figure 8.106 Full Amount of Classification Is Effective in This Posting

We need to run each of these steps at the end of each accounting period, and depending on the reporting structure, we may run these steps at the end of every month or quarter. We will run these steps for each hedging relationship until the maturity or dedesignation of the hedge. The following section will cover the transactions and postings that should occur at the maturity of the hedging relationship.

8.6 Contract Close

The month- or period-end process will happen periodically throughout the hedging relationship, but eventually, the relationship will reach its maturity. When the hedging instrument matures, there are a series of steps that need to occur. First, we need to dedesignate the hedging relationship, and then, we can process and post the contract flows through Transaction TBB1. As in other areas of treasury and risk management, this step will allow for the creation of payment requests that we can initiate in the treasury payment run. After we've exchanged the currencies, we can determine and post the gain or loss in Transaction TPM18. If designation splitting is activated, we should run the hedging relationships through the reclassification to post the OCI reclassification entries on the determined dates. We execute these steps by using the following apps or transactions:

- Release Hedging Business Transactions app or Transaction TPM120 (for both dedesignation and reclassification)

8.6 Contract Close

- Post Flows app or Transaction TBB1
- Post Derived Business Transactions app or Transaction TPM18

8.6.1 Dedesignating Hedging Relationships

We can dedesignate a hedging relationship for different reasons: when the relationship is deemed to not be effective, the hedging instrument has been sold, the hedging relationship is closed, or the underlying transaction is sold. Management generally decides to close the hedging relationship if the exposures change and the hedging strategy is in an overhedged position. If this occurs, management will request dedesignation of the hedging relationship by sending a dedesignation request from the hedge management cockpit. If one of these scenarios occurs or if the contract reaches maturity, then we need to use the Release Hedging Business Transactions app or execute Transaction TPM120 for this purpose. The selection screen and the execution procedure for this transaction are in Section 8.4.5, and the only difference is that we should check the dedesignation box when processing the hedge relation dedesignation.

8.6.2 Posting Cash Flows

As with other areas of treasury and risk management, we need to post the cash flows using the Post Flows app or Transaction TBB1. In an FX forward transaction, we need to settle the nominal amounts of each leg of the transaction, and this will result in two accounting line items for the buy side of the transaction and two for the sell side of the transaction. We covered the selection screen and processing of Post Flows in Chapter 5, Section 5.3.2, and an example of the settled entries is in Figure 8.107.

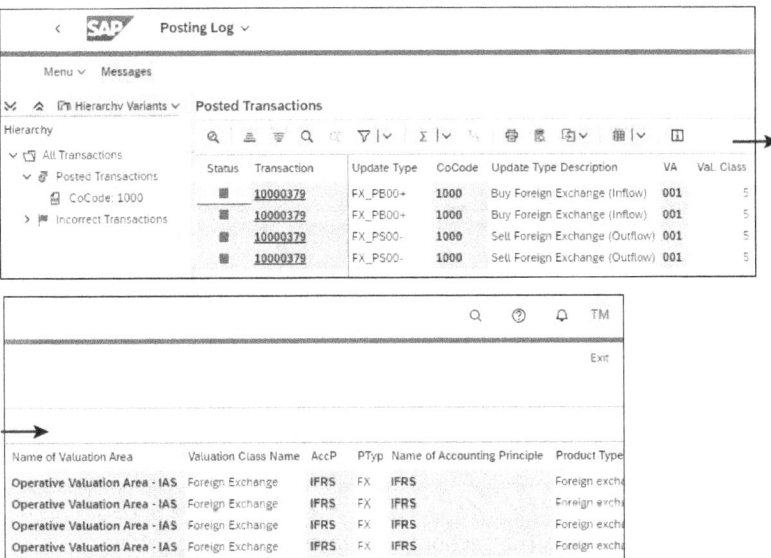

Figure 8.107 Post Flows Posting Both Legs of FX Forward

8 Cash Flow Hedging

8.6.3 Posting Derived Business Transactions

We make the postings we generated following the settlement of the hedge by using the Post Derived Business Transactions app or Transaction TPM18. We execute this transaction after posting the cash flows. We detailed the execution of this transaction in Chapter 6, Section 6.5.5. The transactions shown in Figure 8.108 for hedge accounting that are posted from this transaction are as follows:

- We post each side of the transaction using Transaction TBB1. There's a valuation currency amount that is also reflected in that posting, and this amount is generally the same as the local currency amount. This offset amount of both cash flows is directed into a clearing account that results in an imbalance due to the change in FX rates. The difference between the agreed-upon rate in the FX transaction and the system rate will be in the clearing account after we run Transaction TBB1. Transaction TPM18 clears out this balance from the clearing account and posts the gain or loss from the transaction.

- We also post the OCI valuation by using this transaction. If we determine the hedging relationship to be relevant to document splitting based on the hedging area settings, then we post the OCI amount. We'll reclassify this entry from OCI by using the process defined in Section 8.6.4.

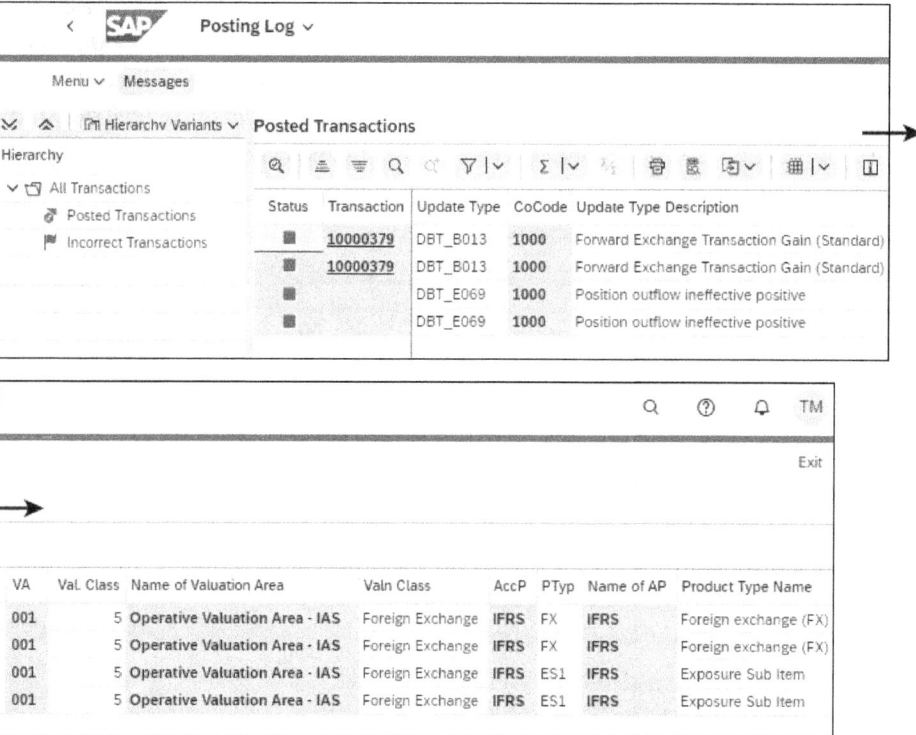

Figure 8.108 Posting Derived Business Transactions Generates Accounting Entries for Maturity of Hedge

8.6.4 Executing Reclassification

When a transaction uses the document splitting function, we post a valuation amount to other comprehensive income at the maturity of the hedging relationship. We can then reclassify this amount from other comprehensive income to a gain or loss account, based on the schedule defined in the **Designation Splitting** section of the hedge management cockpit. The Manage Hedging Relationships app or Transaction TPM100 will show a date when a hedging relationship can be reclassified, and on that date, the Release Hedging Business Transactions app or Transaction TPM120 will be run to initiate the postings.

8.7 Summary

Cash flow hedging in treasury and risk management offers a complete solution for assessing and managing fluctuating risks. The system generates exposures that reflect future cash flow variations, allowing companies to implement various hedging strategies aligned with their overall hedging objectives. This solution enables monitoring and releasing of these exposures, automatically generates hedge requests, and facilitates the creation of accounting entries in accordance with local accounting standards. In the next chapter, we'll examine balance sheet hedging in treasury and risk management and highlight the differences between these two hedging solutions.

Chapter 9
Balance Sheet Hedging

This chapter focuses on balance sheet hedging. It'll cover how to view the balance sheet exposures and automate hedge requests for these exposures. It'll cover the process and the configuration setup and how they differ from cash flow hedging capabilities.

Balance sheet hedging is the way in which we review exposures that come from operational activity. With SAP S/4HANA, we have the constant integration and ability to review the open items or balances from the universal journal. Having real-time access to this information provides clients the ability to understand, analyze, and take action on any balance sheet hedging needs. To start, let's look at the end-to-end process for balance sheet hedging (see Figure 9.1):

1. Analyze balance sheet exposure.
2. Manage balance sheet exposure snapshot.
3. Manage balance sheet exposure hedge request
4. Manage trade request.

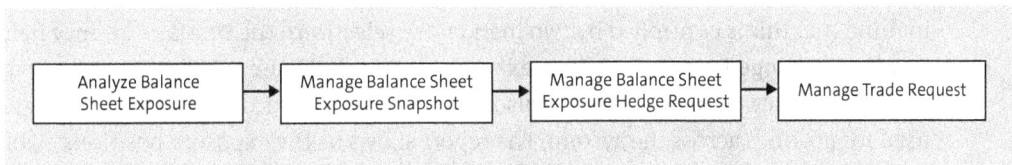

Figure 9.1 Review of Balance Sheet Hedging Process

SAP S/4HANA Finance for treasury and risk management has a fully embedded tool for managing balance sheet risks. Throughout this chapter, we'll explore how to review exposures across the organization, take snapshots of exposure data, review exposure positions, generate hedge requests, send and receive trade requests and trades to and from a trading platform, and discuss the important master data that supports the balance sheet hedging process.

9 Balance Sheet Hedging

9.1 Reviewing Balance Sheet Risk

We'll first analyze our balance sheet exposure, which is composed of general ledger data that's available across SAP S/4HANA. Once we understand those exposures, we'll *take a snapshot*, meaning we'll freeze the data at a point in time so we know what our exposures were at that given time. When we take the snapshot, the system will create hedge requests to carry our exposure data to the trading platform of our choice. In Figure 9.2, we see a first view of our balance sheet hedging report.

Company Code 1710 - BestRun US (USD) Key Date: 03/19/2024					64,694.03 USD
Exposures/Hedges	AUD	CAD	CHF	EUR	GBP
⌄ Exposures	-5,000.00 AUD	-347,021.72 CAD	489,000.00 CHF	-42,130,958.31 EUR	-3,000.00 GBP
› All Accounts Payables	-7,000.00 AUD	-188,310.86 CAD	178,000.00 CHF	-27,755,157.49 EUR	-8,000.00 GBP
› All Accounts Receivables	4,000.00 AUD	2,000.00 CAD	6,000.00 CHF	1,922,728.72 EUR	6,000.00 GBP
› Cash	5,000.00 AUD	-2,400.00 CAD	200,000.00 CHF	5,000.00 EUR	5,000.00 GBP
› Prepaid Expenses	-2,000.00 AUD	-25,000.00 CAD	100,000.00 CHF	-2,000.00 EUR	-2,000.00 GBP
› Taxes Receivables	-4,000.00 AUD	-130,000.00 CAD	5,000.00 CHF	-4,000.00 EUR	-4,000.00 GBP
› In House Bank Balances	-1,000.00 AUD	-3,310.86 CAD		-16,297,529.54 EUR	
⌄ Hedges	5,000.00 AUD	376,158.00 CAD	-450,678.60 CHF	42,130,958.31 EUR	3,000.00 GBP
⌄ Hedges for Current Licence	5,000.00 AUD	376,158.00 CAD	-450,678.60 CHF	42,130,958.31 EUR	3,000.00 GBP
For Hedge	5,000.00 AUD	376,158.00 CAD	-450,678.60 CHF	42,130,958.31 EUR	3,000.00 GBP
Net Exposure (Transaction Currency)	0.00	29,136.28 CAD	38,321.40 CHF	0.00	0.00
Net Exposure (Display Currency)	0.00	21,528.81 USD	43,165.22 USD	0.00	0.00 USD
Hedge Quota	100%	108.4%	92.2%	100%	100%

Figure 9.2 Review of Balance Sheet Hedging Report

Our first step is to analyze balance sheet exposures. To do this, we'll access the Review Balance Sheet FX Risk app, in which we'll review the exposures on a net basis, keeping in mind that this is controlled by two mandatory selections: the **Display Currency** field and the **Exchange Rate Type**. In this example, shown in Figure 9.3, we've selected our display currency as **USD,** and we've also selected the standard SAP **M** rate type that's used for postings across the system. The report shows us the exposure positions available by company code.

Company Code	Company Name	Country/Region Key	Company Code Crcy	Country/Region Name	Absolute Net Exp.		Absolute Hedges		Absolute Exposures	
3010	BestRun AU	AU	AUD	Australia	969,771.83	USD	2,908,229.28	USD	3,878,001.11	USD
1010	BestRun DE	DE	EUR	Germany	112,640.00	USD	0.00	USD	112,640.00	USD
1710	BestRun US	US	USD	USA	64,694.03	USD	46,597,858.50	USD	46,619,494.91	USD

Figure 9.3 Review of Balance Sheet FX Risk by Company Code

9.1 Reviewing Balance Sheet Risk

As we're analyzing the net exposures, we can drill down into the company codes to see what currency we have exposures in. We'll review the exposures for our US company code, 1710, and as we drill down into 1710, we can see that it has balance sheet exposures in the AUD, CAD, CHF, EUR, and GBP currencies. In Figure 9.4, we can see that the report is separated into two key sections:

- Exposures
- Hedges

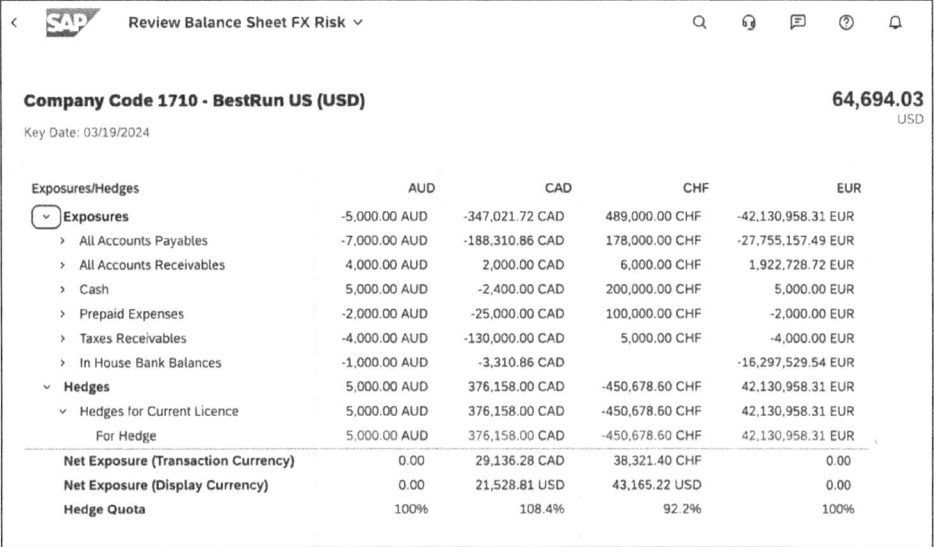

Figure 9.4 Review of Balance Sheet FX Risk: Detailed View

The system reads the exposures from the universal journal, and we can categorized these into whatever types of groupings we'd like to see (e.g., accounts payable, accounts receivable, cash, prepaid expenses). *Hedges* are the trades that we've entered and tagged as appropriate for balance sheet risks. These exposures and hedges are mapped in the Define Key Figures – Balance Sheet Risk app, which we'll discuss in Section 9.6.

As we look at our exposures in Figure 9.5, we see that we've unhedged exposures in AUD, CAD and CHF. The **Hedge Quota** indicates how much of the exposures have been hedged.

> **Note**
>
> SAP assumes that we'll hedge 100% of the open balance sheet risk and propose trade amounts based on that logic. However, we can always adjust the hedge request once it has been generated.

601

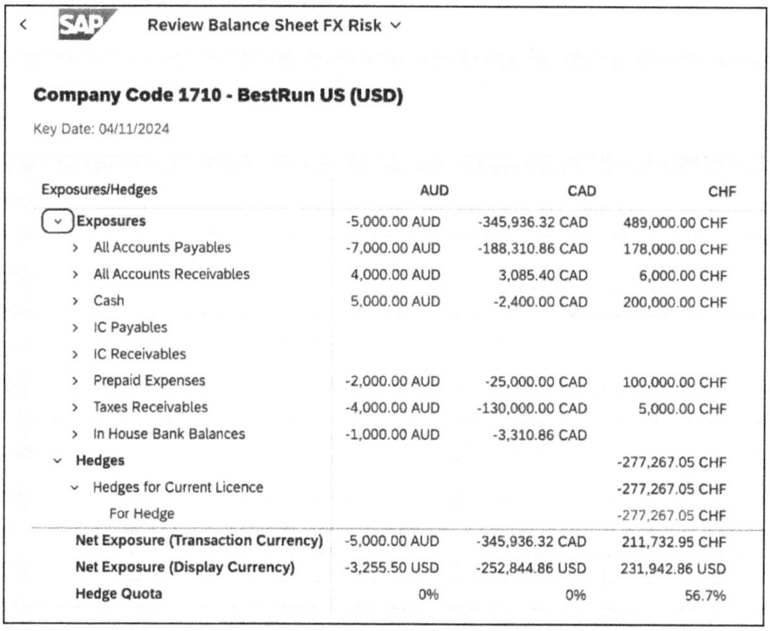

9 Balance Sheet Hedging

Exposures/Hedges	AUD	CAD	CHF
Exposures	-5,000.00 AUD	-345,936.32 CAD	489,000.00 CHF
> All Accounts Payables	-7,000.00 AUD	-188,310.86 CAD	178,000.00 CHF
> All Accounts Receivables	4,000.00 AUD	3,085.40 CAD	6,000.00 CHF
> Cash	5,000.00 AUD	-2,400.00 CAD	200,000.00 CHF
> IC Payables			
> IC Receivables			
> Prepaid Expenses	-2,000.00 AUD	-25,000.00 CAD	100,000.00 CHF
> Taxes Receivables	-4,000.00 AUD	-130,000.00 CAD	5,000.00 CHF
> In House Bank Balances	-1,000.00 AUD	-3,310.86 CAD	
Hedges			-277,267.05 CHF
⌄ Hedges for Current Licence			-277,267.05 CHF
For Hedge			-277,267.05 CHF
Net Exposure (Transaction Currency)	-5,000.00 AUD	-345,936.32 CAD	211,732.95 CHF
Net Exposure (Display Currency)	-3,255.50 USD	-252,844.86 USD	231,942.86 USD
Hedge Quota	0%	0%	56.7%

Company Code 1710 - BestRun US (USD)
Key Date: 04/11/2024

Figure 9.5 Review of Balance Sheet for FX Risk: Hedge Quotas

9.2 Taking a Foreign Exchange Snapshot for Balance Sheet Risk

Now that we've identified the exposures, we'll take a snapshot to lock in the exposure data at this point in time. This is especially important because this report is a direct read from the transactions occurring in the universal journal and deals created across treasury and risk management. You can take a snapshot in one of two ways:

- By using the Take Snapshot for Balance Sheet FX Risk app
- By using the Schedule Treasury Middle Office Jobs app

9.2.1 Taking a Snapshot for Balance Sheet Foreign Exchange Risk

We'll first look at the Take Snapshot for Balance Sheet FX Risk app. This app has three sections to it: **General**, **Dimensions** and **Filters**. First, we'll define the **General** settings of the snapshot, starting with a unique **Snapshot Description**, as shown in Figure 9.6.

> **Note**
> We make the name of each snapshot unique for easy retrieval and historical reference.

9.2 Taking a Foreign Exchange Snapshot for Balance Sheet Risk

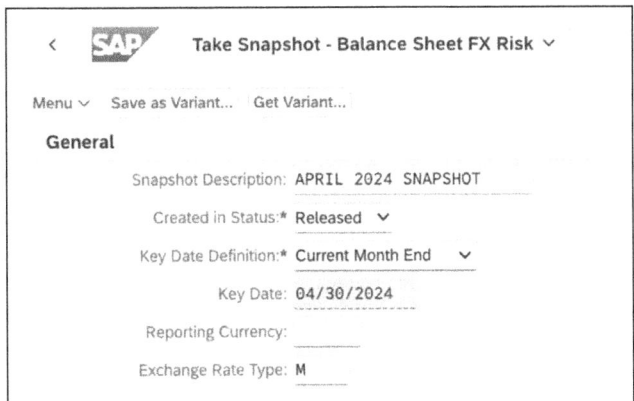

Figure 9.6 Take Snapshot – Balance Sheet FX Risk App: General Section

We'll then determine the **Created in Status**, using one of the following options:

- **Created**
  If we want to have a second user review the snapshot before we process it, we select **Created** so we can release the snapshot in the Process Snapshot – Balance Sheet Risk app.
- **Released**
  If no release is required on the snapshot, we can select **Released**, which will send the snapshot to be available in the Process Hedge Requests – Balance Sheet FX Risk app.

Next, we'll define the **Key Date Definition**, which will be the date that we use for the hedge request date. We have the following options:

- **System Date**
  This is the current date.
- **Current Month End**
  This is the last working day of the month.
- **Current Quarter**
  This is the last working day of the quarter.
- **Fixed Date**
  This is a user-defined date.

Typically, balance sheet hedging is done on a monthly cycle, but we could do it on a more frequent basis, depending on our business process (e.g., daily, weekly, as needed). For this example, we'll select **Current Month End**. Additionally, we can define a **Reporting Currency**, and we've selected the **M** rate as our **Exchange Rate Type**.

As we scroll down through the app screen, we'll find the **Dimensions** section, where we can set additional filters for our snapshot. These optional additional filters include items for **One Exposure**, **Financial Accounting**, **Transaction Manager**, and **Time Pattern**, as shown in Figure 9.7.

603

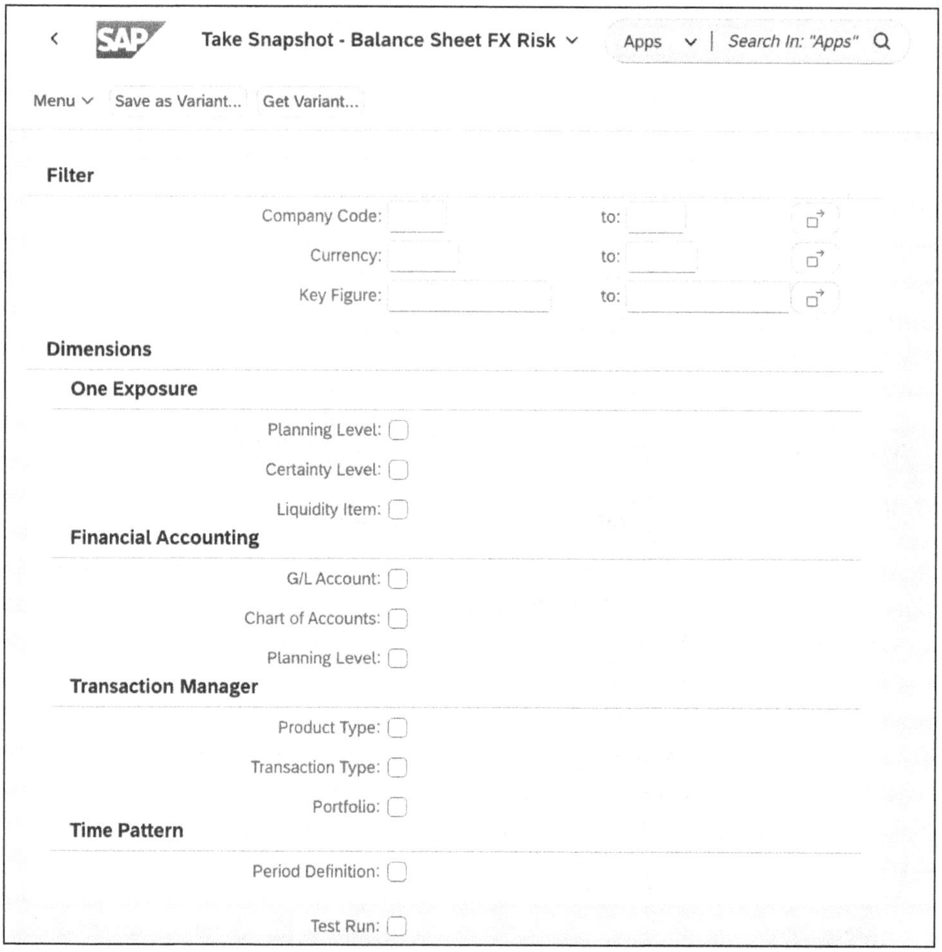

Figure 9.7 Take Snapshot – Balance Sheet FX Risk App

> **Note**
>
> We need to make sure that the **Test Run** box is not checked when we go to generate a snapshot. We can also confirm that the snapshot has successfully processed when we have a snapshot ID displayed on the results page, as shown in Figure 9.8.

Once we've entered all necessary information, we click **Execute**. Then, we'll see a listing of all the exposures in the criteria we've selected (see Figure 9.8). The report is currently summarized by each currency, and we can see the total currency exposures listed at the bottom. Notice that the numerical snapshot ID has also been created at the top of the report.

9.2 Taking a Foreign Exchange Snapshot for Balance Sheet Risk

CoCode	Exp./Hedge	Key Figure Group	Key Figure Identifier	Amount	Currency	Target Amount	Rep. Crcy
				162,400.00-	NZD		
1110	Exposure	Cash	CASH	2,400.00-	SGD	1,772.88-	USD
1110	Exposure	All Accounts Payables	PAYROLL	5,000.00-		3,693.50-	USD
3010	Exposure	In House Bank Balances	IN HOUSE BANK	20,000.00-		14,774.00-	USD
1110	Exposure	Prepaid Expenses	PREPAID EXPENSES	25,000.00-		18,467.50-	USD
1110	Exposure	Taxes Receivables	TAXES REC	130,000.00-		96,031.00-	USD
				182,400.00-	SGD		
1010	Exposure	In House Bank Balances	IN HOUSE BANK	364,383,067.67	USD	364,383,067.67	USD
1010	Exposure	All Accounts Payables	EG4K2	2,670,426.56		2,670,426.56	USD
3010	Hedge	Hedges for Current Licence	HG1K1	1,115,736.96		1,115,736.96	USD
1010	Exposure	All Accounts Receivables	EG1K2	102,248.66		102,248.66	USD
1010	Exposure	All Accounts Receivables	EG2K1	102,248.66		102,248.66	USD
DE20	Exposure	All Accounts Payables	EG1K1	100.00-		100.00-	USD
1010	Exposure	All Accounts Payables	EG1K1	44,000.00-		44,000.00-	USD
DEC1	Exposure	All Accounts Payables	EG1K1	669,313.14-		669,313.14-	USD
AUC1	Exposure	All Accounts Payables	EG1K1	936,298.33-		936,298.33-	USD
3010	Exposure	All Accounts Payables	EG4K2	1,115,736.96-		1,115,736.96-	USD
DEC1	Exposure	In House Bank Balances	IN HOUSE BANK	84,771,222.82-		84,771,222.82-	USD
AUC1	Exposure	In House Bank Balances	IN HOUSE BANK	84,789,721.93-		84,789,721.93-	USD
				196,047,335.33	USD		
				1,571,600.75-	AUD		
				0.00	BRL		
				4,591,450.68	CAD		
				211,732.95	CHF		
				302,287,286.28	EUR		
				2,424,000.00	GBP		
				19,900,000	HUF		

Figure 9.8 Exposures Captured in Balance Sheet Snapshot

9.2.2 Scheduling Treasury Middle Office Jobs

Alternatively, we can elect to use the Schedule Treasury Middle Office Jobs app to generate the hedge request. We can use this app for a variety of actions, such as calculating and deleting market risk key figures, end-of-day processing, taking FX balance sheet snapshots, and generating FX balance sheet hedge requests. While in the Schedule Treasury Middle Office Jobs app, we click **Create**, as shown in Figure 9.9. Much as in Section 9.2.1, we'll need to create a snapshot of the balance sheet exposure data.

Figure 9.9 Schedule Treasury Mid Office Jobs App

9 Balance Sheet Hedging

Upon selecting **Create**, we'll be prompted with a template selection, as shown in Figure 9.10. These are predefined templates that we can then personalize to capture the data we want. We'll first select **Take Snapshot – Balance Sheet FX Risk** and then click the **Step 2** button.

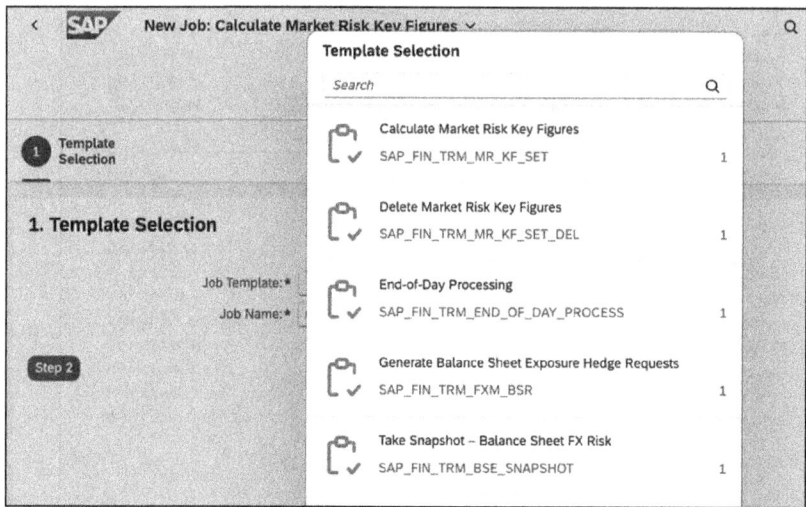

Figure 9.10 Schedule Mid Office Jobs App: Take Snapshot Balance Sheet Risk

Now, we'll determine the scheduling options, as shown in Figure 9.11. Notice that we can run this immediately if we check the **Start Immediately** box, but on the right side of the screen, we can drill down and click **Define Recurrence Pattern**. This allows us to take and save a daily snapshot of the exposures if needed. Once we've made all settings are desired, we click **Step 3**.

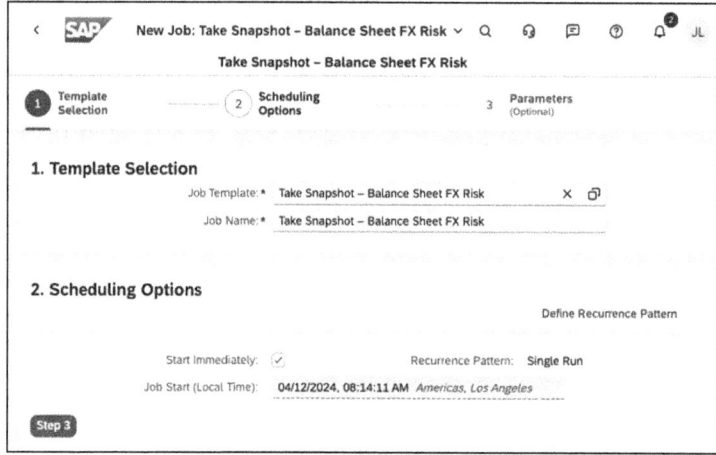

Figure 9.11 Take Snapshot – Balance Sheet FX Risk: Scheduler

9.2 Taking a Foreign Exchange Snapshot for Balance Sheet Risk

In the final section, we'll select our parameters. We'll fill in the required fields with a unique **Snapshot Description**; select the **Created in Status** (**Created** or **Released**); and select our **Key Date Definition,** which we can designate as **System Date, Current Month End, Current Quarter End**, or **Fixed Date**. Once we've entered all the required parameters, we can enter any optional fields as needed and then click **Schedule**, as shown in Figure 9.12.

> **Note**
> We also can set up templates that can be retrieved without needing to enter these values each time.

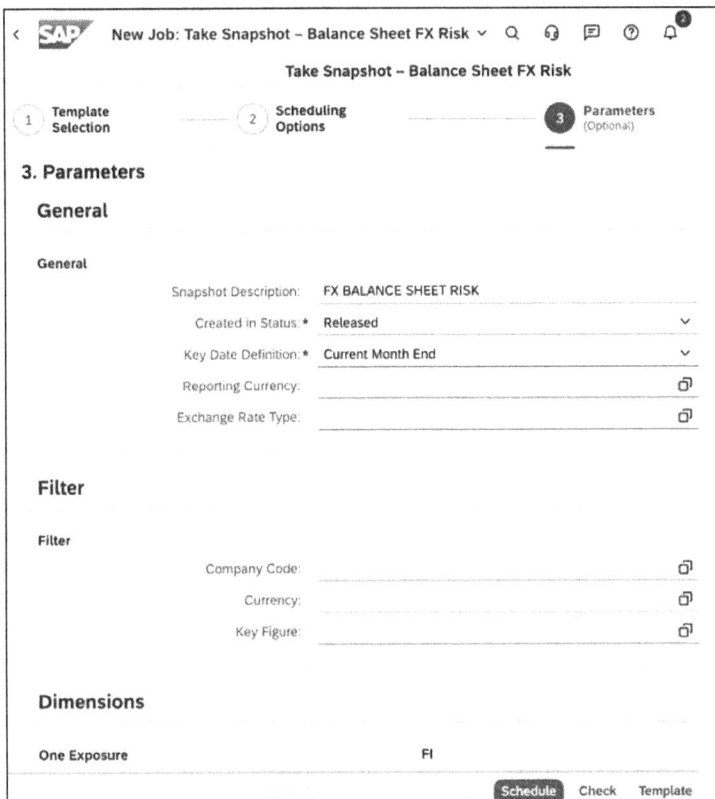

Figure 9.12 Take Snapshot – Balance Sheet FX Risk: Parameters

Upon scheduling and executing the snapshot, we'll see the log shown in Figure 9.13. This log will provide us with the status of the snapshot, with the results and logs listed.

607

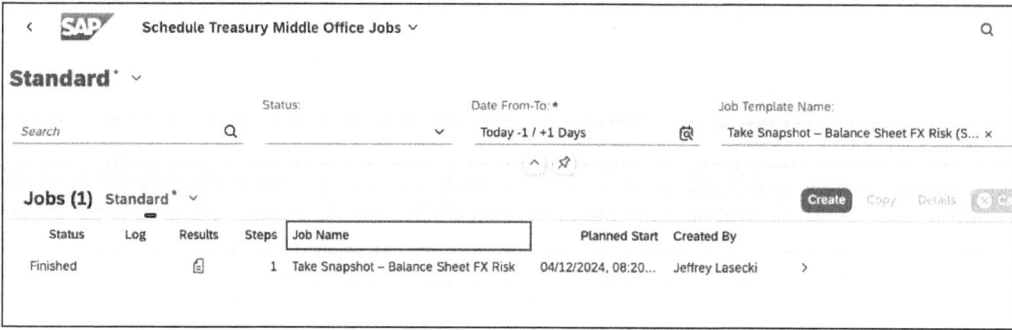

Figure 9.13 Take Snapshot – FX Balance Sheet Job Log

Next, we can click the results icon and see the log of what has been included in our snapshot, as shown in Figure 9.14.

CoCode	Exp./Hedge	Key Figure Group	Key Figure ID	Per. Start	Period End	Amount	Currency	Target Amount	Rep. Crcy
1010	Exposure	In House Bank Balances	IN HOUSE BANK	12/31/9999	12/31/9999	364,383,067.67	USD	364,383,067.67	USD
AUC1	Exposure	In House Bank Balances	IN HOUSE BANK	12/31/9999	12/31/9999	337,999,680.89	EUR	362,538,457.72	USD
USC1	Exposure	In House Bank Balances	IN HOUSE BANK	12/31/9999	12/31/9999	10,893,834.00	EUR	11,684,726.35	USD
3010	Exposure	All Accounts Payables	EG4K2	12/31/9999	12/31/9999	8,027,174.16	EUR	8,609,947.02	USD
UST5	Exposure	All Accounts Payables	EG4K2	12/31/9999	12/31/9999	5,210,000.00	CAD	3,805,905.00	USD
1010	Exposure	All Accounts Payables	EG4K2	12/31/9999	12/31/9999	2,670,426.56	USD	2,670,426.56	USD
1710	Exposure	All Accounts Receivables	EG1K2	12/31/9999	12/31/9999	1,965,370.21	EUR	2,108,056.09	USD
1710	Exposure	All Accounts Receivables	EG2K1	12/31/9999	12/31/9999	1,965,370.21	EUR	2,108,056.09	USD
1010	Exposure	All Accounts Receivables	EG1K2	12/31/9999	12/31/9999	100,000,000	JPY	1,532,801.96	USD
1010	Exposure	All Accounts Receivables	EG2K1	12/31/9999	12/31/9999	100,000,000	JPY	1,532,801.96	USD
1010	Exposure	All Accounts Receivables	EG1K2	12/31/9999	12/31/9999	1,000,000.00	GBP	1,255,350.00	USD
1010	Exposure	All Accounts Receivables	EG2K1	12/31/9999	12/31/9999	1,000,000.00	GBP	1,255,350.00	USD
3010	Hedge	Hedges for Current Licence	HG1K1	12/31/9999	12/31/9999	1,115,736.96	USD	1,115,736.96	USD
1010	Hedge	Hedges for Current Licence	HG1K1	12/31/9999	12/31/9999	500,000.00	GBP	627,675.00	USD

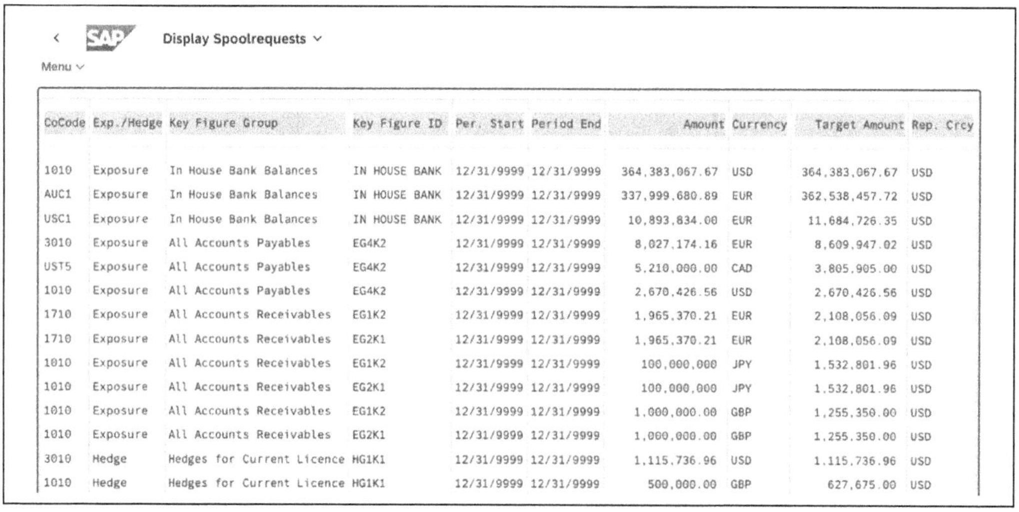

Figure 9.14 Take Snapshot – FX Balance Sheet Job Results

9.3 Balance Sheet Risk Overview Reporting

Once we've successfully captured the snapshot (by using either the Take Snapshot for Balance Sheet FX Risk app or the Schedule Treasury Middle Office Jobs app), we can navigate over to the Balance Sheet FX Risk Overview app, which shows us a view of the information captured during the snapshot process (see Figure 9.15). We'll see a great visualization of our exposures by currency, using SAP S/4HANA embedded analytics. Additionally, we'll also see the snapshot details in the lower portion of the report, where we can start to analyze the differences among snapshots over time.

9.4 Processing Hedge Requests

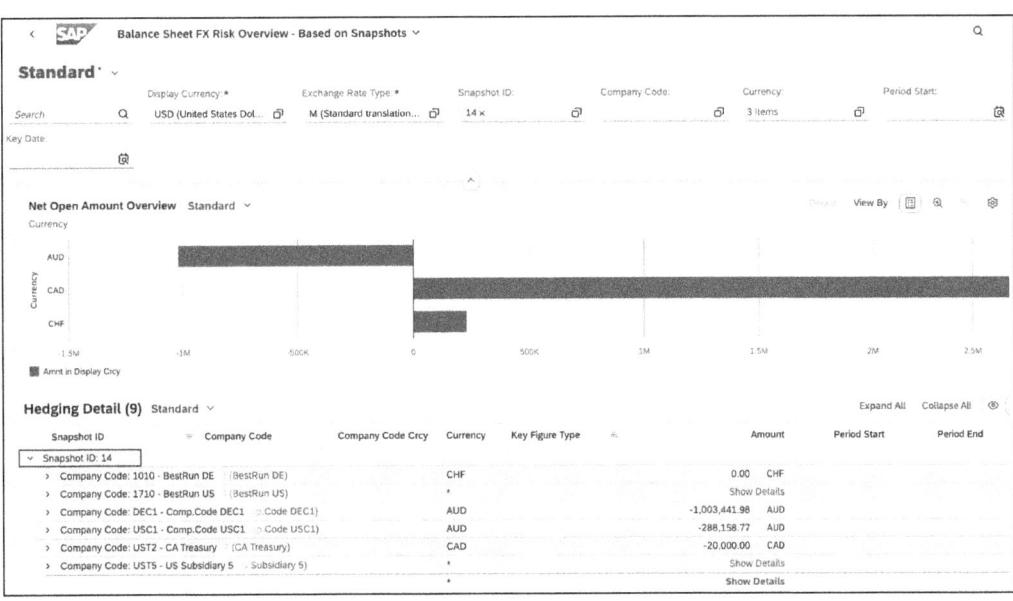

Figure 9.15 Review of Balance Sheet FX Risk Overview – Based on Snapshots

9.4 Processing Hedge Requests

Now, we'll create hedge requests for the exposure data we've locked into the snapshot. Here, we'll convert the snapshot data into individual hedge requests. We have two options for this task, as follows:

- Using the Process Hedge Requests – Balance Sheet FX Risk app
- Using the Schedule Middle Office Jobs app by selecting the **Generate Balance Sheet Exposures Hedge Requests** template

9.4.1 Process Hedge Requests – Balance Sheet FX Risk App

We'll first review the Process Hedge Requests – Balance Sheet FX Risk app. This app will list all of the hedge requests that have been created. As we've just taken the snapshot, we now need to create the hedge requests. To do this, we click **Create** and then select the corresponding **Snapshot ID** from the available list, as shown in Figure 9.16. We'll also have to select a **Request Parameter Group,** which in our case is **HG1K1** (which will be defined in Section 9.6).

9 Balance Sheet Hedging

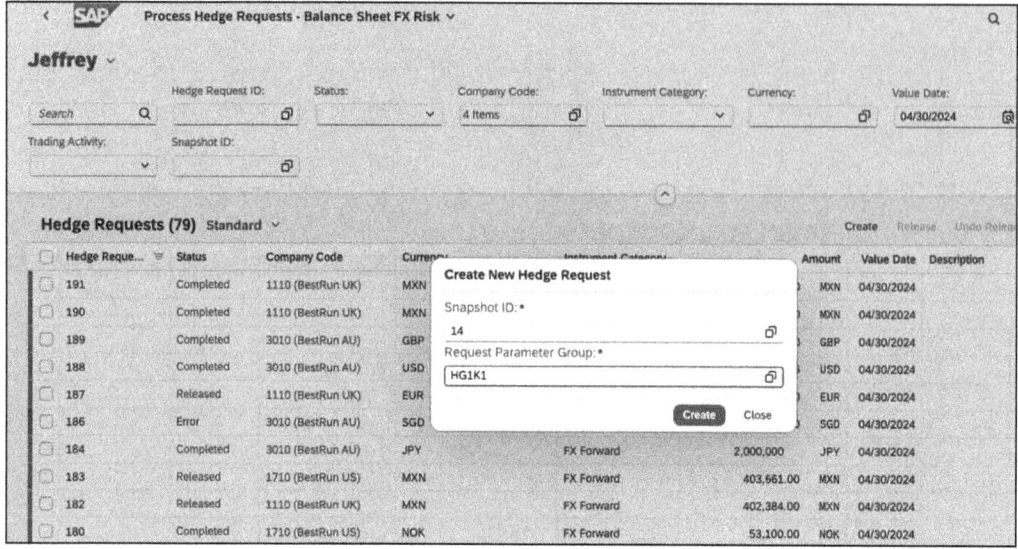

Figure 9.16 Process Hedge Requests – Balance Sheet FX Risk

After we click **Create**, the system displays output containing the newly created hedge requests. We'll see the hedge requests that are in the **Created, Cancelled, Released, Error, Completed**, and/or **In Process** status, as shown in Figure 9.17.

> **Note**
>
> We can add a workflow to the hedge requests to be approved before going to the trading platform for execution.

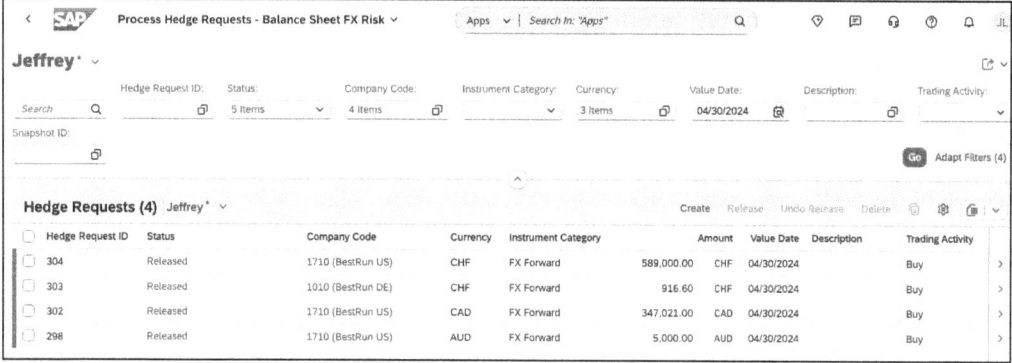

Figure 9.17 Process Hedge Requests Display

9.4.2 Schedule Treasury Middle Office Jobs App

Alternatively, we can elect to use the Schedule Treasury Middle Office Jobs app to generate the hedge request. To do this, in the app, we select the **Generate Balance Sheet**

9.4 Processing Hedge Requests

Exposures Hedge Requests job template and then click **Step 2**, as shown in Figure 9.18. Note that also, this option is sometimes used for background processing of trade requests.

Figure 9.18 Schedule Treasury Middle Office Jobs App: Generate Balance Sheet Exposure Hedge Requests

Next, we'll set the **Scheduling Options** like we did for snapshot generation back in Section 9.2.2, and then, we'll click **Step 3**, as shown in Figure 9.19.

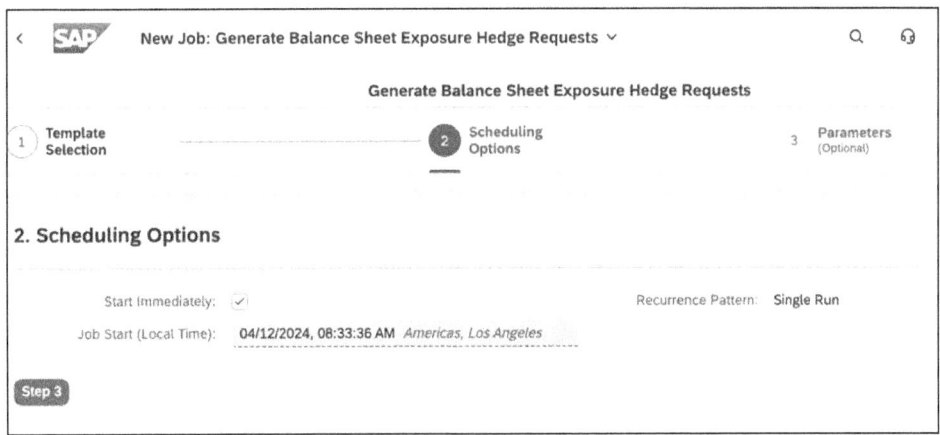

Figure 9.19 Schedule Treasury Mid Office Jobs App: Scheduling Options

Now, we'll enter the **Snapshot Selection**. We'll select the snapshot previously created in the **Snapshot ID** field and then any additional filtering data. Additionally, we'll fill out the **Value Date** section: notice that we've selected a **Fixed Date** in the **Value Date Selection** and input the **Value Date** for the hedge requests. In the **Parameters** section, we can control the type of trade requests to be made (for example, **FX Spots, FX Forwards**, or **FX NDF**) by using the **Instrument Category** field. Once we're satisfied with our entries, we click **Schedule** to generate the FX hedge request, as shown in Figure 9.20.

611

9 Balance Sheet Hedging

> **Note**
> Regardless of which app we choose to use, all hedge requests will be available in the Process Balance Sheet Hedge Request app for review, approval, or display.

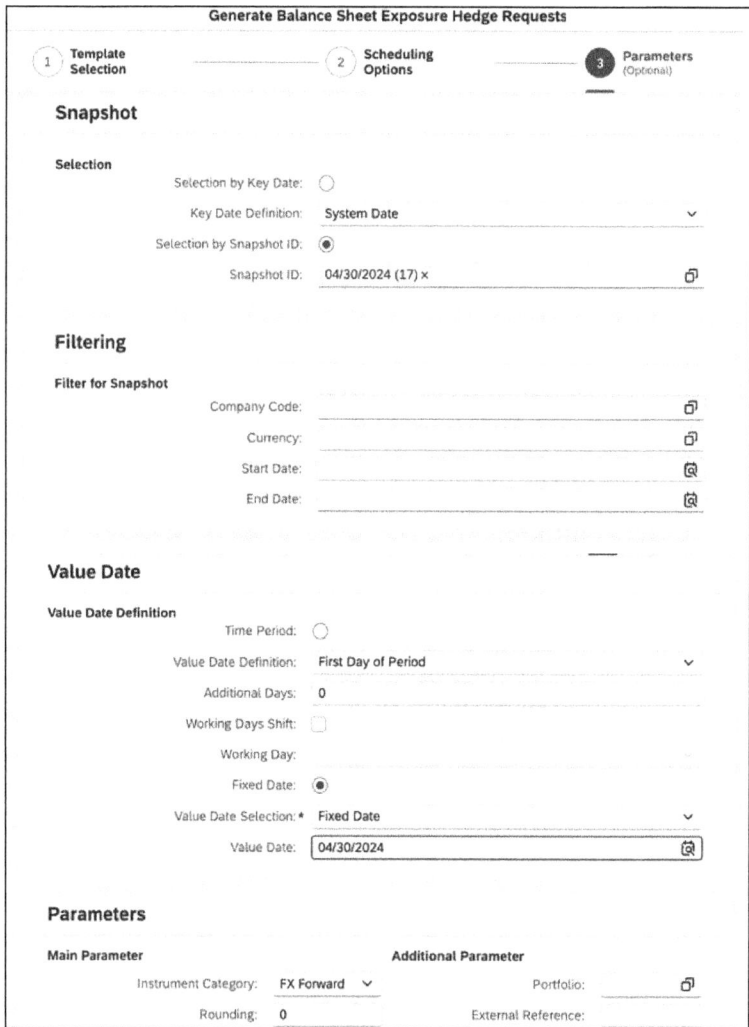

Figure 9.20 Schedule Treasury Middle Office Jobs App: Parameters

Now, we can review the log to see the generated hedge request, as shown in Figure 9.21.

Now that we've successfully created trade requests to be released to a trading platform, there are two options for making the trade request data transfer: we can use SAP trading platform integration (which includes connections to FXall, 360T, and Bloomberg), or we can create a custom secure file transfer protocol interface with our chosen trading platform. In the next section, we'll discuss the best practice of the SAP trading platform integration option.

HReq. ID	CoCode	Company Co	Amount (B/S HR)	Currency	Per. Start	Period End	Snapshot	Buy/Sell	Put/Call	Value Date	Instrument
271	DE20	EUR	100.00	USD			17	B		04/30/2024	FX Forward I
272	DEC1	EUR	1,003,441.98	AUD			17	B		04/30/2024	FX Forward I
273	DEC1	EUR	85,440,535.96	USD			17	B		04/30/2024	FX Forward I
274	1010	EUR	367,213,991.55	USD			17	S		04/30/2024	FX Forward I
275	1010	EUR	0.00	BRL			17	S		04/30/2024	FX Forward I
276	1010	EUR	0.00	CHF			17	S		04/30/2024	FX Forward I
277	1010	EUR	2,427,000.00	GBP			17	S		04/30/2024	FX Forward I
278	1010	EUR	19,900,000	HUF			17	S		04/30/2024	FX Forward I
279	1010	EUR	199,940,000	JPY			17	S		04/30/2024	FX Forward I
280	1710	USD	45,286,875.59	EUR			17	B		04/30/2024	FX Forward I
281	1710	USD	346,336.32	CAD			17	B		04/30/2024	FX Forward I
282	1710	USD	160,831,407,244.04	INR			17	B		04/30/2024	FX Forward I
283	AUC1	AUD	336,772,999.86	EUR			17	S		04/30/2024	FX Forward I
284	AUC1	AUD	85,726,020.26	USD			17	B		04/30/2024	FX Forward I
285	USC1	USD	288,158.77	AUD			17	B		04/30/2024	FX Forward I
286	USC1	USD	10,597,373.29	EUR			17	S		04/30/2024	FX Forward I
287	US20	USD	1,745.00	EUR			17	B		04/30/2024	FX Forward I
288	1710	USD	402,661.00	MXN			17	B		04/30/2024	FX Forward I
289	1710	USD	16.110	JPY			17	B		04/30/2024	FX Forward I

Figure 9.21 Schedule Treasury Middle Office Jobs App: Hedge Request Job Log

9.5 SAP Trading Platform Integration

SAP trading platform integration is a cloud application that connects external trading platforms to our SAP S/4HANA system. SAP trading platform integration covers one-off trades along with additional scenarios, such as subsidiaries trading with their treasury center to transfer FX risk to the treasury center or subsidiaries managing their liquidity. Figure 9.22 provides an overview of the integrated trading process with treasury and risk management.

SAP trading platform integration supports the following business processes:

- **Inbound trading**
  In this process, trades are executed directly with the trading platform and automatically transferred to treasury and risk management.
- **Trading and trade requests**
  In this process, trade requests are created in treasury and risk management and are sent to our external trading platform (outbound from SAP S/4HANA to SAP trading platform integration). The trade is then executed on the trading platform and transferred back automatically to treasury and risk management.

- **Intercompany trading**
  The intercompany trading process handles transactions among subsidiaries, treasury centers, and the trading platform. This can include risk transfer between a subsidiary and a corporation, along with back-to-back trading.

- **Competitive bid capture**
  In this process, competitive bids are captured on each trade. We can use this to evaluate pricing over time.

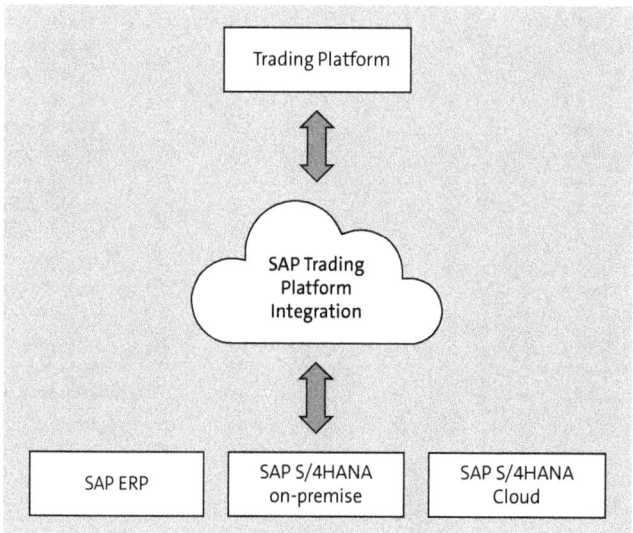

Figure 9.22 SAP Trading Platform Integration Overview

SAP trading platform integration offers preconfigured mapping of fields from the trading platform to key fields in treasury and risk management, along with mapping of key fields in our trading request going to the trading platform. Additionally, the following options are available for connecting to trading platform applications:

- Direct connection based on 360T format
- Connection to other trading platforms (FXall and Bloomberg) using generic format interfaces

To access SAP trading platform integration, we use the Manage Trade Request – Trading Platform app. Once we enter the app, we're welcomed with a dashboard showing key apps for SAP trading platform integration, as shown in Figure 9.23.

In the following sections, we'll walk through the most important apps (Manage Trade Requests, Manage Trades, Manage Block of Trade Requests, and Counterparty Limit Utilization) before diving into some of the related configurations.

9.5 SAP Trading Platform Integration

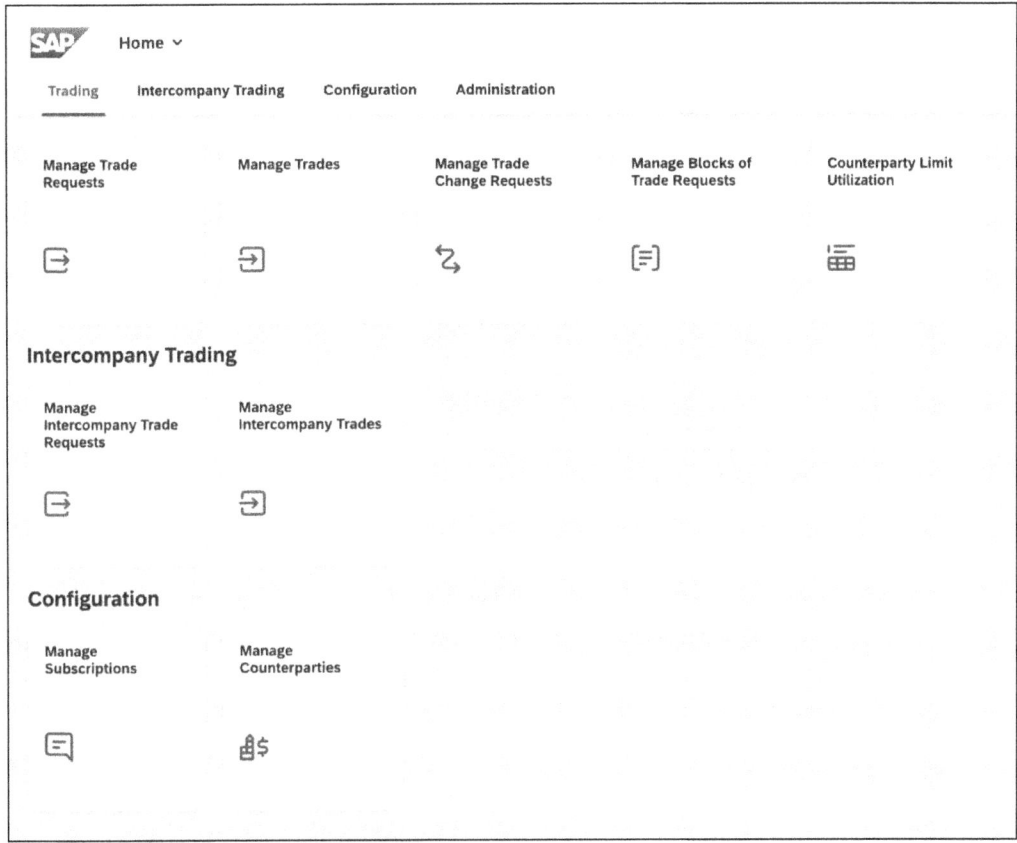

Figure 9.23 SAP Trading Platform Integration Home Screen

9.5.1 Manage Trade Requests App

First, we'll take a look at the Manage Trade Requests app. This app is a dashboard of all the hedge requests and their status, and within it, we can choose to automatically have the hedge requests go to a selected trading platform, or we can allow a user to make the decision on which trading platform it should go to. As shown in Figure 9.24, we've selected our AUD, CAD, and CHF positions, and they're currently in **Initiated** status. There are also some trades that have already been completed in previous hedging cycles.

Once we've selected our hedge requests by checking the boxes to the left of the ones we want, we can select one of the following options:

- Send
- Create Phone Trade
- Split
- Undo Split
- Create Block
- Undo Block
- Assign to Block
- Limit Check

9 Balance Sheet Hedging

Figure 9.24 Manage Trade Requests App

Sending

We'll now select our CHF trade and click **Send**. We'll be prompted with the trading platform options, and we'll select the **360T** option, as shown in Figure 9.25. The trade will then be released from SAP trading platform integration to the 360T trading platform, where a treasury team member will execute the trade.

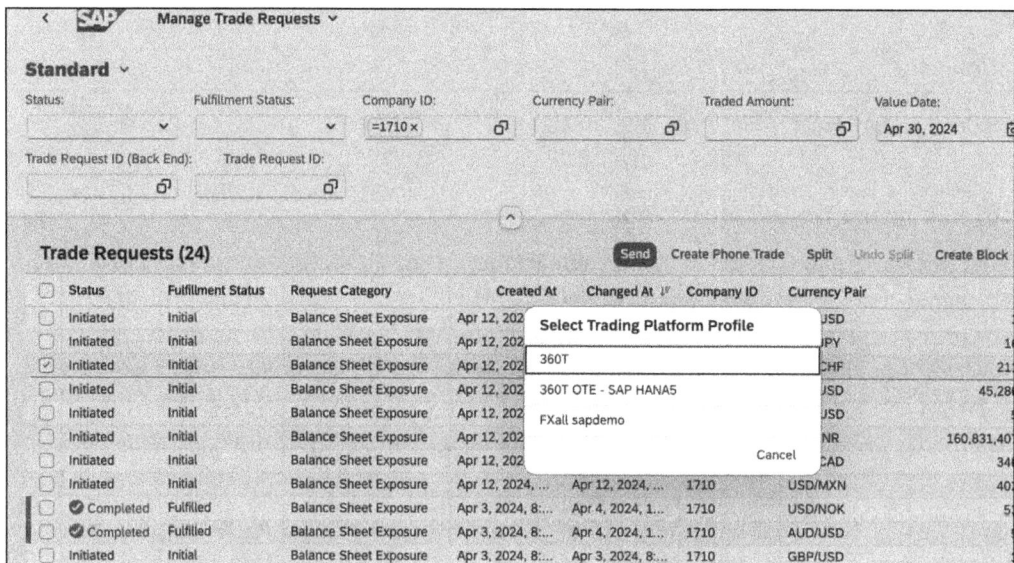

Figure 9.25 Manage Trade Requests App: Select Trading Platform

We'll then see an update message confirming that the trade request was successfully sent to the selected trading platform, as shown in Figure 9.26.

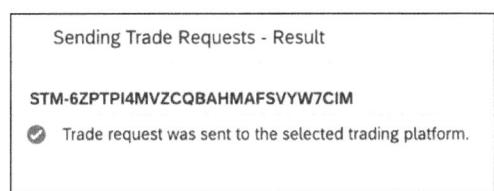

Figure 9.26 Trade Request Successfully Sent to Trading Platform

Now that the trade request has been released, we'll see the status of the request change to **Sent to Trading Platform,** and it will stay in this status until the trade has been executed.

After the trade is executed on the trading platform, we can see the status reflected in the Manage Trade Request app. We'll notice that the status has changed to **Completed** and that the fulfillment status is set to **Fulfilled**, as shown in Figure 9.27.

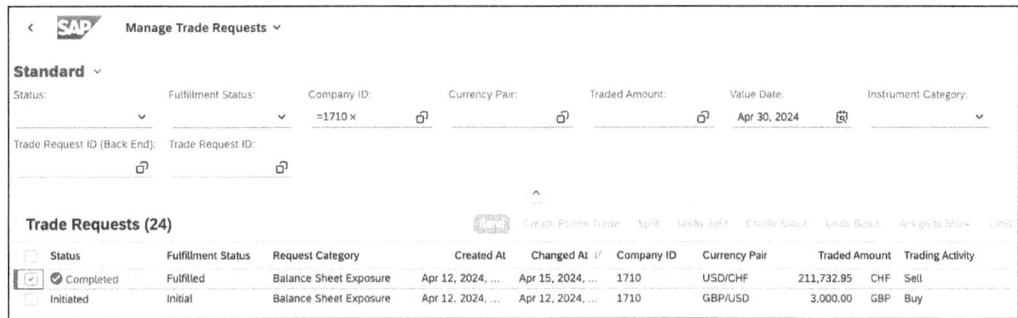

Figure 9.27 Manage Trade Requests App: Status Updated to Completed and Fulfilled

Create Phone Trade App

During our hedging process, there may be times when we'll work directly with the bank on the pricing of trades. Sometimes, we can enter those rates and execute them in the trading platform, but if that's not available, we'll have the option to manually create a phone trade using SAP trading platform integration to populate our trade and complete our hedge request.

In the next example, we have a Swiss franc (CHF) trade, but we've executed it via a phone trade. We'll select the trade request and then click **Create Phone Trade**. In the example in Figure 9.28, we'll enter the required **Trade Date, Counterparty, Forward Rate, and Spot Rate** fields, and we'll validate the contract date. Once we've populated all data fields needed for the trade, we can click **Save**.

9   Balance Sheet Hedging

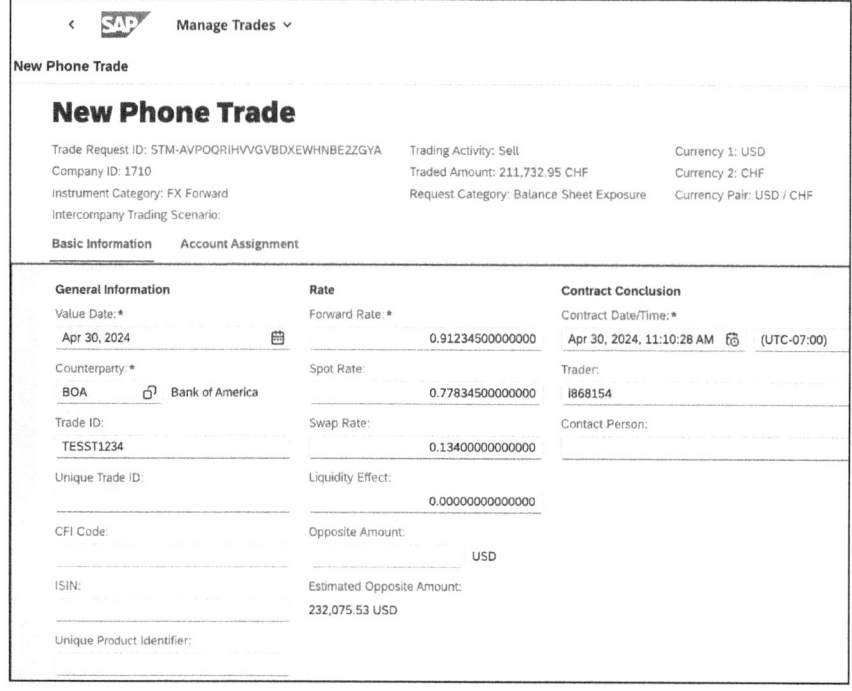

Figure 9.28 Manage Trade Request App: New Phone Trade

Now that we've saved or created the trade, we can navigate to the Manage Trade Request app to see the manually created phone trade as shown in Figure 9.29.

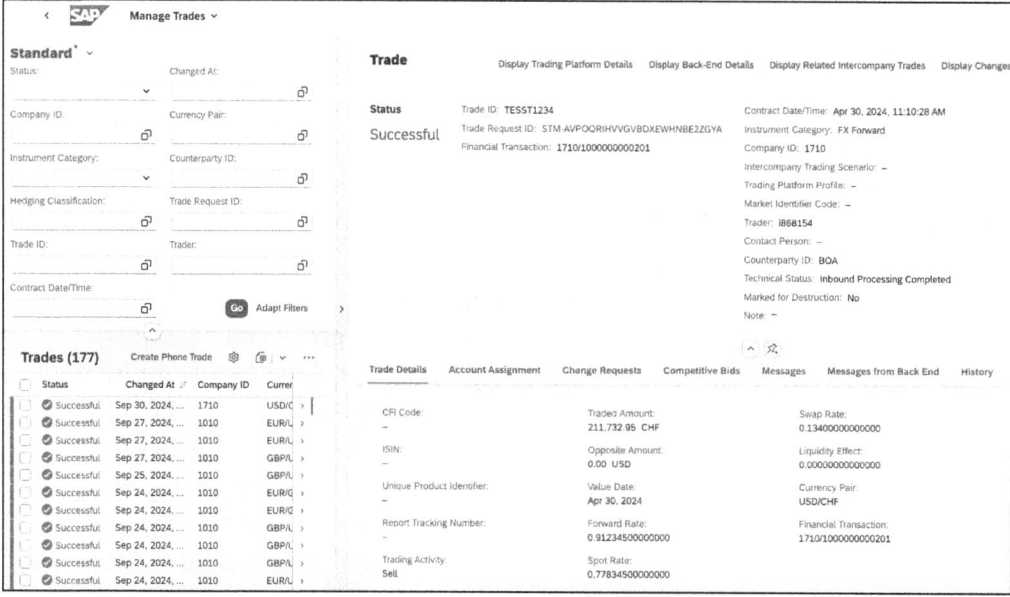

Figure 9.29 Manage Trade Request: Create Phone Trade

9.5 SAP Trading Platform Integration

Now that the **Trade ID** number is present, we can navigate to the Manage Financial Transaction app and review our deal number as it is set up in SAP S/4HANA, as shown in Figure 9.30.

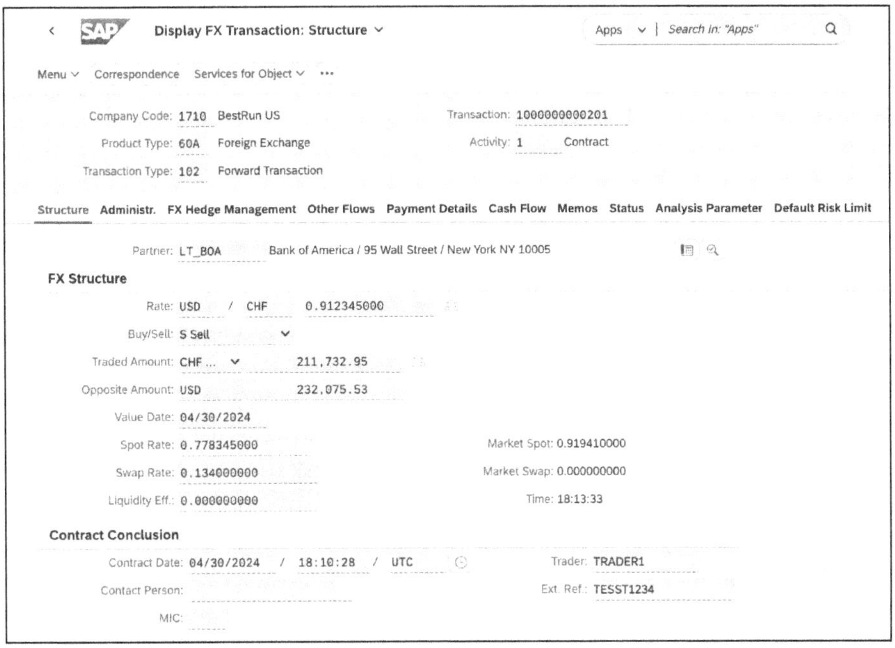

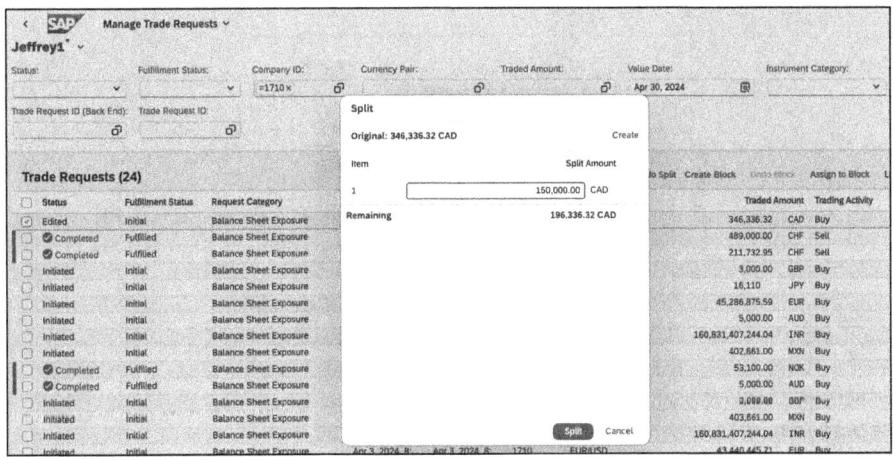

Figure 9.30 Display Trade Created in SAP S/4HANA

Splitting Trades

When we select the **Split** action, we can split trades into different tranches as needed. For our example, we'll leverage our CAD trades and split them into two trades. Once we've entered the amount in the popup shown in Figure 9.31, we'll click **Split**.

Figure 9.31 Manage Trade Request: Create Split Trade

619

9   Balance Sheet Hedging

Once we've completed the split, we'll see two trade requests listed on the dashboard, as shown in Figure 9.32.

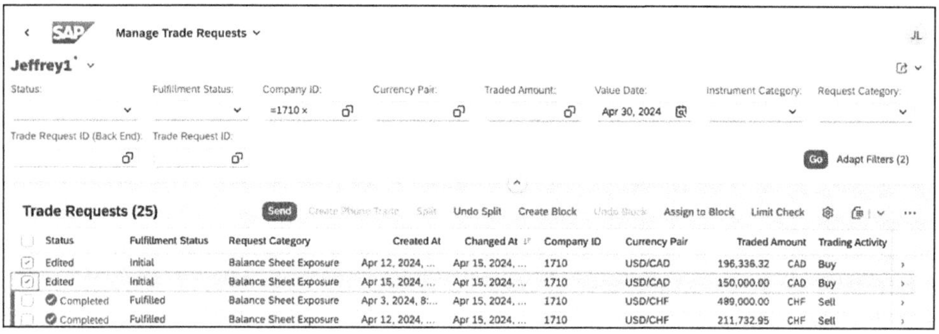

Figure 9.32 Manage Trade Request: Split Trade Requests

Undoing Splits

Clicking **Undo Split** will reverse the previously split trades. When we select the trades in question and click **Undo Split**, a popup will appear and ask us to confirm whether we really want to reverse the split, as shown in Figure 9.33. We confirm this by clicking **Undo Split**, and once the split is undone, we can send the single trade to the trading platform for execution.

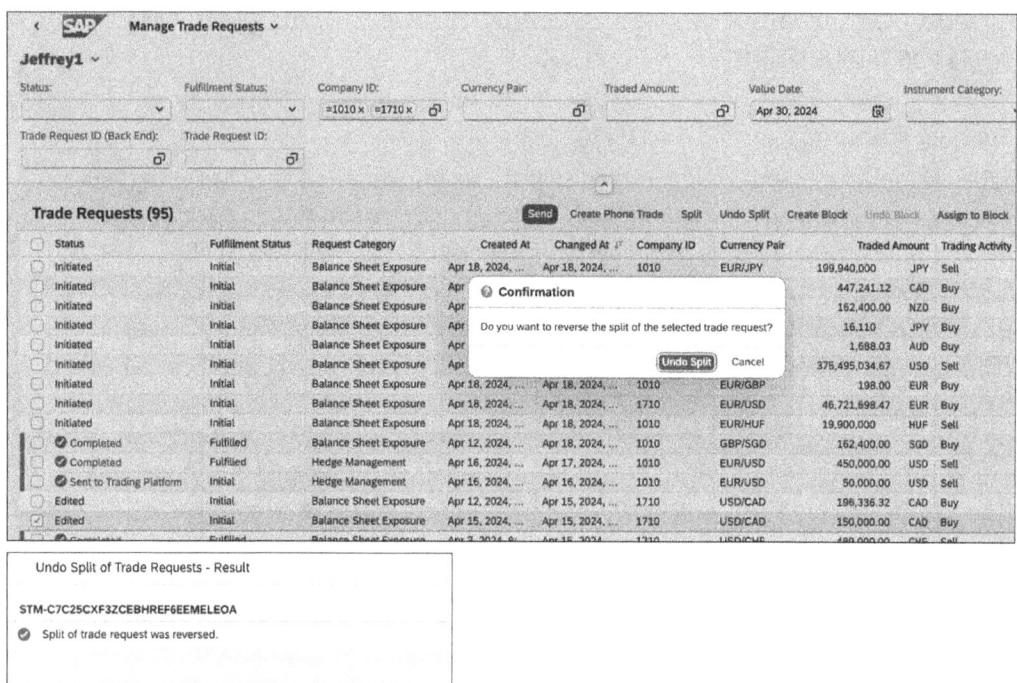

Figure 9.33 Manage Trade Request App: Reversal of Split

Creating Blocks

There may be times when multiple requests for the same currency pair come into SAP trading platform integration. At those times, we may want to trade in bulk as much as possible so we can create blocks of trades. To create blocks of trades, we'll select two or more hedge requests from the Manage Trade Request app and then click **Create Block** (as in Figure 9.34). Once the trades are blocked together, we'll see the confirmation message in Figure 9.35 for the net amount (e.g. 6,000 GBP).

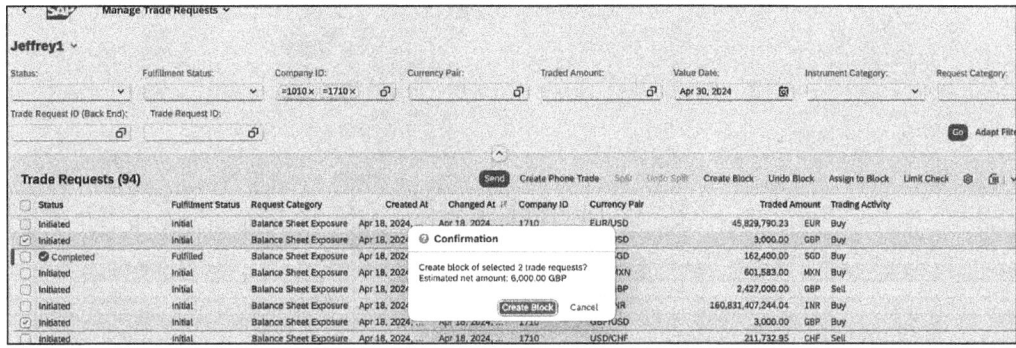

Figure 9.34 Manage Trade Request – Create Block

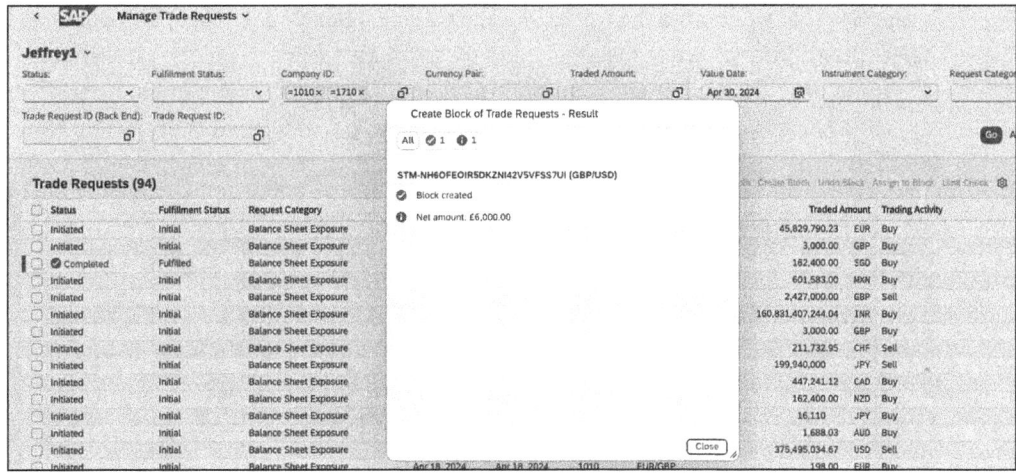

Figure 9.35 Manage Trade Requests: Block Created

Undoing Blocks

The **Undo Block** option will reverse the previously created block of trades. Once we select the trades in question and click **Undo Block**, a popup will appear and ask us to confirm whether we really want to remove the selected trade from the block, as shown in Figure 9.36. We confirm this by clicking **Undo Block**, and once we've undone the block, we'll be able to process each trade separately again.

9 Balance Sheet Hedging

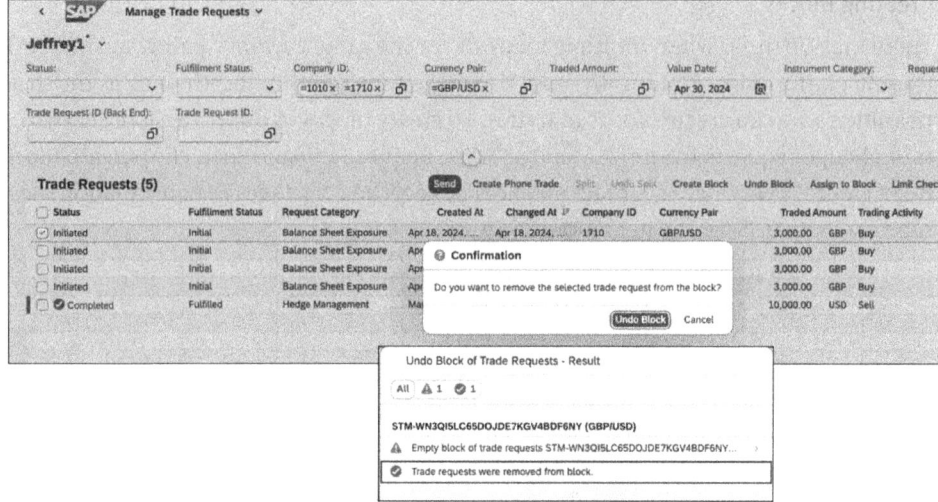

Figure 9.36 Manage Trade Request App: Undo Block

Assign to Block Function

Now that we've created a block of trade requests, we can elect to add more trades to an existing block. We can easily do this by accessing the **Assign to Block** function. In the same way that we created the initial block, we'll select our trade request block and then select the additional hedge request to add to it. After we select the block we wish to add the trade, we click the **Assign to Block** button and the popup shown in Figure 9.37 will appear. We click **Assign to Block** again to confirm, we'll see the results of the additional blocked trade, and our hedge request will be assigned.

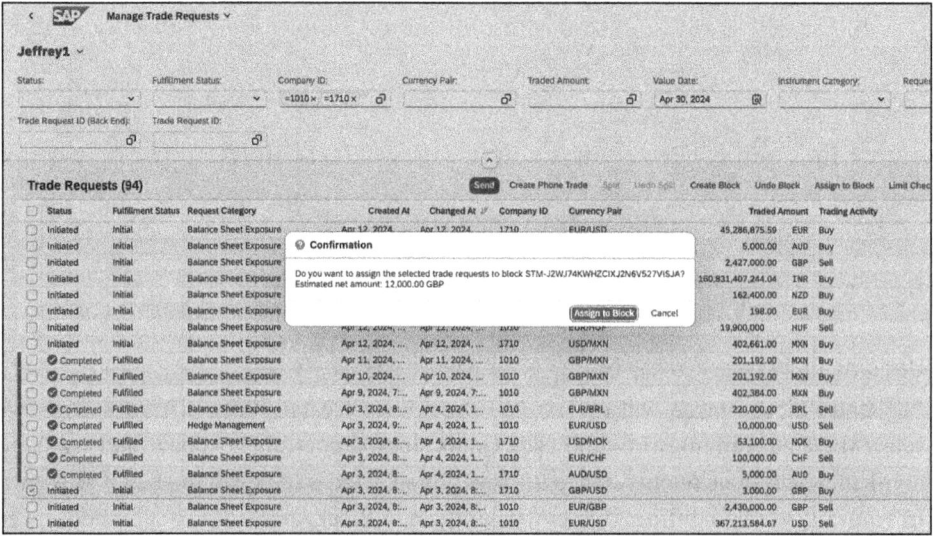

Figure 9.37 Manage Trade Request: Assign to Block Function

> **Note**
> The currency pairs as part of the block must all match. Otherwise, the block will fail.

Limit Checks

In addition to managing our trade requests using SAP trading platform integration, we can review our trade requests against overall limit checks. To perform this, we can select our trade request and select **Limit Check**. Then, the report of all available counterparties will be displayed along with their utilization and limits, as shown in Figure 9.38. The following options are also available for statuses in both **Current Limit Utilization Status** and **Calculated Limit Check Status**:

- Limit Exceeded
- Limit Locked
- No Result
- OK
- Warning Threshold Exceeded

> **Note**
> If any counterparties fail the utilization or limit check, they won't be available for selection in SAP trading platform integration.

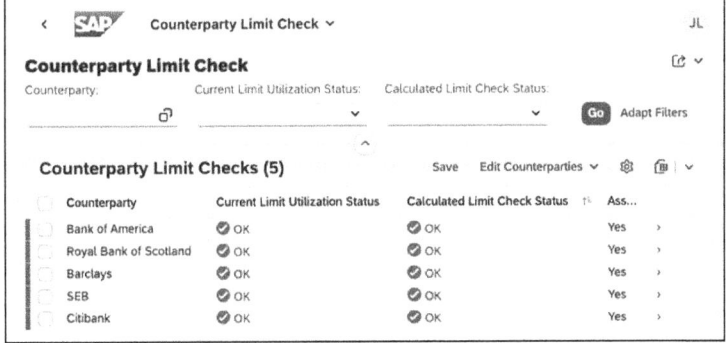

Figure 9.38 Manage Trade Request: Limit Check

Additionally, during this limit check review, we can edit any of the counterparties. We simply select the **Edit Counterparties** function, which will allow us to do the following:

- Remove
- Add
- Discard

9 Balance Sheet Hedging

To act on this, we select the counterparty and select **Remove**, **Add**, or **Discard**, as shown in Figure 9.39. Then, we'll see the list update, and we can click **Save**. At that point, we've verified the business partners that we can use for the trade request.

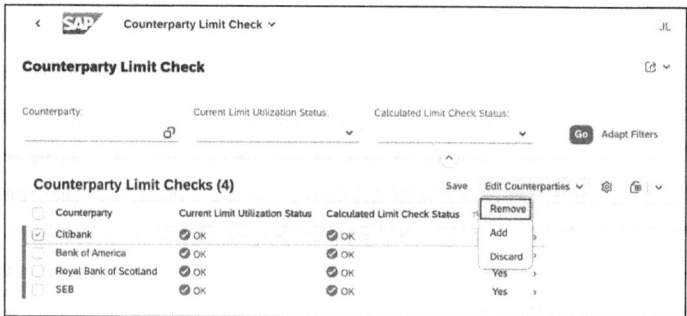

Figure 9.39 Manage Trade Request: Counterparty Modification

Now that we've successfully managed and sent the trade request to the trading platform desired, the trader will execute the trade, and within just a short time after trade execution, the trade will be reflected in the Manage Trades app.

9.5.2 Manage Trades App

The Manage Trades app tracks all the activity that has come in from SAP trading platform integration. As shown in Figure 9.40, this dashboard gives us the list of all the trades with a status for each of the trades.

Notice the only action available in this report is creating phone trades, which we previously showed in Figure 9.29.

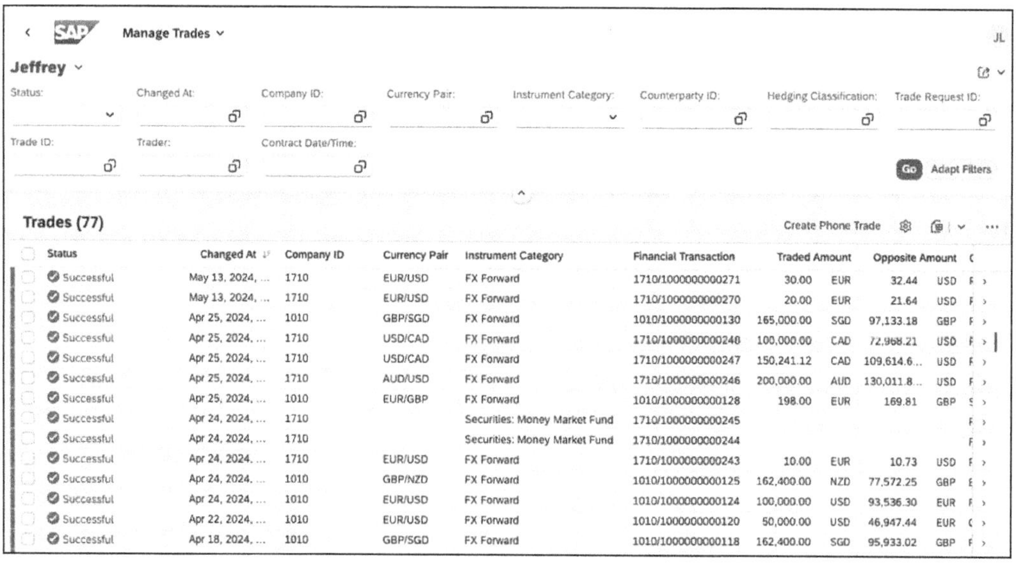

Figure 9.40 Manage Trades Dashboard

9.5 SAP Trading Platform Integration

From this report, we can drill down into the trade and see key information on it. First, we'll note that we can see the **Trade ID**, which is because the program reads the trade from our trading platform and then automatically creates the same entry in SAP S/4HANA. We'll see the following seven tabs available for the trade, as shown in Figure 9.41:

- **Trade Details**
  This gives the key economics of the deal, including rate, dates, and currency pair.

- **Account Assignment**
  This provides us with a view of any key additional attributes we may want to store in the trade for accounting or reporting purposes.

- **Change Request**
  This is the audit log of any changes made to the trade request.

- **Competitive Bids**
  This provides us with a view of all open bids with the counterparties available at the time of deal execution. We can use this to review pricing activities with our counterparties.

- **Messages**
  This provides all messages generated for this trade.

- **Messages from Backend**
  This provides any messages triggered from backend processing.

- **History**
  This provides us with a list of all key activities that have occurred in the trade.

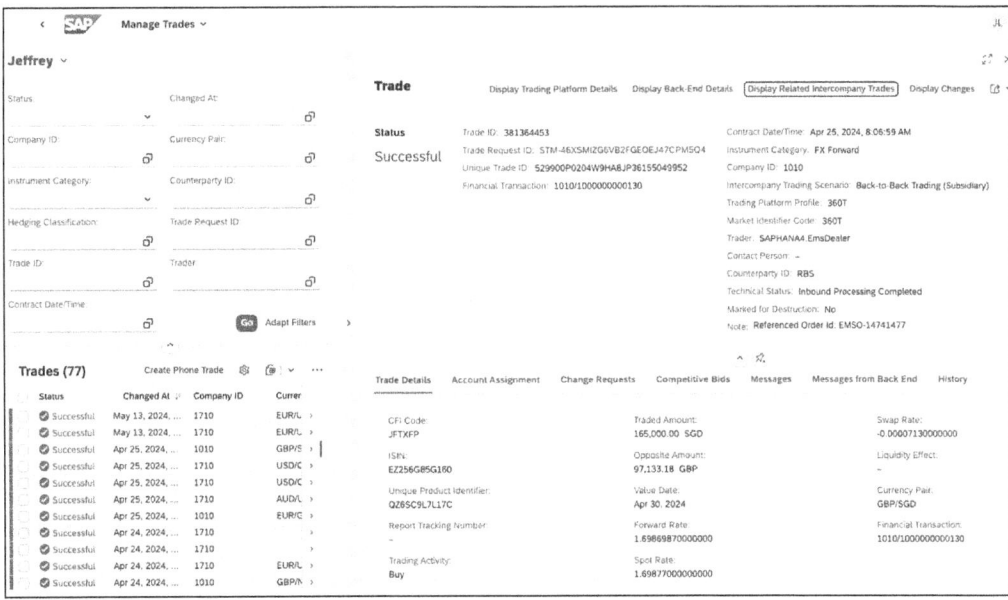

Figure 9.41 Manage Trades Detailed View

625

9 Balance Sheet Hedging

In addition to these tabs, there are some actions that we can review, as follows:

- **Display Trading Platform Detail**
  This is a more technical display of the trade request that was interfaced over from the trading platform. We can see in Figure 9.42 that our example 360T has sent over an XML file.

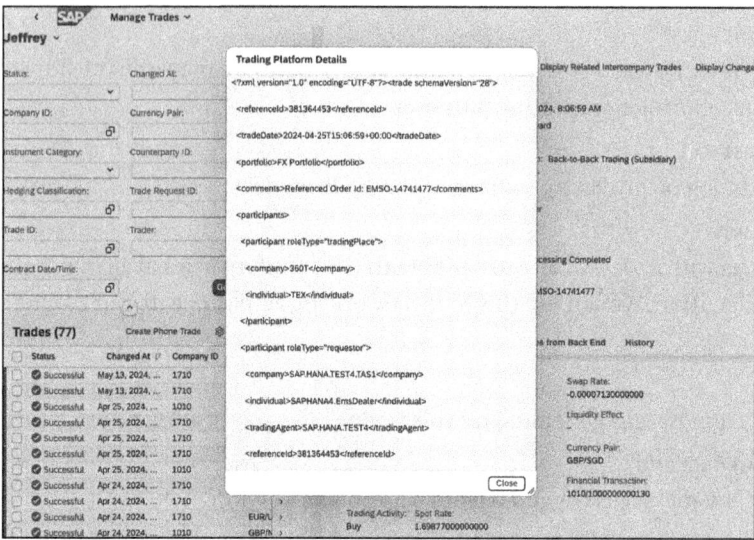

Figure 9.42 Manage Trades File Review

- **Display Back-End Details**
  Here, we'll see how SAP S/4HANA consumed the information from the trading platform, as shown in Figure 9.43. This is a more technical view as well.

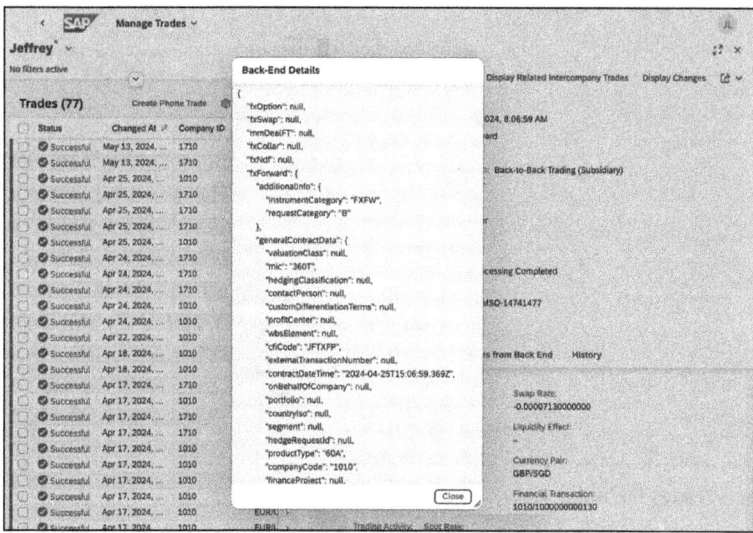

Figure 9.43 Manage Trades Technical Review

- **Display Related Intercompany Trades**
  If our trade is part of back-to-back trading, then using this action will show us all trades (external and internal) in one location so we can easily review each tranche. Once we select the intercompany trade, we'll be directed to the Manage Trades app, as shown in Figure 9.44.

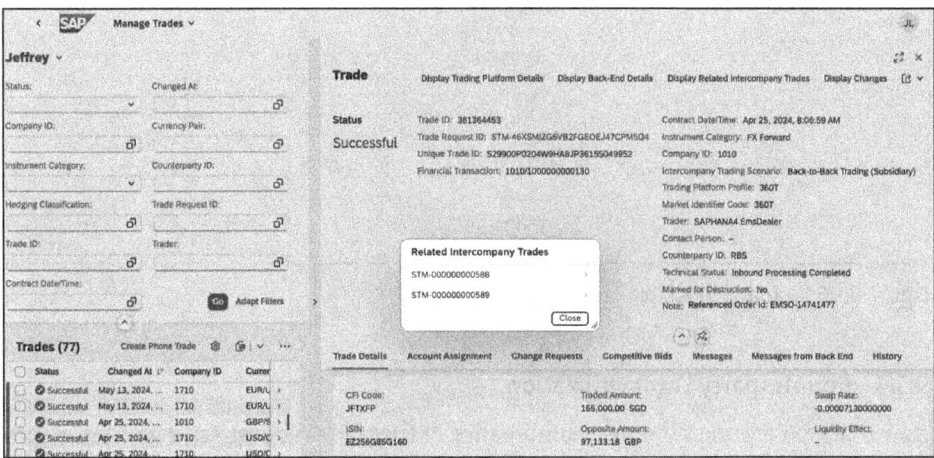

Figure 9.44 Manage Trades: Intercompany Trade Process

- **Display Changes**
  If there were any changes to the trade request, we'll see them noted here, as shown in Figure 9.45.

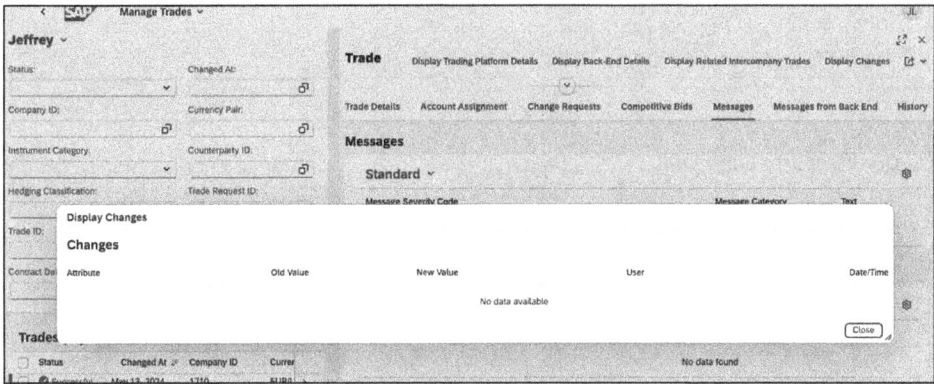

Figure 9.45 Manage Trades: Display Changes

9.5.3 Manage Block of Trade Requests App

Next, we'll discuss the Manage Block of Trade Requests app, which is shown in Figure 9.46. It provides us with an audit trail of all the trades that were sent in a specific block, and it allows us to review the blocks created. However, note that the app doesn't allow us to make any changes to the blocks. To modify any blocks, please refer back to Section 9.5.1.

9 Balance Sheet Hedging

Figure 9.46 Manage Blocks of Trade Requests App

9.5.4 Counterparty Limit Utilization

Now that we've worked on the maintenance of the trade request, we'll look at the pre-trade counterparty limit review. With SAP trading platform integration, we can review our counterparty limits that have been set up in treasury and risk management. This allows us to ensure compliance with our internal and external limits. We can use the Counterparty Limit Utilization app to generate the report shown in Figure 9.47, which shows where we stand with our trades versus our limits.

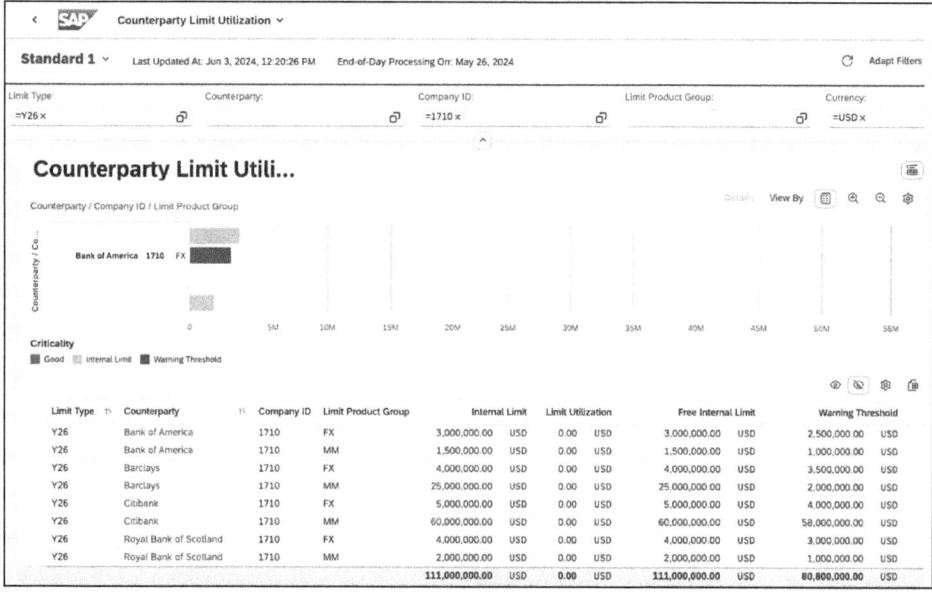

Figure 9.47 Counterparty Limit Utilization Report

Notice that this report is dependent on the end-of-day processing that's done to review all open trades, counterparties, and limits. We'll discuss end-of-day processing in further detail in Chapter 11, Section 11.1.

9.5.5 Configuration

In this section, we'll discuss the configuration settings available for SAP trading platform integration. The key to SAP trading platform integration is mapping both the outbound trade requests and the inbound trade ticket. We'll also walk through the SAP Fiori apps available for this configuration.

Configure Trading Platforms App

First, we'll use the Configure Trading Platforms app to configure the trading platforms that we'll use. This setting will define a unique **Profile ID** that we'll use as we begin to map our fields, as shown in Figure 9.48. From here, we can simply either click **Create** to build a new connection or click on an existing **Profile ID** and make the necessary changes.

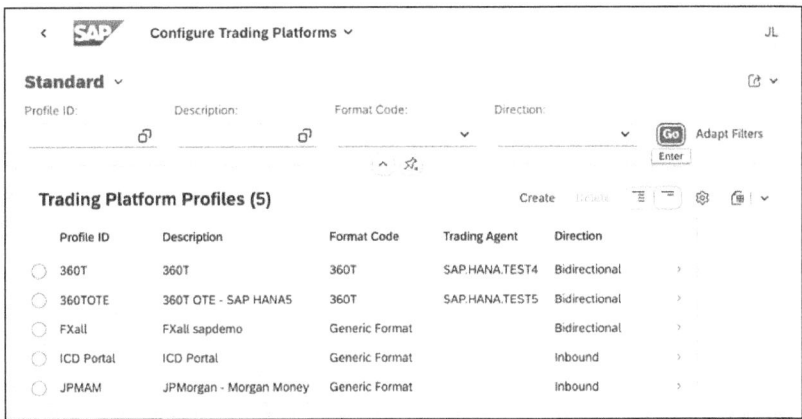

Figure 9.48 Configure Trading Platforms App

For the example in Figure 9.49, we'll review the 360T profile. We can see here that the **Profile ID, Trading Agent, Destination**, and **Description** are free-form fields that will help us identify the connection. Additionally, we'll select the following:

- **Usage Category**
  We can select either **External Trading** or **Intercompany Trading**.
- **Format Code**
  We can select either **360T** or **Generic Format** (which is used for other trading platforms).
- **Directional**
  We will set this to either **Bidirectional** or **Inbound,** depending on our use case.

9 Balance Sheet Hedging

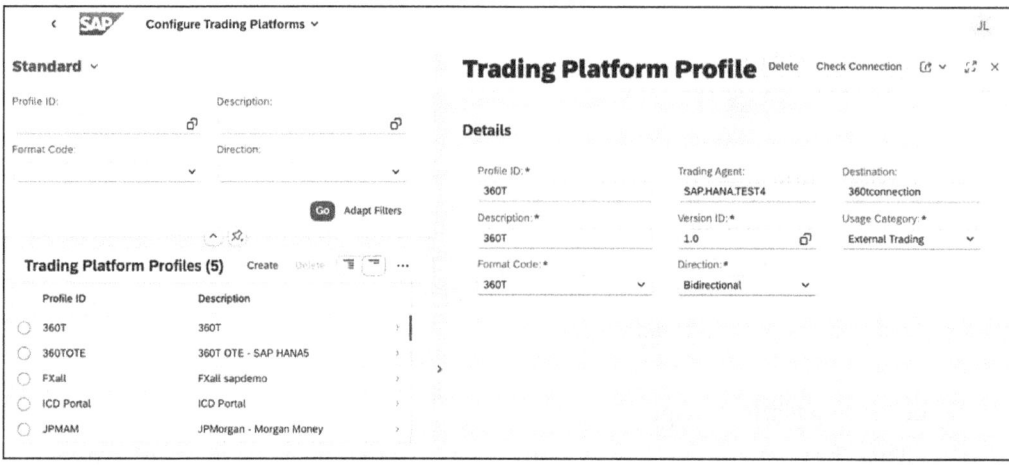

Figure 9.49 Trading Platform Profile

Configure Back-End Systems App

Now that we've identified the profile, we'll also configure the settings for the backend systems in the Configure Back-End Systems app. In this step, we'll create the connection between SAP trading platform integration and our SAP S/4HANA system. We'll populate the **Profile ID** free-form text field, which uniquely identifies our system, and we'll then specify the following settings, as shown in Figure 9.50:

- **Business Systems ID**
  Here, we'll identify the SAP S/4HANA backend system. If we need help finding our business system ID, we run Transaction SLDCHECK on the back-end system, and in the screen output, we check the details section calling function LCR_GET_OWN_BUSINESS_SYSTEM.

- **Direction**
  We'll set this to either **Bidirectional** or **Inbound,** depending on our use case.

- **Business Systems Type**
  Here, we'll select **Cloud Edition** or **On Premise** to match the SAP system we're connecting to.

- **Format Code**
  Here, we'll select **SAP S/4HANA.**

We'll then enter the destination, which will be determined by what we have set up in our SAP BTP cockpit. We can use one of the following types of destinations:

- SAP S/4HANA Cloud backend system: **HTTP**
- SAP S/4HANA backend system: **RFC**

Note that the values entered must be the exact same values as those stored in the SAP BTP cockpit customer account.

9.5 SAP Trading Platform Integration

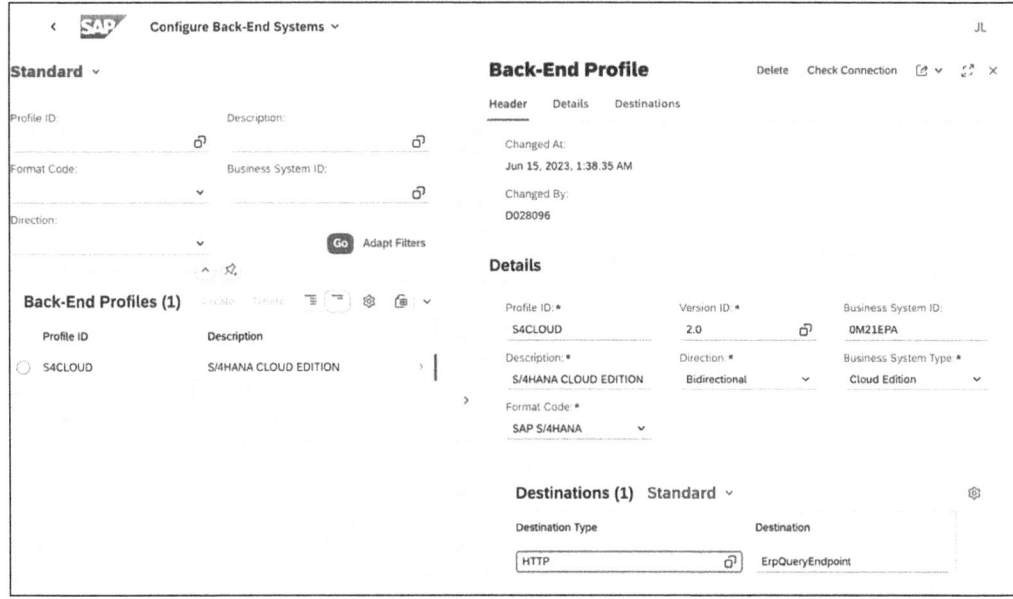

Figure 9.50 Configuring Back-End Profile

Configure Mappings: Trading Platform Format

Now that we've established to which systems the SAP trading platform integration will be connected, we'll review the mapping capabilities to make sure that our trade requests will be mapped appropriately. Using the Configure Mappings – Trading Platform Format app, we'll enter our **Trading Platform Profile** (e.g., **360T**) and the **Direction** (currently, **Inbound** is selected in the example in Figure 9.51).

Once we do this, we'll see the list of available fields and the corresponding status of the field: either **Mapping Provided** or **Reference Table not Provided**, as shown in Figure 9.51. We'll also see sliders that let us toggle on and off whether the fields are required or mapping is enabled. Also note that there are the following three categories of fields:

- Mandatory Fields
- Product Fields
- Custom Fields

This should allow us to map all dimensions needed for the trade request.

Now, let's drill down into the **COMPANY_ID**. We can see that the mapping is already provided, as shown in Figure 9.52, but we want to review the mapping. As we drill down into the entry, we can see that we have four company codes mapped.

631

9   Balance Sheet Hedging

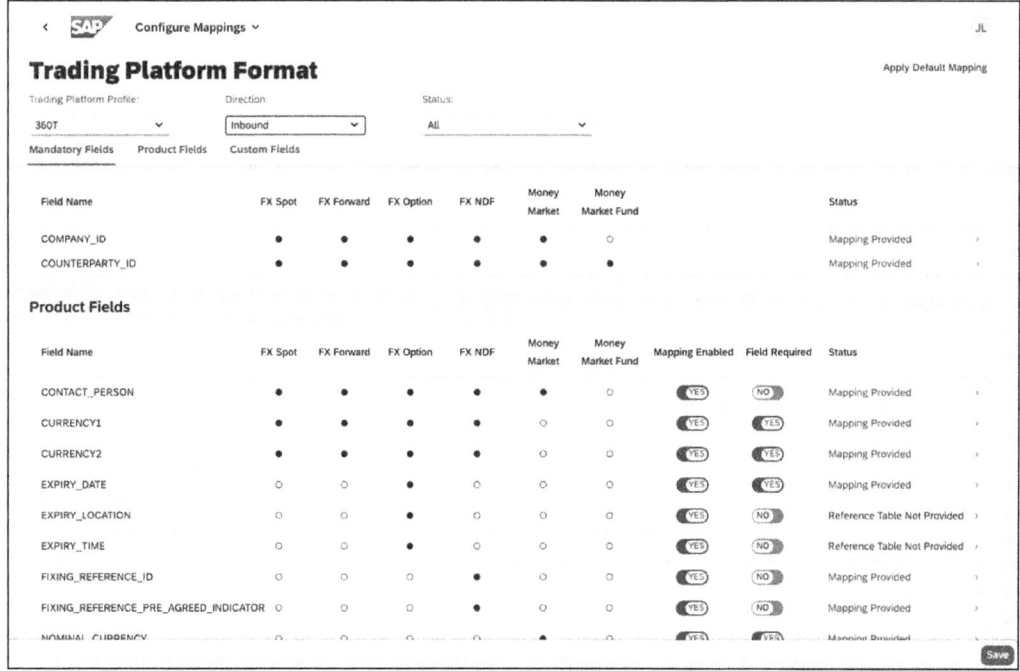

Figure 9.51 Configuring Mappings for Inbound Trading

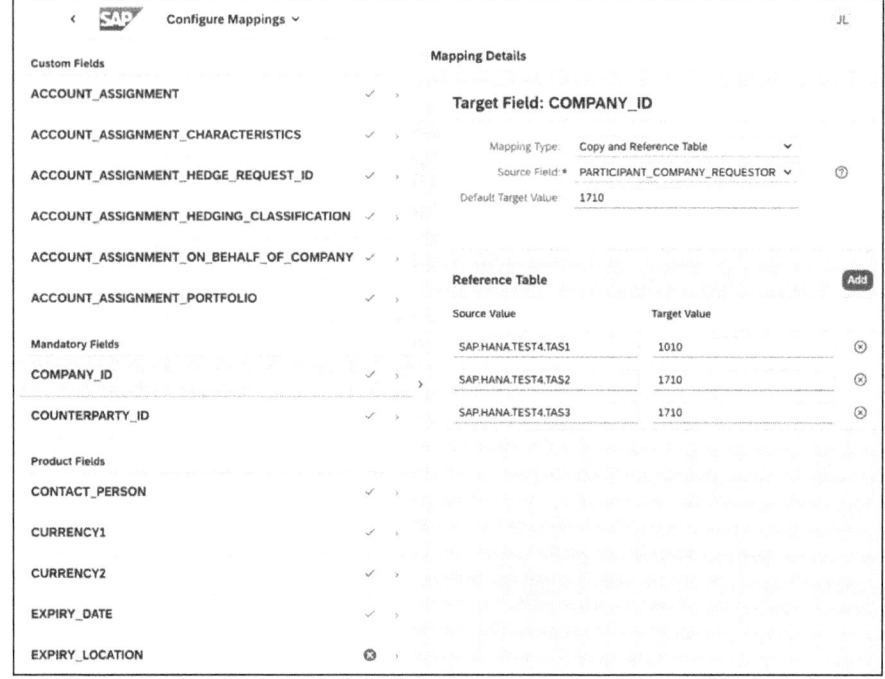

Figure 9.52 Configure Mappings App: Back-End Format

Configure Mappings: Back-End Format

Next, we'll configure the mapping for the inbound trade from the trading platform to treasury and risk management. We'll specify the back-end profile that we established in the Configure Back-End Systems app and the **Direction** (**Inbound** is currently selected, as shown in Figure 9.53).

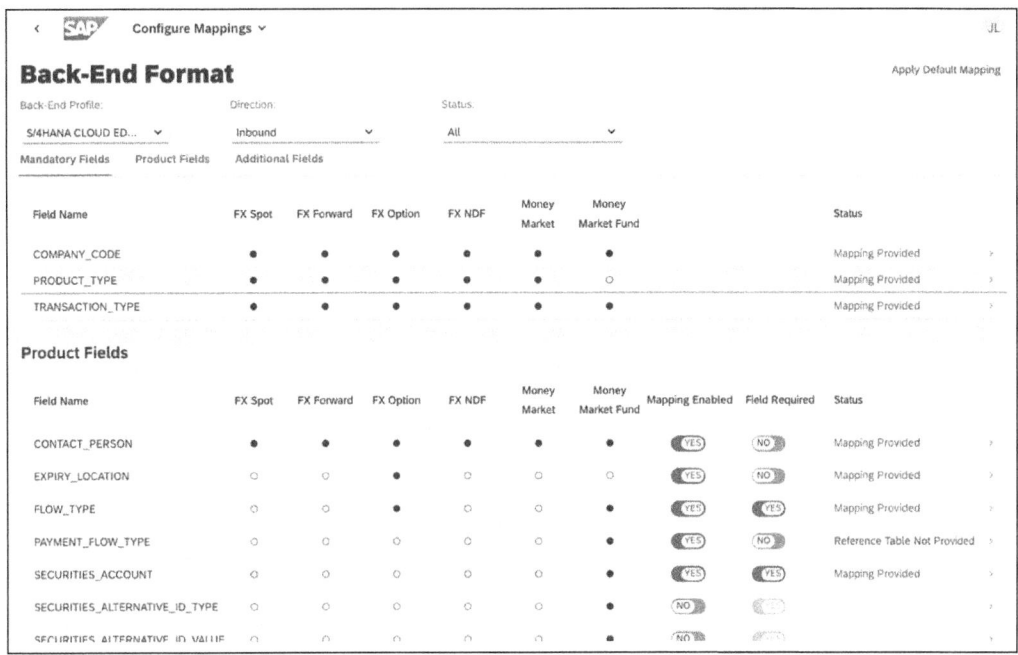

Figure 9.53 Configuring Mappings for Back-End Format: Inbound

For this example, we'll review the mapping for the product type, as shown in Figure 9.54. We see the product type as set up in SAP trading platform integration and what we want to map it to in SAP S/4HANA. In the example, we've used the SAP Best Practices product types, but we can configure them as desired and explained in Chapter 4, Section 4.2:

- 60A: Foreign Exchange External
- 60B: Foreign Exchange Internal
- 76A: FX Option
- 55A: Interest Rate Instrument

9   Balance Sheet Hedging

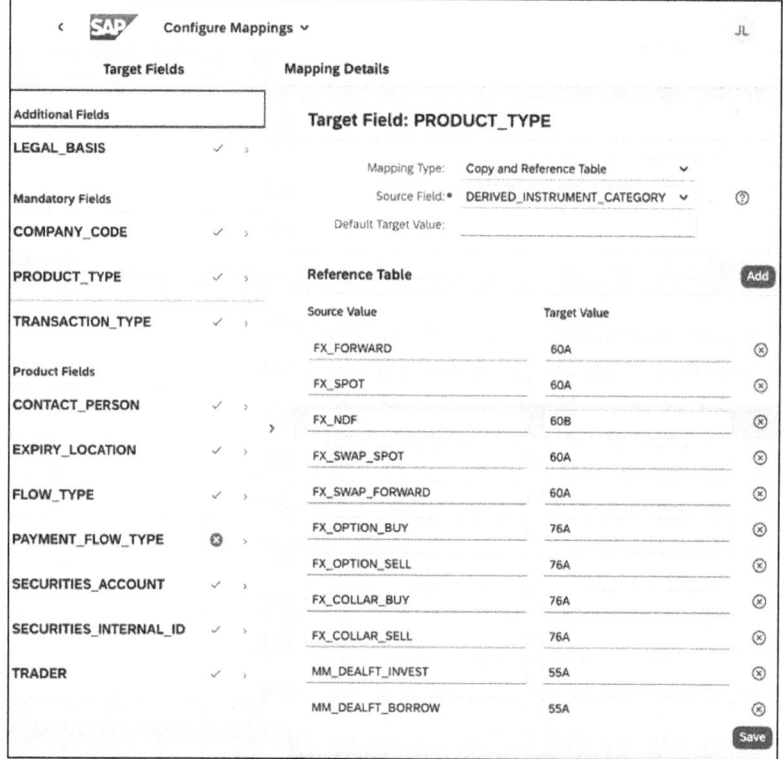

Figure 9.54 Configuring Mappings for Backend Format: Detailed View

Manage Counterparties App

We've set up how the data will be mapped between the trading platform and SAP, so next, we'll map in the master data of the counterparties that we will use. Upon entry into the Manage Counterparties app, we'll find the list of all counterparties that have been set up, as shown in Figure 9.55. We'll see the name of the counterparty, a brief description, and an origin indicator.

We'll then drill down into a counterparty, where we'll see the settings applicable to that counterparty, as shown in Figure 9.56. First, we'll indicate which companies can use the business partner, and if desired, we can select **Unrestricted Access for All Company IDs**, which will enable all company codes for processing. If we don't select this setting, then we'll need to map in each entity individually, under **Authorizations**.

We'll then enter the **Trading Platform Profiles**, which map the trading platform to the counterparty available on that platform. We'll enter the trading platform (for example, "306T") under **Trading Platform Configuration** and map to "Barclays BARX_DEMO" (the name of Barclays on our trading platform) under **Counterparty ID (Trading Platform)**. Notice that this setup can go to multiple trading platforms, so we should have one entry per counterparty that can be mapped to various naming conventions in all the trading platforms.

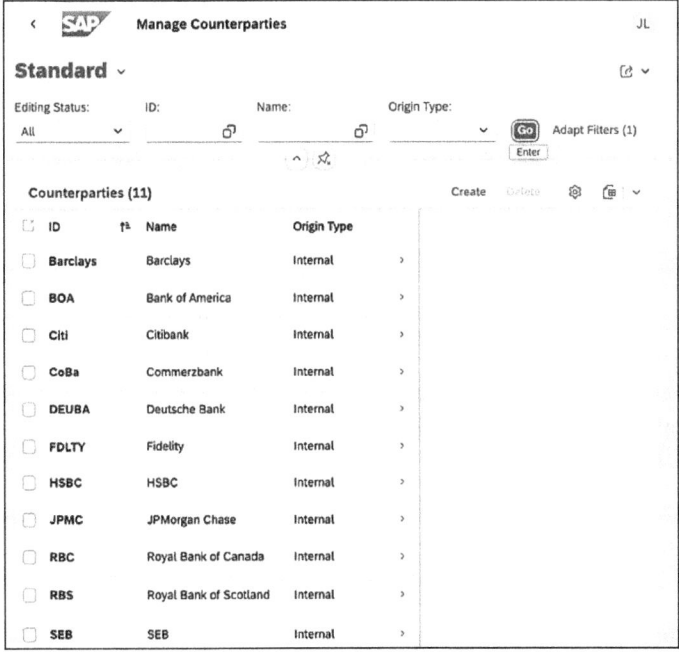

Figure 9.55 Manage Counterparties App Overview

Finally, we'll indicate the business partner that's set up in SAP S/4HANA. We've mapped our "S/4HANA CLOUD EDITION" to our **Business Partner** of "LT_BARCLAY." We must set up the business partner in SAP S/4HANA before configuring this section; instructions are included in Chapter 2, Section 2.5.

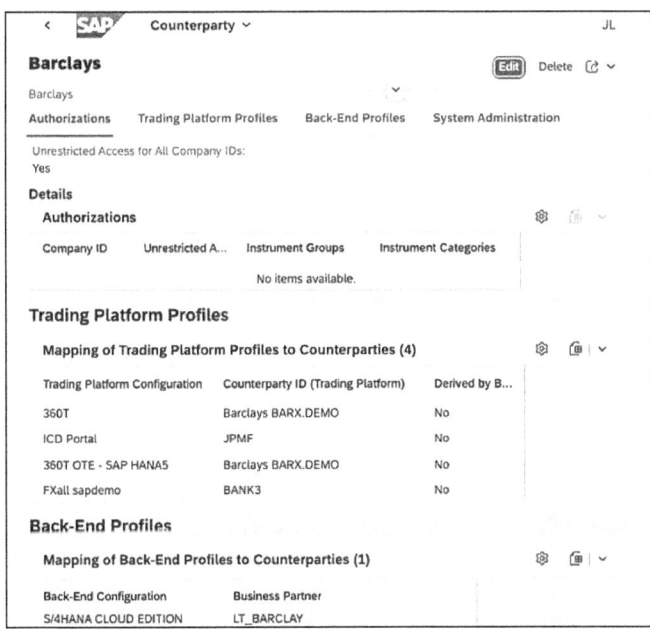

Figure 9.56 Manage Counterparties App: Detailed Mapping

We've reviewed the end-to-end process for collecting our exposures from the universal journal, integrated our hedge requests into the trading platform for execution, and made sure the trade returned complies with our balance sheet hedging program. In the next section, we'll discuss how to setup both the data configuration and the master data that supports the balance sheet hedging process.

9.6 Master Data Setup

As SAP moves toward a self-service model to give treasury teams more agility in making the changes needed instead of a lot of configurations to support the balance sheet process, most of setup is controlled by master data. To support balance sheet hedging, we maintain elements called *key figures*, which help drive the view of the balance sheet hedging report that was shown in Section 9.1.

When defining the key figures, we'll first define the key figure groups, which are the types of categories that we wish to delineate our exposure activity. We can have as few or as many key figure groups as we desire to meet our exposure capture needs.

When defining a key figure group, we'll set up a unique key figure group, provide relevant text to be shown in the Review Balance Sheet FX Risk app, and define the exposure or hedge category.

For our example in Figure 9.57, we'll set up one group called **EXP_GRP1 All Accounts Payables** and another group called **HDG_GRP1 Hedges for Current License**. Notice the two categories that are available: **Hedge** groups and **Exposure** groups.

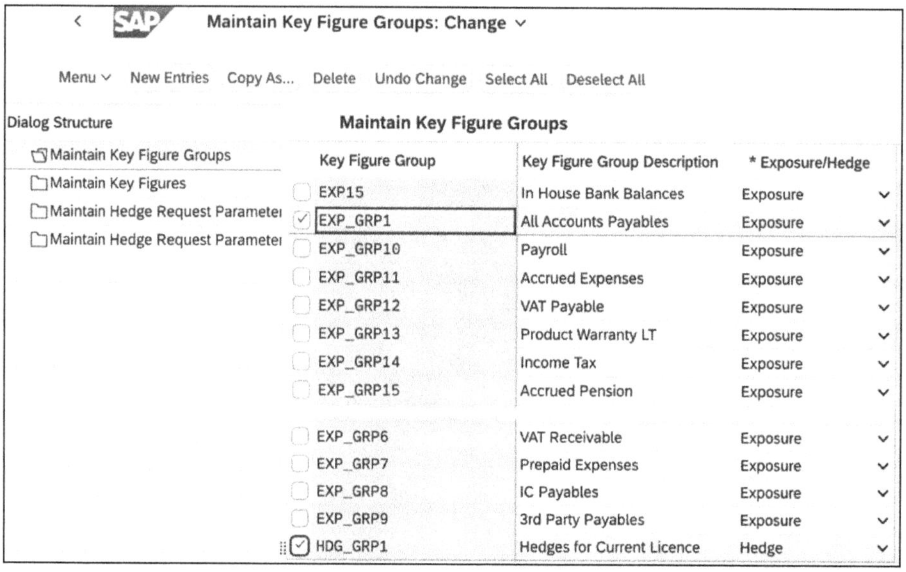

Key Figure Group	Key Figure Group Description	* Exposure/Hedge
EXP15	In House Bank Balances	Exposure
EXP_GRP1	All Accounts Payables	Exposure
EXP_GRP10	Payroll	Exposure
EXP_GRP11	Accrued Expenses	Exposure
EXP_GRP12	VAT Payable	Exposure
EXP_GRP13	Product Warranty LT	Exposure
EXP_GRP14	Income Tax	Exposure
EXP_GRP15	Accrued Pension	Exposure
EXP_GRP6	VAT Receivable	Exposure
EXP_GRP7	Prepaid Expenses	Exposure
EXP_GRP8	IC Payables	Exposure
EXP_GRP9	3rd Party Payables	Exposure
HDG_GRP1	Hedges for Current Licence	Hedge

Figure 9.57 Maintaining Key Figure Groups

9.6 Master Data Setup

Now that we've defined the key figure groups, we'll define the parts of those key figure groups in the **Maintain Key Figures** folder. This is where we'll ultimately define how our report will be displayed in the Review Balance Sheet FX Risk app.

In Figure 9.58, we can see that although we've described **Key Figure Group EXP_GRP1** as **All Accounts Payables**, we've added categories to that key figure group: **Outgoing Payments**, **Payroll**, and **All Accounts Payables**.

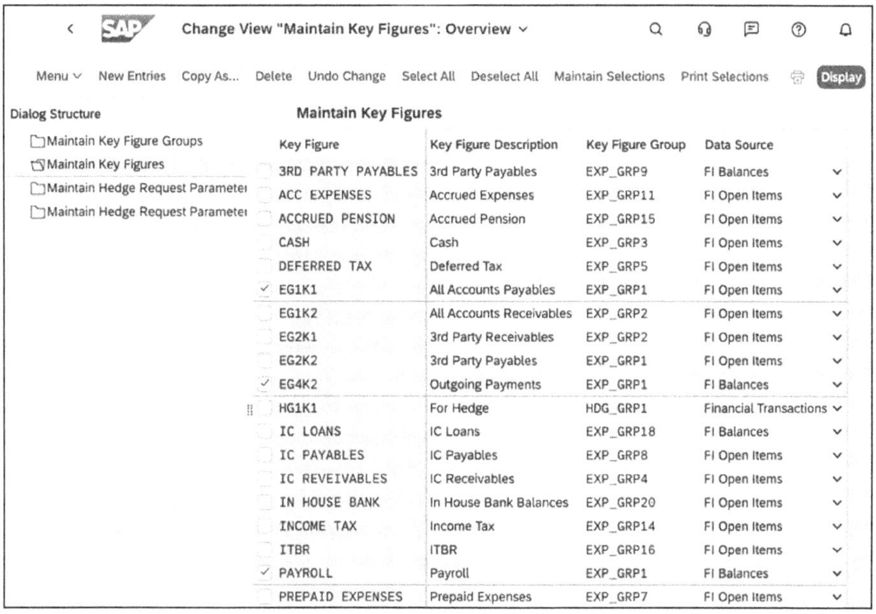

Figure 9.58 Maintaining Key Figures

Notice also that we can set the data source depending on the activity. There are four categories that we can select, as follows:

- **FI Balances**
  We use this for exposure items, and it will show the current balance to be reflected in the report.
- **FI Open Items**
  We use this for exposure items, and it will show all of the detailed transactions.
- **One Exposure**
  We use this for exposure items and to read different certainty levels, planning levels, or liquidity items. We can also use it for manual adjustments (e.g., memo records).
- **Financial Transaction**
  We use this for the hedge items, and it will read directly from treasury and risk management to determine what trades are to be included as part of the balance sheet hedging program.

9 Balance Sheet Hedging

As shown in Figure 9.59, this is now reflected in the reporting where **Payroll**, **Outgoing Payments**, and **All Accounts Payables** are within the grouping of **All Accounts Payable**.

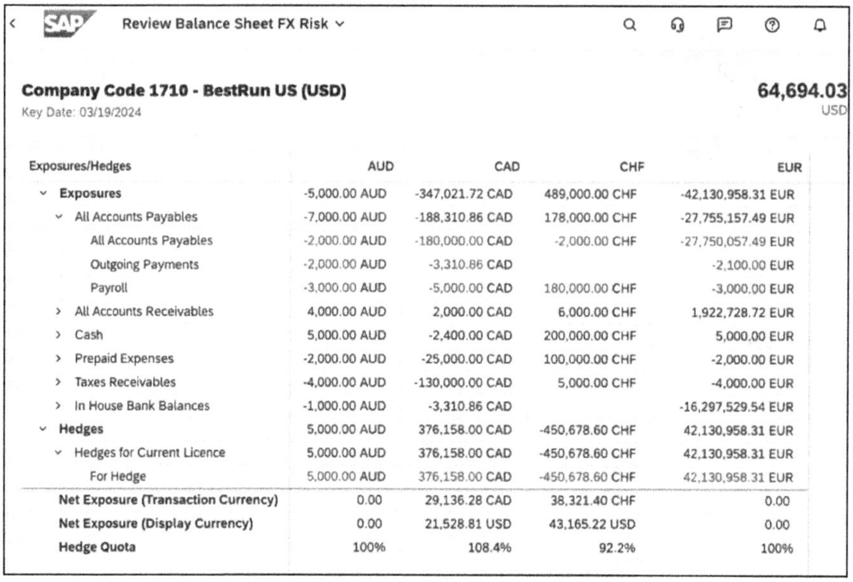

Exposures/Hedges	AUD	CAD	CHF	EUR
Exposures	-5,000.00 AUD	-347,021.72 CAD	489,000.00 CHF	-42,130,958.31 EUR
⌄ All Accounts Payables	-7,000.00 AUD	-188,310.86 CAD	178,000.00 CHF	-27,755,157.49 EUR
All Accounts Payables	-2,000.00 AUD	-180,000.00 CAD	-2,000.00 CHF	-27,750,057.49 EUR
Outgoing Payments	-2,000.00 AUD	-3,310.86 CAD		-2,100.00 EUR
Payroll	-3,000.00 AUD	-5,000.00 CAD	180,000.00 CHF	-3,000.00 EUR
> All Accounts Receivables	4,000.00 AUD	2,000.00 CAD	6,000.00 CHF	1,922,728.72 EUR
> Cash	5,000.00 AUD	-2,400.00 CAD	200,000.00 CHF	5,000.00 EUR
> Prepaid Expenses	-2,000.00 AUD	-25,000.00 CAD	100,000.00 CHF	-2,000.00 EUR
> Taxes Receivables	-4,000.00 AUD	-130,000.00 CAD	5,000.00 CHF	-4,000.00 EUR
> In House Bank Balances	-1,000.00 AUD	-3,310.86 CAD		-16,297,529.54 EUR
⌄ **Hedges**	5,000.00 AUD	376,158.00 CAD	-450,678.60 CHF	42,130,958.31 EUR
⌄ Hedges for Current Licence	5,000.00 AUD	376,158.00 CAD	-450,678.60 CHF	42,130,958.31 EUR
For Hedge	5,000.00 AUD	376,158.00 CAD	-450,678.60 CHF	42,130,958.31 EUR
Net Exposure (Transaction Currency)	0.00	29,136.28 CAD	38,321.40 CHF	0.00
Net Exposure (Display Currency)	0.00	21,528.81 USD	43,165.22 USD	0.00
Hedge Quota	100%	108.4%	92.2%	100%

Company Code 1710 - BestRun US (USD) — Key Date: 03/19/2024 — 64,694.03 USD

Figure 9.59 Review Balance Sheet FX Risk with Rollup of Categories

Now that we've set up the exposure groupings, we want to set up key elements for those groups. To do this, we'll select the **Key Figure** and then click **Maintain Selections**, as shown in Figure 9.60.

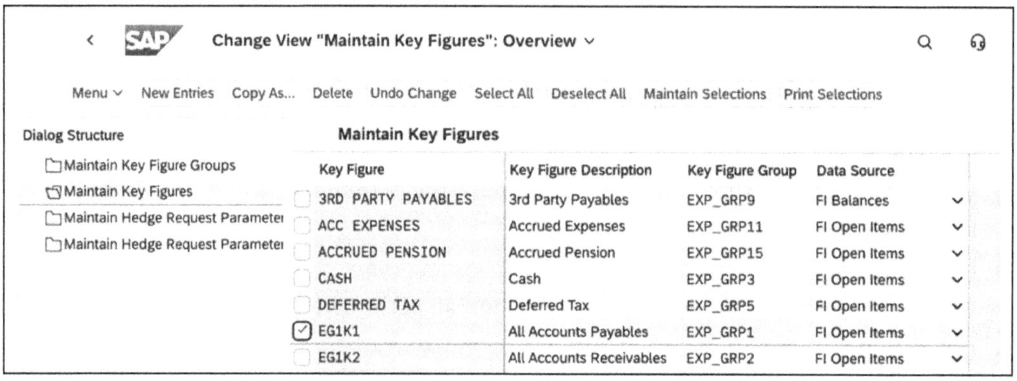

Key Figure	Key Figure Description	Key Figure Group	Data Source
☐ 3RD PARTY PAYABLES	3rd Party Payables	EXP_GRP9	FI Balances
☐ ACC EXPENSES	Accrued Expenses	EXP_GRP11	FI Open Items
☐ ACCRUED PENSION	Accrued Pension	EXP_GRP15	FI Open Items
☐ CASH	Cash	EXP_GRP3	FI Open Items
☐ DEFERRED TAX	Deferred Tax	EXP_GRP5	FI Open Items
☑ EG1K1	All Accounts Payables	EXP_GRP1	FI Open Items
☐ EG1K2	All Accounts Receivables	EXP_GRP2	FI Open Items

Figure 9.60 Maintain Key Figures: Maintain Selections

On the **Maintain Selections** screen, we'll define what information we want to collect for each of the key figures, as shown in Figure 9.61. Continuing our example with **All Accounts Payables**, we want to bring in all the vendor information, so we'll select **Account Type** "K." We can use any of the criteria here to segment exposure information.

9.6 Master Data Setup

Since **All Accounts Payable** is also set up to review **FI Open** items, we'll need to tell SAP S/4HANA what open items we want to see. We've selected "KD" for **Open Items at Key Date,** which will bring in all open items on the key date when the report is executed. Date selection options are listed in Table 9.1.

Date Selection	Description
KD – N	Key date minus number of days
KD	Key day
KD + N	Key day plus number of days
BOM – 1	Start of previous month
BOM	Start of month
BOM + 1	End of previous month
EOM – 1	End of previous month
EOM	End of month
EOM + 1	End of following month

Table 9.1 Available Date Selections for Balance Sheet Hedging

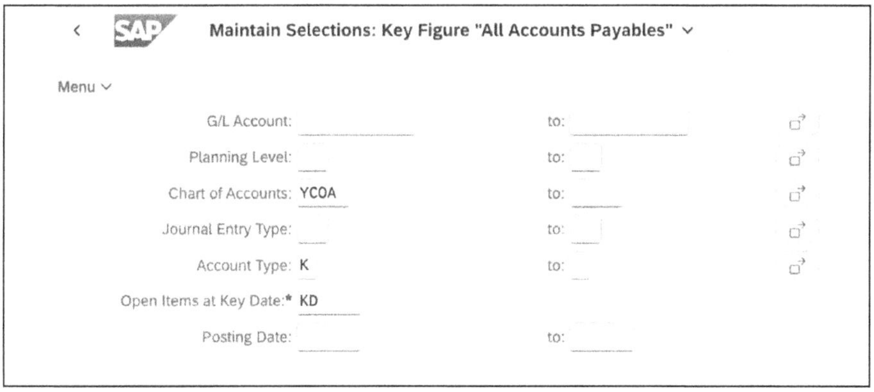

Figure 9.61 Maintaining Key Figures with Maintain Selections App

We may also want to define a key figure based on **FI Balances**. For example, if we want to see the balance for **Outgoing Payments,** we'll indicate the **G/L Account** to be reviewed for exposure collection, as shown in Figure 9.62.

The final way to generate an exposure category is through one exposure for operations. We can use the Review Balance Sheet FX Risk app for cash management flows or even one-off manual adjustments to the report. This can be helpful when there are expected FX exposures that may not be in the system yet. To capture this, we can set up a specific memo record (e.g., by entering "1H" [manual adjustments for planning] in the **Planning Level** field), as shown in Figure 9.63.

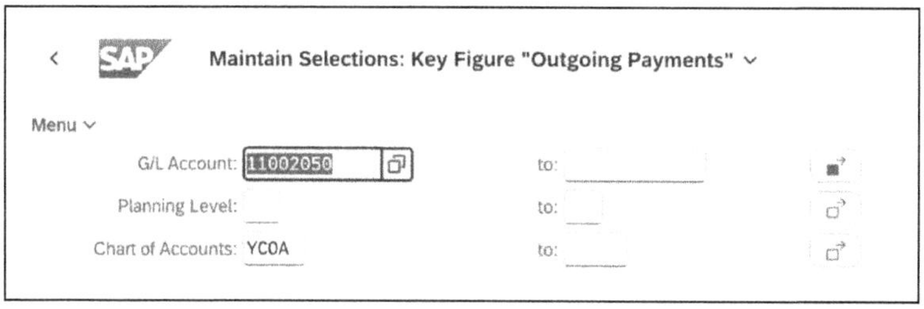

Figure 9.62 Maintaining Key Figures with Maintain Selections App: FI Balances

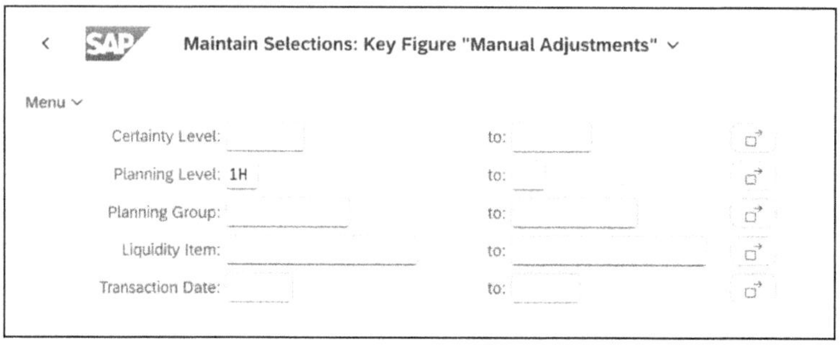

Figure 9.63 Maintaining Key Figures: Maintain Selections—One Exposure for Operations

We've now defined the three options available for an exposure category, but we also need to define a hedge category. In Figure 9.64, we'll define the transactions that will hedge the previously defined exposures. We'll use our standard **Product Type** "60A" (Foreign Exchange) and an **Assignment** of "CLIC" to represent our balance sheet hedge transactions.

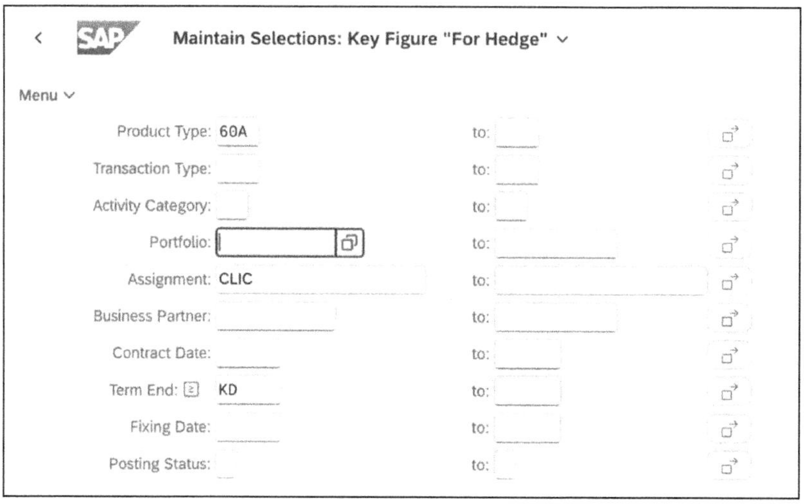

Figure 9.64 Maintaining Key Figures with Maintain Selections App: Hedge Transactions

9.6 Master Data Setup

Note

If you're generating trades for both FX balance sheet hedging and cash flow hedging, it's important to have a defining characteristic as part of the key figures, or all trades (regardless of strategy) could end up in the Review Balance Sheet FX Risk app.

Now that we have all the exposure and hedge categories defined, we'll set up the hedge request groupings by clicking on the **Maintain Hedge Request Parameter Groups** folder. As shown in Figure 9.65, we're currently set up to have one default grouping, but we could have many if desired.

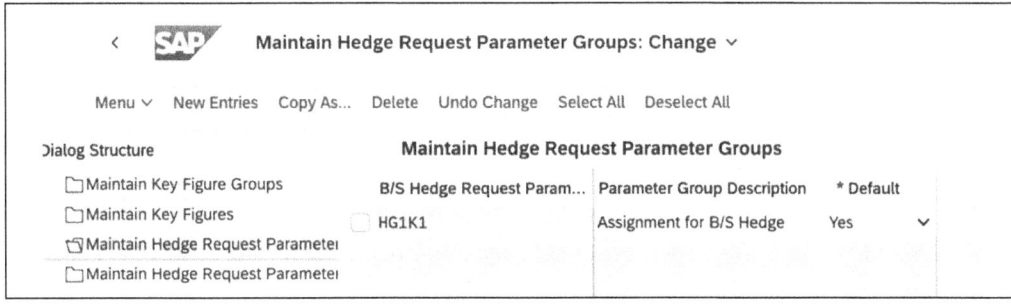

Figure 9.65 Maintaining Hedge Request Parameter Groups

The final items to set up are the hedge request parameters, and we do that by clicking on the **Maintain Hedge Request Parameters** folder. We do this to create the individual hedge requests, and it sets up the parameters that we will use for the hedge request definition (e.g., value date, instrument categories, other additional parameters), as shown in Figure 9.66.

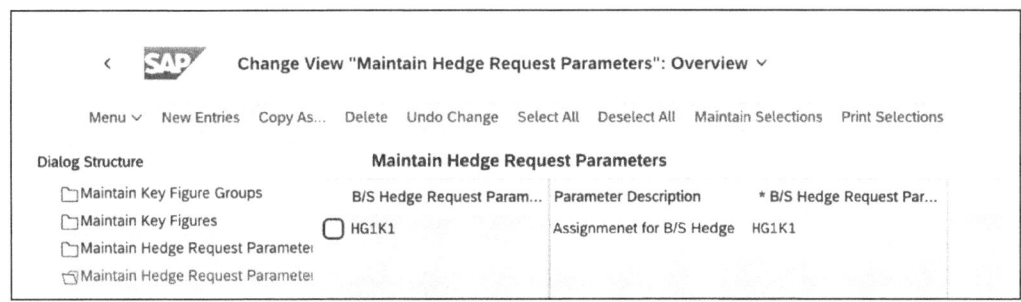

Figure 9.66 Maintaining Hedge Request Parameters

Now that we've defined the parameters, we need to maintain selections for the parameters, as shown in Figure 9.67. To do this, we select our hedge request parameters and then click on **Maintain Selections**. That's where we'll define the parameters we'll use when creating the hedge requests, and we can set up whichever filters for the snapshot

641

9 Balance Sheet Hedging

we want. Our values have been populated for **Company Code** 1010 and 1710; 1010 is listed, and 1710 is in the additional selections box, so when a snapshot is executed, it will only look at exposures in those two entities.

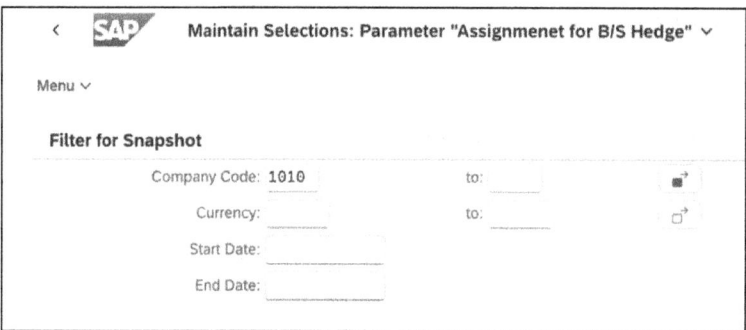

Figure 9.67 Maintain Selections: Assignment for Balance Sheet Hedge Parameter

Next, we'll determine the value dates for when the system will generate hedge requests. Here, there are two options: **Time Period** and **Fixed Date**. If we select **Time Period**, we'll see a few additional fields populate, as shown in Figure 9.68. We'll have the ability to set the value date definition (e.g., first day of period or last day of period). We'll also be able set additional days (+ or –) and add a working day shift including previous-day or next-day scheduling.

> **Note**
> This is important as we don't want to have hedge requests created when trading markets are closed. That will cause the hedge request to fall to error status.

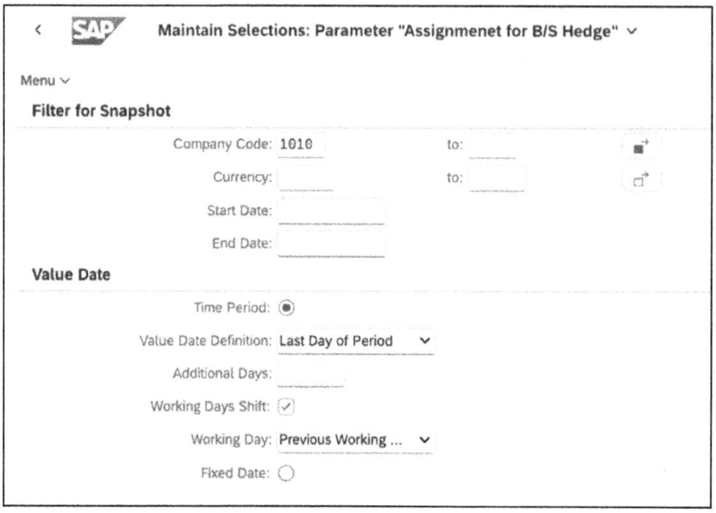

Figure 9.68 Maintain Selections: Assignment for B/S Hedge Parameter—Value Data Settings

9.6 Master Data Setup

The last portion of entry is setting the overall parameters for the hedges. The key items we'll define here are the instrument category, any rounding needs, minimum trade amounts, the release status, and a description of the hedge. We'll notice in the **Instrument Category** that we can create the following:

- FX Spots
- FX Forwards
- FX Collars
- FX Options
- FX Non-Deliverable Forwards

Most commonly, we would select **FX Forwards**, but we do have other options, depending on our balance sheet hedging strategy and the currencies involved.

Additionally, we can define rounding rules and minimum trade amounts, as shown in Figure 9.69. Sometimes, we may prefer to trade in rounded amounts, and with this setting, SAP S/4HANA allows for that. If we simply enter a value like "1,000," this setting will round all the hedge requests to the nearest 1,000 (e.g. 218,275 US dollars would round to a hedge request of 218,000 US dollars).

Figure 9.69 Parameters for Balance Sheet Hedging: Rounding Settings

We can also indicate key minimum amounts of trades. Often, treasury teams may not want to execute low-volume trades, so with this setting, we can have SAP S/4HANA ignore the hedge request until it has an exposure that reaches an acceptable minimum trade value. We'll base this on the minimum amount in a base currency against the exchange rate of our choosing. In Figure 9.70, we've chosen "25,000" US dollars with an exchange rate of "M."

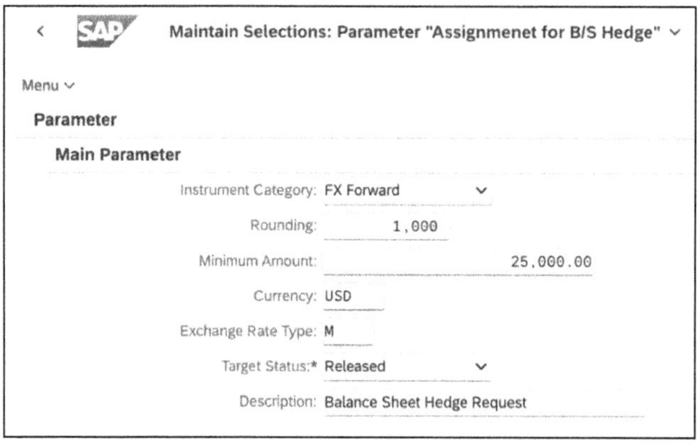

9 Balance Sheet Hedging

Figure 9.70 Parameters for Balance Sheet Hedging: Minimum Trades

We'll then define the **Target Status** and the overall **Description** of the hedge transactions. The **Target Status** defines what will happen after the hedge request is created: will it go into **Created** status, which requires another user to approve the request in the Manage Hedge Requests app, or will it go into **Released** status, where the hedge request will be fully processed and sent to the trading platform for execution? Additionally, it's important to add a description to the hedge transaction so we'll know the source of the request (e.g., balance sheet hedges, cash flow hedges).

> **Note**
>
> There are several additional fields that can be used to further segment balance sheet hedging profiles. This would be decided based on the use cases you have for balance sheet hedging, as shown in Figure 9.71.

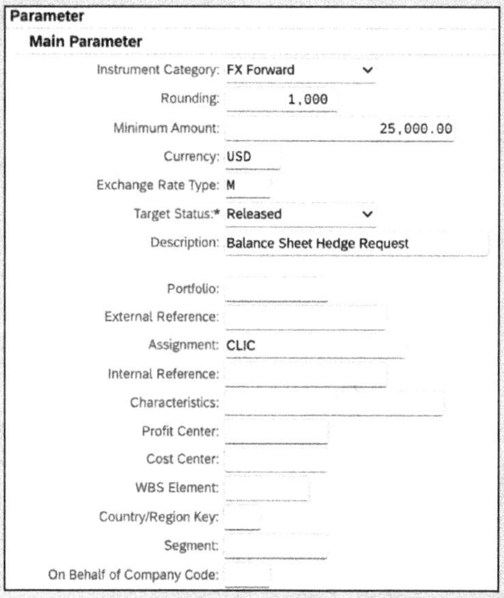

Figure 9.71 Parameters for Balance Sheet Hedging: Additional Fields

9.7 Summary

We've just completed the end-to-end process for balance sheet hedging. Most of this process is completed by the end users, and with SAP S/4HANA, the balance sheet balances are presented to the treasury users, where they control the groupings of how exposures are displayed. Once the risk manager sees the exposures, they can send them to a trading platform for execution and the deal will then populate back in the Review Balance Sheet Risk app. This simplifies the overall process of managing our FX balance sheet risk.

Chapter 10
Correspondence

The correspondence function in treasury and risk management generates and receives messages for both internal and external communication. We can use these messages to integrate with the business processes for the financial transactions.

The correspondence function in SAP S/4HANA Finance for treasury and risk management generates and processes messages related to financial transactions and securities accounts. We can use this function to create internal messages related to financial transactions and to create external messages to send to a counterparty. An example of such a message would be an MT300 or MT320 confirmation of a financial transaction that we send to a business partner. We use the MT300 for foreign exchange (FX) instruments and the MT320 for money market transaction confirmations.

An older version of the correspondence framework was available prior to SAP ERP 6.0 EHP 4, but since that version came out in 2008, we'll cover the more recent correspondence framework in SAP S/4HANA. In this framework, we can generate PDF forms, SAPscript forms, and other file types. On the incoming side, we can process correspondence without changes to ensure we have an audit trail of the transaction documentation.

Overall, the correspondence functionality in treasury and risk management creates a flexible framework for creating a comprehensive solution that delivers and imports messages for financial transactions.

Before jumping into the chapter, let's cover some key terms:

- **Correspondence**
  Correspondence is a generic term to describe the different forms and files we can use in both external and internal communications from the treasury and risk management module. We can also process external incoming messages by using this function.

- **Correspondence framework**
  The *correspondence framework* is the functionality that was built to create and process the correspondence messages. The implementation of this functionality utilizes this framework.

- **Correspondence object**
  When we generate correspondence, records are generated to track all details within the correspondence communication. These include the details of the message and

the data required in SAP to generate and send the message. As the system processes the objects, it updates the status of the correspondence object.

- **Message**
  The actual message that the correspondence function generates and processes is called the *message*. This is the raw form of the message that is generated and sent through a communication channel. The file is created as a printed form, email, fax, or electronic file.

- **Correspondence monitor**
  To track all the correspondence messages, a *correspondence monitor* was created. This monitor has a dashboard that displays the correspondence messages and allows for the actioning of any activities for the messages.

This chapter is a comprehensive guide to the uses of correspondence in SAP S/4HANA. The first few sections of this chapter cover the configuration and master data settings for correspondence. Starting in Section 10.1, we'll cover the step-by-step instructions for the core configuration settings required to create and process correspondence related to financial transactions. Section 10.2 details additional settings required for all inbound and outbound correspondence messages. Section 10.3 dives deeper into the settings, and Section 10.4 continues with more specific settings related to correspondence. These settings include details that are specific to correspondence related to certain company codes, specific product types, and assignment of data for the business partners.

We then cover the business process for correspondence and detail how it supports the business function in treasury and risk management. Section 10.5 starts this by detailing how outbound messages are generated, and Section 10.6 details the inbound processing of messages. Section 10.7 looks into the criteria for how correspondence can be automatically matched when it's imported into SAP S/4HANA, and it also looks at the postprocessing steps for manually matching correspondence. Section 10.8 details how to view the correspondence messages, and it explains the correspondence monitor that serves as the central dashboard for all incoming and outgoing messages. Rules are also determined in configuration to drive alerts for correspondence, so Section 10.9 looks at both the settings and the use of the alerts function.

10.1 Configuration

In this section, we'll cover the details of configuration and the master data changes required to set up correspondence messages. For the purposes of this section, we'll be covering configuration in the **Financial Supply Chain Management • Treasury and Risk Management • Transaction Manager • General Settings • Correspondence** menu path. The configuration in this area creates the core elements of the correspondence function.

Here's a summary of the steps that will be covered in this configuration section. First, the types of communication channels and specific formats are created, and these drive the types of files that are generated and received by the correspondence function. Eventually, the formats will have specific mapping assigned to them to determine how incoming and outgoing files need to be structured.

After these initial settings are completed, the correspondence recipient types determine the type or business partner that is sending or receiving the correspondence message. Then, correspondence classes are determined to detail the types of activities that can generate the messages. The communication profile leverages all the previous settings. In the profile, the format metatypes are mapped to the correspondence class to determine the types of formats that can be generated for each class. Finally, the specific formats are tied to all of the previous settings in the assign formats configuration. This step creates a mapping that includes the communication profile, correspondence class, recipient/sender type, and communication channel to the specific correspondence formats. Each of these details is covered in the following section, which serves as a guide to creating all base configuration settings in the correspondence setup.

10.1.1 Defining Communication Channels

The communication channel configuration is in the **Financial Supply Chain Management • Treasury and Risk Management • Transaction Manager • General Settings • Correspondence • Correspondence Message Interface • Define Communication Channels** menu path. This is an initial configuration that defines the types of files that can be generated as correspondence. The two checkboxes have the following unique functions for the correspondence channels:

- **Response**
  We check this box to indicate that we expect a response to a message. When we're sending a file through the file system or SAP Multi-Bank Connectivity, an acknowledgement or negative acknowledgement can be processed to verify the message was received and that the format was correct.

- **No Canc. Rqrd (no cancellations required)**
  By default, we generate a correspondence message when a financial transaction is created. When the transaction is changed, the existing correspondence messages are invalidated and new messages are generated. As part of this, the system generates cancellation messages for the messages that have already been delivered. We check this box to block the cancellation messages from being generated.

We can see the SAP-delivered correspondence channels in Figure 10.1.

10 Correspondence

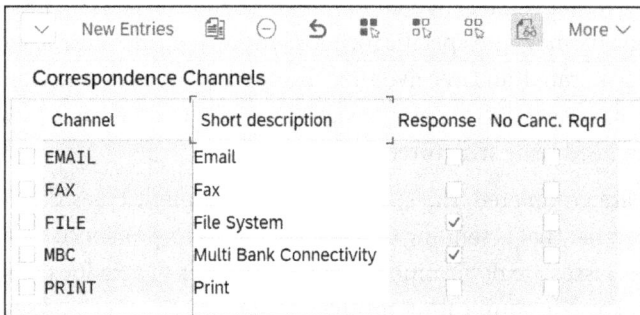

Figure 10.1 SAP-Delivered Correspondence Channels

10.1.2 Defining Format Metatypes

The format metatype configuration is located in the **Financial Supply Chain Management • Treasury and Risk Management • Transaction Manager • General Settings • Correspondence • Correspondence Message Interface • Define Format Metatypes** menu path. This configuration creates the initial settings for the type of form in SAP S/4HANA. Here are some examples of the types of formats available:

- **FORM**
  We use this type of format to generate PDF copies of SAPscript forms.
- **SWIFT**
  We use this type of format for files that are generated in the Society for Worldwide Financial Interbank Telecommunication (SWIFT) format specification.
- **SWIFT MBC**
  We use this type of format for SWIFT messages that are sent through SAP Multi-Bank Connectivity.
- **SWIFT SAP**
  We use this type of format for SWIFT messages that are sent using the SAP integration package 4.

Figure 10.2 shows these formats as they appear onscreen.

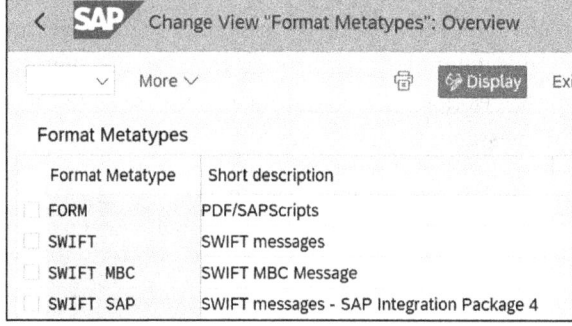

Figure 10.2 Format Metatypes for Forms and Standard SWIFT Files

10.1 Configuration

10.1.3 Defining Formats

The next configuration node is in the **Financial Supply Chain Management • Treasury and Risk Management • Transaction Manager • General Settings • Correspondence • Correspondence Message Interface • Define Formats** menu path. This node, shown in Figure 10.3, covers the specific type of format we'll use for correspondence. The first column determines the type of SWIFT message or PDF form that is generated or processed. The next two columns, for message type and format group, further define the format, as follows:

- **Msg. Type (message type)**
  There are three options for the message type, as follows:
  - **Normal in- or outbound message**
    This is a message going into or out of the correspondence framework.
  - **Acknowledgement message**
    This is a technical message that can be imported to update the status or a correspondence object to say the message was acknowledged.
  - **Negative acknowledgement message**
    This is a technical message that can be imported to update the status or a correspondence object to say it the message was rejected.

- **Frm. Group (format group)**
  This configuration categorizes the correspondence message by type. Possible values to assign in this section are PDF, SAPscript format, and SWIFT MT message. (Since all the SWIFT messages start their name with an MT, they're referred to as *SWIFT MT messages*.)

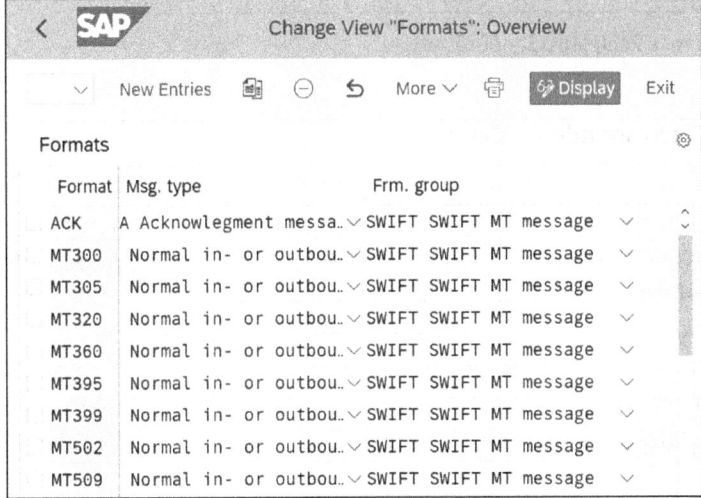

Figure 10.3 Example View of Defined Formats in Configuration

10.1.4 Defining Correspondence Recipient Types

This configuration is the correspondence recipient type located in the **Financial Supply Chain Management • Treasury and Risk Management • Transaction Manager • General Settings • Correspondence • Correspondence Messaging Interface • Define Correspondence Recipient Types** menu path. This node describes options for who is sending or receiving the message. When we have an outgoing message, we'll assign a correspondence recipient that will receive that message. When we receive an incoming message, the correspondence recipient type will be who sent the message. These senders and receivers are tied to the business partner roles, so the applicable business partner roles need to be assigned as in Figure 10.4. The **Recipient/SenderType** is defined in the first column, and the associated business partner role is mapped in the **BP Role** column. Also, the **Internal** column drives whether the messages for this recipient/sender type send and receive internal or external correspondence messages. The default setting assumes the message is meant for external use, so if the message is to be internal within SAP systems, we must check the **Internal** checkbox.

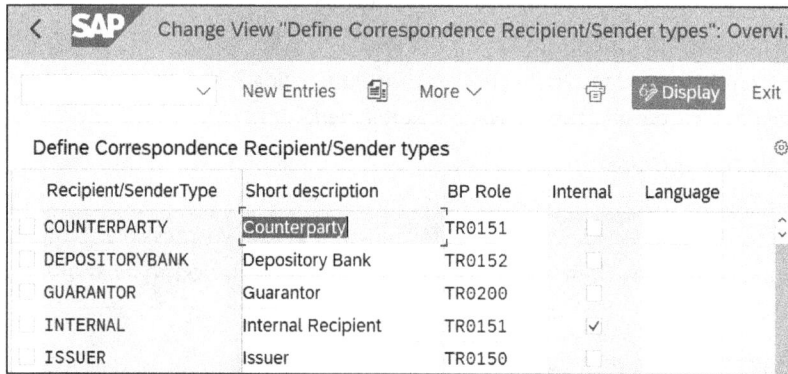

Figure 10.4 Correspondence Recipient/Sender Types

10.1.5 Defining the Correspondence Class

The next step in this process is to determine the activity for which correspondence is sent. We do this in the **Financial Supply Chain Management • Treasury and Risk Management • Transaction Manager • General Settings • Correspondence • Correspondence Messaging Interface • Define Correspondence Classes** menu path. In this node, there are different options for the class, including the DEAL_FX class (see in Figure 10.5). This class generates correspondence when creating an FX transaction. Generally, the standard settings delivered by SAP S/4HANA are adequate, and the correspondence classes don't require additional customizing changes.

Correspondence Class	Short description
DEAL_DER	Deal: Derivatives
DEAL_FREE_TEXT	Free Format Message
DEAL_FX	Deal: Foreign Exchange
DEAL_MM	Deal: Money Market
DEAL_QUERIES	Queries
DEAL_SEC	Deal: Securities
SEC_BUY_SELL	Order to Buy or Sell
SEC_CLIENT_CONF	Client Confirmation of Purchase or Sale

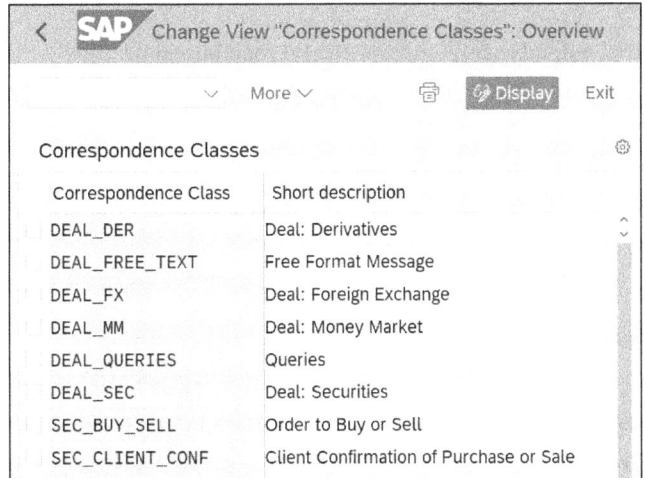

Figure 10.5 Correspondence Classes Determine Why Message Is Sent

10.1.6 Defining the Correspondence Class for Inbound Process

This configuration node starts to map some of the initial settings together: format, recipient/sender type, and correspondence class (see Figure 10.6). These settings are located in the **Financial Supply Chain Management • Treasury and Risk Management • Transaction Manager • General Settings • Correspondence • Correspondence Messaging Interface • Assign Correspondence Class for Inbound Process** menu path.

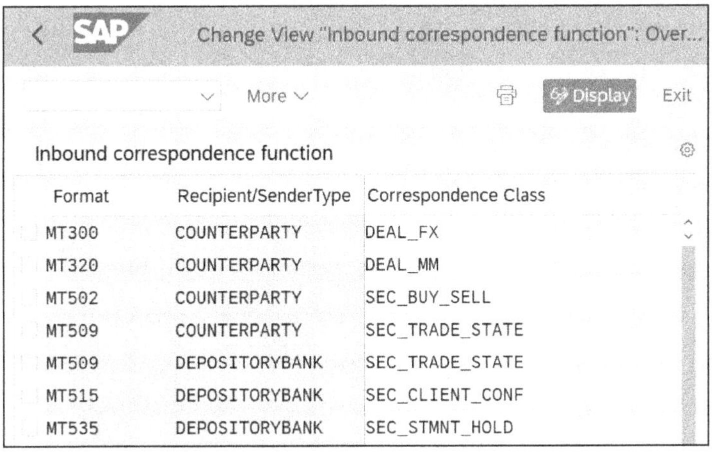

Format	Recipient/SenderType	Correspondence Class
MT300	COUNTERPARTY	DEAL_FX
MT320	COUNTERPARTY	DEAL_MM
MT502	COUNTERPARTY	SEC_BUY_SELL
MT509	COUNTERPARTY	SEC_TRADE_STATE
MT509	DEPOSITORYBANK	SEC_TRADE_STATE
MT515	DEPOSITORYBANK	SEC_CLIENT_CONF
MT535	DEPOSITORYBANK	SEC_STMNT_HOLD

Figure 10.6 Mapping Format, Recipient/Sender Type, and Correspondence Class

This mapping is required specifically for the inbound process since the data is not generated in SAP S/4HANA. These settings need to be defined as follows:

- **Format**
  This is the incoming format that is imported from the business partner.

10 Correspondence

- **Recipient/SenderType**
  Leveraging the settings that already were created in the recipient type configuration, this maps the type of business partner that can send the format mapped in the first column.

- **Correspondence Class**
  This column maps to how the inbound format is processed and determines the channel and additional attributes in the communication.

10.1.7 Defining Communication Profiles

The previous configuration nodes were technical attributes that needed to be created to set up the correspondence. Now that the initial attributes have been created, they can all be assigned to a communication profile. This profile controls details for the creation, sending, and receipt of messages. We need to go through multiple steps to set up this node correctly, as follows:

- Creating a communication profile
- Assigning a channel
- Assigning format metatypes
- Assigning channel-dependent attributes
- Assigning format-dependent attributes

Creating a Communication Profile

The first step in this configuration is to create a communication profile name. We can determine the creation of the communication profile name by first-double clicking on the **Communication Profiles** folder on the left side of the screen. For the example in this chapter, we have created a new profile named **CC_PROFILE**. We created it by clicking on the **Create** button at the top of the screen and determining a **Profile** name and **Short description**. The created profile is shown in Figure 10.7.

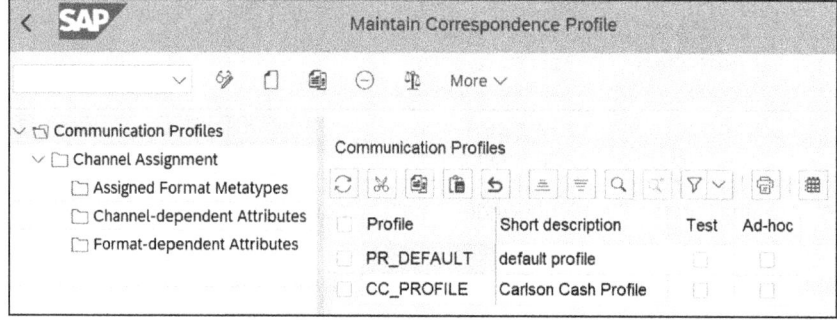

Figure 10.7 Creating Communication Profile

Assigning a Channel

The next step is to assign the correspondence class, recipient/sender type, and communication channel to this profile. To navigate to this function, we highlight the communication profile by clicking on the box to the left of the profile name, and then, we double-click the **Channel Assignment** folder. In Figure 10.8, there are some correspondences that should be sent using a SWIFT file and some that should generate a PDF. This is mapped accordingly in the **Channel** column. The important fields are as follows:

- **Correspondence Class**
  Why is the message being sent?
- **Recipient/SenderType**
  Who is sending or receiving the message?
- **Channel**
  How is the message being sent?

When adding entries to the channel assignment, we need to click the **Create** button each time a new line needs to be added.

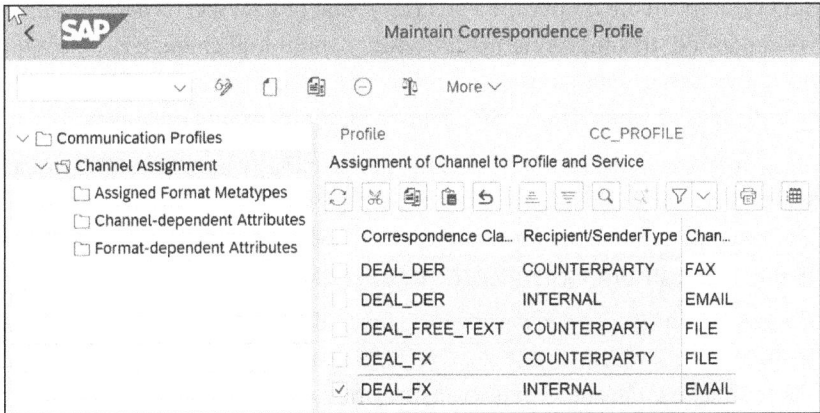

Figure 10.8 Mapping Correspondence Class to Communication Channels

Assigning Format Metatypes

Once we have completed the channel assignment, we can start filling the three dependent folders. For each of the channel assignments, we can highlight the line and then click into the following folder to create the entries. The first folder we'll cover is the format metatypes. We'll assign the created format metatypes from Section 10.1.2 and assign them to the channel. For this example, we'll assign the metatype to the **DEAL_FX**, **COUNTERPARTY**, **FILE** entry (the second-to-last line in Figure 10.8). We'll then send this correspondence by creating a SWIFT file, and we can see in Figure 10.9 that we've assigned the SWIFT **Format Metatype**.

10 Correspondence

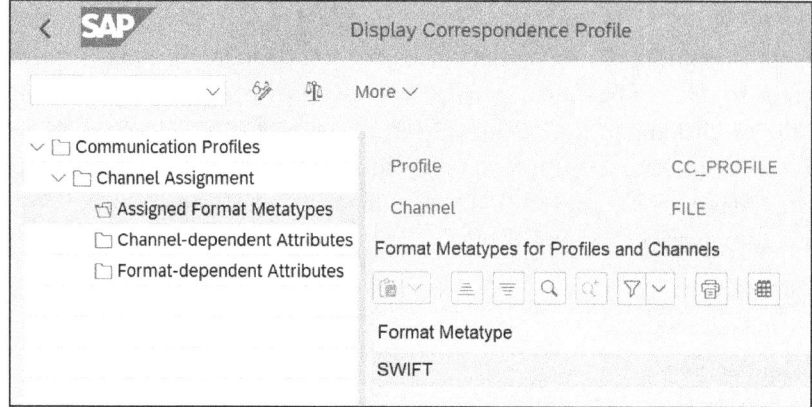

Figure 10.9 Assigning Format Metatype to Correspondence Class/Channel

Assigning Channel-Dependent Attributes

The channel-dependent attributes folder defines details of the format to be sent in this scenario. In the example in this section, we're assigning the MT300 format, and we're assigning the prefix in the next column a custom name of ZMT300. When the correspondence is generated, this prefix name determines the initial characters of the file name. If required, we can use the subsequent columns to define where the file is sent to for outgoing messages or where the file is loaded from for incoming messages.

If we're going to map the logical and physical file paths using Transaction FILE, we can check the box in the **LogicalFld** column. Then, we can define the logical file path in the **FilOut** and **Folder In** columns. This determines the folder where outgoing correspondence is sent and where incoming correspondence should be loaded from.

As with the logical file names, we can determine a file folder for the output and input of messages. In Figure 10.10, we can see that we've directly defined the output and input folders. These should be different folders to ensure one folder is used for the outbound messages and another is defined for the inbound messages. Also, the **LogicalFld** checkbox is not checked in this scenario since we're not using a logical and physical file path.

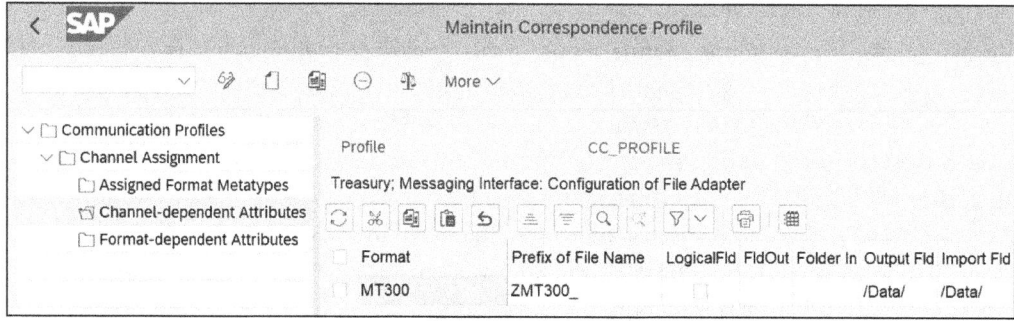

Figure 10.10 Channel-Dependent Attributes with Definition of Output and Import Folders

10.1 Configuration

Assigning Format-Dependent Attributes

The final step in the correspondence profile is to assign the format-dependent attributes for SWIFT messages. When using the SWIFT message types, we need to assign the correct SWIFT number in the first column. We don't include the *MT* in this configuration, so when we're mapping this table for the MT300 format, we only need to enter "300" in the column. The second column is an indicator for Transaction DMEEX, which is used to create different formats within SAP S/4HANA. Generally, we create payment file formats in Transaction DMEEX, but we can map correspondence files in it as well. When the format should be mapped using the DMEEX engine, we can check the **DMEEX Ind.** box. One thing to note is that the appearance of this tab is dependent on the correspondence channel and format. In Figure 10.11, an MT300 format is assigned, so we need to map the 300 into the external format indicator.

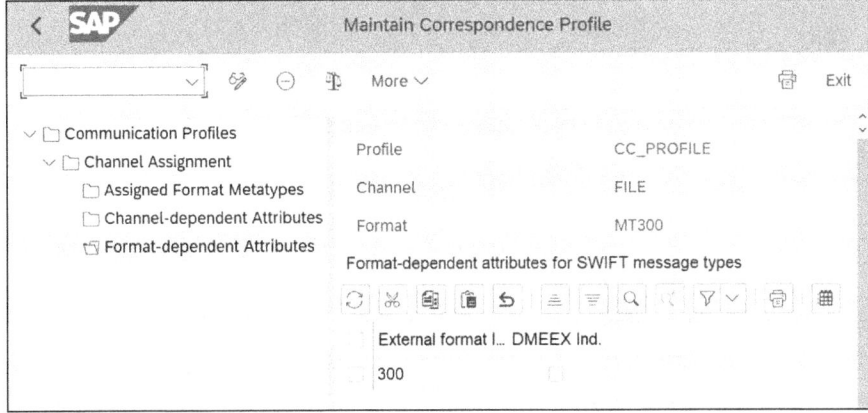

Figure 10.11 Mapping MT300 to Corresponding External Format Indicator

If a PDF form were assigned, the mapping would require us to map a form category and form name like in Figure 10.12.

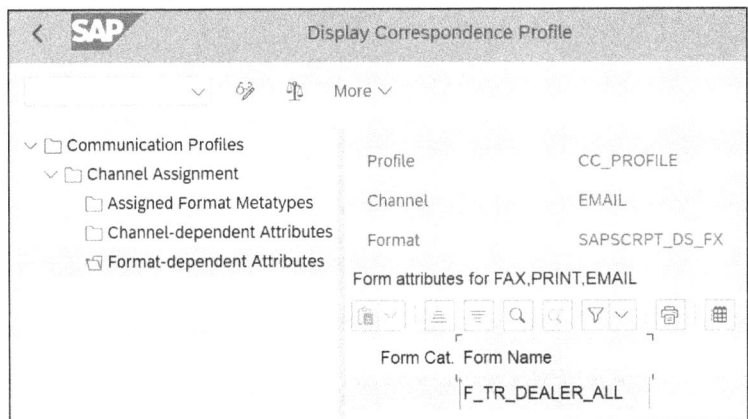

Figure 10.12 Form Category and Form Name Assigned to SAPscript Format

> **Standard Delivered Forms**
>
> There are standard forms available for assignment for both SAPscript and PDF forms. Two SAPscript forms are available, and these are F_TR_CONFIRM_ALL for deal confirmations and F_TR_DEALER_ALL for dealing slips.
>
> Various PDF forms available for assignment for the correspondence process. The following formats are available for assignment in this area:
>
> - **Dealing slips**
>   - TR_F_AFM_CORR_DE (for derivatives)
>   - TR_F_AFM_CORR_FX (for FX transactions)
>   - TR_F_AFM_CORR_MM (for money market transactions)
>   - TR_F_AFM_CORR_SE (for securities)
> - **Confirmations**
>   - TR_F_AFM_CONF_DE (for derivatives)
>   - TR_F_AFM_CONF_FX (for FX transactions)
>   - TR_F_AFM_CONF_MM (for money market transactions)
>   - TR_F_AFM_CONF_SE (for securities)

10.1.8 Assigning Formats

The **Assign Formats** configuration node is an additional mapping table that assigns the formats to a profile, correspondence class, recipient/sender type, and channel. This assignment is done in the **Financial Supply Chain Management • Treasury and Risk Management • Transaction Manager • General Settings • Correspondence • Correspondence Messaging Interface • Assign Formats** menu path. As shown in Figure 10.13, we can assign multiple recipient/sender types and channels to an individual correspondence class to create more than one type of correspondence for an activity.

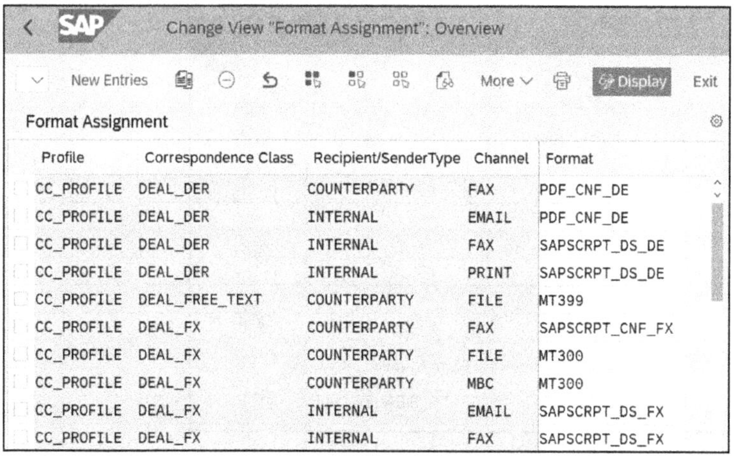

Profile	Correspondence Class	Recipient/SenderType	Channel	Format
CC_PROFILE	DEAL_DER	COUNTERPARTY	FAX	PDF_CNF_DE
CC_PROFILE	DEAL_DER	INTERNAL	EMAIL	PDF_CNF_DE
CC_PROFILE	DEAL_DER	INTERNAL	FAX	SAPSCRPT_DS_DE
CC_PROFILE	DEAL_DER	INTERNAL	PRINT	SAPSCRPT_DS_DE
CC_PROFILE	DEAL_FREE_TEXT	COUNTERPARTY	FILE	MT399
CC_PROFILE	DEAL_FX	COUNTERPARTY	FAX	SAPSCRPT_CNF_FX
CC_PROFILE	DEAL_FX	COUNTERPARTY	FILE	MT300
CC_PROFILE	DEAL_FX	COUNTERPARTY	MBC	MT300
CC_PROFILE	DEAL_FX	INTERNAL	EMAIL	SAPSCRPT_DS_FX
CC_PROFILE	DEAL_FX	INTERNAL	FAX	SAPSCRPT_DS_FX

Figure 10.13 Each Format Needs to Be Mapped Accordingly to Communication Profile

In this transaction, we use the data that was defined in the previous configuration steps. We define the **Profile, Correspondence Class, Recipient/SenderType,** and **Channel** in the first four columns of the configuration. Then, when all of those scenarios match, the **Format** column defines the actual format that should be generated.

10.1.9 Defining the Correspondence Partner

When we're processing correspondence, we need to further define the sender and recipient of the messages. These messages need to have a correspondence partner to determine the recipient of the outgoing messages and the sender of the incoming messages. Here are the steps we need to take to define the correspondence partner:

- Defining internal recipients
- Defining a business partner group
- Assigning settings to business partner groups

Defining Internal Recipients

We define the internal recipients in the **Financial Supply Chain Management • Treasury and Risk Management • Transaction Manager • General Settings • Correspondence • General Settings • Define Internal Recipients** menu path. The purpose of this configuration is to determine the internal recipients within a company. This is very general, and it could indicate a single user, a group of users, or a printer on which to print a correspondence PDF. If a recipient needs to receive correspondence, we can create the entry in this table. For this example, we'll create an entry for an internal recipient called INTERNAL, as shown in Figure 10.14. In this transaction, we create the internal recipient name in the **Int. Recipient** column, and then the second column defines an additional description for this internal recipient (**Text**). The internal recipient will be assigned once we progress further in the configuration.

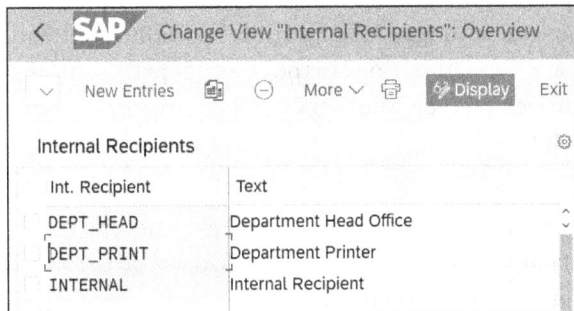

Figure 10.14 Internal Recipients Are Initially Created in This Configuration Node

Defining a Business Partner Group

We can create an additional correspondence partner by defining a business partner group. We create this in the configuration in the **Financial Supply Chain Management •**

Treasury and Risk Management • Transaction Manager • General Settings • Correspondence • General Settings • Define Business Partner Group menu path. We use business partner groups because the business partners are master data, and frequently, the master data is not 100% in sync between systems. Therefore, to avoid any additional dependencies, we create the business partner groups, which allows us to create certain settings for a group of business partners. When we have several business partners that should follow the same correspondence settings, we can assign them to the same business partner group.

There's a default business partner group available, but for this example, we've created a new business partner group named **CC_BPGROUP** (see Figure 10.15), and we'll use in the following configuration. A **Business Partner Group** name is created in the first column, the **Text** column creates an additional description for the business partner group, and the **Active Ind** column contains checkboxes where we can determine whether the business partner is activated.

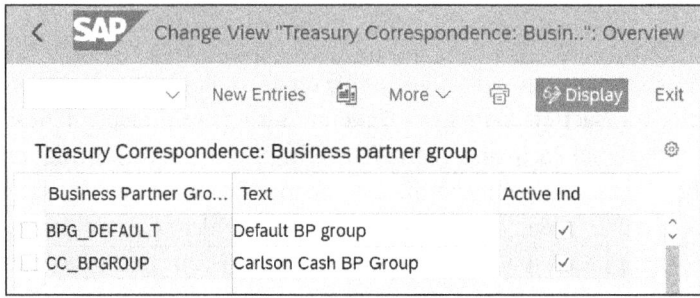

Figure 10.15 Business Partner Groups Are Created and Activated

Assigning Settings to Business Partner Groups

Since the previous configuration only created a name for the business partner group, we still need to add details to the group. We do this in the next configuration, which is located in the **Correspondence • General Settings • Assign Attributes for Business Partner Groups** menu path. In this area, we assign settings to the business partner groups, correspondence class, and recipient/sender type. The details of the settings and checkboxes are as follows (see Figure 10.16):

- **Automatic Correspondence**
  Based on the configuration, the correspondence is set to trigger whenever a certain status for an item is met. By default, the correspondence message will go to the correspondence monitor, and it will be available for us to send it by clicking on the **Create Message and Send** function. If we check this box, the correspondence will skip this step and will automatically be sent.

- **Release Required**
  If we check this box, the correspondence object will need to be released within the workflow process before it can be sent.

10.1 Configuration

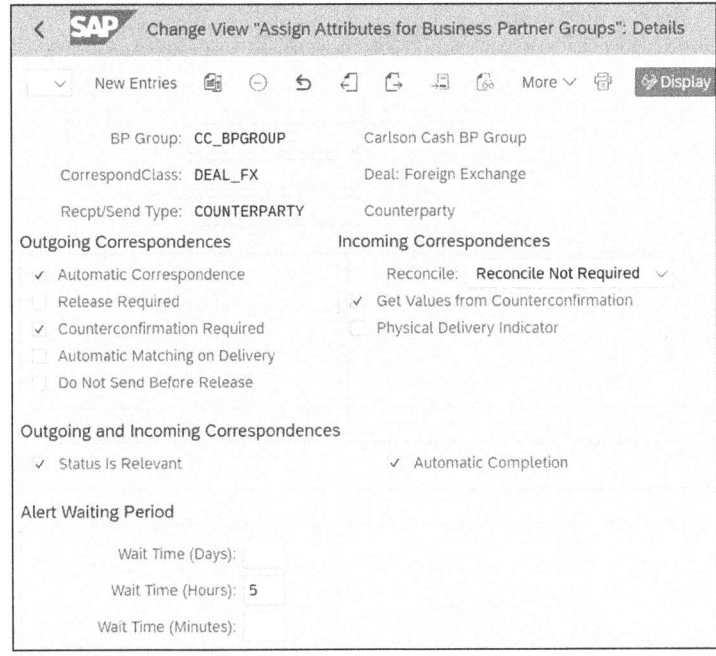

Figure 10.16 Attributes for Business Partner Groups

- **Counterconfirmation Required**
  If we check this box, the system will expect to receive a correspondence to confirm the transaction details. This is required prior to sending the outgoing correspondence. This confirmation should come in with an attribute of **Status Is Relevant**, and once received, the status of the outgoing correspondence can be updated to **Completed**.

- **Automatic Matching on Delivery**
  If we check this box, the system will try to match an incoming correspondence with the outgoing correspondence when it's delivered. This scenario would only apply if we expect to receive the incoming correspondence before the delivery of the outgoing correspondence.

- **Do Not Send Before Release**
  If we check this box, the system will restrict correspondence from being generated for transactions that are not released. If we have a process indicating that transactions need to be released, we can use this indicator to only send correspondence after the release happens. This applies to both manual and automatic correspondence. Automatic correspondence will only be sent once a transaction is released, and similarly, the manual correspondence will show up in the correspondence monitor, but we will not be able to send the correspondence after the transaction is released.

- **Reconcile**
  This indicator is for incoming correspondences. The reconcile feature compares and reconciles the correspondence data with the internal data.

661

- **Get Values from Counterconfirmation**
  When we check this box, we determine that some data from the counterconfirmation should be copied into the confirmation and financial transaction. The data that is pulled in is for the business partner and external reference. (This is determined by the **CONTACT_PERSON** and **EXTERNAL_REFNCE** fields.)

- **Physical Delivery Indicator**
  We should check this box when an incoming correspondence object is related to the physical delivery of a securities transaction. This setting is related to the physical delivery setting of the transaction type for security transactions located in the **Financial Supply Chain Management • Treasury and Risk Management • Transaction Manager • Securities • Transaction Management • Transaction Types • Define Transaction Types – Securities** menu path. This indicator defines that a correspondence indicating the physical delivery should be received.

- **Status Is Relevant (for financial transactions)**
  We should check this box when a correspondence object should influence the status of a financial transaction. When an outgoing correspondence object is marked as **Status Relevant**, then the correspondence needs to be sent to update the transaction status to **Confirmed**. Also, when an outgoing correspondence object is relevant for counterconfirmation and has been matched to a correspondence object that is status relevant, the transaction status should update to **Counterconfirmed**.

- **Automatic Completion**
  When we're looking at correspondence objects in the correspondence monitor, we consider the final status of the objects when they are set to the **Complete** status. If this box isn't checked, then the status of the objects needs to be manually updated to complete, but if this box is checked, then the processing of the incoming and outgoing objects proceeds as follows:
  - Outgoing Correspondences
    - Not Counterconfirmation Relevant
      The outgoing correspondence object is automatically updated to **Complete** when the correspondence is sent.
    - Counterconfirmation Relevant
      In this scenario, the status is updated to **Completed** only when the incoming status-relevant correspondence object is received and matched with the outgoing correspondence object.
  - Incoming Correspondences
    - Not Status Relevant
      These correspondence objects will automatically be set to **Completed**.
    - Status Relevant
      These are only updated to **Completed** once they have been matched to an outgoing correspondence object that is counterconfirmation relevant.

- **Alert Waiting Period**
  We can use the alert monitor to monitor the correspondence. If there's a delay in processing the correspondence, we can send an alert. Examples of when to send an alert are when an outgoing message hasn't been sent, counterconfirmations are pending, or incoming messages haven't been matched. In the **Alert Waiting Period** section, we can determine the waiting period before an alert is sent.

10.1.10 Assigning Format Mapping for Outbound and Inbound Process

There are a couple of configuration nodes we use to alter the inbound and outbound mapping of the correspondence messages. There's also an alternative way to create the correspondence mappings using the Map Format Data for Treasury Correspondence app. The mapping for this transaction is covered in Section 10.4.2. When a mapping exists, it can be assigned in the **Financial Supply Chain Management • Treasury and Risk Management • Transaction Manager • General Settings • Correspondence • Correspondence Messaging Interface • Assign Format Mappings for Outbound Process** and **Assign Format Mappings for Inbound Process** menu paths. In these configuration nodes, we determine the format name and the format mapping name it is being assigned, as shown in Figure 10.17.

This allows us to determine that we want to use the format mappings determined in the Map Format Data app. To define this, we need to assign the original format in the **Format** column, map the applicable **Business Partner Group** in the second column, and map this to the **Format Mapping Name** in the fourth column. This format mapping name is the name that is assigned in the Map Format Data app.

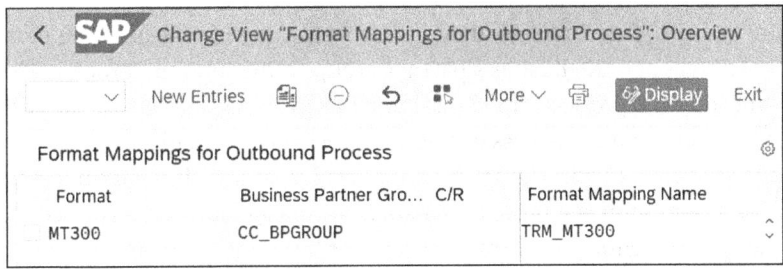

Figure 10.17 Mapping Format Name to Format Mapping in Map Format Data

10.2 Inbound and Outbound Process Settings

In addition to the inbound correspondence configuration we set in the previous section, we should use a couple of user transactions to assign additional information for the inbound processing. As with the inbound settings, we need to assign additional settings to the outbound process to send the correspondence.

10.2.1 Inbound Process

An additional setting that can be determined instead of assigning the correspondence class in configuration. (covered in Section 10.1.6). This setting includes assigning the mapping format, business partner, recipient/sender type, and correspondence class in Transaction FTR_INB_FUNC. Figure 10.18 shows an example of this assignment. The key fields of this assignment are as follows:

- **Format**
  This is the inbound format that is being imported.
- **BPartner**
  This is the business partner that is determined on the inbound format file.
- **Recipient/SenderType**
  This is the business partner type that has sent the format file.
- **Correspondence Class**
  Based on the previous three columns, this correspondence class will be assigned to the inbound file in the correspondence monitor.

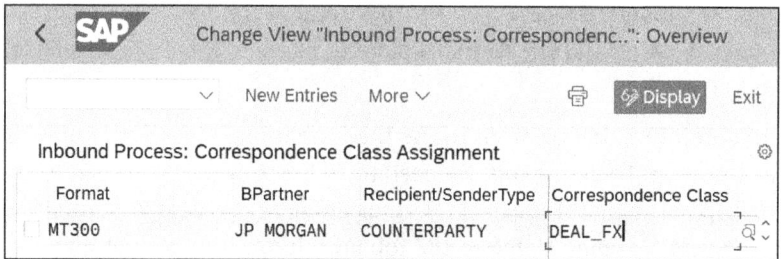

Figure 10.18 Mapping Correspondence Class in FTR_INB_FUNC

Additionally, since a host of settings and attributes were assigned to the business partner group, we need to assign the business partner group to the inbound process. The assignment for the inbound process is in Transaction FTR_INB_ASSIGN. In this transaction, we assign the details to the business partner, transaction information, and the relevant correspondence class like in Figure 10.19. We need to add each scenario to this assignment. A combination of the business partner, company code, recipient/sender type, product category, product type, transaction type, and correspondence class is assigned to the business partner group ID created in Section 10.1.9.

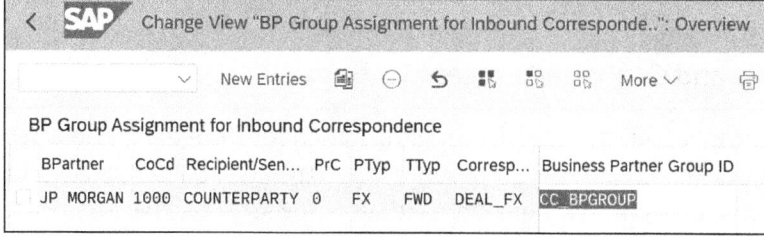

Figure 10.19 Business Partner Group Is Assigned to Inbound Process for Specific Transactions

10.2.2 Outbound Process

When generating the correspondence for the outbound process, we need to assign the correspondence profile to the correct business partner, business partner group, and correspondence recipient type. We complete this assignment in Transaction FTR_EXT_ASSIGN (Assignment of Profiles and BP Groups to External Recipients).

We do this in a couple of different steps. First, we need to create an entry by clicking on the **+** button on the left side of the screen to tie together the desired business partner (**BusPartner**), company code (**CoCd**), and recipient type (**Recipient/SenderType**), as shown in Figure 10.20. Once this initial entry is created, we'll be able to assign further details for the external recipient.

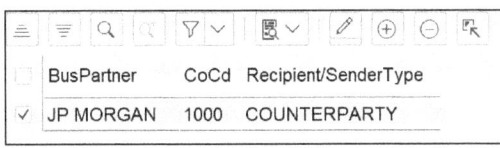

Figure 10.20 Creating Assignment of Profiles for External Recipients

Next, there is a dropdown list at the right side of the screen that lets us drill down into each transaction. This drill down, shown in Figure 10.21, starts with the generic categories for products and then goes into the technical product category, product type, and transaction type.

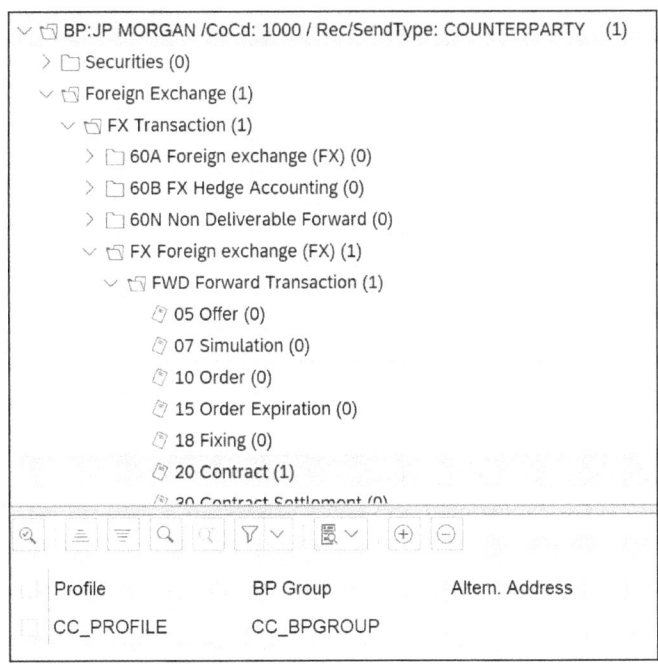

Figure 10.21 Assigning Correspondence Profile and Business Partner Group to Specific Activity for Correspondence Creation

Finally, it lets us see the different statuses for the transactions (contract, settlement, fixing, etc.). Drilling into these statuses, we can assign the correspondence profile and business partner group to the individual status within the product type and transaction type.

In this example, we're assigning the correspondence profile and business partner group to the **Contract** activity for FX forwards. This creates one of the settings required to generate correspondence for the forward when a contract is created.

Finally, we must make the settings for internal recipients, which are similar to those for external recipients. We need to assign the internal recipient to an internal recipient type. The settings for this assignment are located in Transaction FTR_INT_ASSIGN (Assignment of Profiles and BP Groups to Internal Recipients).

As with the previous process for external recipients, there are a few steps involved in assigning internal recipients. First, we need to assign the internal recipient (**Int. Recipient**), company code (**CoCd**), and recipient type (**Recipient/SenderType**), as in Figure 10.22.

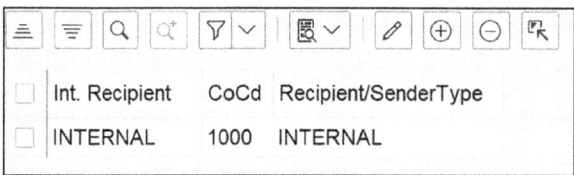

Figure 10.22 Assigning Internal Recipients to Internal Recipient/Sender Type

As in the earlier process, we assign details of the internal recipients to the individual statuses of the product types and contract types. In our example, we're going to send an internal correspondence PDF for the FX contract. Also, as in the external-recipient example, we need to drill down into the desired product type and transaction type to determine which activity should create the correspondence. In this case, we've assigned the CC_PROFILE profile and the CC_BPGROUP BP group.

As we can see in in Figure 10.23, we've defined that the PDF will be sent via email and will be sent to the indicated email address. One thing to note is that there are a couple ways to assign the email address. If we assign a user ID, SAP S/4HANA will send an email to the address assigned to that user ID. If we leave the user ID blank, SAP S/4HANA will send the email to the email address indicated in this configuration.

10.3 Additional Correspondence Settings

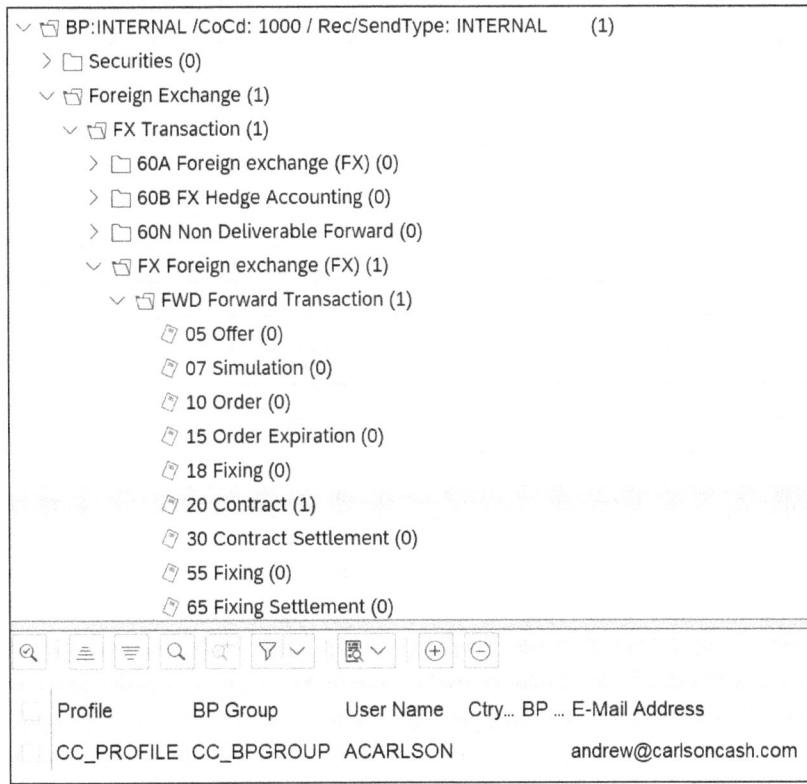

Figure 10.23 Assigning Email Address to Internal Email

10.3 Additional Correspondence Settings

The previous sections cover most of the settings required for the correspondence setup, but this section defines or assigns some additional settings that are required for the end-to-end process, as follows:

- Correspondence activity
- Start and end fields
- Dynamic tables
- BIC codes
- Number ranges

10.3.1 Defining Correspondence Activity

The correspondence activity determines the timing or what triggers the correspondence to be created and sent. The correspondence activity is configured in the **Financial Supply Chain Management • Treasury and Risk Management • Transaction Manager • General**

667

10 Correspondence

Settings • Correspondence • Correspondence Activities • Define Correspondence Activities • Money Market/Foreign Exchange/Derivatives/Securities/Trade Finance menu path. (Note that there are five different nodes that we can use, depending on the type of contract for the correspondence.)

This node specifically assigns the correspondence class and recipient type to the type of transaction being created and the activity that should generate the correspondence. To add any entries, we go into the applicable configuration node. For example, this section is going to cover adding correspondence for an FX transaction, so we'll go into the **Foreign Exchange** node.

First of all, when we enter this configuration, we'll select the company code in the pop-up box. Once we select the company code, the enter checkbox button will let us navigate to the configuration entry screen. We need to select each area in this configuration, so we should enter the product category, product type, transaction type, activity category, recipient/sender type, and correspondence class. As with other areas of SAP S/4HANA, if we leave a field blank, it will be considered a wildcard and be applicable to all values. For example, if we configured this table to be for **Prod. Category** "600" (**FX Transaction**) and left the **Product Type** and **Transactn Type** fields blank, then the configuration would be applicable to all product types and transaction types that use product category "600" (**FX Transaction**). The exception occurs when there's another entry in this table that is more specific. If the entry has the product type and transaction type assigned, that value is be used since it's more specific. The assignment for external recipients is shown in Figure 10.24.

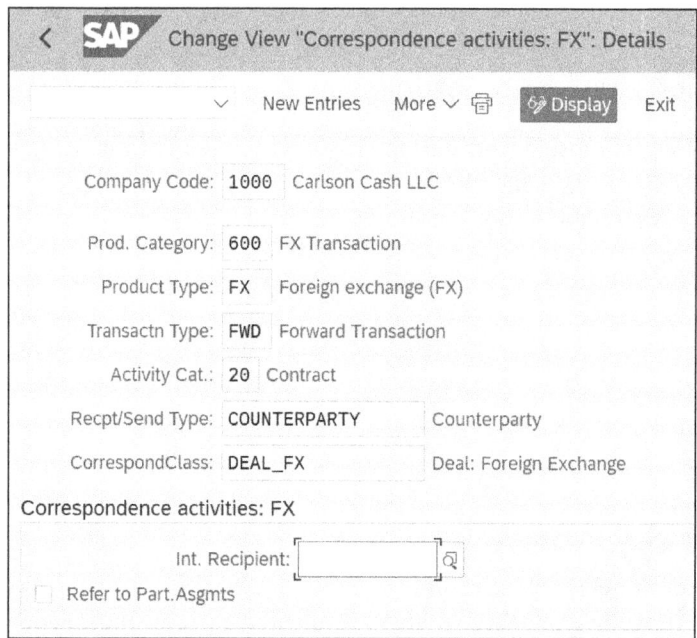

Figure 10.24 Assigning Activity Category to External Recipient

The assignment process for external recipients is identical, with the exception that we should set the **Recpt/Send Type** field and the **Int. Recipient** field to "INTERNAL."

10.3.2 Defining Start and End Fields for Sequences in SWIFT Messages

We can define or review additional settings in the configuration for the start and end fields for SWIFT messages. We can find this configuration in the **Correspondence • Correspondence Messaging Interface • Define Start and End Fields for Sequences in SWIFT Messages** menu path. This helps us define file sequences in the SWIFT format. The standard delivered configuration in this section should be sufficient for the SWIFT messages, but we can configure it if additional SWIFT messages are required. The mapping in this configuration is required for both inbound and outbound SWIFT messages. An example of this mapping for the MT300 messages is in Figure 10.25. It's unlikely that we'll need to make edits in this area, but it's available if need be.

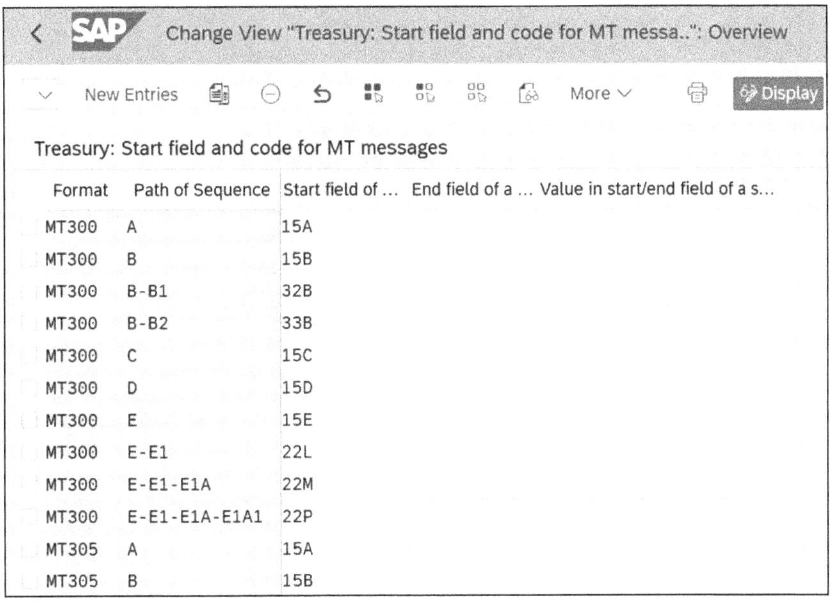

Format	Path of Sequence	Start field of ...	End field of a ...	Value in start/end field of a s...
MT300	A	15A		
MT300	B	15B		
MT300	B-B1	32B		
MT300	B-B2	33B		
MT300	C	15C		
MT300	D	15D		
MT300	E	15E		
MT300	E-E1	22L		
MT300	E-E1-E1A	22M		
MT300	E-E1-E1A-E1A1	22P		
MT305	A	15A		
MT305	B	15B		

Figure 10.25 Standard Mapping for MT300 Messages

10.3.3 Dynamic Table Assignment for Configuration

We can find the dynamic table assignment required for communication channels and formats in the **Correspondence • Correspondence Messaging Interface • Dynamic Table Assignment for Configuration** menu path. This table has standard settings that we should only change if previous customizing decisions require it. The standard settings in this configuration are shown in Figure 10.26: the different types of formats being generated are mapped to a dynamic table for the necessary message generation.

10 Correspondence

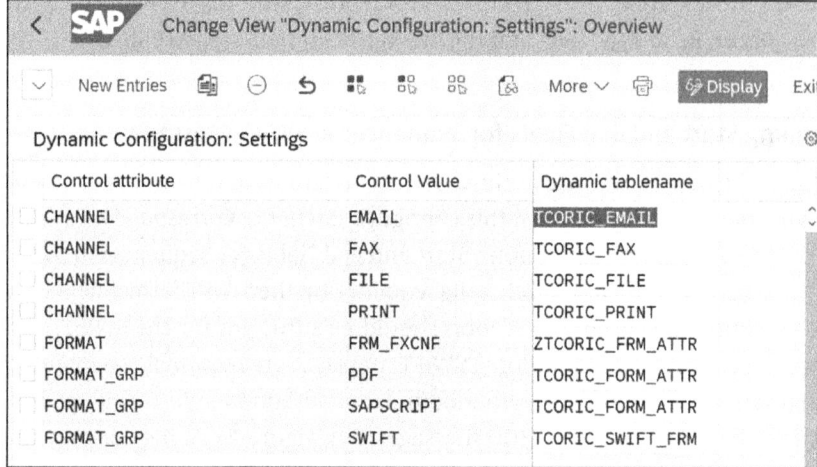

Figure 10.26 Mapping Dynamic Settings for Formats and Channels

10.3.4 Defining BIC Codes and Accounts for Business Partners

When setting up a business partner, we define the business partner's SWIFT BIC code. If this is not maintained directly on a business partner, we can use Transaction FTR_BP_BIC to assign the SWIFT code to the business partner. As we can see in Figure 10.27, this setting is specific to the correspondence class, so we could determine different recipient SWIFT codes for different messages if required. The combination of adding a business partner (**BusPartner**), **Valid From**, recipient/sender type (**Recpt/Send Type**), and correspondence class (**CorrespondClass**) determines the assignment of the SWIFT code. Functionally, this mapping determines how SAP S/4HANA reads the SWIFT code that is delivered on an incoming SWIFT message to determine the business partner to assign in the correspondence monitor.

Figure 10.27 Assigning SWIFT Code to Business Partner for Correspondence

10.3.5 Setting Up Number Ranges

Number ranges are required for all incoming and outgoing correspondence messages. We can set up the relevant number ranges in the following menu paths:

- We set up the number range for match ID via **Correspondence • General Settings • Define Number Ranges for Match ID**, as shown in Figure 10.28.

- We set up the number range for correspondence ID via **Correspondence • General Settings • Define Number Ranges for CO ID**, as shown in Figure 10.29.

We covered configuring number ranges in detail in Chapter 3, Section 3.10.

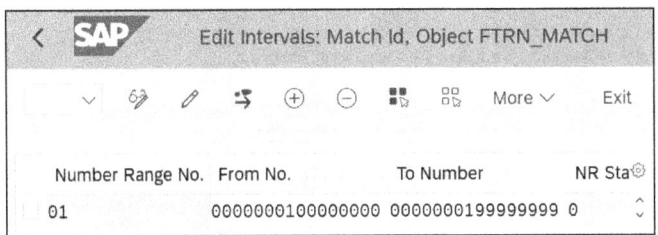

Figure 10.28 Creating Number Ranges for Matching Correspondence

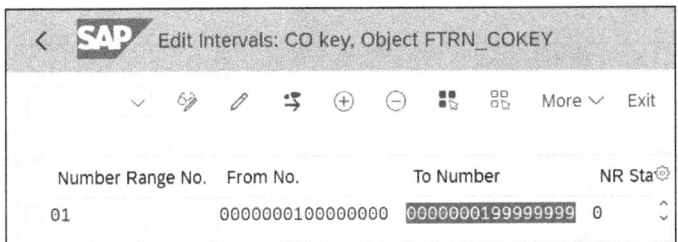

Figure 10.29 Creating Number Ranges for Correspondence Objects

10.4 Correspondence Mapping Rules

Now, you may be wondering how the mapping works for the different file types. In this section, we'll look at how SAP S/4HANA knows how to generate the file, the logic behind the mapping of each field, and how the structure is built in that file. This all occurs in the correspondence mapping rules. In the following sections, we'll walk through how to do this in both the traditional SAP GUI and in the newer SAP Fiori UX.

10.4.1 SAP GUI

To view the mapping settings, we need to navigate to Transaction SM34 (View Cluster Maintenance). To edit this information, we go to the TCORFVC_MAPPING view cluster in Transaction SM34. The initial screen for the mapping table displays the full list of formats available in Figure 10.30.

10 Correspondence

Once in the transaction, we can see the different file formats for both the outgoing and the incoming formats. In this example, we'll look into to the MT300 format by highlighting the format and clicking into the **Mapping rules** folder.

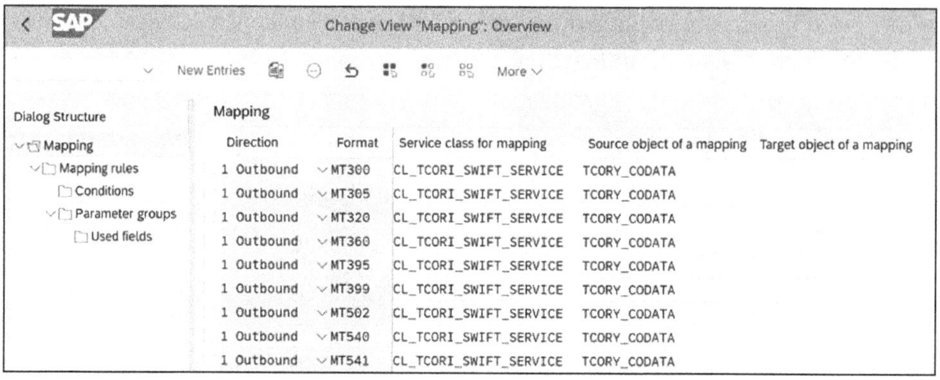

Figure 10.30 Detailed Mapping Settings for SWIFT Messages for Correspondence

The details in this table in Figure 10.31 are used to map each individual value in the SWIFT format. This area is useful if we want to copy a standard format to create our own version of the file mapping.

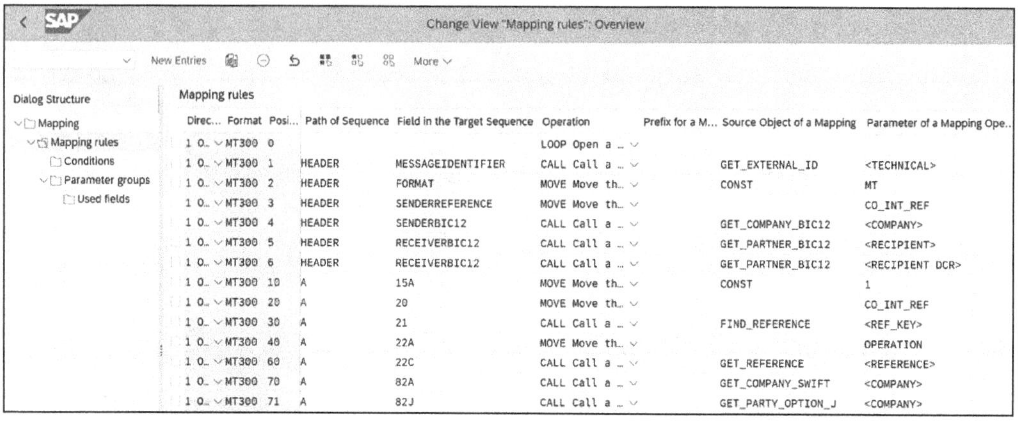

Figure 10.31 Detailed Mapping of Each Position in MT300 Format in Transaction SM34

10.4.2 SAP Fiori

Another function that we can use to map the correspondence is in the SAP Fiori app Map Format Data for Treasury Correspondence. This app allows us to either update format mappings or create our own formats. In our example, we'll examine how an MT300 has been mapped in this app. First of all, the format mapping in this example is called TRM_MT300. As shown in Figure 10.32, a list of correspondence types is already provided in the system.

10.4 Correspondence Mapping Rules

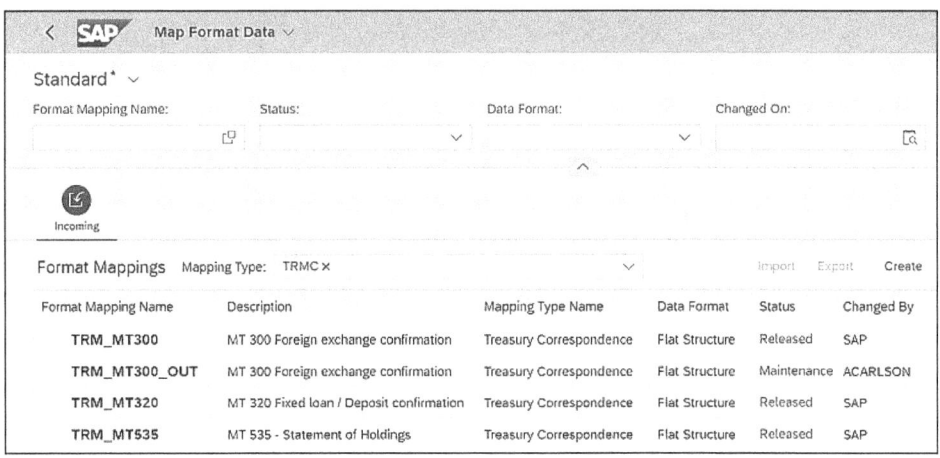

Figure 10.32 Sample and Modified Format Mappings in Map Format Data for Treasury Correspondence

We can drill into each of these formats to see how the format has been mapped. When we go into the format, we first see the initial settings for the format in the **Format Mapping Properties** area. This defines the base settings for the payment format that includes the description, output structures, and definitions of how we delimit the files if there's a space in the mapped data. There are also constants that we can create at the header level and that will be mapped later in the file creation process.

The following section covers a series of settings that we can review and edit when changing the file format mapping. First, we'll cover the header-level properties for the file mapping. Then, we'll cover the different types of nodes that we can create in this file structure. Each of these nodes determines the structure of the file, and there is also a toolbar that determines how we change the nodes when we're editing them. Most importantly, we'll go through the different options for how to define and map the nodes.

Format Mapping Properties

We access the first section of this format mapping by double-clicking into the **Format Mapping Properties** area. This area contains the base settings for the correspondence file format. There are four tabs in these properties that we can edit, as follows:

- **Attributes**
  We can create base attributes in this section to define the name of the format, the structure name that is used for the output, and how fields are separated in the file.
- **Constants**
  We can create constant values that we can reused further in the format mapping.
- **Variables**
  We can define a list of variables in this area and then reuse them in the format mapping. The variables we define here can be assigned in action nodes. The action node

can reference this variable to either create a sum of numbers or concatenate text fields.

- **Enumerations**
We can create a static list of items and assign them in the correspondence format. The name we create in this area will be referenced later in the mapping to assign the defined list of values.

Node Types

Before going into each node and explaining how everything is mapped, we need to describe the different types of nodes available for the mapping. The icons for the node types are shown in Figure 10.33 and are as follows:

- **Record Group**
We use this type of node to group records in the format mapping. We can use this to group similar nodes that we want to map in certain scenarios.

- **Record**
Each record represents a block of information that is being mapped. These records will start with an identifier tag or field name to create the structure of the format.

- **Field Group**
We can group field mapping into a field group. If we want to map multiple data points within a record, we can group the fields using this type of node. We can this if we want to have a certain group of fields that need to be mapped in one scenario but we then want to map a different group of fields in a different scenario. (We can drive this by adding conditions on the field group.)

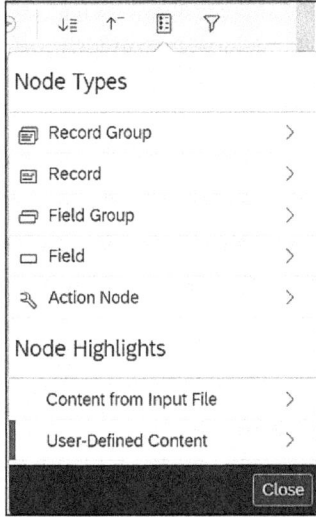

Figure 10.33 Node Types Available to Create and Modify Correspondence Mapping

- **Field**

  This represents a field inside of a record, and we can use this to map individual data points within a record number. We need to map these under either a record or a field group.

- **Action Node**

  The action node allows us to alter the data on the output file by performing an action. An example of this would be if we used an action node to move the data to a different row of the file.

Node Navigation

Now that we've covered the nodes that we can add in the format mapping, we can look at how to alter them in the mapping. The icons in Figure 10.34 appear above the format mapping, and they drive how we can change the mapping tree for the format data. We'll cover each of these mappings in the order they appear:

- **Create Node**

  This creates a new blank node in the location of the tree that we select.

- **Cut**

  This removes the node from the current location of the tree, but we can paste it to a new location with the **Paste** button.

- **Copy**

  This copies a node that we can place in a new location with the **Paste** button. The original node doesn't get deleted.

- **Paste**

  If we cut or copy a node, we can paste the node with this button. It will insert the node in the location that we select in the tree.

- **Duplicate**

  This creates a duplicate of a node.

- **Delete**

  This deletes the node and all the contents in the node.

- **Revert to Original**

  If we've edited a node, we can use this button to revert the settings back to the previous active version of the node.

Figure 10.34 Icons Used to Modify Format Mapping

Using these buttons, we can edit the format mapping of the files. An example of the mapping tree for correspondence is shown in Figure 10.35.

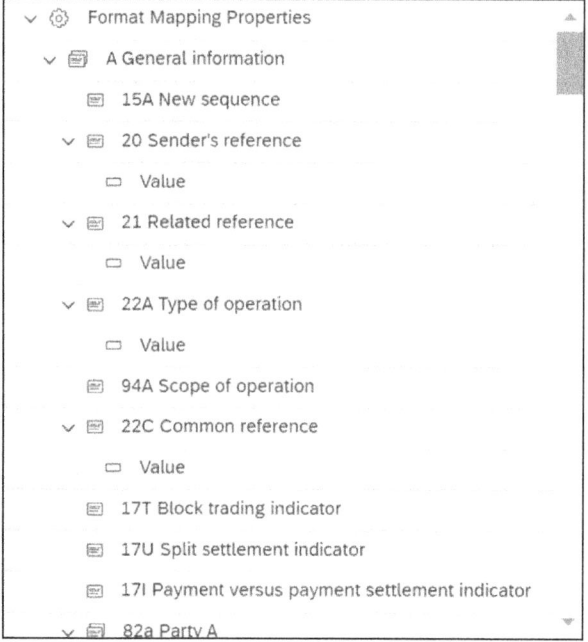

Figure 10.35 Display of Standard Format Mapping Tree for MT300 Format

Node Editing and Mapping

When we're editing the nodes, there are four areas that we can map, and each of them has its own purpose in the mapping:

- Attributes
- Input specification
- Output specification
- Condition

The key details of each of these areas is described in the following sections, along with descriptions of how to map each of them. Additionally, we can create mapping details and logic for the output value adjustment, conditions, and variables via building expressions.

Attributes

The **Attributes** tab has a few details that we use to create base data for the node, as follows:

- **Name**

  This creates a name for the node. When we look at the structure of the nodes on the left-side menu, this name will appear for the node. Typically, we name the node with the MT field name and a description to easily identify the nodes in the tree.

10.4 Correspondence Mapping Rules

- **ID**
  SAP S/4HANA automatically creates a technical ID for each node that we create. This is not editable, but we can reference it within the format mapping hierarchy.

- **Description**
  We can use the description field internally to help define the purpose of the node. The details in this field are not mapped on the file.

- **Mandatory Node**
  When processing a file, the mandatory node determines if the node is absolutely required to be on the file. If a mandatory node is missing from a file that is input into the system, the file will fail processing.

Input Specification

The **Input Specification** tab appears different, depending on whether we are mapping a record or a field in the structure, as follows:

- **Record**
  The record input specification determines the field tag that is being mapped. If the tag reads **:020:**, then that is the corresponding tag that will appear on the file. To actually map the details within the file, we'll need to move on to the field nodes to see how this mapping occurs.

- **Field**
  The mapping of the fields works differently for incoming or outgoing correspondence. When we're looking at the incoming correspondence, we look at the field mapping to first look at the details that are delivered in the file. The output specification will determine what happens with the data that is read from the file. This area also determines the expected details that are delivered in the field, including the data type and structure. These are detailed in Table 10.1, and you'll notice that each data type has a different set of fields to populate. All of these data types are available to enter if we select the **Input Value Matches Custom Criteria** processing criteria.

Data Type	Description	Fields to Populate
Amount	This indicates that the field should have an amount as the value.	■ **Length** This determines how many characters should be in the field. The number of characters needs to match the exact number if the fixed length checkbox is selected. ■ **Decimal Separator** This determines the character that is assigned to separate the decimal places. We can set this to be a period or a comma.

Table 10.1 Different Data Types Available in Format Mapping

Data Type	Description	Fields to Populate
		▪ **Decimal Places** This field allows us to determine how many decimal places are used in the mapping. We can determine an exact number of decimal places, or we can set it up to follow the standard amount of decimal places based on the currency code. In addition to this, the default setting will use the default settings determined in the format mapping properties for the decimal separator.
Currency	This defines that the field should be populated with a currency's ISO code.	No additional settings are required.
Date	This defines that a date should be delivered.	▪ **Date Format** In this field, we define a date format using D, M, and Y to determine the day, month and year (respectively). For example, if we enter MM/DD/YYYY, we are mapping the standard date format for the United States.
Number	This determines that the field should be mapped with a number.	This field uses the same mapping fields that we determined in the **Amount** data type.
Text	We can enter any text in this section.	▪ **Length** This determines the length of the field that should be delivered. ▪ **Fixed Length** This delivers an error if the data in the file is of any other length than that determined in the previous field. ▪ **Starting Value** This determines a fixed starting value that should show up in this field. ▪ **Ending Value** This determines a fixed ending value that should show up in this field.
Time	This indicates that the field will be mapped with a time format.	▪ **Time Format** Similar to the above field, we'll create a time format using H, M, and S to determine the hours, minutes and seconds (respectively) in the mapping. For example, if we wanted to create a format using hours, minutes, and seconds, we would map HH:MM:SS in this field.

Table 10.1 Different Data Types Available in Format Mapping (Cont.)

10.4 Correspondence Mapping Rules

To demonstrate the mapping of an input specification, we have an example of a mandatory date field in Figure 10.36. This **Date Format** field uses a date format of **YYYYMMDD**, so if we're mapping the last day of the year 2024, then the date would appear as 20241231.

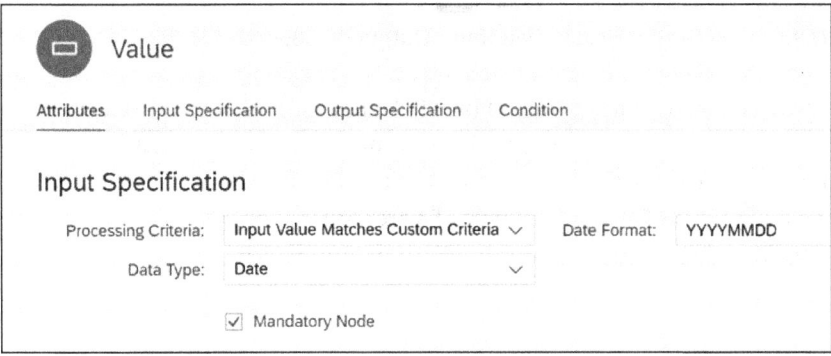

Figure 10.36 Mapping Attributes for Date in Map Format Data

The most detailed processing criteria were covered in Table 10.1, but there are a few other processing criteria that we can map for the input specification, as follows:

- **Input Value Matches Constant Value**
  This determines that a constant value should always be input in the field. If we select the reference mapping, we can have this field use a different field as a reference. If we want to manually type in the constant value, we can determine that in the **Constant Value** field. To use this field, we need to either not populate the reference field or select no reference in that field.

- **Input Value Matches One of Enumerated Values**
  There are fields that we added in the **Format Mapping Properties** area in the **Enumerations** section. If we select this option, we can limit the available values to the values that we determined in the enumerations section.

- **Input Value Matches Referred Node Value**
  In this option, we can reference another node, and the value has to match the reference node. We can only refer to a node that occurred above the node we are mapping.

- **No Input Value Processing**
  We select this option if we want to ignore the value in the input file. We can determine a different output in the output specification section of the setup.

Output Specification

For incoming payment files, the output specification defines where the data from the file gets mapped to in SAP S/4HANA. If we click on the dropdown list in the **Output Specification** section, we can see all the options for the fields that can be mapped. For example, we're going to map the trade date as shown in Figure 10.37.

679

10 Correspondence

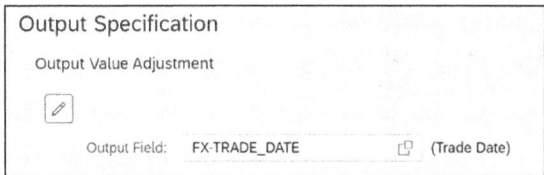

Figure 10.37 Mapping Output Field to Determine Output of Inbound MT300

The dropdown list for the **Output Field**, shown in Figure 10.38, allows us to view various options and sections of information and the information that we can map by selecting the output field. Since we're looking at an MT300 format, the data is relevant for FX, so many of the fields in this **Foreign Exchange - FX** area are mapped.

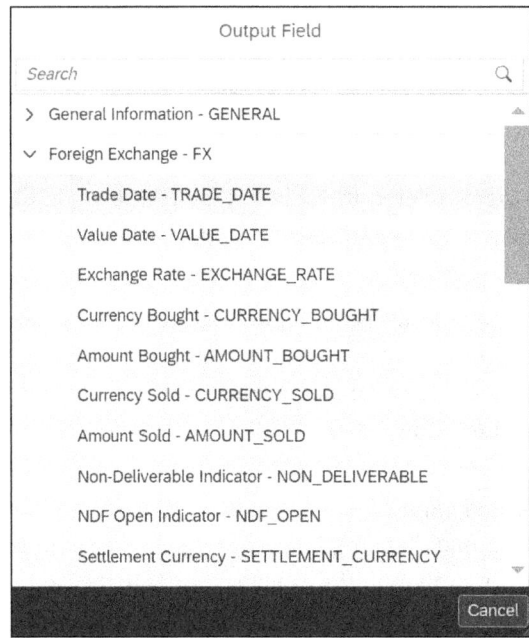

Figure 10.38 Dropdown List in Format Mapping Showing Available Fields

Condition

The **Condition** tab defines when the mapping node should be processed. If the defined conditions are not met, then the node will not be processed. We can add conditions to any type of node and add them to higher nodes in a structure. If we add a condition to a node that has a series of subnodes, the added condition will affect all the subnodes. The options for creating the conditions for mapping are detailed in the following section.

Building Expressions

There are additional mapping details and logic that we can create for the output value adjustment, conditions, and variables. These areas are inherently different from each other, but the mapping of the logic and how the data is derived works very similarly for all of them. This function in SAP S/4HANA is called *building expressions* and is made up of the following elements:

- **Binary operators**
  First of all, the expressions have a series of binary operators that are used to create the logic around the expression (as in Figure 10.39). These values can function differently if we are using numeric or alpha/text characters. If we are mapping numeric characters, then we can use the +, -, *, and / icons to build an equation to calculate the output. If we are using alpha/text characters, then we use the + icon to create a concatenation. It will concatenate the values before and after the + icon that have been mapped.

Figure 10.39 Binary Operators Used to Create Logic in Format Mapping

- **Literal**
  We use the literal section to define a constant value in the expression (either a number or text). Additionally, there are some special characters that we can used in this section. If the literal section is left empty, then we're looking to map an empty character. If we enter a space into this section, then we're mapping a blank space. If we use the Enter key, then we're mapping the expression to go to the next line.

- **Operators**
  The **Operators** function in Figure 10.40 is only available for conditions. Since we generally create some type of logic in a condition, we need to define how the data is being compared to create the logical comparison. These operators are available for us to select when we're creating the condition.

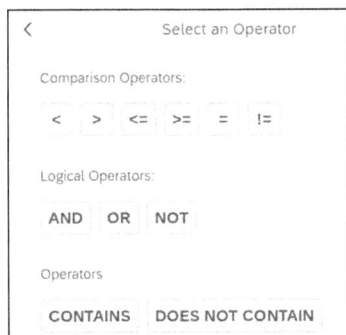

Figure 10.40 Operators for Conditions in Map Format Data

- **Functions**
  When we're defining the output value adjustment or variables, we can use a series of functions to alter the data that is being mapped. These functions are detailed in Table 10.2.

Function	Details
Absolute Value	This function can map an amount that is always a positive value. The expression for the absolute value can be a number that is mapped, either directly or as a reference to a node.
Concatenate	We can concatenate two or more fields together in the output. Values that we concatenate can be constants, variables, or references to nodes.
Default	This mapping references the defined variable in the **Format Mapping Properties** area. The value that is assigned in the variable can be mapped here.
Length	We use this function if we want the system to deliver a count of characters from the input value. If the input value is 12345, then the length that would be output would be 5 because the text is 5 characters long.
Negate	We use this function if we want to exchange positive and negative values. For example, we use this if we have a negative value and we want to make it positive or vice versa. This only applies to mapped numeric values.
Replace	We can use this as a find-and-replace function for certain values. The mapping of this function has three parts to it. The first parameter refers to the input value, the second selects what we want to replace, and the third value determines what we want to map in place of the previous value.
Round	We use this function to determine that a numeric value should be rounded. There are two parts to this function. First, we determine what amount we want to be rounded, and second, we determine how many decimal places should apply. For example, if we have a node that maps an amount of 2.5321 and we map it with "Round(2.5321,2)," then the value of 2.53 will be mapped based on the standard rounding rules.
Round Down	This function works the same as the **Round** function, but it always rounds down regardless of the value of the rounded-off figure. For example, if we are rounding 2.558 to two decimal places, we would output 2.55 and not 2.56 since we're rounding down.
Round Up	This function works the same as the **Round Down** function, except that it always rounds up regardless of the value of the rounded-off figure. For example, if we are rounding 2.552 to two decimal places, we would output 2.56 and not 2.55 since we're rounding up.
Substring	This function allows us to select certain text from a string of text. There are three inputs that we need to map in this function. First, we determine the text we're looking to map. Second, we determine how much we want to offset the text. For example, if we map a 2 in this field, we want to skip the first two characters in the text. Third, we determine how many characters to output. For example, if we map "Substring(1234567,2,4)," we would map 3456 since we skipped the first two characters and mapped the next four.

Table 10.2 List of Functions Available when Building Correspondence Expressions

- **Node references**
  If we want to refer to a node of the mapping tree and use the values from that node, we can create a node reference. As with other areas of this correspondence mapping, we can only refer to nodes that appear higher in the format mapping structure.
- **Constants**
  This section refers to the constants that were defined as part of the **Format Mapping Properties** area.
- **Enumeration**
  We can also reference the enumerations that were defined as part of the **Format Mapping Properties** area. This is only available for conditions and not for mapping the variables and output value adjustment.
- **Variables**
  This section refers to the variables defined in the **Format Mapping Properties** area.
- **System variables**
  In this area, we can refer to system variables such as the date or time.

10.5 Outbound Messaging

Now that we've covered the configuration for the outbound process, we can discuss how the outbound correspondence objects are generated and processed. The correspondence objects can be created automatically, based on the previously defined configuration; or we can create them manually, using a transaction. The process for creating these objects is detailed in the following sections.

10.5.1 Correspondence from a Financial Transaction

Correspondence objects are generated based on the correspondence activity settings that were defined in Transactions FTR_INT_ASSIGN and FTR_EXT_ASSIGN, which were determined in Section 10.2.2. An example of an activity that can be assigned occurs when a contract is created using the Create Financial Transaction app. When a transaction has a contract activity, we can create a correspondence object. Depending on the correspondence settings that were determined in the communication profile, we can generate multiple correspondences with one action. Each business partner recipient also needs to have a business partner group assigned in Transaction FTR_INT_ASSIGN or FTR_EXT_ASSIGN.

The delivery of the correspondence can differ, depending on whether it's being sent to an external recipient or an internal recipient. For external recipients, the delivery of the correspondence was defined in the communication profile. This configuration defined the folder location where the outbound correspondence should be placed when it's successfully created. For internal recipients, the ultimate mapping for the

10 Correspondence

email recipient was defined within Transaction FTR_INT_ASSIGN. The email address either was defined directly or will be derived from the recipient's user ID.

Once the transaction's correspondence is generated, we can view it in the correspondence monitor or directly in the financial transaction by using the Process Financial Transaction app or Transaction FTR_EDIT. We can select and view the desired transaction with the **Display** option. When we're in the transaction, we can click on the **Correspondence** button, and the correspondence objects will be displayed as in Figure 10.41.

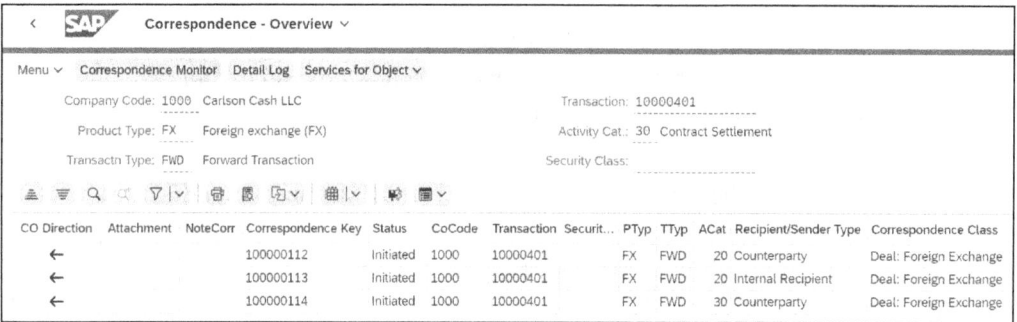

Figure 10.41 Displaying Linked Correspondence Objects from Financial Transaction

10.5.2 Manual Correspondence

We can also generate correspondence objects manually. We can execute this directly in Transaction FTR_COCREATE or by using the **Create** button in the correspondence monitor in Transaction FTR_COMONI. There are two options for creating manual correspondence. We can either enter all details manually in the correspondence object, or we can reference another transaction as a template to copy its data and only update the fields that need to be updated.

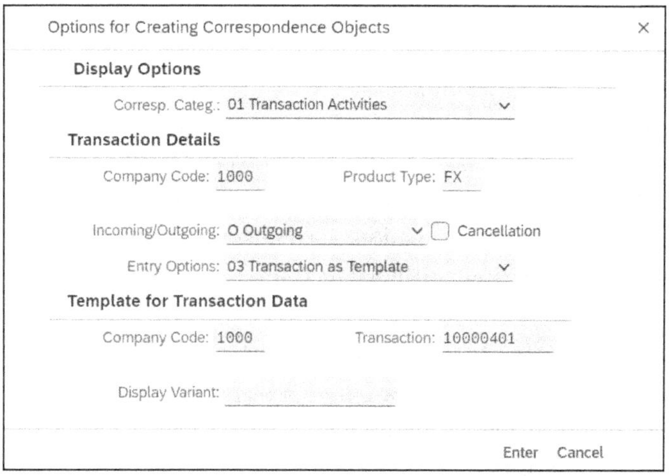

Figure 10.42 Initial Entry Screen for Manually Creating Correspondence Objects

Figure 10.42 shows an example of how to generate the correspondence manually. The options for selection are as follows:

- **Display Options**
  - **Correspondence Category**
    We can choose whether the correspondence should be related to **Transaction Activities** or to a **Securities Account Transfer**.
- **Transaction Details**
  - **Company Code**
    This is the company code in which the correspondence should be created.
  - **Product Type**
    This is the product type to assign to the correspondence object.
  - **Incoming/Outgoing**
    We use this to determine whether the correspondence object is an incoming or outgoing object.
  - **Cancellation**
    We use this to define whether the manual correspondence object is a cancellation object.
  - **Entry Options**
    - **Fast Entry**
      With this option, we can quickly create an incoming correspondence object; this is only available for money market and FX product types.
    - **No Template**
      With this option, no template is used for the entry, and all fields need to be entered manually.
    - **Transaction as Template**
      With this option, another transaction is defined and is used as a reference to populate the correspondence object.
    - **Securities Account as Template**
      With this option, another securities account transfer is defined and is used as a reference to populate the correspondence object.
    - **Correspondence Object as Template**
      With this option, another correspondence object is defined and is used as a reference to populate the correspondence object.
- **Template for Transaction Data**
  This is displayed if we use a template option. The additional fields depend on the type of object we're using as a reference:
  - **Company Code**
    This is the company code that holds the transaction that we'll use for the template creation.

10   Correspondence

- **Transaction**
  This is the transaction number we use to reference the template creation.
- **Securities Account**
  We define securities account transfer in this field for the reference information.
- **Correspondence Key**
  We define another correspondence object here to determine the reference information.

After entering the information, we click on the **Enter** button to move to the next screen.

The screen in Figure 10.43 will be displayed for further entries. In our example, we're using a transaction for a template, and the **BP Group**, **Profile**, and **Recipient/SenderType** all need to be populated. Once these are complete, the rest of the information will populate based on the transaction information. We can update this based on the correspondence object that should be created.

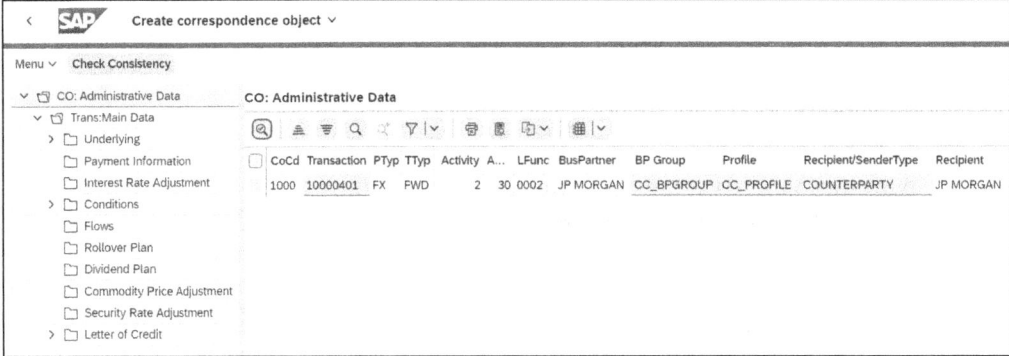

Figure 10.43 Entering Additional Details of Correspondence Object

After the information is updated, we can save the correspondence object and then view it in the correspondence monitor (see Figure 10.44).

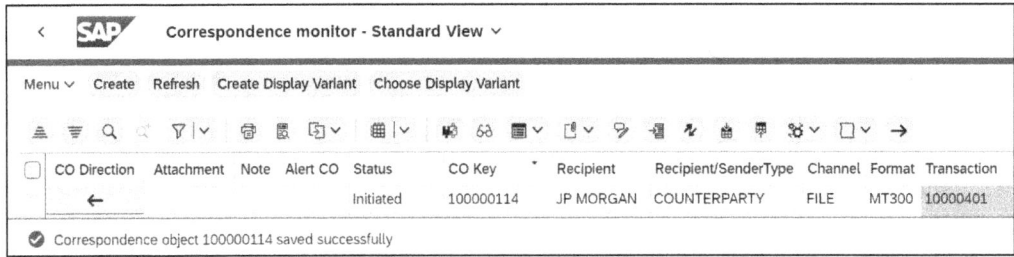

Figure 10.44 Successful Manual Creation of Correspondence Object

10.6 Inbound Messaging

Just like there's an outbound process for the correspondence objects, we have a process for bringing correspondence objects into SAP S/4HANA. When there are electronic correspondence objects, we have a few different processes that we can follow to load them into the system. The main ways we can bring these in are through Transaction FILE, loading through the app server, SAP Multi-Bank Connectivity, and manual uploads, as follows:

- **Transaction FILE and the app server**
  The first two options for loading the incoming correspondence messages are to create a logical and physical file path using Transaction FILE and to map the incoming correspondence directly in a folder path. Both of these options are detailed in the configuration in Section 10.1.5. We will cover this method for loading the correspondence objects further in this section.

- **SAP Multi-Bank Connectivity**
  We can also load messages through SAP Multi-Bank Connectivity. The benefits of using SAP Multi-Bank Connectivity for incoming files is that the files automatically load as they are received and that there's no delay in loading the incoming correspondence messages. However, the settings required to map the incoming files in SAP Multi-Bank Connectivity are outside the scope of this book.

- **Manual**
  An additional transaction is available that lets us manually load incoming correspondence files. If we don't load these files through the previously covered methods, we can load them manually using Transaction FTR_SWIFT_IMPORT.

In the following sections, we'll walk through both automatic and manual inbound processing as well as the different types of incoming messages.

10.6.1 Automatic Inbound Processing

We can start the inbound process for correspondence objects by using Transaction FTR_IMPORT (Import Incoming Messages). For the correspondences to load properly, a file has to be in the location determined in the correspondence profile for the defined correspondence class. It will also only load files with the same prefix that has been defined in the correspondence profile. To run the transaction, we also need to determine the **Profile**, **Correspondence Class**, and **Recipient/SenderType** to import the messages. An example of these settings is in Figure 10.45.

When the transaction is run, all files in the location will be imported and will be viewable in the correspondence monitor. Each file at this location will only be processed once, so SAP S/4HANA won't try to import any file that has already been loaded.

10 Correspondence

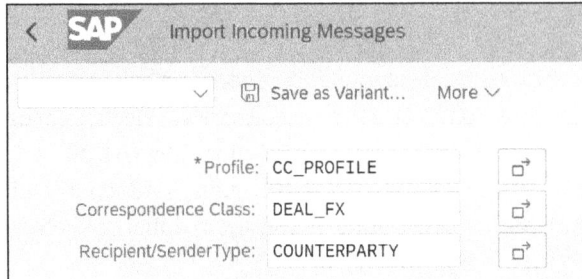

Figure 10.45 Selection Screen for Importing Incoming Correspondence Messages

10.6.2 Manual Inbound Processing

When testing files with correspondence, we may want to simply load a file to test whether it processes correctly. In this scenario, it might be easier to load a file directly from a computer instead of having to place files in a directory. In this case, we can use Transaction FTR_SWIFT_IMPORT (Import SWIFT Messages), in which we have multiple options for import. We can determine a single file or import multiple files in one run, and we can also import these files from either a local PC or the application server. The fields that can be populated in this transaction are as follows:

- **Import Details – Single File**
  - **Input File**
    Here, we define the location and file name to import into SAP S/4HANA.
  - **Format**
    Here, we define the file format that is being imported. This will match the format names that were created in Section 10.1.3.
- **Import Details – Multiple Files**
  - **Application Server/Frontend Server**
    Here, we indicate whether the files to be loaded are on the application server or on our local PC.
  - **Import Folder**
    Here, we define the folder that holds the files to import.
  - **File Name Pattern**
    Here, we can define a pattern for the desired files to load. For example, there may be multiple types of files in the defined folder, and if so, we can determine a common text in the files that should be loaded.
  - **Format**
    This works the same as the import details for a single file. Here, we define the format type that will be loaded in the selected files.
- **Post-Import Details**
  - **Archive Folder**
    Here, we create a copy of the successful file and archive it in the specified folder.

– **Error Folder**
  Here, we create a copy of the file and place it in the specified folder if there's an error.

Checking the display log at the bottom of the screen will result in screen output after running the transaction that shows a report of the processing of the correspondence objects. The screen for the manual inbound processing is detailed in Figure 10.46. Once we have added all the attributes to this screen, we can click the **Execute** button to trigger the loading of the files.

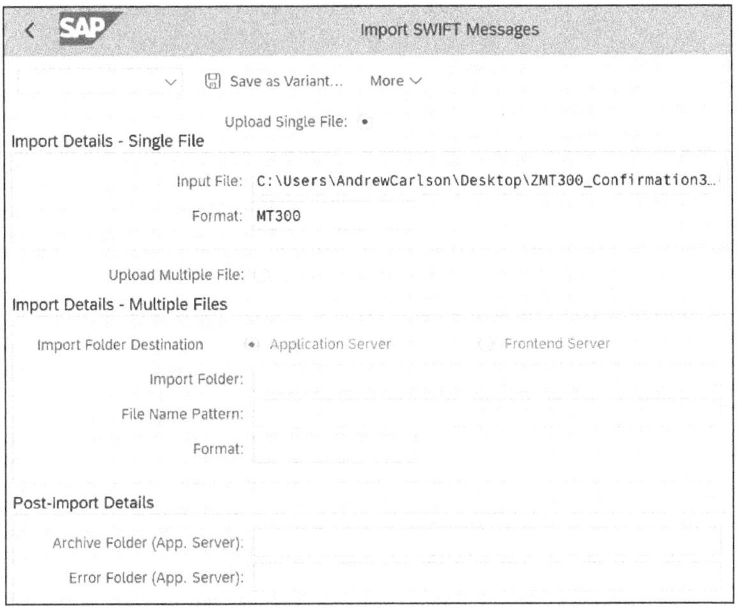

Figure 10.46 FTR_SWIFT_IMPORT Selection Screen

10.6.3 Types of Incoming Messages

There are a few types of incoming messages that can be imported into SAP S/4HANA and that will affect the correspondence monitor. The messages are as follows:

- **Standard messages**
  Standard messages are the same thing as the incoming MT SWIFT messages that we've seen in the previous section. Examples of standard messages are MT300 files for FX contracts and MT320 files for fixed-term loans.

- **Cancellation and amendment messages**
  Cancellation and amendment messages are related to messages that have been previously received. These messages have a reference to a previous message, and they create either a cancellation of that message or an amendment to change the information from the message. Cancellation messages are identified with a **CANC** value in the **Operation** field, and amendment messages are identified with an **AMND** value.

- **Acknowledgement messages**
  When we set up the correspondence channels, we can determine whether a response file is required to tell SAP S/4HANA that the file has been sent. An acknowledgement message is sent to communicate that a message has been successfully sent, and a negative acknowledgement message is sent if a message fails during sending.

  When the correspondence channel is configured to require acknowledgement files, it impacts the outgoing correspondence object status. When the message is generated, the status of the correspondence object will appear as **Acknowledgement Awaited**. This status is then updated to **Delivered** when the acknowledgment is imported.

10.7 Correspondence Matching

Correspondence matching is a process in which we match the outgoing correspondence with a confirmation for a counterparty to document that the correct information was delivered and received. In the following sections, we'll walk through the necessary configuration for correspondence matching and how both automatic and manual matching work.

10.7.1 Configuration

Our main task is to determine in the correspondence matching the fields that we need to compare when we're matching the correspondence objects. In the mapping, we can determine each of the outgoing and incoming correspondence fields that we want to look at, and we can make a determination of whether the fields should match or be different.

We can find the settings for the correspondence matching in the **Correspondence • Match Correspondence Objects • Define Rules for Marching Correspondence Objects** menu path. In Figure 10.47, we can see how the details are mapped in the matching function. The **Outgoing function** and **Incoming function** columns determine the outgoing and incoming correspondence classes, and the table defined in the **Table Name** column determines the data table that is being compared among the correspondence objects. The **Field Name** column determines all the data in that table that is being compared among the correspondence objects for the mapping. Lastly, there are two matching rules that we can assign in the **Matching Rule** column. Either we can determine that the outgoing field needs to equal the incoming field, or we can determine that the outgoing field needs to not equal the incoming field.

The matching rules will compare all the fields that have been defined in this matching definition table, and if everything matches correctly, then the outgoing correspondence will be matched with the incoming correspondence.

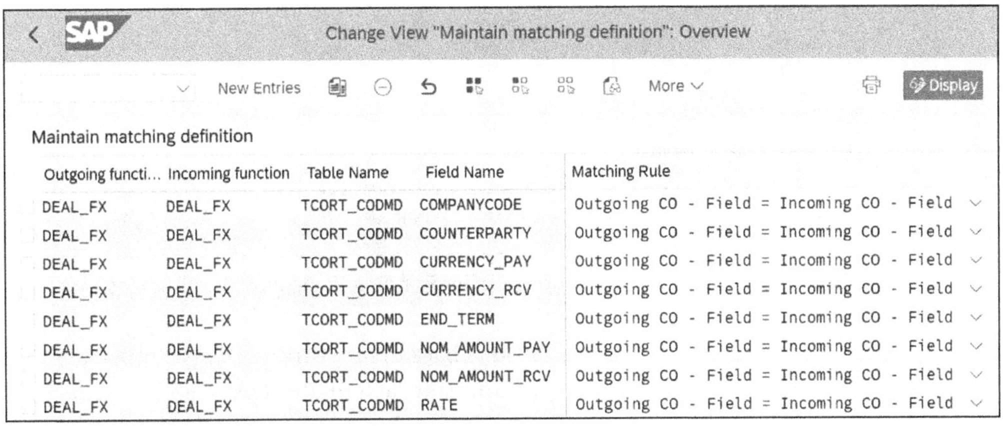

10.7 Correspondence Matching

Outgoing functi...	Incoming function	Table Name	Field Name	Matching Rule
DEAL_FX	DEAL_FX	TCORT_CODMD	COMPANYCODE	Outgoing CO - Field = Incoming CO - Field
DEAL_FX	DEAL_FX	TCORT_CODMD	COUNTERPARTY	Outgoing CO - Field = Incoming CO - Field
DEAL_FX	DEAL_FX	TCORT_CODMD	CURRENCY_PAY	Outgoing CO - Field = Incoming CO - Field
DEAL_FX	DEAL_FX	TCORT_CODMD	CURRENCY_RCV	Outgoing CO - Field = Incoming CO - Field
DEAL_FX	DEAL_FX	TCORT_CODMD	END_TERM	Outgoing CO - Field = Incoming CO - Field
DEAL_FX	DEAL_FX	TCORT_CODMD	NOM_AMOUNT_PAY	Outgoing CO - Field = Incoming CO - Field
DEAL_FX	DEAL_FX	TCORT_CODMD	NOM_AMOUNT_RCV	Outgoing CO - Field = Incoming CO - Field
DEAL_FX	DEAL_FX	TCORT_CODMD	RATE	Outgoing CO - Field = Incoming CO - Field

Figure 10.47 Mapping Outgoing and Incoming Correspondence Fields

10.7.2 Automatic Matching

The automatic matching of correspondence objects follows the logic determined in the configuration or by the available BAdI. We can set up the matches to automatically match, or we can require a user review to determine whether the proposed correspondence objects should be matched together. The automatic matching process can occur at a several different stages in the correspondence process, including the following:

- When incoming correspondence is imported.
- When outgoing correspondence objects are sent (only when the **Automatic Matching** checkbox is indicated on the business partner group).
- We can also trigger the matching process using Transaction FTR_COMATCH. The selection conditions are determined in this transaction, and all applicable correspondence objects will attempt to match based on the configured conditions.
- We can also trigger individual matching in the correspondence monitor. When we run the matching from the correspondence monitor, a manual trigger of the matching occurs, but it will still look to match based on the defined matching rules in customizing.

Let's look at an example of an automatic correspondence match. There's an outgoing MT300 that was created when the FX trade was generated, and when the incoming MT300 is received, SAP S/4HANA compares the data in the outgoing correspondence to the data in the incoming correspondence to create the match. If everything matches based on these rules, we'll see the automatic matching proposal in Figure 10.48. If the transactions are set up to expect a match, and if the allotted time passes as defined in the communication profiles configuration, then an alert message can be generated.

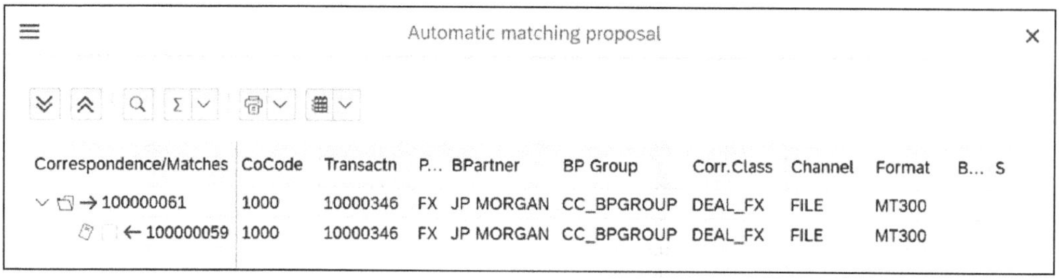

10 Correspondence

Figure 10.48 Automatic Matching Proposal after Bringing In Incoming Correspondence

After these have been matched, we can look back in the correspondence monitor in the Process Correspondence – Monitor app and Transaction FTR_COMONI. In Figure 10.49, we can see that correspondence 100000059 and 100000061 have been successfully matched and that a status of **Completed** is now reflected in the monitor.

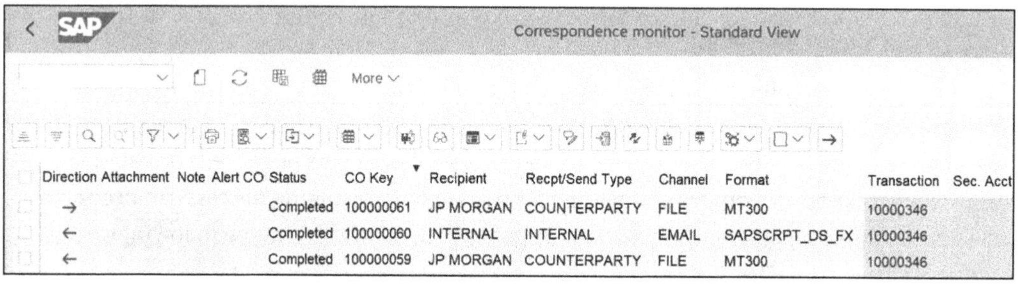

Figure 10.49 View of Completed Correspondence Objects after Matching

> **Automatic Matching**
>
> We can run the automatic matching process with Transaction FTR_COMATCH. This process is run more on an exception basis because it will try match the correspondence objects as they are loaded into SAP S/4HANA. If an error occurs for any reason, we can run the automatic matching transaction to retrigger the matching.
>
> If there are multiple objects that are available to match and that all use the same criteria, the oldest correspondence object will be automatically matched.

10.7.3 Manual Matching

If a correspondence object doesn't automatically match, we also have the ability to manually match correspondence objects. This is a forced match, and the matching configuration doesn't apply in this scenario. This means that the details of the correspondence objects are not reviewed, and we will be reviewing and matching the objects together. Manual matching occurs in the correspondence monitor. To start manually matching the correspondence objects, we need to go to the **Matching View** of the correspondence monitor. The full details of the correspondence monitor are covered in Section 10.8.2, but we'll cover the manual matching portion of the monitor in this section. To get to this view, we go to the Process Correspondence – Monitor app or Transaction FTR_COMONI, and we scroll to the bottom of the initial screen and update the display view. There will be a **Matching View** option in this area, and once we have updated the field, we can execute the report. All correspondence objects that are in a status that can be matched will appear in this matching view. In Figure 10.50, we can see that the status of the outgoing object is **Delivered** and the status of the incoming object is **Received**. We can match these together by first selecting the objects on the left side of the screen and clicking on the **Forced Match** button at the top of the screen.

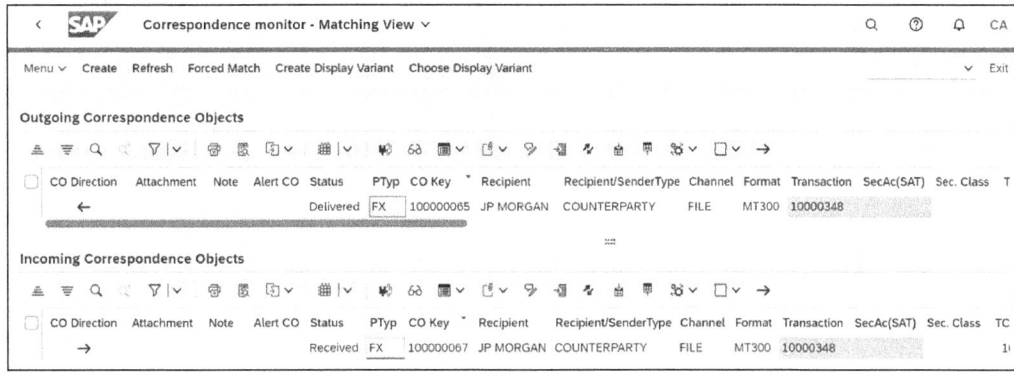

Figure 10.50 Forced Match Option in Correspondence Monitor

The **Matching View** will also ask us to confirm that we want to match the correspondences before the manual matching is complete. This confirmation is shown in Figure 10.51.

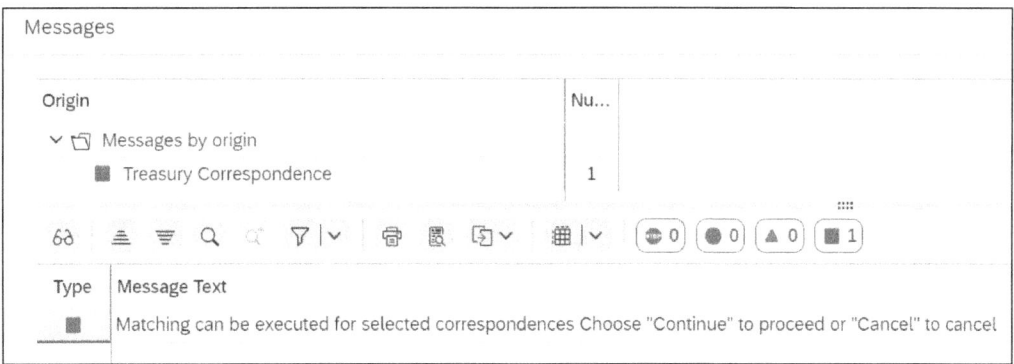

Figure 10.51 Confirmation of Matching Proposal

10.8 Display Correspondence

There are two main ways to display the correspondence details for a transaction. We can do this directly in the financial transaction or in the correspondence monitor. This section will cover both of these methods for viewing and monitoring the correspondence.

10.8.1 Displaying Correspondence in the Financial Transaction

When we process financial transactions, the correspondence will be generated based on the configuration settings and based on the activity type of the transaction. To display the correspondence objects related to the transaction, we can select the **Correspondence** button or use the top menu by following the **Goto • Correspondence** option. All correspondence objects that were generated as part of the financial transaction will appear in this view as shown in Figure 10.52.

One thing to note in this transaction is that any correspondence objects are only created once the changes are saved in the transaction. There's no way to create a "proposal" correspondence object based on changes we're making to the financial transaction. Once the transaction is saved, the relevant correspondence objects will be generated. If there were changes to the transaction, multiple correspondence objects would be generated since the changes would create a cancellation object and a new correspondence object. If we want to look into the reasons why a correspondence object was or was not generated, we can refer to the **Detailed Log** ☐ in the correspondence area. This button will deliver a log that will show details of the correspondence generation and key details of the transaction.

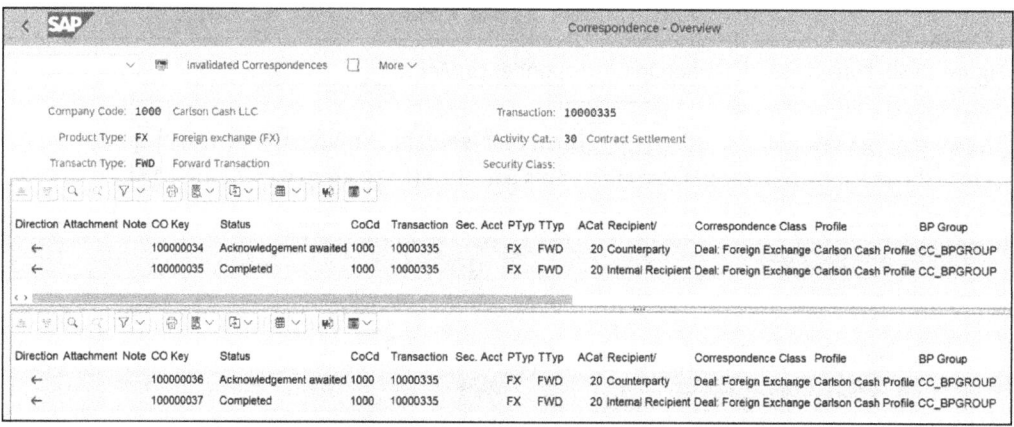

Figure 10.52 Related Correspondence Objects Can Be Viewed in Process Financial Transaction

10.8.2 Correspondence Monitor

In previous sections, we've mentioned the correspondence monitor, but we haven't covered it in detail. We can access the correspondence monitor by going to the Process Correspondence – Monitor app or Transaction FTR_COMONI. We can access most correspondence functions in this transaction. Before we go through the output and functions of the correspondence monitor, we'll walk through the selection screen and cover the different ways we can generate the report.

10.8 Display Correspondence

Selection Screen

The correspondence monitor selection screen is split into three sections. The top section, shown in Figure 10.53, filters the correspondence objects based on administrative data. The pure administrative data of a correspondence object might exist before it's assigned to a financial transaction, so we don't filter by that detailed information at this point. We can see in this top section that the header-level correspondence information is available for filtering, so we see details like messaging status, recipients, direction, and business partner groups in this area. One additional setting in this area that is useful is the checkbox for **All Unassigned Correspondences**. We use this checkbox to select all correspondence objects that haven't been assigned to either a securities account or a financial transaction.

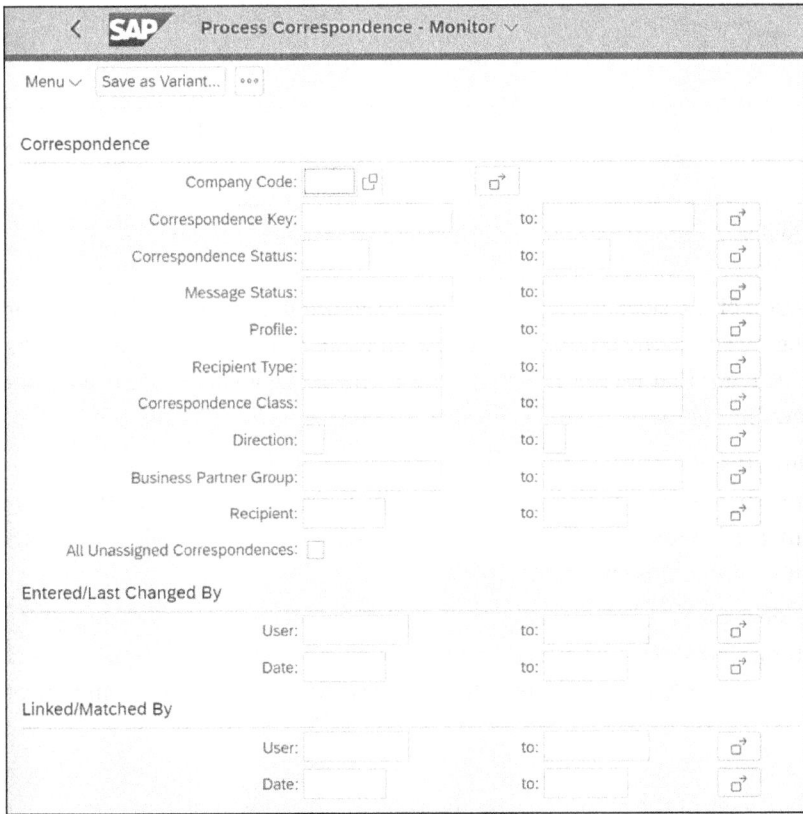

Figure 10.53 First Selection Screen of Correspondence Monitor

The middle section of the selection screen, shown in Figure 10.54, filters correspondence objects that have been assigned to a transaction. We can assign the object to a financial transaction, a securities account, or a settlement instruction. We can filter the correspondence objects by filling out the fields in any of the tabs to ensure we are only viewing the desired correspondence activities.

695

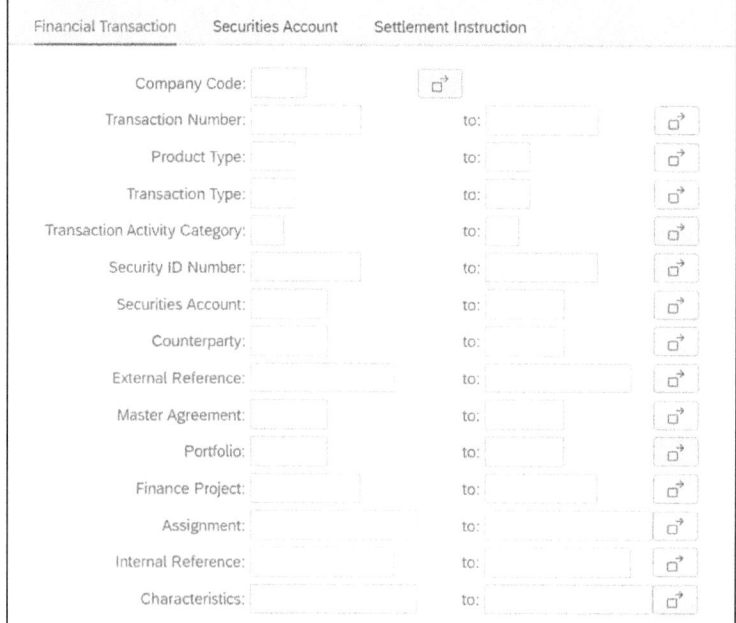

Figure 10.54 Middle Section of Selection Screen of Correspondence Monitor

The third section of the selection screen, shown in Figure 10.55, details how the information will be viewed in the reporting. The key details here are the layout and the display view. The layout that we select will drive how the data appears in the report, and there are a few different views that we can select in the display view:

- **Assignment View**
  The assignment view shows the correspondence objects that should be assigned to either a financial transaction or a securities transaction. We can make the assignment to the transactions either directly or indirectly.

- **Matching View**
  The matching view shows all correspondence objects that have been set up to match with each other. We covered the matching process for correspondence in Section 10.7.

- **Standard View**
  The standard view of the report doesn't filter the information based on assignment or matching, so it will display all correspondence objects based on the filtering.

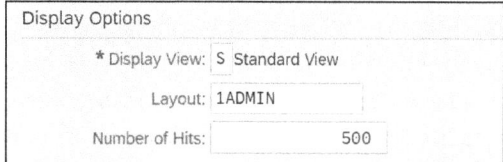

Figure 10.55 Bottom Section of Correspondence Monitor

10.8 Display Correspondence

Correspondence Monitor Report

As mentioned earlier in this section, we can access the correspondence monitor by navigating to the Process Correspondence – Monitor app or Transaction FTR_COMONI. This monitor will show the status of all the incoming and outgoing correspondence messages with all details that are relevant to the correspondence.

As we mentioned in the selection screen section, there are three different views we can use with the correspondence monitor. Figure 10.56 shows the monitor with a few examples of how the incoming and outgoing correspondence can appear in this report. This report shows all details that were described in previous sections, including the matching status, alert details, alert number, and applicable financial information for the correspondence objects.

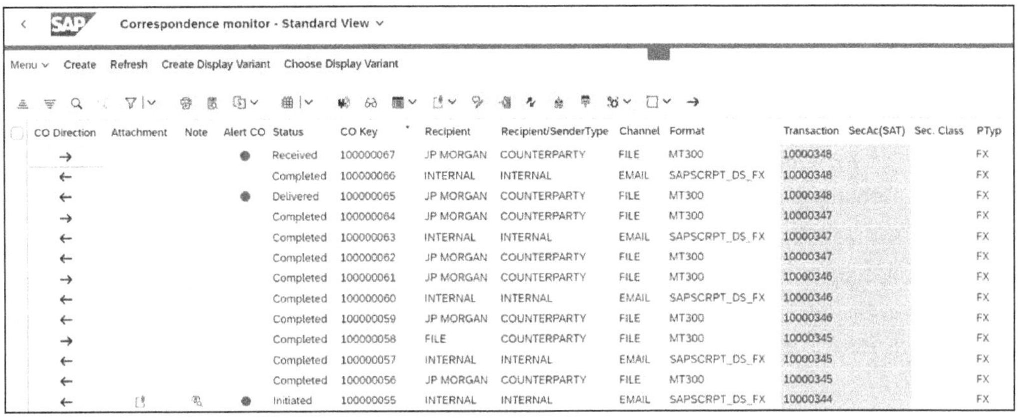

CO Direction	Attachment	Note	Alert CO	Status	CO Key	Recipient	Recipient/SenderType	Channel	Format	Transaction	SecAc(SAT)	Sec. Class	PTyp
→			●	Received	100000067	JP MORGAN	COUNTERPARTY	FILE	MT300	10000348			FX
←				Completed	100000066	INTERNAL	INTERNAL	EMAIL	SAPSCRPT_DS_FX	10000348			FX
←			●	Delivered	100000065	JP MORGAN	COUNTERPARTY	FILE	MT300	10000348			FX
→				Completed	100000064	JP MORGAN	COUNTERPARTY	FILE	MT300	10000347			FX
←				Completed	100000063	INTERNAL	INTERNAL	EMAIL	SAPSCRPT_DS_FX	10000347			FX
←				Completed	100000062	JP MORGAN	COUNTERPARTY	FILE	MT300	10000347			FX
→				Completed	100000061	JP MORGAN	COUNTERPARTY	FILE	MT300	10000346			FX
←				Completed	100000060	INTERNAL	INTERNAL	EMAIL	SAPSCRPT_DS_FX	10000346			FX
←				Completed	100000059	JP MORGAN	COUNTERPARTY	FILE	MT300	10000346			FX
→				Completed	100000058	FILE	COUNTERPARTY	FILE	MT300	10000345			FX
←				Completed	100000057	INTERNAL	INTERNAL	EMAIL	SAPSCRPT_DS_FX	10000345			FX
←				Completed	100000056	JP MORGAN	COUNTERPARTY	FILE	MT300	10000345			FX
←			●	Initiated	100000055	INTERNAL	INTERNAL	EMAIL	SAPSCRPT_DS_FX	10000344			FX

Figure 10.56 Output Screen for Correspondence Monitor

Functions in the Correspondence Monitor

Now that the data has been filtered in the correspondence monitor, we can cover what actions we can take within the monitor. We can use a series of buttons at the top of the screen in the monitor to alter the correspondence objects or view them in more detail. In the following list, we cover the navigation buttons in detail to show the additional functions in the correspondence monitor:

- **Details**
  The correspondence objects are hierarchical-based objects with multiple levels of data. Clicking this button allows us to drill into the other levels of data on the correspondence object.

- **Show Underlying**
  We use this button to view either the financial transaction or the securities account that a correspondence object is assigned to.

- **View Message**
  We use this button to preview or view any correspondence objects in the report. If the message hasn't yet been sent, then we'll use the **Preview Message** function. If the message has been sent, then we'll use the **View Message** function.

- **Attachments**

 As with other areas of this module, we can add and view attachments. We can use this button to add or view attachments for each correspondence object, and if an attachment exists for an object, then an attachment icon will be displayed in the **Attachment** column.

- **Maintain Notes**

 With this button, we can add notes to any of the correspondence objects or edit any previously created notes. If a note has been created for an object, a note icon will be displayed in the **Note** column of the report.

- **Show All Related Correspondence**

 By clicking this button, we can show all correspondence objects that are related to each other. This means that all objects that were created for a single financial transaction or security will be displayed, along with any object that has been referenced by the selected correspondence object.

- **Show Linkages/Messages**

 If a match exists for a correspondence object, we click this button to show the relationship between the matching correspondence objects. We covered further details of the matching process and viewing the matches in Section 10.7.

- **Assign**

 We can use this button to manually assign a correspondence object to a financial transaction or securities account if the assignment doesn't exist yet.

- **Unassign**

 We can use this button to remove an assignment to a financial transaction or securities account. We can only do this if the correspondence object was not generated directly from a financial transaction or securities account.

- **Status Functions**

 We use this button to update the status of a correspondence object. The options that we can select in this function are limited based on what the next available statuses for the correspondence object are.

- **Log**

 Multiple log options are available for each correspondence object. When we click this button, the different options available will be displayed, as follows:

    - **Action Log**

     This displays the actions that have been triggered for a correspondence object and any related messages. If there are issues with the correspondence object, they'll appear in the messages portion of the action log.

    - **Status Log**

     This navigates to status management and will show the current status of the correspondence object.

    - **Release Log**

     We can track the release process by clicking this button.

- **CO Change Documents**
  Any changes to the correspondence object are tracked and can be displayed in the change documents log.
- **Approver's List**
  If correspondence has been configured to require approvers, this list will show all users who can approve the correspondence object.

■ **Send Alert** →
We can configure the correspondence objects to generate an alert after a defined amount of time passes if the object is not updated to a completed status. If this time period passes, the **Alert CO** column will appear with a red indicator ● to highlight that an alert can be generated for that correspondence object. In this scenario, we can highlight the relevant columns we want to send the alerts for, and we can then click the **Send Alert** button. Other ways to send the alerts are described in in the next section.

10.9 Alerts

The correspondence function within SAP S/4HANA is meant to be an automated process, and we should only need to monitor the statuses on an exception basis. With that being said, there are always issues that may arise, and alerts can be configured to notify us when there's an issue with a correspondence object. As we covered in Section 10.1.9, we can define a timeframe for when an alert should occur for a correspondence object. We generally use this when we expect a confirmation or a match of a correspondence object but a certain amount of time passes and we don't receive that confirmation. After the defined time period has been exceeded, an alert can be generated and can be sent using alert management.

In the following sections, we'll walk through how to configure alerts and monitor them.

10.9.1 Configuration

To utilize the alerts function of correspondence, we first need to define the alert categories. The alert categories create the first steps required to define what an alert needs to look like, and they also assign users as the recipients of the alerts. After we create the alert categories, we need to assign them to their applicable transactions.

Defining Alert Categories

We can define alert categories in the **Financial Supply Chain Management • Treasury and Risk Management • Transaction Manager • General Settings • Correspondence • Alert • Define Alert Categories** menu path. To create an alert, we click on the **Display/Change** button and then on the **Create** button once in change mode. The first tab that we need to define is the **Properties** tab. We'll want to assign a **Description** for the treasury alert we're creating and assign the **Classification** of **TRM_ALERTS**, as in Figure 10.57.

10 Correspondence

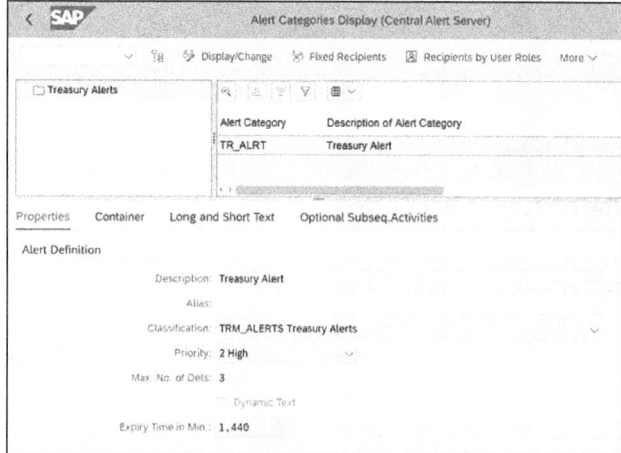

Figure 10.57 Definition of Treasury Alert Category

After this is done, we can navigate to the **Container** tab, where we can start to define additional details for the alert. The first part of the tree that shows up as **ALERT** in Figure 10.58 has additional details behind it. Most importantly, the data type for this container needs to be assigned to VTS_ALERT_MSG_DISP. This is assigned by double clicking on **ALERT** to assign this value as the **ABAP Dictionary Data Type**. Any of the other subnodes in this alert have been assigned to map data in the **Long and Short Text** tab.

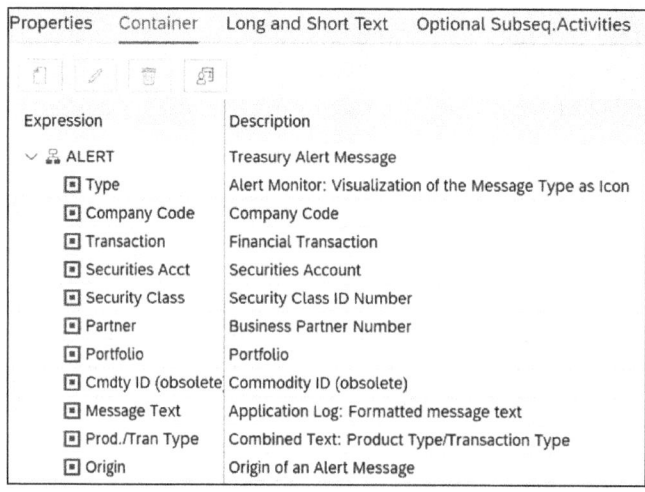

Figure 10.58 Available Mapping Fields in Container Section of Alerts

In the **Long and Short Text** tab, we define the text that is sent when the alert is generated by SAP S/4HANA. As we can see in Figure 10.59, there are defined fields that we can use dynamically to customize each alert to show up with the applicable company code, financial transaction identifier, and any other information that's relevant to the correspondence object.

700

10.9 Alerts

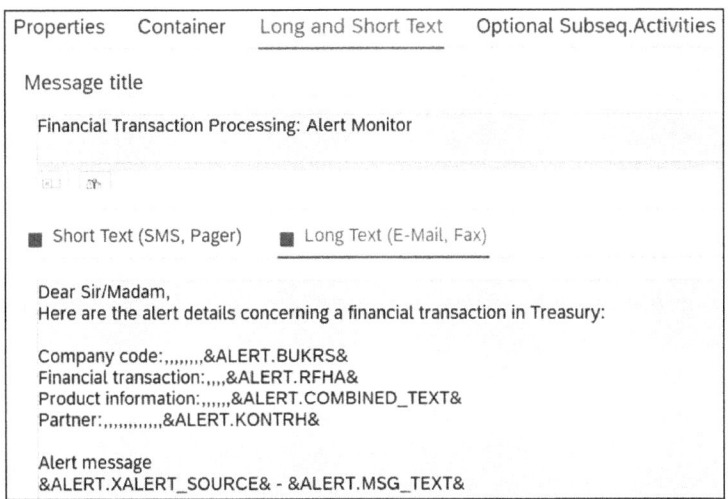

Figure 10.59 Mapping Short and Long Text of Alert Message

After these details are fully defined, we can use the **Fixed Recipients** or **Recipients by User Roles** button to determine who should receive the alert. The alert will be delivered based on the additional details assigned to the user's ID.

Assigning Alert Categories

After we create the alert categories, we need to assign them to their applicable transactions. In the same location as the configuration in **Correspondence • Alert**, there are five different options for assigning the alert categories, depending on the type of transaction. We can assign these for money markets, FX, derivatives, securities, and trade finance transactions. As shown in Figure 10.60, we'll assign each previously defined **Alert Category** to a combination of product types (**PTyp**), transaction types (**TTyp**), and activity categories (**ACat**). This allows us to flexibly assign different alerts to different types of transactions.

Figure 10.60 Mapping Alert Category to Product Types and Transaction Types

10 Correspondence

10.9.2 Monitoring Alerts

The alert monitor in treasury and risk management is not exclusive to correspondence alerts. The monitor allows us to view all types of alerts in this module, including alerts for when transactions should be settled, posted, and paid. In the alert monitor, we can view the correspondence alerts and trigger the alerts to be sent to the configured recipients. This function also is available in the correspondence monitor.

We can find the alert monitor by using Transaction FTR_ALERT. This transaction will show all available alerts in treasury and risk management, not just the correspondence alerts. When we view the correspondence alerts in this transaction (as in Figure 10.61), we can highlight a line and click on the **Send Notification** → button to trigger the sending of the alert. Assuming the customizing has been set up correctly with recipients for the alert, the message defined in the alert category will be successfully sent.

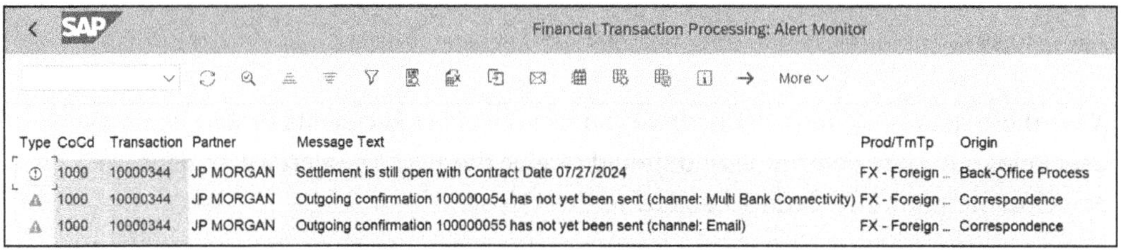

Figure 10.61 Alert Monitor Showing Both Correspondence Objects with Other Treasury Alerts

We can also view the alerts in the SAP Fiori app Display Treasury Alerts – Correspondence, as shown in Figure 10.62. This is only a view transaction, and we can't send the alert messages directly from this app.

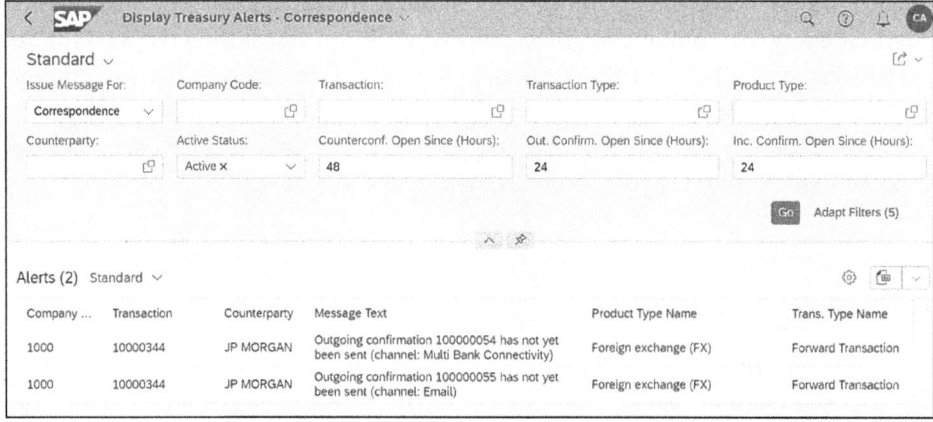

Figure 10.62 Display Treasury Alerts – Correspondence App Showing Applicable Alerts

There's also an alert monitor that we can leverage to send the correspondence alert messages. To do this, we can execute Transaction FTR_ALRT_BTCH. We can use the selection screen to filter the necessary alerts, and any alerts that we choose can all be sent at the same time. Generally, clients will create a variant for this transaction and automatically run the alerts to ensure they are sent to the necessary parties in a timely manner.

We can also trigger the correspondence alerts from the monitor. The transaction for the correspondence monitor is Transaction FTR_COMONI. Once the configured alert waiting period has passed, an icon will appear in the monitor, indicating that we can send an alert. This will appear in the **Alert CO** column as displayed for many of the correspondence objects, as shown in Figure 10.63. We can send multiple correspondence alerts at once, and we do this by selecting multiple lines by clicking the white boxes on the left side of the screen. Once we have selected the boxes, we can trigger the sending of the correspondence alert by clicking the **Send Alert** → button.

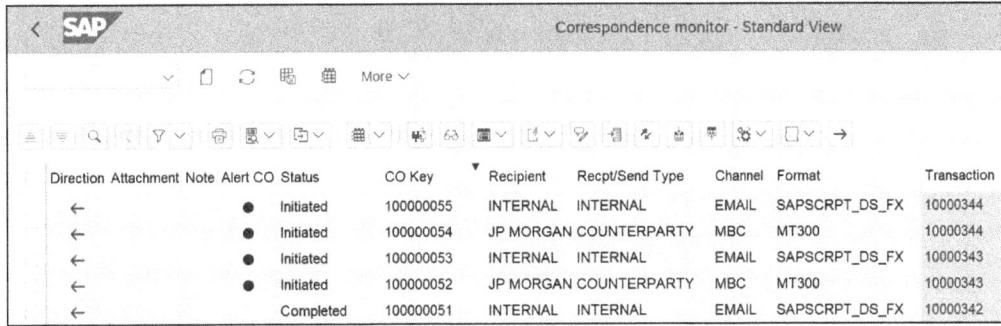

Figure 10.63 Correspondence Monitor Can Send Treasury Alerts Highlighted in Alert CO Column

10.10 Summary

In this chapter, we explored the correspondence framework within treasury and risk management, highlighting its important role in automating and streamlining communication processes within treasury. By configuring and utilizing correspondence, users can automatically generate and distribute key documents such as confirmations, dealing slips, and account statements, thus reducing manual workload and enhancing efficiency. We walked through essential configuration steps, practical examples, and best practices to ensure the smooth operation of incoming and outgoing correspondence. With correspondence, treasury departments can achieve greater transparency, accuracy, and control over their communication processes, making correspondence a vital tool for modern treasury management.

Chapter 11
SAP Treasury Analyzers

SAP S/4HANA Finance for treasury and risk management includes four analyzers that help clients understand key elements surrounding credit risk, market risk, portfolio risk, and accounting risks. The analyzers are not always required for every implementation, but they provide significant value in challenging times to help SAP clients understand their outstanding risks in the market, and they provide quick insights into how clients can react and reposition as necessary.

In this chapter, we'll discuss the use of the SAP analyzers that can be implemented as part of treasury and risk management. SAP has four key analyzers for treasury and risk management, as follows:

- Credit Risk Analyzer
- Market Risk Analyzer
- Portfolio Risk Analyzer
- Accounting Analyzer

While all analyzers can assist organizations, we've found that the most used ones are the Credit Risk Analyzer and the Market Risk Analyzers, so we'll cover them in this chapter. Should you need any information on either the Portfolio Risk Analyzer or the Accounting Analyzer, you can find it at *http://s-prs.co/v590700*.

> **Note**
> All of the analyzers are included in the SAP S/4HANA Finance for treasury and risk management licenses, but note that the Portfolio Risk Analyzer and the Accounting Risk Analyzer are not currently part of SAP S/4HANA Public Cloud Edition.

11.1 Credit Risk Analyzer

The first analyzer we'll discuss is the Credit Risk Analyzer, which allows us to review each trade and aggregate trades against defined limits in the system. This helps treasury organizations ensure compliance with investments policies and manage risk in multiple financial institutions. Throughout this section, we'll explore the setup of limits, how limits are reviewed, and the standard reporting that is available to help us understand limits.

11 SAP Treasury Analyzers

11.1.1 Defining Limits

One of the first activities we'll want to do is set up our limits in the system. We do this as an end user in the Manage Limits app, where we'll see a list of limits that are available. We can set up some common limits by the following parameters:

- Trade type
- Company code
- Business partner
- Traders
- Combinations of these or other attributes

We'll be updating our company code–based limits for a new entity that has been created. Once we find our limit, we'll select it and click **Create Limit**, which brings up the popup shown in Figure 11.1.

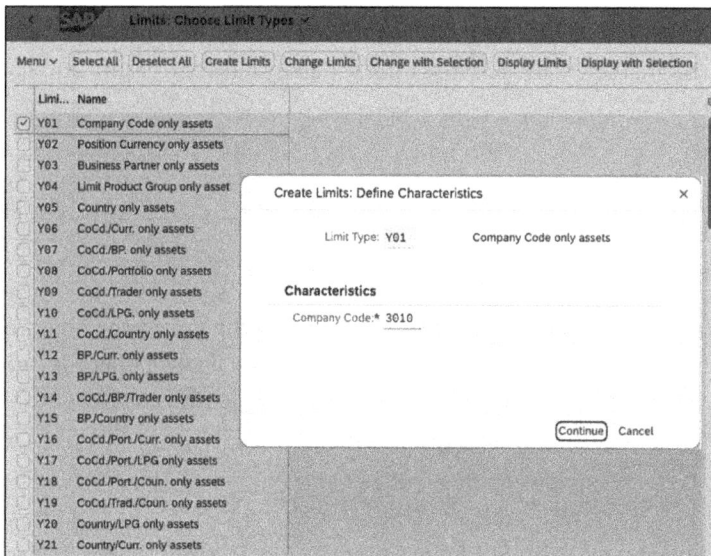

Figure 11.1 Create Limits: Define Characteristics

Since we're using an existing limit, we'll select it from this list. For example, we can select **Y01 Company Code only assets** and then select **Change Limits**, and we'll then see the key characteristics of the limit populate (in this example, **Company Code**). We'll enter our **Company Code** (the example shown here has "3010") and then select **Continue**. On the following screen, since we're setting up a new company code, we'll add in the limits. We'd like to add our company code 3010 to reflect the limits, and we'll set the limit based on our policy. In this example, we have a limit of 80 million Australian dollars, as shown in Figure 11.2. After we add this limit, when trades are created in company code 3010, the Credit Risk Analyzer will do a single transaction check and keep a full listing of all open trades in the company code. The single transaction check is applicable to any traders entering or modifying deals, while traders or management can review the end-of-day report. We'll cover these in more detail later in this chapter.

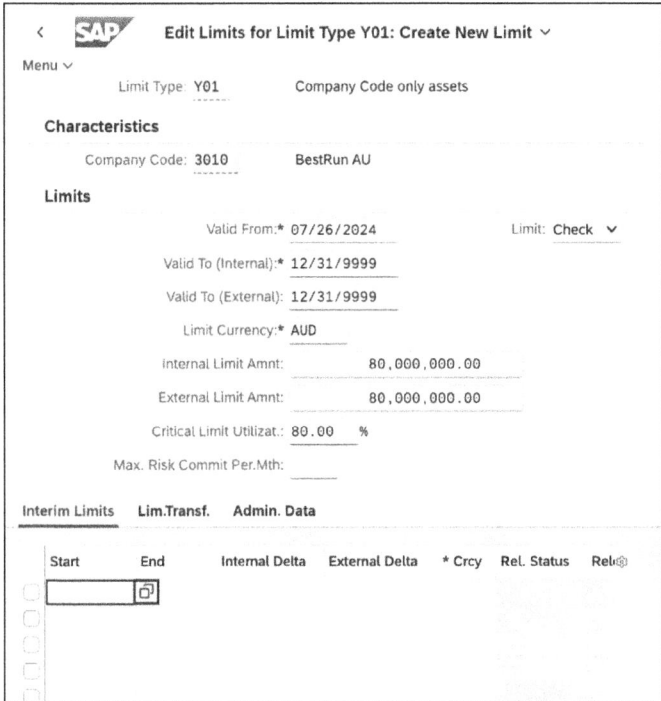

Figure 11.2 Edit Limit Types: Adding Limits

When we create the new limit, we'll enter the **Valid From** and **Valid To** dates, the currency, and the internal and external amounts. We'll enter 80 million Australian dollars, but using credit risk management, we can also input key thresholds of critical utilization. We'll enter "80.00" into the **Critical Limit Utilizat.** percentage field, which will provide alerts to the trader as our credit exposure nears 80 million. This means that when our credit limit is 80% used up, the system will display warnings during the single and aggregate transaction checks.

> **Note**
> If any of our limits are exceeded, SAP S/4HANA won't stop us from entering the trade because this has already been committed in the market.

11.1.2 Single Transaction Checks

During the processing of any deal, if limits are set up, they'll be validated during the entry of that deal. There can be up to three messages displayed, depending on the limit utilization:

11 SAP Treasury Analyzers

- **Limit exceeded**
  This indicates that we're over the limit.

- **Limit warning**
  This indicates that we're over the critical utilization as set up in the limit.

- **Limit successful**
  This indicates that we're within the set-up limits.

There's a view of a trade that has exceeded its limit in Figure 11.3.

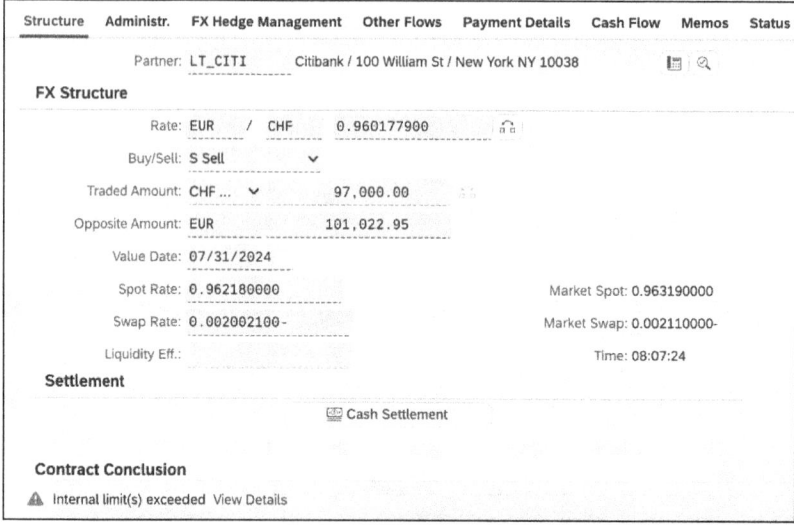

Figure 11.3 Treasury Deal with Limit Exceeded

11.1.3 Interim Limits

Additionally, within the limits, we can set up interim limits, which allow for an increase to our limit for a specific period. An example of an interim limit is shown in Figure 11.4 for an increase of 10 million Australian dollars. To add the 10 million, we can simply click on the interim limits and enter the desired **Start** and **End** dates, the **Internal Delta** and **External Delta** (for example, 10 million), and the **Currency** (which should be in the same currency as the overall limit). Once we've entered this information, we can click the **Interim** button with the green flag to update our limit. This is not a required step in limit setup, but it can be useful to adjust limits if needed.

Once we've completed the limit setup, our limits will be saved and the report in Figure 11.5 will be generated, showing all of the limits that we have based on company code—including our new limit for our 3010 entity. Also, we'll be able to immediately create a deal and have it checked against the limits under a single transaction check.

11.1 Credit Risk Analyzer

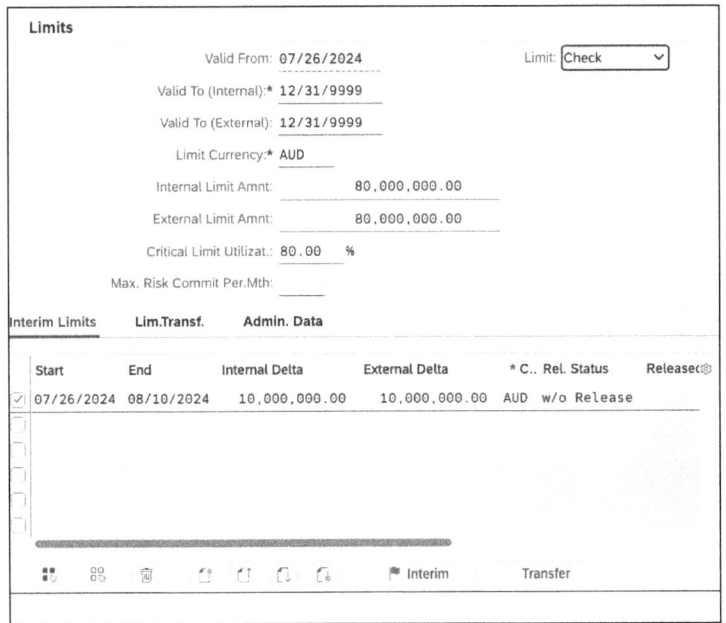

Figure 11.4 Interim Limit Maintenance

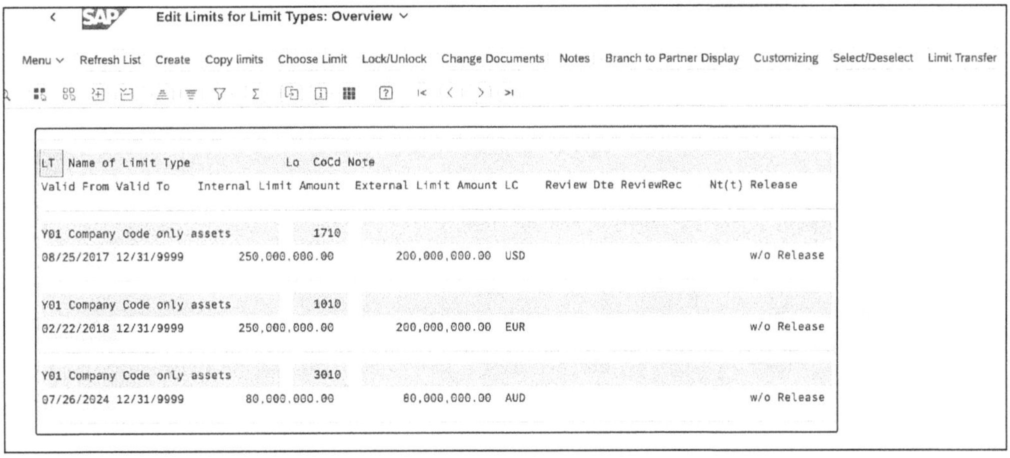

Figure 11.5 Limit Review

11.1.4 End-of-Day Processing

While the limit is active, we can review the single transaction check during trade execution. However, we'll also will want to look at our whole portfolio for all trades and exposures with counterparties. The end-of-day process reviews all transactions in the system for a particular key date (e.g., today) and determines the credit or issuer risk for our portfolio as compared to our limits.

709

To execute the End of Day Processing report shown in Figure 11.6, we must enter the **Valuation Date** and the **Determination Procedure** we wish to use. We'll discuss how to set up these determination procedures in Section 11.1.8. Note that we can set up additional restrictions for company code and control.

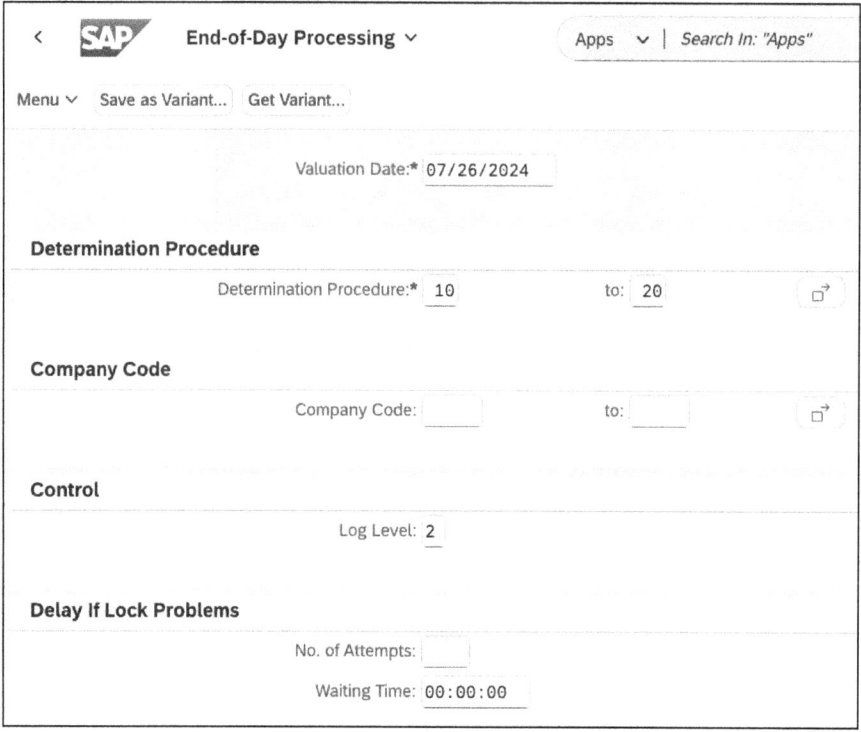

Figure 11.6 End-of-Day Processing Job Selection

> **Note**
> The end-of-day process for credit risk can be run by end users or in the background, but typically, it will be set up as a background job, and the treasury management team will distribute and review its output.

11.1.5 Review Limit Utilizations App

Once we've completed the end-of-day processing, we can navigate to the Review Limit Utilizations app to understand how all of our deals compare with our overall limits.

In this report, we'll notice several settings that we can utilize, as shown in Figure 11.7. First, the **General Access Options** include limiting by **Limit Type**, **Limit Currency**, and **Determination Procedure**, and populating this section is optional. Secondly, there's a **Selection of Utilizations**, in which we can define a **Limit Utilization Base** (current or end-of-day positions) and a **Validity Date for Utilization**, and in which the **Determination Date** field must be populated. Note that the **Determination Date** will automatically populate with the current date, but we can change this to a desired date as needed. We can also activate the **Selection of Limits** as needed, and it contains additional characteristics to help filter our limits. The **Limits Valid From** field will also default to the current date. Once we've populated the required fields and any optional fields, we'll click **Run**.

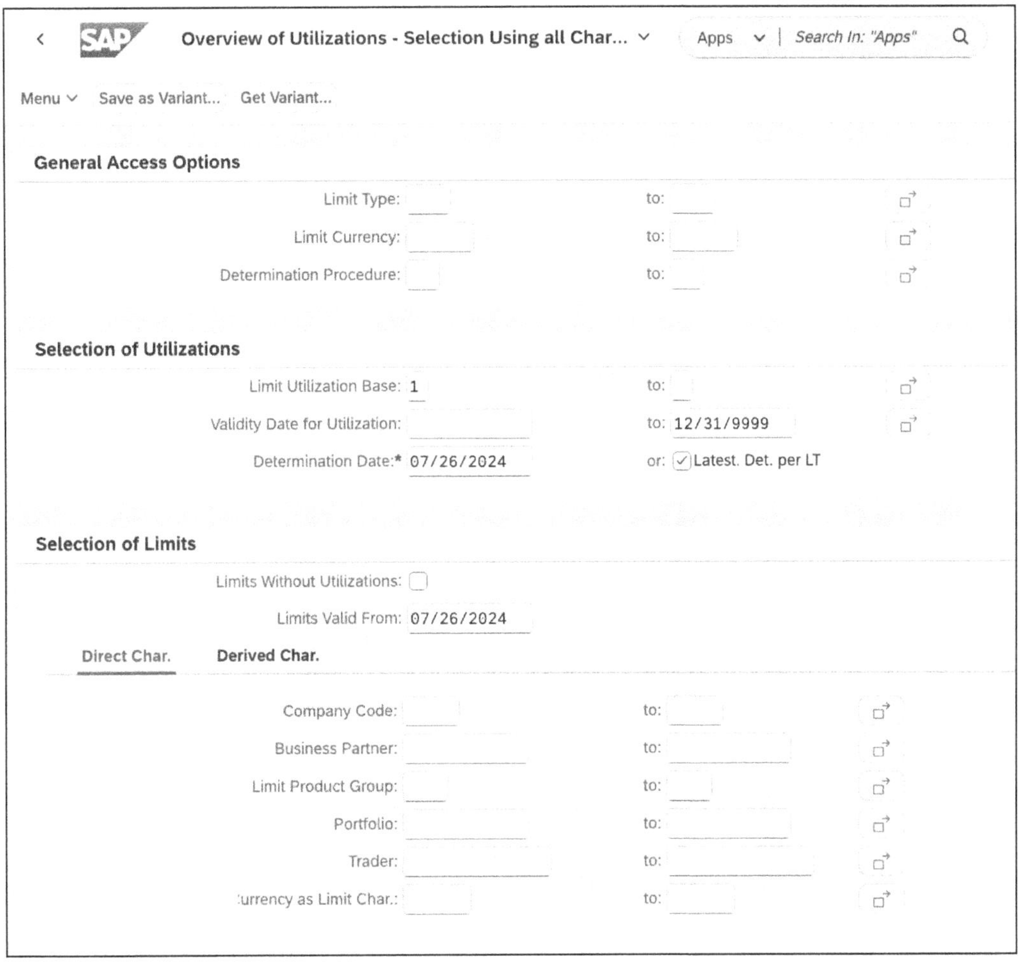

Figure 11.7 Limit Utilization Settings

As we review the report, we'll easily be able to understand our limit utilizations because they'll be color coded by status, as shown in Figure 11.8:

- Green: Within expected range
- Yellow: Within critical utilization as defined in the limit
- Red: Limit has been exceeded

11 SAP Treasury Analyzers

The report shows us the internal and external limits compared to what has been utilized and what is the remaining limit (either **Free (+)** or **Over (−)**).

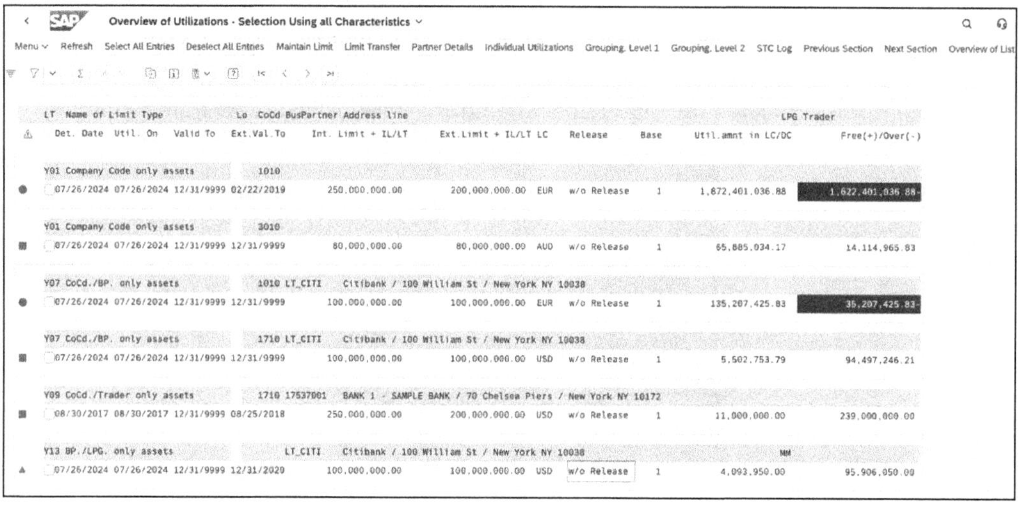

Figure 11.8 Review Limit Utilizations

> **Note**
>
> The utilized amount can be based on the deal value, or it can be derived from the current NPV of the trade (if desired).

While in this report, we can also drill down into the **Free (+)/Over (−)** column to see all the transactions that make up that balance, and the system will identify the trades that caused the limit breach (as shown in Figure 11.9).

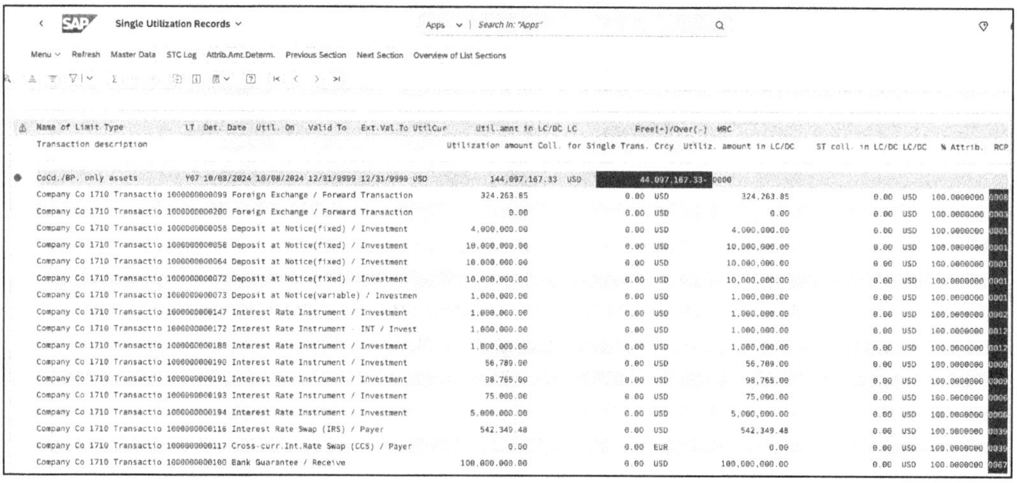

Figure 11.9 Detailed Review of All Trades with Flagging of Trades Violating Limits

We can also drill down one more time to review individual transactions, as shown in Figure 11.10.

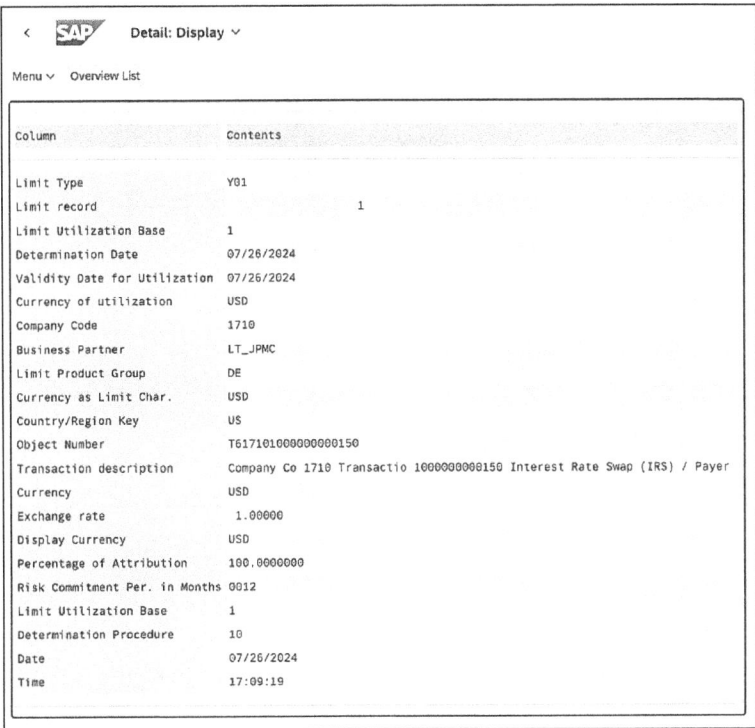

Figure 11.10 Detailed Transaction Review of Trades

11.1.6 Reviewing Bank Risk

In addition to understanding limits, we can review our bank risk using a new app called Bank Risk. This app is a key performance indicator (KPI) and gives end users visibility into how much risk they have at their banks and links that with the bank's credit rating. The Bank Risk app delivers a numerical KPI, shown on the left in Figure 11.11. The KPI shown is a collection of all the banks that have transactions, along with their associated credit rating.

> **Note**
>
> Users can easily control KPI settings, and they may want to change the colors of balances displayed based on values being on target (green), at a warning level (yellow), or critical (red).

We can set the rating threshold in the filter area of the report. This report is currently being viewed **By Low Rating** (see Figure 11.11), but we can also review it by bank or bank and counterparty, using the provided dropdown lists.

11 SAP Treasury Analyzers

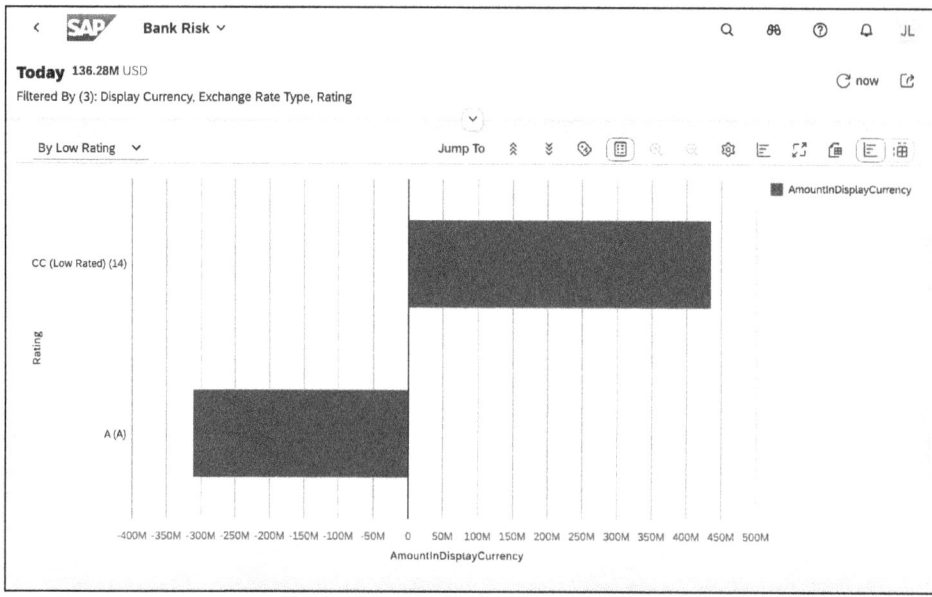

Figure 11.11 Bank Risk Reporting

11.1.7 Reviewing the Deal Default Risk Limit

As part of a deal, we can activate the **Default Risk Limit** tab, as shown in Figure 11.12. (We discuss the activation of this tab in Section 11.1.8). This tab collects all of the risk information (e.g., the **Limit Product Group**, the **Default Risk Rule**) and also displays **Limit Utilization** details.

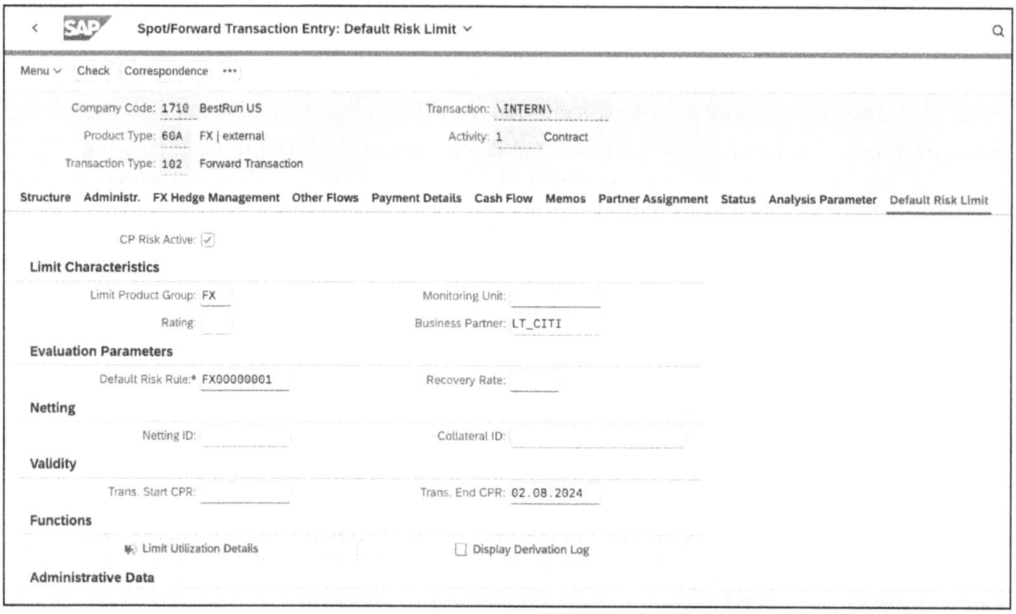

Figure 11.12 Default Risk Limit Tab

11.1 Credit Risk Analyzer

When reviewing a trade, we can select **Limit Utilization** details, and that executes a limit check to see how our trade fits into the limits that have been set up (exceeded, warning, or within limits) as shown in Figure 11.13.

> **Note**
> If we have multiple limits set up, all of them will be checked together during the **Limit Utilization** display.

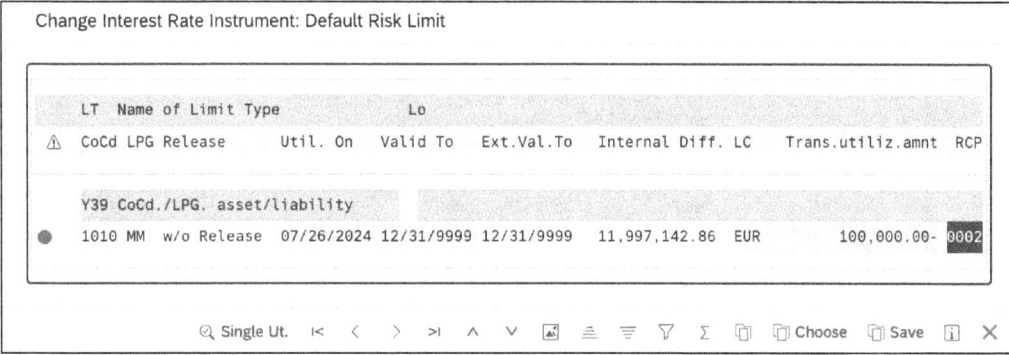

Figure 11.13 Limit Utilization Details Per Trade

11.1.8 Configuration

We now understand how to execute limit reviews from the end-user side, so we'll transition into the system configuration settings that will drive these actions for the users. This is where we'll activate the Credit Risk Analyzer, and we'll also set up and define the limits and determine which trades will automatically group and default the proper limit types.

Defining Global Settings

We navigate to **Financial Supply Chain Management** • **Treasury and Risk Management** • **Credit Risk Analyzer** • **Basic Settings** • **Global Settings for Limit Management** to activate the global settings. As shown in Figure 11.14, we'll set up the key indicators for how limit management will be processed:

- **Default Risk Active**
  If we check this box, we'll enable a tab on the deal entry screen for default risk limit.
- **Deriv. Active**
  If we check this box, we'll automatically check our deals against the limits we've set up.
- **Sec.Acct Pos.**
  If we check this box, the issuer risk of the transaction will be shown in the securities account.

11 SAP Treasury Analyzers

- **Workflow is active**
  If we check this box, a workflow will be automatically generated if a limit has been exceeded. We can find the workflow settings in customizing under **SAP NetWeaver • Application Server • Business Management • SAP Business Workflow**. We can also assign it to a responsible party directly under **Treasury and Risk Management • Credit Risk Analyzer • Basic Settings • Assignment • Assignments to Senders to Recipients**.

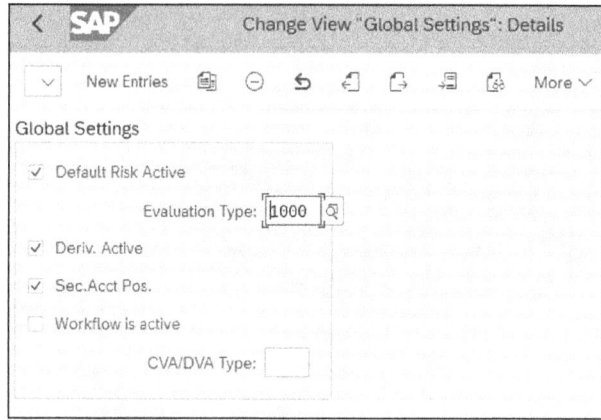

Figure 11.14 Defining Global Setting

Activating Integrated Default Risk Limit

Here we'll configure which applications should be active for trades are applicable across certain company codes. In Figure 11.15, we'll see that we've checked the boxes to select all of the application groups (**Applicat.**): **Loans, Derivatives, Foreign Exchange, Money Market, Securities** and **Trade Finance**. We've set this up for our 1710 company code, but we can do this for all company codes needed. We access this at **Financial Supply Chain Management • Treasury and Risk Management • Credit Risk Analyzer • Basic Settings • Activate Integrated Default Risk Limit Check**.

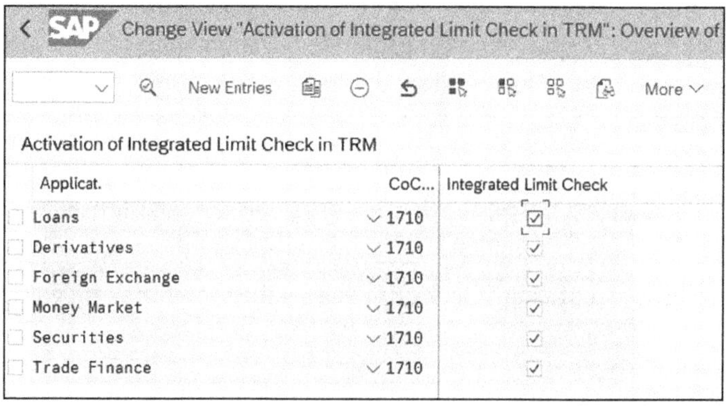

Figure 11.15 Activating Integrated Default Risk Limit Check

Defining Valuation Factors

With this setting, we can define amounts attributable to either counterparty risk or country risk. We will use this later in the setup for defining determination procedures. We find this setup in the **Financial Supply Chain Management • Treasury and Risk Management • Credit Risk Analyzer • Basic Settings • Definitions • Define Valuation Factor** menu path. As shown in Figure 11.16, to set this up, we'll enter a two-digit unique ID, a **Short Test** unique identifier, a **Long Text** unique identifier, and the **CP/Ctry/Reg Risk** indicator. We can set this up with either counterparty/issuer risk or by regions we've selected by counterparty issuer. The final setting in this area defines the **RR Origin**, which is the recovery rate origin. We can set this to the factor in the rating of the business partner or to the country rating of the business partner.

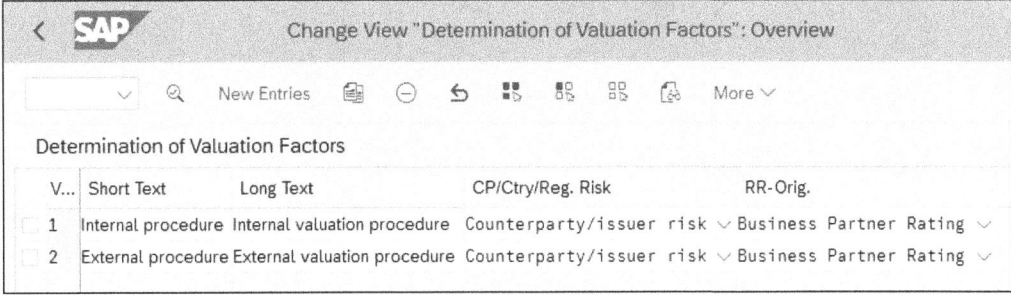

Figure 11.16 Determination of Valuation Factors

Defining Determination Procedures

Now, we'll define the determination procedures in the following menu path: **Financial Supply Chain Management • Treasury and Risk Management • Credit Risk Analyzer • Basic Settings • Definitions • Define Determination Procedures**. We use the determination procedures to control the data, calculation rules, and overall risk categories that we want to generate limit utilizations. There are two determination procedures, as follows:

- 1: Credit risk
- 2: Settlement risk

First, we'll populate enter the determination procedure field (**Determin. Proc.**) with a unique two-digit number (for example, 10). Then, we'll enter a **DP Short Text** and **DP Long Text** that define what this procedure will accomplish, with the goal of providing users with insight into what we're setting up. We also need to set up the control parameters with valuation factors that we defined earlier. For example, we could define the risk category as "1" for **Credit Risk**, and we could set up the **Exposure** as "1" for **Gross** or "2" for **Net**. Finally, we'll set up the **Collateral Valuation Rule**, where we determine whether the primary and secondary risks are economic or political, based on our risk profile as shown in Figure 11.17.

11 SAP Treasury Analyzers

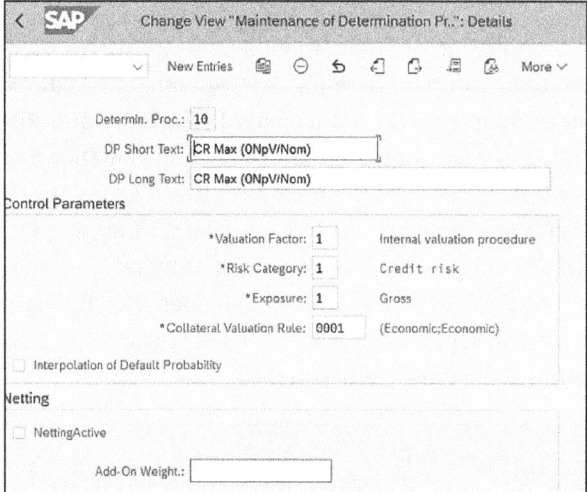

Figure 11.17 Defining Determination Procedures

Defining Limit Types

Next, we'll set up the limit types applicable to the system. These limits will be available to end users, as described in Section 11.1.1, where we defined the limits. We can define the limits as we see fit for our organization. Earlier, we showed how to use the company code limit, so we'll now review the configuration side of how that limit is determined. To define limits, we navigate to **Financial Supply Chain Management • Treasury and Risk Management • Credit Risk Analyzer • Limit Management • Define Limit Types**.

As shown in Figure 11.18, we'll provide a three-digit unique character for the **Limit Type** and assign it a **Name**. This is available to the end users, so the more intuitively we set it up, the better. We'll then assign a **Determination Procedure** (defined in the previous section) and identify the types of **Transactions** to consider.

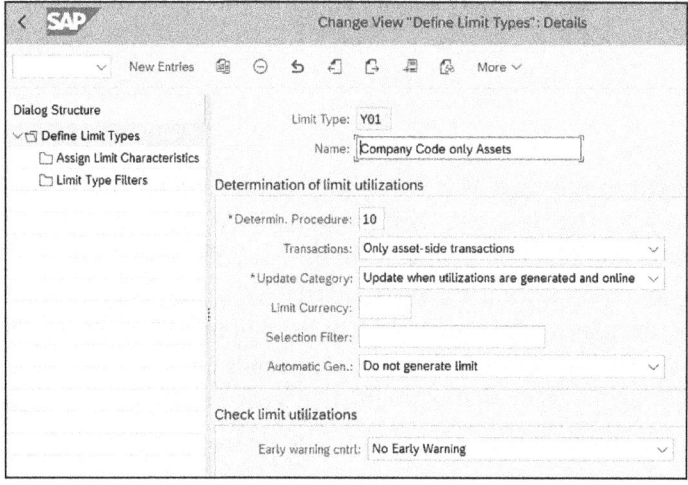

Figure 11.18 Defining Limit Types

Several options exist, including only assets, only liabilities, accumulation of assets/liabilities, and netting of assets/liabilities. We'll also indicate the **Update Category** of either **Update when utilizations are generated** or **Update when utilizations are generated and online**.

We can also have early check warnings that are part of the limit check, and we can set this up with no early limit check, by percentage barrier, or based on an external limit. All these settings set the baseline for the limit, but we also must assign limit characteristics to further segment the different limits we may have. This is outlined in Section 11.1.

One we've made our selections, we click on the **Assign Limit Characteristics** folder in the side menu, as shown in Figure 11.19. Since our example is just a limit based on entity, we'll select **Company Code** from the dropdown list. However, we may elect to have a limit with multiple attributes to it, and we can use the SAP-delivered options or create our own free attributes as well. These settings are reflected in Figure 11.1 from the end user side.

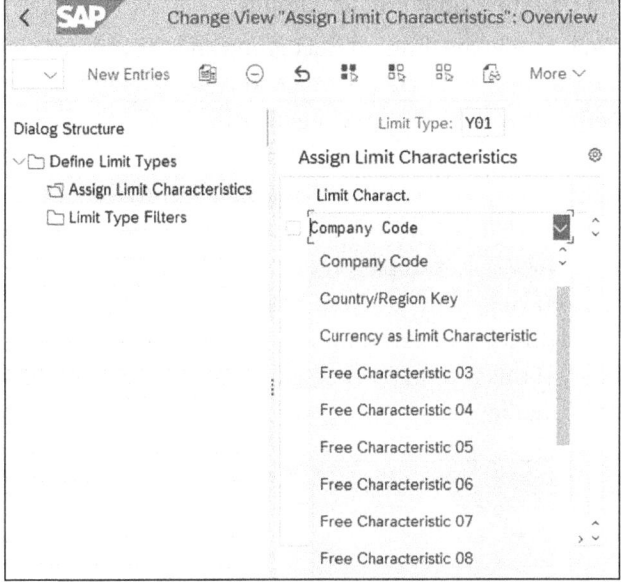

Figure 11.19 Defining Limit Types and Assigning Characteristics

Defining Limit Product Groups

With the limit product group, we define the groupings of transactions that will come together and have similar limits associated with them. Here, we'll set up a limit product group in the **LPG** field (with a maximum of three characters) and a unique name for the limit product group. Typically, this is set up by overall product type as shown in Figure 11.20. The configuration is located at **Treasury and Risk Management • Credit Risk Analyzer • Limit Management • Define Limit Product Groups**.

11 SAP Treasury Analyzers

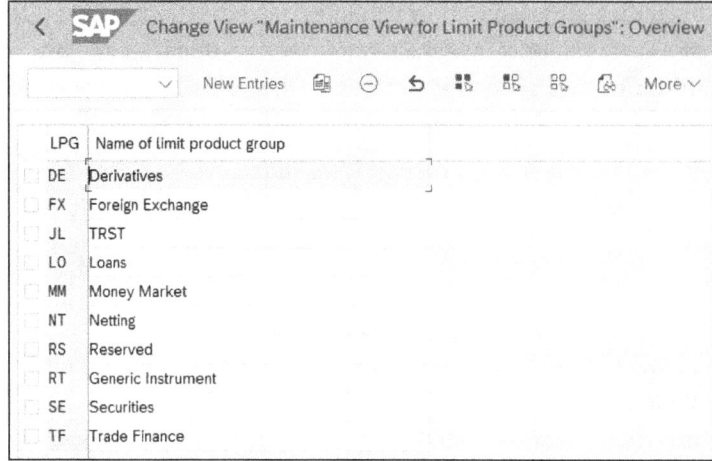

Figure 11.20 Defining Limit Product Groups

Defining Transaction Types and Updating Limit Product Groups

Next, we'll make the connection between the limit product group and the product category via the **Treasury and Risk Management • Transaction Manager • Money Market • Transaction Management • Define Transaction Types** menu path. This will take our product type or transaction type and assign it to the product groups that we just created as in Figure 11.20. The limit product group is mapped as a one-to-one relationship, as shown in Figure 11.21. From here, we'll drill down into the specific product type or transaction type that we want to review (for example, product type **55A** and transaction type **200: Borrowing**). As we drill down into that combination, we'll see the screen in Figure 11.22, which outlines the definition of the transaction type. This will already mostly be set up, but the key part of this is to make the link to the limit group, which we've entered as "MM" for money market.

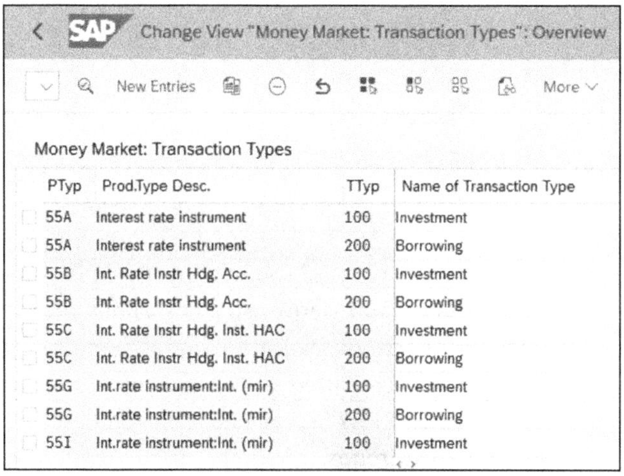

Figure 11.21 Assigning Product Categories to Limit Product Groups

11.1 Credit Risk Analyzer

> **Note**
>
> The implementation guide menu path discussed in this section is located in each of the product areas (e.g., **Money Market, Foreign Exchange, Derivative, Securities, Loans**).

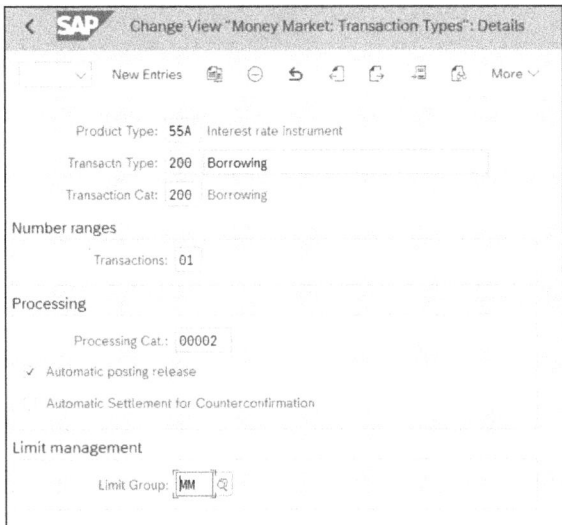

Figure 11.22 Drilldown Settings

Defining the Default Risk Rule

We set up the default risk rule to evaluate the risk of default for the deals that we have set up. We define the default risk rule in the **Treasury and Risk Management • Credit Risk Analyzer • Definitions • Define Default Risk Rule** menu path. To create a new rule, we click **New Entries**, as shown in Figure 11.23.

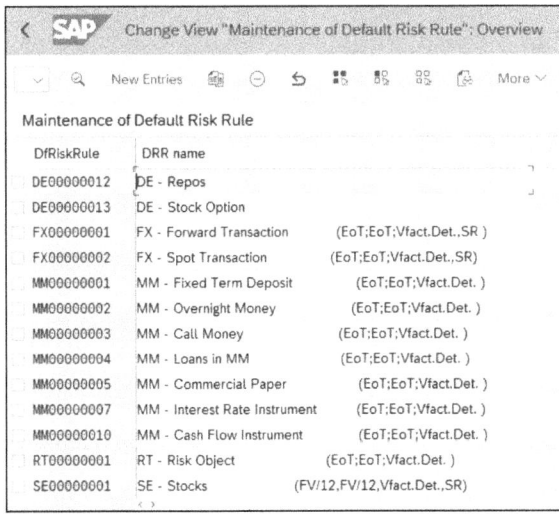

Figure 11.23 Overview of Defining Default Risk Rule

Upon selecting **New Entries,** we'll need to define the following, as shown in Figure 11.24:

- **Market Value Change Period**
  We first enter the market value change period, which defines up to when our market value should be calculated. Options include the following:
  - End of Term
  - Interest Commitment
  - Capital Commitment
  - Fixed Values
  - Ignored
  - Start of Underlying Swaption
- **Risk Commitment Period**
  We then indicate the defined period of the risk, which is closely linked to the market value change period. Our options are as follows:
  - End of Term
  - Interest Commitment
  - Capital Commitment
  - Fixed Values
  - Ignored
- **General Control Parameters**
  In this section, we indicate the recovery rate, which indicates how much we expect to receive if the deal goes into default. We can also check a box indicating whether we would like to consider settlement risk.

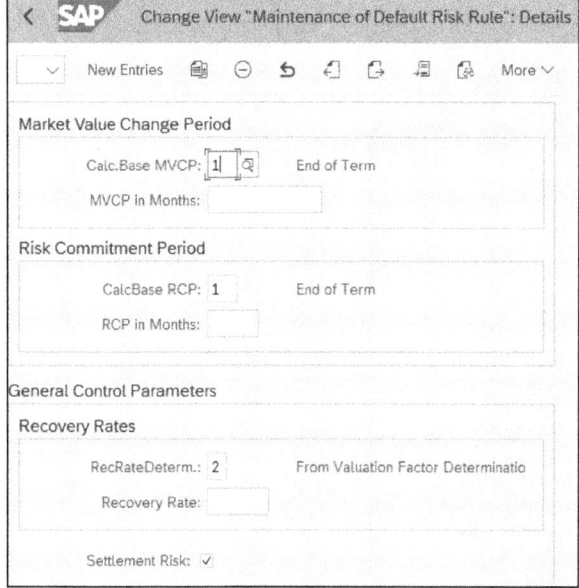

Figure 11.24 Details of Defining Default Risk Rule

11.1 Credit Risk Analyzer

Derivation of Default Risk Rule

The derivation of the default risk rule will take the previously configured risk rule and apply it to specific groups of product types. Additionally, we'll link in the limit product group. We define this in the **Treasury and Risk Management • Basic Analyzer Settings • Automatic Integration of Financial Objects in Master Data • Customize Classic Derivation • Money Market • Derive Default Risk Control Parameters for MM Transactions** menu path.

> **Note**
>
> The menu path in this section will direct us to generic derivations, but we set up the derivation rule in each area (e.g., money market, foreign exchange, OTC derivatives). After we select our area, we'll configure the settings in **Derive Default Risk Control Parameters for XX** transactions, where **XX** identifies the area.

We can then indicate steps to take place as part of our derivation strategy. First, we'll select **Derivation rule**, as shown in Figure 11.25.

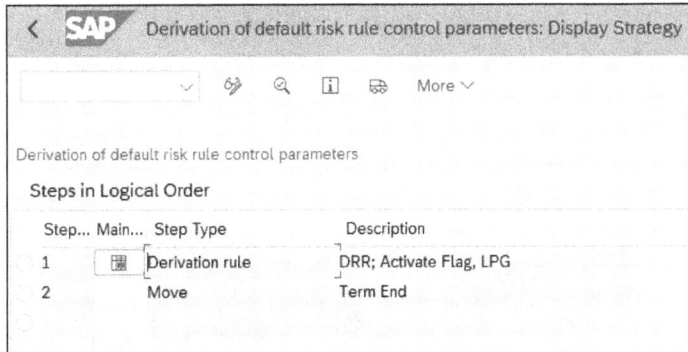

Figure 11.25 Derivation of Default Risk Rule

On the next screen, shown in Figure 11.26, we define the **Source Fields** (for example, **Product Type** and **Financial Transaction Type**). Next, we define the **Target Fields**: the default risk rule, whether the issuer risk is active, and the limit product group. We make these selections based on what we want to use to derive our default risk rules, and we then used this when maintaining rule values as our derivation strategy.

Now, we'll configure the mapping of the data to be derived when we select specific product types and transaction types, as shown in Figure 11.27. Here, we'll enter our **Product Types** (e.g., 60A [Foreign Exchange]), the **Financial Transaction Type** (e.g., 101 [Spot]), and the applicable **Default Risk Rule**, (FX00000002 [FX Spots]). We then select the counterparty risk indicator flag (0 for not active and 1 for active) and add our limit group FX.

11 SAP Treasury Analyzers

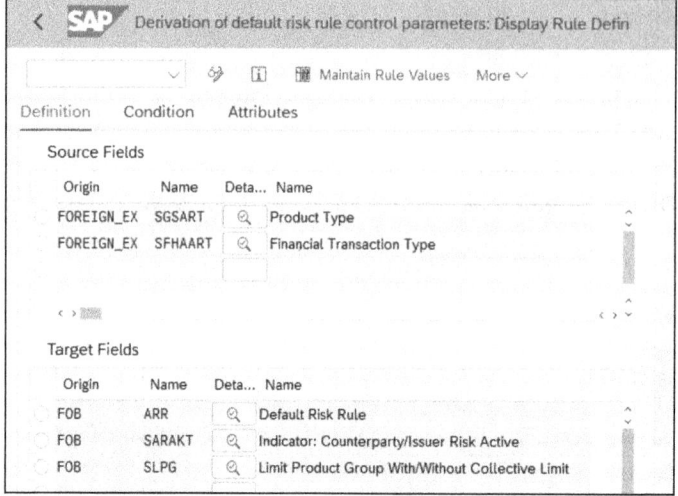

Figure 11.26 Defining Derivation Source Fields and Target Fields

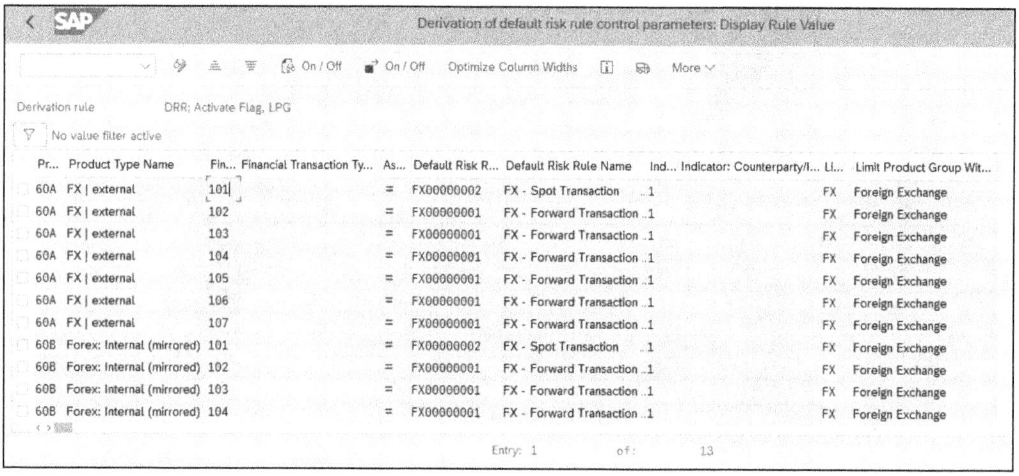

Figure 11.27 Maintaining Rule Values for Default Risk Rule

With the settings we've just reviewed, we'll be able to execute trades and have those trades update to the Credit Risk Analyzer functions for either a single limit check or an end-of-day processing check. This will help us understand our trade exposures in relation to the limits we have setup, and it will also give visibility to outstanding counterparty and settlement risks applicable to our trades.

11.2 Market Risk Analyzer

The next analyzer we'll discuss is the Market Risk Analyzer, which allows us to understand the financial positions we have, compared with market rates. It consumes market rates from our data provider (e.g. Bloomberg, Refinitiv) and prices out each one of the deals to understand that deal's value. Once the value has been computed, we can use it for daily treasury review to help us understand positions, or we can use it at month end to post the impacts of the deal to the general ledger. Having visibility into this information allows treasury the ability to understand and mitigate any impact to the organization.

Here are the four key areas of the Market Risk Analyzer:

- Market data collection
- Mark-to-market valuation
- Sensitivity analysis and simulations
- Value at risk

We'll discuss each of these in the following sections and also provide configuration information.

11.2.1 Market Data Collection

One of the keys to understanding market risks is being able to retrieve overall market data. This data typically is composed of the following market data classes:

- FX spot rates
- Securities
- Interest rates
- FX swap rates
- Exchange rate volatilities
- Interest rate volatilities

As we begin to build out our results database, we need these data points not only on a current basis but on a historical basis. The results database stores and manages all risk-related data and corresponding results from analysis. To bring in the market data, we use the Datafeed: Request Current Market Data app. We can see the selection screen for import in Figure 11.28, and we may set up different **Market Data Class** variants for this in order to process the necessary market data to populate our system. To set this up, we'll select the market data classes that are in our file (for example, **Currencies**), and we will then populate the required **Datafeed Name** field (e.g., with Y001) and any other output controls we wish to select.

Then, we'll click **Execute,** which will show the successful updates of the market data upload. The job log will also be displayed, as shown in Figure 11.29.

Additionally, we can always review rates by using the Current Exchange Rates, Enter FX Swap Rates, Enter Interest Rates, Enter Exchange Rate Volatilities, and Enter Interest Rates Volatilities apps. The Current Exchange Rates app is shown in Figure 11.30.

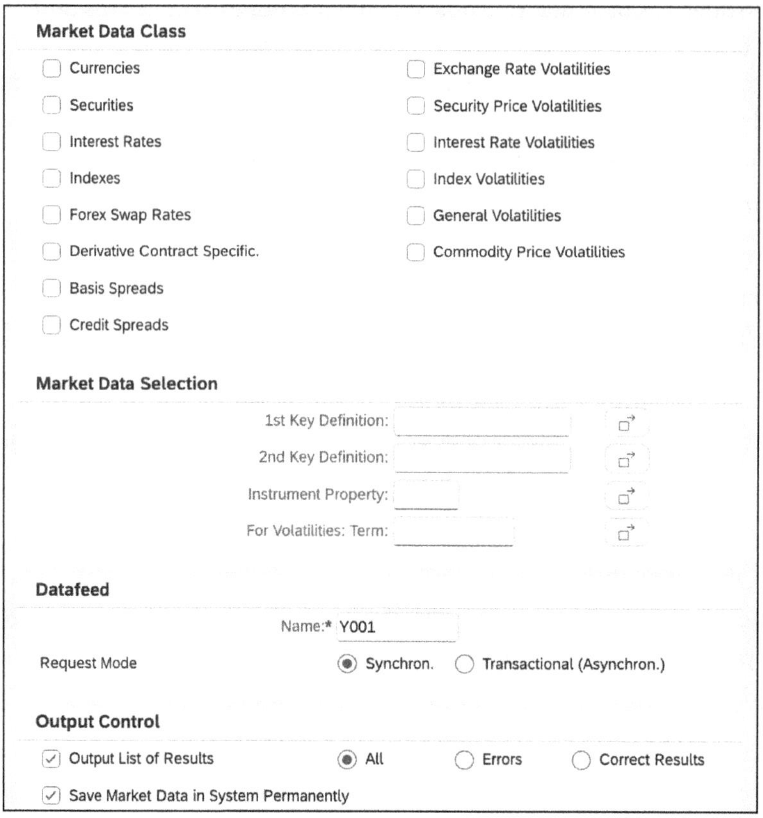

11 SAP Treasury Analyzers

Market Data Class

- [ ] Currencies
- [ ] Securities
- [ ] Interest Rates
- [ ] Indexes
- [ ] Forex Swap Rates
- [ ] Derivative Contract Specific.
- [ ] Basis Spreads
- [ ] Credit Spreads
- [ ] Exchange Rate Volatilities
- [ ] Security Price Volatilities
- [ ] Interest Rate Volatilities
- [ ] Index Volatilities
- [ ] General Volatilities
- [ ] Commodity Price Volatilities

Market Data Selection

1st Key Definition:
2nd Key Definition:
Instrument Property:
For Volatilities: Term:

Datafeed

Name:* Y001

Request Mode   (●) Synchron.   ( ) Transactional (Asynchron.)

Output Control

- [x] Output List of Results   (●) All   ( ) Errors   ( ) Correct Results
- [x] Save Market Data in System Permanently

Figure 11.28 Importing Market Rates

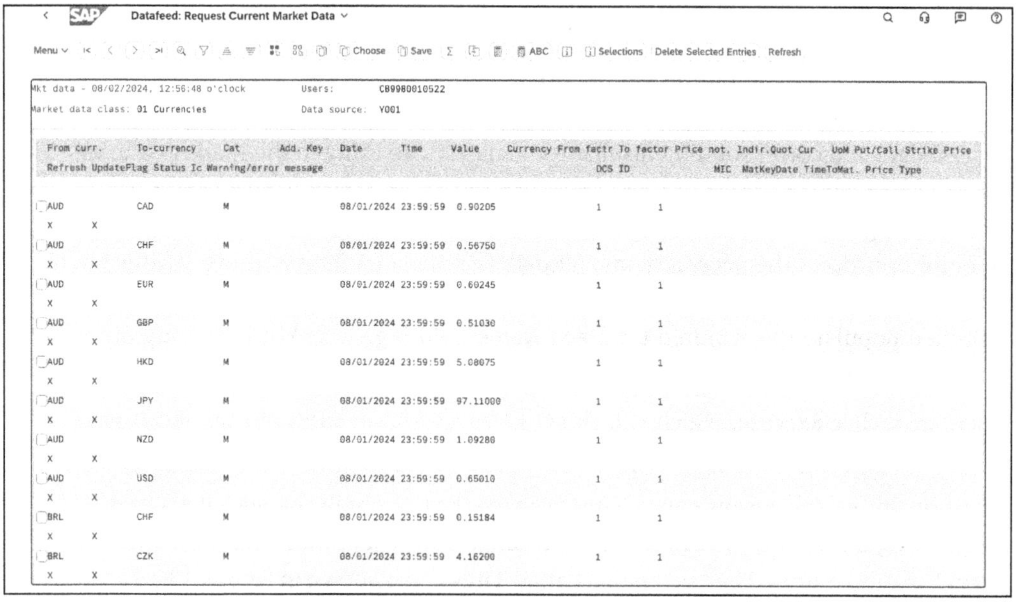

Figure 11.29 Importing Market Rate Data

11.2 Market Risk Analyzer

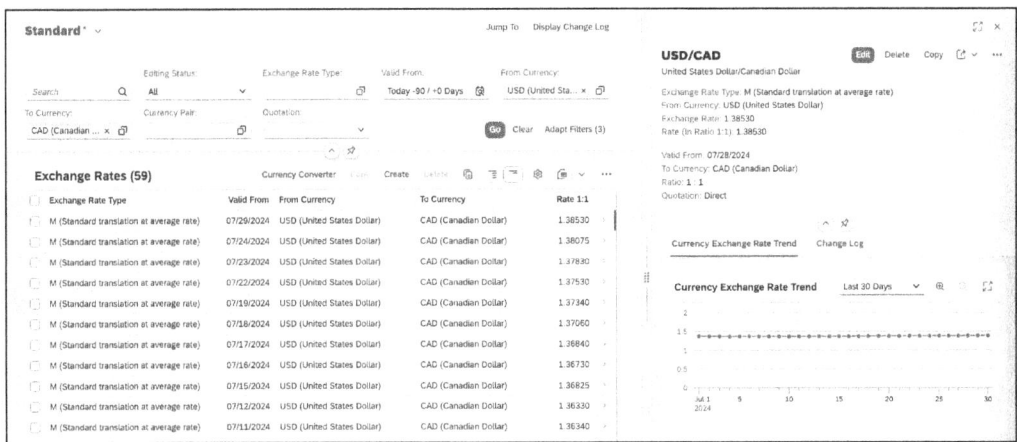

Figure 11.30 Review Currency Rates App Showing Rate History for Thirty Days

11.2.2 Mark-to-Market Valuation

Now that we have our key market data collected, we can start the process of revaluing our transactions. With SAP S/4HANA, we can value our positions as frequently as needed. This helps treasury organizations understand risks as quickly as the market is changing, and they can adjust their positions accordingly. We covered mark-to-market valuation in Chapter 6, Section 6.5. The evaluation type is responsible for bringing in the necessary market data (e.g., yield curves [risky or risk-free], FX rates, volatility types) and correlations that we will use as the bases for calculations in our valuation runs. In Figure 11.31, we can see the output for the mark-to-market valuation.

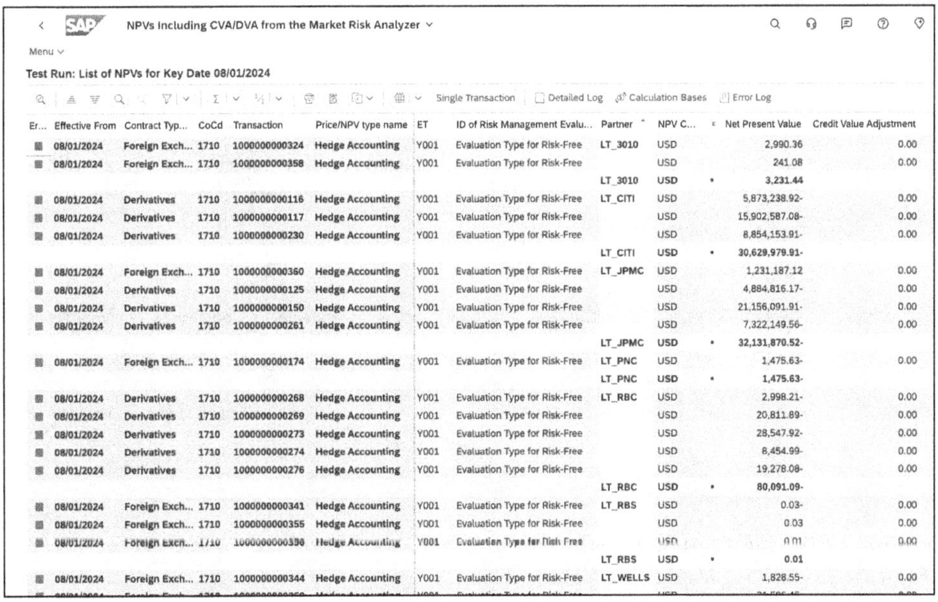

Figure 11.31 Running Mark-to-Market Values

727

11 SAP Treasury Analyzers

11.2.3 Sensitivity Analysis and Simulations

In addition to being able to calculate the mark-to-market valuation of our transactions, we can provide sensitivity analysis and simulations for our deals. Mark-to-market valuation is designed to look at current market data as of a key date, whereas sensitivity analysis is designed to look at the same data points but focus on the possibilities of that market data in the future and the impact it would have on our portfolio.

First, we'll open the Manage Market Data Shifts app, where we'll create, change, display, delete, or copy a shift. To create a new market data shift, we'll enter a unique numerical value in the **Market Data Shift** field and click **Create**, as shown in Figure 11.32.

Figure 11.32 Creating and Changing Market Data Shifts

On the next screen, we'll define both a **Short Name** and a **Long Name** that describe what the shift is designed to do, as shown in Figure 11.33. Then, we'll enter the market data elements that we'd like to have shifted (e.g., exchange rates, FX swap rates, yield curves, credit spreads, indexes, securities, volatilities). Some common thresholds are 2%, 5%, and 10%, but the shift is user defined and could be any percentage value. We may also elect to add a single market data element or a series of elements to each market data shift.

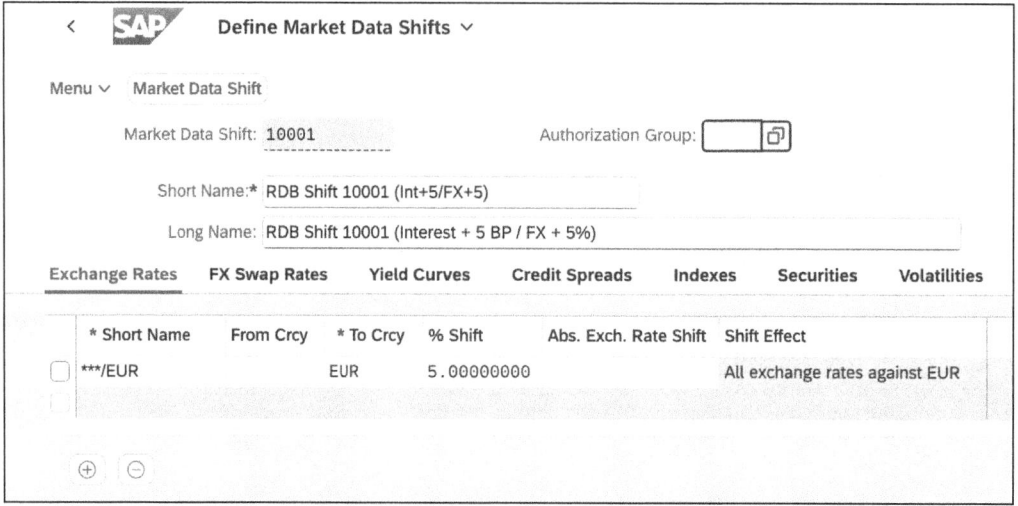

Figure 11.33 Defining Market Data Shift Attributes

In our example, we want to see any 5% shifts of any currency against the euro, so we've entered "***/EUR" in the **Short Name** column, left the **From Crcy** column blank so it will select all currency pairs, entered "EUR" in the **To Crcy** column, and entered "5.000" in the **% Shift** column.

> **Note**
> We may also add authorization objects to our market data shift if we would like some additional control over the create, change, or delete functions.

Now that we've set up the data shifts (and released them, if necessary), to execute a market data shift, we'll use the Analyze NPV app, which requires the following key inputs (see Figure 11.34):

- **Evaluation Currency**
  This is the currency in which we want our values to be displayed.

- **Evaluation Type**
  This is an indicator of what data we want to use. We will define the evaluation type in the configuration settings.

- **Evaluation Date**
  This is the current date.

- **Horizon**
  This is a date in the future we would like our shift to be applicable to.

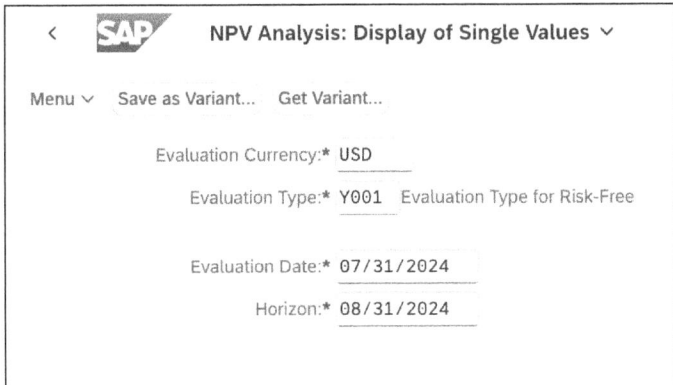

Figure 11.34 Key Inputs into Analyzing NPV

Additionally, we have several tabs that can aid us in displays of data:

- **Gen. Selections**
  As shown in Figure 11.35, we often use this tab to help generate the report in a specific way. For example, we can use the **Portfolio Hierarchy** to sort our report based on **Company code / Position Curr.**, **Company Code/BP**, or **Company Code/Prod.Typ**. The portfolio hierarchy setup is outlined in Section 11.2.5.

11 SAP Treasury Analyzers

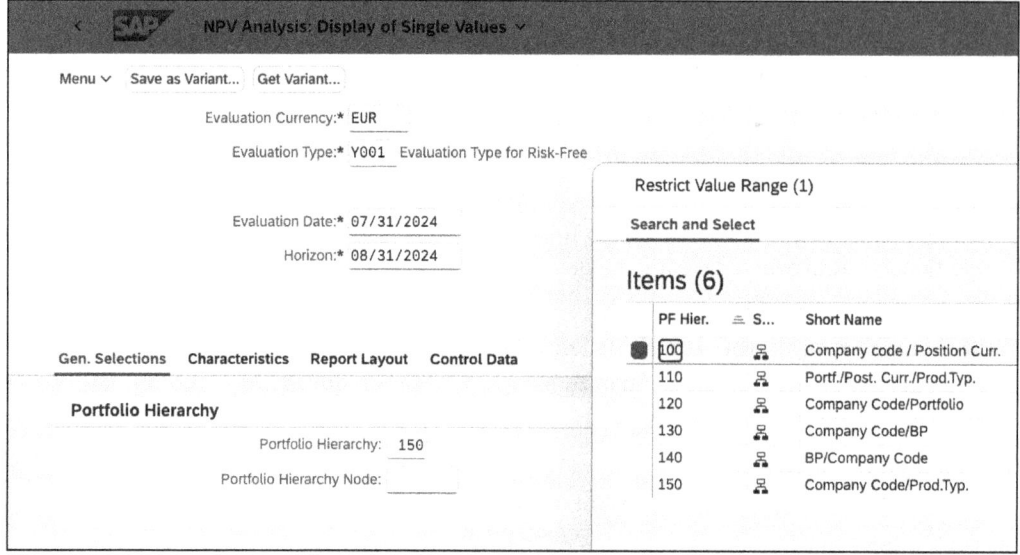

Figure 11.35 Analyzing NPV: General Selections

- **Characteristics**

    As shown in Figure 11.36, this tab is useful if we want to restrict the data further to individual company codes, trades, business partners, etc.

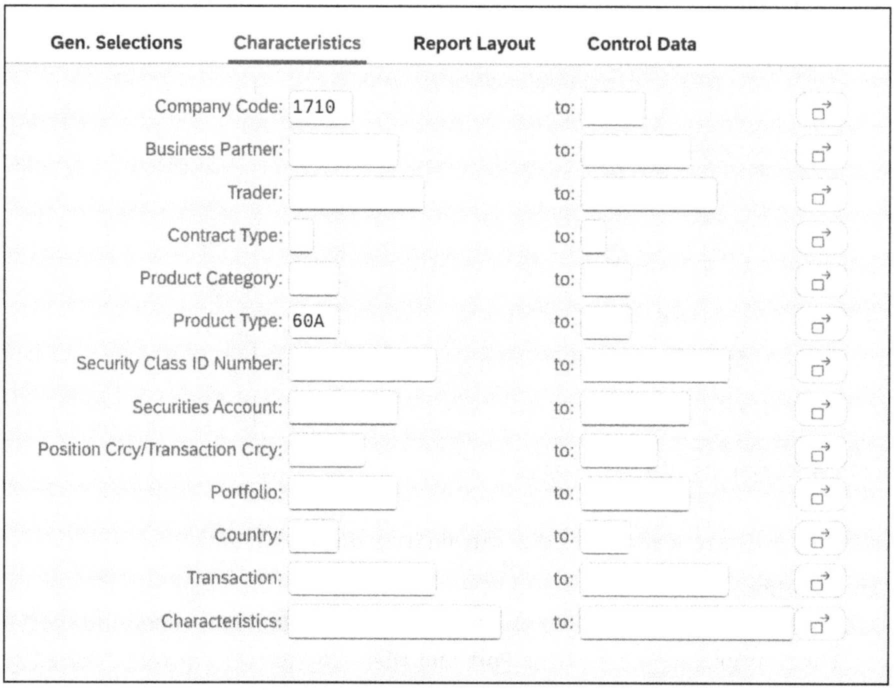

Figure 11.36 Analyzing NPV: Characteristics

- Report Layout

    As shown in Figure 11.37, here we'll define the different market data shifts that we may wish to incorporate into our analysis.

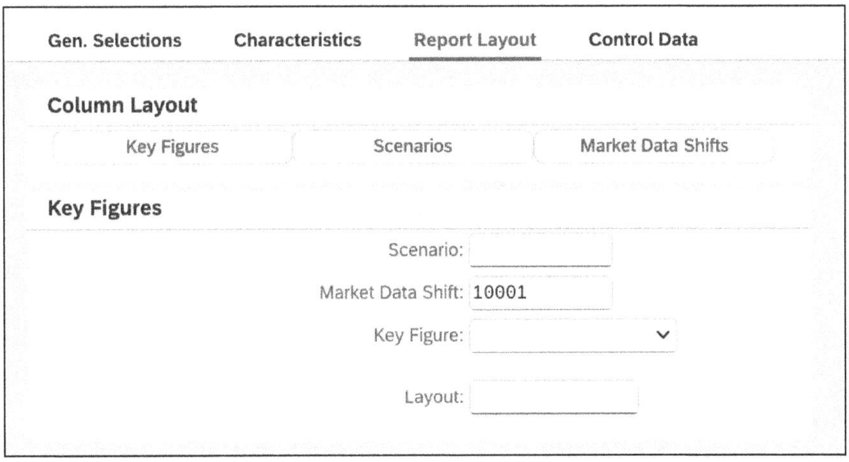

Figure 11.37 Analyzing NPV: Report Layout

- Control Data

    As shown in Figure 11.38, with these settings, we can determine how the report is generated and displayed (e.g., with a detailed log checked), and we'll be able to drill down into each of the transactions and understand the calculations that have been provided.

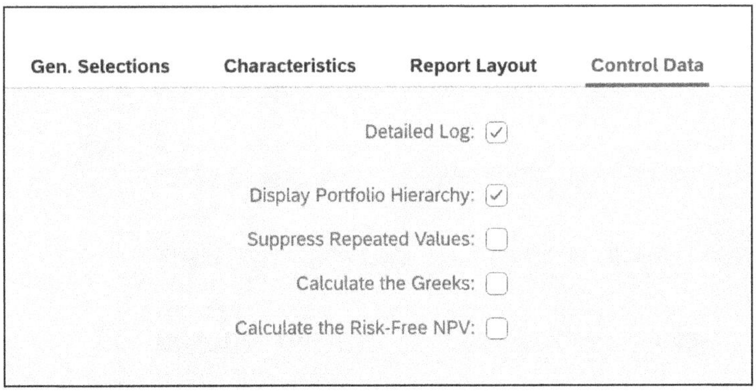

Figure 11.38 Analyzing NPV: Control Data

Upon execution, the report in Figure 11.39 is displayed. We can see that the hierarchy view has been used and that all data is sorted by company code and then product type. If we select a company code, we'll then see all the deals that are associated with that company code. We'll also see the current mark-to-market value with the market data shift that was executed, along with the difference between the two.

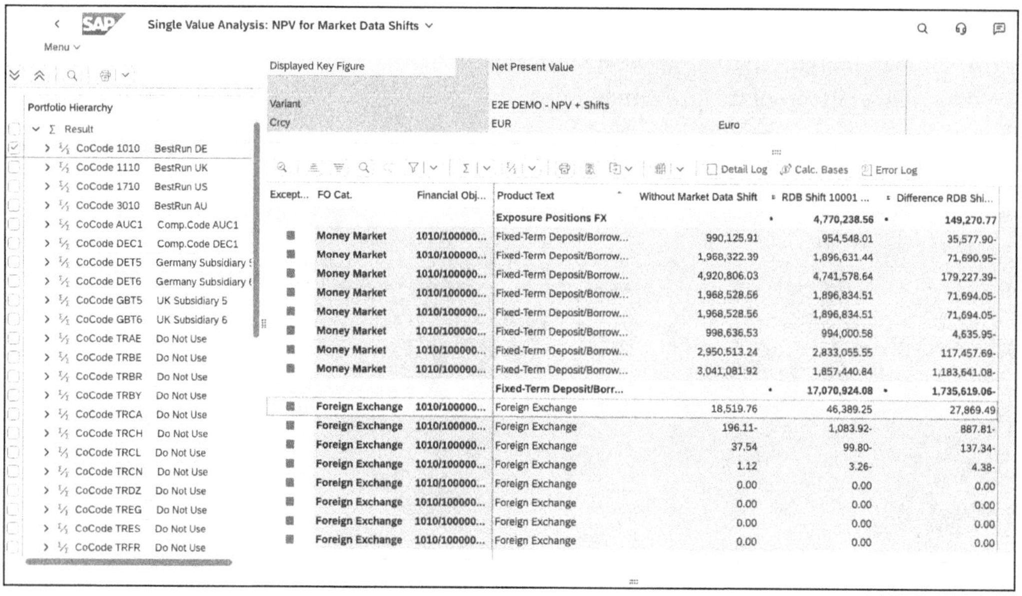

11 SAP Treasury Analyzers

Figure 11.39 Analyzing NPV Report Results with Shifts and Differences

Now that the shift has been calculated, we can drill down to the individual deal and see the detailed log for how the shift was computed, as shown in Figure 11.40. As with the mark-to-market calculation, we can see the calculation basis of what market data was used to value our positions.

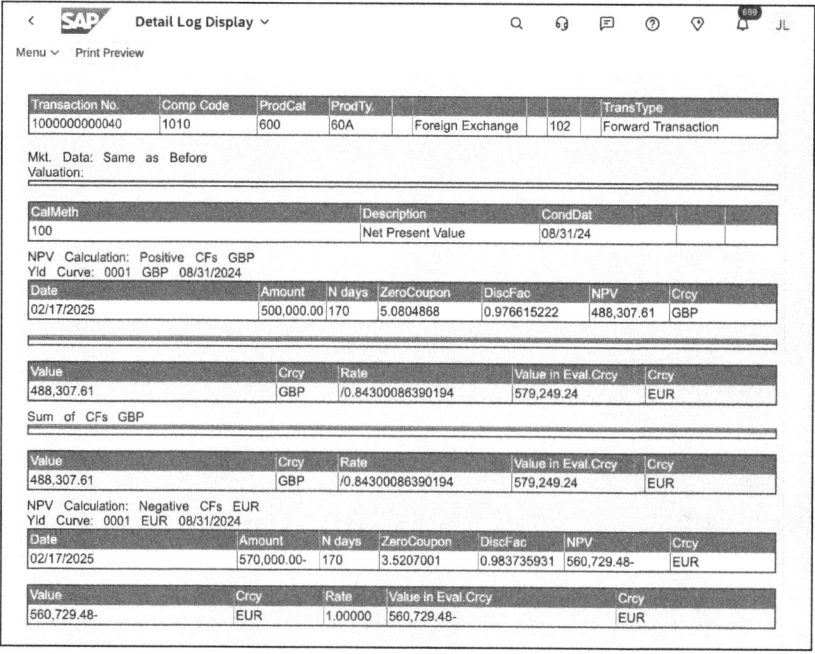

Figure 11.40 Detailed Log of NPV Shift Calculation

11.2.4 Value at Risk

Another feature of the Market Risk Analyzer is its ability to calculate value at risk in our portfolio. Value at risk helps treasury teams understand the levels of financial risk in the portfolio, helps provide a view of the maximum potential losses that could occur within a certain time frame, and provides a confidence level of that happening.

To access the value at risk calculation, we'll use the Single Value Analysis: VAR app. When accessing this app, we'll first need to decide the mode in which we want to execute our value at risk, as follows:

- **VaR Type**
  If we select this mode, we'll use the configuration setup in the **Defining Value at Risk Types** subsection. We can select the value at risk as a historical, Monte Carlo, or risk metrics simulation, and this is the indicator of how the positions will be valued.

- **VaR Key Figure**
  If we select this mode, the evaluation type, valuation area, risk hierarchy, and currency will be taken from the key figure definitions.

- **VaR Key Figure, overwrite settings**
  If we select this mode, we'll be able to enter the settings desired.

In our example in Figure 11.41, we'll select the **VaR Key Figure** mode, input a **VaR Key Figure** value of "10VAR," input our evaluation date, and check the box if we want a detailed log. We'll then click **Execute**.

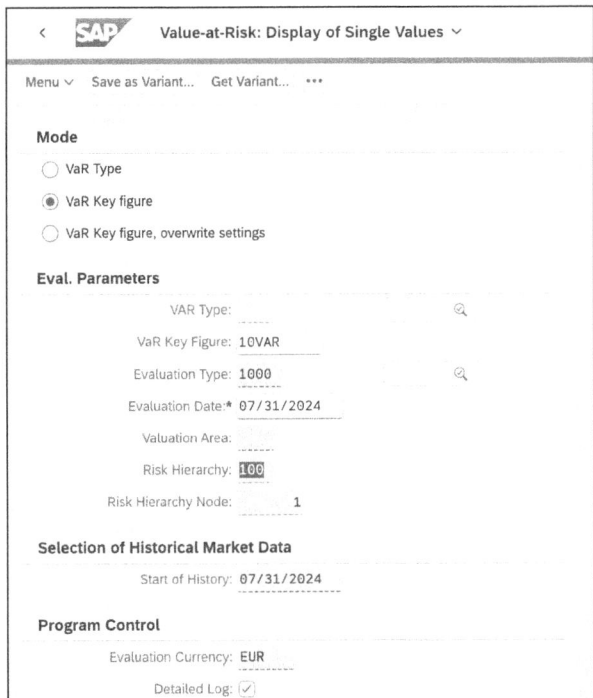

Figure 11.41 Single Value at Risk Selection Screen

11 SAP Treasury Analyzers

Upon execution of the report, we'll see the **Value at Risk** displayed on a deal-by-deal basis, as shown in Figure 11.42. We have the same capabilities available as in the mark-to-market report (e.g., detail log, calculation basis, error log, provide market data shifts).

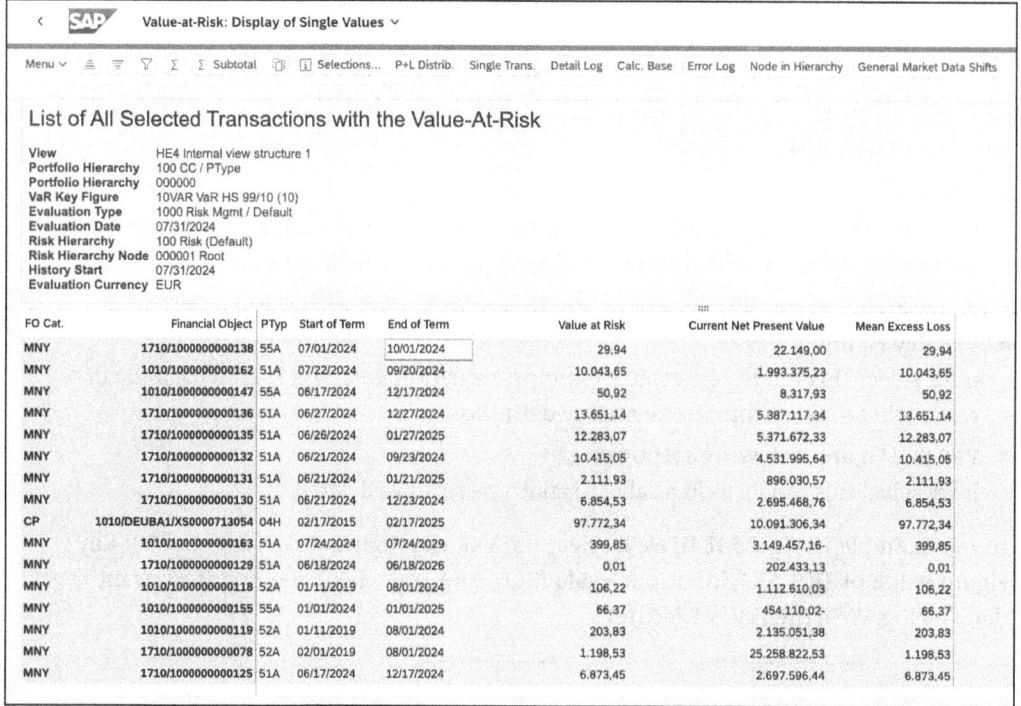

Figure 11.42 Review Value at Risk Output

11.2.5 Configuration

We've seen the generation of reports for the Market Risk Analyzer, and we'll now set up the associated configuration so we can analyze those reports and processes:

- Defining portfolio hierarchies
- Defining value at risk types
- Defining evaluation types

Defining Portfolio Hierarchies

In many of the reports and processes we've reviewed in this section, we saw how we can leverage a portfolio hierarchy that helps control the output of the report. We configure this in the following menu path: **Treasury and Risk Management • Basic Analyzer Settings • Define Portfolio Hierarchy**.

To set up a portfolio, we click **New Entries,** enter a unique three-digit numerical key in the **Portfolio Hierarchy** field, and enter the associated **Short Name** and **Long Name**, as shown in Figure 11.43.

11.2 Market Risk Analyzer

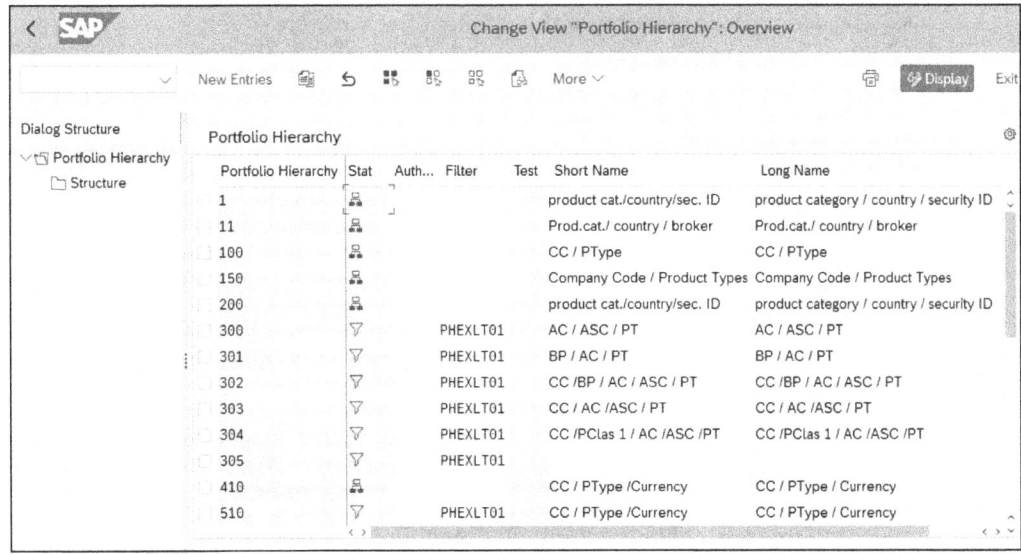

Figure 11.43 Defining Portfolio Hierarchies

Once we've defined the portfolio hierarchy, we click on the **Structure** folder in the menu on the left. The structure will define the sequence of characteristics that we want to see, and we can simply sort the characteristics numerically in the order we desire by entering numbers in the **Sort** field (as shown in Figure 11.44).

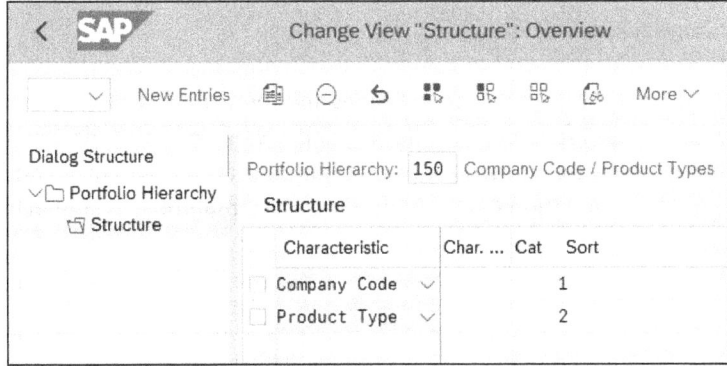

Figure 11.44 Defining Portfolio Hierarchy: Structure

Sections with Complementary Information

We can also see each status of the portfolio hierarchy. One of the following statuses may be displayed:

- Portfolio Hierarchy can be used
- Portfolio Hierarchy with a different tree structure
- Selective Portfolio Hierarchy

735

11 SAP Treasury Analyzers

- Selective portfolio hierarchy with a different tree structure
- Inactive portfolio hierarchy
- Incorrectly defined portfolio hierarchy
- Temporary portfolio

Defining Value at Risk Type

We use the value at risk type to control the calculation of value at risk. This allows us to segment different methods when calculating values. We can indicate a historical, Monte Carlo, or risk metrics simulation method. Figure 11.45 shows the overview of the value at risk types currently set up.

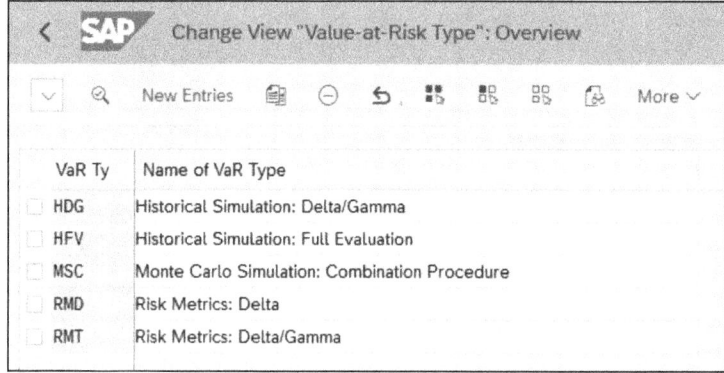

Figure 11.45 Defining Value at Risk Type

To create a new value at risk type, we select **New Entries** and then enter our three-digit alphanumeric value in the **VaR Type** field (e.g., "HDG") and a **Name of VaR Type**. Then, we'll select the **VaR Category**: 0 for simulation or 1 for variance and covariance. Additionally, we'll select the **VaR Simulation Cat.** of 0 for **Historical Simulation**, as shown in Figure 11.46.

> **Note**
> When setting up the VaR simulation categories, keep in mind the following:
> - 0 through 3 are reserved for SAP functions to determine portfolio value distribution.
> - 50 through 53 are reserved for customer-specific categories if necessary.

Additionally, by scrolling down on the page, we can check the boxes to define the **Settings for VaR Determination: Calculate VaR**, **Calculate P+L**, and **Consolidated VaR**. Key settings here are identifying the calendar (e.g., US); the holding period (e.g., 10), which indicates the length of time we plan to hold financial commitments; and a confidence level (e.g., 99), which indicates the probability that a currency won't fall below a certain level in the previously defined holding period (see Figure 11.47).

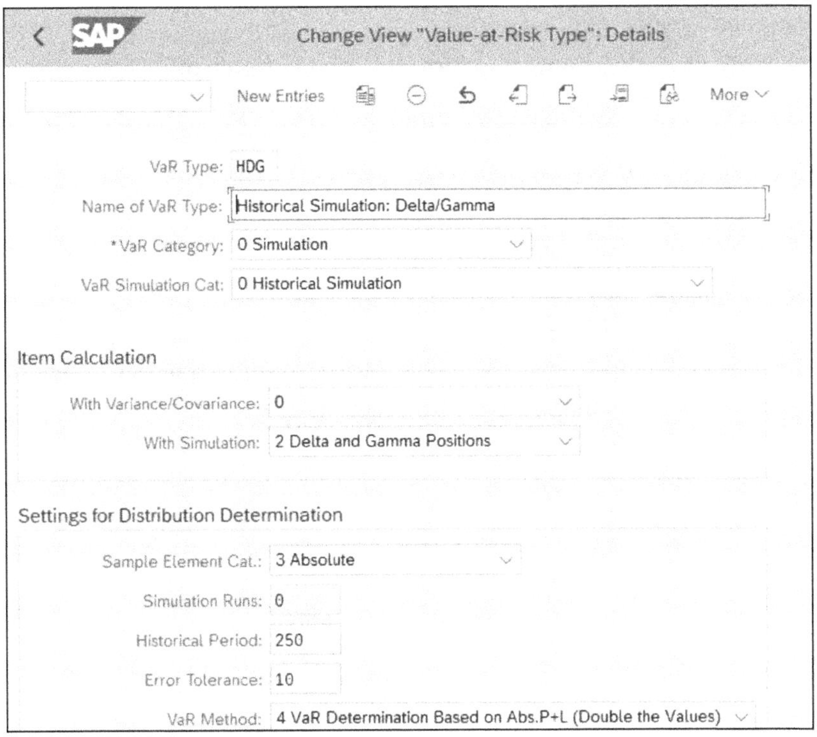

11.2 Market Risk Analyzer

```
VaR Type: HDG
Name of VaR Type: Historical Simulation: Delta/Gamma
*VaR Category: 0 Simulation
VaR Simulation Cat: 0 Historical Simulation

Item Calculation
With Variance/Covariance: 0
With Simulation: 2 Delta and Gamma Positions

Settings for Distribution Determination
Sample Element Cat.: 3 Absolute
Simulation Runs: 0
Historical Period: 250
Error Tolerance: 10
VaR Method: 4 VaR Determination Based on Abs.P+L (Double the Values)
```

Figure 11.46 Defining Value at Risk: Details

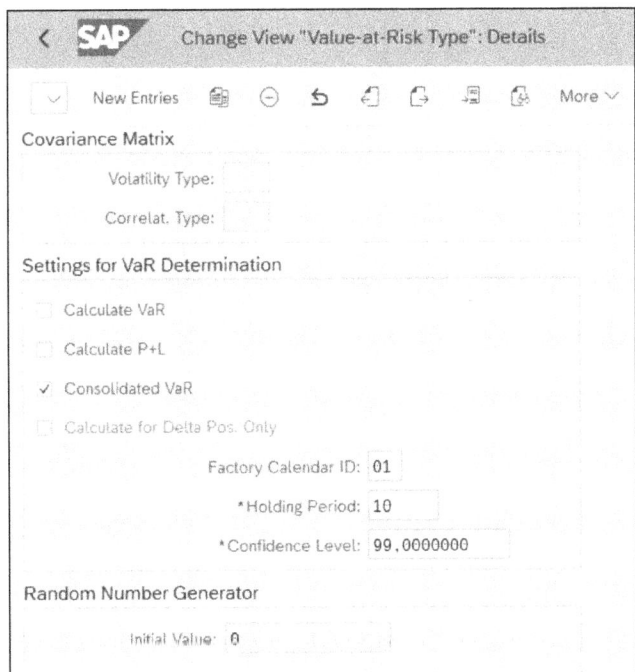

Figure 11.47 Settings for VaR Determination

11 SAP Treasury Analyzers

Defining Evaluation Types

As part of the mark-to-market calculations, we also need to select an evaluation type, which instructs SAP S/4HANA on what data to use for the overall valuation, whether it be for mark to market, market shifts, or value at risk. We configure evaluation types in the **Financial Supply Chain Management • Treasury and Risk Management • Basic Analyzer Settings • Valuation • Define and Setup Evaluation Types** menu path. There are several tabs as part of the evaluation type:

- **Market Data Categories**
  As shown in Figure 11.48, for the varying sources of market data spots, swap rates, yield curves, etc., we'll indicate what rate types we wish to be part of the evaluation type. Note that we've input a price type of 1, which correlates with our yield curve. These settings may differ in your system based on your setup.

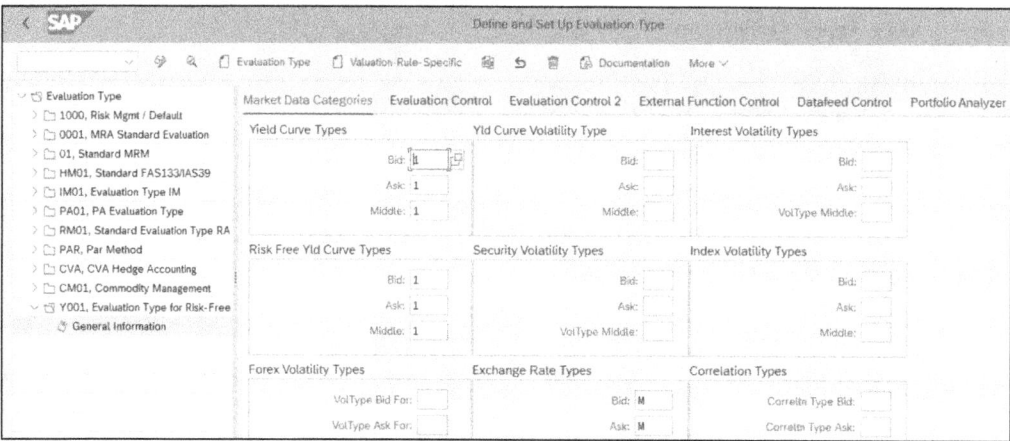

Figure 11.48 Defining and Setting Up Evaluation Types: Market Data Categories

- **Evaluation Control**
  As shown in Figure 11.49, based on the setup of our market data characteristics, we can update these settings if we have the need to bring in specific security prices or take into account netting flows for FX, handling of accrued interest, and defining usage for yield curves.

- **Evaluation Control 2**
  As shown in Figure 11.50, additional attributes are available, including the following:
  - **FX Valuation Method**
    This determines how to impact the price when running the NPV. There are two options:
    - **Discounting Before Currency Conversion**
      If we select this, we'll use the yield curve pricing for our NPV, and then the system will use the effective spot rate to convert the FX.

11.2 Market Risk Analyzer

- **Currency Conversion Before Discounting**
  If we select this, the forward rates will be used for pricing our NPV and then discounted using the yield curve.
- **FX Fixing Details**
  This setting is applicable if we're trading FX non-deliverable forward (NDF) trades and we'll decide to either include the fixing of the FX NDF trade or not. Settings for bank accounts evaluations are as follows:
  - If we plan to do any valuations of our bank account balances, we can identify them here. We'll identify the data source: either **One Exposure** or **Bank Account Balances**. In our example, we've selected **One Exposure**.
  - We'll then define the cash balance type. We can select **Value Date**, **Ledger**, or **Available Balances**.

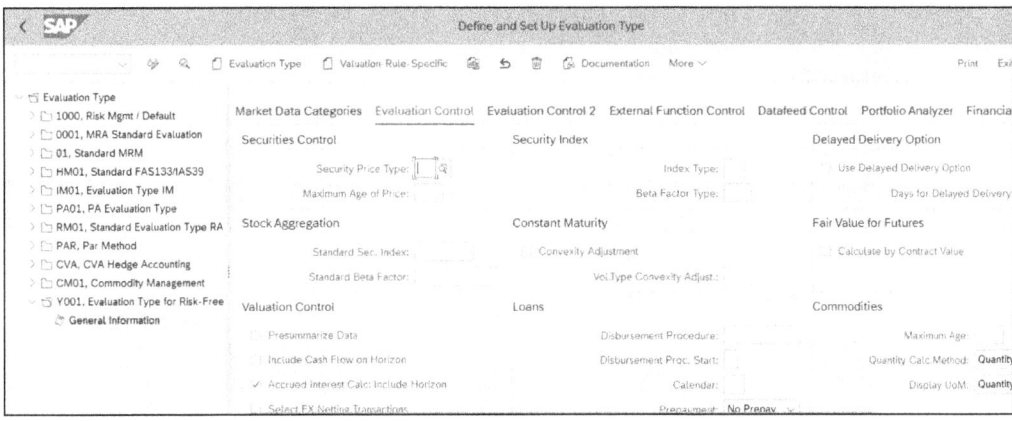

Figure 11.49 Defining and Setting Up Evaluation Types: Evaluation Control

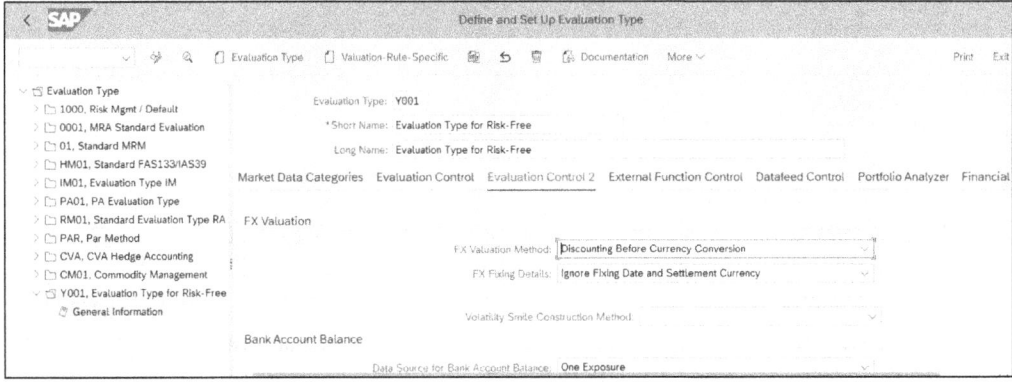

Figure 11.50 Defining and Setting Up Evaluation Types: Evaluation Control 2

- **External Function Control**
  As shown in Figure 11.51, if we're importing any valuations from other systems, we can control remote function calls here.

11 SAP Treasury Analyzers

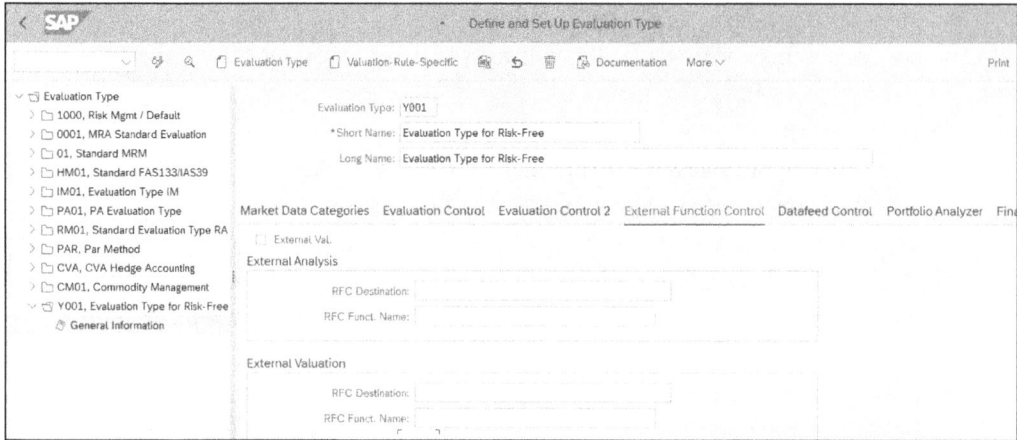

Figure 11.51 Defining and Setting Up Evaluation Types: External Functional Control

- **Datafeed Control**

    As shown in Figure 11.52, we can input settings for market data providers and activate market areas in all our market data (e.g., interest, exchange rates, security prices). Here, we've entered our specific **Data Provider**, "Y001," and we've selected **ER Datafeed,** which relates to exchange rate datafeeds. However, there are several different options we can select: exchange rate volatilities, interest rates, interest rate volatilities, and securities.

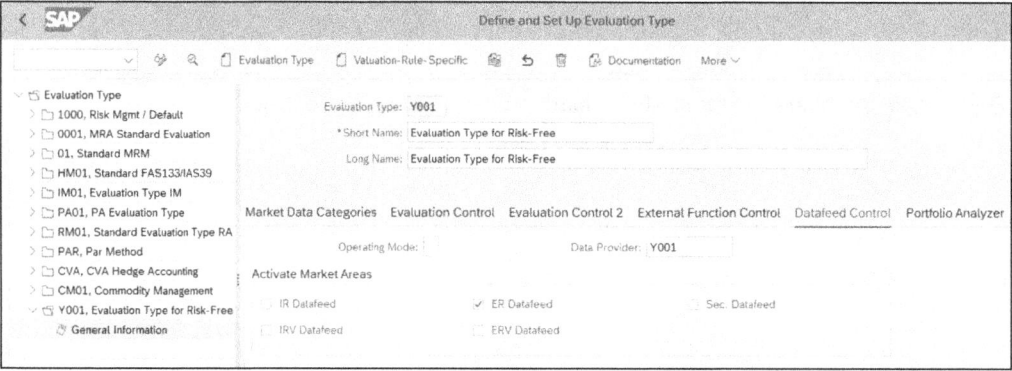

Figure 11.52 Defining and Setting Up Evaluation Types: Datafeed Control

- **Portfolio Analyzer**

    As shown in Figure 11.53, here, we can specify the **Valuation MM** for money market transactions as **NPV (default)**, **Nominal Value**, or **Nominal Value Plus Interest Receivable**.

11.3 Summary

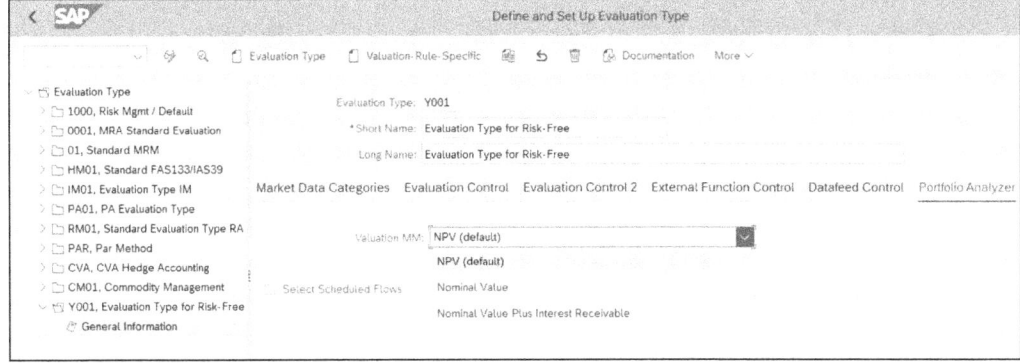

Figure 11.53 Defining and Setting Up Evaluation Types: Portfolio Analyzer

- **Financial Object Selection**
  As shown in Figure 11.54, here, we'll indicate how to handle futures accounts, if applicable. Our two options are **Class Positions in Futures Accounts Have Priority** and **Lot Based Positions in Futures Accounts Have Priority**.

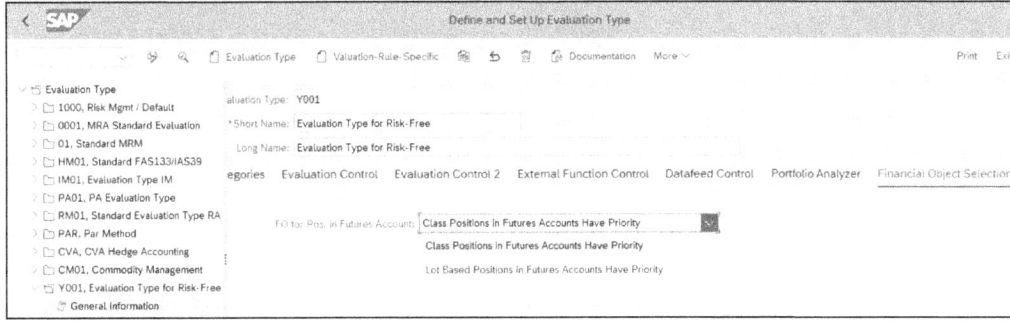

Figure 11.54 Defining and Setting Up Evaluation Types: Financial Object Selection

11.3 Summary

We've just seen how to add the analyzers to our treasury and risk management solution. When activating them, it's key to have good coordination between end users and configurators to get the settings correct. In SAP S/4HANA, credit and market risk reports are available to help us understand the impacts of external elements on the transactions they manage. Once the end users understand the credit or market exposures, they can determine how they need to reallocate, close, or keep the trades they have. This gives users real-time updates to help them make critical decisions.

Chapter 12
Integration with Other Areas

What truly amplifies the value of SAP S/4HANA Finance for treasury and risk management is its native integration with other areas in the SAP S/4HANA ecosystem. By having all treasury activities under the same roof in the broader ERP system, treasury departments can leverage tighter system integration without the need for custom interfaces. In this chapter, we'll examine the key areas of integration between treasury and risk management and the broader SAP S/4HANA ecosystem.

Since SAP S/4HANA is designed as an integrated, end-to-end business suite, its various areas seamlessly integrate with treasury and risk management. This integration allows for more efficient and accurate data handling within treasury processes, ensuring that data flows smoothly across the system, reducing redundancy and eliminating manual reconciliations. Treasury operations don't exist in isolation; they're inherently linked to various financial and operational aspects of a business. Therefore, understanding the key integration points between treasury and risk management and other areas is important for optimizing performance and gaining real-time insights across the S/4HANA landscape.

In this chapter, we'll focus on the key solutions that work in conjunction with treasury and risk management, outlining how they contribute to more effective treasury and risk management. These integrations not only facilitate automated financial postings and payment processing but also enable a unified view of the company's financial health, all within the same ERP system. The key solutions we'll explore include the following:

- **Cash management**
  The integration between treasury and risk management and SAP S/4HANA Finance for cash management enables the automatic and seamless flow of cash flow data from treasury transactions directly into cash management. This native integration means that activities such as payments, receipts, and liquidity movements related to financial instruments in treasury and risk management are immediately reflected in the cash position and liquidity forecasts within cash management.

- **Accounting**
  A key feature of treasury and risk management is its ability to define various activities within a transaction and automate the posting of those activities. Treasury and risk management can automatically process all accounting entries related to

treasury transactions—such as principal payments (incoming and outgoing), amortizations, valuations, accruals, settlements, and other relevant postings—and post them to the general ledger.

- **Payment processing**
  When payments are required for treasury transactions, treasury and risk management integrates with the treasury payments to generate the necessary payment requests. Once the payment requests are created, payment processing facilitates their execution. In this section, we'll demonstrate how this process works and explain how to accurately track these payments and differentiate them within the SAP Bank Communication Management system, ensuring smooth payment handling and proper identification.

- **In-house cash and in-house banking**
  In addition to processing standard payments, treasury and risk management offers flexibility in how to manage payments among internal entities. The fundamental setup for processing treasury and risk management payments through either SAP S/4HANA Finance for in-house cash or in-house banking for SAP S/4HANA Finance for advanced payment management is essentially the same. However, it's important to note that treasury and risk management is capable of integrating with both the legacy in-house cash system and the newer in-house bank introduced in 2022. This flexibility allows organizations to choose the in-house banking solution that best fits their needs while maintaining seamless connectivity with treasury and risk management.

12.1 Cash Management

Cash management is a core functionality of treasury, and it allows us to view all types of cash flows in SAP S/4HANA. First of all, we can define transactions as relevant to cash management, and cash management can reflect cash flows relating to prior-day bank statements, current-day bank statements, projected accounts payable and accounts receivable activity, and cash flows related to treasury and risk management financial transactions. Each of these activities is categorized differently in the cash reporting to assist in the proper reporting of the cash flows and to increase the flexibility to slice and dice the data as required. In cash reporting, the key differentiator to classify the source of the cash flow is the *certainty level*, which we assign at the point of creation to show what application has created the cash flow. Treasury transactions have a designated set of certainty levels, and the following certainty levels detail transactions that originate from bank statements and postings:

- **ACTUAL**
  At this certainty level, the cash postings can be for regular cash activities or for treasury transactions. These transactions are then assigned liquidity items to further categorize the cash activities and show the origin of the cash transaction.

- **INTRAM**
  If the treasury transactions are reported on an intraday bank statement, the cash flows will generate memo records that will have a certainty level of INTRAM.

The following certainty levels are specific to treasury transactions and are generated directly from the transaction cash flows:

- **TRM_D**
  When a treasury contract is created in treasury and risk management, the structure of the contract determines the cash flows. This certainty level is for all treasury and risk management contracts except for options and futures. In a debt or investment contract, the principal flows are on the first day of the contract. Interest calculations and frequency determine the amount of interest that is calculated and forecasted in the future. Additionally, the final repayment of the contract is determined and shows up as a future flow. All of these cash flows are identified as forecasted cash flows for treasury and risk management, and records will be created to write to the `FQM_FLOW` cash management table. Based on the status of the transaction, we can also drive the planning level. These statuses are defined as follows:

  - **Level (bank unknown)**
    Unknown transactions are transactions that have been created in SAP S/4HANA but don't already have a known bank account. If we create a transaction without payment details or if the contract is set up to post directly to an account, then the bank account is unknown. These types of transactions can go to a separate planning level since the bank account is not known at the time the contract was created.

  - **Level (bank known)**
    Known transactions are contracts that have the payment details filled out either in the business partner standing instructions or directly in the contract. Since we know the bank that is sending or receiving the funds, the bank is known and can go to a different planning level.

- **TRM_O**
  This certainty level is also for treasury transactions, but it is reserved for options and futures contracts. The settings and functions for these transactions will be the same as in **TRM_D**, but they'll show up at this certainty level. With an option contract, there's uncertainty about when and if the option will be exercised, so the option contract is separated and assigned a different certainty level.

We can leverage the cash management customizing in treasury and risk management to ensure that the cash flows that originate from cash management flow correctly into a cash position. There are a couple of settings that we need to set up in a certain way to ensure that the flows show up at the correct planning levels. In the following sections, we'll cover how to assign the planning levels for each of the product and transaction types. First, we'll look at how to assign the planning levels in the link to cash management function, and then, we'll look at the exceptions and how we can exclude cash

12 Integration with Other Areas

flows from creating cash management records. We'll also review a newer function in SAP S/4HANA that creates additional flexibility in the assignment planning levels by using additional attributes in the financial transactions. After reviewing all the sections on assignments, we'll review the final output of these settings to show how these cash flows appear in the main cash management SAP Fiori app, Cash Flow Analyzer.

12.1.1 Assigning Planning Levels

The main transaction we use to assign the planning levels for cash management is found in the **Financial Supply Chain Management • Treasury and Risk Management • Transaction Manager • General Settings • Link to Cash Management • Assign Planning Levels** menu path. We set up this transaction specifically for each company code, and the key settings we define in this transaction are as follows:

- **PType**
  This defines the product type we are assigning to this planning level.

- **ACat**
  Each contract has a different set of activity categories, and for the interest rate instruments, activity category 10 is for contract creation and activity category 20 is for contract settlement. Due to this, we can assign different planning levels depending on where the contract is in the processing. An example from Figure 12.1 shows that we can assign a planning level of "T?" to contracts with an activity category of unsettled contracts and a planning level of "TD" to contracts that have been settled.

- **Level (bank known)**
  Accounts with bank details or business partner standing instructions are assigned when we create the treasury and risk management forecast for these flows.

- **Level (bank unknown)**
  Accounts with no bank details assigned will use the planning levels in this column.

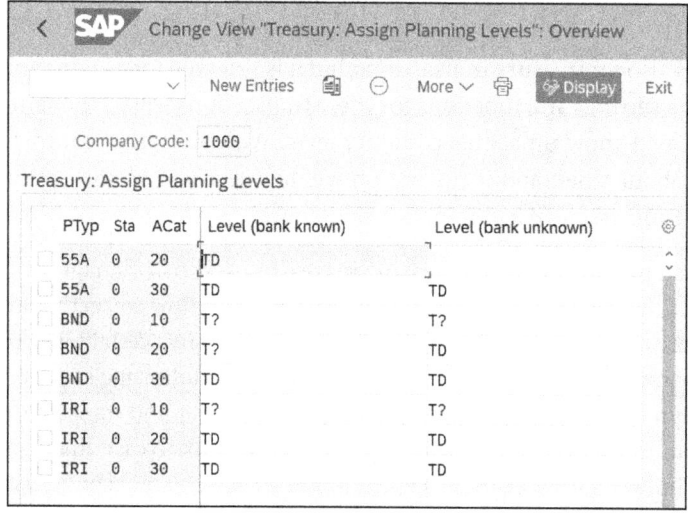

Figure 12.1 Assigning Planning Levels to Each Type of Transaction by Company Code

12.1 Cash Management

12.1.2 Specifying Update Types for Cash Management

There are some cash flows that should not flow through to the cash management reporting. If we need to set up this scenario, you can find the customizing in the **Financial Supply Chain Management • Treasury and Risk Management • Transaction Manager • General Settings • Link to Cash Management • Specify Update Types for Cash Management** menu path. When this is the case, we can drive the cash management record creation by the update type. In this transaction, we don't need to specify all update types, and by default, all the update types will generate cash management records. We should use this transaction on an exception basis, and we can use it to exclude update types from cash management record creation. In Figure 12.2, we have two update types set up in this transaction. In this scenario, update types **MM1100+** and **MM1901–** won't create any cash management records.

This setting is useful when setting up letters of credit. The nominal amounts of the letters of credit are only used for the fee calculation and are not actual cash flows. Since this is the case, we can filter the update types in this configuration to ensure that the forecasted cash flows are accurate.

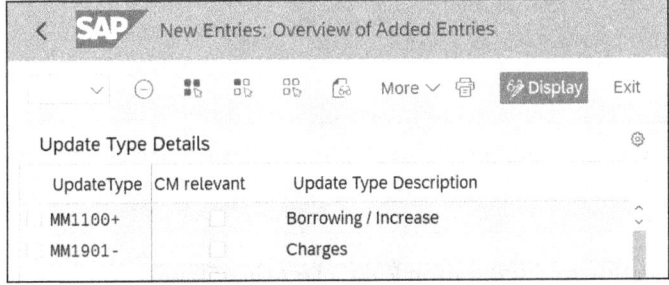

Figure 12.2 Specifying Update Types

12.1.3 Basic Settings for Cash Management Integration

There are some initial settings that we can set up in the **Basic Settings for Cash Management Integration** node. We can find this customizing in the **Financial Supply Chain Management • Treasury and Risk Management • Transaction Manager • General Settings • Link to Cash Management • Basic Settings for Cash Management Integration** menu path. In this configuration, there are a few settings that we can select (see Figure 12.3), as follows:

- **Simplify Flow Generation**
  Each product category has standard flow types that can be assigned to the cash flows. If we don't check this box, then the standard flow types will be assigned to the cash flows, and we can use these to define derivation rules for liquidity items. If we do check this box, then only two flow types will be assigned for treasury and risk management. These are as follows:

12 Integration with Other Areas

- 900100: Incoming Bank Cash (TRM)
- 900101: Outgoing Bank Cash (TRM)

Additionally, if we need to use the enhanced derivation rules for the planning levels, then we need to check this box.

- **Derivation Category for Planning Level – Classic Assignment**
  If we select this option, the planning levels will be assigned using the configuration that was detailed in Section 12.1.1.

- **Derivation Category for Planning Level – Derivation**
  If we select this derivation category, we can use the Substitution Rules for Planning Levels – Treasury Flows app to assign the planning levels.

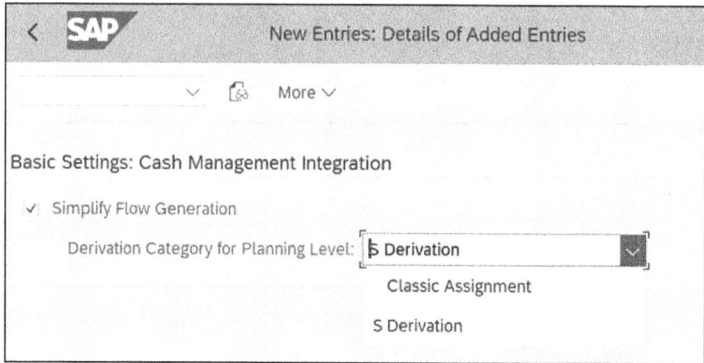

Figure 12.3 Updating this Setting to Derivation Allows for Dynamic Assignment of Planning Levels

12.1.4 Managing Substitution and Validation Rules: Treasury Flows

With SAP S/4HANA, we have an additional way to assign the planning levels to treasury contract flows. Instead of just assigning planning levels directly to a product type and activity type, we have additional attributes we can use to assign the planning levels. To derive the flows, we need to go to the Substitution Rules for Planning Levels – Treasury Flows app, as shown in Figure 12.4.

To derive the flows, we need to set up the following attributes. First, we need to add the **Rule Name** and **Description** for a rule. One thing to note is that the **Rule Name** needs to include only letters, numbers, and underscores (spaces are not allowed).

Second, we need to fill out the **Precondition** to determine the scenarios that the derivation should be triggered for. If we need to set multiple preconditions, we can click the **Add** button to add lines to this area. We can use the following fields to create a precondition:

- **Bank Is Known** (this filters for whether the bank is assigned to a contract or not)
- **Company Code**
- **Activity Category**
- **Product Type**

- Transaction Type
- Hedging Classification
- Portfolio
- Update Type
- Planning Level

Figure 12.4 Setting Up Rules in Substitution Rules for Planning Levels – Treasury Flows App

To select one of the fields or functions, we click on the box to the left of the field. These fields are shown in Figure 12.5.

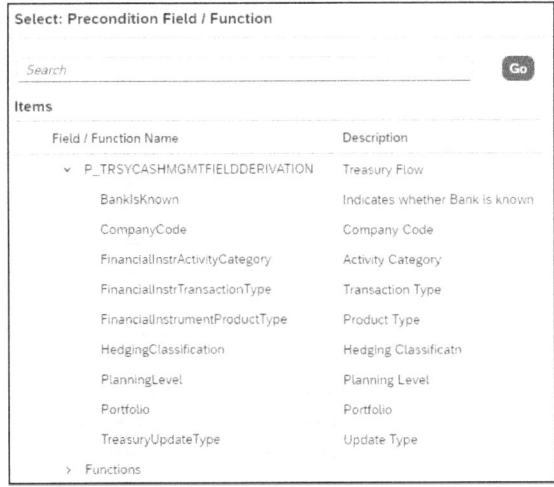

Figure 12.5 Precondition Section Displaying Fields We Can Use to Create Conditions for Planning Level Assignment

12　Integration with Other Areas

Once we've selected the field, the **Value** field will allow us to use the dropdown list to make the appropriate selection. In the example in Figure 12.6, the company code and product type have been selected for the **Precondition**.

Figure 12.6 Precondition Selection

Finally, we can create the substitution, which defines the planning level when the defined precondition is met. In Figure 12.7, we're substituting the planning level TM when the contract has a value of 1000 with a product type of FX.

Figure 12.7 Planning Level TM

12.1.5 Cash Flow Analyzer

Once we've fully assigned the cash management cash flows to the treasury and risk management cash flows, we can view them in the cash management reports. The Cash Flow Analyzer app is the main report we use to view the cash flows. This report can view the forecasted treasury cash flows, and if we want to only view the flows from treasury and risk management, we can either filter them by the specific **Planning Level** defined for treasury and risk management or filter them by **Certainty Level** (certainty levels **TRM_D** and **TRM_O** are specific to treasury and risk management), as shown in Figure 12.8. To view the forecasted treasury flows, select certainty levels **TRM_D** and **TRM_O**.

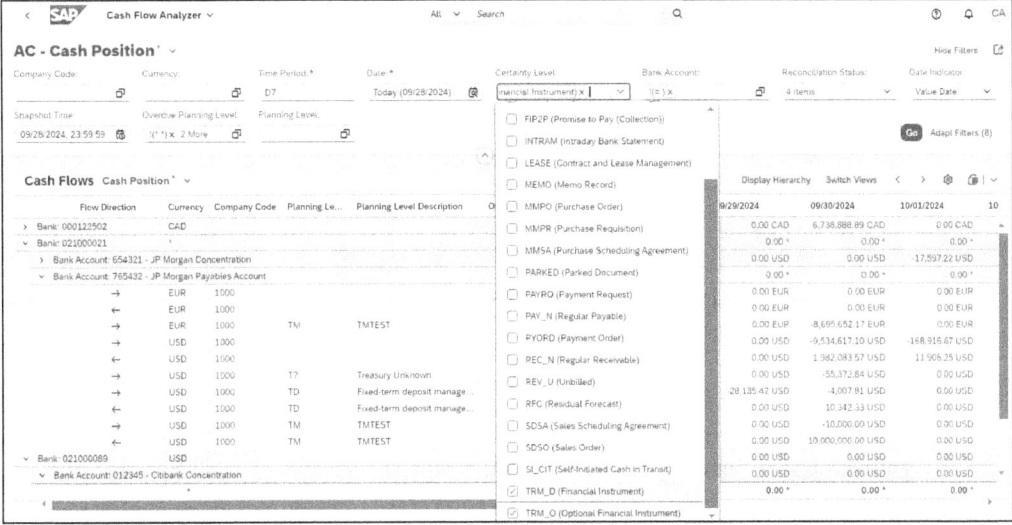

Figure 12.8 Cash Flow Analyzer App

With the **Certainty Level** filtered in the report, we can now see the forecast for all transactions in treasury and risk management, as shown in Figure 12.9.

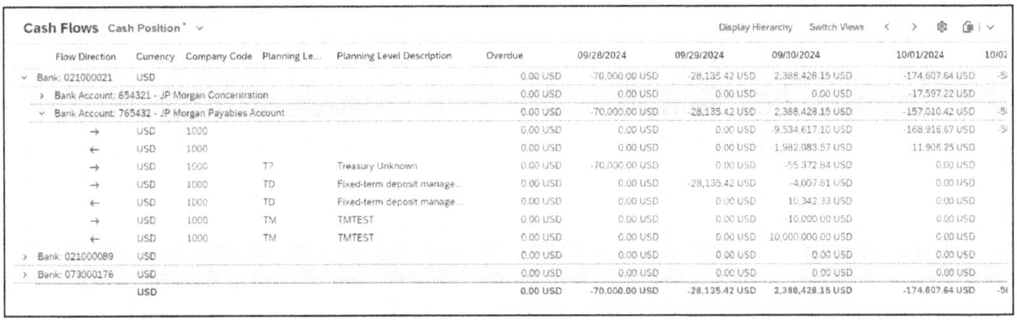

Figure 12.9 Viewing Treasury Forecasted Flows in Cash Flow Analyzer App

12 Integration with Other Areas

Each of the line items in this report has hyperlinks we can use to drill down further into the details. This functionality allows us to drill down all the way into the financial transaction that was created in the Create Financial Transaction app. To achieve this, we click on an amount and select the **Display Cash Flow Items** option. After we drill down further into the specific cash flow, the details will be displayed as in Figure 12.10. From this screen, clicking the **Contract Number** hyperlink will allow us to navigate to the Display Financial Transaction app to view that specific transaction's structure and details.

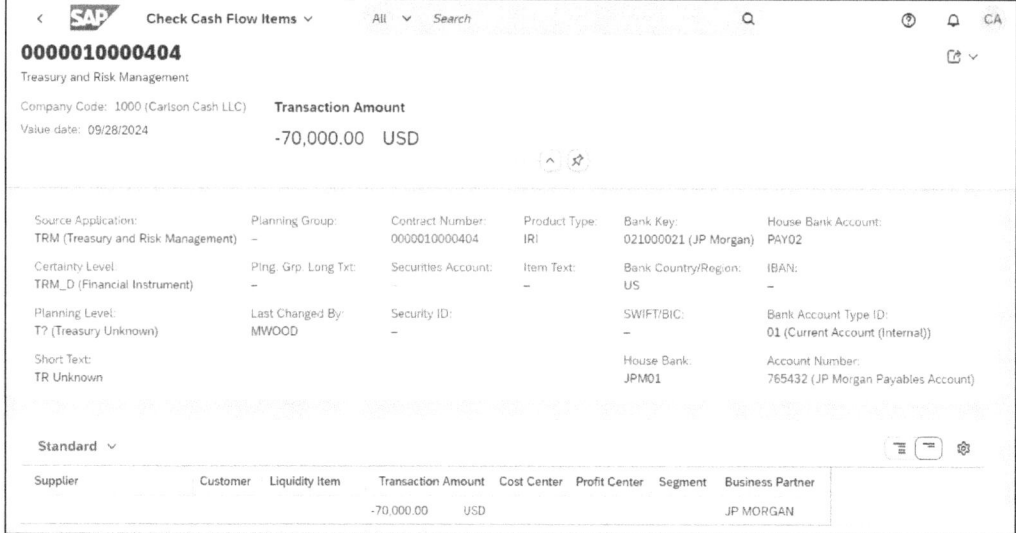

Figure 12.10 View Showing Tie Between This Financial Transaction and Its Related Contract Number

Creating the link between treasury and risk management and cash management in SAP S/4HANA allows for a more comprehensive solution and helps us more effectively manage current cash balances and forecast future cash requirements.

12.2 Accounting

Treasury and risk management works closely with accounting to generate journal entries when the posting programs are run. We covered each of the posting programs in detail in Chapter 5, Section 5.3, so in this section, we'll focus on showing how the accounting process is triggered in treasury and risk management and how the accounting is assigned.

We determine the accounting settings in the following customizing path: **Financial Supply Chain Management • Treasury and Risk Management • Transaction Manager • General Settings • Accounting • Link to Other Accounting Components**. We will not cover all the areas in this section, but we'll cover some of the links between the treasury and risk management contracts and accounting.

First of all, the main driver for all of the accounting treatment in treasury and risk management is the update type. In all scenarios of the accounting postings, an update type has been assigned to the associated flow. There are specific sets of update types for different transactions; general update types for the transaction, key date valuations, accruals and deferrals, and amortizations; and specific update types for other types of postings. Ultimately, all update types go to the accounting settings in **Financial Supply Chain Management • Treasury and Risk Management • Transaction Manager • General Settings • Accounting • Link to Other Accounting Components • Define Account Determination for Treasury and Risk Management**. As shown in Figure 12.11, the **Assignment of Update Types to Posting Specifications** section is where we assign a set of posting specifications to each posting update type.

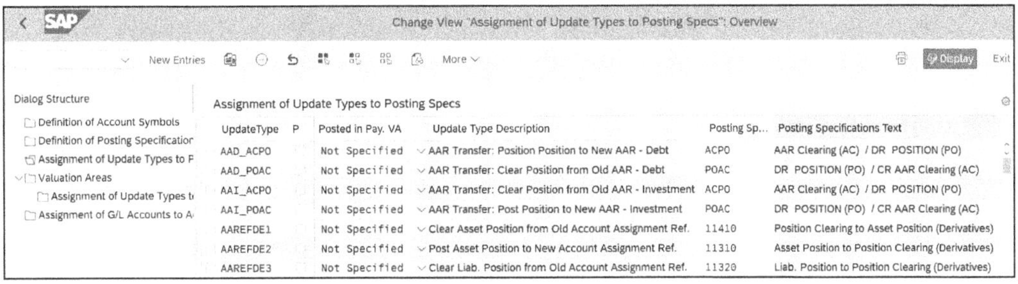

Figure 12.11 Assignment of Update Types to Posting Specifications

We covered these posting specifications in detail in Chapter 3, Section 3.21, but we've highlighted them in this section to give a full picture of the different settings that we use for the integration between treasury and risk management and accounting.

12.3 Payments

Treasury and risk management is deeply integrated into the payment process. Both incoming and outgoing payments can be directed from treasury and risk management to create treasury payment requests. Payment requests are all created using the Post Flows app or Transaction TBB1, and they leverage the settings in the **Payment Details** tab in the Create Financial Transaction app. To fully ensure that a transaction will create an executable payment, we need to make configuration and master data settings. Due to this, before showing how the payments are integrated, we'll show the key areas that we use to determine whether a payment request should be created for a treasury cash flow.

This section will cover the key elements of the payment settings, including setting up the flow types to be relevant to payments, creating payment-relevant posting specifications, assigning a bank general ledger account, and mapping the payment details either in the business partner standing instructions or directly in the transaction.

12 Integration with Other Areas

12.3.1 Flow Types

First of all, we need to set up the flow type to create a payment request. In the flow type setup covered in Chapter 3, Section 3.13, there's a field called **Payment Request**. We can set this payment request to be for incoming payments, outgoing payments, or both. Figure 12.12 shows this flow type set up for both incoming and outgoing payments. In the money market area, we use the principal increase for both an increase of borrowing on a debt contract and an increase in an investment, so depending on when this flow type is called, it could be either an incoming or an outgoing payment.

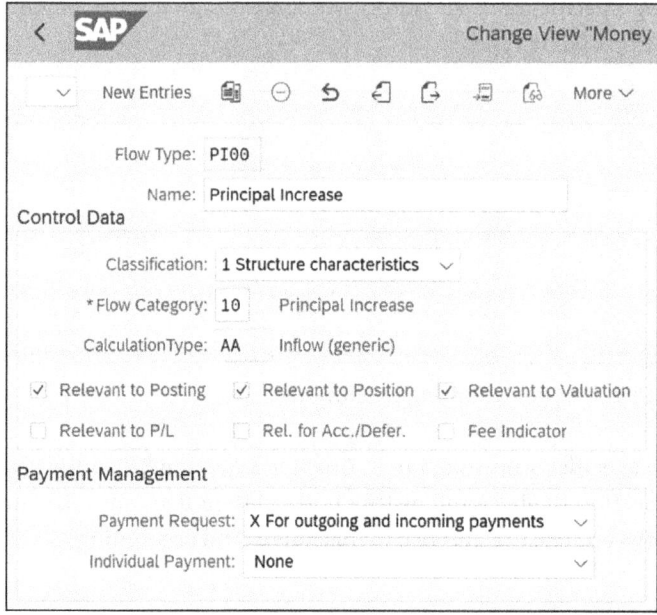

Figure 12.12 Payment Request Setting Determines Direction of Payments Flow Type Can Create

12.3.2 Posting Specifications

To skip forward in the process, we've already created the subsequent configuration to assign the update type to this flow type and assigned a condition type. The next relevant configuration step in this process is in the account assignment step, where we determine the posting specifications. This is located in **Financial Supply Chain Management • Treasury and Risk Management • Transaction Manager • General Settings • Accounting • Link to Other Accounting Components • Define Account Determination for Treasury and Risk Management**. In this area, we'll need to set up the posting specifications to be able to make payments. Figure 12.13 details a posting specification that has an outgoing payment, and we use our **Account Symbol (BK)** to make a **Bank Posting in Payment Currency**. We use this posting category to assign the bank postings in this process.

12.3 Payments

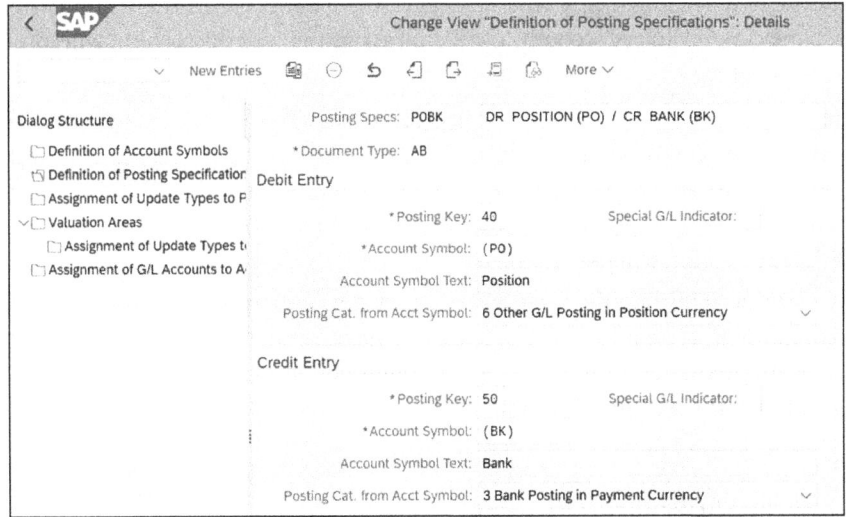

Figure 12.13 Creating Posting Rule to Send Outgoing Payment and Post Credit to Bank Account Symbol

12.3.3 Assigning General Ledger Accounts

The next step is to check the bank account symbol that we used when assigning the posting specifications. As shown in the example in Figure 12.14, we can use masking to assign the general ledger account.

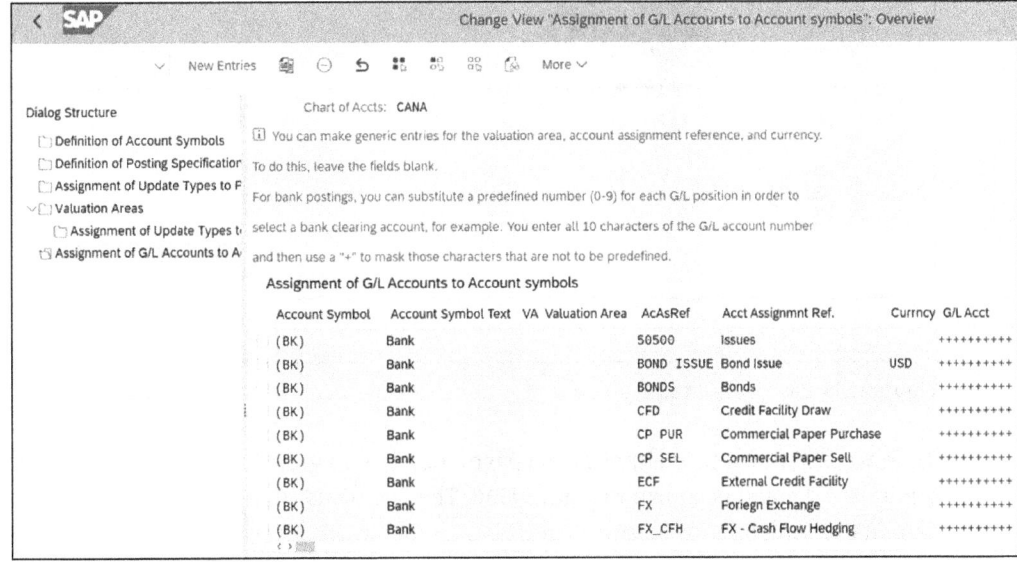

Figure 12.14 General Ledger Account with Wildcard Determines Bank Account Posting for Bank Account Symbol

12 Integration with Other Areas

Ultimately, we're just assigning a wild card here ("++++++++++" in the example) to assign the bank general ledger account. The actual assignment of the account will come in two steps, as follows:

1. The first step happens when the bank posting occurs from the treasury and risk management posting program. Commonly, the Post Flows app will post the cash flows, but there are other apps that will trigger the payment request creation in securities. When the Post Flows app is run, a payment request is created, and the bank side of the transaction will post to the payment request clearing account.

2. The next step occurs when the treasury payment run is executed. This will clear out the payment request clearing account and post the offset to the account assigned in the Automatic Payment Transactions for Payment Requests app or Transaction F111 configuration in the **Account Determination** section.

12.3.4 Contract: Payment Details

Now that we've covered the customizing-specific settings that we need to define for the payments, we'll cover the settings on the contract side. We already covered the business partner standing details in Chapter 2, Section 2.5, so we'll cover how to set up the payments in each contract to create the correct payment request. In Figure 12.15, we have both an incoming and an outgoing side of the payment request in the **Payment Details** tab of the contract entry screen.

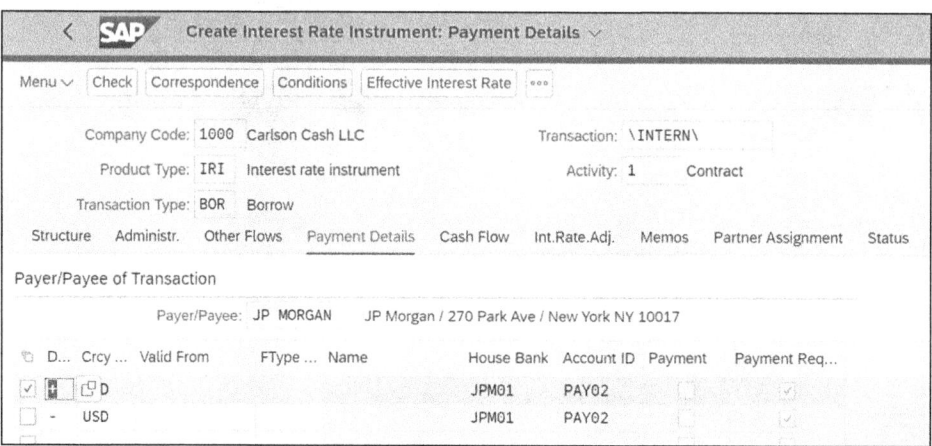

Figure 12.15 Payment Details Tab

Then we drill down into either of the payment flows to view and set up the required additional details, as shown in Figure 12.16. The key fields that we need to enter data into in this screen are as follows:

- **Posting**
  Here, we choose a radio button to have payments posted to the customer or to general ledger accounts. To post the transaction to the cash clearing accounts, we'll select **To GL Accts**.

- **Payment Request**
  Here, if we choose the **With** radio button, we'll create a payment request and queue the transaction to pay through the treasury payment program. If we choose **Without**, the transaction will post directly to an offset general ledger account and no payment request will be created.

- **House Bank and Account ID**
  In these fields, we determine the paying and receiving accounts on the corporate side of the transaction.

- **Repetitive Code**
  Here, we can assign previously defined repetitive codes to the cash flows. Although there are other attributes assigned to the repetitive code, only the reference text will carry forward to the payment request for the treasury and risk management payment.

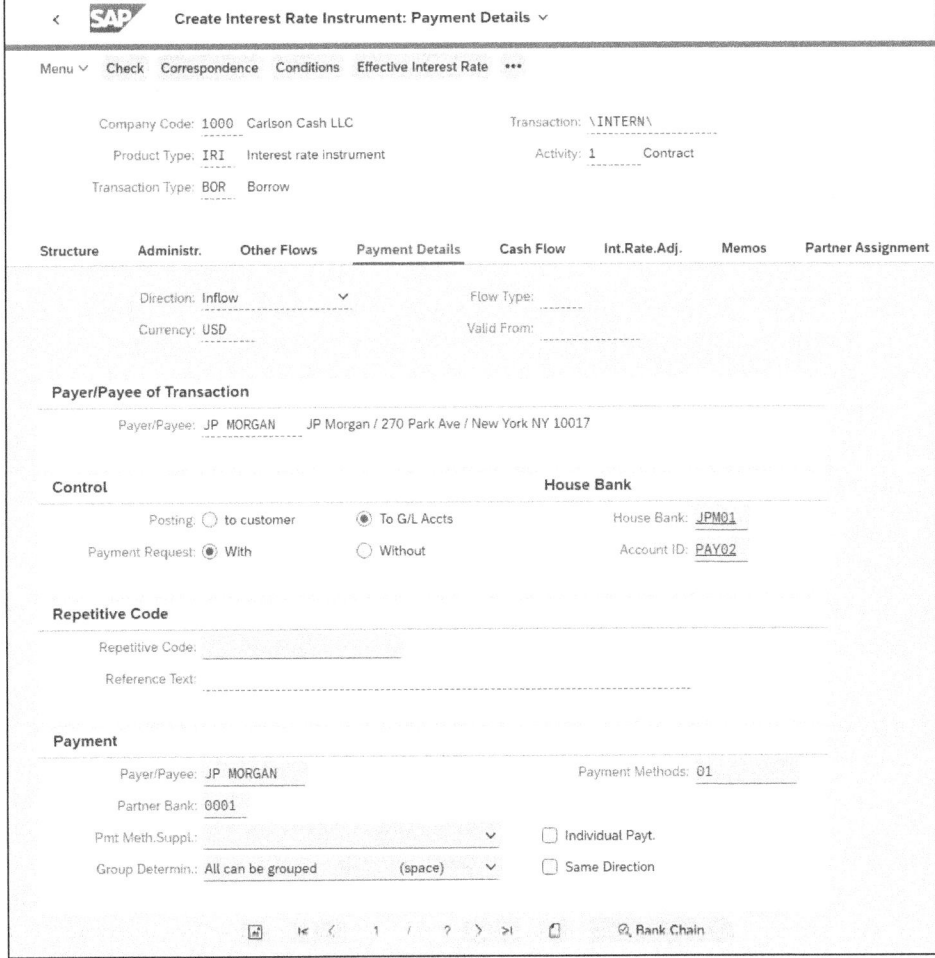

Figure 12.16 Payment Flows: Additional Details

- **Payer/Payee**

  Here, we determine the counterparty we are paying to or from. This is generally the same as the business partner in the transaction, but we can set a different business partner if required.

- **Partner Bank**

  Here, we define the set up payer or payee details we are selecting from the business partner.

- **Payment Methods**

  We can assign multiple payment methods in this field. In this example, we have an outgoing and an incoming payment method defined. The "O" payment method in this case is our incoming payment method, so we will use it for the transaction.

- **Group Determination**

  Here, we define how the treasury payment requests can be grouped in a single treasury payment run.

After we've completed these settings, the Post Flows app and Transaction TBB1 will both create the postings for the cash flows and will also create the payment requests. Once this transaction has been executed, we can pay off the payment requests using the treasury payment program.

12.3.5 Payment Requests and Payment Execution

Once we've created a payment request, we can post and pay it using the treasury payment program. To view the treasury and risk management payment requests that have been created, we can navigate to the Process Free Form Payments app (shown in Figure 12.17). The origin is a classification on the type of payment in SAP S/4HANA, and different payment requests in treasury will have different origins. The main treasury payment origins are as follows:

- **FI-BL**

  Both free-form payments and repetitive treasury payments use this origin.

- **TR-CM-BT**

  This origin is specific to bank-to-bank transfers among a corporation's bank accounts.

- **TR-TM**

  All payment requests created in treasury and risk management will have this origin.

After we review the payment requests created from the Post Flows app, we can initiate the payments by using the treasury payment programs. To complete this step, we execute either the Automatic Payment Transaction for Payment Requests app/Transaction F111 or Transaction F8BX. This will create the required payment postings and will allow for the straight through processing of the treasury payments.

12.4 In-House Cash and In-House Banking

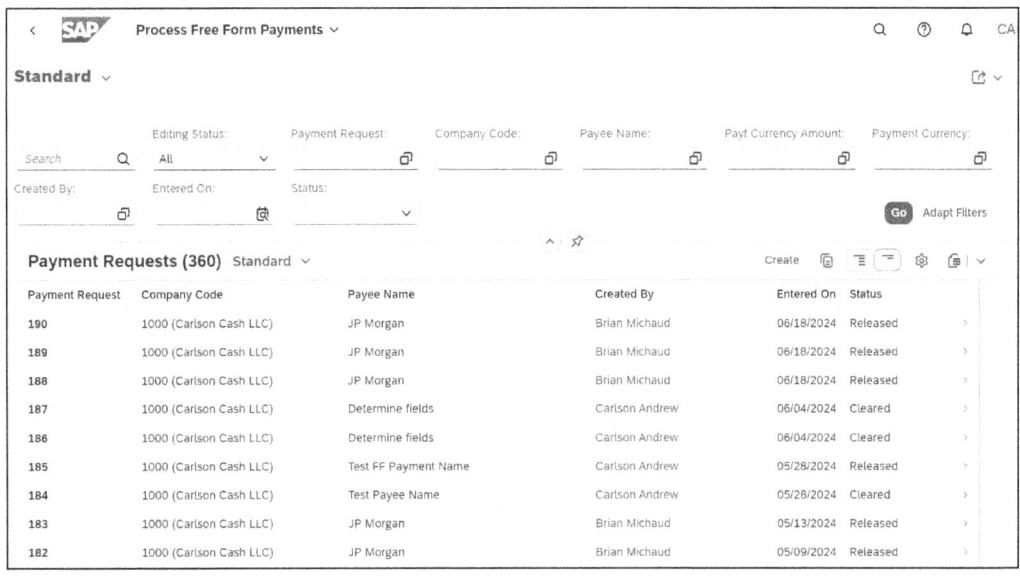

Figure 12.17 Process Free-Form Payments App Displaying All Treasury Payment Requests and Their Statuses

12.4 In-House Cash and In-House Banking

We've already covered how to create and initiate both incoming and outgoing payments within a company to fulfill the contractual obligations of the financial transactions within treasury and risk management. We can also create additional and more complex payment structures by routing payments through in-house cash and in-house banking, which allow for the processing of both internal and external payments. The goals of in-house cash and in-house banking are to reduce a corporation's banking footprint by decreasing the number of external bank accounts and consolidating external payments. This is particularly useful when sending the payments to foreign countries. While the functions of these solutions are very similar, there are distinct differences, so we'll highlight the differences between in-house cash and in-house banking in the following:

- **In-house cash**
  SAP S/4HANA Finance for in-house cash is the successor of the legacy SAP ERP solution that's been available since 1998 and that handles a corporations internal banking solution. An in-house cash center represents the company's internal bank, and subsidiaries hold current accounts with this internal bank to track their account balances. As transactions process through the in-house cash center, internal bank statements are also generated to fully post the transaction flows between the in-house cash center and the current accounts. This solution leverages SAP's intermediate

document messages to fully process and communicate the payment orders and payment items that hold the details of the account activities.

- **In-house banking**
  In-house banking for SAP S/4HANA Finance for advanced payment management is a new solution that was introduced in SAP S/4HANA 2022 and is embedded in the advanced payment management solution. The overall process in in-house banking is similar to that in in-house cash, but the processing has been revamped and these functionalities are all SAP Fiori–based. Many functions of this solution have been transferred to the business users to manage accounts, processing conditions, and batch jobs to increase flexibility for the business instead of relying on IT to make updates to the In-house banking structure. In-house banking also leverages internal XML files for the internal bank statement postings, instead of intermediate document messages.

One function that is similar in in-house cash and in-house banking is the ability to process internal and external payments. The key processes are those for the following functions:

- **Internal payments**
  This function consists of recording internal payments between different subsidiaries.

- **Central payments**
  This function consists of external payments to business partners that are processed centrally by the in-house cash center on behalf of all subsidiaries.

- **Local payments**
  This function consists of payments to external business partners by a designated subsidiary in a foreign country on behalf of the in-house cash center.

- **Central incoming payments**
  This function consists of receiving payment transfers from banking partners and tracking the change in balance in the in-house bank. This structure is generally created for zero-balance accounts for cash pooling.

Treasury and risk management supports the connection to in-house cash and in-house banking with the ability to centrally pay or centrally receive payments in an in-house cash center. The key settings ensure that payment orders are generated from the treasury payments. Additional settings and full configuration are required in in-house cash and in-house banking to generate payment orders from treasury and risk management, but the key integration point from the treasury side is to route payments to in-house cash or in-house banking.

The key integration point is to point the payment instructions to the in-house cash and in-house banking house banks and account IDs. We determine this in the same payment area in the **Payment Details** tab in the Create Financial Transaction app or

Transaction FTR_CREATE. To create the necessary payment orders and to create a payment request within the central paying entity, we need to identify the in-house cash/in-house banking house bank and account ID in the payment details as in Figure 12.18.

Structure	Administr.	Other Flows	Payment Details	Cash Flow	Int.Rate.Adj.	Memos	Partner Assignment	Status

Direction: Inflow   Flow Type:
Currency: USD   Valid From:

Payer/Payee of Transaction

Payer/Payee: JP_MORGAN   JP Morgan / 270 Park Ave / New York NY 10017

Control   **House Bank**

Posting: ○ to customer   ● To G/L Accts   House Bank: IHB01
Payment Request: ● With ○ Without   Account ID: USD1

Figure 12.18 In-House Bank House Bank and Account ID Defined in Payment Details Tab

Once we complete this, we can fully process and release the payment through the central paying entity. This functionality helps consolidate payments for a company and allows for further control of the payment processing.

12.5 Summary

In conclusion, the integration of treasury and risk management within the broader SAP S/4HANA ecosystem significantly enhances the solution's capacity to effectively manage financial activities. By capitalizing on the seamless connections among treasury and risk management and other areas of SAP S/4HANA such as finance, payments, and cash management, businesses gain real-time visibility into their cash positions, streamline processes, and improve overall financial reporting. As companies increasingly prioritize native integration to eliminate costly custom interfaces, treasury and risk management stands out as a logical solution. This integrated framework not only simplifies the management of treasury transactions but also offers organizations a cleaner and more streamlined system architecture.

Chapter 13
Reports, Key Performance Indicators, and Alerts

SAP S/4HANA Finance for treasury and risk management includes tracking and validating many financial instruments that support the treasury operation for a company. To support timely and accurate business decisions, there are standard reporting capabilities that let us view the transactions and their related flows.

While it's important to enter each of the financial transactions for an organization to track its activities, it's equally important to create effective reporting on the financial transactions. Reporting in SAP S/4HANA Finance for treasury and risk management includes consolidating the information in all the transactions to provide summary reports to show totals for all similar transactions. Detailed reports must also be available to so we can drill down into each type of activity to validate that the flows have been both calculated and recorded accurately. Due to these common requirements in all treasury departments, standard reports have been designed to provide flexible reporting for all transactions within treasury and risk management.

While reporting requirements have changed over the years, the technology to create these reports within SAP systems also changed with the introduction of SAP S/4HANA and the related SAP Fiori apps. These apps have enabled additional reports that we will detail in this chapter. The SAP Fiori frontend of SAP enables the frontend view of key performance indicators (KPIs) that are viewable on the home screen and show a summary of key information. Additionally, new SAP Fiori reports have been created to further slice and dice information real time. We detail each of these new reports in this chapter to highlight the new functionalities. The reports that were available in SAP ERP are still available via transaction codes in SAP S/4HANA, and we also detail the key SAP ERP reports in this chapter to ensure all reporting capabilities described.

13.1 Standard SAP Fiori Reports

Treasury and risk management has various reports to assist us in daily operations and help us ensure that it's operating smoothly, and it also has month-end reports that reflect processing and postings. With the introduction of SAP S/4HANA, there are also various new KPI reports that we can view as tiles in the SAP Fiori launchpad.

This section will review all the key reports within treasury and risk management, and it will detail what information we can view in these reports.

13.1.1 Treasury Position History App

The Treasury Position History app allows us to analyze snapshots of the changes to the treasury position values by their book value. We can view the values in this report at a very high level to just view the positions by product type for the company's whole portfolio, or we can add attributes in the settings area to drill down into more specific information. When we've successfully run the report, we can view the data in a chart view, a chart-plus-table view, or a table view (see Figure 13.1). The key fields in this app are as follows:

- **Key Date**
  The key date is the date we're looking for in SAP S/4HANA for the treasury positions. We can use the current date or a future date to view the future positions that will be in treasury and risk management on that date.

- **Period**
  An additional consideration to look at for this report is the period filter. In this field, we specify the period that we want to look at for comparison purposes when running the report, and doing so will greatly change the output. The options for this field are as follows:
  - **Key Date**
    If we select this option, the report will be run for the key date defined in the **Key Date** field. This will only show the positions on that date.
  - **Year to Date**
    If we select this option, the position balance from the last day of the previous year will be shown along with the balance on the defined key date to show how the balances have changed in the current year.
  - **Last**
    Any of the figures that start with **Last** will have a few different outputs. If we select **Last 3 Months**, **Last 6 Months**, or **Last 12 Months**, the report will show the position values at the end of each of the months for either 3, 6 or 12 months, respectively. **Last Quarter** and **Last 4 Quarters** will show a comparison of the start and end of the last full quarters. For example, if we run the report on 06/01/2024, the last full quarter would have ended on 03/31/2024, so the report would compare the position balances from 12/31/2023 to 03/31/2024. **Last 4 Quarters** would show the same information, except that it would show more periods.

- **Display Currency**
  To compare all the positions, we need the system to display a base currency. This helps us compare the positions in multiple currencies and entities. In Figure 13.1, the **Book Value in Display Currency** field shows the value at the given **Treasury Position Ledger Date**, and the **Display Currency** field drives how that field appears in reporting.

- **Exchange Rate Type**
  When the positions are in other foreign currencies, we need to translate them into the display currency. To do this, we use this field to determine the exchange rate type that SAP S/4HANA will use to display the values in the display currency.

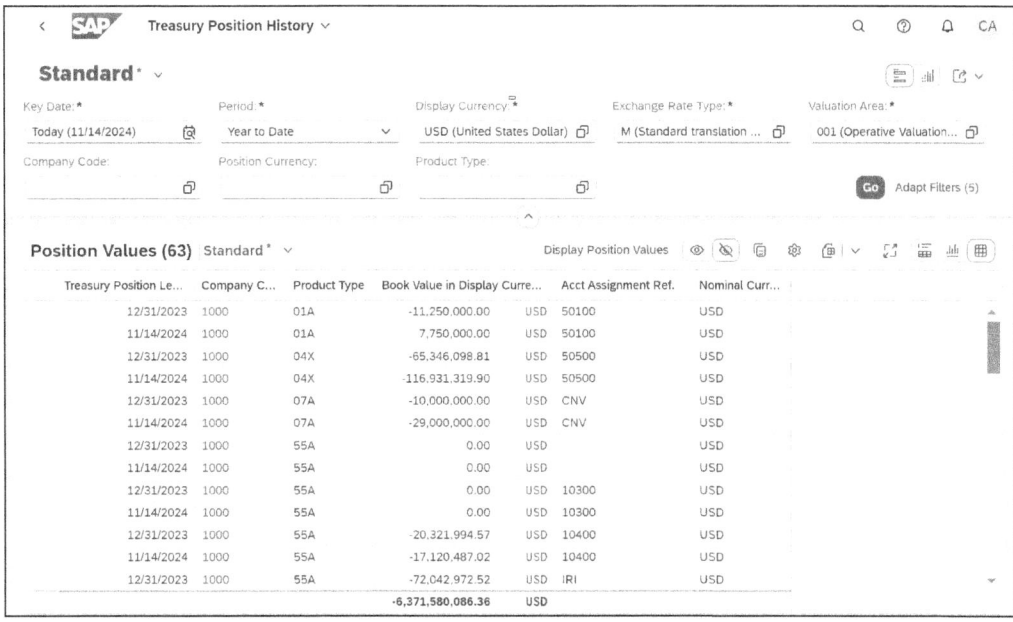

Figure 13.1 Treasury Position History Report Listing All Transactions and Position Changes Over Time

13.1.2 Treasury Position Values App

The next SAP Fiori app we'll cover is the Treasury Position Values app. It displays information similar to that in the Treasury Position History app, but it is meant to look at the positions with a key date and is not meant to be a comparison app. Additionally, this app displays the treasury positions for each individual contract on the key date, so it can give us a snapshot of all treasury positions. As shown in Figure 13.2, we can view all types of contracts in this app, so both the standard treasury contracts and securities contracts are available in it. Figure 13.2 shows an example of the output of this report showing the different values for each transaction at the key date (nominal amount, book value, purchase value, etc.). The key values in this report are as follows:

- **Key Date**
  The key date is the date we're looking for in SAP S/4HANA for the treasury position values. We can set this to the current date or set it to a future date to view the future positions that will be in treasury and risk management on that date.
- **Date Field for Sel.**
  We can view this report by treating the key date in different ways. If we select the

Posting Date, then the report will show the position values based on the posting date. Since the **Treasury Position Ledger Date** can vary from the posting date, because we can edit the posting date when running the posting programs, we can view this report based on the calculated dates for the positions from treasury and risk management.

- **Planned Data Incld.**
  There's also an option to view this report based on planned data or not. We'll use this option used if we set the key date to a future date. If we want to review the report based on the planned data, it doesn't matter if the transaction has been posted or not, and the report will show the project treasury position values. These values won't appear if we select **Exclude Planned Data**.

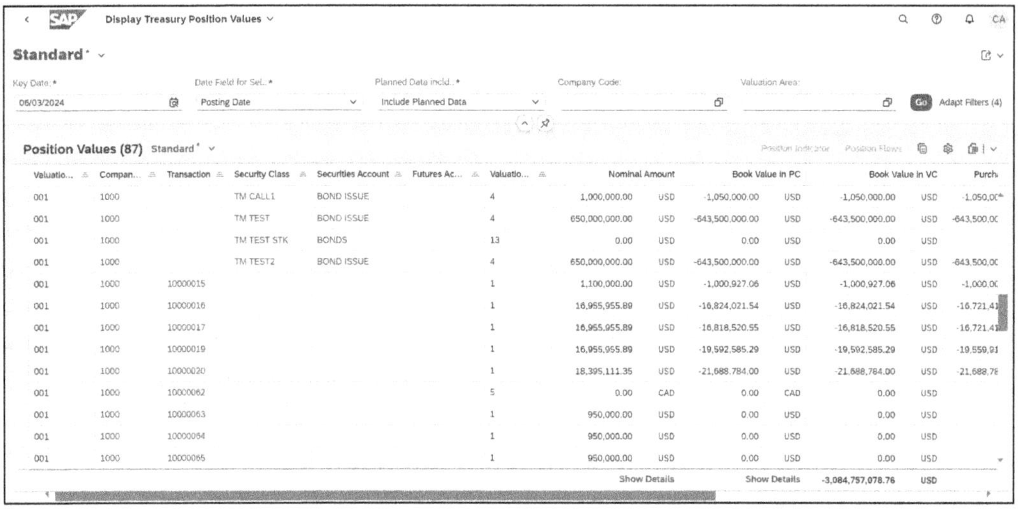

Figure 13.2 Report Output Showing Different Values for Each Transaction on Key Date

13.1.3 Display Treasury Position Flows App

The Display Treasury Position Flows app (see Figure 13.3) drills down further into the transactions to display the individual cash flows related to each contract. This report shows all the relevant information for the flows, including the transaction number, the status of the cash flow, the amount details, and the cash flow date. The key values in this app are as follows:

- **Update Type Name**
  We assign update types to all flows within treasury and risk management, and they ultimately are assigned to specific accounting treatments when the update type is called during the posting programs. The update type name is a longer description of what the flow is used for. In the example report, we can see that the update types include but are not limited to amortizations, principal changes, interest, and final repayments.

- **Business Transaction Status Name**
  This report shows the past position flows that have been posted and the future flows that haven't yet been posted. The posted transactions will have a status of **Fixed**, and the transactions that haven't been posted have a status of **Scheduled**.

> **Navigation to Other Reports**
>
> This report also is connected to other reports in treasury and risk management. We can select any transaction, and we can navigate to the following other reports by clicking on their respective buttons:
>
> - Business Transaction Flows
> - Posting Journal
> - Amortization Log
> - Display Position
> - Payment Journal
> - Position Indicator
> - Original Business Transaction
>
> One thing to note is that not all of these other reports will be applicable to all of the treasury position flows in this report. For example, we can only view the amortization if the flow is an amortization flow, and we can only view the payment journal if the flow is relevant to payments.

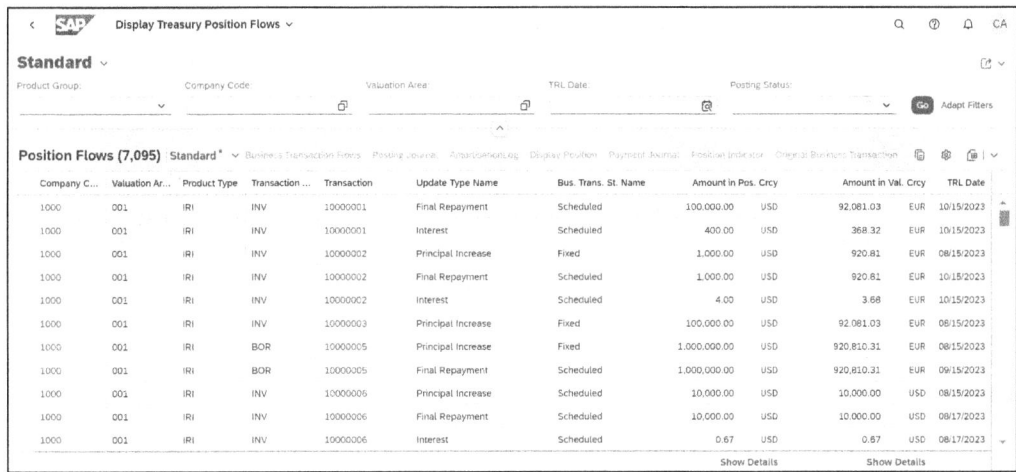

Figure 13.3 Report Detailing Each Position Flow for Transactions

13.1.4 Display Treasury Posting Journal App

The Display Treasury Posting Journal app is an accounting-based SAP Fiori app that details all posted transactions. It details every posting by financial transaction, and we

13 Reports, Key Performance Indicators, and Alerts

can add fields to the display to meet our requirements. This app is a one-stop shop for viewing all treasury postings that have been generated by treasury and risk management posting programs. Figure 13.4 shows an example of how we can run this report to show all postings by financial transaction.

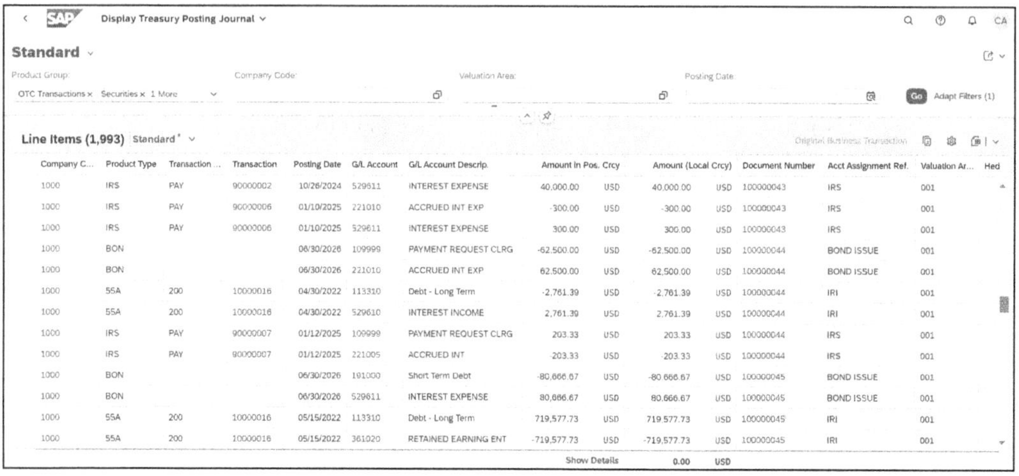

Figure 13.4 Report Showing Journal Entries from Each Financial Transaction

The key values in this report are as follows:

- **Product Group**
  To simply filter this report based on types of transactions, we can filter it by overarching categories called product groups. The product groups that we can use to filter this report are **Listed Derivatives, External Accounts, Loans, OTC Transactions, Securities** and **Exposure Items**.

- **G/L Account**
  This field is a hyperlink that we can use to drill down further into the transactions. By clicking on any of the general ledger accounts in this report, we can navigate to the Display Line Item Entry (see Figure 13.5) and Display Line Items in General Ledger reports, which show additional details on each treasury posting line item.

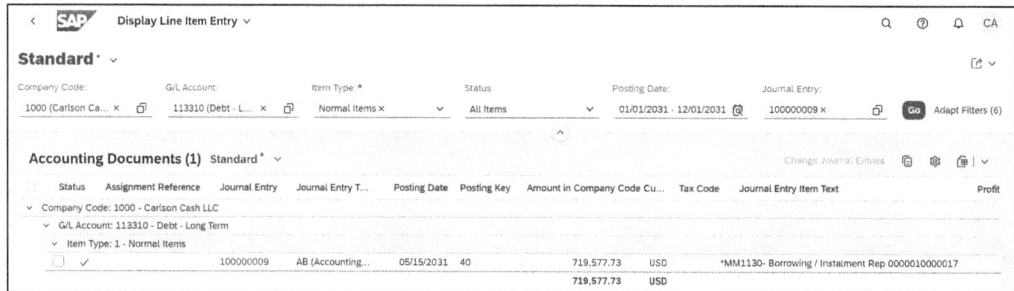

Figure 13.5 Drilling Down into Line Item to Further Review It by Displaying Additional Details

13.1 Standard SAP Fiori Reports

- **Document Number**
  This field is another hyperlink we can use to navigate to a different SAP Fiori app. By clicking the document number, we can navigate to the Manage Journal Entries app (shown in Figure 13.6), which will show us a view of all the line items in the treasury posting.

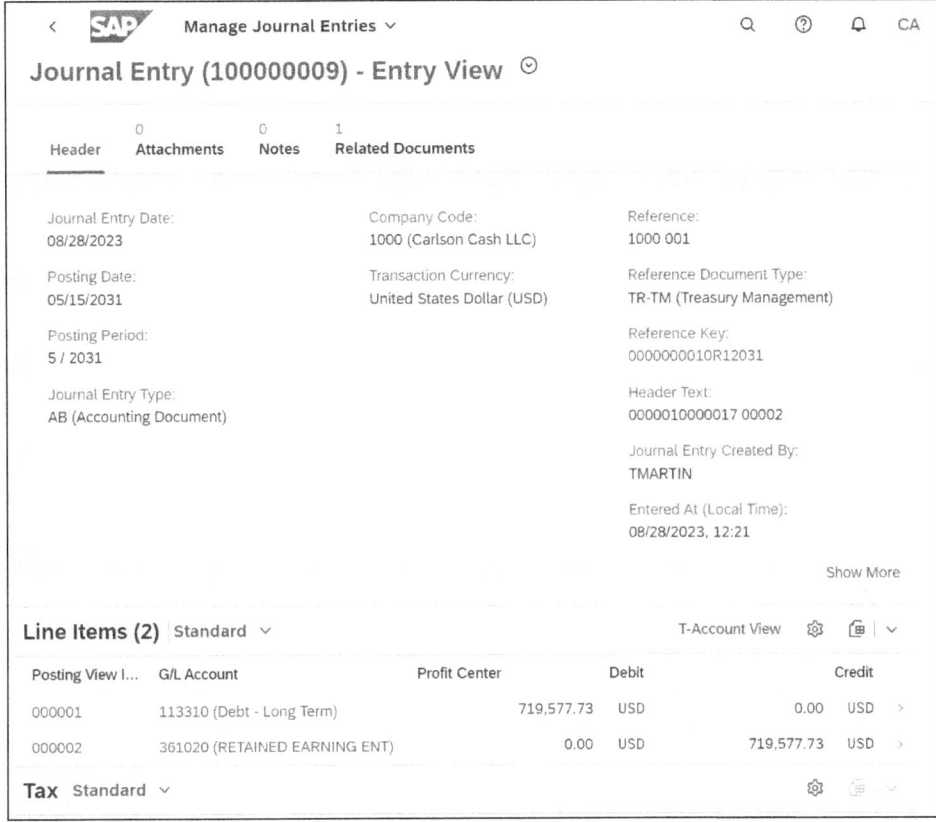

Figure 13.6 Drilling Down into Document Number to Display Full Journal Entry

13.1.5 Interest Rate Overview App

The Interest Rate Overview app (see Figure 13.7) is a useful app that displays many key figures for interest rates and interest-bearing financial transactions. The app is separated into different cards, which are predefined reports that show a high-level view of information. We can edit this screen to meet our requirements, and we can remove cards if they're not applicable.

The key values in this app are as follows:

- **Key Date**
  Here, we enter the key date for this report, which will drive the subsequent dates for forecasting future interest rates.

- **Display Currency**
  When interest rates are in different currencies, the report shows the equivalent in a display currency, which we define in this field.
- **Exchange Rate Type**
  When SAP S/4HANA translates foreign currencies into our display currency, we need to define the exchange rate type to determine the display currency equivalent amount.
- **Number of Years**
  In the **Debt/Investment Maturity Profile** card display, there's a forecast of the future maturity dates for the transactions. In this field, we determine the number of years to forecast for the maturities.
- **Yield Curve Type**
  Here, we define one or more types of yield curves that we want to view in the **Yield Curves** card.

The cards that are available in this app are as follows:

- **Debt/Investment Maturity Profile**
  This card shows a yearly view of the debt and investment maturities.
- **Total Amount of Debts by Key Date**
  This card shows a view of the debt held in the top-five product types for the given key date.
- **Total Amount of Investments by Key Date**
  This card shows a view of the investments held in the top-five product types for the given key date.
- **Debt/Investment by Interest Category**
  This card shows the debt and investments held by their nominal amounts and classifies debt and investments into fixed-rate and variable-rate instruments.
- **Debt/Investment by Reference Interest Rate**
  This card shows the nominal amount of debt and investments that are held with variable interest rates. It also details which interest rates the financial transactions use.
- **Current Reference Interest Rate**
  This card shows the top reference interest rates that are used by financial transactions, and it shows the current interest rate in SAP S/4HANA for the reference interest rates.
- **Historic Reference Interest Rate**
  This card shows reference interest rate trends in the form of a graphical view of how the interest rates have changed over time. We can daily or monthly changes.
- **Yield Curves**
  This card details yield curves by year.

13.1 Standard SAP Fiori Reports

- **IR Swaps by Key Date**
  This card shows the nominal amount of interest rate swaps, and we can view it from either the incoming or the outgoing side of the swap.

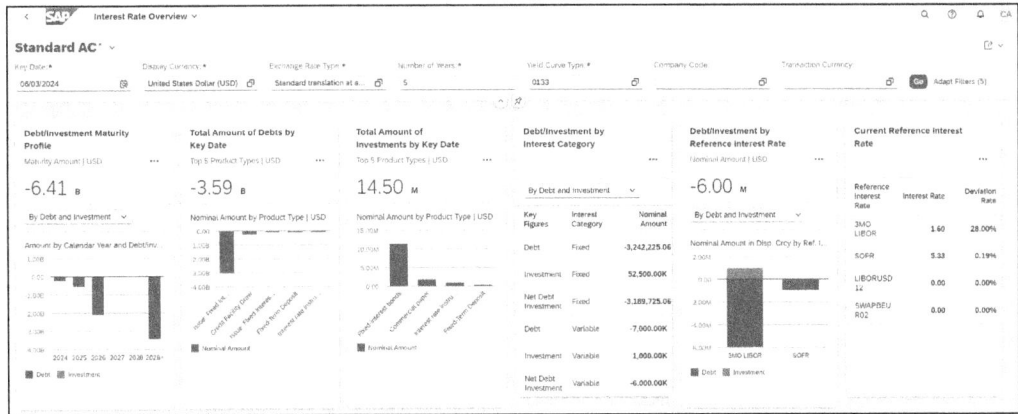

Figure 13.7 Interest Rate Overview SAP Fiori App Displaying Collection of Reports

In the Interest Rate Overview app, we'll always need to populate a few fields. By default, the display currency, exchange rate type, and yield curve type show up as blank, and we'll need to populate them.

If we want to avoid having to populate these fields every time we use the app, we can set default values for them by saving our own view at the top left of the screen or by using the functionality in the settings area. To do the latter, we select our icon at the top right of the screen and navigate to **Settings**, as shown in Figure 13.8.

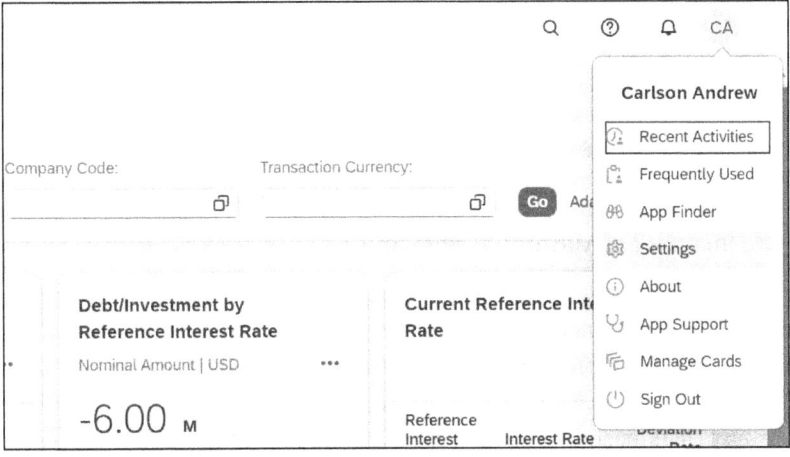

Figure 13.8 Settings Dropdown

13 Reports, Key Performance Indicators, and Alerts

Then, we navigate to the **Default Values** setting and fill out the values we want for this application. In this case, we fill in the three values that don't populate automatically (including **Display Currency** and **Exch. Rate Type**), as shown in Figure 13.9.

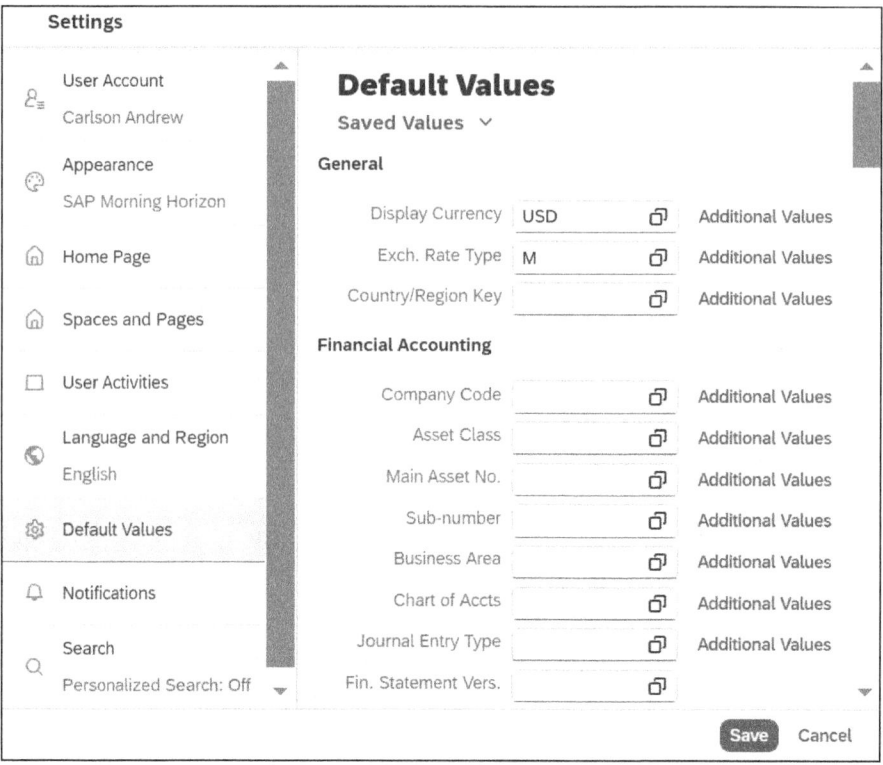

Figure 13.9 Default Values Setting Allows for Auto-Population of Some Settings in Interest Rate Overview Report

Now that we've set the default values for this application, the report will automatically populate when we're running the Interest Rate Overview app, and we won't have to enter the **Display Currency** and **Exchange Rate Type** every time.

13.1.6 Debt and Investment Maturity Profile App

The previous app showed us a very high-level view of the debt and investment maturity profile. To get a more granular view, we can use the Debt and Investment Maturity Profile app (see Figure 13.10). It provides a summary at the top that shows the company codes, product types, and counterparties that have the largest amounts of debt and investments. It also lists each transaction with its respective maturity dates and amounts.

13.1 Standard SAP Fiori Reports

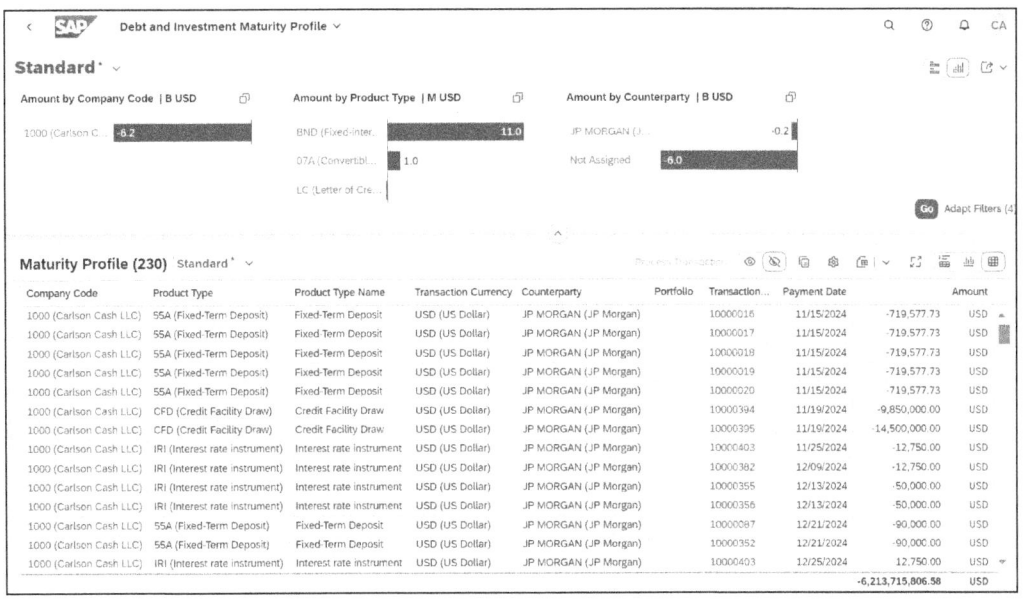

Figure 13.10 Maturity Report Showing Details of Each Transaction with Its Maturity Dates

13.1.7 Debt and Investment Analysis App

The Debt and Investment Analysis app (see Figure 13.11) is another overview app with which we can view all debt and investment contracts by their nominal amount, book value, and NPV. As with the other treasury and risk management apps, we can use this app to create graphical displays to view debt and investments by product type, company code, portfolio, counterparty, etc. We can also view the details of each contract individually by using the table view.

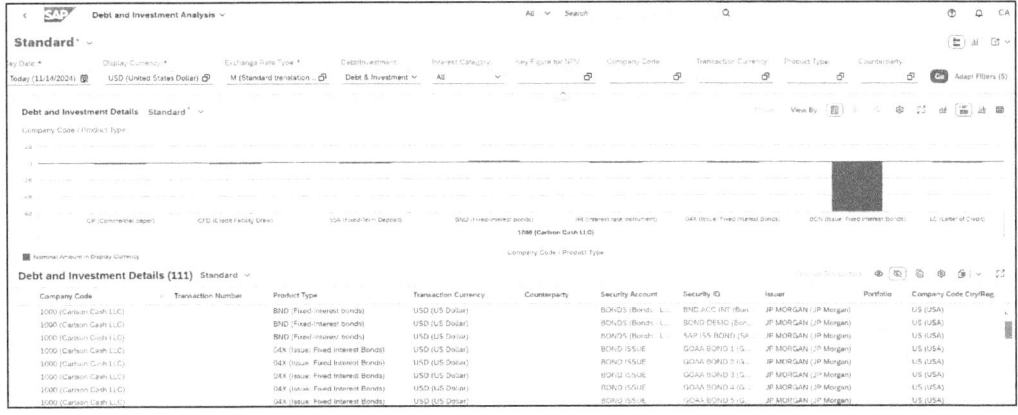

Figure 13.11 Each Debt Item and Investment Is Available in This Report

13.1.8 Foreign Exchange Overview App

Another overview app that displays data in cards is the Foreign Exchange Overview app, which is shown in Figure 13.12. It is customizable like any other overview app, and the available cards in it are as follows:

- **Financial Status in Display Currency**
  This card shows the financial status of treasury positions in the defined display currency.

- **Credit Line Overview in Display Currency**
  This card shows details of the utilization of credit facilities by currency.

- **Cash Position in Display Currency**
  This card displays top currencies in the cash position.

- **Foreign Exchange Rate**
  This card displays the top traded currencies with their current exchange rate. The difference section also compares the current exchange rate to the previous exchange rate and displays it as a percentage to indicate the difference.

- **FX Forwards**
  This card shows the top traded currencies by nominal amount in the transaction currency for FX forwards. The amount is also translated into the defined display currency in the filtering area.

- **FX Options**
  This card shows the top traded currencies by nominal amount in the transaction currency for FX options. The amount is also translated into the defined display currency in the filtering area.

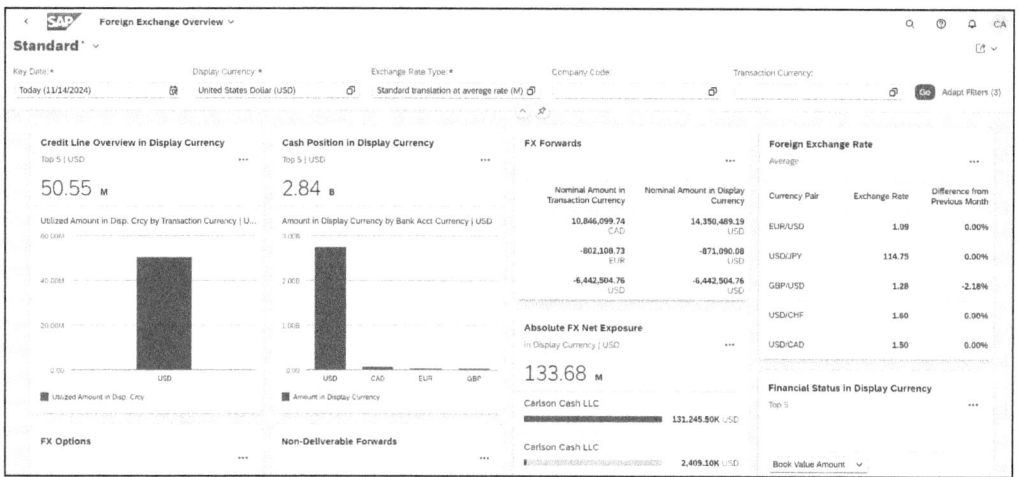

Figure 13.12 Cards Available in Foreign Exchange Overview App

13.1 Standard SAP Fiori Reports

- **Non-Deliverable Forwards**
  This card shows the top traded currencies by nominal amount in the transaction currency for NDFs. The amount is also translated into the defined display currency in the filtering area.

- **Absolute FX Net Exposure**
  This card shows the net FX exposure in the defined display currency by entity and company code.

13.1.9 Market Data Overview App

The Market Data Overview app (see Figure 13.13) also displays key market data points in cards that we can modify to show the specific data points we want to view. Additionally, we can drill down into most of these cards to view the raw data behind the data points displayed in the app. The available cards in this area are as follows:

- **Historic FX Spot Rate**
  This card displays the FX spot rates with a trend analysis of how the rates have changed over the defined time period.

- **FX Spot Rate**
  This card displays the current FX rates along with their percentage change from the previous day.

- **FX Swap Rate**
  This card displays the FX swap rates and forward points by term.

- **Historic Reference Interest Rate**
  This card defines the reference interests in its filtering area. It displays the rates that we have defined, and it shows a trend analysis of how each interest rate has changed over the defined time period.

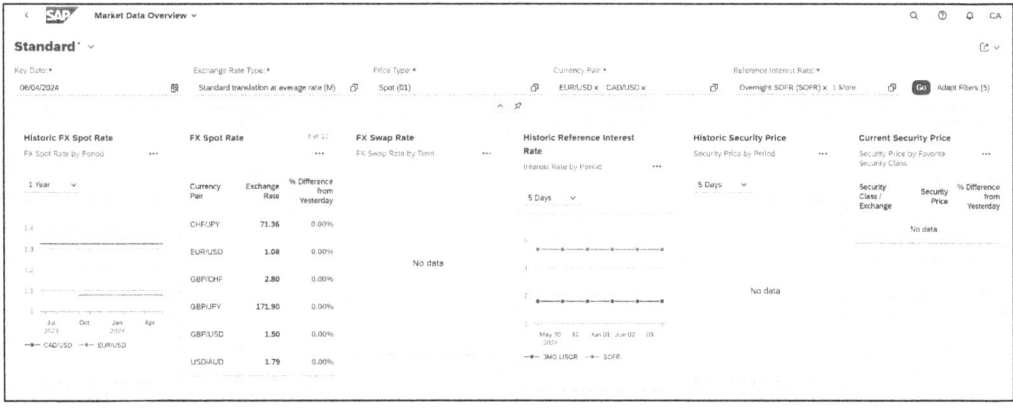

Figure 13.13 Market Data Overview App

- **Historic Security Price**

  We can define security prices in the filtering area, this card displays the security price trend by the defined time period. Available time periods are 5 days, 1 month, 3 months, 6 months and 1 year.

- **Current Security Prices**

  This card displays the defined security prices along with a percentage change that shows the difference between yesterday's security price and today's.

13.1.10 Credit Line Analysis App

The Credit Line Analysis app offers a structured view of all facilities a company holds. As shown in Figure 13.14, the rows and columns of the data analysis are customizable so that we can drive the display of the report. The report can slice and dice the data any way we want it to, and we can view the following credit line data in this report:

- We can view information on the total credit line, available amount, utilized amount, and unutilized amount.
- We can define dates to show a trend of these key figures and how they've changed over time.
- We can view utilized and unutilized rates as percentages.
- If additional information is required for any of the facilities, we can drill down further into the transaction number to display the contract screen for the facility.

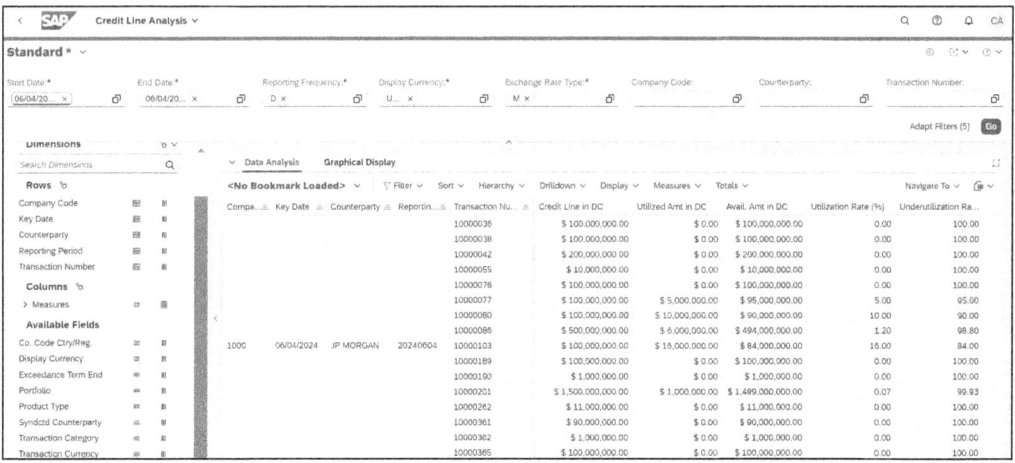

Figure 13.14 Credit Line Analysis Allows for Detailed View of Credit Facilities and Their Utilization

As shown in Figure 13.15, we can display the data as a graph to show the facilities' credit utilization details.

13.1 Standard SAP Fiori Reports

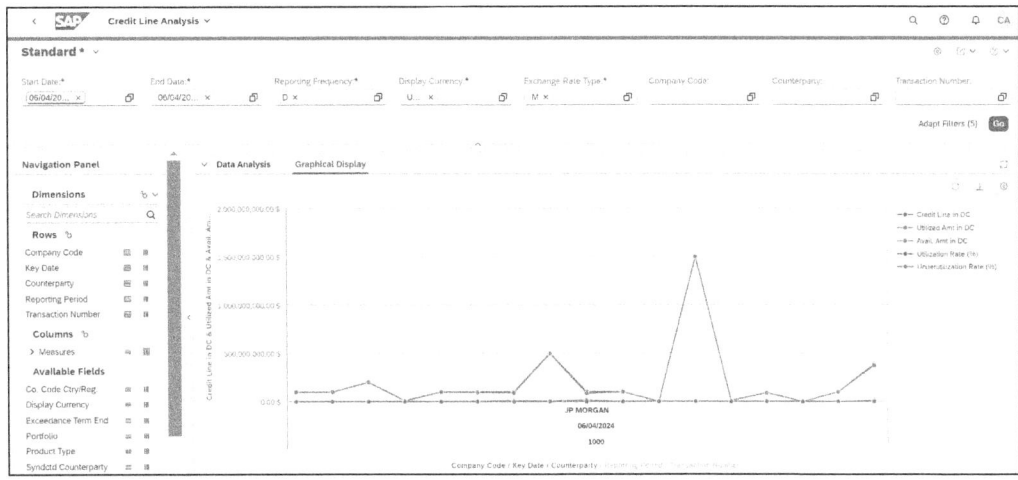

Figure 13.15 Graph Showing Facilities' Credit Utilization Details

13.1.11 Display Treasury Alerts App

Treasury and risk management has a series of alert reports that we can customize to show whether transactions need to be posted, paid, fixed, settled, or released, as shown in Figure 13.16. Technically, we can set up all of these reports in the Display Treasury Alerts app, but we'll cover each of the functions of these reports in the following sections. Also, we can set up these apps as KPIs, and we can save them on the SAP Fiori home screen and quickly reference the KPI number on the app to see if something needs to be actioned in treasury and risk management.

Display Treasury Alerts Posting	Display Treasury Alerts Correspondence	Display Treasury Alerts Payment	Display Treasury Alerts Settlement	Display Treasury Alerts Release	Display Treasury Alerts Interest Rates
1.23 K	1.23 K	1.23 K	1.23 K	1.23 K	1.23 K

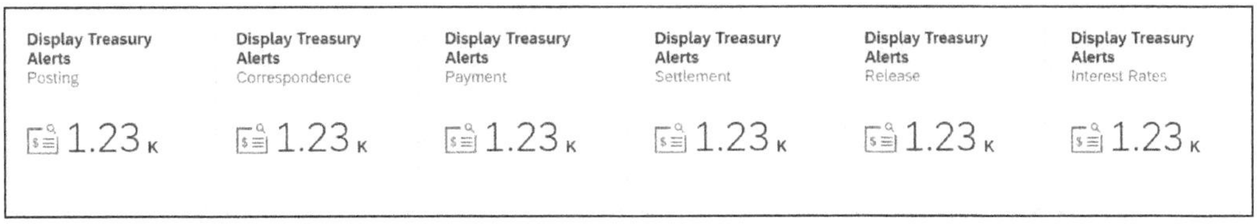

Figure 13.16 Treasury Alerts Detail When Key Activities Are Happening in Treasury Transactions

Display Treasury Alerts – Posting App

The Display Treasury Alerts – Posting app (see Figure 13.17) details each financial transaction and shows whether a posting needs to be executed for the transaction, along with the posting due date. Any transaction that has a posting that is still open will show up in this app, which will detail when the posting should happen.

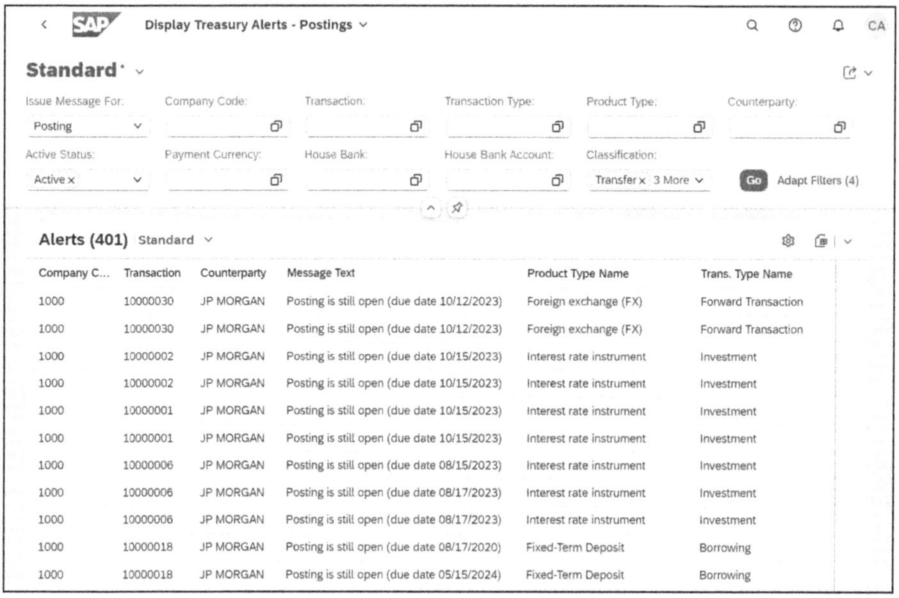

13 Reports, Key Performance Indicators, and Alerts

Figure 13.17 Treasury Alerts Showing When Transactions Need to Be Posted

Display Treasury Alerts – Payment App

The Display Treasury Alerts – Payment app details whether flows need to be paid. As shown in Figure 13.18, it will detail first whether flows have been posted and whether payment runs have been executed. If any flow is relevant to creating a payment and hasn't been processed, then the payment will display as open in the app.

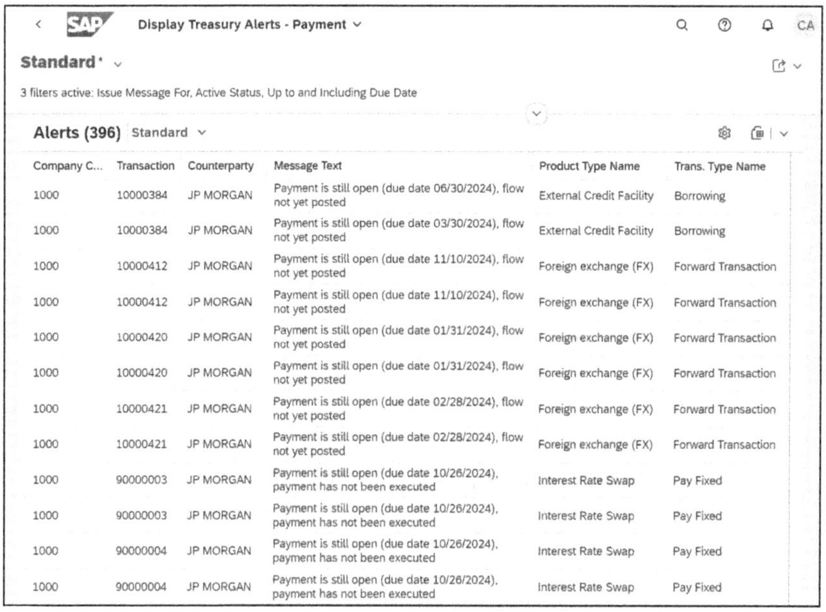

Figure 13.18 Display Treasury Alerts – Payment App

13.1 Standard SAP Fiori Reports

Display Treasury Alerts – Release App

The Display Treasury Alerts – Release app (see Figure 13.19) shows whether a transaction needs to be approved in a workflow. If a transaction has a status of unreleased, the text will show as **Transaction release is open**.

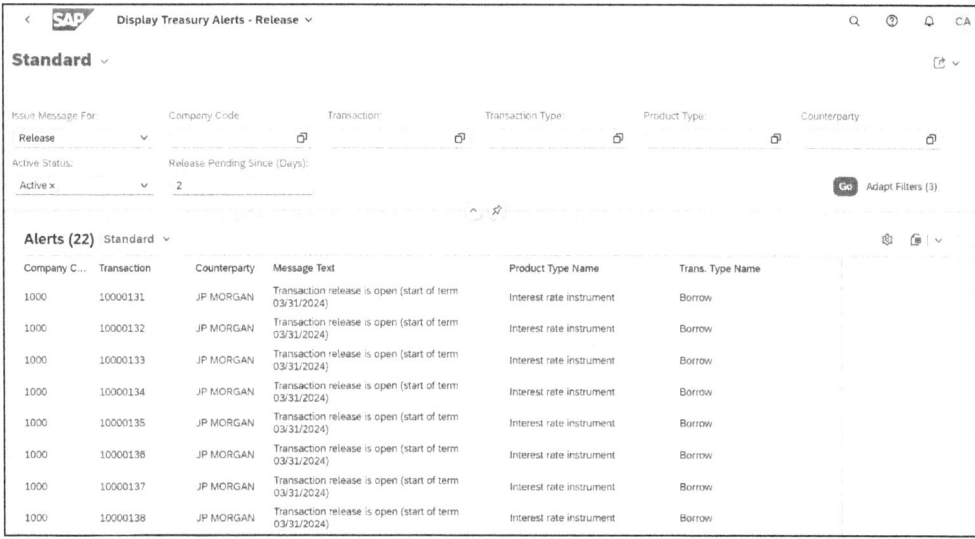

Figure 13.19 Report Displaying Each Transaction That Still Requires Workflow Release

Display Treasury Alerts – Settlement App

Some transactions are set up to require a settlement step to confirm the contract details. If a settlement is required and needs to be executed, the Display Treasury Alerts – Settlement app (see Figure 13.20) will show which transactions still need to be settled.

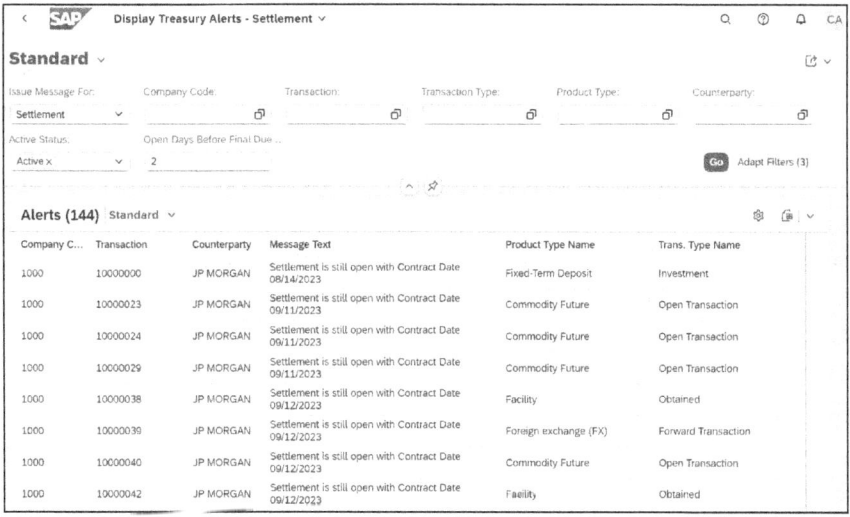

Figure 13.20 Report Showing Transactions That Currently Require Settlement Step

13 Reports, Key Performance Indicators, and Alerts

Display Treasury Alerts – Interest Rates App

Many debt and investment contracts require an interest rate fixing to determine which rate is assigned to an interest flow. The Display Treasury Alerts – Interest Rates app (see Figure 13.21) shows which interest rate fixings we need to execute along with the reference interest rate that we need to fix. The interest rate fixing schedule dictates when we should fix the rates, and this alert will let us know when a rate hasn't been fixed for a transaction.

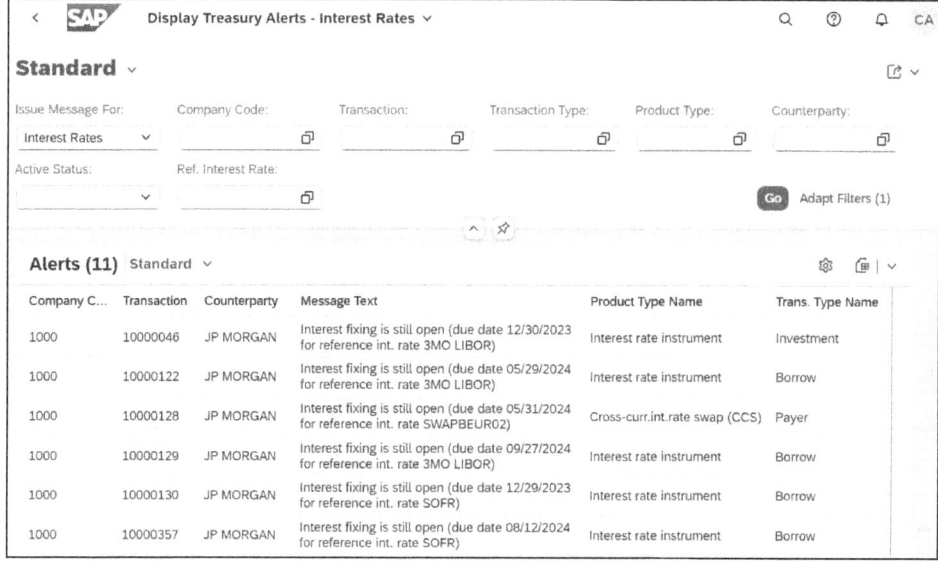

Figure 13.21 Display Treasury Alerts – Interest Rates App

13.1.12 Cash Flow Analyzer App

The Cash Flow Analyzer app is a report in cash management that we use to show cash flows in the past and present, and we can even use it to forecast future cash flows based on data in SAP S/4HANA. Even though this report is housed in cash management, the cash flows from treasury and risk management appear in the Cash Flow Analyzer app. Due to this, we'll cover this report specifically to how the treasury cash flows appear in it.

This report has a couple of different selections that we can used to view the treasury data. The easiest way to identify the treasury cash flows is based on the certainty level. The **TRM_O** certainty level is exclusive to forecasted cash flows based on options product types, and the **TRD_D** certainty level is reserved for all other forecasted cash flows for contracts in treasury and risk management. If we filter the Cash Flow Analyzer app by these certainty levels, we'll see the treasury contract forecast.

Only the cash flows related to treasury transactions are shown in this report (see Figure 13.22). In a real business scenario, we'll likely have a cash position that shows both bank activity and treasury cash flows, but we've filtered this report further to highlight how the treasury transactions appear. If the planning levels have been assigned to the product types, they'll be assigned when the contract is saved. In our example in Figure 13.22, some treasury cash flows don't have a planning level assigned, but we can still view those flows in this report. Additionally, we can see that there's a **T?** row and a **TD** row in the forecast. We assigned **T?** to any treasury contract that has been created but hasn't been settled, and once the contract is settled, it moves to another planning level. **TD** is used for debt contracts in this example, but we can customize the planning levels.

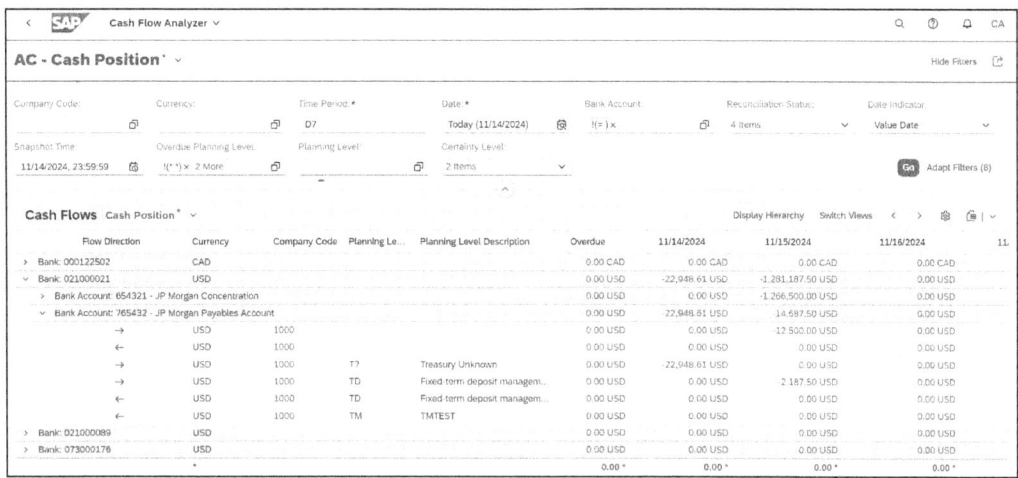

Figure 13.22 Treasury Flows in Cash Flow Analyzer Displaying Cash Impact of Financial Transactions

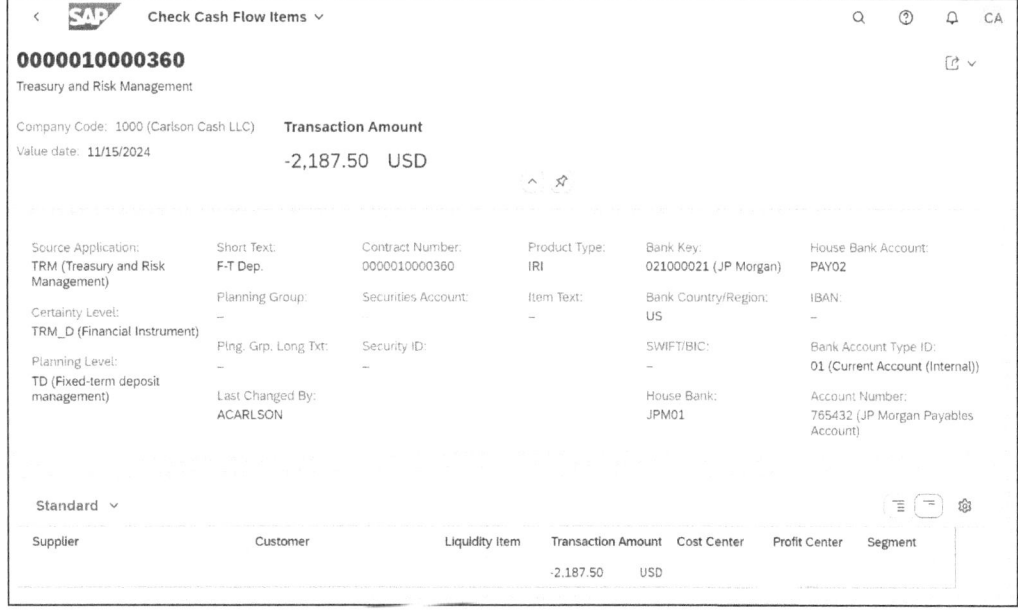

Figure 13.23 Drilling Down into Flow in Cash Flow Analyzer to View Detailed Information

If we drill down into a line item, we can navigate to the Check Cash Flow Items app, and it will show additional details about the cash flow. As shown in Figure 13.23, we can drill down far enough into the transaction to see granular details of the cash flows, including the product type, the paying or receiving bank account, and even the contract number the forecasted cash flow is related to. The contract number is also a hyperlink, so we can navigate straight to the transaction from this report.

13.2 SAP Fiori Analysis Reports

There are a few SAP Fiori reports that appear different from the standard reports that have filters at the top of the screen and have default columns we can view. The analysis reports are based on SAP BusinessObjects Design Studio (more recently part of SAP Lumira, designer edition) and include additional functionalities for viewing data. These reports generally have more data points, we can slice and dice the information in more ways, and we can view the reports like a pivot table. Before going into the Treasury Position Analysis apps, we'll cover how we can alter these reports to meet our needs.

13.2.1 Filtering and Designing Analysis Reports

There are a few different ways to design these reports. We first need to determine the filters at the top of the screen to determine the data that is viewed in these reports. Then, we'll need to define the key attributes, which will alter the rows and columns that are viewed in these reports.

Filters

First of all, we use the top section of this report to filter the data that appears in the **Data Analysis** section. This works like in all the other reports in treasury and risk management, and if we need additional filters, then we click the **Adapt Filters** button to add filters to our reporting, as shown in Figure 13.24. Once we have the desired filters assigned to our reporting, we click the **Go** button to refresh the report results.

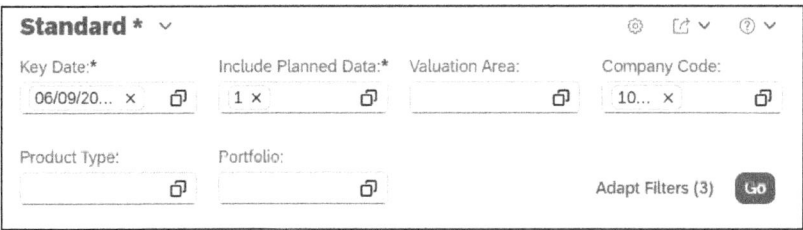

Figure 13.24 Filters at Top of Report Define Key Inputs into Report and Filtering Options

13.2 SAP Fiori Analysis Reports

Navigation Panel

Now that we've filtered the information in the report, we can determine how the information is displayed in the **Navigation Panel**. We'll start with the **Rows** section, which drives the rows of the report, and then we'll cover the columns and filters and show the report output.

Dimensions: Rows

The order of the rows will determine the order that they show up in on the report. For example, the rows on the left show up as **Company Code**, **Product Type**, **Valuation Class**, **Acct Assignment Ref.**, and **Update Type**, as shown in Figure 13.25. This is the order in which we see the report in the **Data Analysis** tab.

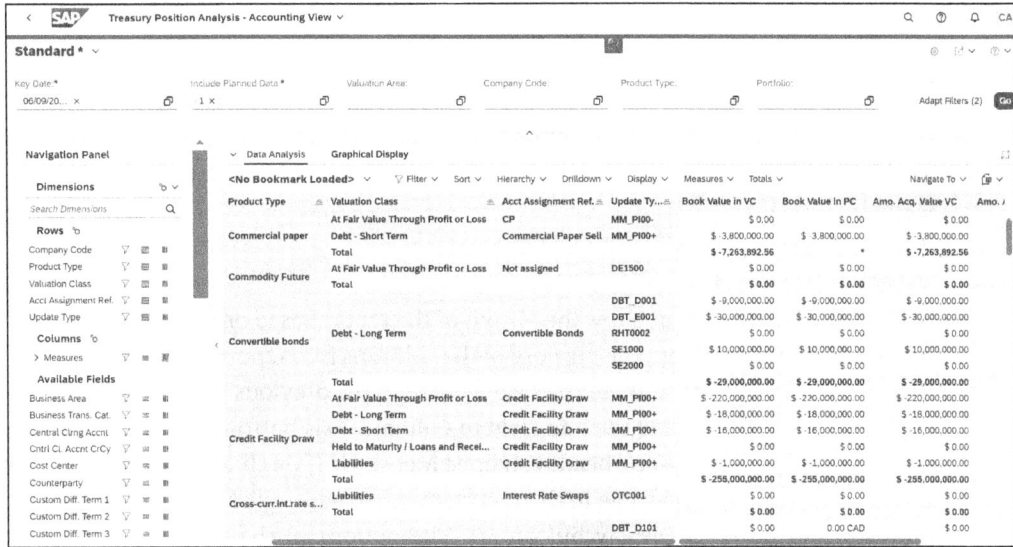

Figure 13.25 Rows Section of Navigation Panel Defines Rows That Appear in Report

We can edit the rows in a couple of different ways. In the **Available Fields** section, we can see all the attributes we can use to influence this report. If we want to add any of these attributes to be a row, we click the **Add Field to Rows Axis** button ≣ , which adds the attribute to the **Rows** section. To reflect the change, we'll need to refresh the report. Similarly, if we want to remove an attribute from the **Rows** section, we click the **Remove Field from Rows Axis** button ≣ . In Figure 13.26, we can see that these icons look slightly different if they're selected and in the **Rows** section or if they're not selected and in the **Available Fields** section.

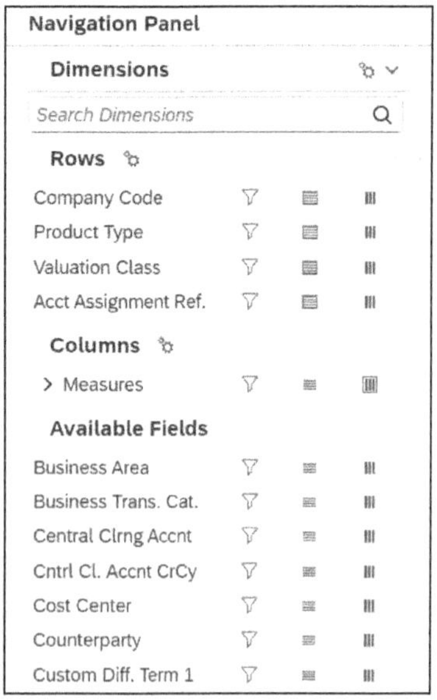

13 Reports, Key Performance Indicators, and Alerts

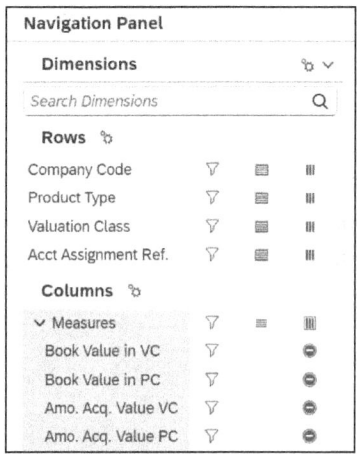

Figure 13.26 Navigation Panel that Drives Data Displayed in Rows and Columns in Report Output

Dimensions: Columns

The columns in this report show the output of the report based on the defined rows. In our example of this report, the purpose of the columns is to report the values based on the filters we've applied and the rows we defined in the previous step. We can add any of the attributes by clicking the **Add Field to Columns Axis** button. Additionally, in the **Columns** section, we can see a heading named **Measures**. If we click it, we can view all of the measures that are available in the report and then add any of them by clicking the **Add Measure Units to Display** button ⊕. The options in the **Columns** section are shown in Figure 13.27.

Figure 13.27 Measures Displayed in Output to Show Different Values for Treasury Positions

Dimensions: Filter

Even though we can filter the report in the top section, it's sometimes easier to alter the report by using the filters in the **Navigation Panel**. To add the ability to filter in this section, we click on the **Panel Settings** button and check the **Show Filter** box (see Figure 13.28). Then, we can click the filter button for any of the **Dimensions** in the **Navigation Panel**.

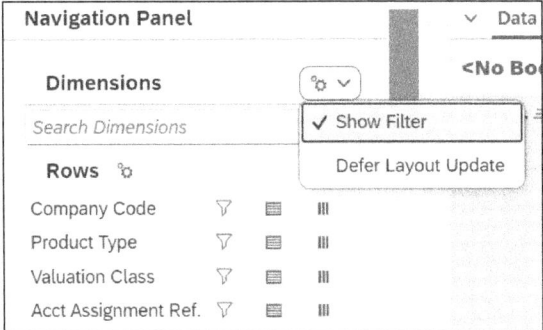

Figure 13.28 Filter Option in Navigation Panel Allows for Filtering by Additional Attributes

When we click the filter button, we're presented with filtering options to include and exclude information as required, as shown in Figure 13.29.

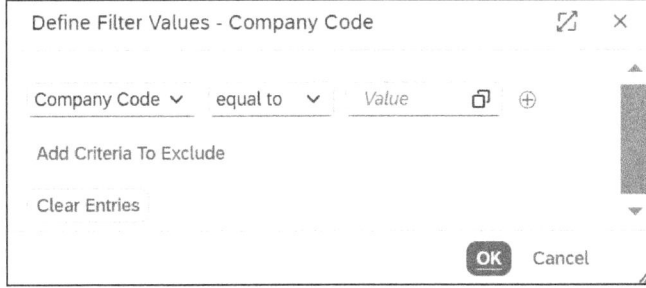

Figure 13.29 Clicking Filter Icon for Company Code Lets Us Define Filtering by Company Codes

Report Output

Now that we've covered how to define the rows and columns of the report, we can review the output (see Figure 13.30). This will show the rows and columns as we defined them in the **Navigation Panel**. We can now look at this report by the **Product Type, Valuation Class,** and the other rows and columns that we defined for this report.

13 Reports, Key Performance Indicators, and Alerts

Product Type	Valuation Class	Acct Assignment Ref.	Update Ty...	Book Value in VC	Book Value in PC	Amo. Acq. Value VC	Amo. Acq. Value PC
		Commercial Paper Sell	MM_PI00+	$ -3,463,892.56	*	$ -3,463,892.56	*
	At Fair Value Through Profit or Loss	CP	MM_PI00+	$ 0.00	$ 0.00	$ 0.00	$ 0.00
Commercial paper			MM_PI00-	$ 0.00	$ 0.00	$ 0.00	$ 0.00
	Debt - Short Term	Commercial Paper Sell	MM_PI00+	$ -3,800,000.00	$ -3,800,000.00	$ -3,800,000.00	$ -3,800,000.00
	Total			$ -7,263,892.56	*	$ -7,263,892.56	*
Commodity Future	At Fair Value Through Profit or Loss	Not assigned	DE1500	$ 0.00	$ 0.00	$ 0.00	$ 0.00
	Total			$ 0.00	$ 0.00	$ 0.00	$ 0.00
			DBT_D001	$ -9,000,000.00	$ -9,000,000.00	$ -9,000,000.00	$ -9,000,000.00
			DBT_E001	$ -30,000,000.00	$ -30,000,000.00	$ -30,000,000.00	$ -30,000,000.00
Convertible bonds	Debt - Long Term	Convertible Bonds	RHT0002	$ 0.00	$ 0.00	$ 0.00	$ 0.00
			SE1000	$ 10,000,000.00	$ 10,000,000.00	$ 10,000,000.00	$ 10,000,000.00
			SE2000	$ 0.00	$ 0.00	$ 0.00	$ 0.00
	Total			$ -29,000,000.00	$ -29,000,000.00	$ -29,000,000.00	$ -29,000,000.00
	At Fair Value Through Profit or Loss	Credit Facility Draw	MM_PI00+	$ 0.00	$ 0.00	$ 0.00	$ 0.00
	Debt - Long Term	Credit Facility Draw	MM_PI00+	$ -18,550,000.00	$ -18,550,000.00	$ -18,550,000.00	$ -18,550,000.00
			DBT_B002	$ 0.00	$ 0.00	$ 0.00	$ 0.00
			DBT_E002	$ 650,000.00	$ 650,000.00	$ 650,000.00	$ 650,000.00
Credit Facility Draw	Debt - Short Term	Credit Facility Draw	MM_PD00-	$ 0.00	$ 0.00	$ 0.00	$ 0.00
			MM_PI00+	$ -41,000,000.00	$ -41,000,000.00	$ -41,000,000.00	$ -41,000,000.00
			NOPOST	$ 0.00	$ 0.00	$ 0.00	$ 0.00

Figure 13.30 Output for Review

Additional Settings

In the **Data Analysis** section, there are a few other settings that are available for us to determine the output of this report. These are shown in Figure 13.31 and are as follows:

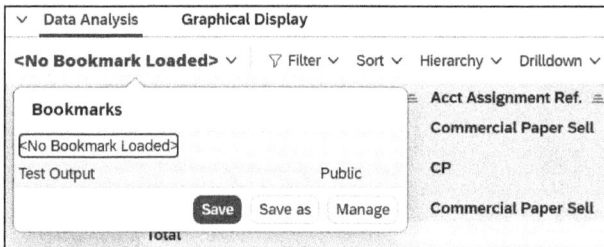

Figure 13.31 Creating Bookmarks to Save Report Settings for Future Use

- **Bookmarks**
  We use the bookmark section to save the settings that we defined in the data analysis. We can save multiple views in this area, and it's easier to find and select the settings if we created a view we want to come back to.

- **Filter**
  This is another filter section available in this report, and it's specific to filtering by the measures that we defined in the **Columns** section. To add this filter, we click into this section, click the **Manage Filter by Measure (Conditions)** button, and click the **+** button to create the condition. Then, we can create rules for how to filter this information as desired, as shown in Figure 13.32.

- **Sort**
  To use the **Sort** function, we need to click into a column first. Then, we click on the dropdown list, we'll see the options for how to sort the information (in ascending or descending order).

13.2 SAP Fiori Analysis Reports

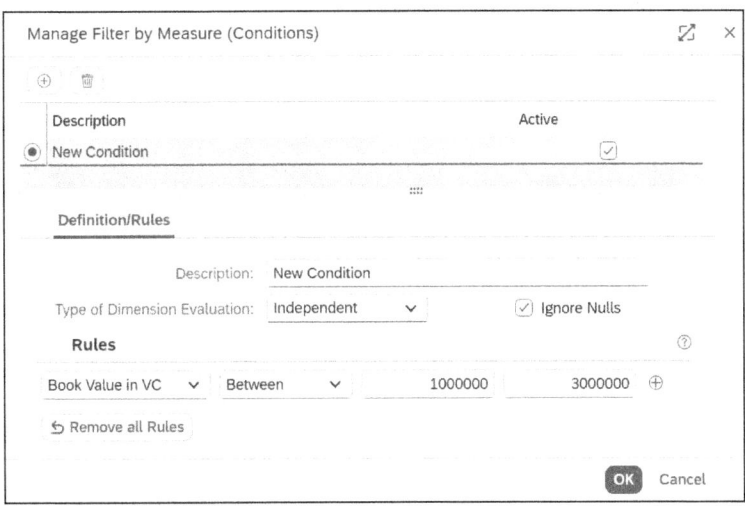

Figure 13.32 Additional Options Available for Filtering

- **Hierarchy**
Depending on the report, a hierarchy of information will be available to view in the output of the report.
- **Drilldown**
This option allows us to drill down to get additional detail.
- **Display**
The display area includes a few different ways in which we can alter the reporting. There are two kinds of properties, as follows:
    - **Grid Properties**
    The grid properties change how the information is shown on the grid in this report. In the initial report, we see that cells have been merged if they have the same values and show up next to each other in the report. We can set the **Cell Merging** field to **No Cell Merging**, and we can ensure that all cells are reported individually. In Figure 13.33, we can now see that the cells are not merged, and they're reported individually. Each product type is displayed in each row without consolidating the cells.
    - **Axis Properties**
    We can use the axis properties to drive additional views in this report. For example, the results column is where we total the information, and we can define whether the totals show up at the top of the rows or at the bottom of the rows. Additionally, we can display the columns or rows as a hierarchy, and this can change how we display this information. Figure 13.34 shows how we can view the information as a hierarchy. Instead of seeing the information as a grid, we can show the totals by company code, then product type, and then all the way down to update type. This is another way that we can view this report to see how these numbers roll up to the totals within an entity. The rollup of the report is displayed in Figure 13.35.

787

13 Reports, Key Performance Indicators, and Alerts

C...	Product Type	Valuation Class	Acct Assignment Ref.	Update Ty...	Book Value in VC	Book Value in PC	Amo. Acq. Value VC	Amo. Acq. Value PC
1000	Commercial paper	At Fair Value Through Profit or Loss	Commercial Paper Sell	MM_PI00+	$ -3,463,892.56	*	$ -3,463,892.56	*
1000	Commercial paper	At Fair Value Through Profit or Loss	CP	MM_PI00+	$ 0.00	$ 0.00	$ 0.00	$ 0.00
1000	Commercial paper	At Fair Value Through Profit or Loss	CP	MM_PI00-	$ 0.00	$ 0.00	$ 0.00	$ 0.00
1000	Commercial paper	Debt - Short Term	Commercial Paper Sell	MM_PI00+	$ -3,800,000.00	$ -3,800,000.00	$ -3,800,000.00	$ -3,800,000.00
1000	Commercial paper	Total	Total	Total	$ -7,263,892.56	*	$ -7,263,892.56	*
1000	Commodity Future	At Fair Value Through Profit or Loss	Not assigned	DE1500	$ 0.00	$ 0.00	$ 0.00	$ 0.00
1000	Commodity Future	Total	Total	Total	$ 0.00	$ 0.00	$ 0.00	$ 0.00
1000	Convertible bonds	Debt - Long Term	Convertible Bonds	DBT_D001	$ -9,000,000.00	$ -9,000,000.00	$ -9,000,000.00	$ -9,000,000.00
1000	Convertible bonds	Debt - Long Term	Convertible Bonds	DBT_E001	$ -30,000,000.00	$ -30,000,000.00	$ -30,000,000.00	$ -30,000,000.00
1000	Convertible bonds	Debt - Long Term	Convertible Bonds	RHT0002	$ 0.00	$ 0.00	$ 0.00	$ 0.00
1000	Convertible bonds	Debt - Long Term	Convertible Bonds	SE1000	$ 10,000,000.00	$ 10,000,000.00	$ 10,000,000.00	$ 10,000,000.00
1000	Convertible bonds	Debt - Long Term	Convertible Bonds	SE2000	$ 0.00	$ 0.00	$ 0.00	$ 0.00
1000	Convertible bonds	Total	Total	Total	$ -29,000,000.00	$ -29,000,000.00	$ -29,000,000.00	$ -29,000,000.00
1000	Credit Facility Draw	At Fair Value Through Profit or Loss	Credit Facility Draw	MM_PI00+	$ 0.00	$ 0.00	$ 0.00	$ 0.00
1000	Credit Facility Draw	Debt - Long Term	Credit Facility Draw	MM_PI00+	$ -18,550,000.00	$ -18,550,000.00	$ -18,550,000.00	$ -18,550,000.00
1000	Credit Facility Draw	Debt - Short Term	Credit Facility Draw	DBT_B002	$ 0.00	$ 0.00	$ 0.00	$ 0.00
1000	Credit Facility Draw	Debt - Short Term	Credit Facility Draw	DBT_E002	$ 650,000.00	$ 650,000.00	$ 650,000.00	$ 650,000.00
1000	Credit Facility Draw	Debt - Short Term	Credit Facility Draw	MM_PD00-	$ 0.00	$ 0.00	$ 0.00	$ 0.00

Figure 13.33 Example of Not Using Cell Merging

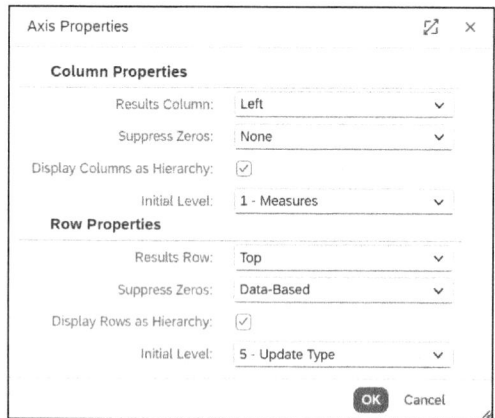

Figure 13.34 Changing Settings for Rows and Columns to Alter How Information Is Displayed

Key Figures	Book Value in VC	Book Value in PC	Amo. Acq. Value VC	Amo. Acq. Value PC
Company Code/Product Type/Valuation Class/Acct Assignment Ref./Up...				
— 1000	$ -5,471,543,105.65	*	$ -5,476,263,546.75	*
— Commercial paper	$ -7,263,892.56	*	$ -7,263,892.56	*
— At Fair Value Through Profit or Loss				
— Commercial Paper Sell				
MM_PI00+	$ -3,463,892.56	*	$ -3,463,892.56	*
— Debt - Short Term				
— Commercial Paper Sell				
MM_PI00+	$ -3,800,000.00	$ -3,800,000.00	$ -3,800,000.00	$ -3,800,000.00
— Convertible bonds	$ -29,000,000.00	$ -29,000,000.00	$ -29,000,000.00	$ -29,000,000.00
— Debt - Long Term				
— Convertible Bonds				
DBT_D001	$ -9,000,000.00	$ -9,000,000.00	$ -9,000,000.00	$ -9,000,000.00
DBT_E001	$ -30,000,000.00	$ -30,000,000.00	$ -30,000,000.00	$ -30,000,000.00
SE1000	$ 10,000,000.00	$ 10,000,000.00	$ 10,000,000.00	$ 10,000,000.00
— Credit Facility Draw	$ -59,900,000.00	$ -59,900,000.00	$ -59,900,000.00	$ -59,900,000.00
— Debt - Long Term				
— Credit Facility Draw				
MM_PI00+	$ -18,550,000.00	$ -18,550,000.00	$ -18,550,000.00	$ -18,550,000.00
— Debt - Short Term				

Figure 13.35 Updating Settings to Display Rows as Hierarchy

- **ID and Description**

  Many of the fields have a short ID and a longer description. By clicking onto any of these columns, we can determine how we want to view this information. For example, the report defaults to showing the product type name, but if we also want to look at the three-character short key of the product type, we can click on the product type column and click on the **ID and Description** option. Then, the report will show both versions of the product type.

- **Suppress Zeros**

  If we only want to view lines that have values, we can click the **Suppress Zeros** option in the report, and then, only rows with values will be displayed. To reverse this option, we click on the report and select the **Display Zeros** option.

  Figure 13.36 shows the report after the zeros have been suppressed.

C...	Product Type	Valuation Class	Acct Assignment Ref.	Update Ty...	Book Value in VC	Book Value in PC	Amo. Acq. Value VC	Amo. Acq. Value PC
1000	Commercial paper	Debt - Short Term	Commercial Paper Sell	MM_PI00+	$ -3,800,000.00	$ -3,800,000.00	$ -3,800,000.00	$ -3,800,000.00
1000	Commercial paper	Total	Total	Total	$ -7,263,892.56	*	$ -7,263,892.56	*
1000	Convertible bonds	Debt - Long Term	Convertible Bonds	DBT_D001	$ -9,000,000.00	$ -9,000,000.00	$ -9,000,000.00	$ -9,000,000.00
1000	Convertible bonds	Debt - Long Term	Convertible Bonds	DBT_E001	$ -30,000,000.00	$ -30,000,000.00	$ -30,000,000.00	$ -30,000,000.00
1000	Convertible bonds	Debt - Long Term	Convertible Bonds	SE1000	$ 10,000,000.00	$ 10,000,000.00	$ 10,000,000.00	$ 10,000,000.00
1000	Convertible bonds	Total	Total	Total	$ -29,000,000.00	$ -29,000,000.00	$ -29,000,000.00	$ -29,000,000.00
1000	Credit Facility Draw	Debt - Long Term	Credit Facility Draw	MM_PI00+	$ -18,550,000.00	$ -18,550,000.00	$ -18,550,000.00	$ -18,550,000.00
1000	Credit Facility Draw	Debt - Short Term	Credit Facility Draw	DBT_E002	$ 650,000.00	$ 650,000.00	$ 650,000.00	$ 650,000.00
1000	Credit Facility Draw	Debt - Short Term	Credit Facility Draw	MM_PI00+	$ -41,000,000.00	$ -41,000,000.00	$ -41,000,000.00	$ -41,000,000.00
1000	Credit Facility Draw	Liabilities	Credit Facility Draw	MM_PI00+	$ -1,000,000.00	$ -1,000,000.00	$ -1,000,000.00	$ -1,000,000.00
1000	Credit Facility Draw	Total	Total	Total	$ -59,900,000.00	$ -59,900,000.00	$ -59,900,000.00	$ -59,900,000.00
1000	Fixed-interest bonds	Held to Maturity / Loans and Receiv...	Bonds	SE_PI00	$ 12,825,000.00	$ 12,825,000.00	$ 12,825,000.00	$ 12,825,000.00
1000	Fixed-interest bonds	Held to Maturity / Loans and Receiv...	Bonds	V301	$ -3,611.44	$ -3,611.44	$ -3,611.44	$ -3,611.44
1000	Fixed-interest bonds	Total	Total	Total	$ 12,821,388.56	$ 12,821,388.56	$ 12,821,388.56	$ 12,821,388.56
1000	Fixed-Term Deposit	At Fair Value Through Profit or Loss	Interest Rate Instrume...	DBT_C004	$ -273,450.85	$ -273,450.85	$ -273,450.85	$ -273,450.85
1000	Fixed-Term Deposit	At Fair Value Through Profit or Loss	Interest Rate Instrume...	DBT_E002	$ 22,707,975.68	$ 22,707,975.68	$ 22,707,975.68	$ 22,707,975.68
1000	Fixed-Term Deposit	At Fair Value Through Profit or Loss	Interest Rate Instrume...	DBT_E018	$ 205,087.19	$ 205,087.19	$ 205,087.19	$ 205,087.19
1000	Fixed-Term Deposit	At Fair Value Through Profit or Loss	Interest Rate Instrume...	MM1105+	$ -88,955,136.00	$ -88,955,136.00	$ -88,955,136.00	$ -88,955,136.00
1000	Fixed-Term Deposit	At Fair Value Through Profit or Loss	Interest Rate Instrume...	V301	$ -295,779.67	$ -295,779.67	$ -295,779.67	$ -295,779.67
1000	Fixed-Term Deposit	At Fair Value Through Profit or Loss	Payables: Other Mon...	DBT_C004	$ -124,393.56	$ -124,393.56	$ -124,393.56	$ -124,393.56

Figure 13.36 Zeroes Have Been Suppressed in this Report to Simplify the View with Fewer Rows

- **Measures**

  In the measures area, we can add highlights to our report under certain conditions. We can add them to only one measure or all defined measures in our report. In this area, we can create a condition for when an alert is created, and the alert highlights cells that match the alert rule condition that we set.

  In Figure 13.37, we're setting an alert that highlights any **Book Value in VC** if the value is under $5,000,000.

13 Reports, Key Performance Indicators, and Alerts

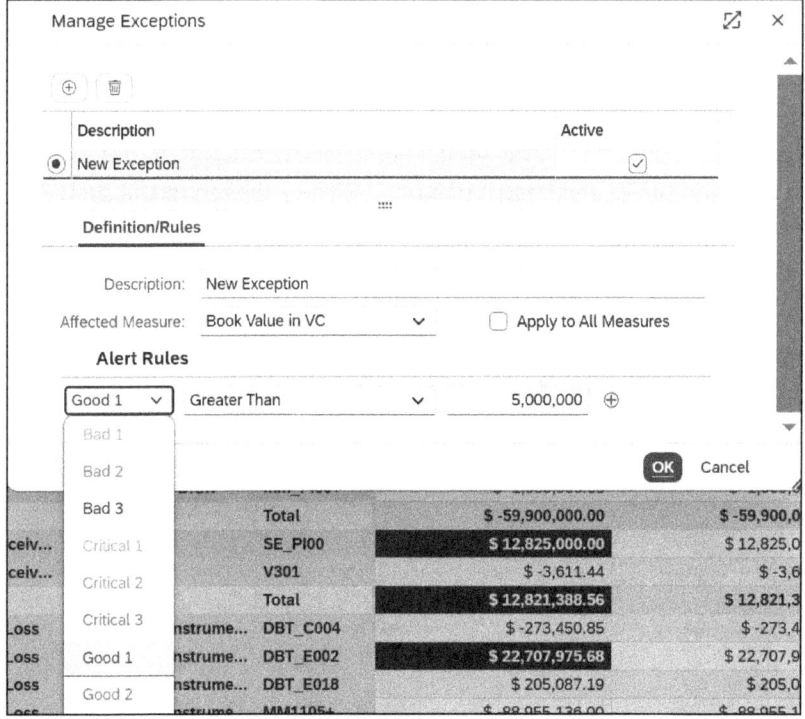

Figure 13.37 Measures Area Is for Creating Scenarios When Cells Should Be Highlighted in Report

Now, in the output, we see that some cells are highlighted based on the alert we created in the **Measures** area, as shown in Figure 13.38.

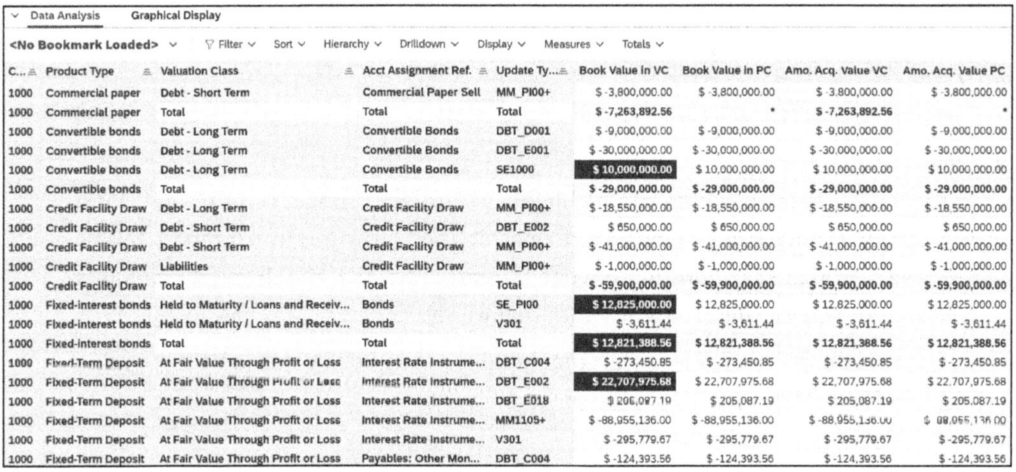

Figure 13.38 Cells with Book Value Under $5,000,000 in Green Based on Settings in Measures Area

- **Totals**
  This area allows us to further define where the totals show up in the final report.

Now that we've covered the basic principles of how to structure these reports and the many ways to edit them, we'll cover the main reports within treasury and risk management that use this functionality.

13.2.2 Treasury Position Analysis Apps

There are a series of reports that all are described as Treasury Position Analysis apps. We'll cover each of these reports because the output of data in these reports varies. The first report we'll view is an overall summary of the positions of all financial transactions for a given key date. In Figure 13.39, every product type available in the system is shown with its total position amounts.

Compa...	Product Type	Units	Nominal Amount	Book Value in VC	Book Value in PC	Amo. Acq. Value VC	Amc
1000	Commercial paper	0.000000000	*	$ -7,263,892.56	*	$ -7,263,892.56	
	Commodity Future	1,041.000000000	$ 0.00	$ 0.00	$ 0.00	$ 0.00	
	Convertible bonds	0.000000000	$ -29,000,000.00	$ -29,000,000.00	$ -29,000,000.00	$ -29,000,000.00	
	Credit Facility Draw	0.000000000	$ 74,900,000.00	$ -59,900,000.00	$ -59,900,000.00	$ -59,900,000.00	
	Cross-curr.int.rate swap (CCS)	1.000000000	$ 0.00	$ 0.00	$ 0.00	$ 0.00	
	Exposure Sub Item	30.000000000	0.00 CAD	$ 0.00	0.00 CAD	$ 0.00	
	Fixed-interest bonds	0.000000000	$ 12,500,000.00	$ 12,821,388.56	$ 12,821,388.56	$ 12,821,388.56	
	Fixed-Term Deposit	0.000000000	$ 83,866,106.80	$ -83,028,968.17	$ -83,028,968.17	$ -83,028,968.17	
	Foreign exchange (FX)	23.000000000	0.00 CAD	$ 4,091,814.70	0.00 CAD	$ 0.00	
	Interest rate instrument	0.000000000	$ 62,687,735.00	$ -60,637,735.00	$ -60,637,735.00	$ -60,637,735.00	
	Interest Rate Swap	6.000000000	$ 0.00	$ 0.00	$ 0.00	$ 0.00	
	Issue: Fixed Interest Bonds	0.000000000	$ 113,325,000.00	$ -116,931,319.90	$ -116,931,319.90	$ -117,439,946.30	
	Issue: Fixed Interest Bonds	0.000000000	$ 5,980,259,000.00	$ -5,139,444,393.28	$ -5,139,444,393.28	$ -5,139,564,393.28	
	Letter of Credit	0.000000000	$ 100,000.00	$ 0.00	$ 0.00	$ 0.00	
	Letter of Credit	0.000000000	$ 27,750,000.00	$ 0.00	$ 0.00	$ 0.00	
	Non Deliverable Forward	0.000000000	0.00 CAD	$ 0.00	0.00 CAD	$ 0.00	
	Stocks	-449,990.500000000	$ 0.00	$ 7,750,000.00	$ 7,750,000.00	$ 7,750,000.00	
	Total	-448,889.500000000	*	$ -5,471,543,105.65	*	$ -5,476,263,546.75	
Grand Total		-448,889.500000000	*	$ -5,471,543,105.65	*	$ -5,476,263,546.75	

Figure 13.39 Treasury Position Analysis App Showing Summary of All Product Types and Their Positions

We'll now jump into the apps for OTC transactions, securities, listed derivatives, and the accounting view.

Treasury Position Analysis – Over the Counter Transactions App

The Treasury Position Analysis – OTC Transactions app uses the same initial report as the regular report, but it filters information by the OTC transactions product group and calculates the subtotal of all the information by product type. Due to this, all money market with debt/investments transactions and FX transactions show up in the reporting. An example of this report is shown in Figure 13.40.

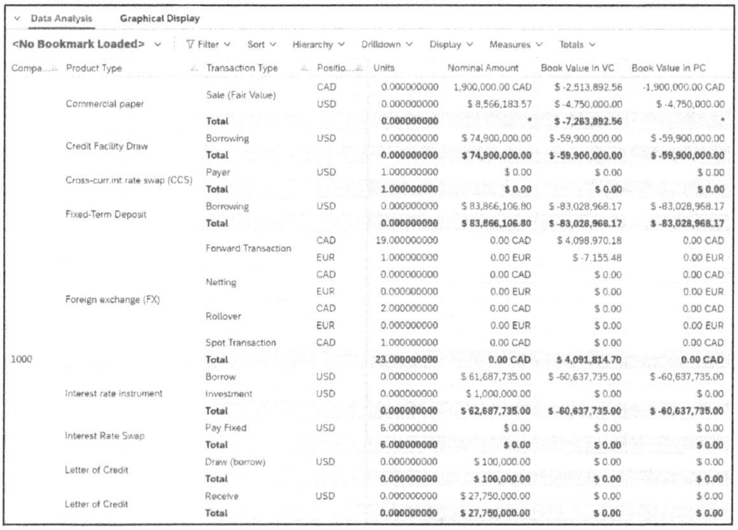

13 Reports, Key Performance Indicators, and Alerts

Compa...	Product Type	Transaction Type	Positio...	Units	Nominal Amount	Book Value in VC	Book Value in PC
	Commercial paper	Sale (Fair Value)		CAD	0.000000000	1,900,000.00 CAD	$ -2,513,892.56	-1,900,000.00 CAD
				USD	0.000000000	$ 8,566,183.57	$ -4,750,000.00	$ -4,750,000.00
		Total			0.000000000	*	$ -7,263,892.56	*
	Credit Facility Draw	Borrowing		USD	0.000000000	$ 74,900,000.00	$ -59,900,000.00	$ -59,900,000.00
		Total			0.000000000	$ 74,900,000.00	$ -59,900,000.00	$ -59,900,000.00
	Cross-curr.int.rate swap (CCS)	Payer		USD	1.000000000	$ 0.00	$ 0.00	$ 0.00
		Total			1.000000000	$ 0.00	$ 0.00	$ 0.00
	Fixed-Term Deposit	Borrowing		USD	0.000000000	$ 83,866,106.90	$ -83,028,968.17	$ -83,028,968.17
		Total			0.000000000	$ 83,866,106.80	$ -83,028,968.17	$ -83,028,968.17
	Foreign exchange (FX)	Forward Transaction		CAD	19.000000000	0.00 CAD	$ 4,098,970.18	0.00 CAD
				EUR	1.000000000	0.00 EUR	$ -7,155.48	0.00 EUR
		Netting		CAD	0.000000000	0.00 CAD	$ 0.00	0.00 CAD
				EUR	0.000000000	0.00 EUR	$ 0.00	0.00 EUR
		Rollover		CAD	2.000000000	0.00 CAD	$ 0.00	0.00 CAD
				EUR	0.000000000	0.00 EUR	$ 0.00	0.00 EUR
		Spot Transaction		CAD	1.000000000	0.00 CAD	$ 0.00	0.00 CAD
1000		Total			23.000000000	0.00 CAD	$ 4,091,814.70	0.00 CAD
	Interest rate instrument	Borrow		USD	0.000000000	$ 61,687,735.00	$ -60,637,735.00	$ -60,637,735.00
		Investment		USD	0.000000000	$ 1,000,000.00	$ 0.00	$ 0.00
		Total			0.000000000	$ 62,687,735.00	$ -60,637,735.00	$ -60,637,735.00
	Interest Rate Swap	Pay Fixed		USD	6.000000000	$ 0.00	$ 0.00	$ 0.00
		Total			6.000000000	$ 0.00	$ 0.00	$ 0.00
	Letter of Credit	Draw (borrow)		USD	0.000000000	$ 100,000.00	$ 0.00	$ 0.00
		Total			0.000000000	$ 100,000.00	$ 0.00	$ 0.00
	Letter of Credit	Receive		USD	0.000000000	$ 27,750,000.00	$ 0.00	$ 0.00
		Total			0.000000000	$ 27,750,000.00	$ 0.00	$ 0.00

Figure 13.40 All OTC Transactions Shown by Product Type

Treasury Position Analysis – Securities App

The Treasury Position Analysis – Securities app (see Figure 13.41) also uses the same basic report, but it filters by securities product group. This app filters the report to only show the product types for securities, and it calculates subtotals for the different product types. One key difference is that the report shows the information by each security class, but we can remove this row if we want to change the output.

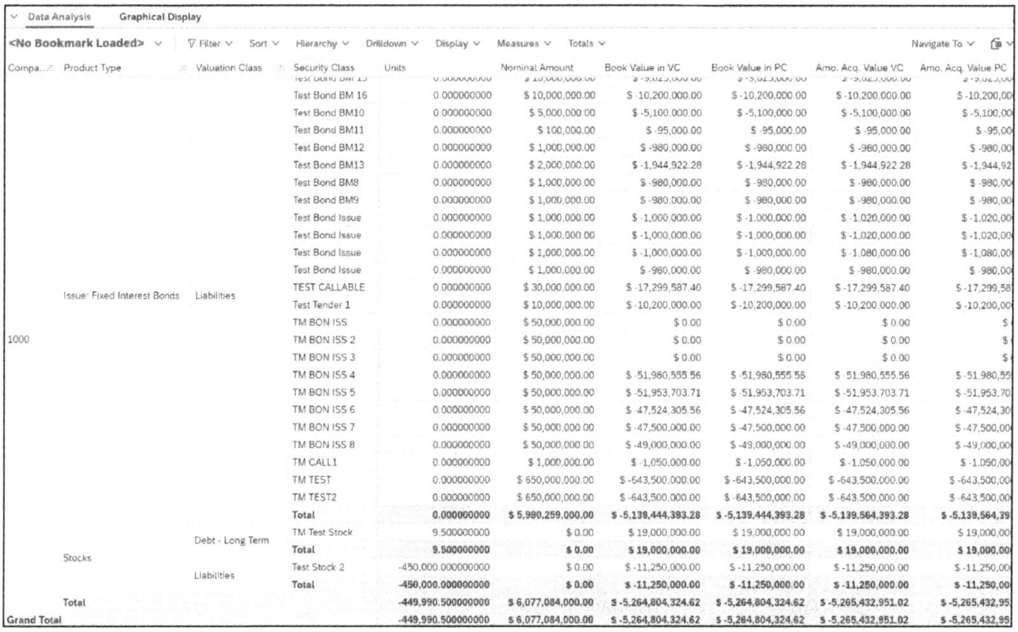

Compa...	Product Type	Valuation Class	Security Class	Units	Nominal Amount	Book Value in VC	Book Value in PC	Amo. Acq. Value VC	Amo. Acq. Value PC
			Test Bond BM 15	0.000000000	$ 10,000,000.00	$ -9,023,000.00	$ -9,023,000.00	$ -9,023,000.00	$ -9,023,000.00
			Test Bond BM 16	0.000000000	$ 10,000,000.00	$ -10,200,000.00	$ -10,200,000.00	$ -10,200,000.00	$ -10,200,000.00
			Test Bond BM10	0.000000000	$ 5,000,000.00	$ -5,100,000.00	$ -5,100,000.00	$ -5,100,000.00	$ -5,100,000.00
			Test Bond BM11	0.000000000	$ 100,000.00	$ -95,000.00	$ -95,000.00	$ -95,000.00	$ -95,000.00
			Test Bond BM12	0.000000000	$ 1,000,000.00	$ -980,000.00	$ -980,000.00	$ -980,000.00	$ -980,000.00
			Test Bond BM13	0.000000000	$ 2,000,000.00	$ -1,944,922.28	$ -1,944,922.28	$ -1,944,922.28	$ -1,944,922.28
			Test Bond BM8	0.000000000	$ 1,000,000.00	$ -980,000.00	$ -980,000.00	$ -980,000.00	$ -980,000.00
			Test Bond BM9	0.000000000	$ 1,000,000.00	$ -980,000.00	$ -980,000.00	$ -980,000.00	$ -980,000.00
			Test Bond Issue	0.000000000	$ 1,000,000.00	$ -1,000,000.00	$ -1,000,000.00	$ -1,020,000.00	$ -1,020,000.00
			Test Bond Issue	0.000000000	$ 1,000,000.00	$ -1,000,000.00	$ -1,000,000.00	$ -1,020,000.00	$ -1,020,000.00
			Test Bond Issue	0.000000000	$ 1,000,000.00	$ -1,000,000.00	$ -1,000,000.00	$ -1,080,000.00	$ -1,080,000.00
			Test Bond Issue	0.000000000	$ 1,000,000.00	$ -980,000.00	$ -980,000.00	$ -980,000.00	$ -980,000.00
	Issue: Fixed Interest Bonds	Liabilities	TEST CALLABLE	0.000000000	$ 30,000,000.00	$ -17,299,587.40	$ -17,299,587.40	$ -17,299,587.40	$ -17,299,587.40
			Test Tender 1	0.000000000	$ 10,000,000.00	$ -10,200,000.00	$ -10,200,000.00	$ -10,200,000.00	$ -10,200,000.00
1000			TM BON ISS	0.000000000	$ 50,000,000.00	$ 0.00	$ 0.00	$ 0.00	$ 0.00
			TM BON ISS 2	0.000000000	$ 50,000,000.00	$ 0.00	$ 0.00	$ 0.00	$ 0.00
			TM BON ISS 3	0.000000000	$ 50,000,000.00	$ 0.00	$ 0.00	$ 0.00	$ 0.00
			TM BON ISS 4	0.000000000	$ 50,000,000.00	$ -51,980,555.56	$ -51,980,555.56	$ -51,980,555.56	$ -51,980,555.56
			TM BON ISS 5	0.000000000	$ 50,000,000.00	$ -51,953,703.71	$ -51,953,703.71	$ -51,953,703.71	$ -51,953,703.71
			TM BON ISS 6	0.000000000	$ 50,000,000.00	$ -47,524,305.56	$ -47,524,305.56	$ -47,524,305.56	$ -47,524,305.56
			TM BON ISS 7	0.000000000	$ 50,000,000.00	$ -47,500,000.00	$ -47,500,000.00	$ -47,500,000.00	$ -47,500,000.00
			TM BON ISS 8	0.000000000	$ 50,000,000.00	$ -49,000,000.00	$ -49,000,000.00	$ -49,000,000.00	$ -49,000,000.00
			TM CALL1	0.000000000	$ 1,000,000.00	$ -1,050,000.00	$ -1,050,000.00	$ -1,050,000.00	$ -1,050,000.00
			TM TEST	0.000000000	$ 650,000,000.00	$ -643,500,000.00	$ -643,500,000.00	$ -643,500,000.00	$ -643,500,000.00
			TM TEST2	0.000000000	$ 650,000,000.00	$ -643,500,000.00	$ -643,500,000.00	$ -643,500,000.00	$ -643,500,000.00
			Total	0.000000000	$ 5,980,259,000.00	$ -5,139,444,393.28	$ -5,139,444,393.28	$ -5,139,564,393.28	$ -5,139,564,393...
	Stocks	Debt - Long Term	TM Test Stock	9.500000000	$ 0.00	$ 19,000,000.00	$ 19,000,000.00	$ 19,000,000.00	$ 19,000,000.00
			Total	9.500000000	$ 0.00	$ 19,000,000.00	$ 19,000,000.00	$ 19,000,000.00	$ 19,000,000.00
		Liabilities	Test Stock 2	-450,000.000000000	$ 0.00	$ -11,250,000.00	$ -11,250,000.00	$ -11,250,000.00	$ -11,250,000.00
			Total	-450,000.000000000	$ 0.00	$ -11,250,000.00	$ -11,250,000.00	$ -11,250,000.00	$ -11,250,000.00
Total				-449,990.500000000	$ 6,077,084,000.00	$ -5,264,804,324.62	$ -5,264,804,324.62	$ -5,265,432,951.02	$ -5,265,432,95...
Grand Total				-449,990.500000000	$ 6,077,084,000.00	$ -5,264,804,324.62	$ -5,264,804,324.62	$ -5,265,432,951.02	$ -5,265,432,95...

Figure 13.41 View of Treasury Position Analysis Only Showing Securities Transactions

Treasury Position Analysis – Listed Derivatives App

The Treasury Position Analysis – Listed Derivatives app shows a report specific to derivatives. It can show different futures and options, and Figure 13.42 shows a report based specifically on derivatives. As we can see in the figure, some additional measures columns have been added to this report—in the previous reports, we were only looking at the nominal amounts and book value. In this report, we're looking at commodity futures, so we've added the variation margin columns to make sure the output is applicable to the report.

Data Analysis | Graphical Display

<No Bookmark Loaded> | Filter | Sort | Hierarchy | Drilldown | Display | Measures | Totals

Compa...	Product Type	Security Class	Futures Account	Units	Book Value in VC	Book Value in PC	Variation Margin VC	Variation Margin PC
1000	Commodity Future	ALI Octob 2023	1000/ALUMINUM	41.000000000	$ 0.00	$ 0.00	$ 0.00	$ 0.00
		ALI Sept 2023	1000/ALUMINUM	1,000.000000000	$ 0.00	$ 0.00	$ 0.00	$ 0.00
	Total			**1,041.000000000**	**$ 0.00**	**$ 0.00**	**$ 0.00**	**$ 0.00**
Grand Total				**1,041.000000000**	**$ 0.00**	**$ 0.00**	**$ 0.00**	**$ 0.00**

Figure 13.42 Position Analysis Displaying Any Derivatives

Treasury Position Analysis – Accounting View App

The Treasury Position Analysis – Accounting View app (see Figure 13.43) is basically the same as the other Treasury Position Analysis reports, but there are some distinct differences. Since valuation class and account assignment reference are some of the main drivers for the accounting treatment of the product types, these values are already defaulted in this report. Another key difference in this report is how we bring in the amortized acquisition value. Since many transactions need to track the amortized acquisition value, it's already delivered in the accounting review of this report to help simplify the reporting.

Data Analysis | Graphical Display

<No Bookmark Loaded> | Filter | Sort | Hierarchy | Drilldown | Display | Measures | Totals

Comp...	Product Type	Valuation Class	Acct Assignment Ref.	Book Value in VC	Book Value in PC	Amo. Acq. Value VC	Amo. Acq. Value PC
1000	Commercial paper	At Fair Value Through Profit or Loss	Commercial Paper Purchase	$ 0.00	$ 0.00	$ 0.00	$ 0.00
1000	Commercial paper	At Fair Value Through Profit or Loss	Commercial Paper Sell	$ -3,463,892.56	*	$ -3,463,892.56	*
1000	Commercial paper	At Fair Value Through Profit or Loss	CP	$ 0.00	$ 0.00	$ 0.00	$ 0.00
1000	Commercial paper	Debt - Short Term	Commercial Paper Sell	$ -3,800,000.00	$ -3,800,000.00	$ -3,800,000.00	$ -3,800,000.00
1000	Commercial paper	Total	Total	**$ -7,263,892.56**	*	**$ -7,263,892.56**	*
1000	Commodity Future	At Fair Value Through Profit or Loss	Not assigned	$ 0.00	$ 0.00	$ 0.00	$ 0.00
1000	Commodity Future	Total	Total	**$ 0.00**	**$ 0.00**	**$ 0.00**	**$ 0.00**
1000	Convertible bonds	Debt - Long Term	Convertible Bonds	$ -29,000,000.00	$ -29,000,000.00	$ -29,000,000.00	$ -29,000,000.00
1000	Convertible bonds	Total	Total	**$ -29,000,000.00**	**$ -29,000,000.00**	**$ -29,000,000.00**	**$ -29,000,000.00**
1000	Credit Facility Draw	At Fair Value Through Profit or Loss	Credit Facility Draw	$ 0.00	$ 0.00	$ 0.00	$ 0.00
1000	Credit Facility Draw	Debt - Long Term	Credit Facility Draw	$ -18,550,000.00	$ -18,550,000.00	$ -18,550,000.00	$ -18,550,000.00
1000	Credit Facility Draw	Debt - Short Term	Credit Facility Draw	$ -40,350,000.00	$ -40,350,000.00	$ -40,350,000.00	$ -40,350,000.00
1000	Credit Facility Draw	Held to Maturity / Loans and Receivabl...	Credit Facility Draw	$ 0.00	$ 0.00	$ 0.00	$ 0.00
1000	Credit Facility Draw	Liabilities	Credit Facility Draw	$ -1,000,000.00	$ -1,000,000.00	$ -1,000,000.00	$ -1,000,000.00
1000	Credit Facility Draw	Total	Total	**$ -59,900,000.00**	**$ -59,900,000.00**	**$ -59,900,000.00**	**$ -59,900,000.00**
1000	Cross-curr.int.rate swap ...	Liabilities	Interest Rate Swaps	$ 0.00	$ 0.00	$ 0.00	$ 0.00
1000	Cross-curr.int.rate swap ...	Total	Total	**$ 0.00**	**$ 0.00**	**$ 0.00**	**$ 0.00**

Figure 13.43 Accounting View with Additional Columns to Further Categorize Book Value Based on Accounting Treatment

13.3 SAP GUI Reports

There are also various legacy reporting transactions in treasury and risk management. These transactions are all GUI based, but we can also run them in SAP Fiori. The following sections walk through the key legacy reporting transactions available to us in SAP S/4HANA.

13.3.1 Treasury: Journal of Financial Transactions Report

Transaction TJ01 brings us to the Treasury: Journal of Financial Transactions report, which gives us a high-level view of the financial transactions. This includes the company code, transaction number, business partner, initial value, and dates to highlight when the contract was created and last changed. We can filter the transaction so that we can view only the transactions we want to view, and if additional details are required, we can click the information icon 📝 to view the details of the transactions. This icon links to Transaction FTR_DISPLAY to show the transaction structure. Figure 13.44 shows how we can view the transactions in Transaction TJ01.

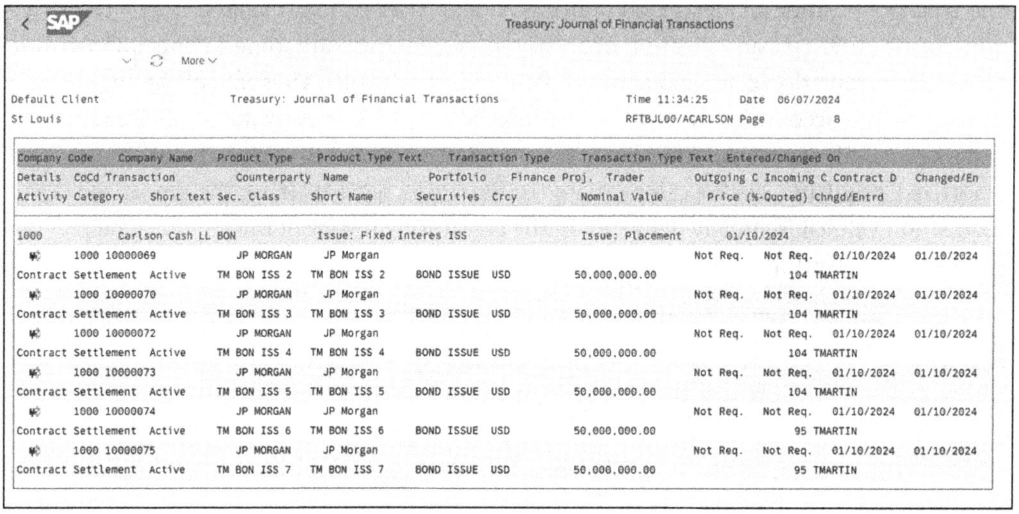

Figure 13.44 List of All Financial Transactions

13.3.2 Transaction Release: Work Item Overview and Status of all Transactions Report

We use Transaction TJ08 to view the Transaction Release: Work Item Overview and Status of all Transactions report, which we use to view the workflow status of any treasury transaction. This can provide a quick view of whether all transactions have been released from workflow or whether there are outstanding transactions that someone hasn't reviewed yet. This helps us ensure that all transactions are up to date and reviewed prior to running the posting programs throughout the month or at month end. We can filter this report by transactional information or based on release information. For example, if we only want to review transactions that are waiting for approval, we could filter the **WI Status** (work item status) by **STARTED** (description: **In Process**) to only view the outstanding contracts that haven't been approved.

Once we've run the transaction, we can see that each change that went to the workflow has its own ID number. This report shows all relevant changes that went to the workflow and shows the work item status (**WI Status**) to reflect whether the workflow is **In Process** or **Completed** (see Figure 13.45).

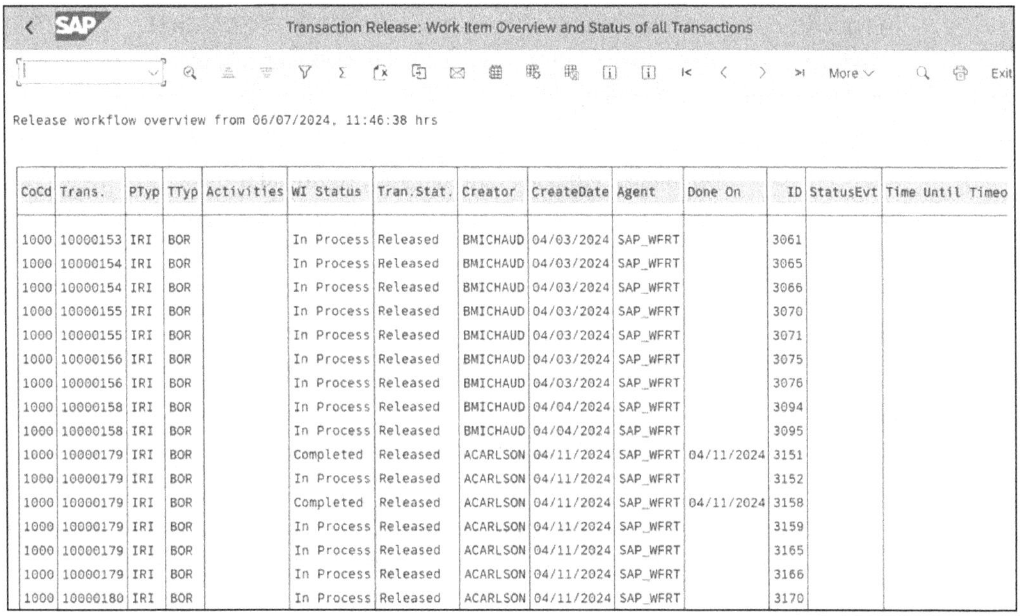
Transaction Release: Work Item Overview and Status of all Transactions

Release workflow overview from 06/07/2024, 11:46:38 hrs

CoCd	Trans.	PTyp	TTyp	Activities	WI Status	Tran.Stat.	Creator	CreateDate	Agent	Done On	ID	StatusEvt	Time Until Timeo
1000	10000153	IRI	BOR		In Process	Released	BMICHAUD	04/03/2024	SAP_WFRT		3061		
1000	10000154	IRI	BOR		In Process	Released	BMICHAUD	04/03/2024	SAP_WFRT		3065		
1000	10000154	IRI	BOR		In Process	Released	BMICHAUD	04/03/2024	SAP_WFRT		3066		
1000	10000155	IRI	BOR		In Process	Released	BMICHAUD	04/03/2024	SAP_WFRT		3070		
1000	10000155	IRI	BOR		In Process	Released	BMICHAUD	04/03/2024	SAP_WFRT		3071		
1000	10000156	IRI	BOR		In Process	Released	BMICHAUD	04/03/2024	SAP_WFRT		3075		
1000	10000156	IRI	BOR		In Process	Released	BMICHAUD	04/03/2024	SAP_WFRT		3076		
1000	10000158	IRI	BOR		In Process	Released	BMICHAUD	04/04/2024	SAP_WFRT		3094		
1000	10000158	IRI	BOR		In Process	Released	BMICHAUD	04/04/2024	SAP_WFRT		3095		
1000	10000179	IRI	BOR		Completed	Released	ACARLSON	04/11/2024	SAP_WFRT	04/11/2024	3151		
1000	10000179	IRI	BOR		In Process	Released	ACARLSON	04/11/2024	SAP_WFRT		3152		
1000	10000179	IRI	BOR		Completed	Released	ACARLSON	04/11/2024	SAP_WFRT	04/11/2024	3158		
1000	10000179	IRI	BOR		In Process	Released	ACARLSON	04/11/2024	SAP_WFRT		3159		
1000	10000179	IRI	BOR		In Process	Released	ACARLSON	04/11/2024	SAP_WFRT		3165		
1000	10000179	IRI	BOR		In Process	Released	ACARLSON	04/11/2024	SAP_WFRT		3166		
1000	10000180	IRI	BOR		In Process	Released	ACARLSON	04/11/2024	SAP_WFRT		3170		

Figure 13.45 Workflow Status of Each Transaction Displayed for All Transactions that Are Relevant to Workflow

13.3.3 Treasury: Change Documents for Transaction Report

Throughout the lifecycle of a financial transaction, there are multiple instances when a contract can change. For example, new interest conditions can be created, flows can be edited, interest rates can be fixed, transactions can be posted, and the whole contract can even be reversed. Transaction TBCD takes us to the Treasury: Change Documents for Transaction report, which tracks and reports any changes to the financial transactions. We can filter this report by company code and transaction number, and if we're trying to view specific changes based on the change date and the user who updated a transaction, we can filter this report to accommodate those requirements.

In the report that is output, there are many fields that we can view, as shown in Figure 13.46. To simplify this, we have filtered this report to only require the relevant information, and we will detail the following key fields to show how this report functions:

- **User ID, First Name, and Last Name**

  These fields are self-explanatory, and they cover who made the change in the financial transaction.

- **Date and Time**

  This field defines exactly when this change occurred.

- **TCode**

  This field defines the transaction that was called for this change. Just to note, the standard Transaction FTR_CREATE is an aggregate transaction that calls many other transaction codes in the background, depending on the type of contract we're creating or editing. Due to this, we won't see Transaction FTR_CREATE in the **TCode** field; instead, we'll see the transaction that was called for this specific transaction type (in this case, Transaction TM_52).

- **Table/Short Description**

  There is a collection of tables that are edited for the financial transactions in treasury and risk management, so the table that was edited from the transaction change is detailed in the log. The short description of the table is shown in the following column.

- **Field/Short Description**

  The field in the table that is edited is called out here along with its description.

- **Old Value/New Value**

  These fields define what the old value of the field was and what it was changed to:

  - **New Entry**

    If the old value was blank, then that value was not populated previously, and the field has been entered and the new entry is shown in the appropriate field.

  - **Delete Entry**

    If the new value is blank, then the entry was completely deleted from the transaction.

  - **Entry Change**

    If both of the previously listed fields are populated, this field will indicate whether a change has occurred.

- **Change Indicator/Short Description**

  There also are fields that show what type of change has occurred. We'll see the following statuses:

  - **I: Insert**

    This indicates that this is a new entry.

  - **D: Delete**

    This indicates that the entry was deleted.

  - **U: Change**

    This indicates that an entry was changed from one value to another.

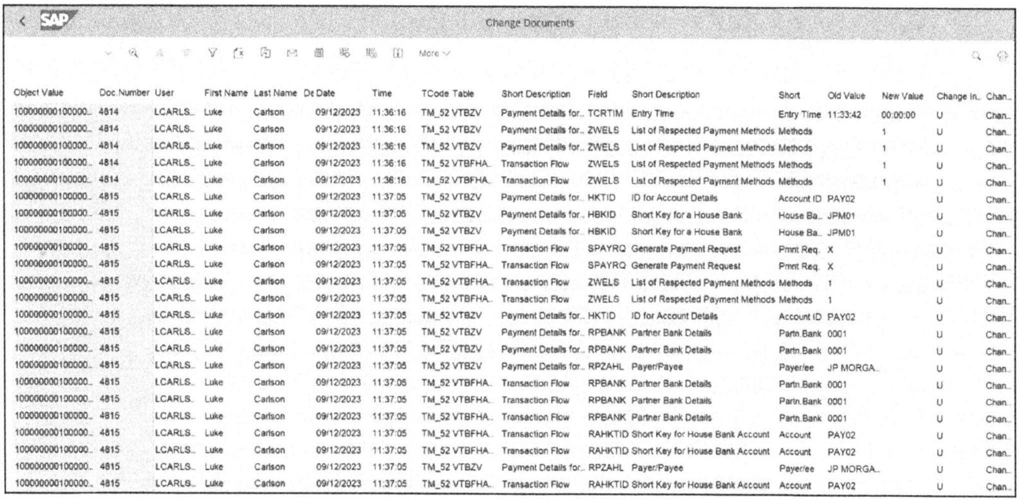

Change Documents

Object Value	Doc.Number	User	First Name	Last Name	De	Date	Time	TCode	Table	Short Description	Field	Short Description	Short	Old Value	New Value	Change In..	Chan..
10000000100000..	4814	LCARLS..	Luke	Carlson		09/12/2023	11:36:16	TM_52	VTBZV	Payment Details for..	TCRTIM	Entry Time	Entry Time	11:33:42	00:00:00	U	Chan..
10000000100000..	4814	LCARLS..	Luke	Carlson		09/12/2023	11:36:16	TM_52	VTBZV	Payment Details for..	ZWELS	List of Respected Payment Methods	Methods		1	U	Chan..
10000000100000..	4814	LCARLS..	Luke	Carlson		09/12/2023	11:36:16	TM_52	VTBZV	Payment Details for..	ZWELS	List of Respected Payment Methods	Methods		1	U	Chan..
10000000100000..	4814	LCARLS..	Luke	Carlson		09/12/2023	11:36:16	TM_52	VTBFHA..	Transaction Flow	ZWELS	List of Respected Payment Methods	Methods		1	U	Chan..
10000000100000..	4814	LCARLS..	Luke	Carlson		09/12/2023	11:36:16	TM_52	VTBFHA..	Transaction Flow	ZWELS	List of Respected Payment Methods	Methods		1	U	Chan..
10000000100000..	4815	LCARLS..	Luke	Carlson		09/12/2023	11:37:05	TM_52	VTBZV	Payment Details for..	HKTID	ID for Account Details	Account ID	PAY02		U	Chan..
10000000100000..	4815	LCARLS..	Luke	Carlson		09/12/2023	11:37:05	TM_52	VTBZV	Payment Details for..	HBKID	Short Key for a House Bank	House Ba..	JPM01		U	Chan..
10000000100000..	4815	LCARLS..	Luke	Carlson		09/12/2023	11:37:05	TM_52	VTBZV	Payment Details for..	HBKID	Short Key for a House Bank	House Ba..	JPM01		U	Chan..
10000000100000..	4815	LCARLS..	Luke	Carlson		09/12/2023	11:37:05	TM_52	VTBFHA..	Transaction Flow	SPAYRQ	Generate Payment Request	Pmnt Req.	X		U	Chan..
10000000100000..	4815	LCARLS..	Luke	Carlson		09/12/2023	11:37:05	TM_52	VTBFHA..	Transaction Flow	SPAYRQ	Generate Payment Request	Pmnt Req.	X		U	Chan..
10000000100000..	4815	LCARLS..	Luke	Carlson		09/12/2023	11:37:05	TM_52	VTBFHA..	Transaction Flow	ZWELS	List of Respected Payment Methods	Methods	1		U	Chan..
10000000100000..	4815	LCARLS..	Luke	Carlson		09/12/2023	11:37:05	TM_52	VTBFHA..	Transaction Flow	ZWELS	List of Respected Payment Methods	Methods	1		U	Chan..
10000000100000..	4815	LCARLS..	Luke	Carlson		09/12/2023	11:37:05	TM_52	VTBZV	Payment Details for..	HKTID	ID for Account Details	Account ID	PAY02		U	Chan..
10000000100000..	4815	LCARLS..	Luke	Carlson		09/12/2023	11:37:05	TM_52	VTBZV	Payment Details for..	RPBANK	Partner Bank Details	Partn.Bank	0001		U	Chan..
10000000100000..	4815	LCARLS..	Luke	Carlson		09/12/2023	11:37:05	TM_52	VTBZV	Payment Details for..	RPBANK	Partner Bank Details	Partn.Bank	0001		U	Chan..
10000000100000..	4815	LCARLS..	Luke	Carlson		09/12/2023	11:37:05	TM_52	VTBZV	Payment Details for..	RPZAHL	Payer/Payee	Payer/ee	JP MORGA..		U	Chan..
10000000100000..	4815	LCARLS..	Luke	Carlson		09/12/2023	11:37:05	TM_52	VTBFHA..	Transaction Flow	RPBANK	Partner Bank Details	Partn.Bank	0001		U	Chan..
10000000100000..	4815	LCARLS..	Luke	Carlson		09/12/2023	11:37:05	TM_52	VTBFHA..	Transaction Flow	RPBANK	Partner Bank Details	Partn.Bank	0001		U	Chan..
10000000100000..	4815	LCARLS..	Luke	Carlson		09/12/2023	11:37:05	TM_52	VTBFHA..	Transaction Flow	RPBANK	Partner Bank Details	Partn.Bank	0001		U	Chan..
10000000100000..	4815	LCARLS..	Luke	Carlson		09/12/2023	11:37:05	TM_52	VTBFHA..	Transaction Flow	RAHKTID	Short Key for House Bank Account	Account	PAY02		U	Chan..
10000000100000..	4815	LCARLS..	Luke	Carlson		09/12/2023	11:37:05	TM_52	VTBFHA..	Transaction Flow	RAHKTID	Short Key for House Bank Account	Account	PAY02		U	Chan..
10000000100000..	4815	LCARLS..	Luke	Carlson		09/12/2023	11:37:05	TM_52	VTBZV	Payment Details for..	RPZAHL	Payer/Payee	Payer/ee	JP MORGA..		U	Chan..
10000000100000..	4815	LCARLS..	Luke	Carlson		09/12/2023	11:37:05	TM_52	VTBFHA..	Transaction Flow	RAHKTID	Short Key for House Bank Account	Account	PAY02		U	Chan..

Figure 13.46 Records Showing Changes to All Fields in Transactions

13.3.4 Journal: Transactions with Cash Flows Report

Transaction TJ12 brings us to the Journal: Transactions with Cash Flows report, which shows all cash-related flows for treasury and risk management transactions (see Figure 13.47). This report shows the dates, amounts, description, and even status of the cash flow. The status will show whether the flow is blocked from posting, carried out, or flagged for posting. When the flow is flagged for posting, it still needs to be carried out when the due date arrives.

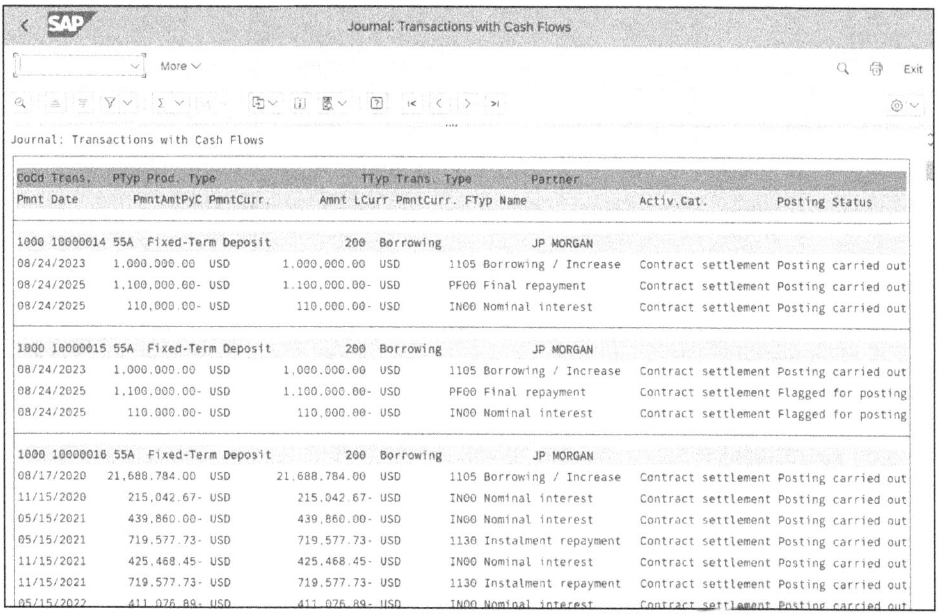

Journal: Transactions with Cash Flows

CoCd Trans. / Pmnt Date	PTyp Prod. Type / PmntAmtPyC	PmntCurr.	Amnt LCurr	PmntCurr.	TTyp Trans. Type / FTyp Name	Activ.Cat.	Posting Status
1000 10000014	55A Fixed-Term Deposit				200 Borrowing	JP MORGAN	
08/24/2023	1,000,000.00	USD	1,000,000.00	USD	1105 Borrowing / Increase	Contract settlement	Posting carried out
08/24/2025	1,100,000.00-	USD	1,100,000.00-	USD	PF00 Final repayment	Contract settlement	Posting carried out
08/24/2025	110,000.00-	USD	110,000.00-	USD	IN00 Nominal interest	Contract settlement	Posting carried out
1000 10000015	55A Fixed-Term Deposit				200 Borrowing	JP MORGAN	
08/24/2023	1,000,000.00	USD	1,000,000.00	USD	1105 Borrowing / Increase	Contract settlement	Posting carried out
08/24/2025	1,100,000.00-	USD	1,100,000.00-	USD	PF00 Final repayment	Contract settlement	Flagged for posting
08/24/2025	110,000.00-	USD	110,000.00-	USD	IN00 Nominal interest	Contract settlement	Flagged for posting
1000 10000016	55A Fixed-Term Deposit				200 Borrowing	JP MORGAN	
08/17/2020	21,688,784.00	USD	21,688,784.00	USD	1105 Borrowing / Increase	Contract settlement	Posting carried out
11/15/2020	215,042.67-	USD	215,042.67-	USD	IN00 Nominal interest	Contract settlement	Posting carried out
05/15/2021	439,860.00-	USD	439,860.00-	USD	IN00 Nominal interest	Contract settlement	Posting carried out
05/15/2021	719,577.73-	USD	719,577.73-	USD	1130 Instalment repayment	Contract settlement	Posting carried out
11/15/2021	425,468.45-	USD	425,468.45-	USD	IN00 Nominal interest	Contract settlement	Posting carried out
11/15/2021	719,577.73-	USD	719,577.73-	USD	1130 Instalment repayment	Contract settlement	Posting carried out
05/15/2022	411,076.89-	USD	411,076.89-	USD	IN00 Nominal interest	Contract settlement	Posting carried out

Figure 13.47 Summary of Status of Cash Flows and Postings

13 Reports, Key Performance Indicators, and Alerts

13.3.5 Rate/Price Adjustment Schedule Report

Various transactions require adjustments of certain market data. For example, the variable rate debt and investment contracts require periodic interest rate adjustments to ensure the interest is calculated correctly. Transaction TJ07 brings us to the Rate/Price Adjustment Schedule report, which allows us to look into the rate/price adjustments for interest rates, commodity prices, security prices, FX rates, and price index adjustments. Once we've run the report, we can view the details of each transaction to see the fixing date, contract details, and status of the adjustment. In Figure 13.48, we can see that the interest rates appear for transactions with a locked **IRA Status**. This shows that the status is fixed, and it shows the determined interest rate from the fixing. If the interest rate hasn't been fixed yet, the rate will be blank and the **IRA Status** will not be locked.

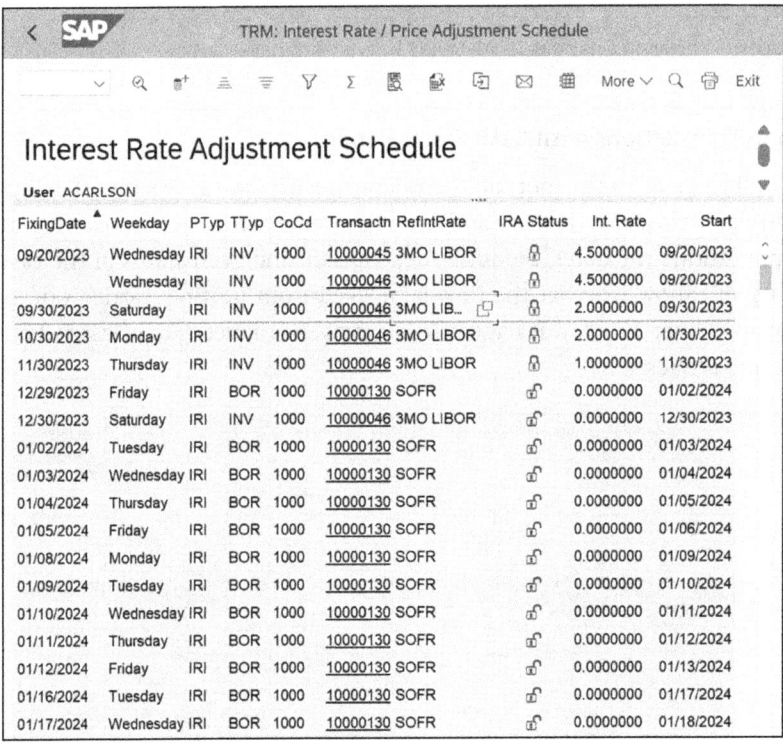

Figure 13.48 Adjustment Schedule for Interest Rates and Whether Rates Have Been Fixed

13.3.6 Facilities: Credit Lines and Utilization Report

The SAP GUI-based transaction we use to view the facility credit line and utilization is Transaction TM_60, which brings us to the Facilities: Credit Lines and Utilization report. This report shows the transaction-level detail of each facility, along with the key utilization information. It shows all types of facilities, so we can view both bilateral and syndicated facilities. An example of this report is in Figure 13.49.

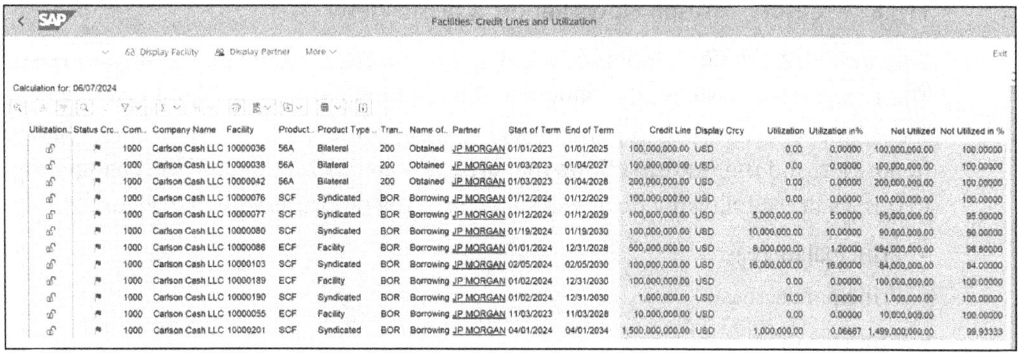

Facilities: Credit Lines and Utilization

Display Facility | Display Partner | More | Exit

Calculation for: 06/07/2024

Utilization..	Status Crc..	Com..	Company Name	Facility	Product..	Product Type ..	Tran..	Name of..	Partner	Start of Term	End of Term	Credit Line	Display Crcy	Utilization	Utilization in%	Not Utilized	Not Utilized in %
		1000	Carlson Cash LLC	10000036	56A	Bilateral	200	Obtained	JP MORGAN	01/01/2023	01/01/2025	100,000,000.00	USD	0.00	0.00000	100,000,000.00	100.00000
		1000	Carlson Cash LLC	10000038	56A	Bilateral	200	Obtained	JP MORGAN	01/03/2023	01/04/2027	100,000,000.00	USD	0.00	0.00000	100,000,000.00	100.00000
		1000	Carlson Cash LLC	10000042	56A	Bilateral	200	Obtained	JP MORGAN	01/03/2023	01/04/2028	200,000,000.00	USD	0.00	0.00000	200,000,000.00	100.00000
		1000	Carlson Cash LLC	10000076	SCF	Syndicated	BOR	Borrowing	JP MORGAN	01/12/2024	01/12/2029	100,000,000.00	USD	0.00	0.00000	100,000,000.00	100.00000
		1000	Carlson Cash LLC	10000077	SCF	Syndicated	BOR	Borrowing	JP MORGAN	01/12/2024	01/12/2029	100,000,000.00	USD	5,000,000.00	5.00000	95,000,000.00	95.00000
		1000	Carlson Cash LLC	10000080	SCF	Syndicated	BOR	Borrowing	JP MORGAN	01/19/2024	01/19/2030	100,000,000.00	USD	10,000,000.00	10.00000	90,000,000.00	90.00000
		1000	Carlson Cash LLC	10000086	ECF	Facility	BOR	Borrowing	JP MORGAN	01/01/2024	12/31/2028	500,000,000.00	USD	6,000,000.00	1.20000	494,000,000.00	98.80000
		1000	Carlson Cash LLC	10000103	SCF	Syndicated	BOR	Borrowing	JP MORGAN	02/05/2024	02/05/2030	100,000,000.00	USD	16,000,000.00	16.00000	84,000,000.00	84.00000
		1000	Carlson Cash LLC	10000189	ECF	Facility	BOR	Borrowing	JP MORGAN	01/02/2024	12/31/2030	100,000,000.00	USD	0.00	0.00000	100,000,000.00	100.00000
		1000	Carlson Cash LLC	10000190	SCF	Syndicated	BOR	Borrowing	JP MORGAN	01/02/2024	12/31/2030	1,000,000.00	USD	0.00	0.00000	1,000,000.00	100.00000
		1000	Carlson Cash LLC	10000055	ECF	Facility	BOR	Borrowing	JP MORGAN	11/03/2023	11/03/2028	10,000,000.00	USD	0.00	0.00000	10,000,000.00	100.00000
		1000	Carlson Cash LLC	10000201	SCF	Syndicated	BOR	Borrowing	JP MORGAN	04/01/2024	04/01/2034	1,500,000,000.00	USD	1,000,000.00	0.06667	1,499,000,000.00	99.93333

Figure 13.49 Display of Each Credit Facility with Its Utilization Details

13.3.7 Facilities: Lines of Credit, Drawing, and Fees Report

The previous report gives a high-level view of the credit facilities, but if we need to look into individual changes to the facility, we can use Transaction TM_60A, which brings us to the Facilities: Lines of Credit, Drawing, and Fees report, shown in Figure 13.50 This report will show any and all activity in the facilities, including drawings from and charges to the facility. One thing to note in this report is that there are nearly eighty columns in it, so we'll likely need to edit the columns to ensure that the report only shows information that's relevant to our needs.

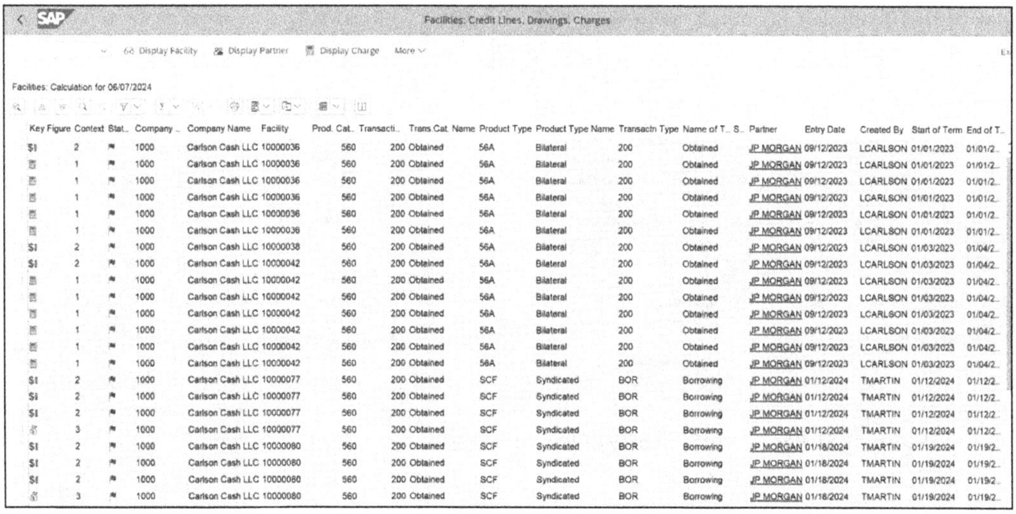

Facilities: Credit Lines, Drawings, Charges

Display Facility | Display Partner | Display Charge | More

Facilities: Calculation for 06/07/2024

Key Figure	Context	Stat..	Company ..	Company Name	Facility	Prod. Cat..	Transacti..	Trans.Cat. Name	Product Type	Product Type Name	Transactn Type	Name of T.. S..	Partner	Entry Date	Created By	Start of Term	End of T..
$I	2		1000	Carlson Cash LLC	10000036	560	200	Obtained	56A	Bilateral	200	Obtained	JP MORGAN	09/12/2023	LCARLSON	01/01/2023	01/01/2..
	1		1000	Carlson Cash LLC	10000036	560	200	Obtained	56A	Bilateral	200	Obtained	JP MORGAN	09/12/2023	LCARLSON	01/01/2023	01/01/2..
	1		1000	Carlson Cash LLC	10000036	560	200	Obtained	56A	Bilateral	200	Obtained	JP MORGAN	09/12/2023	LCARLSON	01/01/2023	01/01/2..
	1		1000	Carlson Cash LLC	10000036	560	200	Obtained	56A	Bilateral	200	Obtained	JP MORGAN	09/12/2023	LCARLSON	01/01/2023	01/01/2..
	1		1000	Carlson Cash LLC	10000036	560	200	Obtained	56A	Bilateral	200	Obtained	JP MORGAN	09/12/2023	LCARLSON	01/01/2023	01/01/2..
	1		1000	Carlson Cash LLC	10000036	560	200	Obtained	56A	Bilateral	200	Obtained	JP MORGAN	09/12/2023	LCARLSON	01/01/2023	01/01/2..
$I	2		1000	Carlson Cash LLC	10000038	560	200	Obtained	56A	Bilateral	200	Obtained	JP MORGAN	09/12/2023	LCARLSON	01/03/2023	01/04/2..
$I	2		1000	Carlson Cash LLC	10000042	560	200	Obtained	56A	Bilateral	200	Obtained	JP MORGAN	09/12/2023	LCARLSON	01/03/2023	01/04/2..
	1		1000	Carlson Cash LLC	10000042	560	200	Obtained	56A	Bilateral	200	Obtained	JP MORGAN	09/12/2023	LCARLSON	01/03/2023	01/04/2..
	1		1000	Carlson Cash LLC	10000042	560	200	Obtained	56A	Bilateral	200	Obtained	JP MORGAN	09/12/2023	LCARLSON	01/03/2023	01/04/2..
	1		1000	Carlson Cash LLC	10000042	560	200	Obtained	56A	Bilateral	200	Obtained	JP MORGAN	09/12/2023	LCARLSON	01/03/2023	01/04/2..
	1		1000	Carlson Cash LLC	10000042	560	200	Obtained	56A	Bilateral	200	Obtained	JP MORGAN	09/12/2023	LCARLSON	01/03/2023	01/04/2..
	1		1000	Carlson Cash LLC	10000042	560	200	Obtained	56A	Bilateral	200	Obtained	JP MORGAN	09/12/2023	LCARLSON	01/03/2023	01/04/2..
	1		1000	Carlson Cash LLC	10000042	560	200	Obtained	56A	Bilateral	200	Obtained	JP MORGAN	09/12/2023	LCARLSON	01/03/2023	01/04/2..
$I	2		1000	Carlson Cash LLC	10000077	560	200	Obtained	SCF	Syndicated	BOR	Borrowing	JP MORGAN	01/12/2024	TMARTIN	01/12/2024	01/12/2..
$I	2		1000	Carlson Cash LLC	10000077	560	200	Obtained	SCF	Syndicated	BOR	Borrowing	JP MORGAN	01/12/2024	TMARTIN	01/12/2024	01/12/2..
$I	2		1000	Carlson Cash LLC	10000077	560	200	Obtained	SCF	Syndicated	BOR	Borrowing	JP MORGAN	01/12/2024	TMARTIN	01/12/2024	01/12/2..
	3		1000	Carlson Cash LLC	10000077	560	200	Obtained	SCF	Syndicated	BOR	Borrowing	JP MORGAN	01/12/2024	TMARTIN	01/12/2024	01/12/2..
$I	2		1000	Carlson Cash LLC	10000080	560	200	Obtained	SCF	Syndicated	BOR	Borrowing	JP MORGAN	01/18/2024	TMARTIN	01/19/2024	01/19/2..
$I	2		1000	Carlson Cash LLC	10000080	560	200	Obtained	SCF	Syndicated	BOR	Borrowing	JP MORGAN	01/18/2024	TMARTIN	01/19/2024	01/19/2..
$I	2		1000	Carlson Cash LLC	10000080	560	200	Obtained	SCF	Syndicated	BOR	Borrowing	JP MORGAN	01/18/2024	TMARTIN	01/19/2024	01/19/2..
	3		1000	Carlson Cash LLC	10000080	560	200	Obtained	SCF	Syndicated	BOR	Borrowing	JP MORGAN	01/18/2024	TMARTIN	01/19/2024	01/19/2..

Figure 13.50 Report with Additional Details to Review Drawings and Charges for Credit Facilities

13.3.8 Treasury Position Flows Report (Classic View)

We can see the classic view of the Treasury Position Flows report in Transaction TPM13. This report is useful since any changes to the position of the transactions can be viewed there. This report actually reports on all types of position flows, so it includes changes in position and the resulting flows that are calculated from the position changes. Due to this, all of the following flows will show up in the report, as shown in Figure 13.51:

- Principal increase
- Principal decrease
- Amortization
- Valuations
- Interest
- Valuation class transfer
- Account assignment reference transfer
- Gains/losses

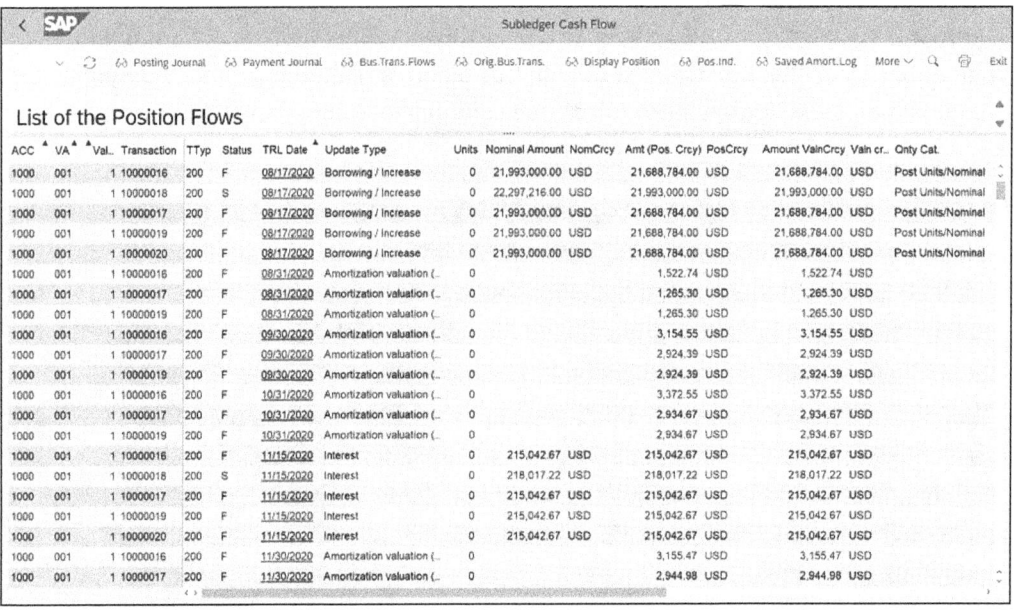

Figure 13.51 Report with Display of Position Flows

13.3.9 Treasury Posting Journal Report (Classic View)

Transaction TPM20 shows us the classic version of the Treasury Posting Journal report, which we use to view any of the postings for the transactions, as shown in Figure 13.52. The posting journal details each posting that has occurred for the defined financial transactions, and we can filter the data in this report to only show the postings for specific transactions or for all transactions with a similar product type.

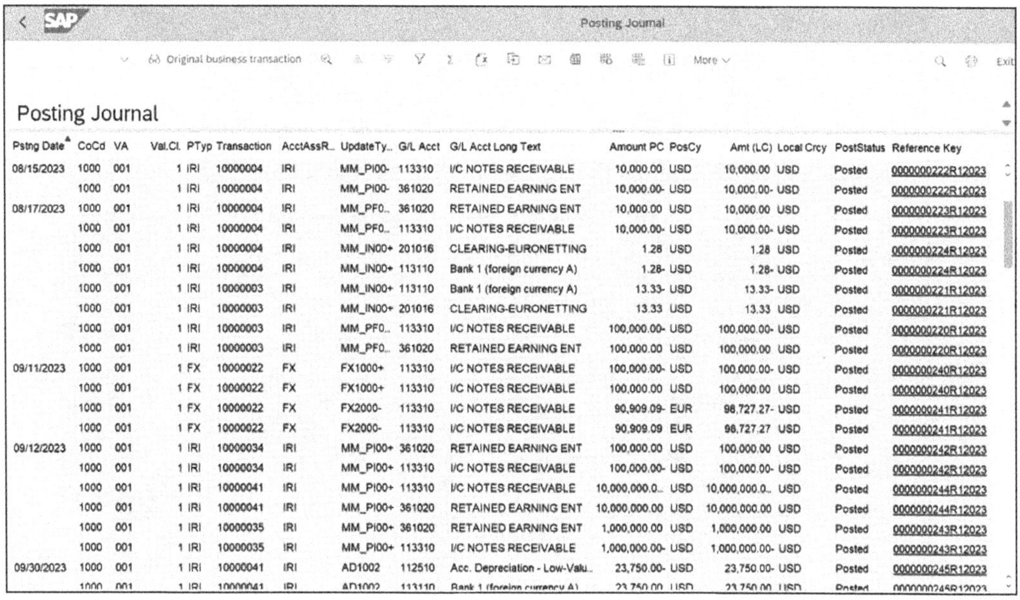

Posting Journal

Pstng Date	CoCd	VA	Val.Cl.	PTyp	Transaction	AcctAssR...	UpdateTy...	G/L Acct	G/L Acct Long Text	Amount PC	PosCy	Amt (LC)	Local Crcy	PostStatus	Reference Key
08/15/2023	1000	001	1	IRI	10000004	IRI	MM_PI00-	113310	I/C NOTES RECEIVABLE	10,000.00	USD	10,000.00	USD	Posted	0000000222R12023
	1000	001	1	IRI	10000004	IRI	MM_PI00-	361020	RETAINED EARNING ENT	10,000.00-	USD	10,000.00-	USD	Posted	0000000222R12023
08/17/2023	1000	001	1	IRI	10000004	IRI	MM_PF0...	361020	RETAINED EARNING ENT	10,000.00	USD	10,000.00	USD	Posted	0000000223R12023
	1000	001	1	IRI	10000004	IRI	MM_PF0...	113310	I/C NOTES RECEIVABLE	10,000.00-	USD	10,000.00-	USD	Posted	0000000223R12023
	1000	001	1	IRI	10000004	IRI	MM_IN00+	201016	CLEARING-EURONETTING	1.28	USD	1.28	USD	Posted	0000000224R12023
	1000	001	1	IRI	10000004	IRI	MM_IN00+	113110	Bank 1 (foreign currency A)	1.28-	USD	1.28-	USD	Posted	0000000224R12023
	1000	001	1	IRI	10000003	IRI	MM_IN00+	113110	Bank 1 (foreign currency A)	13.33-	USD	13.33-	USD	Posted	0000000221R12023
	1000	001	1	IRI	10000003	IRI	MM_IN00+	201016	CLEARING-EURONETTING	13.33	USD	13.33	USD	Posted	0000000221R12023
	1000	001	1	IRI	10000003	IRI	MM_PF0...	113310	I/C NOTES RECEIVABLE	100,000.00-	USD	100,000.00-	USD	Posted	0000000220R12023
	1000	001	1	IRI	10000003	IRI	MM_PF0...	361020	RETAINED EARNING ENT	100,000.00	USD	100,000.00	USD	Posted	0000000220R12023
09/11/2023	1000	001	1	FX	10000022	FX	FX1000+	113310	I/C NOTES RECEIVABLE	100,000.00-	USD	100,000.00-	USD	Posted	0000000240R12023
	1000	001	1	FX	10000022	FX	FX1000+	113310	I/C NOTES RECEIVABLE	100,000.00	USD	100,000.00	USD	Posted	0000000240R12023
	1000	001	1	FX	10000022	FX	FX2000-	113310	I/C NOTES RECEIVABLE	90,909.09-	EUR	98,727.27-	USD	Posted	0000000241R12023
	1000	001	1	FX	10000022	FX	FX2000-	113310	I/C NOTES RECEIVABLE	90,909.09	EUR	98,727.27	USD	Posted	0000000241R12023
09/12/2023	1000	001	1	IRI	10000034	IRI	MM_PI00+	361020	RETAINED EARNING ENT	100,000.00	USD	100,000.00	USD	Posted	0000000242R12023
	1000	001	1	IRI	10000034	IRI	MM_PI00+	113310	I/C NOTES RECEIVABLE	100,000.00-	USD	100,000.00-	USD	Posted	0000000242R12023
	1000	001	1	IRI	10000041	IRI	MM_PI00+	113310	I/C NOTES RECEIVABLE	10,000,000.0...	USD	10,000,000.0...	USD	Posted	0000000244R12023
	1000	001	1	IRI	10000041	IRI	MM_PI00+	361020	RETAINED EARNING ENT	10,000,000.00	USD	10,000,000.00	USD	Posted	0000000244R12023
	1000	001	1	IRI	10000035	IRI	MM_PI00+	361020	RETAINED EARNING ENT	1,000,000.00	USD	1,000,000.00	USD	Posted	0000000243R12023
	1000	001	1	IRI	10000035	IRI	MM_PI00+	113310	I/C NOTES RECEIVABLE	1,000,000.00-	USD	1,000,000.00-	USD	Posted	0000000243R12023
09/30/2023	1000	001	1	IRI	10000041	IRI	AD1002	112510	Acc. Depreciation - Low-Valu...	23,750.00-	USD	23,750.00-	USD	Posted	0000000245R12023
	1000	001	1	IRI	10000041	IRI	AD1002	113110	Bank 1 (foreign currency A)	23,750.00	USD	23,750.00	USD	Posted	0000000245R12023

Figure 13.52 Posting Journal Displaying Details of Each Transaction's Postings

13.3.10 Money Market: Collective Processing Report

The Money Market: Collective Processing report, accessed via Transaction TM00 and shown in Figure 13.53, allows us to view a list of all relevant money market transactions, and it provides us a few extra functionalities. Some useful tools within the collective processing transaction are as follows:

- **Mass Settle**
  We can select multiple transactions and settle them all at once. If the settle function is set up for a company, they generally already have reviewed all the contracts and want to mass-settle their contracts during cutover to production. In this transaction, we can select all relevant lines for settlement.
- **Copy**
  We can copy a transaction, and a new transaction will be created with the exact same details. We frequently use this during the testing phases of a project to determine whether configuration changes were effective. We also use when a company enters into a debt instrument that's similar to one they already have, and it makes it easier to create the contract when we have a frame of reference.
- **History**
  We can view the history of the changes to a transaction to see when it was created and edited.

13   Reports, Key Performance Indicators, and Alerts

- **View or Edit**
  We can also view this transaction in a dashboard and use the glasses and pencil icons to determine which transactions we need to view and edit.

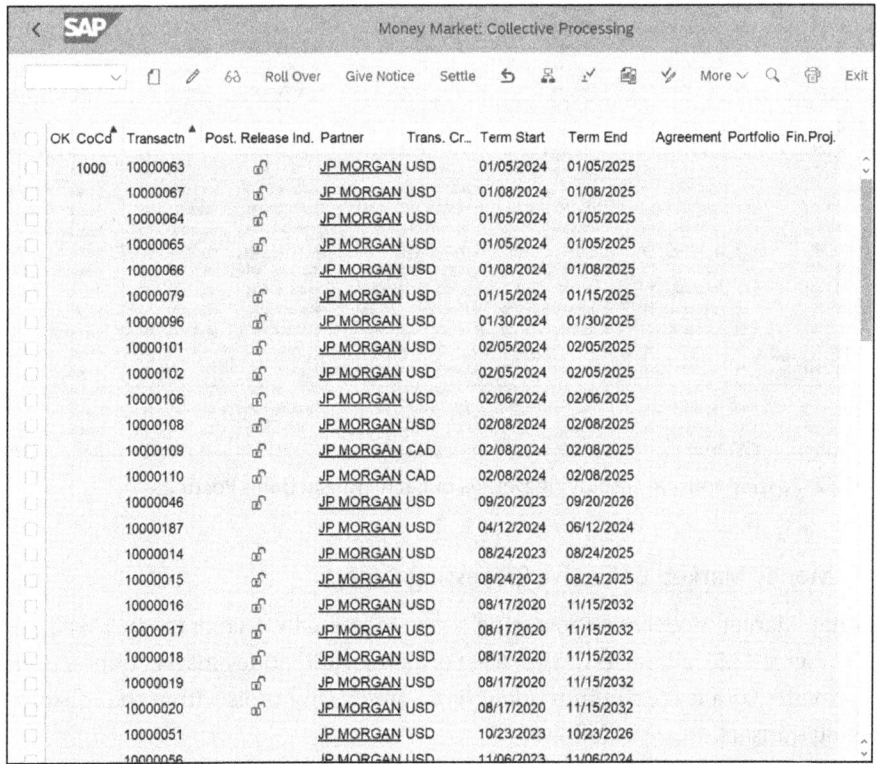

Figure 13.53 Displaying and Managing Money Market Transactions Through Central Collective Processing Transaction

13.3.11   FX: Collective Processing Report

Another collective processing transaction is Transaction TX06, which brings us to the FX: Collective Processing report, shown in Figure 13.54. This report works very much like the Money Market: Collective Processing report, but it includes a few extra features that are not required for the money market contracts. The additional features in this transaction are as follows:

- **Premature Settlement**
  If we need to prematurely settle an FX deal, we can use this function in collective processing. We must configure the contract for premature settlement or it will show an error message when we select this option.

- **Fix**
  We need to fix the rates for the NDFs before we can settle them, and this function allows us to fix the final rates.

- **Terminate**

    If we need to terminate an FX deal early, we use this function. We can also determine the termination amount in the contract once started.

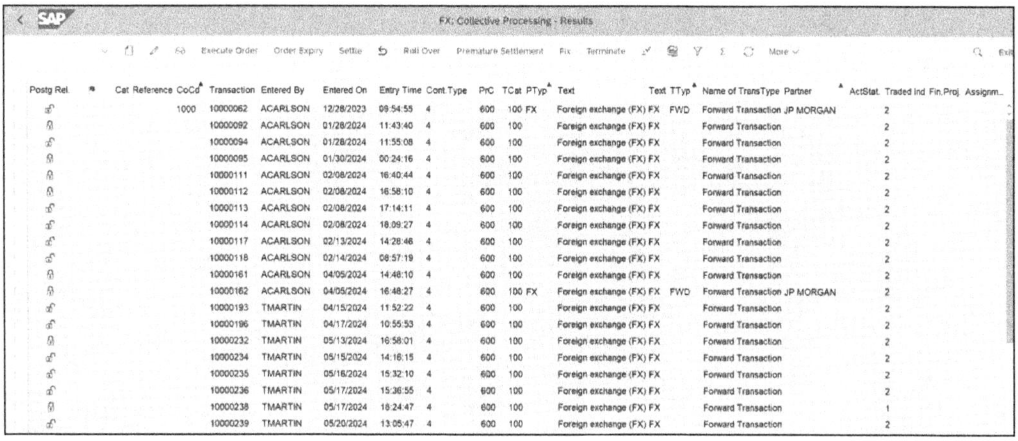

Figure 13.54 Full List of FX Transactions Available in Collective Processing

13.4 Summary

Many different reports are available in treasury and risk management to help us ensure that our reporting needs are met. These include some of the newer KPIs and reports, which add more functionality to the treasury and risk management reporting that did not exist prior to SAP S/4HANA and SAP Fiori. The legacy reports that were available in SAP ERP are still available for SAP S/4HANA Finance for treasury and risk management, but some of them are not available as native SAP Fiori apps. We can either access them using the SAP GUI or map the transaction code as an SAP Fiori app. Being able to review the transaction data ensures that the treasury and risk management solution is complete for end users.

Conclusion

Throughout this book, we've journeyed through the complex world of SAP S/4HANA Finance for treasury and risk management. This book has tried to be a comprehensive guide, shedding light on the most important aspects of treasury and risk management, from the foundational configuration to the end user process in SAP S/4HANA. As we draw this exploration to a close, it is important to reflect on the key insights, reinforce the importance of treasury and risk management in modern treasury operations, and envision the future of treasury management within the SAP ecosystem.

The Evolution and Importance of Treasury and Risk Management

From its inception in the early 1990's, treasury and risk management has continually evolved to meet the ever-changing demands of corporate treasury departments. Its longstanding presence in the market is a testament to its robustness, adaptability, and commitment to continuous improvement. Treasury and risk management has earned a reputation as a reliable and comprehensive solution, offering unparalleled support for managing a wide array of financial instruments. Its integration with the broader SAP ecosystem further enhances its value and provides a seamless solution for organizations that already use SAP.

Over the years, we've seen significant advancements in the usability and functionality of treasury and risk management. particularly with the rollout of SAP S/4HANA and the SAP Fiori user interface. These innovations have made the solution much more intuitive and user friendly, making the learning curve less steep and enabling treasury professionals to harness the solution's full potential. The real-time availability of data, coupled with the enhanced reporting capabilities of treasury and risk management, has empowered treasury departments to make informed decisions with greater accuracy than in the past.

Addressing Common Scenarios and Challenges

In this book, we've focused on the most common scenarios and financial instruments encountered within treasury management. In doing so, we've aimed to provide practical and relevant insights that align with the day-to-day operations of treasury departments. From money market instruments to derivatives, FX, securities, and trade finance, we have delved into the specifics of each instrument, offering step-by-step instructions, practical examples, and detailed walkthroughs.

We've also addressed the complexities of configuring and managing these instruments within treasury and risk management, highlighting foundational setup requirements and instrument-specific configurations. By bridging the gap between system configuration and practical applications, we have ensured that users can efficiently manage their treasury operations and leverage treasury and risk management to its full potential.

Practical Applications and the User Experience

A significant portion of this book has been dedicated to the practical applications of treasury and risk management. We've walked through the steps of creating and processing contracts, managing daily operational tasks, and handling common issues, and we've provided troubleshooting tips. By providing detailed instructions, practical examples, and screenshots, we've aimed to make the user experience as seamless and intuitive as possible.

The importance of practical applications cannot be overstated. Understanding the theoretical aspects of treasury and risk management is essential, but the real value lies in the ability to apply this knowledge effectively in real-world scenarios. By focusing on the user side of the process, we've equipped you with the tools and understanding you'll need to put the system into practice and use it on a daily basis to manage treasury operations.

Integration

One of the key strengths of treasury and risk management is its ability to integrate seamlessly with other areas of SAP S/4HANA, such as SAP S/4HANA Finance, the general ledger, and payment processing. This native integration ensures cohesive functionality across the enterprise without relying on a custom interface to connect treasury operations to the ERP. Treasury and risk management is embedded within the SAP S/4HANA solution, which creates natural efficiencies and a much smoother user experience.

The enhanced reporting capabilities of treasury and risk management have revolutionized corporate treasury management over the past decade. Modern treasury management systems offer advanced reporting tools that provide detailed insights into various aspects of treasury operations, and the ability to customize reports and dashboards has empowered treasury departments to present data in a more meaningful way, facilitating better communication with senior management and other key stakeholders. This has led to more strategic discussions about financial health and risk management, ultimately contributing to more effective treasury operations.

Looking Ahead

As we look to the future, it's clear that the role of treasury management will continue to evolve, driven by technological advancements and the increasing complexity of financial markets. Treasury and risk management is well positioned to meet these challenges, offering a robust and flexible solution that can adapt to the changing landscape. The ongoing development and enhancement of treasury and risk management, particularly with the integration of emerging technologies such as artificial intelligence (AI) and machine learning, promise to further elevate treasury and risk management's capabilities, enabling treasury departments to operate more efficiently and strategically.

The treasury departments of the future will expect to have real-time, accurate information at their fingertips. Enhanced reporting capabilities will continue to revolutionize corporate treasury management, providing advanced tools that offer detailed insights into cash flow, liquidity, and risk exposures. These tools will enable treasurers to generate comprehensive reports that meet the needs of internal stakeholders and comply with regulatory requirements. The ability to customize reports and dashboards will empower treasury departments to present data in a more meaningful way, facilitating better communication with senior management and other key stakeholders.

Final Thoughts

In conclusion, this book has aimed to provide a comprehensive and practical guide to treasury and risk management. By focusing on the most commonly used treasury instruments, addressing real-world challenges, and offering detailed instructions and practical examples, we hope to have equipped you with the knowledge to successfully implement treasury and risk management.

Writing this book has been both a challenging and a rewarding experience. Despite the considerable effort involved, it was satisfying to take our many years' worth of experience in treasury and risk management and see it published within this book. Our passion for treasury and risk management has driven us throughout this project, and we hope that this book will contribute to a greater understanding and broader adoption of treasury and risk management in the marketplace. Each chapter, section, and example we've provided in these pages is a reflection of our deep commitment to helping professionals navigate the intricacies of treasury and risk management. This book is not just a collection of technical instructions but rather a distillation of our experience and insights gained over many years of consulting and project delivery in the treasury and risk management space.

Conclusion

Our dedication to this project stems from our belief in the transformative power of treasury and risk management. We've seen firsthand how effective use of this solution can streamline operations, enhance risk management, and drive financial performance. It's our hope that through this book, more organizations will be able to harness these benefits and achieve greater efficiency and effectiveness in their treasury functions.

Thank you for embarking on this journey with us. We hope that this book serves as a valuable resource, guiding you through the complexities of treasury and risk management and empowering you to achieve excellence in your treasury management endeavors. In closing, we want to express our gratitude for your trust and engagement with this material. It has been our privilege to share our knowledge and experience with you, and we look forward to seeing the positive impact that treasury and risk management can have on your treasury operations. Here's to your continued success and the bright future of treasury and risk management.

The Authors

Luke Carlson began his career in financial services before transitioning into SAP treasury consulting. With more than 18 years of expertise with SAP Treasury and Risk Management, he has successfully delivered SAP treasury projects for more than 25 companies worldwide. A recognized thought leader in the SAP treasury space, Luke is a frequent speaker at SAP Financials and SAP Treasury conferences. He has also collaborated with SAP to help design and provide valuable insights on numerous treasury and cash management features introduced into SAP S/4HANA. In 2017, Luke cofounded Carlson Cash with a team of SAP treasury experts, driven by the vision of addressing a market need for a firm exclusively focused on the seamless implementation of SAP Treasury and Risk Management projects.

Andrew Carlson began working with SAP treasury solutions in 2009, first as a business user and soon after as a consultant. Prior to cofounding Carlson Cash, his SAP treasury consulting experience ranged from a boutique consulting firm, independent consulting, as well as a Big 4 accounting firm. He is a certified treasury professional and has delivered more than 20 successful SAP Treasury and Risk Management implementations throughout his career. Within SAP Treasury and Risk Management, he has focused on implementing solutions in all financial instruments, with a particular focus on exposure management and hedge management.

Jeffrey Lasecki has specialized in SAP Treasury and Risk Management since 2006 and has worked with clients across North America and Europe in the areas of cash management and treasury. Jeffrey spent nine years consulting with a boutique firm specializing in treasury and assisted more than 20 clients with their implementations. Jeffrey was an independent consultant for four years, initially working with a Fortune 500 spin-off and then remained with the parent company, supporting their treasury operations. In 2018, Jeffrey joined SAP America and is currently leading all treasury presales efforts in North America, including electronic banking, cash management, treasury and risk management, and in-house banking.

Index

A

Acceptance payment 243
Account assignment reference 184, 345
   allocate 193
   assign general ledger accounts to posting
      keys 192
   create derivation rule 186
   define 185
   derivation rule attributes 188
   derivation rule conditions 187
   determine 186
   maintain values 189
   transfer 358
   transfer process 199
Account IDs 59
Account symbols 176, 199
Accounting 752
Accounting codes 106
Accruals and deferrals 165, 169
Administration fields 171
Alerts 699, 763
   assign categories 701
   categories 699
   configuration 699
   monitor 702
Amount to hedge 570
Analyze NPV app 729
App Finder 38
Authorization groups 522
Automatic Debit Position and Posting –
   Security Account app 432
Automatic Debit Position and Posting
   app 433
Automatic Payment Transactions for
   Payment Requests app 365, 756

B

Balance Sheet FX Risk Overview app 608
Balance sheet hedging 599
Balance sheet risk 600
Bank guarantees 50, 141
Bank master 55
Bank Risk app 713
BAPI BUS5990 495
Bilateral facilities 136, 397
Binary operators 681

Bond warrant 141
Bonds 48, 140
Book and reset 166
Building expressions 681
Business partners 60, 325
   addresses 64
   assign authorizations 71
   assign payment details 67
   maintenance 62
   payment transactions 65

C

Calculate Net Present Values – With CVA and
   DVA app 442, 448, 466, 591
Calendars 84
Caps and floors 49, 137
Cash Flow Analyzer app 751, 780
Cash flow hedging 515
   process 558
   take snapshot 559
Cash flow transaction 136
Cash flows 465
Cash management 744
   basic settings for integration 747
Cash settlement 314
Central incoming payments 760
Central payments 760
Certainty level 744
Charges 208, 243
Check Cash Flow Items app 782
Commercial paper 47, 133, 135
Commodities 306
Commodity contracts 476
   cash settlement 480
   contract creation 477
   contract settlement 483
   process flow 476
Commodity derivatives 305
   assign exchange to MIC 309
   define exchange 308
   flow types 312
   position updates 317
   product types 310
   transaction types 311
   update types 315
   update types relevant to posting 318

Index

Commodity forwards	139
Commodity prices	98
Commodity swaps	139
Commodity types	305
Communication channels	649
Communication profiles	654
assign channel	655
assign format metatypes	655
channel-dependent attributes	656
create	654
format-dependent attributes	657
Company codes	54, 324, 545, 548
additional data	104
Competitive bid capture	614
Condition category	163
Condition types	161
assign transaction types	164
define	162
Configure Back-End Systems app	630, 633
Configure Trading Platforms app	629
Contract close	594
Contract settlement	342
dual control	342
workflow	343
Contracts	321
Core configuration	103
Correspondence	647
additional settings	667
BIC code	670
configuration	648
define activity	667
display	693
dynamic table assignment	669
inbound messaging	687
inbound process	664
manual	684
mapping rules	671
number ranges	671
outbound messaging	683
outbound process	665
SWIFT message sequence	669
Correspondence class	652
inbound	653
Correspondence framework	647
Correspondence matching	690
automatic	691
configuration	690
manual	692
Correspondence monitor	648, 694
functions	697
report	697
selection screen	695

Correspondence objects	647, 683
Correspondence partners	659
business partner groups	659
internal recipients	659
Correspondence recipient types	652
Counterparty Limit Utilization app	628
Create Adjustment – Rates/Prices app	384
Create Financial Transaction app	88, 322, 399, 422, 580, 752–753, 760
administration	332
borrowing	327
cash flow	337
cash flow calculation dates	338
contract conclusion	331
header	325
interest flows	338
interest structure	330
intial screen	323
memos	341
other flows	332
payment details	334
repayment structure	330
reverse flows	340
status	341
structure	326
term	329
Create Manual Posting app	436
Create Phone Trade app	617
Credit facilities	47, 226
condition types	235
define partner rank	238
flow types	228
product types	226
transaction types	228
update types	233
Credit Line Analysis app	776
Credit Risk Analyzer	30, 705
configuration	715
default risk rule	721
derivation of default risk rule	723
determination procedure	717
global settings	715
integrated default risk limit check	716
limit product groups	719
limit types	718
transaction types	720
valuation factors	717
Credit value adjustment (CVA)	442, 526
Currency barrier options	49
Currency warrant	140

Index

D

Data feeds 100
   Request Current Market Data app 725
Deal default risk limit 714
Debit value adjustment (DVA) 442, 526
Debt and Investment Analysis app 773
Debt and Investment Maturity Profile
   app 772
Debt and investments 201, 369
   cash flows 382
   conditions for interest structure 375
   interest rate adjustment 380
   interest rate adjustment condition 376
   interest structure 372
   process flow 370
   repayment structure 377
   variable interest rate 373
Define Hedging Area app 541
Define Time Pattern app 520
Deposits at notice 47, 135
Derivation rules 186
Derivative contract specification 74
   basic data 75
   contract-specific data 78
   MIC 77
   month codes 83
   periods 80
   quantity/UoM 79
   release procedure 82
Derivatives 49, 293
   assign condition types 300
   assign flow types 299
   define condition types 300
   define flow types 296
   derived business transactions 303
   position management procedure 301
   product types 293
   transaction types 295
   valuation 302
Designation activation 555
Designation level 554
Designation splitting 556
Designation types 523
Differentiation criteria 546
Dirty prices 122
Display Financial Transaction app 752
Display Journal Entries – In T-Account View
   app 367
Display Line Item Entry report 768
Display Treasury Alerts – Correspondence
   app 702
Display Treasury Alerts – Interest Rates
   app 780
Display Treasury Alerts – Payment app 778
Display Treasury Alerts – Posting app 777
Display Treasury Alerts – Release app 779
Display Treasury Alerts – Settlement app 779
Display Treasury Alerts app 777
Display Treasury Position Flows app 766
Display Treasury Posting Journal app 767

E

Effectiveness test 531
End of Day Processing report 710
Enter Book Values for Manual Valuation
   app 446
Equities 48
Equity warrant 140
Exchange rates 105
Exchange-traded option 138
Execute Classification app 592
Execute Debit Position – Manual Debit
   Position app 431
Execute Valuation Class Transfer app 361
Exercise Rights app 434
Expected exposures 443
Exposure management 31, 485
   configuration 497
   define derivation strategy 510
   exposure origin 513
   exposure position types 510
   exposure types 499
   free attributes 502
   global settings 498
   periods 498
   product types 509
   release procedure 503
   users and roles 504
External number range 145
External underlying options 138

F

Facilities 136, 390
   charges 229, 235, 392
   Credit Lines and Utilization report 798
   dates 395
   Lines of Credit, Drawing, and Fees report 799
   process flow 390
   profiles 395
   rules 396
   structure 391

813

Index

Facility drawings 399
Factory calendar 84
Fee on available amount 245
Fee on overdraft amount 245
Fee on presented amount 244
Fee on total amount 244
Field selection 148
Final repayment 208
Financial transactions 321
Fixed-term deposit 135
Fixing date 530
Fixing spread 261
Flow categories 153
Flow types 151, 754
   assign transaction types 154
   define 152
Foreign exchange (FX) 47, 256, 438
   assign company code 274
   assign fixing spreads 261
   assign flow types 264
   assign FX attributes 259
   assign update types for valuation ... 265
   book and reset procedure 451
   contract creation 440
   contract posting 441
   derived business transactions ... 266, 450
   difference procedure 451
   flow types 262
   hedging classifications 275
   manual valuation 446
   mirroring 270
   mirroring mapping 273
   mirroring product types 272
   monthly processing 442
   nondeliverable currencies 269
   position management procedure ... 264
   position-related update types 265
   posting flows 450
   posting valuations 448
   process flow 439
   processing incoming data 274
   product types 257
   rollovers 456
   transaction settlement 450
   transaction types 258
   update types for position outflows ... 268
Foreign Exchange Overview app 774
Format metatype 650
Formats 651, 658
   mapping 663
Forward contracts 47, 50
Forward loans 139
Forward rate agreement 49, 137
Forward securities 138
Forwards 138
From-currency 530
Functions 681
Futures 50
Futures contract 137
FX options 49
FX rates 90
   enter rates 91
   leading/following currencies 90
   rate types 91
FX swap requests 575
FX valuations 116

G

General valuation classes 109, 526
   assign 111
   define 109
   valuation class transfer 110
Generally Accepted Accounting Principles
(GAAP) 32

H

Hedge accounting calculation types 526
Hedge management 32, 517
Hedge management cockpit 546, 561
   automated request creation 573
   dedesignate hedge 576
   differentiation criteria 565
   exposure totals 568
   filter 567
   hedge targets 570
   hedges 569
   initial settings 561
   key figures 566
   layout details 564
   layout IDs 562
   manual request creation 574
   new layout ID 563
   overwrite target quota 571
   period 566
   report layout 568
Hedge Management Cockpit app 561
Hedge relationship grouping 540
Hedge request reasons 522
Hedge requests 609
   groupings 641
   log 612
   parameters 641

Index

Hedges 601
Hedging areas 522, 541
   currencies 547
   filters for exposures 548
   filters for hedges 550
   financial transactions 544
   FX hedge request 552
   general settings 545
   hedge accounting 554, 557
   hedge request settings 553
   integration 543
   main data 542
   value date 553
Hedging business transactions 539
Hedging classifications 518, 546
Hedging profiles 523
   define 534
Hedging relationship scenario 534
Hedging relationships 582
   dedesignation 595
   details 584
   documentation 586
   efectiveness test 587
   header 583
   hedged items 585
   hedging instrument 585
Holiday calendar 84
House Bank Account Connectivity app 59
House bank accounts 59
House banks 57
Hypothetical derivative 529

I

Import Raw Exposures – Spreadsheet
   app 494
Inbound trading 613
Incremental accrual/deferral method 166
Index warrant 140
In-house banking for SAP S/4HANA Finance
   for advanced payment management 760
Installment bonds 140
Integration 743
Intercompany FX contracts 48
Intercompany loans 46, 213, 402
   assign company codes 225
   cash flows 405, 410
   changes 413
   condition types 219
   contract creation 404
   contract types for mirroring 222
   flow types 217

Intercompany loans (Cont.)
   mapping types 223
   mirroring 220, 406
   mirroring mode 221
   object links 408
   process flow 402
   processing incoming data 224
   product types 214
   settlement 411
   transaction types 215
Intercompany trading 614
Interest calculation method 143
Interest rate adjustments 383, 464
   automatic 386
   manual 384
Interest rate instruments 46, 136
Interest Rate Overview app 769
Interest rate swaps 50, 137, 293, 459
   entry 461
   interest condition settings 463
   interest structure 462
   process flow 460
Interim limits 708
Internal number range 145
Internal payments 760
International Financial Reporting Standards
   (IFRS) 32
Investment funds 49, 139
Issue amount 128
Issue spread 128

J

Journal: Transactions with Cash Flows
   report 797

K

Key figures 636
Key performance indicators (KPIs) 763

L

Letters of credit 50, 141, 239
   assign condition types 251
   bank guarantee types 255
   condition types 250
   conditions 475
   conditions for payment 254
   contract creation 468
   document 474
   document templates 252

Letters of credit (Cont.)
  document types 251
  flow types 242
  goods and shipping 472
  main data 471
  payment 473
  product types 239
  reasons for rejection 255
  structure 469
  transaction types 240
  update types 248
Limits 706
Literals 681
Local payments 760
London Interbank Offered Rate (LIBOR) 379

M

Maintain Business Partner app 62
Manage Bank Accounts app 59
Manage Banks – Cash Management app 58
Manage Banks – Master Data app 55
Manage Block of Trade Requests app 627
Manage Counterparties app 634
Manage Hedge Requests app 644
Manage Hedging Relationships app 540, 581–582, 597
Manage Journal Entries app 769
Manage Limits app 706
Manage Market Data Shifts app 728
Manage Securities Accounts app 417
Manage Securities Classes app 419
Manage Trade Request – Trading Platform app 614
Manage Trade Requests app 615
Manage Trades app 624
Map Format Data for Treasury Correspondence app 672
  attributes 676
  building expressions 680
  condition 680
  data types 677
  input specification 677
  mapping properties 673
  node editing 676
  node navigation 675
  node types 674
  output specification 679
Market data 89
Market data collection 725
Market data interfaces 99
Market Data Overview app 775

Market identifier code (MIC) 307, 331
Market Risk Analyzer 30, 724
  configuration 734
  evaluation types 738
  portfolio hierarchies 734
  value at risk type 736
Market value component calculation 528
Mark-to-market valuation 465, 727
Master data 53, 636
Messages 648
Minimum trades 643
Money market 46
Money market contracts 203
  assign flow types 212
  condition types 211
  flow types 206
  product types 203
  transaction types 204
Money Market: Collective Processing report 801
MT300 647
MT320 647
My Inbox app 343

N

Negotiation spread 128
Nominal decrease 242
Nominal increase 242
Nominal interest 208
Non-deliverable forwards (NDFs) 47, 438, 452
  accounting 455
  entry screen 453
  fixing 453
Number ranges 145

O

Operators 681
OTC derivatives 117, 459
Other comprehensive income (OCI) 596
Other flows 332
Overnight SOFR 379
Over-the-counter (OTC) options 138

P

Payment flows 756
Payment management 154
Payment obligation 246
Payment processing 364
Payment requests 232

Index

Payment runs 365
Payments 753
Period end closing 591
Planned records 387
Planning levels 746
Portfolios 170
Position component 128
Position management procedure 113, 117
   amortization 126
   categories 113
   component for valuation 120
   current portion transfer procedure 130
   customizing valuations 118
   define and assign 113
   foreign currency valuation 119
   impairment procedure 129
   one-step price valuation 123
   rate valuation for FX transactions 124
   security valuation 121
   valuation steps 115
Position transfers 194
   post 197
Post Derived Business Transactions app 596
Post Flows app 346, 427, 595, 753, 756
   application 346
   general selections 346
   messages 349
   payment log 349
   posting control 348
Posting keys 179
Posting specifications 178, 754
   assign update types 183
   define 178
Precondition selection 750
Principal decrease 207
Principal increase 207
Process Balance Sheet Hedge Request app 612
Process Business Transactions app 340
Process Correspondence – Monitor app 692–694
Process Financial Transaction app 343, 346, 407, 413
Process Free Form Payments app 758
Process Hedge Requests – Balance Sheet FX Risk app 603, 609
Process Hedge Requests app 578
Process Raw Exposure app 486, 495
Process Raw Exposures – Collective Processing app 491, 496
Product categories 134, 459
   derivatives 137
   foreign exchange 136

Product categories (Cont.)
   money market 135
   securities 139
   trade finance 141
Product types 141, 324
   assign update types 525
   exposure subitems 524
Purchase commodity forward 313

R

Rate/Price Adjustment Schedule report 798
Raw exposures 486
   header 488
   line item data 489
   release 495
   user data 490
Reference interest rates 94
   configuration 94
   entry 95
Release Hedging Business Transactions app 589, 595, 597
Remaining credit amount 246
Reporting 763
Reporting time pattern 543
Reprocess Transactions – Automated Designation app 581
Repurchase agreements 49, 138
Review Balance Sheet FX Risk app 600, 637, 639
Review Limit Utilizations app 710
Risk-free currency 542
Risk-free interest rates 379
Rounded amounts 643
Rules 504
   apply 505
   responsibilities 506
Run Accrual/Deferral app 350, 430
Run Automatic Adjustments – Rates/Prices app 386
Run Valuation app 264, 355, 429, 448, 592

S

Sale commodity forward 314
SAP Bank Communication Management ... 367
SAP BTP cockpit 630
SAP Fiori 32
   add/remove apps 38
   analysis report 782
   axis properties 787
   columns 784

Index

SAP Fiori (Cont.)
  custom tiles 43
  custom user groups 39
  custom views 41
  default values 36, 772
  filters 782, 785–786
  grid properties 787
  home screen 33
  measures 789
  navigation 783
  output 785
  rows 783
  settings 34, 771
  share views 45
  spaces 35
  standard reports 763
  user profile 33
  user screen 37
  zeroes 789
SAP Fiori apps reference library 34
SAP GUI reports 794
SAP Multi-Bank Connectivity 687
SAP S/4HANA Finance for in-house cash 759
SAP trading platform integration 613, 623
  backend mapping 633
  configuration 629
  mapping 631
SAPscript 658
Schedule Treasury Middle Office Jobs
  app 605, 610
Secured Overnight Financing Rate
  (SOFR) 379
Securities 48, 275, 415
  account transfer 291
  accounting postings 428
  accounts 417
  accrued interest 424
  amortization 429
  assign condition types 278
  assign flow types 285
  assign repayment types 280
  assign update types 289
  automatic debit position 432
  basic data 420
  cash flows 427
  classes 418
  company code-dependent settings 281
  condition types 276
  conditions 420
  configuration 276
  contract creation 422
  contracts 422

Securities (Cont.)
  date rules 284
  define general classification 282
  define groups 277
  define update types 287
  derivation rules for tax flows 291
  flow types 284
  interest accrual 430
  manual debit position 431
  notices 421
  one-off postings 434
  options 434
  other flows 436
  position management 286
  position management procedure 286
  posting flows 429
  process flow 416
  product types 279
  rights category 292
  search terms 419
  specify update types 288
  transaction management 283
  transaction reversals 438
  transaction types 283
Securities lending 138
Security prices 93
Single transaction checks 707
Single Value Analysis: VAR app 733
Snapshots 600
  selection 611
Spot contracts 47
Spot designation 527
Stocks 139
Subscription rights 140
Substitution Rules for Planning Levels –
  Treasury Flows app 748
Swap rates 92
Swaps 49
SWIFT 650
Syndicated facilities 136, 238, 398

T

T-accounts 351
Take Snapshot app 559
Take Snapshot for Balance Sheet FX Risk
  app 602
Target quota types 519
Target quotas 545, 551
Term SOFR 379
Time deposits 46
Total return swap 137

818

Index

Trade finance 468
   instruments 50
   process flow 468
Trade requests 613
Traders 86
   assign 88
   user data 86
Trades 615
   assign to block 622
   creating blocks 621
   limit checks 623
   sending 616
   splitting 619
   undoing blocks 621
   undoing splits 620
Transaction
   F111 .... 365, 756
   FDCS17 98
   FI01 55
   FI06 57
   FI12_HBANK 57
   FILE 687
   FTR_ALERT 702
   FTR_ALRT_BTCH 702
   FTR_BP_BIC 670
   FTR_COMATCH 691
   FTR_COMONI 692–694
   FTR_CREATE 323, 399, 761
   FTR_DISPLAY 794
   FTR_EDIT 343
   FTR_EXT_ASSIGN 665, 683
   FTR_IMPORT 687
   FTR_INB_FUNC 664
   FTR_INT_ASSIGN 666, 683
   FTR_SWIFT_IMPORT 688
   FW18 94
   FWBS 436
   FWER 434
   FWSO 432
   FWZE 431
   FWZZ 419
   JBIRM 95
   SBWP 343
   SLDCHECK 630
   SM34 671
   TAC1 145
   TBB1 .... 346, 427, 595, 753
   TBCD 795
   TBD4 .... 94, 99
   TBT1 88
   TI10 384
   TJ01 794
   TJ05 386
   TJ07 798

Transaction (Cont.)
   TJ08 794
   TJ09 388
   TJ12 797
   TM_52 796
   TM_60 798
   TM_60A 799
   TM00 801
   TOE_HEDGING_AREA 541
   TOE_TIME_PATTERN 520
   TOEHREQO 578
   TOENE 561
   TOESNAP 559
   TPM1 .... 264, 355, 429, 448, 592
   TPM10 340
   TPM100 .... 540, 581–582, 597
   TPM101 592
   TPM104 581
   TPM120 .... 589, 595, 597
   TPM13 800
   TPM15M 111, 361
   TPM18 .... 594, 596
   TPM20 800
   TPM28 359
   TPM44 .... 350, 430
   TPM60CVA 115, 442, 591
   TPM74 446
   TRS_SEC_ACC 417
   TX06 802
Transaction manager 29
Transaction posting 344
Transaction Release: Work Item Overview and Status of all Transactions report 794
Transaction types 146, 324
Transfer Account Assignment Reference app 359
Transfer book value 358
Treasury Position Analysis – Accounting View app 793
Treasury Position Analysis – Listed Derivatives app 793
Treasury Position Analysis – OTC Transactions app 791
Treasury Position Analysis – Securities app 792
Treasury Position Flows report 800
Treasury Position History app 764
Treasury Position Values app 765
Treasury Posting Journal report ... 800
Treasury: Change Documents for Transaction report 795
Treasury: Journal of Financial Transactions report 794

819

U

Update Planned Records app 388
Update types 156, 538
   assign flow types 159
   assign usages 158
   cash management 747
   define 156
   relevant to posting 160
User data 131
User profiles 133

V

Valuation areas 107
Valuation class transfer 361
Valuation classes 108, 361

Value at risk 733
Variable commodity prices 105
Variable interest rates 105
Variable security prices 106

W

Warrant bonds 140
Warrants 48
Workflows 172, 343
   adjust template 174
   release procedures 173

Y

Yield curves 96

- Set up financial accounting and controlling processes in SAP S/4HANA
- Configure your system with step-by-step instructions
- Prepare for testing, go-live, and production support

Stoil Jotev

Configuring SAP S/4HANA Finance

Starting a new SAP S/4HANA Finance implementation? Get it right the first time! From setting up an organizational structure to defining master data, this comprehensive guide to configuring SAP S/4HANA Finance walks you through each key task. Follow step-by-step instructions organized by functional area: general ledger, accounts payable and receivable, margin analysis, group reporting, and more. Customize SAP S/4HANA to meet your FI/CO needs!

approx. 744 pp., 3rd edition, avail. 01/2025
E-Book: $84.99 | **Print:** $89.95 | **Bundle:** $99.99

www.sap-press.com/5920

Aylin Korkmaz

Financial Reporting with SAP S/4HANA

Having access to clear, accurate financial reports is key for any organization! In this comprehensive guide, you'll walk step by step through creating such high-quality reports in SAP S/4HANA. You'll learn about key reporting requirements for finance and how to guide system design to produce the best reports for your company. Coverage includes core financial areas like general ledger, AP/AR, assets, cash management, overhead costing, margin analysis, and more. Make sense of your financial data with this expert guide!

707 pages, pub. 02/2022
E-Book: $84.99 | **Print:** $89.95 | **Bundle:** $99.99

www.sap-press.com/5416

- Configure your cash management processes in SAP S/4HANA
- Run bank relationship management, cash positioning and operations, and liquidity management
- Migrate your cash data into SAP S/4HANA

Dirk Neumann, Lawrence Liang

Cash Management with SAP S/4HANA

Managing your cash is critical—so master cash management in SAP S/4HANA! Follow step-by-step instructions to run bank relationship management, cash positioning and operations, and liquidity management, and then tailor each process to your system. Walk through the One Exposure from Operations data model, including integration scenarios, transactions, and configuration. Discover extensibility options for bank relationship management and key SAP Fiori apps. Get equipped for cash management!

561 pages, 2nd edition, pub. 11/2020
E-Book: $84.99 | **Print:** $89.95 | **Bundle:** $99.99

www.sap-press.com/5169

- Manage your financial consolidation with SAP S/4HANA Finance
- Configure and run currency translation, intercompany elimination, consolidation of investments, and more
- Create consolidation reports and use new reporting tools

Ryan, Bala, Raghav, Mohammed, Kukreja

Group Reporting with SAP S/4HANA

The Financial Consolidation Guide

Get financial consolidation up and running! With this comprehensive guide, learn how to set up and use SAP S/4HANA Finance for group reporting. Organize your financials information by configuring global settings, master data, and transaction data. Walk through each group reporting process—intercompany elimination, consolidation of investments, and matrix consolidation—following step-by-step instructions for setup and execution. With details on consolidation reporting, this is the only book you need for a successful corporate close!

484 pages, 2nd edition, pub. 09/2024
E-Book: $94.99 | **Print:** $99.95 | **Bundle:** $109.99

www.sap-press.com/5872

Interested in reading more?

Please visit our website for all new book
and e-book releases from SAP PRESS.

www.sap-press.com